Notes on Numerical Fluid Mechanics and Multidisciplinary Design

Volume 140

Notes on Numerical Fluid Mechanics and Multidisciplinary Design publishes state-of-art methods (including high performance methods) for numerical fluid mechanics, numerical simulation and multidisciplinary design optimization. The series includes proceedings of specialized conferences and workshops, as well as relevant project reports and monographs.

More information about this series at http://www.springer.com/series/4629

Charles Hirsch · Dirk Wunsch
Jacek Szumbarski · Łukasz Łaniewski-Wołłk
Jordi Pons-Prats
Editors

Uncertainty Management for Robust Industrial Design in Aeronautics

Findings and Best Practice Collected During UMRIDA, a Collaborative Research Project (2013–2016) Funded by the European Union

Springer

Editors
Charles Hirsch
NUMECA International S.A.
Brussels
Belgium

Dirk Wunsch
NUMECA International S.A.
Brussels
Belgium

Jacek Szumbarski
Institute of Aeronautics
and Applied Mechanics
Warsaw University of Technology
Warsaw
Poland

Łukasz Łaniewski-Wołłk
Institute of Aeronautics
and Applied Mechanics
Warsaw University of Technology
Warsaw
Poland

Jordi Pons-Prats
CIMNE International Centre
for Numerical Methods
in Engineering
Barcelona
Spain

ISSN 1612-2909 ISSN 1860-0824 (electronic)
Notes on Numerical Fluid Mechanics and Multidisciplinary Design
ISBN 978-3-030-08534-6 ISBN 978-3-319-77767-2 (eBook)
https://doi.org/10.1007/978-3-319-77767-2

Printed on acid-free paper

This Springer imprint is published by the registered company Springer International Publishing AG part of Springer Nature
The registered company address is: Gewerbestrasse 11, 6330 Cham, Switzerland

Contents

Part I
The UMRIDA Project

Vision, Objectives and Research Activities

Charles Hirsch and Dirk Wunsch

Vision, Concepts and Main Challenges

The ultimate objective of aircraft design methodologies, within a fully computerized design environment based on the concept of virtual prototyping (VP), is to rely essentially on analysis: computer-aided design (CAD) definitions for the geometry, computational fluid dynamics (CFD) for the aerodynamic loads and noise sources; computational structural mechanics (CSM) for stresses and lifetime evaluations, including combined fluid–structure interactions (FSI); conjugate heat transfer (CHT) for heat fluxes and heat transfer; combustion modelling for emissions; computational aeroacoustics (CAA) for noise estimations. This objective should lead to a shorter design cycle, with fewer costly tests and as importantly, fewer redesigns due to unanticipated test results.

An important limitation of the extensive application of analytical tools, at the basis of virtual prototyping, is connected to the **level of uncertainty** they introduce **in the analysis and design process**.

For example, when performing 3-D viscous flow analysis of engine components, there is uncertainty in the definition of the boundary conditions representing the operational environment; in the discrepancy between the CAD geometry and the real geometry resulting from the manufacturing tolerances and assembly process; in the true deformed geometry of the parts being analysed at engine operating conditions. In addition, modelling uncertainties are present, resulting from imperfect models for turbulence, multi-species reaction rates or combustion, as well as

C. Hirsch (✉) · D. Wunsch
Numeca International, Brussels, Belgium
e-mail: charles.hirsch@numeca.be

D. Wunsch
e-mail: dirk.wunsch@numeca.be

C. Hirsch et al. (eds.), *Uncertainty Management for Robust Industrial Design in Aeronautics*, Notes on Numerical Fluid Mechanics and Multidisciplinary Design 140, https://doi.org/10.1007/978-3-319-77767-2_1

numerical errors from the set-up of the simulations (such as grid dependencies or convergence levels).

This leads to a **global uncertainty on the results of the analysis**, on which design decisions have to be taken. The ability to **quantify the impact of these uncertainties on the predicted behaviour** of aeronautical components and to **account for these uncertainties in the design process** is crucial for a **reliable risk management** and better estimates of safety margins. Managing the design process in the light of these uncertainties is therefore the key to robust design.

Introducing the probabilistic nature of uncertainties in simulation software systems is a highly challenging undertaking, as the whole process transforms the resolution of deterministic physical conservation laws, to **non-deterministic methods**, governed by stochastic partial differential equations (SPDE). As a consequence, predicted quantities, such as loads, lift, drag, efficiencies, emissions, noise, …, are not represented anymore by single numbers, but by a *Probability Density Function* (PDF), providing a domain of confidence, associated to the considered uncertainties, introducing hereby a **fundamental shift in paradigm** for the whole of the VP methodology.

The main concept is summarized in Fig. 1. The current deterministic approach is to set single-valued computational conditions, leading to a single value for the output quantity η (in red). When the uncertainty is introduced by a PDF, the output quantity is also transformed into a probabilistic quantity. Figure 1 illustrates this fundamental change, which can be strongly dependent on how the input uncertainty parameter is defined. From the output PDF, one can derive the C*umulative Distribution Function* (CDF), leading to the definition of a domain of variation for a given level of confidence, say 95%. This is shown on the right side of Fig. 1, which shows the predicted PDF and CDF for the drag on an airfoil with geometrical uncertainties. The vertical bar represents the deterministic prediction.

A very important property, shown intentionally on this Fig. 1, **is that the deterministic output**, **corresponding to the mean value of the input uncertain variables, is <u>not equal</u> to the mean value of the output PDF**.

The importance of assessing and quantifying the various uncertainties affecting the virtual prototyping process, in particular in aeronautics, has been growing significantly in the last few years in Europe and in the USA, and in particular as an

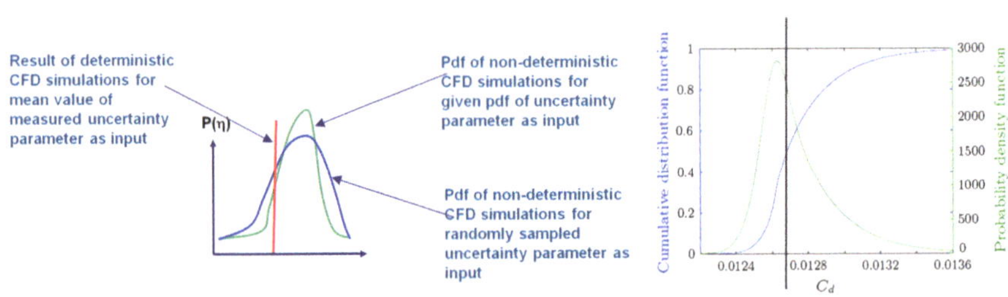

Fig. 1 Main concept of non-deterministic simulations

outcome of the FP6 project NODESIM-CFD (Non-Deterministic Simulations for CFD-based Design Methodologies), precursor of this UMRIDA project.

The incorporation, up to an advanced industrial level of uncertainties in the VP methodology, has become a major necessity, in order to reduce the risks associated to the design decisions based on numerical simulations and virtual prototyping.

Traditional Approach to Risk Management: Safety Factors

A traditional approach to risk management is the introduction of safety factors, or safety margins, comparing the resistance of a system (which we can term as its ***capacity***), compared to the estimated loads (termed in general as ***requirement***). In the traditional and still current practice, left part of Fig. 2, a deterministic value is estimated for the load x_R and a value is provided, as best as possible, for the maximum capacity x_C, on basis of which a ***safety margin*** $k = x_C/x_R$ is imposed on the system.

Taking the safety margin at a sufficiently high value minimizes risk, but this will generally have detrimental consequences on cost and performance. On the other hand, taking into account uncertainties and their PDFs (right figures) provides ranges (under the form of PDFs) for requirements (loads) and capacity (resistance) in a rational way, allowing to define a failure region where the two PDFs overlap. Full safety, taking into account the known uncertainties, is obtained for $k > 1$, upper right figure; while when $k \cong 1$, a certain risk factor will exist, due to the small

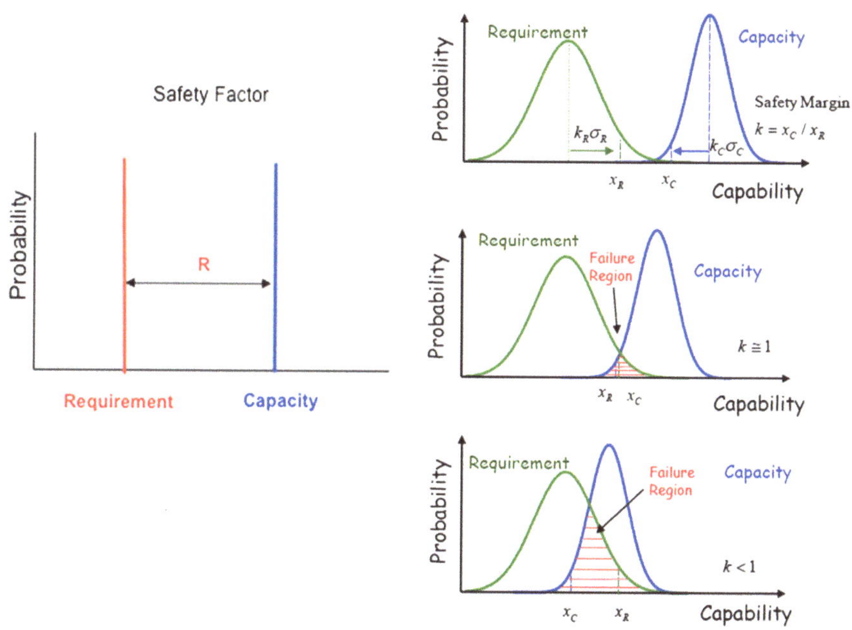

Fig. 2 Rational approach to safety factors, based on UQ methodology (taken from Green [2])

overlapping failure region. In case k < 1, a large failure region exists, which would lead to a catastrophic design, of course to be rejected.

An important element in risk of failure assessment is the so-called ***inverse design*** problem, namely trying to determine a level of acceptable uncertainty of the input quantities, in order to keep the output uncertainties below an accepted level of risk of failure.

Uncertainty Categorization

It is important to separate different types of uncertainties, based on their nature and origin. It has now become standard practice in the Uncertainty Quantification (UQ) literature (see, for instance, the recent overview book of Oberkampf and Roy [4], to distinguish between ***epistemic***, also called ***reducible***, and ***aleatory***, also called ***irreducible***, uncertainties).

The former are globally generated by numerical errors due to discretization approximations and grid dependences, as well as lack of knowledge associated to the imperfect physical models, such as turbulence, combustion or multiphase models. They are considered as *reducible* uncertainties, since they could be reduced through increased understanding and research, or more relevant physical data, and are globally related to the lack of knowledge about the appropriate value to use for the considered quantity.

The important consequence is that ***epistemic uncertainties have a fixed, but poorly known, value in the analysis***. For instance, the elastic modulus for a composite material in a specific component is fixed but its value can be unknown or poorly known; the turbulent viscosity in a CFD simulation is known to be subject to the many approximations attached to turbulence models.

On the other hand, ***aleatory*** uncertainties are related to the ***inherent randomness*** of the system being analysed, such as variability of operational conditions, geometrical randomness from the manufacturing process, which cannot be reduced by further data.

Hence, epistemic uncertainties are a property of the models applied in the analysis, including the choices made by the modeller, while the aleatory uncertainties are a strict property of the system being analysed.

The main reason for this distinction is the different ways of treatment and quantification. The epistemic uncertainties will essentially be handled through probabilistic techniques, while aleatory uncertainties, which form the main objective of the UMRIDA project, require more sophisticated and innovative techniques in the framework of non-deterministic methods, such as Monte Carlo methods or sensitivity-based methodologies, or Polynomial Chaos methods.

The methods for handling epistemic uncertainties, see, e.g. Helton and Oberkampf [3] or Eldred et al. [1], place some type of bounds on the resulting output uncertainty, largely based on subjective estimates of error and model uncertainty levels. It is indeed difficult to provide objective estimates of the numerical errors, or of the error associated with the weaknesses of a given turbulence model.

As both types of uncertainties are always present, their highly unknown and nonlinear interactions are of importance.

Scientific and Technical Objectives

The UMRIDA project focuses on uncertainty management at all levels of the analysis and design process, and it is believed to offer a significant potential for innovative and safer designs. The technical objectives of UMRIDA can be summarized as follows:

Objective 1: Develop innovative methods for UQ and Robust Design (RD) to respond to the main challenges mentioned above.
Objective 2: Develop new methods for the large-scale introduction of the UQ methodologies in robust design methods, in order to produce designs incorporating the major uncertainties.
Objective 3: Apply the developed UQ and RD methods to complex systems of particular interest for the aeronautical industry. To achieve this objective, industry has pre-defined Industrial application Challenges (IC's) ***with prescribed uncertainties***, including multi-physics applications, demonstrating the industries strong needs for UQ and RD methods in their day-to-day work. This leads to a ***new generation of database***, including whenever possible, new experimental data with controlled uncertainties, on which the methods will be tested and *validated*, as well as **best practices** in the application of UQ and RD methods to industrial cases.
Objective 4: Advance UQ and RDM to the Technology Readiness Level 5–6 from a presently estimated TRL of 2–3. Figure 3 illustrates this in relation to the precursor project NODESIM-CFD. A quantifiable objective is set by the industrial partners as: **handling at least ten simultaneous uncertainties, in a turn-over time of no more than 10 h on a 100 core parallel processor**.

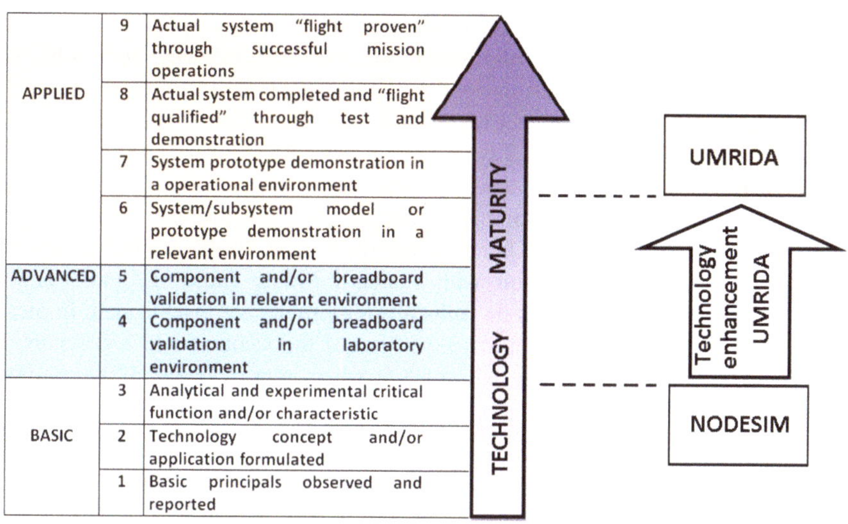

Fig. 3 Progress in Technology Readiness Level (TRL) during the UMRIDA project

Objective 5: Facilitate cooperation and dissemination of UQ and RDM awareness towards European industries, research establishments and universities and foster cooperation between different industries as airframe, turbo-engines, helicopters, and sea and ground transportation, including the European CleanSky project.

Reaching these objectives will enable industry and all partners involved:

- To strengthen the competitiveness of the European Aeronautical Industry, since UQ and RDM are not yet systematically used in industry (apart from some very few attempts on research level). It is expected that, at the outcome of the project, the innovative UQ and RD methodologies, to be developed in UMRIDA, will be ready to being applied at a wider scale in the aeronautical industry.
- To secure global leadership by promoting the novel UQ and RD methods in day-to-day industrial practice, beyond aeronautics, towards other transport areas, such as land and sea. The downstream objective of the UMRIDA project is to contribute to the integration of generalized risk analysis into all design practices whereby safety margins, and the associated risks are identified at each level of the product development.

Progress Beyond the State of the Art

In order to reach these objectives, a significant progress beyond the state of the art is needed. The activities of UMRIDA project partners can be grouped into three main fields:

I. **Advance methods for uncertainty quantification**

Three different approaches for uncertainty propagation are investigated. First, **method of moments (perturbation method) and adjoint-based methods**; the principle of these methods is the use of a Taylor series to expand an output quantity around its mean. The key element here is the evaluation of first- and second-order sensitivity derivatives in this Taylor series. Second, **Monte Carlo methods (multi-level Monte Carlo)** are studied by several partners. The basic idea is to perform a high number of deterministic simulations with sampled parameters and evaluate output quantities on different levels in order to reduce significantly the computation time in comparison with standard Monte Carlo methods. Third, non-intrusive **polynomial chaos or collocation methods** are investigated. In these methods, the principle resides in the expansion of the solution into a polynomial chaos or an interpolating polynomial, which forms an approximation of the solution around the uncertainties.

II. **Characterization of most influential uncertainties and dimension reduction**

A second approach to tackle many simultaneous uncertainties is to determine the most influential uncertainties in the problem and solve only for these, in order to reduce the dimensionality of the problem. This implies that the input uncertainties are well known, which represents a challenge in itself for many industrial problems.

Thus, a first line of action is the accurate **quantification of input uncertainties.** Output PDFs are dependent on the shape of input PDFs, and it is necessary to develop a methodology that allows to derive correct information for input PDFs from generally scarce experimental data or knowledge. A second line of action is the **identification of the most important uncertainties** in the problem description. A third line of action focuses on **surrogate models or reduced order models** for the description of aerodynamic or turbomachinery test cases.

III. **Advances in robust design and optimization methodologies**

The last year of the UMRIDA project will focus on Robust Design Optimization techniques. The objective of robust design is to produce **designs which are less sensitive to variations of conditions/parameters** due to uncertainties. In general, it must be noted that **design under uncertainties is a new field of research** and it is unclear so far how to enforce robustness, robust objective/constraint formulations and dependence on used optimizers within an industrial design environment. These issues are addressed within UMRIDA.

Research Consortium

The research consortium formed to address the above challenges consists of the following 21 partners, NUMECA International (Coordinator), DASSAULT Aviation, EASN-ITS, LEONARDO Aircraft, MAN Diesel & Turbo Schweiz, TURBOMECA, NPO-SATURN, ESTECO, ONERA, DLR, INRIA-Sophia, CIRA, CIMNE, CERFACS, TU Delft, Vrije Universiteit Brussel, Warsaw University of Technology, EPFL, Linkoping University, AIRBUS Group Innovation, TU Dresden. The logos are shown in Fig. 4.

Research Activities: UMRIDA Work Plan

One of the main objectives, namely the creation of a new generation of database, focused on UQ and RDM, is further detailed, as it is at the heart of the research work in UMRIDA. This database is in first priority constructed around the industrial challenges (IC), submitted by the industrial partners of the UMRIDA project. This is part of the top-down approach taken in this project, guided by the industrial requirements of the aeronautical industry towards safer and more optimal designs,

Fig. 4 UMRIDA consortium

taken into account large numbers of uncertainties. This new database, with prescribed uncertainties, will be the backbone of the two workshops, which are expected to become landmarks in the validation and evaluation of UQ and RD methods. It is intended to open up the workshops to worldwide participation with the aim of gathering the best worldwide experts and to confront the UMRIDA results and progress with related activities worldwide. This translates into the following work package structure.

WP1: General management and coordination

WP2: Improvement of methods for Uncertainty Quantification towards industrial readiness

Objectives: Extension of UQ methods able to handle the UQ challenges identified in Chapter "UMRIDA Test Case Database with Prescribed Uncertainties", paving the road towards a TRL 5–6, with a ***quantified objective to handle at least 10 simultaneous uncertainties within a turn-around time of the order of 10 h on 100 cores***.

- Task 2.1: UQ methods for efficient handling of large number of uncertainties
- Task 2.2: Development of efficient UQ methods for general geometrical uncertainties
- Task 2.3: Impact of numerical properties of CFD codes, numerical errors, including issues of shock discontinuities on assessment and validation of UQ methods
- Task 2.4: Methods for identification and quantification of input experimental uncertainties, including methods to define tolerances in input parameters to satisfy tolerance in output (inverse robust design).

WP3: Validation and evaluation of UQ methods for industrial test cases (industrial challenges)

Objectives: Develop the methodology of UQ towards industrial readiness, by applying and assessing the methods developed in WP2, in view of the quantitative objectives relevant to a TRL 5-6 objective.

- Task 3.1: Detailed specification of test cases, from basic to industrial challenge level, including multi-physics, with preset uncertainties, including the creation of new experimental data, with 'controlled' uncertainties, to establish an innovative database for UQ validations
- Task 3.2: Application of methods of WP2 to the selected test cases
- Task 3.3: Efficient UQ methods for multidisciplinary applications.

WP4: **Robust design methodologies and applications**

<u>**Objectives**</u>: Bring robust design methods to industrial readiness levels, covering large number of uncertainties, including geometrical uncertainties, with applications to Multidisciplinary Design Optimization (MDO).

- Task 4.1: Analysis of different methodologies for uncertainty incorporation in objective/constraint formulations of the optimization process
- Task 4.2: Innovative methods for robust design under uncertainties
- Task 4.3: Application of robust design methods to selected industrial relevant test cases, including general geometrical uncertainties.

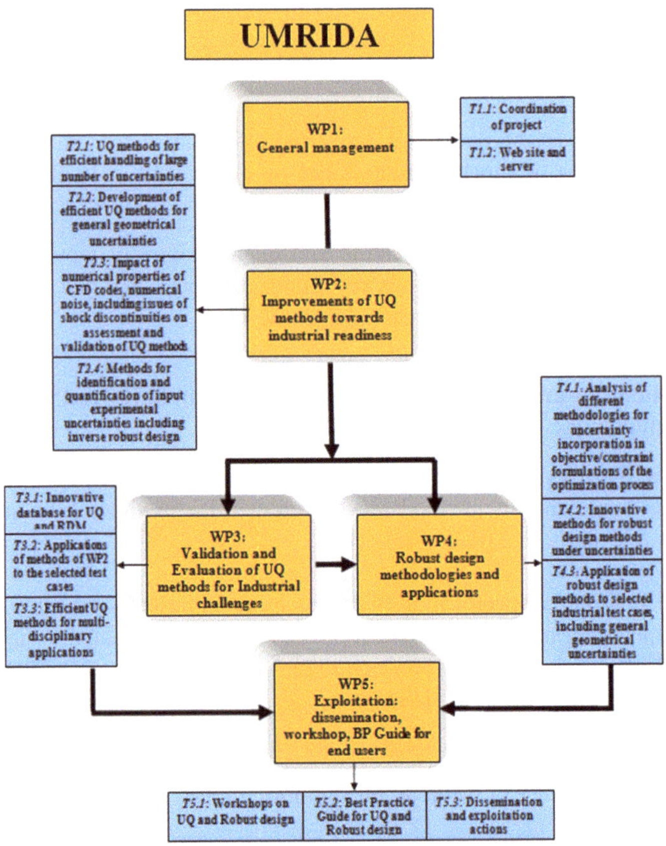

Fig. 5 UMRIDA work plan

WP5: Workshop, BP guideline for end-users, exploitation: dissemination,

Objectives: Set up two workshops for validation and evaluation of UQ and RD methods at the level of the IC test cases. Dissemination and exploitation of the UMRIDA methodologies.

- Task 5.1: Workshops on UQ and robust design at mid-term and at end of project
- Task 5.2: Best Practice Guide (BPG) for UQ and robust design methodologies
- Task 5.3: Dissemination and exploitation.

Figure 5 shows a diagram of the interactions between the different work packages and tasks.

References

1. Eldred, M.S., Swiler, L.P., Tang, G.: Mixed aleatory-epistemic uncertainty quantification with stochastic expansions and optimization-based interval estimation. Reliab. Eng. Syst. Saf. **96**(9), 1092–1113 (2011)
2. Green, L.L.: Lecture on Advanced Uncertainty Analysis. NASA/NIA Summer Design Institute on Uncertainty (2011)
3. Helton, J., Oberkampf, W. (eds.): Special issue on representations of epistemic uncertainty. Relia. Eng. Syst. Saf. **85**(1–3) (2004)
4. Oberkampf, W.L., Roy, C.J.: Verification and Validation in Scientific Computing. Cambridge University Press (2010)

UMRIDA Test Case Database with Prescribed Uncertainties

Sönke Klostermann

The UMRIDA Database for Uncertainty Quantification and Robust Design Methods comprises different test cases with prescribed uncertainties which have been compiled using a common way of describing the different test cases. Generally, the database describes the test case regarding the geometry, the mesh (if prescribed), and the flow conditions. Furthermore details for the considered uncertainties are given separating between geometrical uncertainties, operational uncertainties, and modeling uncertainties which may be caused by epistemic uncertainties, for example. It is differentiated between »Basic Challenges« (Table 1) and »Industrial Challenges« (Table 2) of increased complexity regarding the model complexity itself and the considered uncertainties. The industrial test cases have been submitted by the UMRIDA project partners.

The purpose of the generic methodology description is to be (partly) independent from the specific use case, allowing for a common view by all partners on the different cases. A common language of all partner's use case descriptions enables for easier collaboration and better means of comparability of different UQ and RDM methods.

In the section at hand we will briefly describe the basic and industrial challenges given in Tables 1 and 2. More detailed test case descriptions of specific properties will be given in the respective sections when necessary.

S. Klostermann (✉)
Airbus Group Innovations, Hein-Saß-Weg 22, 21129 Hamburg, Germany
e-mail: soenke.klostermann@airbus.com

C. Hirsch et al. (eds.), *Uncertainty Management for Robust Industrial Design in Aeronautics*, Notes on Numerical Fluid Mechanics and Multidisciplinary Design 140, https://doi.org/10.1007/978-3-319-77767-2_2

Table 1 Overview of basic challenges

Reference	Test case name	Provider
BC-01	NASA Rotor 37	NUMECA
BC-02	RAE 2822 airfoil	NUMECA
BC-03	DLR-F6	DLR
BC-05	Shock–boundary layer interactions	Stanford University (CIRA)

Table 2 Overview of industrial challenges

Reference	Test case name	Provider
IC-01	Helicopter engine combustor	Turbomeca
IC-02	3D DLLM wing	Airbus Group Innovations
IC-03	Falcon jet	Dassault Aviation
IC-04	NPO Saturn industrial fan blade	NPO Saturn
IC-05	NPO Saturn gas turbine HPT blade	NPO Saturn
IC-06	Acoustic liner	Alenia Aermacchi
IC-07	Industrial compressor stage	MAN
IC-08	Supersonic laminar flow business jet	CIRA/Dassault Aviation
IC-09	High-pressure compressor blade design	Rolls-Royce Deutschland

BC-01: NASA Rotor 37

The Rotor 37 test case was designed and initially tested as part of a research program involving four related axial-flow compressor stages. Then, the rotor was retested as an isolated component, which is the geometry described here and referred to as Rotor 37. Detailed description of the geometry, the experimental setup, and a series of simulations cross-plotting the predictions can be found in [1].

The geometry of the Rotor 37 is given schematically in Fig. 1. The mesh is of HOH-type and has shown a satisfying insensitivity to refinement with a final of 4.7 million mesh points. Further, a mesh with approximately 700,000 mesh points is available. Data available for the Rotor 37 were obtained at the design point value of an equivalent rotational speed of 17188.7 rpm (1800 rad/s). All data are available in [1].

The following geometrical uncertainties are of importance and are result of manufacturing and assembling tolerances:

- Tip clearance: Uncertainty in the tip clearance results from different sources, such as tolerances in the casing or temperature differences between casing and flow field (symmetric beta PDF, $a = 0.5 * M_{tip}$, $b = 1.02 * M_{tip}$, $M_{tip} = 0.356$ mm)

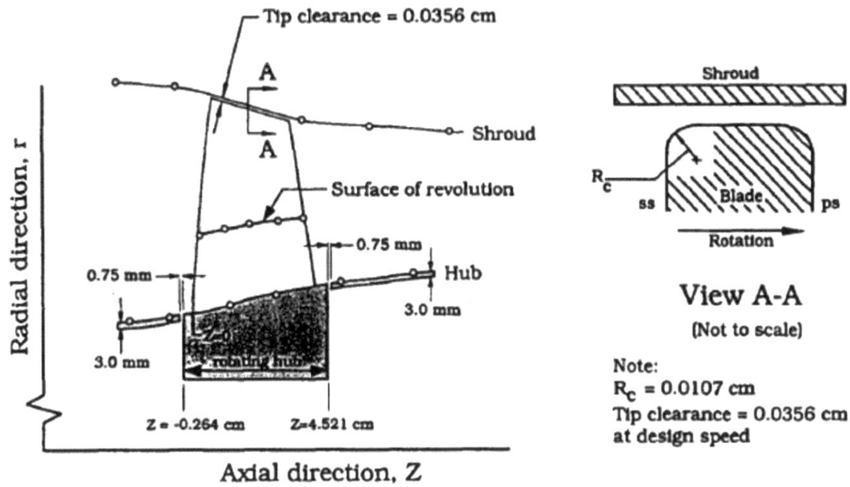

Fig. 1 Blade and flow path coordinates of BC-01 (from [1])

- Camber angles: (symmetric beta PDF, $\beta_1 = 5°$, $\beta_2 = -70°$, Min. value (a): 95% β_1, 95% β_2, Max. value (b): 105% β_1, 105% β_2)
- Leading and trailing edge radii: Deviations in leading and trailing edge radii result from manufacturing tolerances. The nominal values vary along spanwise direction (symmetric beta PDF, a = 0.90 of Nominal, b = 1.10 of Nominal)
- Non-uniformity of blade spacing: Non-uniformity in the circumferential blade spacing can be accounted for if more than one blade is simulated. Each inter-blade spacing should be independent from the others, such that the circumferential displacement of the next blade is not linked to the displacement of its neighbor (symmetric beta PDF, a = –1% of blade spacing, b = +1% of blade spacing)
- Blade surface roughness: The RMS blade surface roughness is given in [1] to lie between 0.5 and 1.25 μm (symmetric beta PDF).

The following operational uncertainties are of importance (values given in [1]):

- Inlet total pressure (symmetric beta PDF, a = 0.98, b = 1.02 of Nominal values)
- Static outlet pressure (symmetric beta PDF, a = 0.98, b = 1.02 of Nominal values)
- Radial distribution of the inlet flow angle (cubic polynomial law with normal distribution of random coefficients, Min. value −5%, Max. value +5%).

No epistemic uncertainties are considered.

BC-02: RAE 2822 Airfoil

The test case corresponds to cases 6 and 9 in 2. Figure 2 shows the airfoil geometry. The experimental data cover a range of conditions from fully subcritical flow to conditions where a comparatively strong shock wave exists in the flow above the upper surface of the airfoil. In total, 14 configurations have been measured, for which the free-stream Mach number, angle of attack, and Reynolds Number are given in [2].

The experimental flow conditions of these test cases must be corrected to eliminate wall interference, since the experiments are performed with wind tunnel walls. Several corrections are used for this test case, but no final convergence on the correction to apply can be identified in available studies on this test case.

The following geometrical uncertainties are of importance and are result of manufacturing and assembling tolerances:

- Thickness-to-chord ration: An uncertainty on the thickness-to-chord ratio is imposed. The proposed uncertainty PDF is of symmetric beta PDF shape. It can be varied for different control points (a = 0.97 Nominal, b = 1.03 Nominal)
- Camber line: The camber line is modified following a symmetric beta PDF at several control points (a = Nominal −0.01%, b = Nominal +0.01%).

The following operational uncertainties are of importance:

- Free-stream Mach number (symmetric beta PDF, a = 0.95, b = 1.05)
- Angle of attack (symmetric beta PDF, a = 0.98, b = 1.02)

No epistemic uncertainties are considered.

Fig. 2 RAE 2822 airfoil

BC-03: DLR F6

DLR-F6 wing-body configuration (DLR-F6 WB) is the geometry used for AIAA CFD Drag Prediction Workshop (DPW) 2 and 3 in 2003 and 2006, respectively, focusing on drag, lift, and pitching-moment predictions (Fig. 3). The design condition is that $M_\infty = 0.75$, $C_L = 0.5$, $Re = 5 * 10^6$, which was used for the single-point grid-sensitivity study on three grids (coarse, medium, and fine) for two kinds of configurations, DLR-F6 WB without and with fairing (FX2B).

In DPW-3, a test case by changing angle of attack (AoA), which is one of the operational uncertainties, in the range [−3.0°, −2.0°, −1.0°, −0.5°, 0.0°, 0.5°, 1.0°, 1.5°] was also discussed. In the fixed C_L design condition, AoA is in the range (−1.2°, 0.3°) for without fairing and (−0.9°, 0.4°) for with fairing, and CD is (261, 332 counts) for without fairing and (253, 293 counts), respectively, by twenty of data by the participants [3]. Here, 1 count means 0.0001. Note that the AoA in the design condition is in the range (−0.2°, 0.0°) by the DLR data [4].

The FX2B (DLR-F6 WB with fairing) has a fairing at the trailing edge (TE) of the wing–fuselage junction to suppress the flow separation. This point is one of the most sensitive points in grid densities for C_P in the configuration without fairing [3]. All necessary data for CFD calculations, which was used in DPW-3, are available in the DPW-3 official Web site [5].

In UMRIDA, the wing thickness and leading edge shape for both DLR-F6 WB with and without fairing are considered as the geometrical uncertainties. In UMRIDA, the Mach number, angle of attack (AoA), and Reynolds number are considered as the operational uncertainties. In DPW-3, some data were provided by the participants by changing AoA in the range [−3.0°, −2.0°, −1.0°, −0.5°, 0.0°, 0.5°, 1.0°, 1.5°] for both DLR-F6 WB with and without fairing. In DPW-2, constant C_L ($C_L = 0.5$) that drags rise in the range of M_∞ [0.50, 0.60, 0.70, 0.72, 0.74, 0.75, 0.76, 0.77] was studied for DLR-F6 WB without fairing [6, 7].

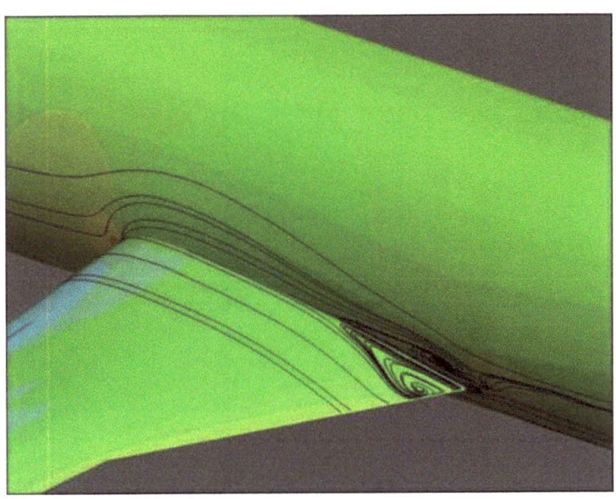

Fig. 3 View of wing-body configuration DLR-F6

BC-05: Shock–Boundary Layer Interactions

The objective of this test case is to study the shock–boundary layer interactions under free-stream and geometrical uncertainties of a 24° compression wedge in a Mach 2 flow. An experimental data set is provided that is well suited to be used for the validation of CFD codes which are intended to be used in design and analysis of systems with stochastic inputs.

The experimental and computational setups are aimed to validate and quantify of aleatory and epistemic uncertainties, determining the dominant contributions to changes in the streamwise location of the shock-crossing point. Aleatory uncertainties considered include geometric perturbations and variation of the inflow conditions. The test section configuration and geometry is described in full detail in [8–11], and it is sketched in Fig. 4.

A continuously operated Mach 2.05 wind tunnel is used in the experiment which is fed from a 2D converging/diverging nozzle followed by a constant-area development section of 45 × 47.5 mm. A contraction is produced by a 3-mm-long, 20°-angle compression wedge that spans the top wall of the duct and is responsible for generating the oblique shock that impinges and reflects at the bottom wall. Another constant-area section follows. The turbulent incoming boundary layers had an average thickness of 5.4 mm at a streamwise location 21 mm upstream of the foot of the wedge.

Small geometric perturbations in the form of a bump of height h_b ($<0.2\,\delta_0$) spanning the bottom wall are introduced at several streamwise locations, upstream of the foot of the wedge. The quantity of interest is chosen as the streamwise location of the shock-crossing point of the first bottom-wall STBLI measured at the z = 21 mm (near-center) plane.

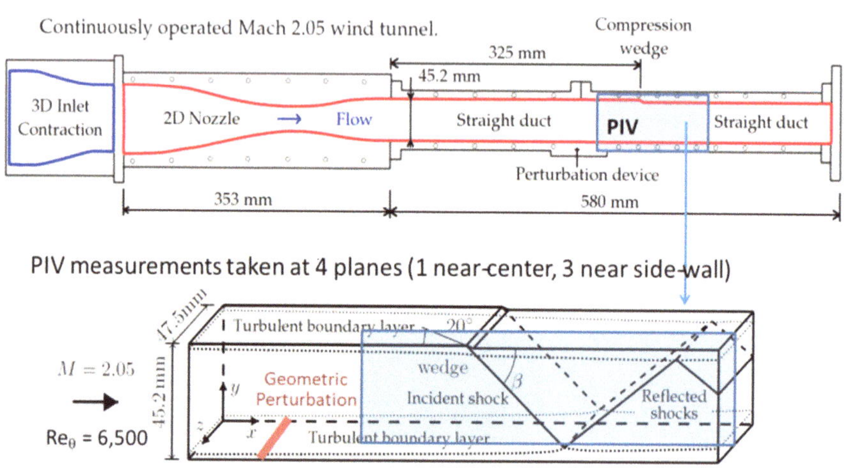

Fig. 4 Shock–boundary layer interaction experimental setup

Besides geometric perturbations, we also consider aleatory uncertainties in the inflow conditions in terms of the Mach number and the thickness of the turbulent boundary layers. The experimental uncertainty in the measurement of streamwise velocity was estimated as 5 m/s, which translates into Mach number variations of $\pm 1\%$. For the boundary layer thickness, we consider variations of $\pm 5\%$.

A possible approach to turbulent model uncertainties is described in [11] where the epistemic uncertainty is associated with the Boussinesq eddy viscosity hypothesis.

IC-01: Helicopter Engine Combustor

The test case concerns a database of pressure signal records taken in the combustor casing of a helicopter engine. Such a database comprises operating conditions for which thermo-acoustic instabilities appear. These instabilities are revealed by the high amplitude of pressure fluctuations. Their occurrence is promoted at given thermodynamics conditions such as pressure and temperature at the compressor exit, but is very dependent on the combustor design. The combustion chamber comprises a casing, breathed by the compressor, the combustor itself, which is perforated by primary holes, dilution holes and cooling films, and injectors.

The combustor mesh covers the whole combustion chamber domain to account for the effect of all components on combustion instabilities. This is particularly important since the occurrence of these phenomena is highly sensitive to fine geometrical features of the combustor. Several discretizations can be considered depending on the simulation tool to be used. For instance, a Helmholtz solver needs less than 1 million elements to give access to eigen-frequencies of the whole combustor, while a large Eddy simulation code requires several tens of millions of elements to resolve acoustics/flame interactions (Fig. 5). The considered operating

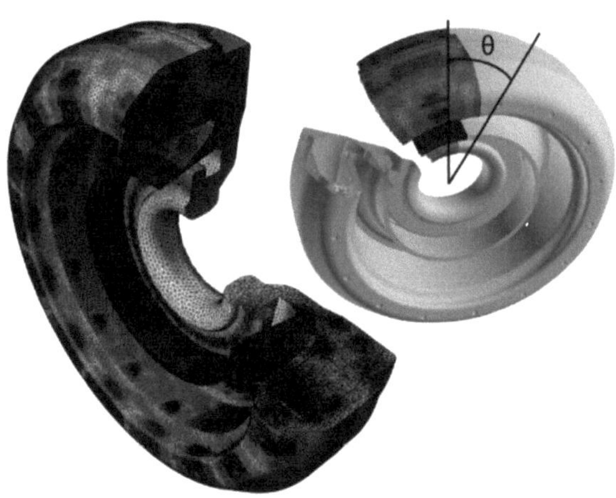

Fig. 5 LES mesh of the Turbomeca combustor

conditions are compressor outlet pressure ~8 bars and compressor outlet temperature ~500 K.

Uncertainties concern the primary and dilution holes' diameters and shapes. Such uncertainties are of the order of a few percent but can modify the flow distribution in the combustor. Other uncertainties are linked to the precise injection point position in the swirlier both azimuthally and longitudinally which scatter less than 1% of the reference value. Some of the combustor walls are also multiperforated where the diameter of the holes can scatter about 10% of the reference value. In this study, large uncertainties on thermodynamic conditions should be considered around the nominal point for which experimental data are available.

For the LES and the Helmholtz solvers, modeling uncertainties are mainly related to boundary conditions. Notably the impedance of upstream (compressor stages) and downstream (turbines stages) components are poorly characterized. The same problem arises concerning the multiperforated walls' impedance. Other modeling uncertainties also concern combustion processes inside the combustor.

IC-02: 3D DLLM Wing

For the robust optimization of a 3D wing, based on a differentiated lifting line method (DLLM), the objective is to achieve an optimal robust design of the wing regarding the wing's lift over drag ratio. For preliminary drafts of the aircraft configuration that is needed in early design phases, a full 3D simulation is too costly and—even more importantly—the detailed design of the aircraft needed for this approach normally does not yet exist.

The DLLM (based on Prandtl lifting line theory) is a one-dimensional simulation process that relies on information from two-dimensional information on airfoil level. This model proposes an improvement with regard to a typical nonlinear lifting line model. In particular, it proposes not to start the resolution from a priori known lift and circulation distribution, but to deduce it from the series of airfoils composing the wing. Thus, the DLLM relies on a spanwise geometric description of the wing where the plan form geometry is specified by several parameters (Fig. 6).

The 3D wing is defined by the shape of six airfoils and a smooth interpolation procedure between the different airfoils. The shape of each of the airfoils is defined by 11 PARSEC parameters as proposed in [12]. The optimization of the lift over drag ratio is constrained by a minimum value of the lift (accounting for aircraft mass) and a minimum volume of the wing (accounting for required mission fuel mass).

We consider that the design variables may exhibit some kind of uncertainty which represents manufacturing tolerances or assembly effects, for example. Taking into account uncertainties of design variables—we restrict the design space to the

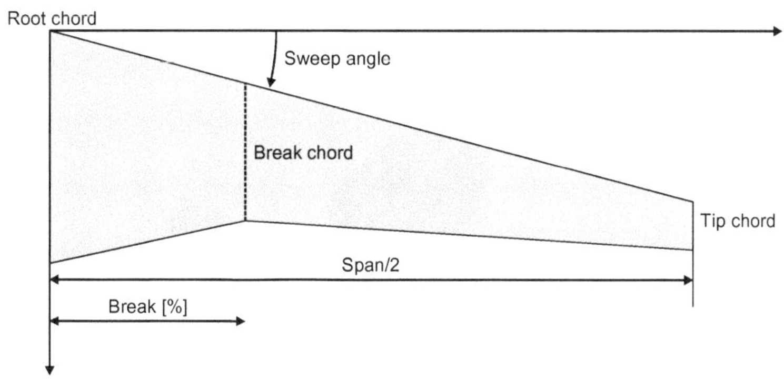

Fig. 6 Illustration of wing parameters for definition of simplified wing geometry

three chord lengths and the break value along the wing's longitudinal axis—the uncertainties are modeled as noise vector $\xi = (\xi_1, \xi_2, \xi_3, \xi_4)$ given by a joint probability density of normally distributed ξ_1, ξ_2, ξ_3, ξ_4 for a set of standard deviations $\sigma = [0.1, \ldots, 0.5]$ to illustrate the effect of uncertainty on the location of the optimal design point. We can assume that in addition to the uncertainties of the design variables the operating conditions exhibit some form of uncertainty as well. For example, the environmental conditions underlie a natural form of physical scatter (temperature, pressure, humidity) or the operating conditions are controlled by some kind of control loop that relies on measured data. For this reason, we impose some normally distributed noise on the operating conditions.

IC-03: Falcon Jet

Industrial challenge IC-03 represents an industrial relevant external flow test case of a Falcon business jet (Fig. 7) exhibiting several forms of uncertainty for the geometry, the operating conditions, and the modeling. Table 3 summarizes the test case parameters.

Geometric uncertainties are given by the spanwise twist angle distribution of the wing (x/c = 0.25; rotation/ control section plane) on eight control sections modeled by bounded, asymmetric beta PDF distributions (Holland approach) with a delta twist angle minimum of –0.5° and a delta twist angle maximum of 0.2° with the most likely value for the delta twist angle of 0.01°. The free-stream Mach number and the angle of attack are modeled by the same type of distribution with AoA Min of 1.97°, AoA Max of 2.1° and a free-stream Mach number minimum of 0.795 and 0.807 for the maximum value. Turbulence is modeled by Spalart–Allmaras, k-ε SST two layers, k-ω SST Menter.

Fig. 7 Falcon business-jet geometry

Table 3 General test case description of the Falcon jet test case

Mach number	0.8
Angle of attack	2°
Altitude	40,000 ft
Reynolds number	14.512 Million
Flow regime	Fully turbulent
Wall treatment	Adiabatic no-slip wall
Ref. density	0.30132 kg/m^3
Ref. pressure	18820.15 Pa
Ref. temperature	216.65 K
Ref. velocity	236.57 m/s
Ref. length (mean aerodynamic chord)	2.888 m
Ref. area	49 m^2
Moment reference point	x = 9.0355 m, y = 0 m, z = 0 m

IC-04: NPO Saturn Industrial Fan Blade

A modern civil aircraft fan blade is considered as an investigation object. This blade must provide a high level of the aerodynamic efficiency (adiabatic coefficient of efficiency) and necessary strength properties. Another important characteristic is the fan blade flutter phenomenon sensitivity.

A fan blade is a complicated object, and obviously, it is subjected to geometrical uncertainties from manufacturing tolerances and other production deviations. In spite of all uncertainties, the fan blade should provide stable aerodynamic efficiency and strength properties. That is why it is considered to solve multidimensional and multidisciplinary optimization task (aerodynamic and strength) in robust statement under uncertainties.

In aerodynamic calculations only the blade profile is modeled. In the strength calculations, full blade (with foot) is considered. The blade foot has no geometrical uncertainties (nominal geometry). Total pressure and total temperature in stationary frame were used as inlet boundary conditions. The flow direction is set by dimensionless angle components. In the proposed test case, geometrical uncertainties from fan blade manufacture tolerances and deviations are considered (Fig. 8). Probability density function was obtained as a result of statistical operation upon the results of blades' coordinate measurements.

The main uncertainties considered as it is shown in Fig. 8 are blade thicknesses in different profile locations and maximal blade thickness. Profile angle of incidence is also taken into consideration. In such a case, the total number of the considered uncertainties is 54. Most of experimental stochastic parameters can be described by means of Gaussian PDF distribution law with acceptable accuracy level. Some parameters will have to be described by means of lognormal distribution and beta distribution.

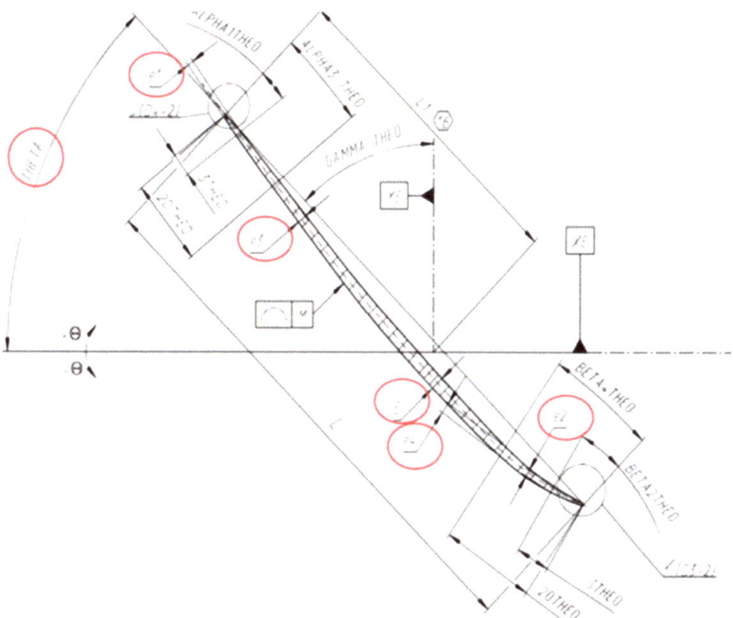

Fig. 8 Main geometrical uncertainties considered in blade section

In this test case, operational uncertainties are not considered. Aerodynamic calculation will have to be carried out in one point on operating line. Epistemic uncertainties are not considered. Standard approach for fan blades aerodynamic computations must be used (RANS modeling with using k-ε turbulence model).

IC-05: NPO Saturn Gas Turbine HPT Blade

A high-pressure turbine blade from an industrial gas turbine engine used for electric power generation is considered as test case. The considered blade must provide necessary level of aerodynamic efficiency and necessary temperature distribution of the blade. The HPT blade at the real gas turbine engines shows strong influence of the operational uncertainties on blade temperature distribution and aerodynamic efficiency. Deviations in operational parameters result in oscillations in temperature and pressure distributions at the inlet of the blade hot gas channel and in cooling air pressure under the blade. At the same time, deviations of the tip gap values have a strong influence on the aerodynamic efficiency and cooling effectiveness of the blade tip. The turbine blade should provide stable cooling effectiveness of the leading edge and high aerodynamic effectiveness; therefore, it is considered to solve multidimensional optimization task (aerodynamic and heat transfer) (Fig. 9).

Total pressure and total temperature radial distribution in stationary frame were used as hot gas path inlet boundary conditions. The blade tip gap is the only considered geometrical uncertainty. Based on several experimental investigations on prototype engines, it is considered to use theoretical Gaussian distribution for tip gap PDF with a mean value of 0.5 mm and a standard deviation of 0.1 mm.

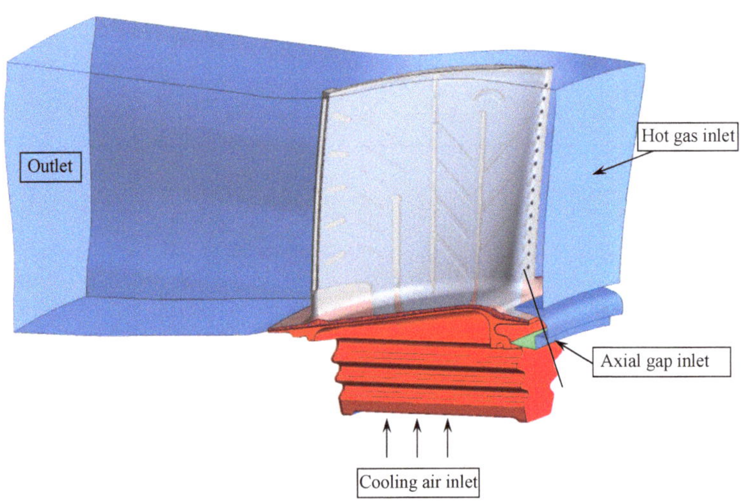

Fig. 9 Industrial HPT blade computational model

Table 4 Operational uncertainties of IC-05 NPO Saturn gas turbine HPT blade

Uncertainty	Mean	Standard deviation	Distribution
ΔT^* (K)	8.244	21.43	Normal
ΔP^* (bar)	−0.1165	0.236	Normal
ΔP^*_{rel} (bar)	−0.1165	0.236	Normal

Hot gas total pressure and total temperature mean values T^*, P^* in the channel inlet and cooling air total pressure under the blade in relative coordinate frame P^*_{rel} are operational uncertainties. Hot gas temperature and pressure inlet values in the channel were obtained from 90 industrial gas turbine engines experimental investigations and modeled by normal distributions (Table 4). Epistemic uncertainties are not considered. Standard and verified NPO Saturn approach for turbine blades CHT computations was used (RANS modeling with SST turbulence model).

IC-06: Acoustic Liner

Test case IC-06 deals with the optimal design of acoustic panels installed in typical business-jet aero engine intakes. A dedicated acoustic panel construction will be selected for multiobjective optimization. The objective functions are represented by noise attenuations computed at two certification flight conditions in the far field surrounding the nacelle. Acoustic requirements are prescribed by customers and aviation agencies for the three typical flight conditions approach, side-line (or takeoff) and flyover (or cutback).

As these requirements can be in contrast to each other and an acoustic liner providing the best attenuation in one flight condition not necessarily performs well in the other conditions, hence liners shall be designed through a multiobjective optimization procedure. Specifically, a number of samples will be manufactured and tested to detect the steady (or DC) flow resistance for a set of given impinging air velocities by means of the AAEM DC-Flow rig (Fig. 10).

Geometrical parameters tolerances considered are influenced by the following main factors: Honeycomb height (h), Facing sheet thickness (t_{fs}), Facing sheet effective open area (*POAeff*) and Holes diameter (d). Flight conditions uncertainties are not being considered. Proprietary impedance model is not being investigated as a source of (epistemic) uncertainties.

IC-07: Industrial Compressor Stage

The test case comprises the static components of industrial, radial compressor stages as the main focus in radial compressor development has been put on the impellers over the last years. Static parts have been less addressed. Therefore it is of

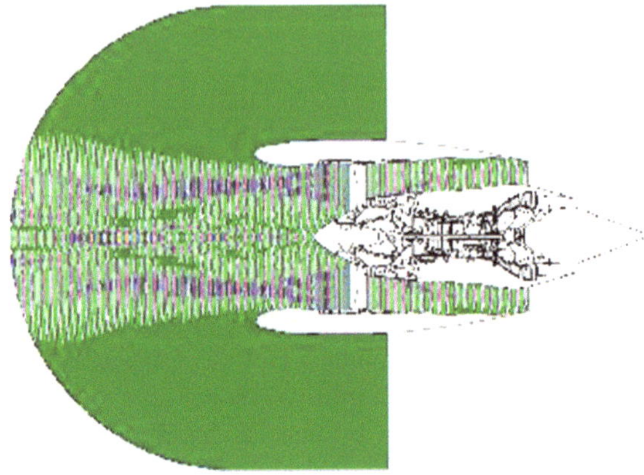

Fig. 10 Illustration of IC-06 acoustic liner

Fig. 11 IC-07: Industrial compressor stage

interest to analyze these parts in more detail, especially since less data are available and uncertainties have been less addressed (Fig. 11).

For the flow conditions gases with real gas properties will be used for operating conditions with high Reynolds numbers due to the elevated operating pressures of industrial compressor stages. The test case exhibits a high operating range that means a high incidence variation may occur. The design flow for this application is in the subsonic range.

For the consideration of uncertainties a special emphasis is laid on the entire product life cycle of the industrial compressor due to the typically harsh operating environment. The uncertainties take into account design adaptations, manufacturing tolerances, assembly tolerances, operating conditions and service requirements. Experimental reference data are available for validation.

IC-08: Supersonic Laminar Flow Business Jet

IC-08 deals with the design of a swept wing with a large extension of laminar flow in working conditions for a supersonic laminar flow design of a business-jet class aircraft (Fig. 12). The reference configuration is the optimized wing-body shape produced by CIRA within the SUPERTRAC EU project. This shape, optimized for natural laminar flow, used as baseline a configuration produced by Dassault Aviation within the framework of the Supersonic Business Jet project, and made available within SUPERTRAC project. Wing section airfoils and twist angle were optimized in order to maximize the laminar flow region while monitoring and controlling the pressure (vortex and wave) drag. A redesign of the wing using robust or reliability based optimization tools is proposed for the UMRIDA project.

The flow conditions are those defined by the main cruise design point with a Mach number of 1.6, a Reynolds number of $51.8 * 10^6$, reference length of 6.27 m, Angle of Attack of 3.65° and a lift coefficient of 0.182. The detailed description of the original optimization problem is reported in Reference [13].

Natural laminar flow is mostly sensitive to the shape of the leading edge region defined by the radius of the leading edge and the airfoil section thickness at ten percent of the chord. The nominal range of variation for these parameters is 15% of the radius of the leading edge of the initial configuration and 10% of the thickness in the assigned position. A uniform probability distribution should be considered for both parameters. However, an inverse approach may be considered where these uncertain parameters are regarded as unknowns to be determined within the design process.

Operational uncertainties are the classical ones related to Mach number and lift coefficient. Mach and C_L are modeled as four parameter beta distributions (Table 5). A significant uncertainty in the determination of transition location is

Fig. 12 Supersonic laminar flow design of a business-jet class aircraft

Table 5 Operational uncertainties of IC-04 supersonic laminar flow business jet

Parameter	α	β	a	b
Mach	4	4	1.55	1.65
C_L	2.5	2.5	0.180	0.184

inherent to the methods for numerical transition prediction and in particular to the e^N method. A robust design approach should take into account this epistemic uncertainty source. It will be modeled by a uniform distribution in the interval [16, 20].

IC-09: High-Pressure Compressor Blade Design

IC-09 is the aerodynamic axial high-pressure compressor rotor blade design in the presence of measured and identified manufacturing uncertainties. One important step for aleatory uncertainty quantification is to analyze and quantify the variability of input data for a physical system such as an aero engine. This test case covers a one and a half stage axial high-pressure compressor with one rotor blade and two adjacent stator rows as physical system (Fig. 13) and a probabilistic model which reflects real manufacturing uncertainties for the transonic rotor blade.

For one operating point the aerodynamic boundary conditions are part of the test case. Upstream the first stator the inlet boundary conditions for the total pressure, the total temperature and the absolute flow angle in radial direction are defined as well as the radial static pressure distribution downstream the second stator at the outlet of the computational domain.

The test case concentrates on realistic manufacturing tolerances, i.e. geometrical uncertainties, thus no operational, modeling or other uncertainties are defined. For a parametric model according to [14] probability density functions for each uncertain variable and a positive definite correlation matrix are defined (Fig. 14). The

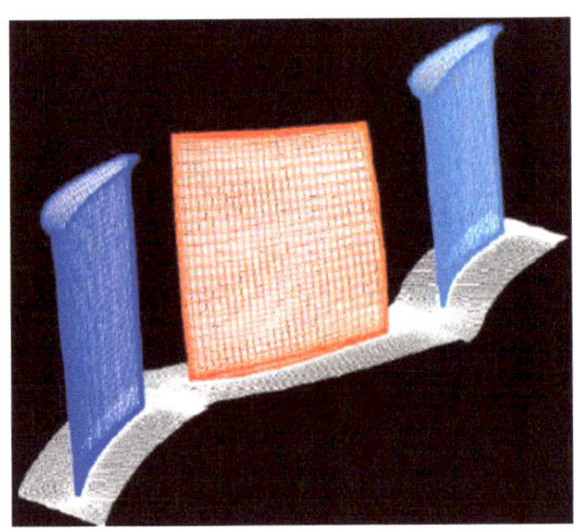

Fig. 13 One and a half stage axial high-pressure compressor

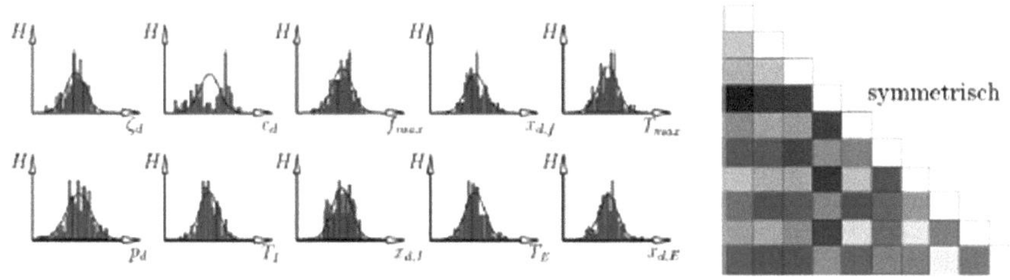

Fig. 14 Qualitative illustration of ten uncertainties with correlation structure [16]

probabilistic model has been derived from a sufficient number of optical surface measurements of manufactured rotor blades. The resulting sets of data have been reduced, discussed and statistically quantified [15].

In total the uncertain design space consists of 14 parameters. For the rotor reference geometry the associated values for each uncertain parameter are known. Based on the probabilistic model random deviations can be added to the mean/reference values. These deviations as a consequence lead to a disturbed rotor geometry and hence a varying aerodynamic performance.

References

1. Dunham, J.: CFD Validation for Propulsion System Components, AGARD-AR-355 (1998)
2. Various Authors: Experimental data base for computer program assessment. In: Report of the Fluid Dynamics Panel Working Group, AGARD-AR-138 (1979). ISBN 92-835-1323-1
3. Vassberg, J.C., Tinoco, E.N., Mani, M., Brodersen, O.P., Eisfeld, B., Wahls, R.A., Morrison, J.H., Zickuhr, T., Laflin, K.R., Mavriplis, D.J.: Abridged summary of the third AIAA computational fluid dynamics drag prediction workshop. J. Aircr. **45**(3), 781–798 (2008)
4. Brodersen, O., Eisfeld, B., Raddatz, J, Frohnapfel, P.: DLR results from the third AIAA computational fluid dynamics drag prediction worksop. J. Aircr. **45**(3), 823–836 (2008)
5. 3rd AIAA CFD Drag Prediction Workshop. http://aaac.larc.nasa.gov/tsab/cfdlarc/aiaa-dpw/Workshop3/workshop3.html
6. Laflin, K.R., Klausmeyer, S.M., Zickuhr, T., Vassberg, J.C., Wahls, R.A., Morrison J.H., Brodersen, O.P., Rakowitz, M.E., Tinoco, E.N., Godard, J.: Data summary from second AIAA computational fluid dynamics drag prediction workshop. J. Aircr. **42**(5), 1165–1178 (2005)
7. Brodersen, O., Rakowitz, M., Amant, S., Larrieu, P., Destarac, D., Sutcliffe, M.: Airbus, ONERA, and DLR results from the 2nd AIAA drag prediction workshop. J. Aircr. **42**(4), 932–940 (2005)
8. Campo, L.M.: Effects of Shock Strength, Confinement, and Geometric Perturbations on Shock Boundary Layer Interactions. Stanford University (2014)
9. Campo, L., Helmer, D., Eaton, J.: Validation experiment for shock boundary layer interactions: sensitivity to upstream geometric perturbations. In: 53rd AIAA/ASME/ASCE/AHS/ASC Structures, Structural Dynamics and Materials Conference. American Institute of Aeronautics and Astronautics (2012). https://doi.org/10.2514/6.2012-1440

10. Helmer, D., Campo, L.M., Eaton, J.K.: Three-dimensional features of a Mach 2.1 shock/boundary layer interaction. Exp. Fluids **53**(5), 1347–1368 (2012). http://dx.doi.org/10.1007/s00348-012-1363-8
11. Bermejo-Moreno, I., Larsson, J., Campo, L., Bodart, J., Emory, M., Palacios, F., Helmer, D., Iaccarino, G., Eaton, J.: Multi-fidelity numerical simulations of shock/turbulent-boundary-layer interaction in a duct with uncertainty quantification. In: Center for Turbulence Research—Annual Research Briefs 2012 (2012)
12. Sobieczky, H.: Parametric airfoils and wing. In: Notes on Numerical Fluid Mechanics (1998)
13. Iuliano, E., Quagliarella, D., Donelli, R.S., Salah El Din, I., Arnal, D.: Design of a supersonic natural laminar flow wing-body. J. Aircr. **48**, 1147–1162 (2011). https://doi.org/10.2514/1.C031039
14. Lange, A., Vogeler, K., Gümmer, V., Schrapp, H., Clemen, C.: Introduction of a parameter based compressor blade model for considering measured geometry uncertainties in numerical simulation. In: Proceedings of ASME Turbo Expo, GT2009-59937, Orlando (2009)
15. Lange, A., Voigt, M., Vogeler, K., Schrapp, H., Johann, E., Gümmer, V.: Probabilistic CFD simulation of a high-pressure compressor stage taking manufacturing variability into account. In: Proceedings of ASME Turbo Expo, GT2010-22484, Glasgow (2010)
16. Flassig, P.: Unterstützende Optimierungsstrategien zur robusten aerodynamischen Verdichterschaufelauslegung. Dissertation, BTU Cottbus, Aachen: Shaker (2011)

Part II
Uncertainty Quantification (UQ) and Efficient Handling of a Large Number of Uncertainties

Uncertainties in Compressor and Aircraft Design

Dirk Büche, Sönke Klostermann, G. Rogé and X. Loyatho

Introduction

The tremendous increase in computational power over the last decades allows for complex numerical simulations of internal and external flows ranging from single components such as compressor blades up to entire airplanes and aircraft engines [1]. Furthermore, it allows increased modelling complexity: early 2D Euler solvers have been replaced by Reynolds-averaged Navier–Stokes (3D RANS) simulations. Nowadays dynamic simulations such as unsteady RANS and detached eddy simulation (DES) become more and more feasible. Besides the higher modelling complexity, the computational power in combination with automated execution of simulations (e.g. by scripting and queuing systems) allows for variation analysis as well as automated design optimization.

Nevertheless, simulations will not be able to capture real-life machine behaviour precisely as insufficient information is typically available. There is a lack of information about the actual geometrical scattering of the manufactured and assembled parts. Furthermore, limited knowledge exists about operation, for example about operating conditions and machine degradation by fouling, wear and corrosion. Finally, assumptions and simplifications are required for the computational modelling of non-resolvable effects such as unsteady flow or small turbulent length scales.

D. Büche (✉)
MAN Diesel & Turbo Schweiz AG, Hardstr. 319, 8005 Zurich, Switzerland
e-mail: dirk.bueche@man.eu

S. Klostermann
Airbus Group Innovations, Hein-Saß-Weg 22, 21129 Hamburg, Germany

G. Rogé · X. Loyatho
Dassault Aviation, 78 Quai Marcel Dassault, 92552 Saint-Cloud, France

C. Hirsch et al. (eds.), *Uncertainty Management for Robust Industrial Design in Aeronautics*, Notes on Numerical Fluid Mechanics and Multidisciplinary Design 140, https://doi.org/10.1007/978-3-319-77767-2_3

Most often simulations are deterministic using the nominal values for all inputs, and by doing so, they ignore uncertainties. In contrast, the real performance of a machine is non-deterministic and could be best described by probability distributions reflecting the scattering in geometry, operation and modelling uncertainties.

Several actions have been started to reduce this gap of uncertainty by capturing information about production scattering as well as operational data and to extend the simulation software to non-deterministic frameworks. For example, Garzon [2] showed that the mean loss coefficient of an airfoil in a non-deterministic simulation can be several percent higher compared to the deterministic simulation with the nominal geometry.

This chapter focuses on how geometrical and operational uncertainties can be captured and modelled. These models serve as input for the uncertainty quantification and robust design optimization in the following chapters of this book. Due to the different manufacturing techniques, geometrical uncertainties in massive turbomachinery parts of industrial compressors differ to aircraft's sheet metal structures. Uncertainties in the turbulence modelling and grid generation are handled in chapters "Estimation of Model Error Using Bayesian Model-Scenario Averaging with *Maximum a Posteriori*-Estimates and Numerical Uncertainties Estimation and Mitigation by Mesh Adaption".

Geometrical Uncertainties in Compressors

The focus is put on geometrical uncertainties that are relevant for the aerodynamic performance of a turbomachine. The main interest lies in the main flow path which comprises elements such as compressor blades, radial impellers, stator vanes, bends and hub/shroud contours.

Geometrical uncertainties exist for the entire design and manufacturing chain beginning with geometrical parameterization reaching over to the individual machining steps and to final assembly activities of the turbomachine. Preparations for the production of the compressor blades or impellers typically require conversion of a parametrized geometrical model into a CAD system and further to CAM data. This inevitably leads to some deviations due to different surface representation models such as splines, Bézier curves, NURBS, multiple circular arcs or STL, just to name a few. For axial compressors and turbines, the parameterization may consist of several profiles along a stacking line. The stacked profiles are converted into 3D surfaces within the CAD environment. Smoothing is often applied in order to minimize geometric oscillations. The conversion of the CAD data into the CAM environment may require further changes in the geometric representation. In the following, we exclude the effects of these conversions and assume that there are adequate tools to analyse and precisely control deviations originating from the second versions.

The manufacturing as such typically requires a sequence of different steps, e.g. casting, forging, welding, heat treatments, bending and machining. All steps in a manufacturing sequence are relevant for the final geometrical uncertainties; e.g., residual stresses may be present after forging, heat treatments or welding. These may lead to geometric deformations upon machining.

Often, a variety of alternative manufacturing techniques can be applied, which differ in cost and geometrical precision. The goal is to find an optimal trade-off between cost and scattering. Only sensitive surface areas should be manufactured with high precision, as the additional effort obviously has consequences for the overall machining time and cost. Finding the required precision can be formulated as optimization task as described in [3].

Classification by Size

DIN 4760 [4] defines six different orders of surface deviations. The first four orders are relevant for aerodynamic performance, while the fifth and sixth order relate to the microstructure of the material surface and are aerodynamically irrelevant. These orders differ in the magnitude of the geometric deviations:

The first-order deviations affect the characteristic dimensions of the part. For example, the total length, the leading edge thickness or an angle may deviate from the nominal value as illustrated in Fig. 1. The deviations may be caused by the inherent imprecision of the milling machine, an unprecise clamping or by wear of the milling tool.

For the second-order deviations, only a small fraction of the surface is analysed. There, the surface may be wavy; i.e., it may contain periodical surface structures with very high aspect ratios. The aspect ratio between wavelength in surface direction and perpendicular deviations is about 100:1–1000:1. The causes typically are attributed to vibrations, to part deflection under machining forces or to shape deviations of the milling tool itself.

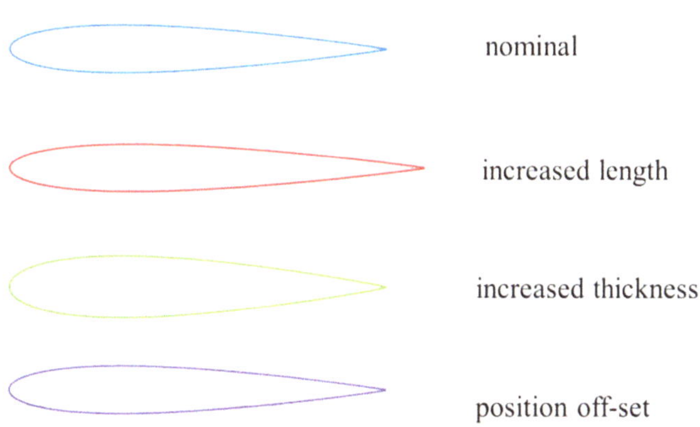

Fig. 1 First-order geometrical uncertainties in shape and position

The third-order deviations affect even smaller portions of the part surface. They may be periodic or non-periodic deviations and are typically machining scores. The milling scores depend on the tool-feeding motion. The ratio between wavelength and deviation depths lies in the range of about 100:1–5:1.

The surface roughness is an example of the fourth-order deviation and is in the range of a single micrometre.

On manufacturing drawings, tolerance intervals are often specified as illustrated in Fig. 2. These intervals limit the maximum deviation. Geometrical deviations can be classified by their actual shape. Offsets shift the entire shape and are typically first-order deviations. Waviness is a typically second- or third-order deviation. Another example is a kink or step. Kinks may be present at the interface of two mating parts that differ with regard to their surface slopes or when surfaces are machined from both ends. Steps may result from mating of parts at their interfaces. Deviations of this kind may also be attributed to interrupted milling or turning processes, e.g. after an exchange of the cutting tool or when parts are machined from two directions.

Quantifying Geometrical Uncertainties

In the following sections, examples are given for how to quantify the first four orders of shape deviations defined in the previous chapter. Different measurement devices may be required.

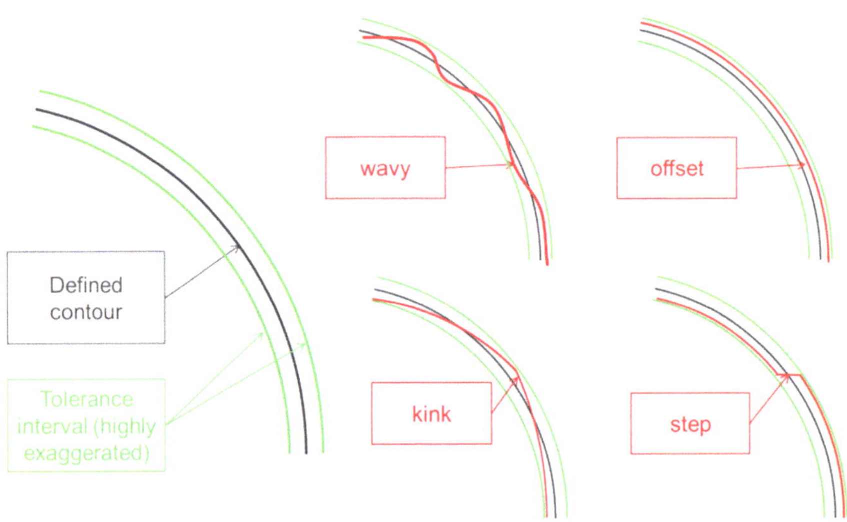

Fig. 2 Possible shapes of geometrical deviations

First-Order Shape Deviations
Maximum values for the first-order shape deviations are defined as geometrical tolerances on the manufacturing drawings. These tolerances are either explicitly given for a certain measure or can be given as general tolerances that are applicable for all measures that are not specifically tolerated. A well-established general tolerance is the standard ISO 2768 [5]. Part I of this standard defines tolerances for linear and angular dimensions. Part 2 adds shape and position tolerances.

ISO 2768 specifies the four different tolerance classes: *very coarse*, *coarse*, *medium* and *fine*. These classes reflect typical workshop capabilities. For class *medium*, Fig. 3 shows the tolerances for linear dimensions as a function of the actual measured length. As expected, the absolute tolerance increases in steps as the part grows in size. The relative tolerance however decreases substantially, as larger dimensions can be manufactured with higher relative precision compared to small ones.

Assuming that an aerodynamic profile has a length of 100 mm and trailing edge thickness of 1 mm, the absolute tolerance is 0.3 mm and 0.1 mm, respectively. This means that the relative tolerance for the length is 0.3%, whereas the tolerance for the trailing edge thickness is 10%. This is an important fact to be carefully considered when performing scale-ups or scale-downs of machines, especially when small-sized prototypes are manufactured and tested and scaling to the final product size is performed afterwards.

Figure 4 shows an example for a surface measurement of a compressor blade. The deviation of the measured profile from the nominal contour is almost constant along the measured surface, illustrating that this deviation represents the first-order deviation. Neither waviness nor bumps are present in this example. The deviation could result from a limited positioning of the part before milling. Clamping forces that cause part deflection or deflections caused by milling forces could also be

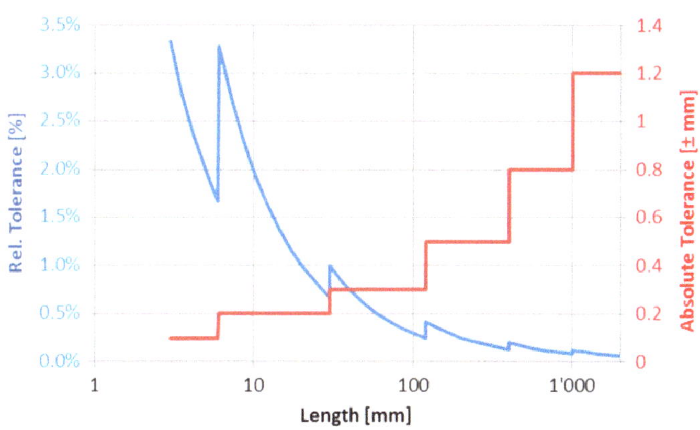

Fig. 3 Relative and absolute tolerances for linear measures from DIN ISO 2768 part I

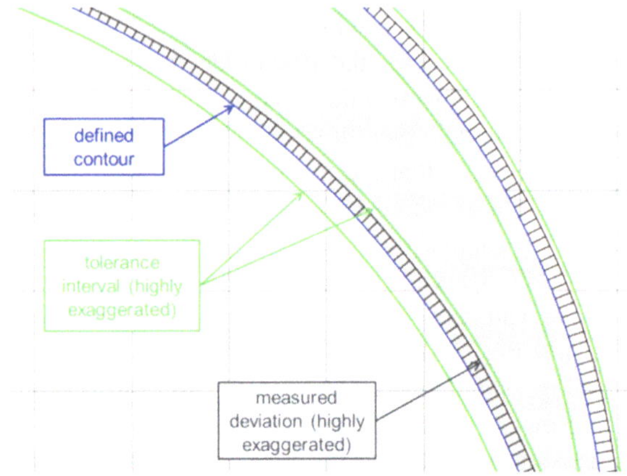

Fig. 4 Measured first-order deviation (black) between a stator part (blue) and given tolerance interval (green). The deviation and tolerance interval are highly exaggerated

considered as being causal to this deviation. The deviation is highly exaggerated in the graphic representation as the actual one could hardly be seen by eye.

As a general rule of thumb, first-order shape deviations are typically in the order of 0.1–1 mm.

Systematic measurements of actual compressor blades were reported by Garzon [2] and Schmidt et al. [6] for 150 and 400 blades, respectively. Today, capturing 3D geometric measurement data becomes more cost and time efficient as it can be done in an automated manner using tools such as optical scanners which generate complete surface information in one measurement.

Extracting statistical information from such measurements, e.g. probability distributions, is only appropriate if a sound statistical basis exists; i.e., a sufficiently high number of parts need to be manufactured with the same or similar manufacturing tools before being measured and analysed. If only a low number of parts are manufactured and even different suppliers are assigned, then very limited statistical information can be reasonably extracted.

Second- and Third-Order Shape Deviations

Since the first-order shape deviations are typically larger than 0.1 mm, comparatively simple tools such as sliding callipers can be used. The second- and third-order shape deviations are smaller, typically in the order of 0.001–0.1 mm. They require special measurement devices to be assessable.

Figure 5 shows the result of a 3D measuring device that measured points along a line every 0.01 mm. The waviness of the surface is difficult to see with the naked eye as the wavelength is about 0.6 mm and the amplitude of the deviation is only in the order 0.005 mm. Therefore, the surface normal component is magnified in the figure by a factor of 16. The waves result from the feed rate of the milling tool. Due to the ratio of amplitude to wavelength, the measured property is located somewhere between the second- and third-order shape deviations.

Fig. 5 Exaggerated actual surface measurement with high resolution for the second- and third-order shape deviations. Only by exaggerating the deviation in the normal direction of the surface, these deviations from machining turn visible

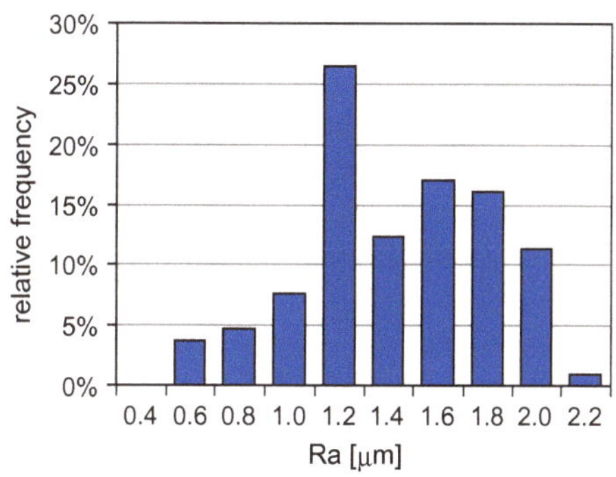

Fig. 6 Histogram generated from surface roughness measurements on a large set of radial compressor impellers. Taken from Büche et al. [7]

Fourth-Order Shape Deviations The fourth-order deviation is surface roughness and is typically expressed in Ra or Rz. Definitions are given in DIN EN ISO 4287. The surface roughness is in the range of 0.1–10 μm, again one order of magnitude smaller than the third-order deviations.

Büche et al. [7] describe the measurement of the surface roughness for a large set of radial compressor impellers. The measured roughness values are shown in Fig. 6 with the aid of a histogram. The difficulty is to transfer these measured technical roughness values into equivalent sand roughness values, which is typically used for fluid mechanical calculations.

Geometrical Uncertainties from Compressor Operation

By experience, machine operation causes surface degradation due to corrosion, erosion, particle impact and fouling. This increases surface roughness as well as leakage flows in labyrinth seals. The extent can be quantified during service inspections. These deviations are often larger than the acceptable manufacturing tolerances.

Reducing the effect of this kind of uncertainties during operation is possible, but it comes at an additional cost: parts can be made of higher-grade materials which are more resistant against wear. Furthermore, in some applications better filter devices can be implemented with the effect that fouling and erosion are reduced. Finally, shorter inspection and replacement intervals as well as frequent cleaning can reduce the degradation level further.

Integration of Geometrical Uncertainties into Uncertainty Quantification

The number of possible geometrical shape deviations is unlimited. For example, the impact of a large particle can alter a blade surface very locally. Surface measurements, in general, generate large amounts of data, which must be condensed in some way before this data can be analysed further. This reduction can be done by extracting only first-order deviations. Information about second- and third-order deviation is typically suppressed.

One possible approach by means of random fields is described in chapter "Estimation of Model Error Using Bayesian Model-Scenario Averaging with *Maximum a Posteriori*-Estimates". Another method is presented by Garzon [2], who measured the surfaces of a large set of blades and computed the mean and standard deviation for the geometrical deviations at each blade position. From this local deviation information, covariance matrices were deducted. The main information was then extracted with the use of an eigenvalue decomposition that revealed the largest eigenvectors. These eigenvectors were used for the subsequent robustness analysis.

A third approach is given by Schmidt et al. [6] who also measured a large set of compressor blades. A standard profile parameterization was fitted through each measured blade surface. The resulting parameter values allow for more intuitive interpretation of the geometrical variation. Again, statistical information was extracted from the parameterization.

In all approaches, the statistical information about the blade is condensed to a set of eigenvectors as for Garzon [2] or to the blade parameters as for Schmidt et al. [6].

Operational Uncertainties for Compressors

Exact operating conditions are often unknown a priori and reasonable assumptions become necessary in order to layout compressors. When, for example, an industrial compressor is designed for an oil and gas production facility, the gas composition of the future production can only be roughly forecasted as it changes with the depletion of the field. Hence, the layout of industrial compressors must typically fulfil a wide range of operating points with different pressure ratios and volumetric flows. Uncertainty with regard to the suction flow such as gas composition, pressure and temperature as well as operating point characterized by speed and pressure ratio is typically represented by three non-dimensional parameters which are Mach number, Reynolds number and flow coefficient.

Geometrical Uncertainties in the Aircraft Outer Geometry

In order to treat geometrical uncertainties in aircraft design, we propose a methodology to simulate random geometries based on measurements (e.g. by photogrammetry [8]) by means of random fields. This allows us to take into account uncertainties of the geometry for the computation of the aerodynamic performance of an aircraft.

The proposed methodology is based on a probabilistic modelling of the manufactured geometry as a random perturbation of a nominal geometry $\mathcal{G}$, the random perturbation being modelled as a random field using an extension of the Box and Jenkins method [9]. The methodology involves the following steps that we will further detail in the following section:

1. Data preparation step for obtaining a meaningful statistical representation of the measured geometrical uncertainties.
2. Data standardization step to make the data more suitable for the random field modelling (Box-Cox transformation).
3. Identification of a deterministic trend in the data: the random geometry will be a combination of this deterministic transformation and an additional random perturbation of the nominal geometry.
4. Modelling of the random perturbation by a random field using its Karhunen–Loève expansion.
5. The composition of all the previous steps in order to get the final random field model.
6. The simulation of a random field in order to produce virtual prototypes of our probabilistic model that can be used as input geometry for a subsequent CFD performance calculation.

The information available to build a random geometry generator is made of two distinct sequential steps: the representation of a nominal geometry (usually a discretized mesh) and a perturbation of the geometry based on a set of actual measurements (usually coordinates of points). To begin with, we cannot directly use these empirical data sets for statistical inference as there is no meaningful link between the nominal geometry and the data sets. The data preparation step consists of a data alignment step and a data pairing step.

For the *data alignment* step, we conduct a rigid body transformation of the raw measurements in order to match the measurements as closely as possible to the mesh vertices of the nominal geometry. This can be done by mapping the centres of gravity for the mesh and the measurement cloud and by aligning the principle axes of these two by means of a principal component analysis [10]. One measurement cloud represents one measured component such as an aircraft wing.

Once the measurements and the vertices have been aligned, the most challenging task is the *pairing* between the measurements and the vertices of the nominal geometry. If we suppose that the measurements are a perturbation of the mesh characterized by a small amplitude, it can be done in an efficient way by means of nearest neighbour search: each vertex x_i is associated with the measurement ς_i^k (being the geometrical coordinates of the ith point of the measurement cloud k) such that

$$\varsigma_i^{k*} = \underset{\varsigma_j^k}{\operatorname{argmin}} \left\| \boldsymbol{x}_i - \varsigma_j^k \right\|. \tag{1}$$

With the displacement $\boldsymbol{d}_i^k = \varsigma_i^{k*} - \boldsymbol{x}_i$ associated with $\boldsymbol{x}_i$, the set of measurements is converted into a corresponding displacement field $\left(\boldsymbol{x}_i,\ \boldsymbol{d}_i^k\right)$ as shown in Fig. 7. The collection of these fields forms a process sample that allows for building a probabilistic model by means of a random field.

A random field is a generalization of a stochastic process that describes the stochastic function of a point in an n-dimensional topological space [11]. The random field is said to be Gaussian if all finite-dimensional distributions are multivariate normal distributions and fully defined by its mean and covariance function [12].

Since we cannot expect this normally distributed property to hold for all geometrical uncertainties that we want to take into account, we apply a Box-Cox transformation in order to transform the raw data into a more suitable shape (i.e. normally distributed shape) [13]. The transformation is defined by:

$$\forall d > \delta,\ T_{\lambda,\delta}(d) = \begin{cases} \frac{(d+\delta)^{\lambda}-1}{\lambda} & \text{if } \lambda > 0 \\ \log(d+\delta) & \text{if } \lambda = 0 \end{cases} \tag{2}$$

Fig. 7 Displacement field (amplitude)

where $\lambda \geq 0$ is a shape parameter and $\delta \in \mathbb{R}$ is a shift parameter. In the case of a multidimensional random process, the transformation is applied component by component. Given the value of δ^k (a typical choice is $\delta^k = 0$ for symmetric distributions), the optimal value of λ^k is obtained by maximizing the following log-likelihood function:

$$L\left(\lambda^k; \delta^k, (X_i^k)_{i=1,\dots,N}\right) = \begin{array}{c} -\frac{N}{2}\log\left(\frac{N-1}{N} Var\left(T_{\lambda^k,\delta^k}(X^k)\right)\right) \\ +\left(\lambda^k - 1\right) \sum_{i=1}^{N} \log\left(X_i^k\right) \end{array} \tag{3}$$

where N is the total number of measurement points for each geometrical uncertainty. Given a discretized nominal geometry $\mathcal{G}$ represented by a mesh with vertices $(\boldsymbol{x}_i)_{i=1,\dots,n}$, the restriction to a random field $X: \Omega \times \mathcal{G} \to \mathbb{R}^3$ is the collection of random vectors $(\boldsymbol{X}_{x_i})_{i=1,\dots,n}$. If we stack these random vectors into a large random vector $\boldsymbol{X}_n: \Omega \to \mathbb{R}^{3n}$, it has a Gaussian distribution with mean vector

$$\boldsymbol{\mu}_n = \begin{pmatrix} \boldsymbol{\mu}(x_1) \\ \vdots \\ \boldsymbol{\mu}(x_n) \end{pmatrix} \tag{4}$$

and covariance matrix

$$\boldsymbol{\Sigma}_n = \begin{pmatrix} \boldsymbol{C}(\boldsymbol{x}_1, \boldsymbol{x}_1) & \cdots & \boldsymbol{C}(\boldsymbol{x}_1, \boldsymbol{x}_n) \\ \vdots & \ddots & \vdots \\ \boldsymbol{C}(\boldsymbol{x}_n, \boldsymbol{x}_1) & \cdots & \boldsymbol{C}(\boldsymbol{x}_n, \boldsymbol{x}_n) \end{pmatrix}. \tag{5}$$

This representation is called the trend/covariance representation of the random field.

For a discretized geometry represented by a mesh, the numerical representation of the Gaussian random field by its mean and covariance vector is restricted to small- and medium-sized problems due to its computational costs. For large-sized problem, some kind of compression technique is required.

The Karhunen–Loève expansion allows representing the random field $\boldsymbol{X}$ by a set of random variables $(Z_l)_{l \in \mathbb{N}}$ (that are uncorrelated for Gaussian random fields)

$$\forall (\omega, \boldsymbol{x}) \in \Omega \times \mathcal{G}, \boldsymbol{X}(\omega, \boldsymbol{x}) = \sum_{l \in \mathbb{N}} Z_l(\omega) \psi_l(\boldsymbol{x}) \tag{6}$$

and deterministic basic functions $(\psi_l)_{l \in \mathbb{N}}$ where, for each $l \in \mathbb{N}$, $\psi_l: \mathcal{G} \to \mathbb{R}^3$ such that

$$\forall l \in \mathbb{N}^2, \int_{\mathcal{G}} \psi_l(\varsigma) C(\boldsymbol{x}, \varsigma) d\varsigma = \lambda_l^2 \psi_l(x) \tag{7}$$

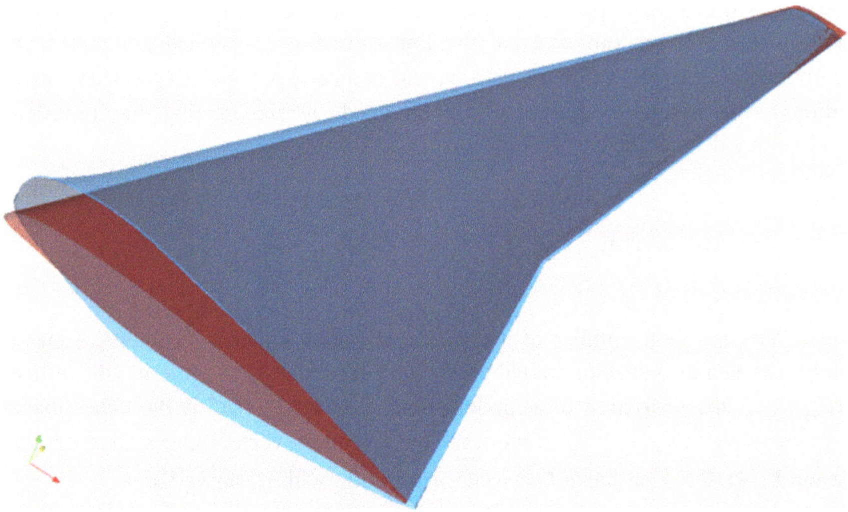

Fig. 8 Illustration of the first basic function (blue) compared to the nominal geometry (red)

where λ_l^2 is a decreasing summable sequence of positive real numbers (Fig. 8). The coefficients' distribution is given by $Z_l \approx \mathcal{N}(0,\ \lambda_l^2)$. The basis is orthonormal with respect to the covariance function [14]. In many cases, the decrease in the sequence λ_l^2 allows for a truncation of the sum in (2) leading to an effective compression. The resulting covariance function reads

$$\forall(\boldsymbol{x},\varsigma)\in\mathcal{G}^2, C(\boldsymbol{x},\varsigma)=\sum_{l\in\mathbb{N}}\lambda_l^2\psi_l(\boldsymbol{x})\psi_l(\varsigma). \tag{8}$$

The randomness of the probabilistic model is concentrated in the random variables $(Z_l)_{l\in\mathbb{N}}$. The Karhunen–Loève expansion possesses some favourable properties if all the random variables are normally distributed, which is why we applied the Box-Cox transformation on the raw data.

The simulation of a random field in order to produce virtual prototypes of our probabilistic model is straightforward.

1. Simulate a realization of the random vector $(Z_l)_{l\in\mathbb{N}}$ according to its joint distribution function. In the case of the Karhunen–Loève representation of a Gaussian random field, the random variables $(Z_l)_{l\in\mathbb{N}}$ are Gaussian and independent.
2. Evaluate the $(l+1)$ deterministic functions $\psi_0,\ \ldots,\ \psi_l$ at the vertices of the reference mesh $(\boldsymbol{x}_1,\ \ldots,\ \boldsymbol{x}_n)$. *These computations are common to all the realizations of the random field and can then be stored and reused.*

3. Compute the deterministic trend function at the vertices of the reference mesh $(\boldsymbol{x}_1, \ldots, \boldsymbol{x}_n)$. *These computations are common to all the realizations of the random field and can then be stored and reused.*
4. Apply the inverse Box-Cox transformation to each value of the random field realization.
5. Convert the displacement realization into a random mesh realization by moving each vertex $\boldsymbol{x}_i$ by its associated displacement $\boldsymbol{d}_i^k$. Each displacement realization represents one non-nominal geometry that can be used as input geometry for a subsequent CFD performance calculation.

Geometrical Uncertainties from Aircraft Operation

The change of geometry during the lifetime of the aircraft may occur for several reasons. As for compressor operation, surface degradation due to corrosion, erosion, impact and fouling have to be considered for aircraft operation as well as aeroelastic deformation as illustrated in Fig. 9. These deviations are also often larger than the acceptable manufacturing tolerances.

Concerning geometry uncertainties, we have to take into account the details such as discontinuities, bleed flows and protuberances like antennas as shown in Fig. 10 that are not included in an ideal CAD.

Last but not least, icing can change dramatically the shape, especially in the locations highlighted in Fig. 11.

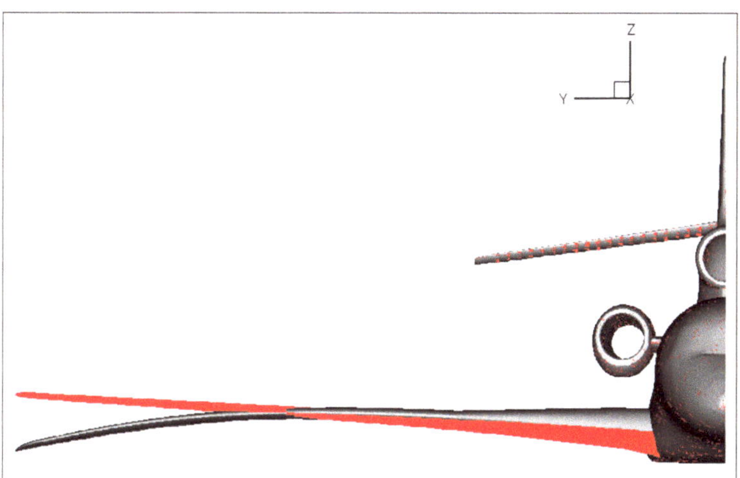

Fig. 9 Aeroelasticity effects on the aircraft

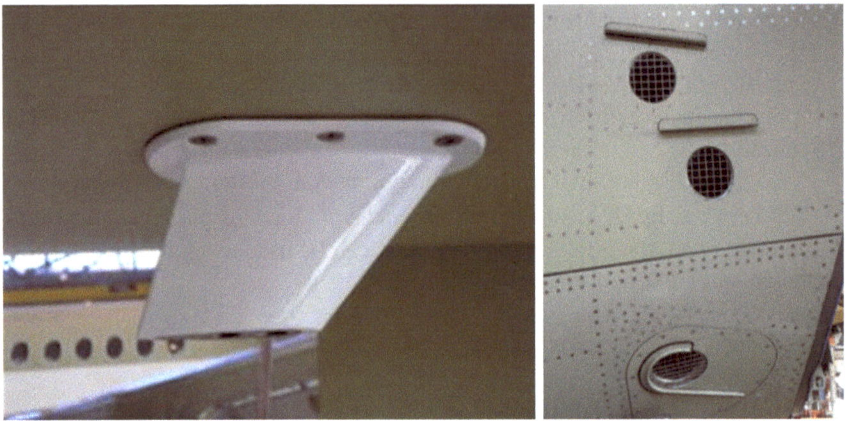

Fig. 10 Typical elements that are missing in CAD models: antenna and bleed

Fig. 11 Typical locations for icing

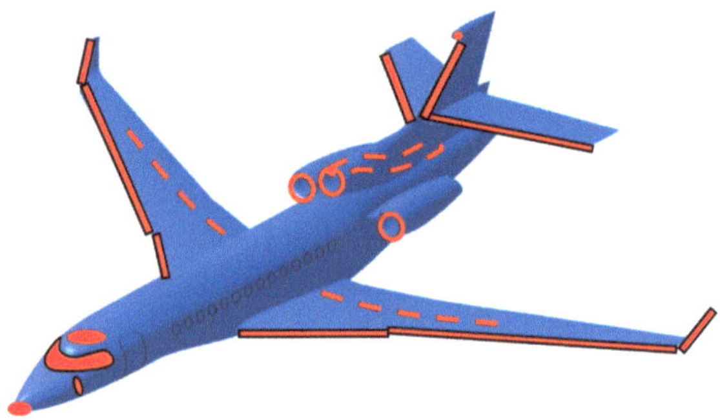

Operational Uncertainties for Aircraft

As for compressor operation, aircraft operating conditions are often assumed and not precisely known. For example, we can compare temperature, pressure versus altitude for both standard and real atmospheres (database DLR ERA40/ECMWF, www.ecmwf.int) as shown in Fig. 12 and Fig. 13, respectively.

In the same DLR database (a vertical profile measured somewhere between Ireland and England, on 08 July 1992), we can observe the evolution of the module and the direction of the wind velocity with the altitude (Fig. 14).

Despite non-homogeneous behaviour of the atmosphere with respect to location and time, engineers use standard atmospheres for CFD computation. Moreover, uniform pressure, temperature and velocity are imposed along the artificial boundary of the computational domain (so-called infinity boundary).

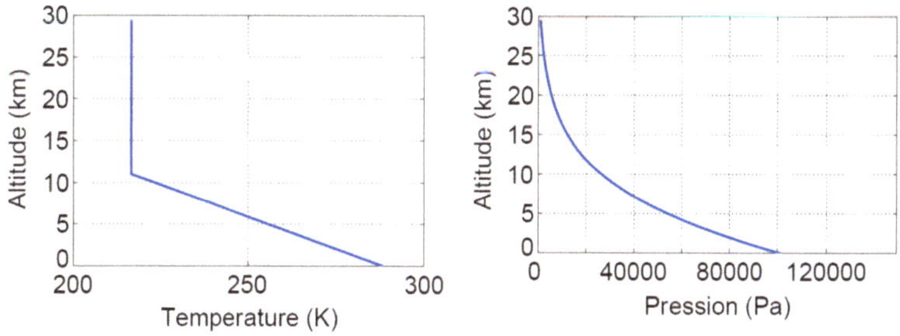

Fig. 12 Standard atmosphere—temperature and pressure distribution

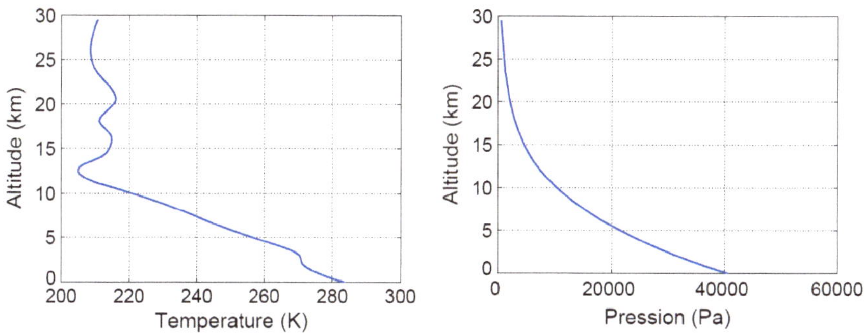

Fig. 13. Real atmosphere—temperature and pressure distribution

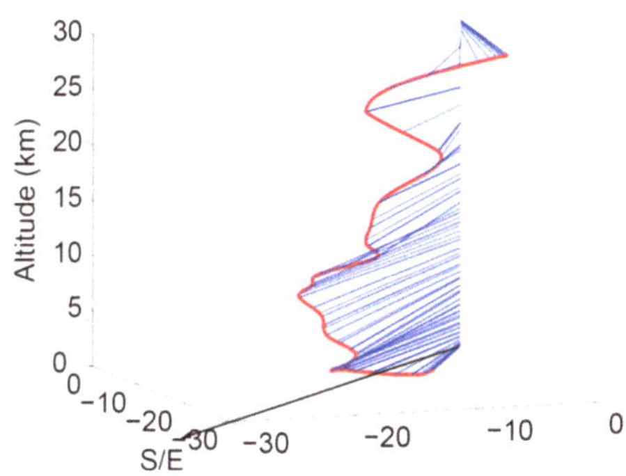

Fig. 14. Real atmosphere—wind vector

Of course, unsteady flows, like gust or aircraft manoeuvre effects, are not considered for basic CFD calculations.

The same considerations can be made for engine nozzle boundary conditions. For example, engineers consider basically uniform total temperature, total pressure and Mach number. Sometimes, data are provided on engine exit plane, but not for all the considered flight points.

Fuel consumption yields weight decrease during the flight (equilibrium of forces imply modification of the lift coefficient). Despite fuel management during flight, centre of gravity changes and trim modifications are needed.

Conclusions

Geometrical uncertainties exist in all physical parts due to manufacturing and assembly tolerances as well as thermo-mechanical deformation and degradation during operation. Reducing the level of uncertainty is in most cases possible but typically increases cost. Tightening manufacturing tolerances should only be accepted where the influence level justifies this.

Geometrical uncertainties are difficult to quantify and model as they can vary over several orders of magnitude in size, beginning from surface roughness up to length variations of the entire machine. Furthermore, geometrical deviations can be arbitrary in shape, possibly manifested as kinks, bumps, local defects or surface waviness, just to name a few. Geometrical uncertainties are also highly dependent on the underlying manufacturing process(es) such as forging, casting or machining, with or without finishing for massive compressor parts or assemblies of sheet metals for aircraft wings. Geometrical uncertainties from aircraft operation and operational uncertainties are often larger than the acceptable manufacturing tolerances.

Methodologic approaches, nevertheless, exist that are able to reduce the complexity by extracting the main features of a geometrical uncertainty. This allows for integration of the variation in numerical simulations for uncertainty quantification and robust design optimization. These simulations help in understanding uncertainties and developing designs that are insensitive to uncertainties.

References

1. Schlüter, J.U., Shankaran, S., Kim, S., Pitsch, H.A.J.J., Moin, P.: Towards multi-component analysis of gas turbines by CFD: integration of RANS and LES flow solvers. In: ASME Turbo Expo 2003, GT2003-38350 (2003)
2. Garzon, V.E.: Probabilistic Aerothermal Design of Compressor Airfoils (2003)
3. Javed, A., Pecnik, R., van Buijtenen, J.: Optimization of a centrifugal compressor impeller design for robustness to manufacturing uncertainties. In: ASME Turbo Expo 2013: GT2013-95515, San Antonio, Texas, USA
4. DIN 4760, Gestaltabweichungen; Begriffe, Ordnungssystem
5. ISO 2768-1, General Tolerances—Part 1: Tolerances for Linear and Angular Dimensions Without Individual Tolerance Indications
6. Schmidt, R., Voigt, M., Vogeler, K., Meyer, M.: Comparison of two methods for sensitivity analysis of compressor blades. In: Proceedings of ASME Turbo Expo 2016, GT2016-57378 (2016)

7. Büche, D., Beetz, M., Ribi, B.: Uncertainty analysis for large-scale industrial radial compressors. In: ASME Turbo Expo 2010, GT2010-22918 (2010)
8. Elishakoff, I., et al.: Optimization and antioptimization of buckling load for composite cylindrical shells under uncertainties. AIAA J. **50**(7), 1513–1524 (2012)
9. Box, G., Jenkins, G.: Time Series Analysis: Forecasting and Control. Holden-Day, San Francisco (1970)
10. Jolliffe, I.T.: Principal Component Analysis. Springer, New York (2002)
11. Adler, R.J., Taylor, J.E.: Random Fields and Geometry. Springer, New York (2007)
12. Adler, R.J.: The Geometry of Random Fields. Wiley, Chichester (1981)
13. Box, G., Cox, D.: An analysis of transformations. J. Roy. Stat. Soc. B **26**(2), 211–252 (1964)
14. Schenk, C.A., Schuëller, G.I.: Uncertainty Assessment of Large Finite Element Systems. Lecture Notes in Applied and Computational Mechanics p. 24 Springer, Berlin (2005)

Estimation of Model Error Using Bayesian Model-Scenario Averaging with *Maximum a Posterori*-Estimates

Martin Schmelzer, Richard P. Dwight, Wouter Edeling and Paola Cinnella

Introduction

The understanding of turbulence is one of the key challenges in classical mechanics. A turbulent flow has a three-dimensional, time-dependent and random velocity field [1], which is composed of a wide range of scales varying from the level close to molecular dissipation of heat to the scales determined by the boundary conditions of the flow domain. Despite great developments in the field of high-fidelity turbulence modelling, such as improvements of the large eddy simulation (LES) approach and hybrid methods combining Reynolds-averaged Navier–Stokes (RANS) and LES, RANS 'continues to be the standard approach used to predict a wide range of flows for very complex configurations across virtually all aerospace product categories' [2]. However, using the less-computationally demanding RANS approach comes at the price of uncertainty due to approximate physical modelling such as turbulence modelling.

M. Schmelzer (✉) · R. P. Dwight
Faculty of Aerospace Engineering, Delft University of Technology,
Kluyverweg 2, Delft, The Netherlands
e-mail: m.schmelzer@tudelft.nl

R. P. Dwight
e-mail: r.p.dwight@tudelft.nl

W. Edeling
Center for Turbulence Research, Stanford University, 488 Escondido mall,
Stanford 94305, USA
e-mail: wedeling@stanford.edu

P. Cinnella
DynFluid laboratory, Arts et Métiers ParisTech, 151 Boulevard de l'Hospital,
75013 Paris, France
e-mail: paola.cinnella@ensam.eu

C. Hirsch et al. (eds.), *Uncertainty Management for Robust Industrial Design in Aeronautics*, Notes on Numerical Fluid Mechanics and Multidisciplinary Design 140, https://doi.org/10.1007/978-3-319-77767-2_4

In recent years, research has focussed on two types of epistemic uncertainties in relation to turbulence modelling, namely parameter uncertainty and model-form uncertainty. Parameter uncertainty arises due to the fact that the closure coefficients of the models are determined via calibration against simple flow configurations using experimental data or scale-resolving simulations. A Bayesian calibration incorporating both the error of the experimental data and model inadequacy revealed how strongly the posterior distributions of the closure coefficients vary even for a simple flow scenario of flow over a flat plate, with different pressure gradients ranging from favourable to strongly adverse [3]. This observation makes the generalisation of the coefficient unjustifiable. Similarly, a Bayesian analysis dealing with the predictive performance of commonly used turbulence models, i.e. Launder-Sharma $k-\epsilon$, Wilcox $k-\omega$, Spalart-Allmaras, Baldwin-Lomax and Stress-ω showed that no superior model could be identified either for the given flow configurations [3].

Bayesian Model-Scenario Averaging (BMSA) uses a set of different closure models to compute both an ensemble prediction as well as a-posteriori measures of uncertainty due to the choice of closure model. In addition, the variability of the closure coefficients over different calibration scenarios is included to inject uncertainty caused by applying a set of coefficients to a predictive case for which they were not specifically calibrated [4]. The full BMSA approach requires the propagation of various posterior distributions through a CFD code, an expensive endeavour when the underlying flow case is computationally expensive. The costs could be mitigated by using surrogate models per turbulence model in order to propagate the distributions more efficiently [5]. However, also the construction of the surrogate can become expensive for models with many closure coefficients.

In this work, we use a major simplification of the full BMSA approach in order to make the technique accessible for industrial flow cases. The reduction of the costs is based on using *maximum a posteriori* (MAP) estimates of the posterior distributions of the closure coefficients, which means that only a single set of closure coefficients per scenario and per model needs to be propagated through the code [6]. Furthermore, the BMSA method was developed using data of flat plate boundary-layer flows, for which the Bayesian inference of the posterior closure coefficient was conducted with a cheap boundary-layer code. The method has recently been applied to airfoil and wing cases as examples in order to assess the predictive capabilities of the BMSA framework for wall-bounded flows in external aerodynamics [6], which represents a natural next step in terms of flow complexity. In this work, we apply the method to other flow cases such as flow in a turbulent pipe at $Re = 44{,}000$ and flow over periodic hills at $Re_H = 5600$ in order to assess the robustness of the method also for flow configurations outside of this natural domain. Finally, we also show results for one of the industrial challenges of the UMRIDA project: a generic Falcon Jet (IC-03).

Bayesian Predictive Methodology

Calibration

The BMSA framework is based on the Bayesian calibration of the coefficients of closure models. Treating calibration as a stochastic problem, we obtain posterior probability distributions for the coefficients, which serve as our uncertain estimates of the coefficients under the measurement error of the reference data and the modelling error. The data used for the calibration consists of boundary-layer data sets from the 1968 AFOSR-IFP-Stanford conference proceedings [7], which are both highly resolved and subject to low measurement noise. The set contains a collection of wide range of favourable and adverse pressure gradients.

A *scenario*, denoted S, is a particular flow set-up, including boundary conditions, material parameters, and all other physical properties needed to define the flow, with corresponding experimental data $\mathbf{z}$. The CFD code $m_{\text{CFD}}(S; M, \boldsymbol{\theta})$ takes as arguments the scenario $S \in \mathscr{S} = \{S_1, \dots, S_K\}$, a turbulence model $M \in \mathscr{M} = \{M_1, \dots, M_I\}$ and its closure coefficients $\boldsymbol{\theta}$. Given an operator $H_z(\cdot)$ that maps the state to the measured quantities $\mathbf{z}$ we can define a statistical model to relate $\mathbf{z}$ and $\boldsymbol{\theta}$:

$$\mathbf{z} = \delta \left[H_z \circ m_{\text{CFD}}(S; M, \boldsymbol{\theta}) \right] + \epsilon, \tag{1}$$

where ϵ, δ are random variables (RVs) representing measurement noise and multiplicative model error respectively. The error of the measurements is modelled as zero-mean additive Gaussian noise and the RV δ is a representation of the model error, which following the approach of Cheung et al. [8] specifies δ as a correlated Gaussian process, see also [3]. The Gaussian choice for both ϵ and δ yields a Gaussian likelihood function, i.e. the pdf describing the probability of observing the data given a realisation of $\boldsymbol{\theta}$. Finally, an application of Bayes theorem [9] yields the expression for the posterior distribution of $\boldsymbol{\theta}$

$$p(\boldsymbol{\theta}|z, M, S) = \frac{p(z|\boldsymbol{\theta}, M, S)\, p(\boldsymbol{\theta}|M, S)}{p(z|M, S)} \propto p(z|\boldsymbol{\theta}, M, S)\, p(\boldsymbol{\theta}|M, S). \tag{2}$$

Here, $p(\boldsymbol{\theta}|M, S)$ is the prior distribution which represents our knowledge of $\boldsymbol{\theta}$ prior to training M on z. Since the denominator in (2) does not depend on $\boldsymbol{\theta}$ it is omitted from consideration.

To obtain samples from the posterior distribution $p(\boldsymbol{\theta}|\mathbf{z}, M, S)$, we employ the Markov-Chain Monte Carlo method [10]. To reach convergence of the Markov-chain, we observed that roughly 40,000 code samples were required [3]. Ordinarily, this would constitute an excessive strain on available computational resources in a CFD context. However, as our experimental data consisted of boundary-layer quantities, we were in a position to use a fast boundary-layer code. As such, no real computational bottleneck exists during the calibration phase.

Prediction

Let Δ be a quantity of interest (QoI) in a particular scenario, which may be a scalar, vector, or functional quantity derived from the flow state, which can be expressed as

$$\Delta \simeq H_{\Delta} \circ m_{\mathrm{CFD}}(\tilde{S}; M, \boldsymbol{\theta}), \tag{3}$$

in which $\tilde{S}$ represents a flow scenario outside of the set of scenarios used within the calibration phase.

We wish to obtain a stochastic estimate of Δ conditional on a set of models $\mathscr{M}$ and a set of training scenarios $\mathscr{S}$ for the predictive scenario $\tilde{S} \notin \mathscr{S}$. The BMSA methodology offers the evaluation of the posterior predictive distribution (ppd) for Δ in case $\tilde{S}$ conditional on all training data:

$$p(\Delta \mid \tilde{S}, \mathbf{z}) = \sum_{i=1}^{I} \sum_{k=1}^{K} \int_{\boldsymbol{\Theta}_i} p(\Delta \mid \tilde{S}, M_i, \boldsymbol{\theta})\, p(\boldsymbol{\theta} \mid S_k, M_i, z_k)\, \mathbb{P}(M_i \mid S_k, z_k)\, \mathbb{P}(S_k)\, \mathrm{d}\boldsymbol{\theta}. \tag{4}$$

The first term on the right-hand side inside the integral represents the probabilistic equivalent of the simulation results for the QoI given the flow scenario $\tilde{S}$, a turbulence model M_i and closure coefficients $\boldsymbol{\theta}$. The second and third terms are the posterior probability density distribution of the closure coefficients and the posterior model probabilities respectively, which are the output of the previously conducted calibration procedure summarised in section "Calibration" and detailed in Edeling et al. [3, 4]. The last term represents the scenario probabilities. The solution of the multi-dimensional integral over the closure coefficients is the expensive part of the method, because it requires as many code calls as the entries in the Markov-chain, in this case 40, 000.

Therefore to obtain a practical estimate of $p(\Delta \mid \tilde{S}, \mathbf{z})$, we propose to approximate the marginal posterior probability distributions $p(\boldsymbol{\theta} \mid S_k, M_i, z_k)$ with Dirac-δ functions at their *maximum a posteriori* (MAP) values[1]

$$\boldsymbol{\theta}_{i,k}^{\mathrm{MAP}} := \underset{\theta \in \boldsymbol{\Theta}_i}{\operatorname{argmax}}\; p(\boldsymbol{\theta} \mid S_k, M_i, z_k) \tag{5}$$

so that

$$p(\boldsymbol{\theta} \mid S_k, M_i, z_k) \simeq \delta\left(\boldsymbol{\theta} - \boldsymbol{\theta}_{i,k}^{\mathrm{MAP}}\right). \tag{6}$$

The effect of this approxmation is to neglect the effect of *within-model within-scenario* variance on the ppd. Thus the ppd variance will be reduced, but still include the effect of multiple models and scenarios. Note that if perfectly plentiful data were available in the training scenarios (and the models were able to fit the data exactly for

[1]The MAP estimates are available online at [11].

some values of the closure coefficients), then $p(\boldsymbol{\theta} \mid S_k, M_i, z_k)$ would be δ-functions. So one way to think of this approximation is as neglecting the effect of imperfect training.

Substituting (6) into (4) leads to an approximation of the posterior predictive distribution $p(\Delta \mid \mathbf{z}) \simeq \hat{p}(\Delta \mid \mathbf{z})$

$$\begin{aligned}\hat{p}(\Delta \mid \tilde{S}, \mathbf{z}) &= \sum_{i=1}^{I} \sum_{k=1}^{K} \mathbb{P}(M_i \mid S_k, z_k)\, \mathbb{P}(S_k) \int_{\boldsymbol{\Theta}_i} p(\Delta \mid \tilde{S}, M_i, \boldsymbol{\theta})\, \delta\left(\boldsymbol{\theta} - \boldsymbol{\theta}_{i,k}^{\mathrm{MAP}}\right) \mathrm{d}\boldsymbol{\theta} \\ &= \sum_{i=1}^{I} \sum_{k=1}^{K} \mathbb{P}(M_i \mid S_k, z_k)\, \mathbb{P}(S_k)\, p(\Delta \mid \tilde{S}, M_i, \boldsymbol{\theta}_{i,k}^{\mathrm{MAP}}) \\ &\overset{(a)}{=} \sum_{i=1}^{I} \sum_{k=1}^{K} \mathbb{P}(M_i \mid S_k, z_k)\, \mathbb{P}(S_k)\, \delta\big(\Delta - m_{\mathrm{CFD}}\left(\tilde{S}; M_i, \boldsymbol{\theta}_{i,k}^{\mathrm{MAP}}\right)\big). \end{aligned} \tag{7}$$

Equality (a) follows from the fact that the prediction of m_{CFD} is deterministic for deterministic $\boldsymbol{\theta}$. The approximate ppd is therefore a weighted-sum of $I \times K$ δ-functions, one at each prediction of $m_{\mathrm{CFD}}(\tilde{S})$ for each model, and each scenario's MAP-estimate of $\boldsymbol{\theta}$. The cost of evaluating the ppd is $I \times K$ runs of $m_{\mathrm{CFD}}(\tilde{S})$.

The first moment of $\hat{p}(\Delta \mid \tilde{S}, \mathbf{z})$ can be derived directly from (7). The expectation is

$$\begin{aligned}\mathbb{E}[\Delta \mid \tilde{S}, \mathbf{z}] &= \int \Delta \cdot \hat{p}(\Delta \mid \tilde{S}, \mathbf{z})\, \mathrm{d}\Delta \\ &\overset{(a)}{=} \sum_{i=1}^{I} \sum_{k=1}^{K} \mathbb{P}(M_i \mid S_k, z_k)\, \mathbb{P}(S_k) \int \Delta \cdot \delta\left[\Delta - m_{\mathrm{CFD}}\left(\tilde{S}; M_i, \boldsymbol{\theta}_{i,k}^{\mathrm{MAP}}\right)\right] \mathrm{d}\Delta \\ &\overset{(b)}{=} \sum_{i=1}^{I} \sum_{k=1}^{K} \mathbb{P}(M_i \mid S_k, z_k)\, \mathbb{P}(S_k)\, m_{\mathrm{CFD}}\left(\tilde{S}; M_i, \boldsymbol{\theta}_{i,k}^{\mathrm{MAP}}\right), \end{aligned} \tag{8}$$

where (a) follows from (7) and (b) is the integral-identity encoding the statement that the mean of a deterministic quantity is the quantity itself.

Unlike the posterior model probabilities $\mathbb{P}(M_i \mid S_k, z_k)$, which are informed using the reference data during the training phase [4], the scenario probabilities $\mathbb{P}(S_k)$ are of a predictive nature. If reference data for the scenario $\tilde{S}$ would be available, $\mathbb{P}(S_k)$ could be optimised accordingly. However, in a predictive setting this data isn't always available. Therefore, $\mathbb{P}(S_k)$ is defined based on model agreement per scenario S_k: If the models show a high level of agreement regarding the value of Δ under a specific scenario S_k, this scenario receives a higher weight compared to other scenarios, in which the models rather disagree. For that we exploit the principle, that if $\tilde{S}$ is similar to S_k the models are expected to give accordingly similar predictions, because $\boldsymbol{\theta}_{i,k}^{\mathrm{MAP}}$ has been calibrated under the same data z_k. As introduced in [4], this principle is modelled by

$$\mathbb{P}(S_k) := \frac{\xi_k^{-p}}{\sum_{j=1}^{K} \xi_j^{-p}}, \quad \xi_k = \sum_{i=1}^{I} \| m_{\mathrm{CFD}}\left(\tilde{S}; M_i, \theta_{i,k}^{\mathrm{MAP}}\right) - \mathbb{E}[\Delta \mid \tilde{S}, z_k] \|_2, \tag{9}$$

in which p serves as a tuning parameter scaling the weighting procedure, i.e. for $p = 0$ all scenarios are equally weighted but for $p \to \infty$ a single scenario is preferred.

Given all terms in (7), we now want to deduce uncertainty estimates from the ppd. If the full range of $\mathbb{P}(M_i \mid S_k, z_k)\, \mathbb{P}(S_k)$ is used, the uncertainty bound is largest and determined by the extreme predictions of the entire ensemble. However, a more reasonable approach is to draw samples from the ppd and use percentiles as min/max levels to obtain a confidence interval of the ppd leading potentially to tight uncertainty bounds. This is done in the following.

Incompressible Pipe Flow at *Re* = 44,000

We apply the BMSA method to turbulent flow in a straight pipe at $Re = 44{,}000$ with validation data from a DNS simulation [12]. This flow case is, together with turbulent flow over a flat plate and channel flow, one of the three canonical wall-bounded flow types [1]. The main difference between flow in a pipe or channel and over a flat plate lies in the development of the boundary layer. Assuming a uniform inflow into a circular straight pipe a boundary layer at the wall develops and its thickness increases downstream similarly to the flat plate case. Forced by the geometry of the pipe the boundary layers from all sides will eventually merge in the middle of the pipe. The distance from the entrance of the pipe to the point, where the boundary layer fills the entire diameter, is known as the entry length. Beyond the entry length, the flow is homogeneous in stream-wise direction. In contrast, the boundary-layer thickness of a flat plate flow with zero pressure gradient is not bounded. However, for all these wall-bounded flow types the physics close to the wall can be assumed to be similar. Therefore, with the application of BMSA based on flat plate flow to pipe flow is a proof-of-concept in order to assess the predictive capabilities of BMSA for wall-bounded flows in general.

Simulation Set-up

The validation of the BMSA method is based on a comparison with averaged velocity profiles from a DNS simulation [12]. Therefore, the steady-state RANS simulation mimics the time-averaged DNS simulation by using appropriate boundary conditions given in Table 1. In the DNS simulation, the mass flow rate is kept constant via a time-varying adjustment of the pressure gradient in the stream-wise direction [12]. For the RANS simulation, the mass flow rate is kept constant via a fixed inlet velocity and the pressure gradient is set to zero at the inlet. At the outlet, both

Table 1 The Inlet conditions of velocity U, pressure P/ρ and turbulent quantities [13]

	U (m/s)	P/ρ (m²/s²)	k	ϵ	ω	$\tilde{\nu}$
Inlet	$(1,0,0)^T$	$\mathbf{n}\cdot\nabla P=0$	0.00375	0.00835	25.0516	0.00027
Outlet	$\mathbf{n}\cdot\nabla\mathbf{U}=0$	$(0,0,0)^T$	$\mathbf{n}\cdot\nabla k=0$	$\mathbf{n}\cdot\nabla\epsilon=0$	$\mathbf{n}\cdot\nabla\omega=0$	$\mathbf{n}\cdot\nabla\tilde{\nu}=0$
Wall	$(0,0,0)^T$	$(0,0,0)^T$	$(0,0,0)^T$	$(0,0,0)^T$	8367238.00803	$(0,0,0)^T$

the velocity gradients are set to zero and pressure is set to zero. In this way, the pressure drop is computed according to the fixed mass flow. For the forward simulations $m_{CFD}\left(M_i, \theta_{i,k}^{MAP}\right)$, the results of simulations with nominal values of the coefficients $m_{CFD}(M_i, \theta_{i,k}^{o})$ are used as an initial condition in order to reduce iteration counts.[2]

The geometry of a straight pipe is determined by its diameter and length. While the diameter D is set in order to achieve the target Reynolds number, the length of the pipe needs to be chosen sufficiently longer than the entry length in order to overcome all effects at the inlet, such as the usage of different closure coefficients or the uniform inflow velocity profile. The entry length is estimated by an empirical relation based on the Reynolds number [13]

$$L_{turb} = 1.359 \cdot D \cdot Re_D^{1/4} \Rightarrow L_{turb} \sim 20D. \tag{10}$$

For all simulations, a conservative value $L = 200D$ was chosen. Due to the symmetry of the case a wedge-shaped mesh with symmetric boundary conditions orthogonal to the stream-wise direction was chosen [14]. Different meshes were initially studied, with a total number of 2900, 5900 and 9900 cells, respectively. The differences for the velocity profiles between the meshes were negligible, but in order to keep the error small the finest mesh was chosen for all forward simulations. In order to study the effect of the turbulence modelling and not additional modelling due to wall functions, the thickness of the smallest cells close to the wall was defined according to $y^+ \leq 1.0$ and Low-*Re* turbulence models were used.

BMSA Prediction for Radial Velocity Profile

Applying the BMSA framework to the set of simulations leads to the expectation of the radial velocity $\mathbb{E}(U/U_{bulk}|\mathbf{z})$ as shown in Fig. 1a–c for different tuning parameters of the smart-weighting method $p = [0, 1, 2]$. For the confidence interval, 10,000 samples were drawn from the posterior predictive mass function given in Eq. 7 for each x/R position. The 5th and 95th percentile of the population was used as the lower and upper limit, respectively, containing 90% of the samples. For $p = 0$, i.e. uniform

[2]The template-cases for each turbulence model are available on GitHub: https://github.com/shmlzr/UQOpenFOAM.

weighting of the scenarios, the impact of outliers on the confidence interval is large leading to an over-prediction of the modelling error. Outliers are caused by sensitivities of the models with respect to certain coefficients leading to predictions very different compared to the ensemble mean. The effect is already reduced for $p = 1$, for which the confidence interval is covering a tight bound around the expectation. The shape matches with the shape of the majority of the ensemble of forward simulations. For increasing p, the confidence interval becomes tighter, since the smart weighting consecutively increases the weight of $S_k = 1300$ as shown in Fig. 2b and decreases the influence of the others, reducing the overall variance. Interestingly, the MAP estimates for scenario $S_k = 1300$ are acquired based on a flow over a flat plate with a moderately favourable pressure gradient and a boundary layer close to an equilibrium state [4], which is very similar to the conditions for the flow in a pipe. A comparison of the DNS data and the expectation for different p via the L^2-norm

$$||\mathbb{E}[U/U_{bulk}|\tilde{\mathrm{S}}, \mathbf{z}] - U_o||_2 \tag{11}$$

given in Fig. 2a, where the high-resolution DNS data is interpolated by a cubic spline to enable a point-by-point comparison, shows a minimal error for $0 < p < 2$, but also an increase of the error for higher values of p. Note that due to the use of MAP estimates, the variance is missing the *within-model within—scenario* contribution. Therefore, a reduction of the variance with smart weighting should be done with care in order to not reduce the already underestimated variance even further. Thus, the purpose of the smart weighting for the BMSA-with-MAP-approach should be to exclude outliers, which can be achieved with relatively low p values. In such a way an optimum of the width of the confidence interval and the correctness of the expectation can be achieved.

BMSA Prediction for Pressure Drop

Another relevant quantity of interest for the turbulent pipe flow is the pressure drop ΔP over the pipe length. Table 2 gives the values for the pressure drop $\Delta P/\rho$ normalised by the fluid density ρ and the corresponding Darcy friction coefficient f_{Darcy}

$$f_{Darcy} = \frac{2\ \Delta P\ D}{\rho\ L\ U^2} \tag{12}$$

per model. The empirical reference value for the pressure drop is calculated from the Colebrook–White equation, which is an approximation of the empirical Moody-diagram,

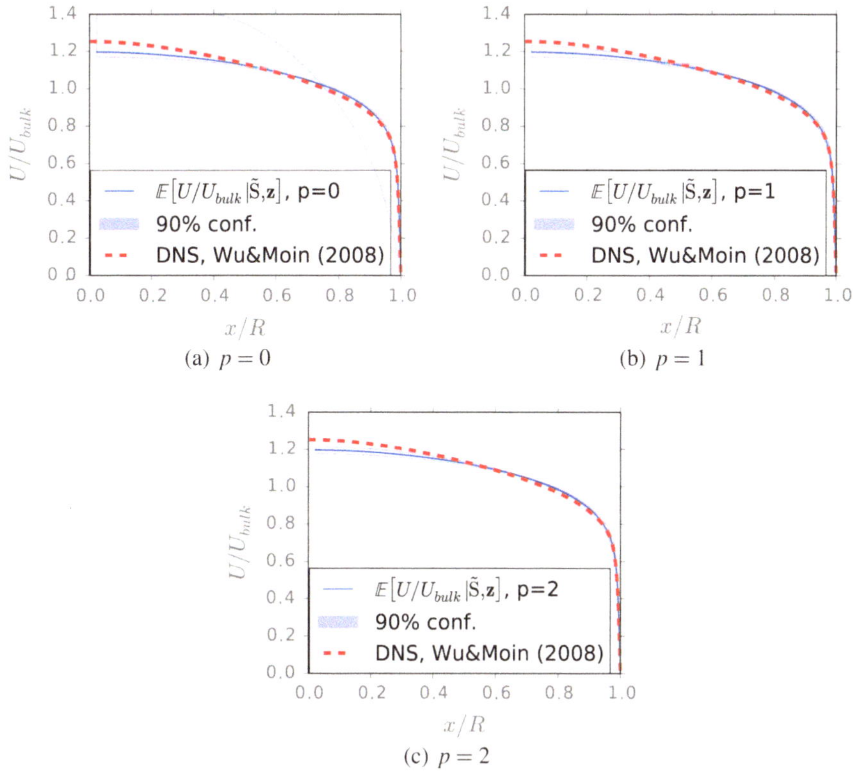

Fig. 1 BMSA prediction using different smart-weighting tuning parameter p

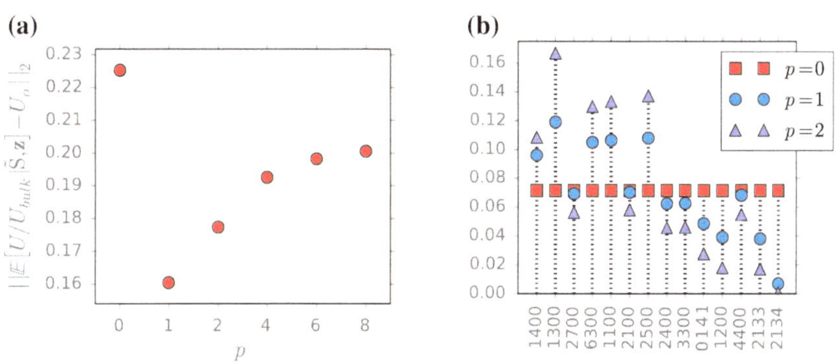

Fig. 2 **a** L^2 error norm of difference between DNS data and BMSA expectation against p—value. **b** Scenario-weights for several p-values. Description of scenarios in [3]

Table 2 Pressure drop over pipe length for Launder-Sharma $k-\epsilon$, Wilcox (2006) $k-\omega$ and Spalart-Allmaras using nominal closure coefficient [4]. The corresponding Darcy friction coefficient f_{Darcy} is calculated using the Darcy–Weisbach equation. The empirical value for f_{Darcy} is based on the Colebrook equation for $Re = 44{,}000$ and the corresponding $\Delta P/\rho$ is calculated using the Darcy–Weisbach equation

Model	$\Delta P/\rho\,(\frac{m^2}{s^2})$	f_{Darcy}
Wilcox (2006) $k-\omega$	2.1427	0.02144
Launder-Sharma $k-\epsilon$	2.0478	0.02049
Spalart-Allmaras	2.2074	0.02208
Empirical value	2.1499	0.021499

$$\frac{1}{\sqrt{f}} = -2\log_{10}\left(\frac{2.51}{Re\sqrt{f}}\right) \tag{13}$$

with zero roughness at $Re = 44{,}000$. For several tuning parameters of the smart-weighting methods, i.e. $p \in [0, 8]$, the BMSA expectation of the pressure drop $\mathbb{E}[\Delta P \mid \tilde{S}, \mathbf{z}]$ is shown in Fig. 3 as well as the empirical reference data calculated based on the Colebrook–White equation. Up to $p = 4$ the confidence interval still covers a range including also the empirical reference value, which suggests that the method successfully gives a reasonable uncertainty range. For larger tuning parameters, the confidence interval reduces drastically and puts all weight on one single scenario, so that the confidence interval is not visible anymore. In line with the results for the velocity profile (see section "BMSA Prediction for Radial Velocity Profile"), also for the pressure drop p = 2 is a reasonable choice. The expectation consistently underestimates the true value and is therefore worse than the baseline predictions, see Table 2. However, since the BMSA method was not trained on pressure drop data this is not unexpected and the main gain of BMSA is the confidence bounds instead of replacing the baseline prediction with the point-estimate of the expectation.

Incompressible Flow over Periodic Hills at $Re_H = 5600$

The periodic hill test case is based on the channel flow case, but with a lower surface modified by a series of periodically arranged hills, and deals with flow separation on the curved surface of the hills and reattachment on the flat plate region between the hills [15]. Being computationally relatively cheap but still challenging in terms of flow physics it has been used in several workshops as a benchmark test case [16]. We use DNS data of the mean flow field from Breuer et al. [15] for $Re_H = 5600$, defined using the stream-wise bulk velocity U_b between the hills crest and the upper surface and the hill height H. The test case is known to be especially challenging for linear eddy-viscosity models, which are not able to predict the mean effect of the

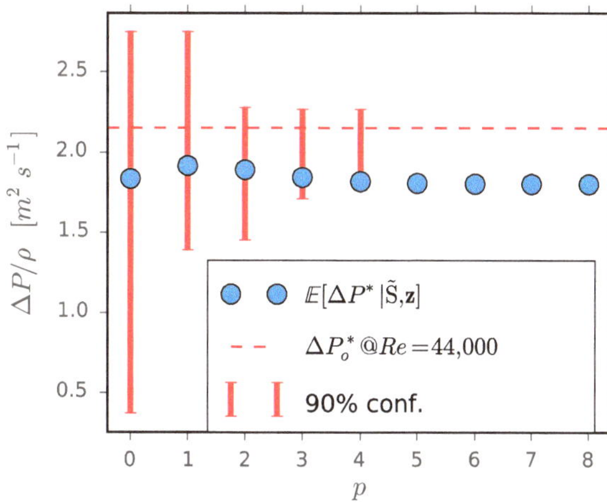

Fig. 3 BMSA prediction of pressure drop including confidence interval for several tuning parameter p of smart-weighting method

unsteady fluctuation of the separation and reattachment points correctly [17]. Thus, the application of BMSA to this challenging flow configuration is a test of the limits of the framework utilising linear eddy-viscosity models being calibrated for flat plate boundary-layer flows.

Simulation Set-up

In order to mimic the periodicity of the hill-geometry, periodic boundary conditions were applied at the inlet and outlet and no-slip conditions at the walls. A volume forcing is applied to each cell, which maintains a bulk velocity of $U_b = 1.0$ between the hill's crest and the upper surface. The functional form of the lower surface is defined according to the ERCOFTAC test case description.[3] In order to study the effect of the turbulence modelling and not additional modelling due to wall functions, the thickness of the smallest cells close to the wall were defined according to $y^+ \leq 1.0$ and Low-*Re* turbulence models were used: Spalart-Allmaras, Launder-Sharma $k - \epsilon$ and Wilcox (2006) $k - \omega$ (for details of these models see [4]). Furthermore, a mesh convergence study was conducted for each model using baseline coefficients and the meshes in Table 3 were used for the application of the BMSA method.

Similar to the pipe flow case in section "Incompressible Pipe Flow at Re = 44,000", the results of the forward simulations $m_{CFD}\left(M_i, \boldsymbol{\theta}_{i,k}^{MAP}\right)$ with nominal values of the coefficients $m_{CFD}(M_i, \boldsymbol{\theta}_{i,k}^{o})$ are used as an initial condition in order to reduce iteration counts. For $S_1 4$, the simulation using Wilcox (2006) $k - \omega$ didn't converge so that this scenario was excluded from the set leading to 13 used scenarios in total.

[3]Underlying flow regime 3–30, 2D Periodic Hill Flow: http://qnet-ercoftac.cfms.org.uk.

Table 3 Mesh size per closure model for flow of periodic hills at $Re_H = 5600$

	Mesh ($n_x \times n_y$)
Spalart-Allmaras	100×110
Launder-Sharma $k - \epsilon$	150×140
Wilcox (2006) $k - \omega$	100×110

BMSA Prediction for Velocity Field

The baseline simulations for the three models show the expected behaviour as reported in the literature for a larger *Re*-number [17]: while both Spalart-Allmaras and the Wilcox (2006) $k - \omega$ over-predict the size of the recirculation zone, characterised by a zero velocity component close to the wall, the Launder-Sharma $k - \epsilon$ under-predicts this flow feature.

The BMSA approach using MAP estimates of the posterior probability distributions outputs both an expectation of the velocity $\mathbb{E}[\Delta|\tilde{\mathrm{S}}, \mathbf{z}]$, for which $\Delta = U_x$, and a confidence interval. For the latter two types were chosen: one obtained by sampling from the posterior predictive distribution and using the range from 1.0 to 99.0 percentile, i.e. 98% confidence interval, and another one by using the min/max range of the ppd. We have chosen both a larger confidence interval and the entire range of the distribution in order to take the expected low performance of the linear eddy-viscosity models for this test case into account. Especially, the latter can be seen as an approach trying to envelop the true process by all models and for all scenarios. The smart-weighting technique to obtain the scenario probabilities as described in section "Prediction" was applied along the y/H-direction for each stream-wise location for the 98% confidence interval. In that manner, the predictive similarity for each scenario per model is evaluated locally for the stream-wise direction x.

The BMSA expectation $\mathbb{E}[\Delta|\tilde{\mathrm{S}}, \mathbf{z}]$, as shown in Fig. 4, over-predicts the recirculation zone, gives similar velocity profiles for $x = 3.0$ and 4.0 close to the lower surface, but is completely off for other locations. Especially, for $x = 0.0$ the expectation does not capture the local maximum of the velocity close to $y/H = 1.0$, i.e. at the hills crest, and over-predicts the velocity within the channel for every other location. Throughout the different locations the expectation shows large differences compared to the DNS for the upper part of the velocity profile.

The entire range of the posterior predictive distribution (ppd) and the 98% confidence interval show large differences over the entire domain as shown in Fig. 4. The ppd captures the DNS data for $x \geq 3$, but does not capture the local velocity maxima in the area of the free shear layer in the leeward region of $x =$ 0–2.0. Interestingly, for $x \leq 3.0$ when the range of the ppd shrinks locally in y/H-direction also the DNS is still inside. However, this pattern is not the same for every y/H-position, e.g. at $x = 2.0$ the expectation and the DNS match for $0.5 \leq y/H \leq 1$, but the error is high. For the relatively large confidence value 98%, the intervals shrink drastically, so that

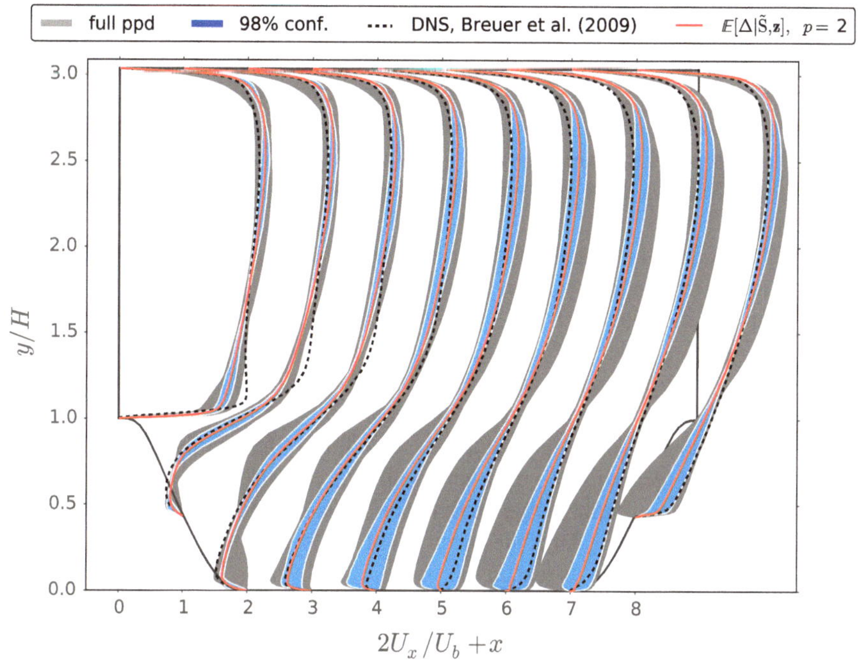

Fig. 4 BMSA for stream-wise velocity component U_x at several x-locations, $\mathbb{E}[\Delta|\tilde{S},\mathbf{z}]$ (red), full posterior predictive distribution (ppd) range (shaded grey) and confidence interval of 98% (shaded blue) using $p=2$. DNS data of Breuer et al. [15] (black)

the DNS data is only captured in a small lower band $y/H \leq 1.0$ for $3 \leq x \leq 7$ and for the part of the velocity profile in the middle of the channel.

Interestingly, in the reattachment region between the two hills, where the case is similar to a flat plate, the BMSA approach in the present form with a confidence of 98% is able to capture the DNS data. However, the interval for the upper surface is always negligible but the DNS data does not match with the expectation.

The large effect of the min/max profiles, which determine the min/max range of the ppd, to capture the DNS data proves the aforementioned fact that the used linear eddy-viscosity models suffer from restrictions which inhibit the reproduction of the true flow state for this test case. However, posterior probability distributions of the closure coefficients of the models inferred on other flow scenarios more equal to the one here might lead to a different picture.

Industrial Test Case: Generic Dassault Falcon

The final test case we consider is the generic Falcon jet—a business jet including tail, engine nacelles and winglets—at transonic conditions (Industrial challenge IC-

03 of UMRIDA). The geometry is far more complex than any considered so far, with a resulting spectrum of primary and secondary flows. We provide this case as a demonstration of the applicability of our approach to problems of relevance in the aerospace industry—unfortunately without a reference solution. LES is impractically expensive here, and experimental data is not available. We attempt to justify the results of BMSA based on the expected response of the simulation to the closure modelling.

The CFD code used is Petrov–Galerkin finite-element RANS solver *AETHER* used within Dassault Aviation. As a preliminary step, it was verified that the closure models implemented in *AETHER* were identical to the models for which closure coefficients were calibrated. The implementations of Spalart-Allmaras and Wilcox $k-\omega$ were found to be sufficiently similar to the calibrated models, and in addition are regarded by Dassault Aviation as suitable for this test case. Of the 14 sets of closure coefficients computed with each model, only five cases were able to be successfully converged with $k-\omega$, whereas all S-A cases converged without issue. BMSA requires a minimum of two models per scenario, and as such only five scenarios could be considered. Thus the spectrum of model results is substantially more limited than for both previous test cases.

The resulting uncertainty in the pressure distribution at a cut on the wing (with $p=0$) can be seen in Fig. 5a, b. The uncertainty is concentrated around the shock on the suction side, and even there is barely visible. Increasing p reduces the uncertainty further. The very low variance can be attributed to two effects: (a) the real lack of sensitivity of the pressure distribution to the turbulent boundary layer in an essentially attached flow, for which an inviscid solution is already satisfactory, and (b) the limited range of models and coefficients used in the study. Indeed by eliminating exactly those coefficients that cause convergence problems (for $k-\omega$), we are introducing additional bias into the sampling, likely biasing against extreme results. The only alternative—of including unconverged solutions—is even less attractive however.

More insight can be gained by looking at force coefficients: Fig. 6 shows total, pressure and viscous drag coefficients, all evaluated by surface integration, with mean and 95% confidence intervals, as a function of p. While pressure and viscous drag have approximately the same magnitude, the viscous part completely dominates the uncertainty—with $C_{D,p}$ varying at most 3 counts, and $C_{D,v}$ varying between 15 and 25 counts, depending on p. Thus the uncertainty in total drag is driven entirely by $C_{D,v}$. Given the observed lack of uncertainty in C_p-profiles, and the high sensitivity of friction coefficients to closure modelling, this is not surprising—though it does indicate that varying the closure model does not appear to have a significant effect on separation behaviour in this case. Examining the relationship of uncertainty against p reveals which quantities are dominated by model-differences (where uncertainty is approximately constant with p), and for which quantities scenario differences are significant. Here $C_{D,p}$ belongs to the former class, and $C_{D,v}$ to the latter.

In summary, the limited number of turbulence models and range of coefficients lead to limited uncertainty in this case, for quantities of interest related to the pressure, which is essentially governed by the inviscid flow behaviour. Significant uncer-

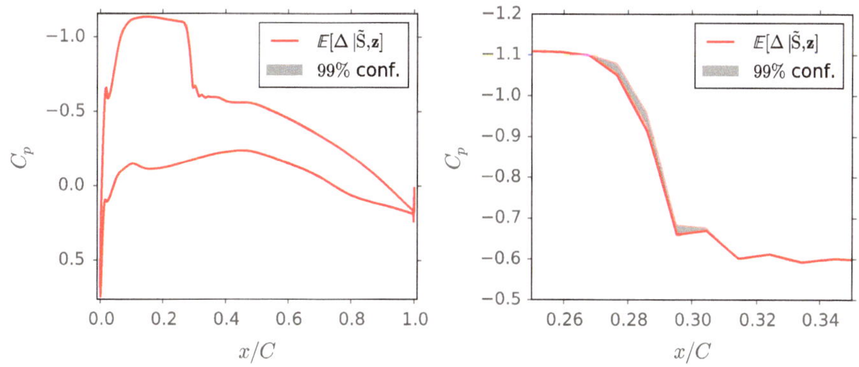

(a) Full x/C domain.
(b) Zoom at shock location.

Fig. 5 Pressure distribution at 30% span of the Falcon wing, $y = 3848$ mm

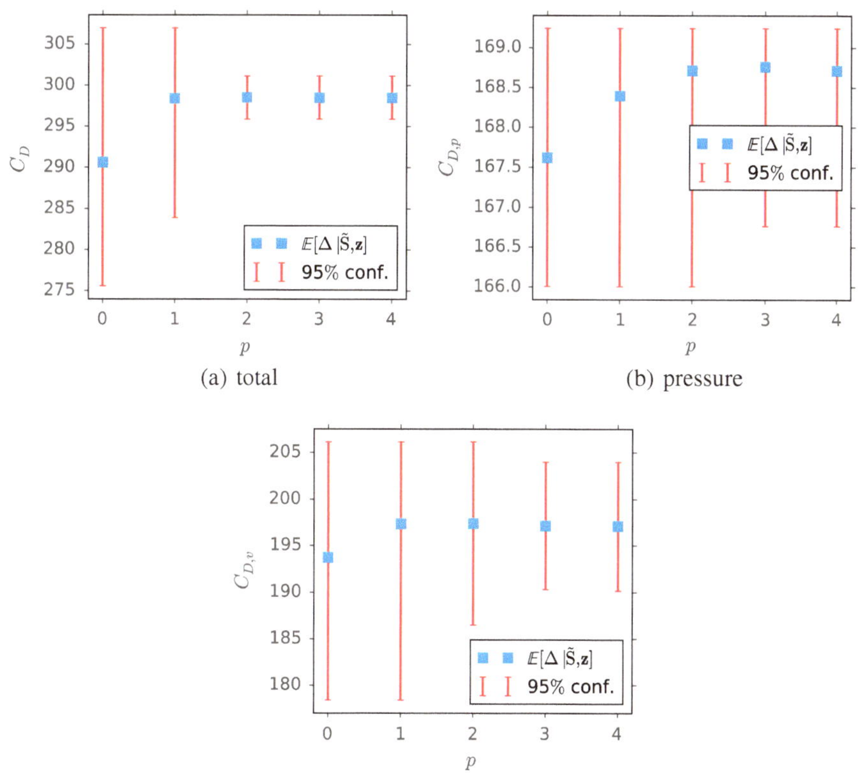

(a) total
(b) pressure
(c) viscous

Fig. 6 Total, pressure and viscous drag coefficients for the full Falcon tainty is observed for viscous quantities such as friction drag. Resolving this in practical applications is the subject of ongoing work.

Conclusion

The BMSA approach based on MAP estimates of the posterior probability distributions of the closure coefficient has been applied to three different test cases.

For the flow in a straight pipe at $Re = 44{,}000$, the results for the velocity profile and the pressure drop are in line with the validation data for this case, which shows in general that the approach can successfully be applied to other test cases outside of the set used for calibration.

The application of BMSA to the periodic hill test case at $Re_H = 5600$ proves that this test case is challenging for the linear eddy-viscosity assumption and also shows the limits of the BMSA framework in the present form, provided that all models in the chosen set employ the Boussinesq hypothesis. However, the resulting ppd is able to capture most of the validation data successfully. A more suitable test case will be evaluated for BMSA in the future, such as the flow over a backward-facing step, for which the separation is forced by the geometry and not result of the simulation itself. Furthermore, the question of how to incorporate model-form error within a stochastic framework for UQ purposes is the topic of ongoing research.

Finally, the Falcon Jet test case (IC-03) revealed the open problem of how to deal with many non-converged solutions within the BMSA methodology and the successive artificial reduction of uncertainty by excluding these simulations from the set of simulations, which is also a topic for further research.

Acknowledgements We thank Dassault Aviation and especially Gilbert Roge for the collaboration on the test case of the generic falcon jet.

References

1. Pope, S.B.: Turbulent Flows. Cambridge University Press (2000)
2. Slotnick, J., Khodadoust, A.: CFD Vision 2030 Study: A Path to Revolutionary Computational Aerosciences. Technical report, NASA—National Aeronautics and Space Administration, Langley Research Center Hampton, Virginia 23681–2199 (2014)
3. Edeling, W.N., Cinnella, P., Dwight, R.P., Bijl, H.: Bayesian estimates of parameter variability in the k-epsilon turbulence model. J. Comput. Phys. **258**, 73–94 (2014)
4. Edeling, W.N., Cinnella, P., Dwight, R.P.: Predictive RANS simulations via bayesian model-scenario averaging. J. Comput. Phys. **275**, 65–91 (2014)
5. Edeling, W.N.: Quantification of modelling uncertainties in turbulent flow simulations. PhD thesis, Technical University Delft, Aerospace Department (2015)
6. Edeling, W.N., Schmelzer, M., Dwight, R.P., Cinnella, P.: Bayesian predictions of reynolds-averaged Navier–Stokes uncertainties using maximum a posteriori estimates. AIAA J. (Article in Advance) (2018)

7. Coles, D., Hirst, E.: Computation of turbulent boundary layers. In: Proceedings of AFOSR-IFP Stanford Conference, vol. 2 (1968)
8. Cheung, S.H., Oliver, T.A., Prudencio, E.E., Prudhomme, S., Moser, R.D.: Bayesian uncertainty analysis with applications to turbulence modeling. Reliab. Eng. Syst. Safety **96**(9), 1137–1149 (2011)
9. Gelman, A., Carlin, J., Stern, H., Rubin, D.: Bayesian data analysis, vol. 2. Chapman & Hall/CRC Boca Raton, FL, USA (2014)
10. Hastings, W.K.: Monte carlo sampling methods using markov chains and their applications. Biometrika **57**(1), 97–109 (1970)
11. Edeling, W.N., Cinnella, P., Dwight, R.: MCMC Traces of Posterior Distributions of Closure Coefficients for 6 Turbulence Models [online database]. https://www.researchgate.net/publication/308609583_MCMC_traces_of_posterior_distributions_of_closure_coefficients_for_6_turbulence_models_ASCII_format (2016). Accessed 25 Sept 2016
12. Wu, X., Moin, P.: A direct numerical simulation study on the mean velocity characteristics in turbulent pipe flow. J. Fluid Mech. **608**, 81–112 (2008)
13. Venkateshan, S.: Mechanical Measurements. 2 edn. Wiley (2015)
14. OpenFOAM: OpenFOAM User Guide. OpenFOAM Foundation (Feb 2014)
15. Breuer, M., Peller, N., Rapp, C., Manhart, M.: Flow over periodic hills—numerical and experimental study in a wide range of Reynolds numbers. Comput. Fluids **38**(2), 433–457 (2009)
16. Manceau, R.: Report on the 10th joint ERCOFTAC (SIG-15)/ IAHR / QNET-CFD Workshop on Refined Turbulence Modelling. ERCOFTAC Series **57** (2002)
17. Jakirlic, S.: Extended Excerpt Related to the Test Case: Flow Over a Periodical Arrangement of 2D Hills. Technical Report June (2012)

Uncertainties for Thermoacoustics: A First Analysis

A. Ndiaye and F. Nicoud

Introduction to the Thermoacoustic Framework

Despite decades of intense research, combustion instabilities remain a challenging topic in a range of engineering applications [1–6]. In particular, thermoacoustic instabilities arise from the coupling between the combustor acoustics (with related waves propagation and reflection) and flame dynamics (and associated heat release fluctuations). These heat release fluctuations are generally delayed with respect to incident disturbances (noise, modulation of mixture fluctuations, convection of hydrodynamic processes, etc.) and give rise to an unstable growth of pressure oscillations if they happen to be in phase with the acoustic pressure fluctuations [7]. In extreme cases, these oscillations lead to irreversible damages which can destroy the combustor or at least decrease its lifetime. Therefore, the prediction and control of all acoustic modes of the configuration at the design stage are required to avoid these instabilities and their negative impact on the combustion system of interest.

Several approaches are available to model and compute combustion instabilities. For example, large eddy simulation techniques appeared to be tremendously powerful to capture both the combustion dynamics and even unstable and various types of acoustic modes in complex gas turbines' combustors [8–11]. Nevertheless, LES approaches are known to be CPU expensive because they require solving the full 3D Navier–Stokes equations at high Reynolds number by taking into account many complex physical processes in addition to incidental effects related to acoustics and combustion phenomena. Moreover, if LES offers a nice picture of the flow structure in a combustor, they do not explain the mechanisms by which unstable modes appear.

A. Ndiaye
CERFACS, 42 avenue Gaspard Coriolis, 31052 Toulouse, France
e-mail: ndiaye@cerfacs.fr

F. Nicoud (✉)
Université de Montpellier, Place Bataillon, 34095 Montpellier, France
e-mail: franck.nicoud@umontpellier.fr

C. Hirsch et al. (eds.), *Uncertainty Management for Robust Industrial Design in Aeronautics*, Notes on Numerical Fluid Mechanics and Multidisciplinary Design 140, https://doi.org/10.1007/978-3-319-77767-2_5

These difficulties have encouraged the widespread development and use of low-order modelling techniques such as analytical tools or Helmholtz solvers. The advantage of low-order modelling tools [12–14] lies in their ability to determine at reasonable CPU cost both the eigenfrequencies and the growth or the decay rate of thermoacoustic modes. In this study, a 3D parallel in-house Helmholtz solver is used to represent thermoacoustic instabilities [15]. Such Helmholtz solver is based on an approximate linear wave equation for small pressure disturbances, $p_1(x,t)$, which is derived from the Navier–Stokes equations for reactive flows [6]:

$$\frac{\partial^2 p_1(\mathbf{x},t)}{\partial t^2} - \nabla \cdot c_0^2(\mathbf{x}) \nabla p_1(\mathbf{x},t) = (\gamma - 1)\frac{\partial q_1(\mathbf{x},t)}{\partial t} \tag{1}$$

where $q_1(\mathbf{x},t)$ is the heat release fluctuation. In the framework of linear acoustics, the pressure fluctuations as well as the heat release fluctuations have the form of an harmonic waves: $p_1 = \Re[\hat{p}(\mathbf{x})e^{-i\omega t}]$ and $q_1 = \Re[\hat{q}(\mathbf{x})e^{-i\omega t}]$. Therefore, using the harmonic assumption for the fluctuating variables, the wave equation in the time domain, Eq. (1), becomes the following Helmholtz equation in the frequency domain:

$$\nabla \cdot c_0^2(\mathbf{x}) \nabla \hat{p}(\mathbf{x}) + \omega^2 \hat{p}(\mathbf{x}) = i\omega(\gamma - 1)\hat{q}(\mathbf{x}) \tag{2}$$

where c_0 is the speed of sound of the baseline flow, ω is the complex-valued pulsation ($\omega = 2\pi f$), ρ_0 corresponds to the mean density, and $\hat{q}(\mathbf{x})$ represents the unsteady heat released from the flame: $q'(\mathbf{x},t) = \hat{q}(\mathbf{x})e^{-i\omega t}$.

The right-hand side term of Eq. (2) represents the flame response to acoustic perturbations. In this study, it is modelled thanks to the $n - \tau$ formalism initially introduced by [16] and [17] for confined flames. This formalism is also related to the Flame Transfer Function (FTF) formulation:

$$\mathscr{F}(\omega) = \frac{\hat{Q}}{\hat{u}} = n(\omega)e^{i\omega\tau(\omega)} \tag{3}$$

where n is the gain of the flame response and τ the time delay between the overall unsteady heat release produced by the flame $\hat{Q} = \int_{V_f} \hat{q}\, dV$ and the velocity fluctuation $\hat{u}$ measured at the burner mouth, in the cold gas region. The latter parameters can be determined either experimentally [18], analytically [19] or numerically [20].

Once the flame response is modelled, Eq. (2) corresponds to an inhomogeneous Helmholtz equation which is then solved as a nonlinear eigenvalue problem in the frequency domain. The resolution of the eigenvalue problem gives access to the pressure field $\hat{p}$ and the complex angular frequency ω of the mode. The real part of ω, ω_r, is related to the frequency of oscillation of the mode while its imaginary part, ω_i, is the growth rate of the acoustic disturbances. Therefore, from the sign of ω_i, it is possible to build a stability map which gathers all the thermoacoustic modes as shown in Fig. 1.

The stability of a combustion chamber may be drastically impacted by many effects ranging from combustion chemistry, impedance boundary conditions,

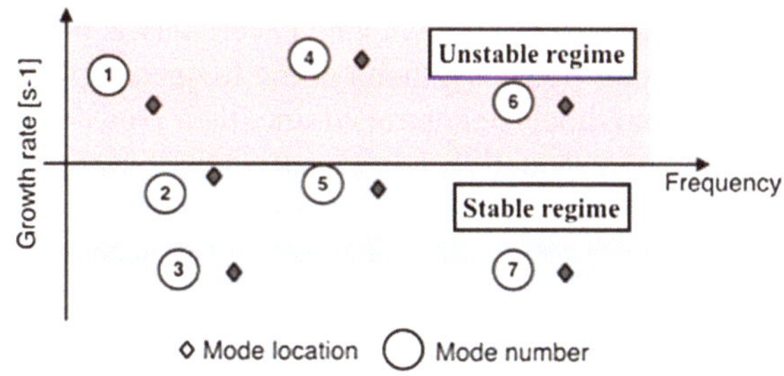

Fig. 1 Location of the first six thermoacoustic modes in a typical combustor

geometry, swirler and combustor design, wall heat transfer, inlet temperatures to spray characteristics. Aside from impedance boundary conditions and chamber design away from the flame, all the above-mentioned factors are embodied in the Flame Transfer Formulation that accounts for the Flame/Acoustic modelling response. As of today, a key challenge remains in the development of accurate and predictive combustion response models to detect potential combustor instability. Effective modelling of the flame dynamics would certainly improve the understanding of processes such as nonlinear phenomena responsible for limit-cycle oscillations, the flame–acoustic coupling in industrial geometries, flame–vortices interactions and the interaction of flames with distributed reaction zones or well-stirred reactors. Due to the limited knowledge on all the aforesaid phenomena and the well-known sensitivity of the flame response to any details (surface condition of the injection system, manufacturing tolerances, fuel chemical composition, inlet conditions, thermal conditions of the mainframe, etc), introducing Uncertainty Quantification in the context of deterministic mathematical modelling to analyse the probabilistic aspects of the simulation of thermoacoustic instabilities is an appealing perspective.

One of the overriding concern of the present work is to perform Uncertainty Quantification to address the sensitivity of thermoacoustic results with respect to the flame modelling response parameters. Until very recently, Uncertainty Quantification had virtually never been applied to thermoacoustics. This task will shed more light towards reliable predictions of unstable modes beyond allowing to get insight on the likely risk of a mode to be amplified or damped in gas turbine combustors. Figure 2 represents a stability map from a thermoacoustic analysis under Uncertainty

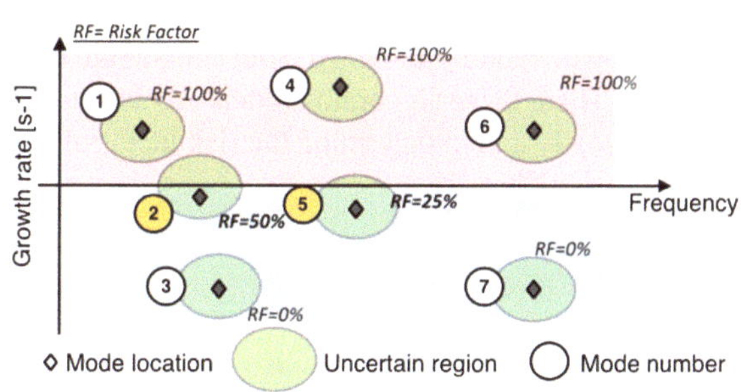

Fig. 2 Location of the first six thermoacoustic modes in a typical combustor on top of their uncertain regions

Quantification analysis. When no uncertainty is present, each mode corresponds to a single point (black symbols) in the frequency plane. Here, modes 1, 4 and 6 are unstable and should be controlled since their growth rate ω_i is positive. If uncertainties are present, each mode belongs to an admissible region of the frequency plane. Mode 2 and 5 are now potentially dangerous and should be controlled too. Therefore, Uncertainty Quantification analysis will contribute to study how the uncertainties on the input parameters n and τ propagate into uncertainties on the growth rate ω_i. Introducing Uncertainty Quantification naturally gives rise to a new concept in thermoacoustics, namely the risk factor of the acoustic mode, i.e. the probability for a mode to be unstable ($\omega_i > 0$):

$$\textbf{Risk Factor}\,(\%) = 100 \int_0^{\infty} PDF(\omega_i)\, d\omega_i \qquad (4)$$

where $PDF(\omega_i)$ stands for the probability density function of the growth rate of the acoustic disturbances. This quantity allows a continuous classification of the thermoacoustic modes while the classical analysis is only binary (stable vs. unstable). Note that to fairly assess the risk factor, it is necessary to have a realistic statistical distribution of the input parameters n and τ, given by experimental data or early numerical results.

The remaining of this section is devoted to a first UQ analysis of a swirled stabilized combustion chamber, developed and built at EM2C laboratory in Paris (France) [18, 21]. Two input parameters are considered: the amplitude (n) and time delay (τ) of the Flame Transfer Function formulation. At first, taking advantage of the affordable numerical resources associated to the Helmholtz solver, a brute force Monte Carlo approach is used to propagate random perturbations on the FTF input parameters and compute probabilistically the growth rate of the acoustic mode. As the literature does not confer a clear accurate analysis on the uncertainty range of the FTF input parameters, an uncertainty range based on experimental measurements of the flame response (from experimentalists at EM2C (Paris) and IMFT (Toulouse)) was used. Besides, in the absence of more information regarding the probability density functions, two different distributions were considered: a uniform distribution and a β-distribution with the same mean and variance. The ranges of the uniform distributions are directly deduced from the experimental data of the FTF amplitude and time delay. Then, the UQ analysis is extended by using a reduced two-step UQ strategy to deal efficiently with thermoacoustic phenomena in such a system. First, three surrogate models are tuned from a moderate number of Helmholtz solutions (a few tens). Then, these algebraic models are used to perform a Monte Carlo analysis even less costly and to determine the risk factor of the acoustic mode.

Experimental Set-up Description

The laboratory-scale experiment used in this study corresponds to a single swirled stabilized combustor designed and built by Palies et al. [18, 21, 22] at the EM2C laboratory. Initially, this academic system was used to investigate the nonlinear mechanisms involved in the flame dynamics of complex systems. As illustrated in Fig. 3, the system features a confined swirled flame, an upstream manifold, an injection unit equipped with a swirler and a cylindrical flame tube. The fuel/oxidizer is injected through the sidewalls located at the bottom of the upstream manifold. Once formed, the mixture flows through the honeycomb grid to wreck large-scale turbulent structures. Then, the gas stream is accelerating into the convergent tube to decrease the boundary layer thickness. The motion stabilizes the turbulent flame within the combustion chamber, despite the absence of bluff-body.

This experiment is handy and practical because it was thought and conceived in such a way that both the upstream manifold and the combustion chamber may take, respectively, three and four different lengths. Hence, this simple system leads to twelve possible geometries as summarized in Table 1.

To measure the flame response, a loudspeaker is placed at the back end of the system. Moreover, two experimental conditions corresponding to two different air flow rates were experimentally tested corresponding to flames **A** and **B**, with larger power in the latter than in the former. Thus, from twelve possible geometries, the system offers the advantage to investigate finally 24 different operating conditions. Also, acoustic losses of the system were measured during the experimental phase. This has been realized by sending an acoustic wave through the combustion chamber

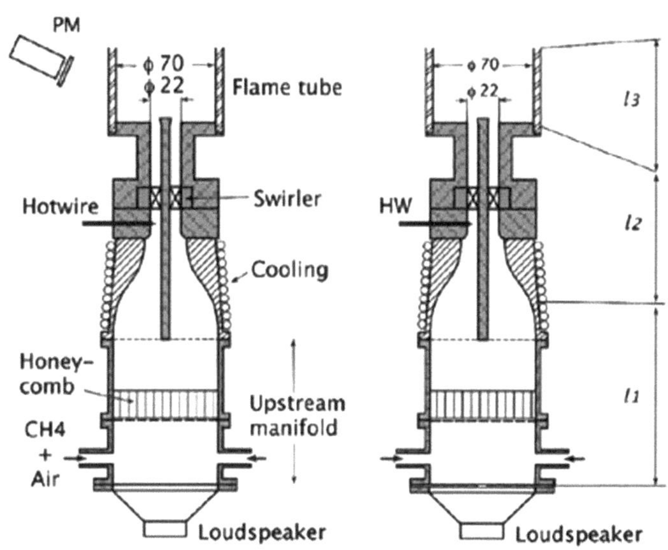

Fig. 3 Numerical/experimental configuration. From Ref. [21]

Table 1 Twelve different configurations explored: l_1 indicates the upstream manifold length and l_3 corresponds to the combustion chamber length. Dimensions are given in millimetres. From Ref. [23]

Cases studied		$l_3=100$	$l_3=150$	$l_3=200$	$l_3=400$
Expe./Simu.	$l_1=96.0$	C01	C02	C03	C04
Expe./Simu.	$l_1=160.0$	C05	C06	C07	C08
Expe./Simu.	$l_1=224.0$	C09	C10	C11	C12

to measure the response of the flame for a range of frequencies around resonance. These losses are expressed for both types of flames: $\alpha_A = 82\ \text{s}^{-1}$ for flame A and $\alpha_B = 125\ \text{s}^{-1}$ for flame B with an uncertainty of $\Delta\alpha = \pm 10\ \text{s}^{-1}$.

The numerical acoustic modelling of the swirled combustor and its associated linear stability analysis has been realized [23] via a Helmholtz solver developed at CERFACS (the AVSP solver [15]). Since no damping is contained in the Helmholtz equation, the system is considered to be stable when the growth rate ω_i is smaller than the damping rate α and similarly, when the computed growth rate is larger than the damping rate, the system is considered to be unstable. Moreover, accounting for the error $\Delta\alpha$ leads to the subsequent classification:

- Stable **S**: $\omega_i < \alpha - \Delta\alpha$
- Unstable **U**: $\omega_i > \alpha + \Delta\alpha$
- Marginal **S/U**: $\alpha - \Delta\alpha < \omega_i < \alpha + \Delta\alpha$.

Experimentally, a mode is denoted **S/U** when a low amplitude frequency of oscillation is detected, **S** if no fluctuation appears and **U** if a large amplitude limit cycle is observed. The global comparative study between the experimental and numerical stability results [23] is displayed in Fig. 4 and sum up in Table 2.

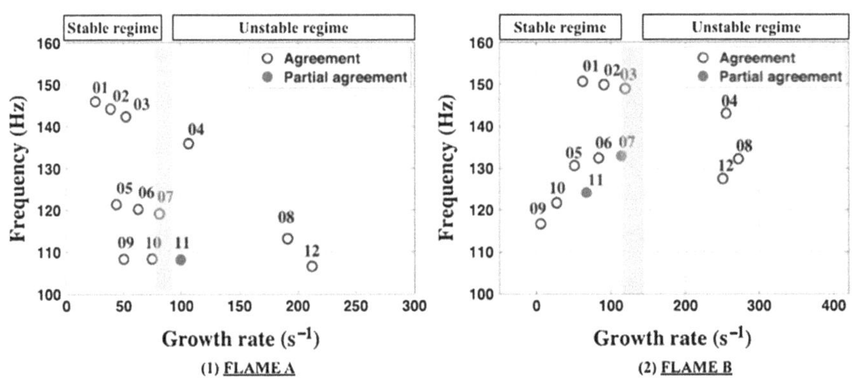

Fig. 4 Linearized stability prediction. The grey bounds indicate the marginally stable region defined by $\Delta\alpha = \pm 10\,\text{s}^{-1}$. Empty symbols indicate agreement with experimental results while filled symbols represent partial agreement. Adapted from [23]

Table 2 Linear stability analysis of flame A and flame B. Comparison between experimental and numerical results. (**S**) Stable, (**S/U**) marginally stable/unstable, (**U**) unstable. The geometrical configurations C01–C12 are defined in Table 1

Case	Flame A				Flame B			
	C01	*C02*	*C03*	*C04*	*C01*	*C02*	*C03*	*C04*
Experiment	**S**	**S**	**S**	**U**	**S**	**S**	**S-U**	**U**
Simulation	**S**	**S**	**S**	**U**	**S**	**S**	**S-U**	**U**
	C05	*C06*	*C07*	*C08*	*C05*	*C06*	*C07*	*C08*
Experiment	**S**	**S**	**S-U**	**U**	**S**	**S**	**S**	**UU**
Simulation	**S**	**S**	**S-U**	**U**	**S**	**S**	**S-U**	**U**
	C09	*C10*	*C11*	*C12*	*C09*	*C10*	*C11*	*C12*
Experiment	**S**	**S**	**S-U**	**U**	**S**	**S**	**S-U**	**U**
Simulation	**S**	**S**	**U**	**U**	**S**	**S**	**S**	**U**

An overall agreement in most of the cases is observed for the numerical and experimental stability analysis. Only three partial disagreements are observed because the experiment predicts marginal stability (**S**/**U**) while the computation gives an instability or conversely. Uncertainty Quantification is thus used to further analyse one of these operating points with partial disagreement and investigate if the difference between the computational and experimental results could be explained by the lack of knowledge of the input parameters: configuration 07 for the flame B was selected for this purpose.

Uncertainty Quantification Analysis

Force Monte Carlo Method

The Monte Carlo method is a conventional Uncertainty Quantification algorithm frequently used due to its conceptual simplicity and straightforward implementation. It is used here to propagate the flame response uncertainties through the system with the aim to forecast the PDF of the growth rate and subsequently the probability of the first acoustic mode of the system to be unstable ($f_0 = \frac{\omega_0}{2\pi} = 133$ Hz), i.e. the risk factor of the first acoustic mode.

Having only limited information regarding the actual range of uncertainty relevant to the Flame Transfer Function parameters, quantitative data from two independent experimentalists groups at (i) EM2C (Paris) and (ii) IMFT (Toulouse) was collected. They provide statistically valid and reliable estimation of the uncertainties on n and τ parameters which is fixed to $\frac{\sigma_n}{\bar{n}} = \frac{\sigma_\tau}{\bar{\tau}} = 10\%$. This range of uncertainty is therefore adopted and applied to the following nominal experimental value: $\bar{n} = 1080$ J/m and $\bar{\tau} = 4.73$ ms. Also, the type of distribution followed by the FTF parameters is not

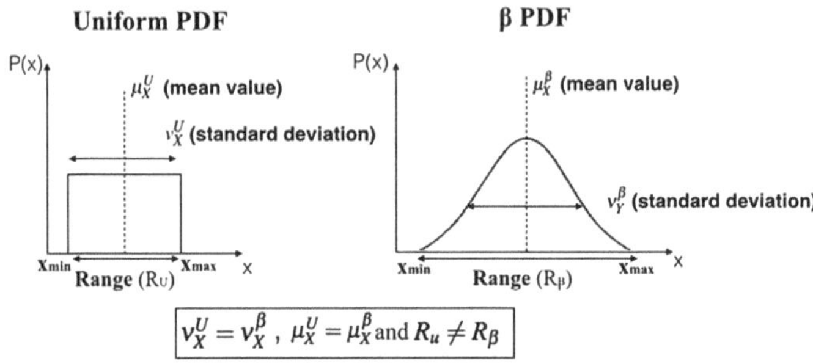

Fig. 5 Uniform and β-PDF of an arbitrary random variable X with similar mean (μ) and standard deviation (σ), but with different range (R)

known and one needs to make sure that the shape of the PDF has only a limited impact on the risk factor value. This is done by considering two typical distributions, namely (i) a uniform distribution and (ii) a β-distribution (Fig. 5).

- **The uniform distribution**: The ranges of the uniform distributions are directly deduced from the experimental values of the amplitude and time delay, i.e 10% of the mean values (Fig. 5). The uniform PDF reads:

$$f_X^U = \frac{1}{||x_{max} - x_{min}||} \quad \text{for} \quad x_{min} \leq x \leq x_{max} \tag{5}$$

Therefore, the mean μ_X^U and the variance υ_X^U are:

$$\mu_X^U = \frac{x_{min} + x_{max}}{2} \quad \text{and} \quad \upsilon_X^U = \frac{1}{12}(R_U \mu_X^U)^2 \tag{6}$$

where R_U represents the normalized range $\frac{x_{max} - x_{min}}{\mu_X^U}$ of the uniform distribution: here $R_U = 10\%$.

- **The β-distribution**: The β-distribution is characterized by its density function:

$$f_Y^\zeta = B(\alpha, \zeta)^{-1} y^{\alpha-1}(1-y)^{\zeta-1} \quad \text{for} \quad 0 \leq y \leq 1 \tag{7}$$

where $B(\alpha, \zeta) = \frac{\Gamma(\alpha)\Gamma(\zeta)}{\Gamma(\alpha+\zeta)}$ denotes the beta function, $\Gamma(.)$ is the gamma function, and α and ζ are two free parameters. Note that f_Y^ζ is only defined for a reduced random variable Y on [0, 1]. The parameters α and ζ which characterize the β-PDF are deduced from the desired mean μ_Y^ζ and variance υ_Y^ζ of this reduced variable Y:

$$\alpha = \mu_Y^\zeta \left(\frac{\mu_Y^\zeta(1 - \mu_Y^\zeta}{\upsilon_Y^\zeta} - 1 \right) \tag{8}$$

and

$$\zeta = (1 - \mu_Y^\zeta \left(\frac{\mu_Y^\zeta(1 - \mu_Y^\zeta}{\upsilon_Y^\zeta} - 1 \right) \tag{9}$$

To close the problem, the reduced variable Y in [0, 1] is related to the desired random variable X in $[x_{min}, x_{max}]$:

$$X = \mu_X^\zeta(1 + R_\zeta[2Y - 1]) \tag{10}$$

Taking the mean and variance of the previous equation leads to the following relations between characteristics of X and Y:

$$\mu_Y^\zeta = 1/2 \text{ and } \upsilon_Y^\zeta = \frac{\nu_X^\zeta}{4R_\zeta^2}(\mu_X^\zeta)^2 \tag{11}$$

Consequently, the mean value of Y is fixed and its variance can be deduced by imposing that the beta and uniform PDFs have the same characteristics, i.e. $\mu_X^\zeta = \mu_X^U$ and $\nu_X^\zeta = \nu_X^U$. Note, however, that the range of the β-PDF appears in (μ_X^ζ) (Eq. 11). If this range is chosen equal to the range of the previous uniform PDF (i.e. $R_\zeta = R_U = 10\%$), then the ζ-distribution degenerates to the previous uniform PDF. Consequently, the range R_ζ is an additional free parameter. For this study, this range is fixed to $R_\zeta = 30\%$ leading to the characteristic values $\alpha = \zeta = 2.87$.

The Monte Carlo sampling realized for the configuration 07 of the Flame B (see Table 2) using the uniform distribution is presented in Fig. 6.

In Fig. 6a, each point corresponds to a Helmholtz simulation in the complex domain. The horizontal solid lines denote the acoustic losses α: 115 $s^{-1} < \alpha_B <$ 135 s^{-1}. The stable or unstable regions are evaluated using the difference $\omega_i - \alpha$:

1. $\omega_i - 115\ s^{-1} < 0$ corresponds to a stable system (**S**).
2. $\omega_i - 135\ s^{-1} > 0$ corresponds to a unstable system (**U**).
3. $115\ s^{-1} < \omega_i < 135\ s^{-1}$ corresponds to a situation where the system is marginal (neither stable nor unstable) (**S/U**).

The 4000 samples are then classified into three types: stable regime (**S**), unstable regime (**U**) and marginal regime (**S/U**). In Fig. 6b, the PDF of the growth rate (ω_i) is presented and shows that most of the thermoacoustic modes found by the Helmholtz solver are in the stable regime. This leads to a risk factor close to 24%.

Following a similar methodology as for the uniform distribution, 4000 runs have been performed using the Helmholtz solver by considering a β-distribution for the input parameters n and τ. It has been found that the risk factor obtained from the β-distribution is close to the one obtained by the uniform distribution: 24% for the uniform distribution against 22% for the β-distribution. This shows that UQ results are weakly affected by the distributions chosen for the input parameters n and τ for the study of such academic cases which suggests that assessing the risk factor of a mode without a clear knowledge of the uncertainties on the input data is relevant. In the rest

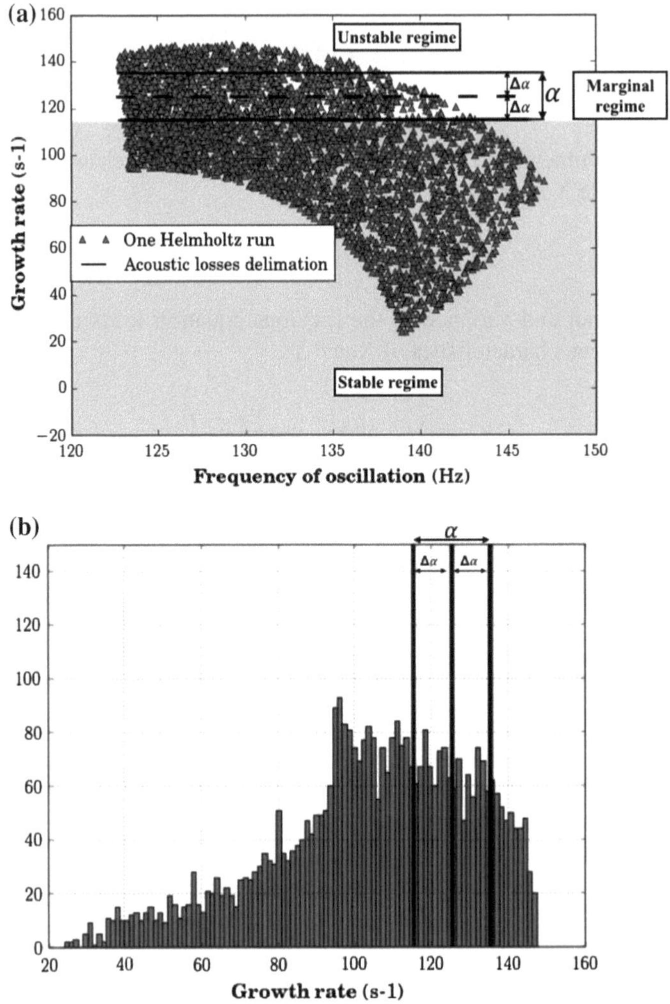

Fig. 6 **a** Uncertainty region for the first acoustic mode for a uniform PDF with 10% uncertainty on both the amplitude and time delay of the flame response. **b** Histogram of the growth rate of acoustic disturbance for 4000 samples using the uniform distribution for the parameters n and τ

of the study, only the uniform distribution is kept. Moreover, the risk factor being 22–24%, this simple UQ analysis shows that the computation is actually consistent with the experimental data. Indeed, accounted for a realistic 10% uncertainty in the flame response, this risk factor value means that the mode of interest is computationally found stable in approximately 76–78% of the cases (recall that the mode of Case 07—Flame B was observed stable in the experiment; see Table 1).

Multiple Linear Regression

Because Eq. (2) is an eigenvalue problem which is nonlinear in ω_i, the response surface $\omega_i = \omega_i(n, \tau)$ is implicit and nonlinear. To speed up the UQ analysis, it is worth investigating if this response surface can be approximated by explicit surrogate models. Linear and quadratic models based on the uncertainties on the Flame Transfer Function parameters n and τ are investigated in this section:

1. $\mathbf{LM}_{n-\tau}$: a linear model based on the parameters n and τ of the Flame Transfer function:
$$\omega_i^{model_1} = \zeta_0 + \zeta_1 n + \zeta_2 \tau \tag{12}$$

2. $\mathbf{LM}_{FTF}$: based on the Flame Transfer Function evaluated at $\omega = \omega_0$, where ω_0 corresponds to the mode without flame coupling (corresponding to n=0). The Flame Transfer Functions incorporate here physical nonlinearities into the model:
$$\omega_i^{model_2} = \zeta_0 + \zeta_1 \Re(ne^{j\omega_0\tau}) + \zeta_2 \Im(ne^{j\omega_0\tau}) \tag{13}$$

3. $\mathbf{QM}_{FTF}$: is a quadratic model based on the Flame Transfer Function also evaluated at $\omega = \omega_0$. Here, the physical nonlinearities are taken into account into the model.

$$\omega_i^{model_3} = \zeta_0 + \zeta_1 \Re(ne^{j\omega_0\tau}) + \zeta_2 \Im(ne^{j\omega_0\tau}) + \zeta_3 \Re(ne^{j\omega_0\tau})^2 \tag{14}$$
$$+\zeta_4 \Im(ne^{j\omega_0\tau})^2 + \zeta_5(\Re(ne^{j\omega_0\tau}) \times \Im(ne^{j\omega_0\tau})) \tag{15}$$

The models $\mathbf{LM}_{n-\tau}$, $\mathbf{LM}_{FTF}$ and $\mathbf{QM}_{FTF}$ can be written in linear algebra notation as follows:
$$\omega_i = X\zeta + \varepsilon = \omega_i^{model} + \varepsilon \tag{16}$$

where $X\zeta$ is the matrix-vector product, and $\zeta = [\zeta_0, \zeta_1, \zeta_2, \zeta_3, \zeta_4, \zeta_5]^T$ corresponds to the regression coefficients of the model. These coefficients represent the mean change in the response variable for one unit of change in the predictor variable. ω_i is considered to be a N $\times$ 1-dimensional vector containing the growth rate ω_i determined from N Helmholtz computations, and X is the matrix containing 1, n and τ when using $\mathbf{LM}_{n-\tau}$ or 1, $\Re(ne^{j\omega_0\tau})$, $\Im(ne^{j\omega_0\tau})$ with $\mathbf{LM}_{FTF}$ or 1,$\Re(ne^{j\omega_0\tau})$, $\Im(ne^{j\omega_0\tau})$, $\Re(ne^{j\omega_0\tau})^2$, $\Im(ne^{j\omega_0\tau})^2$ and $(\Re(ne^{j\omega_0\tau}) \times \Im(ne^{j\omega_0\tau}))$ with $\mathbf{QM}_{FTF}$. This for each samples and ε the N $\times$ 1 vector of residuals:

$$\omega_i = \begin{bmatrix} \omega_{i_1} \\ \omega_{i_2} \\ \vdots \\ \omega_{i_N} \end{bmatrix}, X = \begin{bmatrix} 1 & n_1 & \tau_1 \\ 1 & n_2 & \tau_2 \\ \vdots & \ddots & \vdots \\ 1 & n_N & \tau_N \end{bmatrix}, \zeta = \begin{bmatrix} \zeta_0 \\ \zeta_1 \\ \zeta_2 \\ \zeta_3 \\ \zeta_4 \\ \zeta_5 \end{bmatrix} \text{ and } \varepsilon = \begin{bmatrix} \varepsilon_1 \\ \varepsilon_2 \\ \vdots \\ \varepsilon_N \end{bmatrix}$$

A least squares methodology is used to determine the coefficients ζ of the three models which minimize the error ε:

$$\tilde{\zeta} = \left(X^t X\right)^{-1} X^t \omega_i \tag{17}$$

where $\tilde{\zeta}$ corresponds to the estimated parameters from the least squares, $(X^t X)^{-1}$ is called the "information matrix", and X^t corresponds to the transpose of the X matrix. The predicted values $\tilde{\omega_i}$ for the mean of ω_i are then determined as follows:

$$\tilde{\omega_i} = X\tilde{\zeta} = X\left(X^t X\right)^{-1} X^t \omega_i \tag{18}$$

The objective is to use $\mathbf{LM}_{n-\tau}$, $\mathbf{LM}_{FTF}$ and $\mathbf{QM}_{FTF}$ to estimate accurately and at low cost the risk factor of the first acoustic mode of the system. To achieve this, a reduced two-step UQ strategy is proposed:

1. **Step 1**: Find, at reasonable cost, the regression coefficients associated to surrogate models, $\mathbf{LM}_{n-\tau}$, $\mathbf{LM}_{FTF}$ and $\mathbf{QM}_{FTF}$, using only a few samples of Helmholtz simulations instead of 4000 Helmholtz simulations as performed in the section "Conclusions" using a brute force Monte Carlo method.
2. **Step 2**: Apply a Monte Carlo on these surrogate models to assess the risk factor of the mode.

At first, to find the regression coefficients of the surrogate models, the 4000 Helmholtz simulations of the Monte Carlo database obtained in the section "Conclusions" are used. These coefficients are computed using Eq. (17), and the correlation between the surrogate models and the reference Monte Carlo database is evaluated.

$$R = \frac{\mathbb{E}[(\omega_i - \mathbb{E}(\omega_i))(\omega_i^{model} - \mathbb{E}(\omega_i^{model}))]}{\sigma_{\omega_i}\sigma_{\omega_i^{model}}} \tag{19}$$

In Eq. (19), $\mathbb{E}$ is the expectation, ω_i corresponds to the reference growth rate, ω_i^{model} is the growth rate issued from linear least squares fitting, and σ corresponds to the standard deviation from the reference growth rate and the estimated growth rate from linear least squares fitting. Results of the model fitting are shown in Fig. 7 and their corresponding correlations to the full Monte Carlo database are merged in Table 3.

Results of the model fitting showed that $\mathbf{LM}_{FTF}$ (Eq. 13) and the quadratic model $\mathbf{QM}_{FTF}$ (Eq. 15) are able to reproduce, respectively, 98% and almost 100% of the growth rate variation, whereas the model $\mathbf{LM}_{n-\tau}$ reproduced 95% correlation of the growth rate variations. For the rest of this study, the linear model $\mathbf{LM}_{FTF}$ and the quadratic model $\mathbf{QM}_{FTF}$ are kept to approximate the risk factor of the mode at low cost, i.e. relying on much less than 4000 Helmholtz computations.

Initially, to assess the number of Helmholtz computations required to fit these surrogate models, several tuning of the ζ coefficients were performed. Typically, a set of 3, 5, 6, 10, 20, 40 and 100 Helmholtz computations randomly selected from the full Monte Carlo database are used. Once the surrogate models tuned, they are

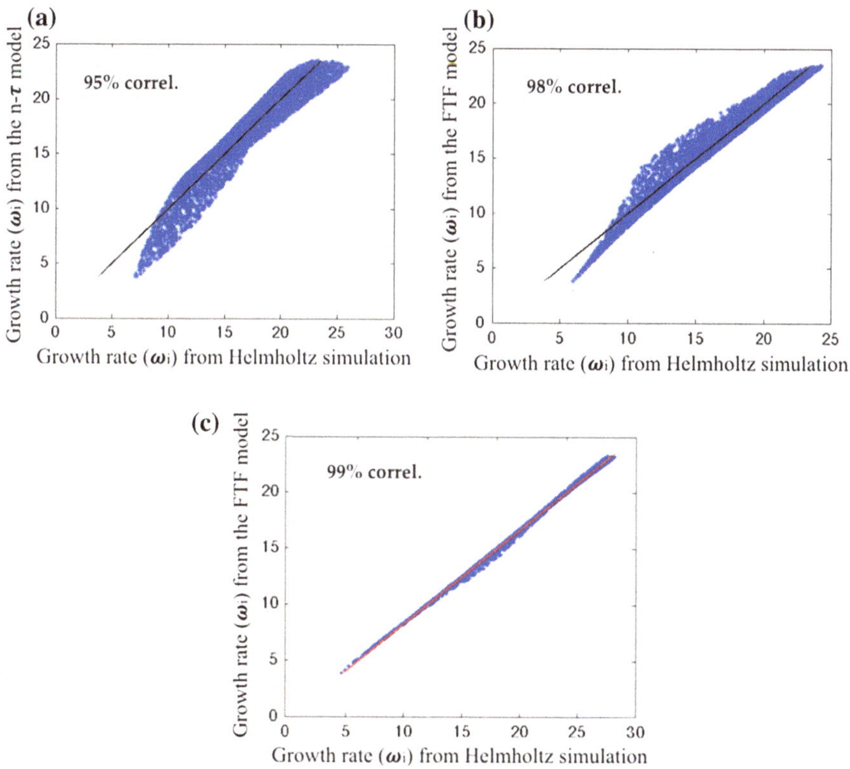

Fig. 7 Multiple linear regression computation: **a** using the surrogate model $\mathbf{LM}_{n-\tau}$, **b** using the surrogate model $\mathbf{LM}_{FTF}$ and **c** using the surrogate model $\mathbf{QM}_{FTF}$

Table 3 Correlations coefficients of the surrogate models and the full Monte Carlo database computed from AVSP.

Models	Correlations
$\mathbf{LM}_{n-\tau}$	0.9481
$\mathbf{LM}_{FTF}$	0.9761
$\mathbf{QM}_{FTF}$	0.9990

used to perform an affordable Monte Carlo to estimate the corresponding risk factor of the mode. Besides, to evaluate the variability of the risk factor, the Monte Carlo analysis based on the surrogate models $\mathbf{LM}_{FTF}$ and $\mathbf{QM}_{FTF}$ repeated **100 times** for each chosen set of Helmholtz simulations (from 3 to 100 Helmholtz simulations). The results are displayed in Fig. 8 when using the linear model $\mathbf{LM}_{FTF}$ and in Fig. 9 when using the quadratic model $\mathbf{QM}_{FTF}$.

In Figs. 8 and 9, the dashed lines represent the reference risk factor (≈24%), obtained by the force Monte Carlo analysis over 4000 Helmholtz computations. The full

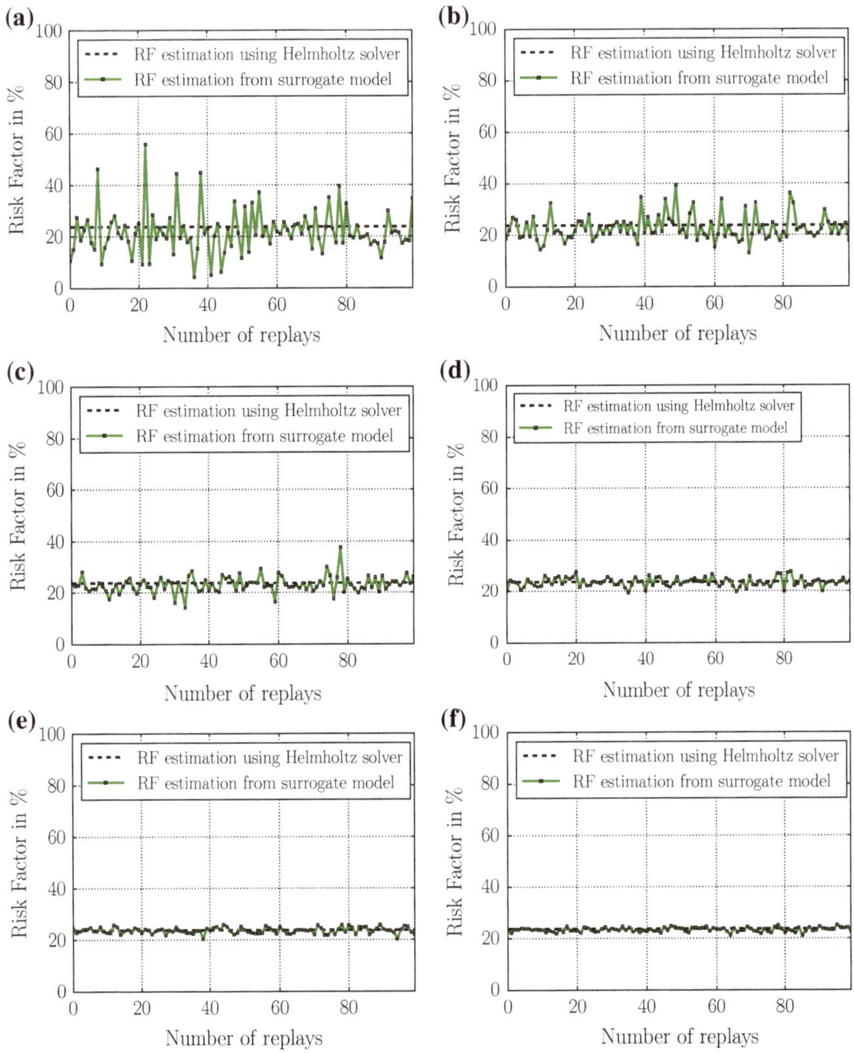

Fig. 8 Risk factor estimated from a Monte Carlo analysis using the linear model $\mathbf{LM}_{FTF}$: **a** with 3 arbitrary samples from the full Monte Carlo database, **b** with 5 arbitrary samples from the full Monte Carlo database, **c** with 10 arbitrary samples from the full Monte Carlo database, **d** with 40 arbitrary samples from the full Monte Carlo database, **e** with 70 arbitrary samples from the full Monte Carlo database and **f** using 100 arbitrary samples from the full Monte Carlo database

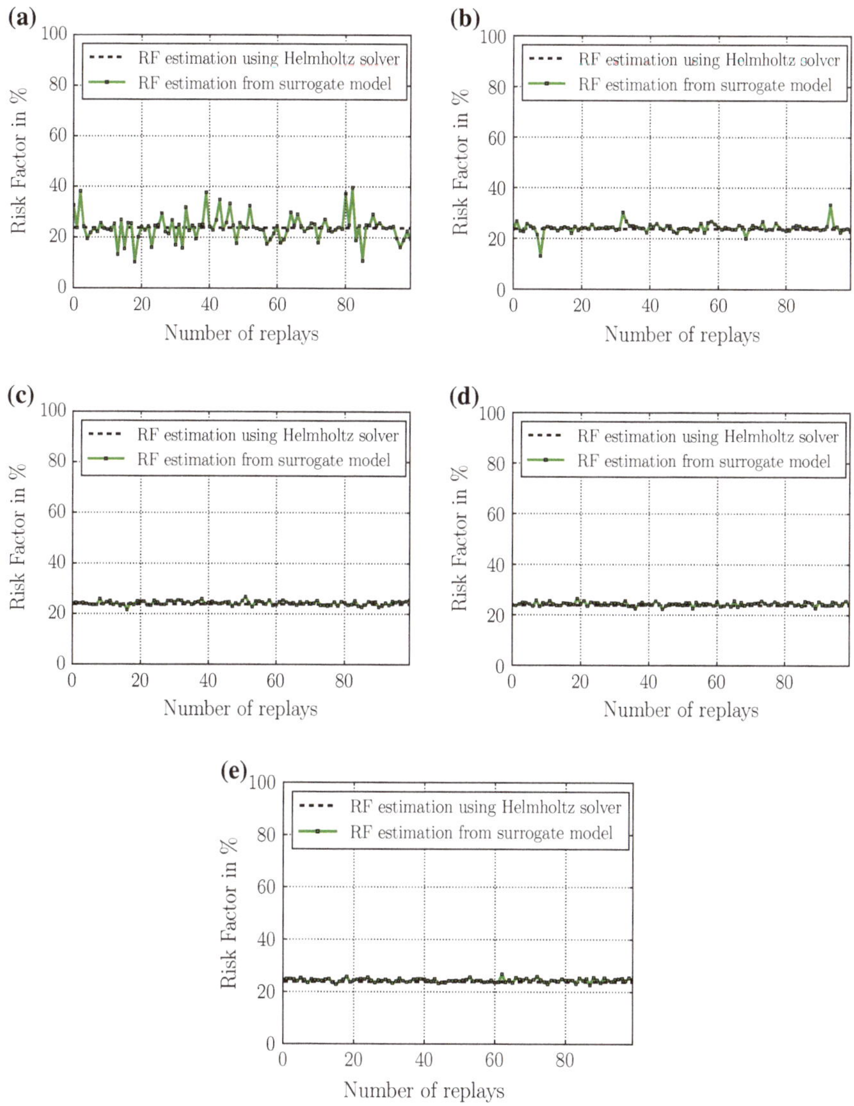

Fig. 9 Risk factor estimated from a Monte Carlo analysis using the linear model $\mathbf{QM}_{FTF}$: **a** with 6 arbitrary samples from the full Monte Carlo database, **b** with 10 arbitrary samples from the full Monte Carlo database, (**i**) with 20 arbitrary samples from the full Monte Carlo database, **c** with 40 arbitrary samples from the full Monte Carlo database, **d** with 70 arbitrary samples from the full Monte Carlo database and **e** using 100 arbitrary samples from the full Monte Carlo database

Table 4 Risk factors and their associated standard deviations computed by the Monte Carlo and surrogate models $\mathbf{LM}_{FTF}$ and $\mathbf{QM}_{FTF}$ using a different number of Helmholtz simulations from the full MC database

	Mean risk factors (in %)	Standard deviations
Number of samples for the MC study using $\mathbf{LM}_{FTF}$		
3	21.45	8.92
5	22.88	4.93
10	23.13	3.18
20	23.54	1.80
40	23.59	1.20
100	23.32	0.83
Number of samples for the MC study using $\mathbf{QM}_{FTF}$		
6	23.69	6.95
10	24.19	1.95
20	24.24	0.81
40	24.31	0.73
100	24.40	0.69

lines correspond to the risk factor assessed by the surrogate models. Results show that the discrepancies between the reference risk factor ($\approx$24%) and the risk factor estimated from the surrogate models decrease when the size of the samples increases, as expected. A reliable estimation of the risk factor is reached on the basis of 10–100 Helmholtz simulation.

To ensure these observations, the mean risk factors and associated standard deviations for each set of samples (from 3 to 100 Helmholtz simulations) were investigated to approximate the number of Helmholtz simulations required to tune the ζ-coefficients and to accurately estimate the risk factor of the mode. The results are shown in Table 4, and they show that a few tens of Helmholtz simulations are enough to get an accurate risk factor estimation with surrogate models. These results suggest that a purely algebraic model is able to fairly assess at reduced cost the risk factor of thermoacoustic modes and using about 20 Helmholtz simulations is sufficient to accurately estimate the risk factor of the mode.

Conclusions

An Uncertainty Quantification analysis has been applied in the context of thermoacoustic instabilities in a single swirled combustor experiment. All eigenmodes of the combustor have been assessed by means of a parallel Helmholtz solver. The Flame

Transfer Function measured experimentally has been used as a flame model to feed the Helmholtz solver. The frequency of oscillation as well as the growth rate of the first thermoacoustic mode was computed in 24 different operating points, and the stability analysis of the system has been realized by [23]. Numerical predictions are coherent with the experimental observations of the combustor, except in 3 cases (out of 24) where the agreement is only partial. Introducing Uncertainty Quantification allows a more accurate mode classification than the usual binary one (stable or unstable), and thus a more reliable comparison between experimental observations and numerical predictions. This leads to a continuous classification of the thermoacoustic modes based on their probability to be unstable given the uncertainties on the flame response, also called their risk factor. The risk factor associated to the first acoustic mode of the combustor was first assessed using a Monte Carlo approach based on 4000 Helmholtz simulations of a single experimental operating point but with random perturbations on the Flame Transfer Function parameters. Then, a two-step UQ strategy was used to deal with thermoacoustics in such a system: (i) First, three surrogate models were tuned from a moderate number of Helmholtz solutions. (ii) Then, these algebraic models were used to perform a Monte Carlo analysis affordably and to approximate the risk factor of the mode. The study proves that analytical surrogate models can be used to predict the risk factors at reduced cost.

Acknowledgements This study was performed within the UMRIDA project funded by the European Commission (FP7-AAT-2013-RTD-1-605036). The authors also thank the Center for Turbulence Research for its support during the 2014 Summer Program. Professor G. Iaccarino (Stanford University) and P. Constantine (UMINES Colorado) are acknowledged for their help regarding the UQ strategy.

References

1. Putnam, A.A.: American Elsevier (1971)
2. Dowling, A.P., Stow, S.R.: J. Prop. Power **19**, 751–764 (2003)
3. Culick, F.E.C.: AGARD 72B PEP meeting
4. Lieuwen, T., Torres, H., Johnson, C., Zinn, B.: J. Eng. Gas Turbines Power **123**, 182–189 (2001)
5. Lieuwen, T., Banaszuk, A.: J. Prop. Power **21**, 25–31 (2005)
6. Poinsot, T., Veynante, D.: Theoretical and Numerical Combustion, 3rd edn. www.cerfacs.fr/elearning (2011)
7. Rayleigh, L.: Nature July 18 (1878) 319–321
8. Wolf, P., Staffelbach, G., Gicquel, L., Muller, J.-D., Poinsot, T.: Combust. Flame **159**, 3398–3413 (2012)
9. Staffelbach, G., Gicquel, L., Poinsot, T.: Lecture Notes in Computational Science and Engineering. In: Complex effects in Large Eddy Simulation, vol. 56, pp. 326–336. Springer
10. Schmitt, P., Poinsot, T., Schuermans, B., Geigle, K.P.: J. Fluid Mech. **570**, 17–46 (2007)
11. Motheau, S.: NICOUD, J. Sound Vib. 1–26 (2012)
12. Bauerheim, M., Cazalens, M., Poinsot, T.: Proc. Combust, Inst (2014)
13. Bauerheim, M., Parmentier, J., Salas, P., Nicoud, F., Poinsot, T.: Combust. Flame 1374–1389 (2014)

14. Bauerheim, M., Ndiaye, A., Magri, L., Moreau, S., Poinsot, T., Nicoud, F.: J. Fluid Mech. (2016)
15. Nicoud, F., Benoit, L., Sensiau, C., Poinsot, T.: AIAA J. **45**, 426–441 (2007)
16. Crocco, L.: J. Am. Rocket Soc. **21**, 163–178 (1951)
17. Crocco, L.: J. Am. Rocket Soc. **22**, 7–16 (1952)
18. Palies, P., Durox, D., Schuller, T., Candel, S.: Combust. Flame **157**, 1698–1717 (2010)
19. Schuller, T., Durox, D., Candel, S.: Combust. Flame **134**, 21–34 (2003)
20. Giauque, A., Selle, L., Poinsot, T., Buechner, H., Kaufmann, P., Krebs, W.: J. Turb. **6**, 1–20 (2005)
21. Palies, P.: Dynamique et instabilités de combustion de flammes swirlées. Ph.D. Thesis, Ecole Centrale Paris (2010)
22. Schuller, T., Durox, D., Palies, P., Candel, S.: Combust. Flame **159**, 1921–1931 (2012)
23. Silva, C.F., Nicoud, F., Schuller, T., Durox, D., Candel, S.: Combust. Flame **160**, 1743–1754 (2013)

Numerical Uncertainties Estimation and Mitigation by Mesh Adaptation

Frédéric Alauzet, Alain Dervieux, Loïc Frazza and Adrien Loseille

Introduction

The numerical approximation involved in simulation Navier–Stokes software carries certainly a systemic error, since this error can be reduced by applying some supplementary effort. But the deviation between exact solution and approximated one remains in practice an uncertain factor. The common strategy for the engineer is to get convinced that numerical error is small. Of course if it is not enough small, the simulation output cannot be efficiently used. A rational standpoint is to combine (i) a strategy for obtaining a small error with (ii) an estimation of it. In this chapter, we propose to address (i) with a sophisticated mesh adaptation method and to address (ii) with the computation of a corrector, approximating the actual error.

The approximation or the estimate of the actual numerical error is a difficult task addressed by many ways. Let us refer to the works [1] which also introduce the error in the adaptive process. Our approach is close to the method of functional correction of Giles, see for example [2].

Adaptive methods in aeronautics have been used for many different purposes. The first one is generally to improve the prediction of complex phenomena (sonic-boom prediction, drag prediction, high-lift configuration, blast, vortices, ...) while minimizing the CPU cost. Then, it may be used to guarantee the optimal (second) order

F. Alauzet (✉) · L. Frazza · A. Loseille
Gamma3, INRIA, 1 rue Estienne d'Orves, 91126 Palaiseau, France
e-mail: frederic.alauzet@inria.fr

L. Frazza
e-mail: loic.frazza@inria.fr

A. Loseille
e-mail: adrien.loseille@inria.fr

A. Dervieux
Ecuador, INRIA, 2002 route des lucioles, 06902 Sophia Antipolis, France
e-mail: alain.dervieux@inria.fr

C. Hirsch et al. (eds.), *Uncertainty Management for Robust Industrial Design in Aeronautics*, Notes on Numerical Fluid Mechanics and Multidisciplinary Design 140, https://doi.org/10.1007/978-3-319-77767-2_6

of convergence of the numerical scheme, especially when discontinuities (such as shocks waves) are present in the flow field [3]. In addition, adaptivity is also concerned with the assessment of the numerical solution. We distinguish the following class of adaptive methods according to these purposes.

A first set of methods is based on the *minimization of the interpolation error* of one or several sensors depending on the CFD solution [4–8]. Given a numerical solution W_h, a solution of higher regularity $R_h(W_h)$ is recovered, so that the following interpolation error estimate [9, 10] holds:

$$\|R_h(W_h) - \Pi_h R_h(W_h)\|_{L^p} \leq N^{-\frac{2}{3}} \left(\int_\Omega \det \left(|H_{R_h(W_h)}| \right)^{\frac{p}{2p+3}} \right)^{\frac{2p+3}{3p}}$$

where $H_{R_h(W_h)}$ is the Hessian of the recovered solution and N an estimate of the desired number of vertices. If an anisotropic mesh prescription is naturally deduced in this context, interpolation-based methods do not take into account the features of the PDE. Note that from a practical point of view, $R_h(W_h)$ is never recovered, only its first and second derivatives are estimated. Standard recovery techniques include least-square, L^2-projection, green formula, or the Zienkiewicz–Zhu recovery operator.

A second set of methods tends to couple adaptivity with the assessment of the numerical prediction of one important scalar output computed from the flow. *Goal-oriented methods* [2, 11–14] aim at minimizing the error

$$\delta j = |j(W) - j_h(W_h)|$$

committed on the evaluation of a scalar functional j depending on the approximate field. An usual functional is the observation of the pressure field on an observation surface γ:

$$j(W) = \int_\gamma \left(\frac{p - p_\infty}{p_\infty} \right)^2,$$

where W and W_h are the solution and the numerical solution of the compressible Euler equation, respectively. Goal-oriented methods do take into account the features of the PDE, through the use of an adjoint state that gives the sensitivity of W to the observed functional j. In order to solve the goal-oriented anisotropic mesh optimization problem, an a priori analysis has been introduced [12, 15] which restricts to the main asymptotic term of the local error. While a super-convergence of $|j(W) - j_h(W_h)|$ is observed in some cases [16, 17], goal-oriented optimal methods are specialized for a given output, and in particular do not systematically provide a convergent solution field. In other words, the convergence of $\|W - W_h\|$ (in any norm) is not ensured. In addition, if the observation of multiple functionals is possible (by means of multiple adjoint states), the optimality of the mesh and the convergence properties of the approximation error may be lost.

In each case, the aforementioned adaptive strategies address specifically one goal. Consequently, it is still a challenge to find an adaptive framework that encompass all the desired requirements: anisotropic mesh prescription, asymptotic optimal order of convergence, assessment of the convergence of the numerical solution to the continuous one, control of multiple functionals of interest, ... This paper is a contribution with a first attempt to formally predict all the different requirements. Our approach is based on the design of a *norm-oriented optimal method*, which takes into account the PDE features, and produces an approximate solution field which does converge to the exact one in the norm chosen by the user. To do so, we derive a corrector that estimates the approximation error. Contrary to the goal-oriented mesh adaptation, the functional may be now any function of the approximation error. A particular case involves multiple functionals of interest to be minimized simultaneously. For instance, instead of the above δj, we can minimize the semi-norm-like functional:

$$j(W_h) = (\text{drag}(W) - \text{drag}(W_h))^2 + (\text{lift}(W) - \text{lift}(W_h))^2$$

while the goal-oriented should use two functionals, one for drag and one for lift or specify a combination of them.

The paper is organized as follows. Section "Flow Solver Models" briefly recalls the considered PDE and the numerical discretization. In section "Formal Error Analysis Within the Continuous Mesh Framework," the Hessian-based multiscale and the goal-oriented error estimates are recalled, then a new norm-oriented error analysis is derived formally. The norm-oriented mesh adaptation uses correctors to estimate the approximation error, and section "Correctors for the Compressible Navier–Stokes Equations" proposes two approaches for the case of the compressible Euler equations within a linear and non linear setting. Finally, section "Numerical Experiments: A Turbulent Transonic Falcon" compares different adaptation methods and evaluates the corrector for 3D CFD problems.

Flow Solver Models

The methods which we propose rely on estimates applicable to $\mathbb{P}^1$-exact approximations, i.e., approximations which are exact for affine analytical solutions. In particular they apply directly to CFD solvers which are inspired by $\mathbb{P}^1$-continuous finite-element approximation or, equivalently, based on a vertex-centered approximation like, N3S [18], SU2 [19], AERO [20, 21], or FUN3D [11]. In the present paper, we have used the vertex-centered research software Wolf [4, 22].

Flow Equations

The Navier–Stokes equations for mass, momentum, and energy conservation read:

$$\begin{cases} \dfrac{\partial \rho}{\partial t} + \nabla \cdot (\rho \mathbf{u}) = 0 \\ \dfrac{\partial (\rho \mathbf{u})}{\partial t} + \nabla \cdot (\rho \mathbf{u} \otimes \mathbf{u}) + \nabla p = \nabla \cdot \mathbf{F}^v_{\mathbf{u}}, \\ \dfrac{\partial (\rho E)}{\partial t} + \nabla \cdot ((\rho E + p)\mathbf{u}) = \nabla \cdot F^v_E, \end{cases}$$

where ρ denotes the density, $\mathbf{u}$ the velocity, E the total energy per unit mass, and p the pressure. Symbols $\mathbf{F}^v_{\mathbf{u}}, F^v_E, \mathbf{F}^v$ are used for viscous fluxes which we shall not detail for simplicity, see [22]. This system can be rewritten under vectorial form:

$$W_t + F^c_1(W)_x + F^c_2(W)_y + F^c_3(W)_z = \nabla \cdot \mathbf{F}^v,$$

where W is the non-dimensioned conservative variables vector:

$$W = (\rho, \rho u, \rho v, \rho w, \rho E)^T$$

and $\mathbf{F}^c(W) = (F^c_1(W), F^c_2(W), F^c_3(W))$ are the convective (Euler) flux functions:

$$\begin{aligned} F^c_1(W) &= (\rho u,\ \rho u^2 + p,\ \rho uv,\ \rho uw,\ u(\rho E + p))^T \\ F^c_2(W) &= (\rho v,\ \rho uv,\ \rho v^2 + p,\ \rho vw,\ v(\rho E + p))^T \\ F^c_3(W) &= (\rho w,\ \rho uw,\ \rho vw,\ \rho w^2 + p,\ w(\rho E + p))^T. \end{aligned}$$

A weak formulation of this system writes for $W \in V = \left[H^1(\Omega)\right]^5$ as follows:

$$\forall \phi \in V, \quad (\Psi(W), \phi) = \int_\Omega \nabla \phi \cdot \mathscr{F}(W)\, \mathrm{d}\Omega + \int_\Gamma \phi\, \bar{\mathscr{F}}(W) \cdot \mathbf{n}\, \mathrm{d}\Gamma \;=\; 0, \qquad (1)$$

with

$$\mathscr{F}(W) = (F^c_1(W), F^c_2(W), F^c_3(W))^T + (F^v_1(W), F^v_2(W), F^v_3(W))^T$$

and where Γ is the boundary of the computational domain Ω, $\mathbf{n}$ the outward normal to Γ, and the boundary flux $\bar{\mathscr{F}}$ contains the boundary conditions.

In the case of a turbulent flow simulation, which we address with the Spalart–Allmaras model, an equation for the turbulent viscosity ν_t is added to a slight modification of the previous system.

Spatial Discretization

Equation (1) is discretized by a vertex-centered upwind finite-volume formulation applied to unstructured tetrahedra meshes. The interested reader is invited to find a detailed presentation in [4, 22]. To carry out the variational analysis, it is interesting to present the finite-volume formulation as a stabilization of the Galerkin approximation.

Let $\mathscr{H}$ be a mesh of Ω composed of tetrahedra. We denote by Ω_h and Γ_h the linear approximate of Ω and Γ defined by $\mathscr{H}$. Let us introduce the following approximation space:

$$V_h \;=\; \left\{\phi_h \in V \cap \mathscr{C}^0 \mid \phi_{h|K} \text{ is affine } \forall K \text{ element of } \mathscr{H}\right\}.$$

The interpolation operator of the previous section is chosen as the usual $\mathbb{P}^1$ operator:

$$\Pi_h \,:\, V \cap \mathscr{C}^0 \to V_h \;\text{ we have }\; \Pi_h\varphi(\mathbf{x}_i) = \varphi(\mathbf{x}_i).$$

The weak discrete formulation writes:

$$\begin{gathered}\forall \phi_h \in V_h,\; \left(\Psi_h(W_h)\,,\,\phi_h\right) = 0,\\ \left(\Psi_h(W_h)\,,\,\phi_h\right) = \int_{\Omega_h} \nabla\phi_h \cdot \mathscr{F}_h(W_h)\,\mathrm{d}\Omega_h + \int_{\Gamma_h} \phi_h \bar{\mathscr{F}}_h(W_h)\cdot\mathbf{n}\,\mathrm{d}\Gamma_h \;=\; 0\\ \mathscr{F}_h = \Pi_h \mathscr{F}\;;\; \bar{\mathscr{F}}_h = \Pi_h \bar{\mathscr{F}}.\end{gathered} \tag{2}$$

Taking as in Relation (2) the $\mathbb{P}^1$-interpolation of the fluxes $\mathscr{F}_h$ as a discretization principle produces a finite-element scheme which is identical to the central-differenced finite-volume scheme built on the so-called median dual cells. In practice, this family of Mixed-Element-Volume schemes cannot be used in a non-dissipative purely centered version. In [23, 24], MUSCL versions are described and analyzed. For our analysis, we consider that the scheme under study is a Galerkin formulation enriched with artificial stabilization terms under the form of numerical diffusion. We write this as follows:

$$\forall \phi_h \in V_h,\quad \int_{\Omega_h} \nabla\phi_h \cdot \mathscr{F}_h(W_h)\,\mathrm{d}\Omega_h + \int_{\Gamma_h} \phi_h \bar{\mathscr{F}}_h(W_h)\cdot\mathbf{n}\,\mathrm{d}\Gamma_h = -\int_{\Omega_h} \phi_h\, D_h(W_h)\,\mathrm{d}\Omega_h \tag{3}$$

According to [24], the numerical diffusion term is of higher order as soon as it is applied to the interpolation of a smooth enough field W on a sufficiently regular mesh:

$$\left|\int_{\Omega_h} \phi_h\, D_h(W_h)\mathrm{d}\Omega_h\right| \;\le\; h^3 K(W) |\phi_h|_{L^2}\,.$$

As a result, the numerical diffusion term is neglected in the remaining analysis.

It is now useful to introduce the linearized operators $\mathscr{A}$ (resp. $\mathscr{A}_h$) expressed in terms of Jacobians of $\mathscr{F}$ and $\bar{\mathscr{F}}$ computed at W (resp. $\mathscr{F}_h$ and $\bar{\mathscr{F}}_h$ computed at W_h):

$$\forall W \in V, \quad \forall \delta W \in V, \quad \mathscr{A}(W)\delta W \in (V)' \quad \text{and} \quad \forall \phi \in V,$$

$$(\mathscr{A}(W)\delta W\,,\,\phi) = \int_\Omega \nabla\phi \cdot \frac{\partial\mathscr{F}}{\partial W}(W)(\delta W)\,\mathrm{d}\Omega + \int_\Gamma \phi \frac{\partial\bar{\mathscr{F}}(W)}{\partial W}(\delta W)\cdot\mathbf{n}\,\mathrm{d}\Gamma = 0 \quad (4)$$

$$\forall W_h \in V_h, \quad \forall \delta W_h \in V_h, \quad \mathscr{A}_h(W_h)\delta W_h \in (V_h)' \quad \text{and} \quad \forall \phi_h \in V_h,$$

$$\left(\mathscr{A}_h(W_h)\delta W_h\,,\,\phi_h\right) = \int_{\Omega_h} \nabla\phi_h \cdot \frac{\partial\mathscr{F}_h}{\partial W}(W_h)(\delta W_h)\,\mathrm{d}\Omega_h + \int_{\Gamma_h} \phi_h \frac{\partial\bar{\mathscr{F}}_h(W_h)}{\partial W}(\delta W_h)\cdot\mathbf{n}\,\mathrm{d}\Gamma_h = 0\,. \quad (5)$$

$\mathscr{A}$ and $\mathscr{A}_h$ are assumed to be invertible. We use in the sequel the notations $\mathscr{A}^{-1}RHS$ and $\mathscr{A}_h^{-1}RHS_h$ for the results of solving the corresponding systems with RHS and RHS_h as right-hand sides.

Formal Error Analysis Within the Continuous Mesh Framework

The norm-oriented approach is an extension of the previous developments on anisotropic (Hessian-based) and goal-oriented mesh adaptation. In these latter methods, the anisotropic mesh prescription (orientations and sizes) is given in a close form. Each of them are tightly related to interpolation error measured on the solution field for Hessian-based methods, and the various fluxes for the goal-oriented approach. We first recall formally the derivation of these estimates in the continuous mesh framework. It leads to the definition of two *kernels* (interpolation and goal-oriented) providing the optimal mesh. The norm-oriented approach is a combination of these kernels with the derivation of a solution corrector. In this section, we focus on controlling the implicit error $\Pi_h u - u_h$. Controlling the approximation error will consist in controlling the implicit error (corrector) and the interpolation error terms simultaneously as:

$$u - u_h = u - \Pi_h u + \Pi_h u - u_h.$$

Note that the implicit error can be seen as a vertex-wise error between the exact solution and the numerical one, whereas the interpolation error can be seen as geometric deviation between the continuous solution and its $\mathbb{P}^1$ representation on the mesh.

Continuous Mesh Framework Formalism

In [25], we prove that any mesh can be represented by a continuous Riemannian metric field $\mathcal{M}$. The link between continuous mesh and discrete mesh is based on the unit mesh concept [26]: given a Riemannian metric field $\mathcal{M}$, a unit mesh is a mesh having:

$$\begin{aligned}
&\text{for all edges } \mathbf{e} = AB, \quad \ell_{\mathcal{M}}(\mathbf{e}) = \int_0^1 \sqrt{{}^t AB \mathcal{M}((1-t)A + tB) AB}\, \mathrm{d}t \in \left[\frac{1}{\sqrt{2}}, \sqrt{2}\right], \\
&\text{for all elements } K, \quad |K|_{\mathcal{M}} \approx \frac{\sqrt{2}}{12}.
\end{aligned} \tag{6}$$

From a practical point of view, generating an anisotropic unit mesh $\mathcal{H}$ with respect to $\mathcal{M}$ requires to use any anisotropic mesh generators, see [8, 11, 27–34] in 3D. Conversely, given a mesh $\mathcal{H}$, the following metric field is a continuous representative of $\mathcal{H}$:

$$\mathcal{M}_P = \exp\left(\frac{\sum_{K \ni P} |K| \ln(\mathcal{M}_K)}{\sum_{K \ni P} |K|}\right),$$

where P is a vertex of $\mathcal{H}$ and $\mathcal{M}_K$ is the unique metric representing element K, and $|K|$ is the volume of K. Consequently, if u_h denotes a discrete quantity computed on a given mesh, we use equivalently the notation $u_{\mathcal{M}}$ that represents the same quantities represented on any unit mesh with respect to $\mathcal{M}$. In the case of the interpolation error, there is a strict equivalence between continuous $u - \pi_{\mathcal{M}} u$ and discrete interpolation error $u - \Pi_h u$, see [10]. The parametrization of a mesh by $\mathcal{M}$ instead of h is advantageous for a priori analysis with anisotropic mesh. Indeed, it exists also quantities of interest as the density, anisotropic ratios, differentiation that are well defined on $\mathcal{M}$.

Hessian-Based Multiscale Adaptation

Let us consider a mesh $\mathcal{H}$ which is unit for a metric $\mathcal{M}$, in other words, $\mathcal{M}$ is a continuous model of $\mathcal{H}$. A Hessian-based adaptation relies on the choice of a *sensor* u depending on the state variable W. According to the continuous mesh theory [10, 25], the $\mathbb{P}^1$ interpolation error $u - \Pi_h u$ can be approximated in terms of second derivatives of u, *i.e.*, the Hessian H_u of u, and of metric $\mathcal{M}$:

$$|u - \Pi_h u| \approx |u - \pi_{\mathcal{M}} u| \stackrel{\Delta}{=} \operatorname{trace}(\mathcal{M}^{-\frac{1}{2}} |H_u| \mathcal{M}^{-\frac{1}{2}}), \tag{7}$$

where $|H_u|$ is derived from H_u by taking the absolute value of the eigenvalues. The above expression is the continuous approximation of the $\mathbb{P}^1$ interpolation error. Minimizing $\|u - \Pi_h u\|_{L^1(\Omega_h)}$ for a given number N of vertices can be recast in the continuous setting as minimizing $\|u - \pi_{\mathcal{M}} u\|_{L^1(\Omega)}$ for a complexity $\mathcal{C}(\mathcal{M}) = N$ where calculus of variation is available. The complexity $\mathcal{C}$ of $\mathcal{M}$ is the continuous counter part of the number of vertices.

$$\mathcal{C}(\mathcal{M}) = \int_{\Omega} \sqrt{\det(\mathcal{M}(\mathbf{x}))}\,\mathrm{d}\mathbf{x}. \tag{8}$$

Solving this optimization problem provides an optimal interpolation-based metric (continuous mesh):

$$\mathcal{M}_{L^1}^{opt}(u) = \underset{\mathcal{C}(\mathcal{M})=N}{\arg\min}\ \mathrm{trace}(\mathcal{M}^{-\frac{1}{2}} H_u \mathcal{M}^{-\frac{1}{2}}).$$

The expression of the optimal continuous mesh in 3D is:

$$\mathcal{M}_{L^1}^{opt}(u) \;=\; \frac{N^{\frac{2}{3}}}{\left(\int_{\Omega} (\det|H_u|)^{\frac{1}{5}}\right)^{\frac{2}{3}}} (\det|H_u|)^{-\frac{1}{5}}\, |H_u|\,. \tag{9}$$

The first factor of the RHS is a global normalization term set to obtain a continuous mesh with complexity N, and $(\det|H_u|)^{-\frac{1}{5}}$ is a local normalization taking into account the sensitivity of the L^1 norm. Note that expressing the continuous interpolation error for the optimal metric, Relation (9) shows that second-order convergence is obtained for a smooth sensor [10]. The approach can be extended to non-smooth sensor and still recovers the second-order convergence [3, 35].

In practice, computing the optimal metric is done approximatively, *i.e.*, in a discrete context with a couple (mesh, solution) denoted $(\mathcal{H}, W_h)$, and iteratively through the following fixed point algorithm. The Hessian of sensor u is replaced by a numerical sensor of higher regularity $R_h(u_h)$ computed from the numerical sensor u_h using any recovery techniques.

Algorithm 1 Hessian-based multiscale adaptation

1. Compute state W_h on mesh $\mathcal{H}$
2. Compute sensor $u_h = u(W_h)$ and $R_h(u_h)$
3. Compute optimal metric $\mathcal{M}_{L^1}^{opt}(R_h(u_h))$
4. Generate a new adapted mesh $\mathcal{H}$ which is unit for metric $\mathcal{M}_{L^1}^{opt}(R_h(u_h))$
5. If not converge, goto 1.

For the remaining analysis, it is useful to introduce the *kernel* function $\mathcal{K}_{L^1}$ that gives the optimal metric for the $\mathbb{P}^1$ interpolation error in L^1 norm as a function of the Hessian of u instead of u directly:

$$\mathcal{M}_{L^1}^{opt}(u) \;=\; \mathcal{K}_{L^1}(|H_u|)\,. \tag{10}$$

Indeed, for goal-oriented and norm-oriented analysis, the kernel will be applied to more complex Hessian-like functions.

Goal-Oriented Adaptation

The Hessian-based multiscale adaptation is geometric thus generic and does not take into account the PDE from which W is obtained. On the contrary, the goal-oriented analysis relies on the considered PDE. According to Relations (1) and (3), we assume that solution W and numerical solution W_h verify:

$$\int_{\Omega} \phi \nabla \cdot \mathcal{F}(W) + BI = 0 \quad \text{and} \quad \int_{\Omega_h} \phi_h \nabla \cdot \mathcal{F}_h(W_h) + BI_h = 0, \tag{11}$$

where BI and BI_h are boundary integrals, and the discrete fluxes are simply:

$$\mathcal{F}_h(\cdot) = \Pi_h \mathcal{F}(\Pi_h(\cdot)).$$

The goal-oriented analysis relies on the minimization of the error committed on a scalar output functional j. We assume j to be smooth enough such that j can be observed through its Taylor expansion:

$$j(W) \approx j(W_h) + (\frac{\partial j}{\partial W}(W), W - W_h)$$

which leads to:

$$j(W) - j(W_h) \approx (\frac{\partial j}{\partial W}(W), W - W_h) = (g_{go}, W - W_h). \tag{12}$$

We recall in short the main result of the analysis given in Loseille et al. [12]. Introducing the adjoint state W^*_{go} defined by $W^*_{go} = (\mathcal{A}^{-1})^* g_{go} \equiv \mathcal{A}^{-*} g_{go}$, we have:

$$(g_{go}, W_h - W) \leq$$
$$\int_{\Omega_h} |\nabla W^*_{go}| \, |\mathcal{F}(W) - \Pi_h \mathcal{F}(W)| \, d\Omega_h + \int_{\Gamma_h} |W^*_{go}| \, |(\bar{\mathcal{F}}(W) - \Pi_h \bar{\mathcal{F}}(W)) \cdot \mathbf{n}| \, d\Gamma_h.$$

If the boundary terms are neglected, it simplifies to

$$|j(W) - j(W_h)| \approx (|\nabla W^*_{go}|, |\mathcal{F}(W) - \Pi_h \mathcal{F}(W)|). \tag{13}$$

Similarly to the previous section, from the continuous mesh theory, we get:

$$
\begin{aligned}
(|\nabla W^*_{go}|, |\mathcal{F}(W) - \Pi_h \mathcal{F}(W)|) &\equiv (|\nabla W^*_{go}|, |\mathcal{F}(W) - \pi_{\mathcal{M}} \mathcal{F}(W)|) \\
&= \operatorname{trace}(\mathcal{M}^{-1/2} \, |\nabla W^*_{go}| \cdot |H_{\mathcal{F}(W)}| \, \mathcal{M}^{-1/2}) \,,
\end{aligned}
$$

where $\mathcal{M}$ is a metric field representing the current mesh. Then, minimizing the approximation error on functional j in L^1 norm is equivalent to solve the optimization problem:

$$
\mathcal{M}^{opt}_{go} = \underset{\mathcal{C}(\mathcal{M})=N}{\arg\min} \; \operatorname{trace}(\mathcal{M}^{-1/2} \, |\nabla W^*_{go}| \cdot |H_{\mathcal{F}(W)}| \, \mathcal{M}^{-1/2}) \,,
$$

and $|\nabla W^*_{go}| \cdot |H_{\mathcal{F}(W)}|$ is a positive combination of symmetric matrices. Similarly to the Hessian-based error analysis where the optimal metric is given by Relation (9) and the kernel definition (10), we get:

$$
\mathcal{M}^{opt}_{go}(W) = \mathcal{K}_{L^1}(|\nabla W^*_{go}| \cdot |H_{\mathcal{F}(W)}|) = \mathcal{K}_{L^1}(|\nabla(\mathcal{A}^{-*} g_{go})| \cdot |H_{\mathcal{F}(W)}|). \qquad (14)
$$

Note that if we want to observe many output functionals, as many adjoints must be evaluated. In addition, if we want to minimize the norm of the approximation error, the above analysis cannot be applied directly.

Relation (14) gives a continuous expression of the optimal continuous mesh. In a discrete context, all the continuous quantities are evaluated on the current mesh. We use iterative Algorithm 2 to converge to the optimal solution.

Algorithm 2 Goal-oriented adaptation

1. Compute state W_h on mesh $\mathcal{H}$
2. Compute adjoint state $W^*_{go,h}$ using $g_{go,h} = \dfrac{\partial j}{\partial W}(W_h)$ and $R_h(\mathcal{F}_h(W_h))$
3. Compute optimal metric $\mathcal{M}^{opt}_{go}(W_h) = \mathcal{K}_{L^1}(|\nabla W^*_{go,h}| \cdot |H_{R_h(\mathcal{F}_h(W_h))}|)$
4. Generate a new adapted mesh $\mathcal{H}$ which is unit for metric $\mathcal{M}^{opt}_{go}(W_h)$
5. If not converge, goto 1.

Norm-Oriented Adaptation

We are now interested in the minimization of a semi-norm such as:

$$
\|L(W) - L(W_h)\|^2_{L^2(\Omega_h)}
$$

where L is a given vector- or scalar-valued operator. Linearizing L using a Taylor expansion, we have:

$$
\Big(L(W) - L(W_h), L(W) - L(W_h)\Big) \approx \Big(\frac{\partial L}{\partial W}(W)(W - W_h), \frac{\partial L}{\partial W}(W)(W - W_h)\Big) \,.
$$

Introducing the adjoint operator $\left(\dfrac{\partial L}{\partial W}\right)^*$, we can rewrite:

$$\left(L(W)-L(W_h),L(W)-L(W_h)\right)\approx\left(\left(\frac{\partial L}{\partial W}(W)\right)^*\left(\frac{\partial L}{\partial W}(W)\right)(W-W_h),(W-W_h)\right)$$

then choosing:

$$g_{no}=\left(\frac{\partial L}{\partial W}(W)\right)^*\left(\frac{\partial L}{\partial W}(W)\right)(W-W_h) \tag{15}$$

we get:

$$\|L(W)-L(W_h)\|^2_{L^2(\Omega_h)}\approx\left(g_{no},(W-W_h)\right)\,.$$

Now, we apply the goal-oriented analysis to minimize $\left(g_{no},(W-W_h)\right)$ that leads to the optimal norm-oriented metric:

$$\mathcal{M}^{opt}_{no}(W)=\mathcal{K}_{L^1}(|\nabla(\mathcal{A}^{-*}g_{no})|\cdot|H_{\mathcal{F}(W)}|)=\mathcal{K}_{L^1}(|\nabla W^*_{no}|\cdot|H_{\mathcal{F}(W)}|)\,. \tag{16}$$

where $W^*_{no}=\mathcal{A}^{-*}g_{no}$. The main difficulty is thus to evaluate g_{no}. Indeed, contrary to the goal-oriented case, g_{no} depends on the approximation error $W-W_h$ and not only on the solution W. Consequently, it is necessary to derive an estimate of $W-W_h$, this is done by the computation of a corrector, see section "Correctors for the Compressible Navier–Stokes Equations".

Now, let us give two examples. The simplest one is to consider the L^2 norm of the approximation error:

$$\|W-W_h\|^2_{L^2(\Omega_h)}\,.$$

In this particular case, $\dfrac{\partial L}{\partial W}(W)$ is the identity and g_{no} reduces to the corrector itself $(W-W_h)$. A second example is controlling the approximation on the pressure p in L^2 norm:

$$\|p-p_h\|^2_{L^2(\Omega_h)}\,,$$

and we have $\dfrac{\partial p}{\partial W}(W)=(\gamma-1)\left(\dfrac{\mathbf{u}^2}{2},-u,-v,-w,1\right)$ and

$$\left(\frac{\partial p}{\partial W}(W)\right)^*\left(\frac{\partial p}{\partial W}(W)\right)=(\gamma-1)^2\begin{pmatrix}\frac{\mathbf{u}^4}{4} & -u\frac{\mathbf{u}^2}{2} & -v\frac{\mathbf{u}^2}{2} & -w\frac{\mathbf{u}^2}{2} & \frac{\mathbf{u}^2}{2}\\ -u\frac{\mathbf{u}^2}{2} & u^2 & -uv & -uw & -u\\ -v\frac{\mathbf{u}^2}{2} & -uv & v^2 & -vw & -v\\ -w\frac{\mathbf{u}^2}{2} & -uw & -vw & w^2 & -w\\ \frac{\mathbf{u}^2}{2} & -u & -v & -w & 1\end{pmatrix}.$$

In practice, all the continuous quantities are evaluated on the current mesh, so does the corrector. Iterative Algorithm 3 is considered to converge to the optimal solution.

Algorithm 3 Norm-oriented adaptation

1. Compute state W_h on mesh $\mathcal{H}$
2. Compute an approximation $g_{no,h}$ of $W - W_h$
3. Compute adjoint state $W^*_{no,h}$ and $R_h(\mathcal{F}_h(W_h))$
4. Compute optimal metric $\mathcal{M}^{opt}_{no}(W_h) = \mathcal{K}_{L^1}(|\nabla W^*_{no,h}| \cdot |H_{R_h(\mathcal{F}_h(W_h))}|)$
5. Generate a new adapted mesh $\mathcal{H}$ which is unit for metric $\mathcal{M}^{opt}_{no}(W_h)$
6. If not converge, goto 1.

Correctors for the Compressible Navier–Stokes Equations

The norm-oriented error estimate presented in the previous section requires an estimate of the approximation error $W - W_h$. The approximation error can be decomposed into the interpolation error and the implicit error:

$$W - W_h = W - \Pi_h W \; + \; \Pi_h W - W_h \, .$$

Then, the Hessian-based mesh adaptation theory (section "Hessian-Based Multiscale Adaptation") provides an estimate of the interpolation error, and the implicit error is estimated by computing a corrector.

Given a numerical solution on a mesh $\mathcal{H}_h$ of size h, we intend to control the implicit error. To this end, we geometrically divide the mesh size by two and split all the elements with patterns leading to a new embedded mesh $\mathcal{H}_{h/2}$ of size $h/2$. Then, if we compute a solution of this new refined mesh, then it will have two different effects:

- Add new degrees of freedom, so that every function (solution, fluxes, source terms, …) is better represented. This will reduce the interpolation error.
- Change the nodal value of the solution even for vertices issued from the initial mesh of size h. For example, this process may increase the solution at every vertices, so that $\Pi_h \left(W_{h/2}\right) - W_h > 0$. This is implicit error that we try to control here.

To do so, we first need to relate the implicit error on the solution to the mesh as it has already been done for the interpolation error [10, 25]. In other words, we need to figure out how adding/removing/moving a node and leaving the other unchanged will globally affect the solution. As we do not know a priori W, we reconstruct a corrected solution W_c closer to the exact one.

As we do not have access to $\Pi_h W$, we choose to approach it with $\Pi_h W_{h/2}$. However, $W_{h/2}$ is expensive to compute and once we have it, it is worthless to work with W_h. Therefore, we are seeking for a way to mimic $\Pi_h W_{h/2}$ without computing $W_{h/2}$. Note that we want a solution on $\mathcal{H}_h$ and not $\mathcal{H}_{h/2}$. If $W_{h/2}$ or $\nabla \cdot \mathcal{F}_h \left(\Pi_h W_{h/2}\right)$ are known, we can compute a corrector W_c by solving

$$\int_{\Omega_h} \phi_h \nabla \cdot \mathcal{F}_h \left(W_c\right) = \int_{\Omega_h} \phi_h \nabla \cdot \mathcal{F}_h \left(\Pi_h W_{h/2}\right),$$

where the right-hand side is a fixed source term. In that case, we have $W_c = \Pi_h W_{h/2}$ and the implicit error is given by $W_c - W_h$. As we do not know $W_{h/2}$, we have to clarify how to compute an approximation of the right-hand side term. In fact, we are interested in the global modification of the solution at vertices induced by the mesh refinement via the flux operator and not only in local modifications obtained by a better representation through solution interpolation. Indeed, the residual of the interpolated solution on the refined mesh is different from the residual of the solution on the current mesh:

$$\begin{aligned}\int_{\Omega_h} \phi_h \;\; \nabla \cdot \mathcal{F}_{h/2} \left(I_{h/2} W_h\right) &= \int_{\Omega_{h/2}} I_{h/2} \phi_h \;\; \nabla \cdot \mathcal{F}_{h/2} \left(I_{h/2} W_h\right) \\ &\neq \int_{\Omega_h} \phi_h \; \nabla \cdot \mathcal{F}_h \left(W_h\right) = 0.\end{aligned}$$

where $I_{h/2}$ is an interpolation operator from mesh $\mathcal{H}_h$ to mesh $\mathcal{H}_{h/2}$. The corrector is thus given by the solution of the following system:

$$\int_{\Omega_h} \phi_h \nabla \cdot \mathcal{F}_h \left(W_c\right) = \int_{\Omega_h} \phi_h \; \nabla \cdot \mathcal{F}_{h/2} \left(I_{h/2} W_h\right).$$

This consists in computing the residual of W_h with a higher precision, similarly to what can be done with a higher order scheme. Another advantage is that the same numerical scheme is considered to compute the corrector, only a source term is added. Using multigrid tools, we can also compute $\mathcal{F}_{h/2} \left(I_{h/2} W_h\right)$ on the finer grid and assemble the source term by accumulating this defect on the coarse grid.

Generally speaking, we can see that, if $\mathcal{F}_{h/2} \left(I_{h/2} W_h\right)$ has locally a mean value close to 0, it means that $W_{h/2}$ will have small local adjustments related to the new degrees of freedom. On the contrary, if $\mathcal{F}_{h/2} \left(I_{h/2} W_h\right)$ is always positive, it means that $W_{h/2}$ will have a global change which will be reflected by the corrector.

From a practical point of view, the finer grids are never generated to avoid memory and CPU time overhead. Indeed, we can solve local problem in vertices neighborhood by virtually refining the elements to compute the source terms. The flow solver is then used to inverse the error equation directly, and provides the corrected solution W_c.

In the numerical experiments section, we have only considered the implicit error $W_c - W_h$ to represent the approximation error in the computation of g_{no} given by Relation (15). The interpolation error term has been neglected.

Numerical Experiments: A Turbulent Transonic Falcon

Test Case Description

The method is applied to the UMRIDA test case IC-03, Falcon jet, described in this book. The flow is transonic with a Mach number of 0.8, an angle of attack $\alpha = 2°$, and a Reynolds of 14.512 million. The viscosity is computed using the Sutherland law with a reference static temperature of 216.65 K. The Spalart–Allmaras RANS turbulence model with no trip is used.

RANS Mesh Adaptation Simulation

For the RANS mesh adaptation, the structured boundary layer mesh is kept frozen up to $y^+ = 500$ and the mesh is adapted in the upper boundary layer region and the outer field. We choose to control the interpolation error on the local Mach number in L^2 norm. Fifteen mesh adaptation iterations are performed. We split the adaptation loop into three steps with an increasing theoretical complexity (outside of the boundary layer region) specification at each step ranging from 100 000 to 400 000. Within each step, the adapted mesh at a fixed theoretical complexity is converged in five iterations. The final adapted meshes for each step contain:

- 2 298 958 vertices and 13 407 595 tetrahedra for a theoretical complexity of 100 000
- 6 168 815 vertices and 36 614 955 tetrahedra for a theoretical complexity of 200 000
- 10 337 483 vertices and 61 629 069 tetrahedra for a theoretical complexity of 400 000.

The final adapted mesh for the largest theoretical complexity is illustrated in Fig. 1.

Such adapted meshes considerably enhance the efficiency of the flow solver and the solution accuracy. We first notice that the wake is highly resolved and the wing tip vortices are well captured. Second, mesh refinements along the shock on the upper surface of the wing lead to an accurate computation of the shock–boundary layer interaction. We also observe a nice transition between the boundary layer-structured mesh and the adapted anisotropic mesh.

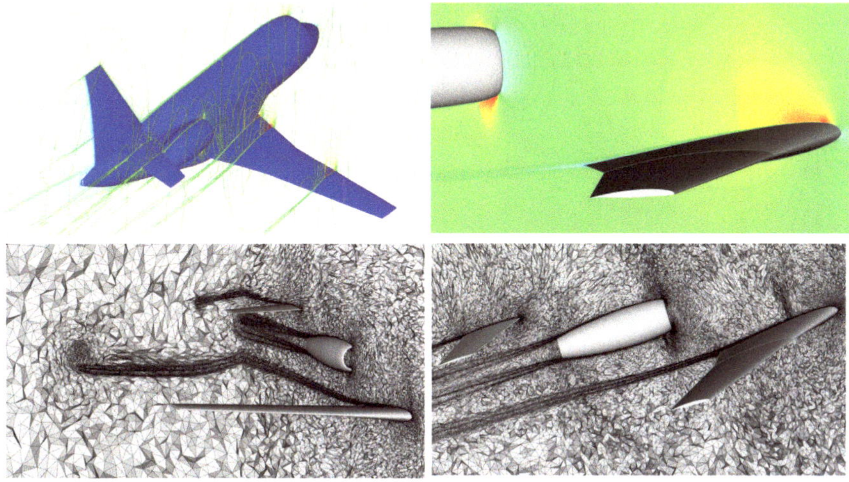

Fig. 1 Transonic Falcon business jet. Top, Mach solution field. Bottom, final adapted meshes for several cut planes in the volume

Discretization Uncertainties Quantification and Reduction

In this section, we are presenting how the corrector described in section "Correctors for the Compressible Navier–Stokes Equations" can be used to quantify uncertainties due to the mesh discretization and how mesh adaptation reduces these uncertainties. They are analyzed on the pressure contours on the wing section at chord $y = 3.84861$.

The reference solution is the solution on the finest adapted mesh at iteration 15 containing more than 10 million vertices and 60 million tetrahedra. For the first and the second adaptation steps, we compare the reference solution to the solutions obtained at the first and the last adaptation of the step at a fixed complexity, see Figs. 2 and 3. The pressure solutions are presented with error bars which are given by plus or minus the solution correction. First, we observe the convergence of the couple mesh-solution within the adaptive process at a fixed complexity, and also the convergence of the corrector with a reduction of the error bars. This points out that our formulation takes into account the algorithmic errors, i.e., the lack of iterative convergence of the mesh adaptation process. Second, we observe the reduction of the discretization uncertainties through the adaptive process when the complexity is increased. Indeed, for the lowest complexity we notice large uncertainties for the shock region due to its wrong position while these uncertainties are drastically reduced for the medium complexity because its position is now well captured. Note that the corrector is always computed on the current mesh and not on the finer grid, and thus is able to accurately detect the regions of higher discretization error and numerical error.

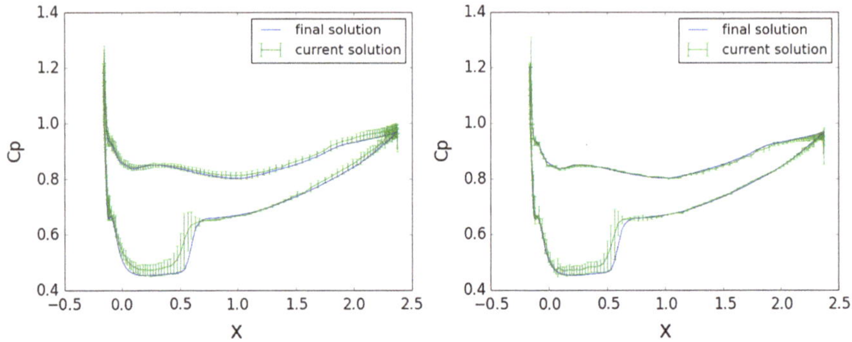

Fig. 2 Transonic Falcon business jet. Pressure extraction on the wing section for the first (iteration 1) and the last (iteration 5) adapted meshes at theoretical complexity 100 000. In blue, the reference pressure contour on the finest adapted mesh. In green, the current adapted mesh pressure contour with the error bars provided by the corrector

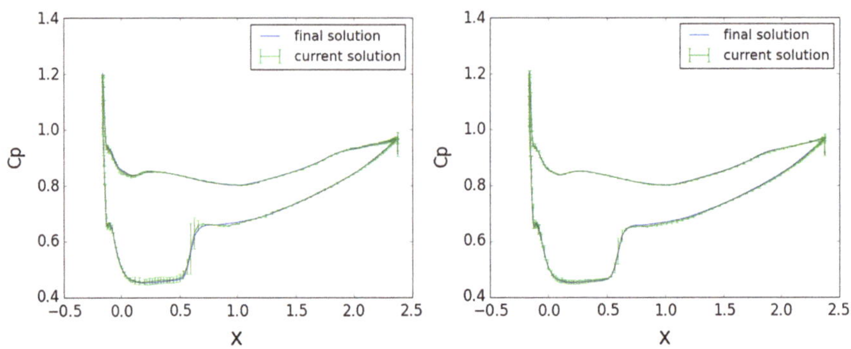

Fig. 3 Transonic Falcon business jet. Pressure extraction on the wing section for the first (iteration 6) and the last (iteration 10) adapted meshes at theoretical complexity 200 000. In blue, the reference pressure contour on the finest adapted mesh. In green, the current adapted mesh pressure contour with the error bars provided by the corrector

Conclusions and Perspectives

We have described a method combining into a single formalism mesh adaptation and solution correction strategies. Not only it helps obtaining more surely mesh convergence, but also the corrector provides an estimate of the final error. The method is based on a priori or a posteriori analysis of the different component of the error arising when discretizing a PDE: interpolation error and implicit error, the approximation error being the sum of these two errors. The interpolation error is controlled by the second derivatives of the solution, and its estimation is given by the feature-based mesh adaptation theory. The implicit error is estimated by computing a corrector. This corrector is computed by solving the state equation with an added source term. This source term relies on the residual defect of the current numerical solution by locally computing the solution residual on a subdivided grid. We show in the numerical results that our formulation also takes into account the algorithmic errors (lack of iterative convergence). Contrary to standard goal-oriented approaches [2, 11–14], a corrector W_c of the whole flow field is computed. Consequently, this approach allows us to correct any set of functional of interests simultaneously. The corrected functionals are simply evaluated with respect to W_c.

For mesh adaptation, the corrector is used to compute the right-hand side term of the adjoint state linear system. Then, the standard goal-oriented mesh adaptation theory is used to derive the optimal adapted mesh [12].

If this method gives a track for better controlling numerical error, controlling the second component of simulation error, namely model error, is not solved. While in some future, we can hope to control the model error with very-high-fidelity DNS-like calculations, the next decades will address it only with very large database comparing accurate simulations and measurements. We refer, for instance, to [36–40].

References

1. Ponsin, J., Fraysse, F., Gomez, M., Cordero-Gracia, M.: An adjoint-truncation error based approach for goal-oriented mesh adaptation. Aerospace Sci. Technol
2. Giles, M.B., Suli, E.: Adjoint methods for PDEs: a posteriori error analysis and postprocessing by duality. In: Acta Numerica, pp. 145–236. Cambridge University Press (2002)
3. Loseille, A., Dervieux, A., Frey, P.J., Alauzet, F.: Achievement of global second-order mesh convergence for discontinuous flows with adapted unstructured meshes. In: 37th AIAA Fluid Dynamics Conference, AIAA Paper 2007-4186. Miami, FL, USA, Jun (2007)
4. Alauzet, F., Loseille, A.: High order sonic boom modeling by adaptive methods. J. Comp. Phys. **229**, 561–593 (2010)
5. Castro-Díaz, M.J., Hecht, F., Mohammadi, B., Pironneau, O.: Anisotropic unstructured mesh adaptation for flow simulations. Int. J. Numer. Meth. Fluids **25**, 475–491 (1997)
6. Dompierre, J., Vallet, M.G., Fortin, M., Bourgault, Y., Habashi, W.G.: Anisotropic mesh adaptation: towards a solver and user independent CFD. In: AIAA 35th Aerospace Sciences Meeting and Exhibit, AIAA-1997-0861. Reno, NV, USA, Jan (1997)
7. Frey, P.J., Alauzet, F.: Anisotropic mesh adaptation for CFD computations. Comput. Methods Appl. Mech. Eng. **194**(48–49), 5068–5082 (2005)
8. Loseille, A., Löhner, R.: Adaptive anisotropic simulations in aerodynamics. In: 48th AIAA Aerospace Sciences Meeting, AIAA Paper 2010-169. Orlando, FL, USA, Jan (2010)
9. Chen, L., Sun, P., Xu, J.: Optimal anisotropic meshes for minimizing interpolation errors in L^p-norm. Math. Comp. **76**(257), 179–204 (2007)
10. Loseille, A., Alauzet, F.: Continuous mesh framework. Part II: validations and applications. SIAM J. Numer. Anal. **49**(1), 61–86 (2011)
11. Jones, W.T., Nielsen, E.J., Park, M.A.: Validation of 3D adjoint based error estimation and mesh adaptation for sonic boom reduction. In: 44th AIAA Aerospace Sciences Meeting and Exhibit, AIAA-2006-1150. Reno, NV, USA, Jan (2006)
12. Loseille, A., Dervieux, A., Alauzet, F.: Fully anisotropic goal-oriented mesh adaptation for 3D steady Euler equations. J. Comp. Phys. **229**, 2866–2897 (2010)
13. Power, P.W., Pain, C.C., Piggott, M.D., Fang, F., Gorman, G.J., Umpleby, A.P., Goddard, A.J.H.: Adjoint a posteriori error measures for anisotropic mesh optimization. Comput. Math. Appl. **52**, 1213–1242 (2006)
14. Venditti, D.A., Darmofal, D.L.: Grid adaptation for functional outputs: application to two-dimensional inviscid flows. J. Comp. Phys. **176**(1), 40–69 (2002)

15. Belme, A., Dervieux, A., Alauzet, F.: Time accurate anisotropic goal-oriented mesh adaptation for unsteady flows. J. Comp. Phys. **231**, 6323–6348 (2012)
16. Giles, M.B.: On adjoint equations for error analysis and optimal grid adaptation in CFD. Technical Report NA-97/11, Oxford (1997)
17. Giles, M.B., Pierce, N.: Improved lift and drag estimates using adjoint Euler equations. AIAA Paper, pp. 1999–3293 (1999)
18. Degrez, G., Giraud, L., Loriot, M., Micelotta, A., Nitrosso, B., Stoessel A.: Parallel industrial CFD calculations with n3s. In: Proceedings of the International Conference and Exhibition on High-Performance Computing and Networking, HPCN Europe '95, pp. 820–825. Springer, London, UK (1995)
19. Palacios, F., Colonno, M.R., Aranake, A.C., Campos, A., Copeland, S.R., Economon, T.D., Lonkar, A.K., Lukaczyk, T.W., Taylor, W.R., Alonso, J.J.: Stanford University unstructured (su2): an open-source integrated computational environment for multi-physics simulation and design. In: 51st AIAA Aerospace Sciences Meeting including the New Horizons Forum and Aerospace Exposition 07–10 Jan 2013. Grapevine (Dallas/Ft. Worth Region), Texas, AIAA 2013-0287 (2013)
20. Geuzaine, P., Farhat, C., Brown, G.: Application of a three-field nonlinear fluid-structure formulation to the prediction of the aeroelastic parameters of an f-16 fighter. Comput. Fluids **32**, 3–29 (2003)
21. Geuzaine, P., Brown, G., Harris, C., Farhat, C.: Aeroelastic dynamic analysis of a full f-16 configuration for various flight conditions. AIAA J. **41**, 363–371 (2003)
22. Menier, V., Loseille, A., Alauzet, F.: CFD validation and adaptivity for viscous flow simulations. In: 44th AIAA Fluid Dynamics Conference, AIAA Paper 2014-2925. Atlanta, GA, USA, Jun (2014)
23. Cournède, P.-H., Koobus, B., Dervieux, A.: Positivity statements for a Mixed-element-volume scheme on fixed and moving grids. Eur. J. Comput. Mech. **15**(7–8), 767–798 (2006)
24. Mer, K.: Variational analysis of a mixed element/volume scheme with fourth-order viscosity on general triangulations. Comput. Methods Appl. Mech. Eng. **153**, 45–62 (1998)
25. Loseille, A., Alauzet, F.: Continuous mesh framework. Part I: well-posed continuous interpolation error. SIAM J. Numer. Anal. **49**(1), 38–60 (2011)
26. Frey, P.J., George, P.-L.: Mesh generation. Application to finite elements, 2nd edn. ISTE Ltd and Wiley (2008)
27. Bottasso, C.L.: Anisotropic mesh adaption by metric-driven optimization. Int. J. Numer. Meth. Eng. **60**, 597–639 (2004)
28. Compère, G., Remacle, J.-F., Jansson, J., Hoffman, J.: A mesh adaptation framework for dealing with large deforming meshes. Int. J. Numer. Meth. Eng. **82**, 843–867 (2010)
29. Coupez, T.: Génération de maillages et adaptation de maillage par optimisation locale. Revue Européenne des Éléments Finis **9**, 403–423 (2000)
30. Dobrzynski, C., Frey, P.J.: Anisotropic Delaunay mesh adaptation for unsteady simulations. In: Proceedings of 17th International Meshing Rountable, pp. 177–194. Springer (2008)
31. George, P.L.: Gamanic3d, adaptive anisotropic tetrahedral mesh generator. Technical Note, INRIA (2003)
32. Li, X.L., Shephard, M.S., Beall, M.W.: 3D anisotropic mesh adaptation by mesh modification. Comput. Methods Appl. Mech. Eng. **194**(48–49), 4915–4950 (2005)
33. Pain, C.C., Umpleby, A.P., de Oliveira, C.R.E., Goddard, A.J.H.: Tetrahedral mesh optimisation and adaptivity for steady-state and transient finite element calculations. Comput. Methods Appl. Mech. Eng. **190**, 3771–3796 (2001)
34. Tam, A., Ait-Ali-Yahia, D., Robichaud, M.P., Moore, M., Kozel, V., Habashi, W.G.: Anisotropic mesh adaptation for 3D flows on structured and unstructured grids. Comput. Methods Appl. Mech. Eng. **189**, 1205–1230 (2000)
35. Dervieux, A., Loseille, A., Alauzet, F.: High-order adaptive method applied to high speed flows. In: Proceedings of WEHSFF (2007)
36. Edeling, W.N., Cinnella, P., Dwight, R.P., Bijl, H.: Bayesian estimates of parameter variability in the k-epsilon turbulence model. J. Comput. Phys. **258**, 73–94 (2014)

37. Esfahanian, V., Rahbari, I.: Epistemic uncertainty quantification of Rans turbulence models. 2013. In: 15th Conference on Fluid Dynamics, The University of Hormozgan, Bandar Abbas, Iran
38. Gorlé, C., Rossi, R., Iaccarino, G.: Uncertainty quantification for RANS simulation of flow over a wavy wall. In: 6th European and African Conference on Wind Engineering, Cambridge, UK
39. Oliver, T.A., Moser, R.D.: Bayesian uncertainty quantification applied to Rans turbulence models. J. Phys.: Conf. Ser. 318 042032 (2011)
40. Schmidt, T., Mockett, C., Thiele, F.: Minimization and quantification of errors and uncertainties in Rans modeling. In: Eisfeld, B. et al. (eds.), Management and Minimisation of Uncertainity and Errors, NNFM 122. Springer, Berlin, Heidelberg

General Introduction to Polynomial Chaos and Collocation Methods

Chris Lacor and Éric Savin

Polynomial Chaos Methodology

The polynomial chaos methodology (PCM) is a rather recent approach, which offers a large potential for computational fluid dynamics (CFD) related non-deterministic simulations, as it allows the treatment of a large variety of stochastic variables and properties that can be described by probability density functions (PDFs). The method is based on a spectral representation of the uncertainty where the basis polynomials contain the randomness, described by random variables $\boldsymbol{\xi}$ with values in a set Γ, and the unknown expansion coefficients are deterministic, resulting in deterministic equations. More specifically, if u is a random variable indexed by a spatial variable $\mathbf{x} \in \mathscr{D} \subseteq \mathbb{R}^d$ (typically, $d = 3$ in physical space) and time $t \geq 0$, the so-called **polynomial chaos expansion** (PCE) reads:

$$u(\mathbf{x}, t, \boldsymbol{\xi}) \simeq \mathbb{P}^P[u](\mathbf{x}, t, \boldsymbol{\xi}) = \sum_{i=0}^{P} u^i(\mathbf{x}, t)\psi_i(\boldsymbol{\xi}) \,. \tag{1}$$

In the above, u^i are the deterministic unknown expansion coefficients and represent the random mode i of the random variable u. ψ_i are N-variate polynomials which are functions of $\boldsymbol{\xi} = (\xi_1, \xi_2, \dots, \xi_N)$ where ξ_j is a random variable with values in a set Γ_j. N is the number of input uncertainties which is also the **number of random**

C. Lacor (✉)
Mechanical Engineering Department, Vrije Universiteit Brussel,
Pleinlaan 2, 1050 Brussels, Belgium
e-mail: clacor@vub.ac.be

É. Savin
Computational Fluid Dynamics Department, Onera–The French Aerospace Lab,
29 avenue de la Division Leclerc, 92322 Châtillon, France
e-mail: eric.savin@onera.fr

C. Hirsch et al. (eds.), *Uncertainty Management for Robust Industrial Design in Aeronautics*, Notes on Numerical Fluid Mechanics and Multidisciplinary Design 140, https://doi.org/10.1007/978-3-319-77767-2_7

dimensions. It is assumed that these variables are **independent** and are real valued, and hence $\Gamma = \Gamma_1 \times \Gamma_2 \times \cdots \times \Gamma_N \subseteq \mathbb{R}^N$. Input uncertainties could, e.g., be associated with uncertain operational conditions or uncertainty in the geometry. For an external flow around an airplane, the inlet Mach number, angle of attack, inlet pressure, etc., are examples of operational conditions. Geometrical uncertainties are then uncertainties on the shape of the shape of the plane due to manufacturing tolerances. It is clear that because of the uncertain input, any flow variable, say u becomes also uncertain, and can therefore be described as in Eq. (1). The total number of terms $P + 1$ used in (1) depends on the highest order of the polynomial that is used (denoted by p) and on the number of random dimensions. One has, see [1]:

$$P + 1 = \frac{(N + p)!}{N!p!} . \tag{2}$$

The methodology was originally formulated by Wiener [2] and was much later rediscovered and used for CFD applications by several groups, e.g., Xiu and Karniadakis [3], Lucor et al. [4], Le Maître et al. [5], Mathelin et al. [6], and Walters and Huyse [7] among others.

In the original method of Wiener [2], the projection basis ψ_i is constituted by **Hermite** polynomials. These are optimal for random variables with **Gaussian** distribution. Optimal means that, for increasing polynomial order, the expansion will quickly converge in the mean-square sense. The condition for optimality is that the polynomials are orthogonal with a weighting function $\boldsymbol{\xi} \mapsto W_N(\boldsymbol{\xi})$ which is exactly the PDF of the set of random variables, i.e.:

$$\int_{\Gamma} \psi_i(\boldsymbol{\xi})\psi_j(\boldsymbol{\xi})W_N(\boldsymbol{\xi})d\boldsymbol{\xi} := \langle \psi_i, \psi_j \rangle = \gamma_j \delta_{ij} , \tag{3}$$

where δ_{ij} is the Kronecker symbol, and γ_j is a normalization constant. With a proper scaling though, one can always normalize the polynomial basis such that $\gamma_j := \langle \psi_j, \psi_j \rangle = 1 \ \forall j$. In the case of a multivariate Gaussian distribution, the Hermite polynomials satisfy the condition above with W_N given by:

$$W_N(\boldsymbol{\xi}) \equiv \frac{1}{\sqrt{(2\pi)^N}} \exp(-\frac{1}{2}\boldsymbol{\xi} \cdot \boldsymbol{\xi}) , \tag{4}$$

where $\boldsymbol{\xi} \cdot \boldsymbol{\xi} = \sum_{j=1}^{N} \xi_j^2$ is the standard Euclidian scalar product in $\mathbb{R}^N$. Note that because of the independence of the uncertainties, the PDF is the product of the PDF of each of the uncertainties, i.e., $W_N(\boldsymbol{\xi}) = \prod_{j=1}^{N} W_1(\xi_j)$ as defined above for Gaussian uncertainties.

For uncertainties with other distributions, the orthogonality condition (3) gives adapted polynomials, see e.g. [3], leading to the so-called **Askey scheme**; for example, as already mentioned Hermite polynomials for Gaussian distributions, and further Charlier polynomials for Poisson distributions, Laguerre polynomials for Gamma distributions, Jacobi polynomials for Beta distributions, etc. In case of less common distributions, an optimal PCM can always be found by constructing the polynomials via a Gram-Schmidt procedure; see Witteveen and Bijl [8], in order to satisfy (3). It should be noted that, if the optimal polynomials are not used, the PCM will also converge (with increasing order) in the mean-square sense but much slower than the exponential convergence with optimal polynomials; see [1].

In cases where the response of the system shows a localized sharp variation or a discontinuous change, local expansions may be more efficient than expansions with global polynomials, whose convergence will deteriorate due to the Gibbs phenomenon. This has led to developments using wavelet expansions [9] and to multi-element polynomial chaos [10, 11]. In the latter case, the random space is subdivided in smaller elements in which new random variables are defined with associated orthogonal polynomials that are constructed numerically.

As already mentioned, the dimension of the problem N is determined by the number of independent random input variables. In case of a random process (as opposed to a random variable), a Karhunen-Loève expansion (also known as Principal Component Analysis or Proper Orthogonal Decomposition) [12, 13] can be applied to the correlation function $R(\mathbf{x}, \mathbf{y})$ of the random process $u(\mathbf{x})$ indexed by $\mathbf{x} \in \mathscr{D}$, to decompose the random input process in a set of uncorrelated random variables. Assuming $\int_{\mathscr{D}} R(\mathbf{x}, \mathbf{x}) d\mathbf{x} < +\infty$ (which is untrue for a stationary process with $\mathscr{D} \equiv \mathbb{R}^d$) and solving the eigenvalue problem:

$$\int_{\mathscr{D}} R(\mathbf{x}, \mathbf{y}) \phi_i(\mathbf{y}) d\mathbf{y} = \lambda_i \phi_i(\mathbf{y}) \tag{5}$$

with $\phi_i(\mathbf{x})$ the eigenfunctions and λ_i the eigenvalues, the Karhunen-Loève expansion of the random field $u(\mathbf{x})$ becomes:

$$u(\mathbf{x}) - \overline{u(\mathbf{x})} = \sum_i \sqrt{\lambda_i} \xi_i \phi_i(\mathbf{x}), \tag{6}$$

where the ξ_is are uncorrelated random variables, and $\overline{u(\mathbf{x})}$ is the mean value at the indexation point $\mathbf{x}$. Note that if the process u is Gaussian, the random variables ξ_i are Gaussian as well, and hence, they are mutually independent.

A geometrical uncertainty is typically a random process where the coordinates of a geometry are uncertain with some specific correlation length. Depending on the correlation length of the process, the eigenvalues λ_i become quickly very small, so that only few terms in the summation above have to be kept. This is not the case however for a very short correlation length (e.g., white noise) resulting in a high-dimensional chaos expansion for such processes. Non-Gaussian random processes are much more difficult to treat than Gaussian [14]. In the former case, mean and covariance are far from sufficient to completely specify the process. This remains an active area of research.

The PCM can be implemented either in an **intrusive** or in a **non-intrusive** way as follows.

Intrusive Polynomial Chaos

In an intrusive PCM, the polynomial expansion of the unknown variables, Eq. (1), is introduced in the model, e.g., for CFD applications, the Navier–Stokes equations. Each unknown u is therefore replaced with its expansion coefficients u^i. The number of unknowns is therefore basically multiplied with a factor $P + 1$, which can be quite high for high stochastic dimensions and/or high polynomial order. In addition, the model, e.g., CFD code, has to be adapted. The required effort for extending a deterministic CFD code with the intrusive PCM depends on the characteristics of the code: computer language, structured/unstructured, handling of data storage, etc. In the framework of the NODESIM-CFD EU project, an intrusive PCM was implemented in the commercial code Fine/Turbo of NUMECA. This has led to one of the first applications of intrusive PCM to three-dimensional turbulent Navier–Stokes flows [15]. The number of additional lines of code is very limited, compared to the length of the original, deterministic code. However, changes are not restricted to a local part of the code. This increases the risk of introducing bugs and requires someone who is very familiar with all aspects of the code. This is a big disadvantage compared to non-intrusive PC and the main reason why the application of intrusive PCM in commercial codes is very limited.

Nonetheless intrusive methods are more flexible and in general more precise than non-intrusive methods; see Aleksev et al. [16]. This is also confirmed by Xiu [14], who mentions that the intrusive method offers the most accurate solutions involving the least number of equations in multi-dimensional random spaces, even though the resulting equations are coupled.

It is to be noted that the treatment of geometrical uncertainties needs a different approach compared to operational uncertainties. A possibility is to use a transformation such that the deterministic problem in a stochastic domain becomes a stochastic problem in a deterministic domain, e.g., Xiu and Tartakovsky [17]. An alternative is the use of a so-called fictitious domain method [18, 19], or by introducing the uncertainty directly in the surface normals within a control volume approach [20, 21].

Non-intrusive Polynomial Chaos

In the UMRIDA EU project, all PCM contributions relate to non-intrusive approaches. Basically, two different classes of approaches have been formulated: (i) the so-called **projection method**, which is based on a numerical evaluation of the Galerkin integrals; see Le Maître et al. [5, 22, 23] and Nobile et al. [24]; (ii) **regression methods** based on a selected set of sample points; see Berveiller et al. [25], and Hosder et al. [26–28].

In the **projection methods**, starting from Eq. (1), the projection on ψ_j yields:

$$\begin{aligned}\int_{\Gamma} u(\mathbf{x},t,\boldsymbol{\xi})\psi_j(\boldsymbol{\xi})W_N(\boldsymbol{\xi})d\boldsymbol{\xi} &= \sum_{i=0}^{P} u^i(\mathbf{x},t) \int_{\Gamma} \psi_i(\boldsymbol{\xi})\psi_j(\boldsymbol{\xi})W_N(\boldsymbol{\xi})d\boldsymbol{\xi} \\ &= \gamma_j u^j(\mathbf{x},t)\end{aligned} \tag{7}$$

The last equation results from the orthogonality condition (3) and can be considered as an equation for the unknown expansion coefficient u^j. It requires the evaluation of the integral in the left-hand side. A numerical quadrature formula is used. For a single variable parameter, it reads:

$$\int_{\Gamma_1} u(\mathbf{x},t,\xi)\psi_j(\xi)W_1(\xi)d\xi \simeq \sum_{l=1}^{q} w^l u(\mathbf{x},t,\xi^l)\psi_j(\xi^l)\,. \tag{8}$$

The evaluation of the sum in the right-hand side requires an evaluation of the unknown u in q sample points $\{\xi^l\}_{1\leq l\leq q}$ in Γ_1 associated to q weights $\{w^l\}_{1\leq l\leq q}$. Depending on the weighting function (PDF) W_1, adapted Gaussian quadrature formulations exist for an accurate evaluation: With q sample points, a polynomial of order $2q-1$ is integrated exactly in one dimension. Examples are the Gauss-Legendre quadrature ($W_1 = 1/2$ corresponding to a uniform distribution), the Gauss-Hermite quadrature (W_1 given by Eq. (4) in one dimension), etc. For a PCM of order p, one takes $q = p+1$. This guarantees exact quadrature if $u(\mathbf{x},t,\xi)$ can be described by a polynomial of maximum order $p+1$.

This extends to multiple stochastic dimensions by using a full-tensor product quadrature with $Q = q^N$ sample points. This approach quickly becomes very expensive for high-order and high stochastic dimensions. This has led to the use of **sparse grid sampling** techniques, avoiding the full-tensorial sampling, e.g., the Smolyak scheme [29]. Sparse grid schemes can be combined with the non-nested Gaussian quadratures invoked above, as well as with nested quadratures, e.g., Clenshaw-Curtis, Gauss-Patterson [30–33]. More recently, adaptive algorithms have been developed that further reduce the cost [34–36]. The choice of quadrature sets is discussed further on in section "Choices of Interpolation Set" in relation with the stochastic collocation method. Alternatively, the numerical quadrature can also be achieved using Monte Carlo simulation [37, 38], or Latin Hypercube sampling [39]. All in all, the evaluation of the left-hand side of Eq. (7) using Q sampling points $\{\boldsymbol{\xi}^l\}_{1\leq l\leq Q}$ in Γ associated to Q weights $\{w^l\}_{1\leq l\leq Q}$ yields:

$$u(\mathbf{x},t,\boldsymbol{\xi}) \simeq \mathbb{P}_Q^P[u](\mathbf{x},t,\boldsymbol{\xi}) = \sum_{i=0}^{P}\left(\frac{1}{\gamma^i}\sum_{l=1}^{Q} w^l u(\mathbf{x},t,\boldsymbol{\xi}^l)\psi_i(\boldsymbol{\xi}^l)\right)\psi_i(\boldsymbol{\xi})\,. \tag{9}$$

In linear **regression methods**, the stochastic problem is solved in S samples in stochastic space. For each sample s, Eq. (1) can be written as:

$$u(\mathbf{x}, t, \boldsymbol{\xi}^s) = \sum_{i=0}^{P} u^i(\mathbf{x}, t)\psi_i(\boldsymbol{\xi}^s)\,. \tag{10}$$

This leads to S equations for the $P+1$ unknowns u^i. Note that this forms a linear system. In order to make the solution less dependent on the choice of the samples, oversampling is used and the system is solved with regression (i.e., the least squares method); see Berveiller et al. [25] and Hosder et al. [26–28]. As a rule of thumb, $S = 2(P+1)$ is a good choice; see [27]. Different sampling techniques can be used such as Random, Latin Hypercube, Hammersley [27], roots of Hermite polynomials of order $p+1$ (for PCM of order p with Gaussian uncertainties) [25], Sobol' quasi-random sampling [40], etc.

In case of geometrical uncertainties, each of the different samples–both in the projection and the regression method–will correspond to a different geometry. Geometrical uncertainties therefore require no special treatment in contrast with the intrusive method.

The Collocation Method

The stochastic collocation (SC) method based on Lagrange interpolation has been introduced in [41] and developed further on in e.g. [24, 42–45]. Examples of applications can be found in [46–51] among others. Along the same lines as Eq. (1), the SC expansion is formed as a sum of multi-dimensional Lagrange interpolation polynomials with respect to the N–dimensional random input variable $\boldsymbol{\xi}$. Lagrange polynomials interpolate a set of points in one dimension $\{\xi_1^l\}_{1\leq l\leq q_1}$ in a bounded interval Γ_1 by the following functional form:

$$L_l(\xi) = \prod_{\substack{k=1\\k\neq l}}^{q_1} \frac{\xi - \xi_1^k}{\xi_1^l - \xi_1^k}\,, \tag{11}$$

such that $L_l(\xi_1^k) = \delta_{kl}$, $1 \leq k, l \leq q_1$; in addition, all L_l's have order $q_1 - 1$. For interpolation in multiple dimensions, the tensor product of one-dimensional Lagrange polynomials can be formed. Eventually at this stage, it is assumed that the interpolation set is formed by tensorization of one-dimensional sets. In other words, structured interpolation sets are considered, for multivariate Lagrange interpolation on unstructured, arbitrary sets of nodes still raises numerous theoretical and practical difficulties. Letting $\mathbf{l} = (l_1, l_2 \ldots l_N)$ be a multi-index in $\mathbb{N}^N \setminus \{\mathbf{0}\}$, the multi-dimensional Lagrange polynomial $L_\mathbf{l}$ reads:

$$L_\mathbf{l}(\boldsymbol{\xi}) = L_{l_1}(\xi_1) \otimes L_{l_2}(\xi_2) \otimes \cdots \otimes L_{l_N}(\xi_N)\,, \tag{12}$$

where different interpolation sets $\{\xi_j^l\}_{1\leq l\leq q_j}$ in different intervals Γ_j may possibly be used for each different dimension j. If Q is now the total number of such multi-dimensional interpolation points counted by a single index l, $\{\boldsymbol{\xi}^l\}_{1\leq l\leq Q}$, the SC expansion of the random field u reads:

$$u(\mathbf{x},t,\boldsymbol{\xi}) \simeq \mathbb{I}^Q[u](\mathbf{x},t,\boldsymbol{\xi}) = \sum_{l=1}^{Q} u(\mathbf{x},t,\boldsymbol{\xi}^l)L_l(\boldsymbol{\xi}), \tag{13}$$

where the expansion coefficients are the random field evaluated at $\boldsymbol{\xi}^l$.

Choices of Interpolation Set

The key issue of the SC method is the choice of appropriate interpolation sets. A natural, straightforward choice is quadrature nodes and weights as in Eq. (8). Multi-dimensional quadrature sets $\boldsymbol{\Theta}(N,Q) = \{\boldsymbol{\xi}^l, w^l\}_{1\leq l\leq Q}$, where $\boldsymbol{\xi}^l$ is the l-th node in $\Gamma = \prod_{j=1}^{N}\Gamma_j$ and w^l is the corresponding weight, may be constructed from one-dimensional (univariate) quadrature sets by full tensorization or sparse tensorization, using Smolyak's algorithm [29] as already invoked above.

Univariate Gauss quadratures $\Theta(1,q_1)$ based on q_1 integration points are tailored to integrate on $\Gamma_1 \equiv [a,b]$ a smooth function $\xi \mapsto f(\xi)$:

$$\int_{\Gamma_1} f(\xi)W_1(\xi)d\xi \simeq \sum_{l=1}^{q_1-r} w^l f(\xi^l) + \sum_{m=1}^{r} w^{q_1-r+m} f(\xi^{q_1-r+m}), \tag{14}$$

such that this rule turns to be exact for univariate polynomials up to the order $2q_1 - 1 - r$. Here, r is the number of fixed nodes of the rule, typically the bounds a, b. Depending on the choice of r, different terminologies are used:

- $r = 0$ is the classical Gauss rule;
- $r = 1$ is the Gauss-Radau (GR) rule, choosing $\xi^{q_1} = a$ or $\xi^{q_1} = b$ for instance;
- $r = 2$ is the Gauss-Lobatto (GL) rule, choosing $\xi^{q_1-1} = a$ and $\xi^{q_1} = b$ for instance.

Multivariate quadratures may subsequently be obtained by full or sparse tensorization of these one-dimensional rules. Firstly, a fully tensorized grid is obtained by the straightforward product rule:

$$\boldsymbol{\Theta}(N,Q) = \bigotimes_{j=1}^{N} \Theta(1,q_j), \tag{15}$$

which contains $Q = \prod_{j=1}^{N} q_j$ grid points in Γ. Secondly, a sparse quadrature rule can be derived thank to the Smolyak algorithm [29]. The so-called k–th level, N-dimensional Smolyak sparse grid $\widehat{\boldsymbol{\Theta}}(N,k)$ is obtained by the following linear combination of product formulas [52]:

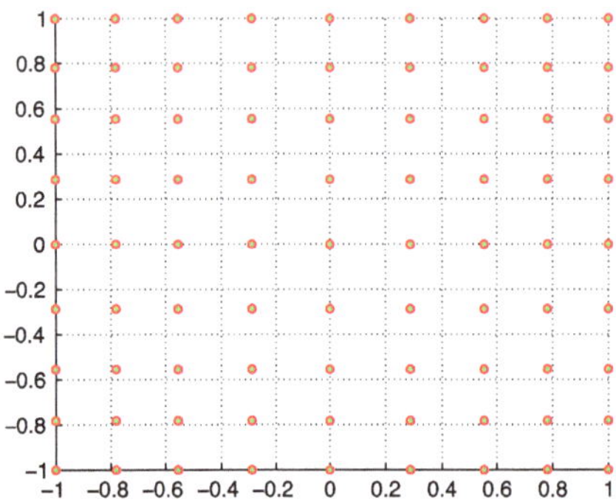

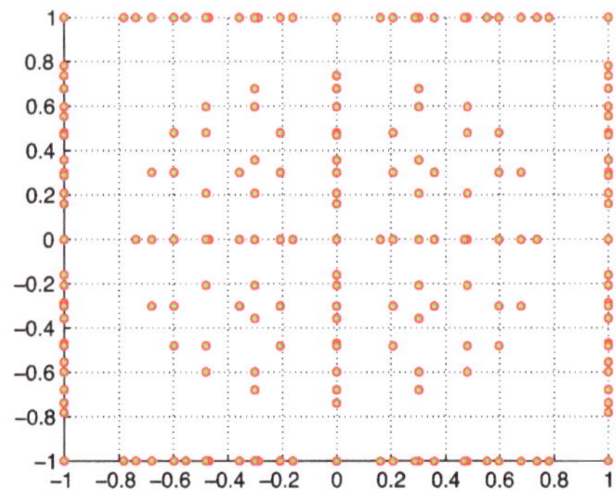

Fig. 1 Two-dimensional ($N = 2$) nodes based on a non-nested, one-dimensional Gauss-Lobatto quadrature rule with $q = 9$ nodes. Left: fully tensorized grid ($Q = 81$). Right: sparse tensorized grid from Smolyak's algorithm with $k = 9$ ($Q = 193$)

$$\widehat{\boldsymbol{\Theta}}(N,k) = \sum_{l=k-N}^{k-1} \sum_{q_1+\cdots+q_N=N+l} \Theta(1,q_1) \otimes \cdots \otimes \Theta(1,q_N). \tag{16}$$

Clearly, the above sparse grid is a subset of the full-tensor product grids. It typically contains $Q \sim (2N)^{k-1}/k-1!$ nodes in Γ whenever $N \gg 1$ and k is fixed. By a direct extension of the arguments divised in [31, 33], it can be shown that provided the univariate quadrature rules $\Theta(1,q)$ are exact for all univariate polynomials of order up to $2q-1$ (Gauss rules) or $2q-3$ (GL rules), the foregoing rule is exact for all N–variate polynomials of total order up to $2k-1$ or $2k-3$, respectively. Figure 1 displays for example the two-dimensional full and sparse rules for an underlying univariate GL quadrature (14) with $q = 9$ nodes and $W_1(\xi) = (1-\xi^2)^3$, $\Gamma_1 = [-1,1]$. For this example:

$$\begin{aligned}\widehat{\boldsymbol{\Theta}}(2,9) &= \Theta(1,2)\otimes\Theta(1,7) + \Theta(1,3)\otimes\Theta(1,6) + \Theta(1,4)\otimes\Theta(1,5)\\ &\quad + \Theta(1,2)\otimes\Theta(1,8) + \Theta(1,3)\otimes\Theta(1,7) + \Theta(1,4)\otimes\Theta(1,6)\\ &\quad + \Theta(1,5)\otimes\Theta(1,5) + \text{perm.}\end{aligned}$$

Here, $Q = 193$, compared to $Q = 81$ with the fully tensorized rule (15). In [53], it has been observed that sparse quadratures outperform fully tensorized quadratures with non-nested underlying one-dimensional rules whenever $N \geq 4$, though. If $\Theta(1,q_i)$ is now Clenshaw-Curtis (CC) univariate quadrature of i-th level for $i > 1$, such that:

$$\xi^l = -\cos\frac{(l-1)\pi}{q_i-1}, \quad 1 \leq l \leq q_i = 2^{i-1}+1,$$

then the associated third-level bivariate sparse rule as constructed in, e.g., [32] for, say, $q = 9$ is:

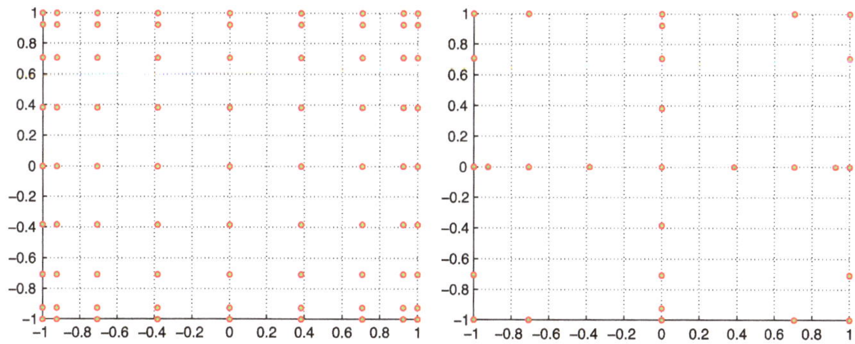

Fig. 2 Two-dimensional ($N = 2$) nodes based on a nested, one-dimensional Clenshaw-Curtis quadrature rule with $q = 9$ nodes. Left: fully tensorized grid ($Q = 81$). Right: sparse tensorized grid from Smolyak's algorithm with $k = 3$ ($Q = 29$)

$$\begin{aligned}\widehat{\boldsymbol{\Theta}}(2,3) &= \Theta(1,1)\otimes\Theta(1,5)+\Theta(1,3)\otimes\Theta(1,3)\\ &\quad+\Theta(1,1)\otimes\Theta(1,9)+\Theta(1,3)\otimes\Theta(1,5)+\text{perm.}\end{aligned} \tag{17}$$

The underlying univariate CC rules $\Theta(1,q_i)$ are nested, that is, $\Theta(1,q_i)\subset\Theta(1,q_{i+1})$, and consequently, the multivariate rules are nested as well, $\widehat{\boldsymbol{\Theta}}(N,k)\subset\widehat{\boldsymbol{\Theta}}(N,k+1)$. They are in addition exact at least for all multivariate polynomials of total order k [32]. Figure 2 displays the two-dimensional full rule (15) and third-level sparse rule (17) corresponding to the univariate CC quadrature with $q = 9$ nodes. The total number of nodes is significantly reduced with such a nested rule.

Link with Polynomial Chaos

The multi-dimensional Lagrange polynomials may be expanded on the multi-dimensional polynomial chaos basis $\{\psi_i\}_{0\leq i\leq P}$ as in Eq. (1):

$$L_l(\boldsymbol{\xi}) = \sum_{i=0}^{P}\langle L_l,\psi_i\rangle\psi_i(\boldsymbol{\xi}),\quad 1\leq l\leq Q,$$

where P is given by Eq. (2) with polynomial total order $p = \sum_{j=1}^{N} q_j - N$. The expansion coefficients $\ell_{li} := \langle L_l,\psi_i\rangle$ can be evaluated with the quadrature rule $\{\boldsymbol{\xi}^l, w^l\}_{1\leq l\leq Q}$ also used as the interpolation set:

$$\ell_{li}\simeq\frac{1}{\gamma_i^Q}\sum_{m=1}^{Q} w^m L_l(\boldsymbol{\xi}^m)\psi_i(\boldsymbol{\xi}^m) = \frac{1}{\gamma_i^Q}w^l\psi_i(\boldsymbol{\xi}^l),$$

where the second equality stems from the very definition of Lagrange polynomials. Here, $\gamma_i^Q = \sum_{l=1}^{Q} w^l(\psi_i(\boldsymbol{\xi}^l))^2$ is the normalization constant for the polynomial chaos, which is simply $\gamma_i^Q = \gamma_i$ if the quadrature rule integrates exactly polynomials of total order $2p$. Consequently, the SC expansion (13) of the random field u reads:

$$\begin{aligned}\mathbb{I}^Q[u](\mathbf{x},t,\boldsymbol{\xi}) \simeq \mathbb{I}_P^Q[u](\mathbf{x},t,\boldsymbol{\xi}) &= \sum_{l=1}^{Q} u(\mathbf{x},t,\boldsymbol{\xi}^l) \sum_{i=0}^{P} \frac{1}{\gamma_i^Q} w^l \psi_i(\boldsymbol{\xi}^l)\psi_i(\boldsymbol{\xi}) \\ &= \sum_{i=0}^{P} \left(\frac{1}{\gamma_i^Q} \sum_{l=1}^{Q} w^l u(\mathbf{x},t,\boldsymbol{\xi}^l)\psi_i(\boldsymbol{\xi}^l) \right) \psi_i(\boldsymbol{\xi}) \,.\end{aligned} \tag{18}$$

The bracketed sum above is the evaluation of the PC expansion coefficients u^i by the quadrature rule at hand. Hence, both PC and SC expansions are mathematically equivalent, $\mathbb{I}_P^Q \equiv \mathbb{P}_Q^P$, though they are numerically slightly different [54].

Application to Uncertainty Quantification (UQ)

Once the polynomial expansion (1) or (13) has been derived, the first moments and/or cumulants of the random field u can be computed using a quadrature rule $\boldsymbol{\Theta}(N, Q)$ and associated evaluations $u(\mathbf{x}, t, \boldsymbol{\xi}^l)$, $1 \leq l \leq Q$. Indeed, for a regular function $u \mapsto f(u)$, one can estimate a mean output functional by:

$$\mathbb{E}\{f(u)\}(\mathbf{x},t) = \int_{\Gamma} f(u(\mathbf{x},t,\boldsymbol{\xi}))W_N(\boldsymbol{\xi})d\boldsymbol{\xi} \simeq \sum_{l=1}^{Q} w^l f(u(\mathbf{x},t,\boldsymbol{\xi}^l)) \,.$$

The mean μ is obtained for $f(u) = u$, the variance σ^2 is obtained for $f(u) = (u - \mu)^2$, the skewness β_1 for $f(u) = (\frac{u-\mu}{\sigma})^3$, the kurtosis β_2 for $f(u) = (\frac{u-\mu}{\sigma})^4$, etc. More generally, the j-th moment m_j is obtained for $f(u) = u^j$ and may be used to compute the characteristic function $\boldsymbol{\Phi}_U$:

$$\boldsymbol{\Phi}_U(V) = \int \mathrm{e}^{\mathrm{i}U \cdot V} W_U(dU) = \sum_{j=0}^{+\infty} \frac{m_j}{j!} (\mathrm{i}U)^j \,,$$

where by the causality principle (or transport of PDFs) for the random variable $U \sim u(\cdot, \boldsymbol{\xi})$ one has:

$$W_U(dU) = \left| \frac{du^{-1}}{dU} \right| W_N(u^{-1}(dU)) \,.$$

Sobol' sensitivity indices or global sensitivity indices may be computed alike; see [14, 53, 55–57] and references therein. Denoting by $\mathscr{I}_j$ the set of indices corresponding to the polynomials ψ_k depending only on the j-th variable parameter ξ_j, the main-effect PCE-based Sobol' indices are given by (see e.g. Sudret [57]):

$$S_j(\mathbf{x},t) = \frac{1}{\sigma^2} \sum_{k \in \mathscr{I}_j} \gamma_k (u^k(\mathbf{x},t))^2 ,$$

owing to the normalization condition (3). More generally, if $\mathscr{I}_{j_1 j_2 \dots j_s}$ is the set of indices corresponding to the polynomials ψ_k depending only on the parameters $\xi_{j_1}, \xi_{j_2}, \dots \xi_{j_s}$, the s-fold joint PCE-based Sobol' indices are:

$$S_{j_1 j_2 \dots j_s}(\mathbf{x},t) = \frac{1}{\sigma^2} \sum_{k \in \mathscr{I}_{j_1 j_2 \dots j_s}} \gamma_k (u^k(\mathbf{x},t))^2 .$$

Conclusions

In this chapter, we have outlined the main ingredients of polynomial expansion methods for the pseudo-spectral analysis of random variables and fields, using either projections on orthonormal polynomials–the generalized polynomial chaos method, or interpolations on Lagrange polynomials–the stochastic collocation method. We have also shown how both approaches are actually intimately connected by a proper choice of the integration/interpolation nodal sets used to compute the polynomial expansion coefficients. However, alternative strategies have been recently considered in order to evaluate them, which are detailed in the following chapters "Generalized Polynomial Chaos for Non-intrusive Uncertainty Quantification in Computational Fluid Dynamics" through "Screening Analysis and Adaptive Aparse Collocation Methods". Applications to uncertainty quantification and robust design optimization for industrial challenges are given in parts III and IV.

References

1. Xiu, D., Karniadakis, G.: The Wiener-Askey polynomial chaos for stochastic differential equations. SIAM J. Sci. Comput. **24**(2), 619–644 (2002)
2. Wiener, N.: The homogeneous chaos. Am. J. Math. **60**(4), 897–936 (1938)
3. Xiu, D., Karniadakis, G.: Modeling uncertainty in flow simulations via generalized polynomial chaos. J. Comput. Phys. **187**(1), 137–167 (2003)
4. Lucor, D., Xiu, D., Su, C.H., Karniadakis, G.: Predictability and uncertainty in CFD. Int. J. Numer. Methods Fluids **43**(5), 483–505 (2003)
5. Le Maître, O., Knio, O., Najm, H., Ghanem, R.: A stochastic projection method for fluid flow. I. Basic formulation. J. Comput. Phys. **173**(2), 481–511 (2001)
6. Mathelin, L., Hussaini, M., Zang, T.: Stochastic approaches to uncertainty quantification in CFD simulations. Numer. Algorithms **38**(1), 209–236 (2005)
7. Walters, R., Huyse, L.: Uncertainty analysis for fluid mechanics with applications. ICASE Report no. 2002-1 (2002)
8. Witteveen, J., Bijl, H.: Efficient quantification of the effect of uncertainties in advection-diffusion problems using polynomial chaos. Numer Heat Transf Part B **53**, 437–465 (2008)
9. Le Maître, O., Knio, O.M., Najm, H.N., Ghanem, R.G.: Uncertainty propagation using Wiener-Haar expansions. J. Comput. Phys. **197**(1), 28–57 (2004)

10. Wan, X., Karniadakis, G.: An adaptive multi-element generalized polynomial chaos method for stochastic differential equations. J. Comput. Phys. **209**(2), 617–642 (2005)
11. Wan, X., Karniadakis, G.: Beyond Wiener-Askey expansions: handling arbitrary PDFs. J. Sci. Comput. **27**(1), 455–464 (2006)
12. Karhunen, K.: Zur spektraltheorie stochastischer prozesse. Ann. Acad. Sci. Fennicae **34**, 1–7 (1946)
13. Loève, M.: Fonctions aleatoires du seconde ordre. In: Levy, P. (ed.) Processus Stochastiques et Movement Brownien, Paris (1948)
14. Xiu, D.: Fast numerical methods for stochastic computations: a review. Commun. Comput. Phys. **5**(2–4), 242–272 (2009)
15. Dinescu, C., Smirnov, S., Hirsch, C., Lacor, C.: Assessment of intrusive and non-intrusive non-deterministic CFD methodologies based on polynomial chaos expansions. Int. J. Eng. Syst. Model. Simul. **2**(1–2), 87–98 (2010)
16. Alekseev, A.K., Navon, I.M., Zelentsov, M.E.: The estimation of functional uncertainty using polynomial chaos and adjoint equations. Int. J. Numer. Methods Fluids **67**(3), 328–341 (2011)
17. Xiu, D., Tartakovsky, D.: Numerical methods for differential equations in random domains. SIAM J. Sci. Comput. **28**(3), 1167–1185 (2006)
18. Parussini, L., Pediroda, V.: Investigation of multi geometric uncertainties by different polynomial chaos methodologies using a fictitious domain solver. Comput. Model. Eng. Sci. **23**(1), 29–52 (2008)
19. Parussini, L.: Fictitious domain approach via Lagrange multipliers with least squares spectral element method. J. Sci. Comput. **37**(3), 316–335 (2008)
20. Perez, R., Walters, R.: An implicit polynomial chaos formulation for the Euler equations. In: AIAA 2005-1406 (2005) 43rd AIAA Aerospace Sciences Meeting and Exhibit, 10–13 Jan 2005, Reno, NV
21. Perez, R.: Uncertainty analysis of computational fluid dynamics via polynomial chaos. Ph.D. thesis, Virginia Polytechnic Institute and State University (2008)
22. Le Maître, O., Reagan, M., Najm, H., Ghanem, R., Knio, O.: A stochastic projection method for fluid flow. II. Random process. J. Comput. Phys. **181**(1), 9–44 (2002)
23. Reagan, M.T., Najm, H.N., Ghanem, R.G., Knio, O.M.: Uncertainty quantification in reacting-flow simulations through non-intrusive spectral projection. Combust. Flame **132**(3), 545–555 (2003)
24. Nobile, F., Tempone, R., Webster, C.G.: A sparse grid stochastic collocation method for partial differential equations with random input data. SIAM J. Numer. Anal. **46**(5), 2309–2345 (2008)
25. Berveiller, M., Sudret, B., Lemaire, M.: Stochastic finite element: a non-intrusive approach by regression. Rev. Eur. Mécanique Numér. **15**(1–3), 81–92 (2006)
26. Hosder, S., Walters, R.W., Perez, R.: A non-intrusive polynomial chaos method for uncertainty propagation in CFD simulations. In: AIAA paper 2006-891 (2006) 44th AIAA Aerospace Sciences Meeting and Exhibit, 9–12 Jan 2006, Reno, NV
27. Hosder, S., Walters, R.W., Balch, M.: Efficient sampling for non-intrusive polynomial chaos applications with multiple uncertain input variables. In: AIAA paper 2007-1939 (2007) 48th AIAA/ASME/ASCE/AHS/ASC Structures, Structural Dynamics, and Materials Conference, 23–26 April 2007, Honolulu, HI
28. Hosder, S., Walters, R.W., Balch, M.: Efficient uncertainty quantification applied to the aeroelastic analysis of a transonic wing. In: AIAA paper 2008-729 (2008) 46th AIAA Aerospace Sciences Meeting and Exhibit, 7–10 Jan 2008, Reno, NV
29. Smolyak, S.: Quadrature and interpolation formulas for tensor products of certain classes of functions. Sov. Math. Dokl. **4**, 240–243 (1963)
30. Gerstner, T., Griebel, M.: Numerical integration using sparse grids. Numer. Algorithms **18**(3–4), 209–232 (1998)
31. Novak, E., Ritter, K.: Simple cubature formulas with high polynomial exactness. Constr. Approx. **15**(4), 499–522 (1999)
32. Barthelmann, V., Novak, E., Ritter, K.: High dimensional polynomial interpolation on sparse grids. Adv. Comput. Math. **12**(4), 273–288 (2000)

33. Heiss, F., Winschel, V.: Likelihood approximation by numerical integration on sparse grids. J. Econ. **144**(1), 62–80 (2008)
34. Gerstner, T., Griebel, M.: Dimension-adaptive tensor-product quadrature. Computing **71**(7), 65–87 (2003)
35. Ma, X., Zabaras, N.: An adaptive hierarchical sparse grid collocation algorithm for the solution of stochastic differential equations. J. Comput. Phys. **228**(8), 3084–3113 (2009)
36. Resmini, A., Peter, J., Lucor, D.: Sparse grids-based stochastic approximations with applications to aerodynamics sensitivity analysis. Int. J. Numer. Methods Eng. **106**(1), 32–57 (2016)
37. Ghanem, R.: Hybrid stochastic finite elements and generalized monte carlo simulation. ASME J. Appl. Mech. **65**(4), 1004–1009 (1998)
38. Field, R.V.: Numerical methods to estimate the coefficients of the polynomial chaos expansion. In: Proceedings of the 15th ASCE Engineering mechanics Conference (2002)
39. Ghiocel, D., Ghanem, R.: Stochastic finite-element analysis of seismic soil-structure interaction. ASCE J. Eng. Mech. **128**(1), 66–77 (2002)
40. Raisee, M., Kumar, D., Lacor, C.: A non-intrusive model reduction approach for polynomial chaos expansion using proper orthogonal decomposition. Int. J. Numer. Methods Eng. **103**(4), 293–312 (2015)
41. Tatang, M.A., Pan, W., Prinn, R.G., McRae, G.J.: An efficient method for parametric uncertainty analysis of numerical geophysical models. J. Geophys. Res. **102**(D18), 21925–21931 (1997)
42. Babuška, I., Nobile, F., Tempone, R.: A stochastic collocation method for elliptic partial differential equations with random input data. SIAM J. Numer. Anal. **45**(3), 1005–1034 (2007)
43. Ganapathysubramanian, B., Zabaras, N.: Sparse grid collocation schemes for stochastic natural convection problems. J. Comput. Phys. **225**(1), 652–685 (2007)
44. Xiu, D., Hesthaven, J.S.: High-order collocation methods for differential equations with random inputs. SIAM J. Sci. Comput. **27**(3), 1118–1139 (2005)
45. Xiu, D.: Numerical Methods for Stochastic Computations: A Spectral Method Approach. Princeton University Press, Princeton, NJ (2010)
46. Eldred, M.S., Burkardt, J.: Comparison of non-intrusive polynomial chaos and stochastic collocation methods for uncertainty quantification. In: AIAA paper 2009-0976 (2009) 47th AIAA Aerospace Sciences Meeting, 5–8 Jan 2009, Orlando, FL
47. Loeven, G.J.A., Witteveen, J.A.S., Bijl, H.: Probabilistic collocation: An efficient non-intrusive approach for arbitrarily distributed parametric uncertainties. In: AIAA paper 2007-317 (2007) 45th AIAA Aerospace Sciences Meeting and Exhibit, 8–11 Jan 2007, Reno, NV
48. Loeven, G.J.A., Bijl, H.: Airfoil analysis with uncertain geometry using the probabilistic collocation method. In: AIAA paper 2008-2070 (2008) 49th AIAA/ASME/ASCE/AHS/ASC Structures, Structural Dynamics, and Materials Conference, 7–10 April 2008, Schaumburg, IL
49. Preston, J.S., Tasdizen, T., Terry, C.M., Cheung, A.K., Kirby, R.M.: Using the stochastic collocation method for the uncertainty quantification of drug concentration due to depot shape variability. IEEE Trans. Biomed. Eng. **56**(3), 609–620 (2009)
50. Xiu, D.: Efficient collocational approach for parametric uncertainty analysis. Commun. Comput. Phys. **2**(2), 293–309 (2007)
51. DeGennaro, A.M., Rowley, C.W., Martinelli, L.: Uncertainty quantification for airfoil icing using polynomial chaos expansion. AIAA J. Aircr. **52**(5), 1404–1411 (2015)
52. Wasilkowski, G.W., Woźniakowski, H.: Explicit cost bounds of algorithms for multivariate tensor product problems. J. Complex. **11**(1), 1–56 (1995)
53. Savin, É., Resmini, A., Peter, J.: Sparse polynomial surrogates for aerodynamic computations with random inputs. In: AIAA paper 2016-0433 (2016) 18th AIAA Non-Deterministic Approaches Conference, 4–8 Jan 2016, San Diego, CA
54. Hesthaven, J.S., Gottlieb, S., Gottlieb, D.: Spectral Methods for Time-Dependent Problems. Cambridge Monographs on Applied and Computational Mathematics, Vol. 21. Cambridge University Press, Cambridge (2007)
55. Arnst, M., Ponthot, J.P.: An overview of nonintrusive characterization, propagation, and sensitivity analysis of uncertainties in computational mechanics. Int. J. Uncertain. Quantif. **4**(5), 387–421 (2014)

56. Eldred, M.S.: Design under uncertainty employing stochastic expansion methods. Int. J. Uncertain. Quantif. **1**(2), 119–146 (2011)
57. Sudret, B.: Global sensitivity analysis using polynomial chaos expansions. Reliab. Eng. Syst. Saf. **93**(7), 964–979 (2008)

Generalized Polynomial Chaos for Non-intrusive Uncertainty Quantification in Computational Fluid Dynamics

Vincent Couaillier and Éric Savin

Introduction

Complex aerodynamic analysis and design of aircraft use high-fidelity computational fluid dynamics (CFD) tools for shape optimization, for example, whereby some robustness is achieved by considering uncertain operational, environmental, or manufacturing parameters. Cruise flight conditions are generally transonic, as the flow may become locally supersonic depending on the wing profiles. Hence, high-fidelity simulations must be carried out in order to obtain a detailed description of the flow structure for optimization purposes. When it comes to consider variable parameters for sensitivity and robustness analyses, non-intrusive methods for uncertainty quantification (UQ) are typically considered in CFD. Indeed, the complex flow solvers are preferably treated as black boxes in order to compute the output quantities of interest that are required to evaluate the objective function of an optimization process. The latter is often expressed in terms of moments of the quantities of interest, such as the mean, standard deviation, or even higher-order moments (skewness, kurtosis…). Together with the Monte Carlo method or the method of moments, the stochastic collocation and non-intrusive polynomial chaos expansion methods introduced in the previous chapter "General Introduction to Polynomial Chaos and Collocation Methods" are widely used for evaluating stochastic objective functions. We more particularly focus on the latter approach in this chapter.

The polynomial chaos (PC) or homogeneous chaos expansion defined as the span of Hermite polynomial functionals of a Gaussian random variable has been introduced by Wiener [1] for stochastic processes. Mean-square convergence is guar-

V. Couaillier · É. Savin (✉)
Computational Fluid Dynamics Department, Onera–The French Aerospace Lab,
29 Avenue de la Division Leclerc, 92322 Châtillon, France
e-mail: eric.savin@onera.fr; savin@onera.fr

V. Couaillier
e-mail: vincent.couaillier@onera.fr

C. Hirsch et al. (eds.), *Uncertainty Management for Robust Industrial Design in Aeronautics*, Notes on Numerical Fluid Mechanics and Multidisciplinary Design 140, https://doi.org/10.1007/978-3-319-77767-2_8

anteed by the Cameron-Martin theorem [2] and is optimal (i.e., exponential) for Gaussian processes. For arbitrary random processes, the numerical study in [3] has shown that the convergence rates are not optimal. This observation has prompted the development of generalized chaos expansions (gPC) involving other families of polynomials [3, 4]. They consist in expanding any function of random variables into a linear combination of orthogonal polynomials with respect to the probability density functions (PDFs) of these underlying random variables. PC and gPC expansions have recently received a broad attention in engineering sciences, where they are extensively used as a constructive tool for representing random vectors, matrices, tensors, or fields for the purpose of quantifying uncertainty in complex systems. Several implementation issues and applications are described in, e.g., [5–13] and references therein.

The intrusive PC expansion originally introduced in [7, 14, 15] is based on a Galerkin-type projection formulation of the model equations, typically the incompressible or compressible Navier–Stokes equations in CFD, to derive the governing equations for the spectral expansion coefficients of the output quantities of interest. More precisely, the PC expansions of the model parameters and variables are substituted in the model equations, which in turn yield the evolution equations for the outputs from Galerkin projections using the orthogonal polynomials of the PC expansions [6, 7, 10]. The projection coefficients are thus obtained by solving ordinary differential equations in time. Regarding non-intrusive PC expansions, two approaches for computing the PC coefficients of the output quantities of interest have usually been considered: (i) the projection approach, in which they are computed by structured quadratures, i.e., Gauss quadratures, or unstructured quadratures, i.e., Monte Carlo or quasi Monte Carlo sampling; and (ii) the regression approach, minimizing some error measure or truncation tolerance of the PC expansion for some particular values of the inputs (which can be the quadrature sets invoked just above, for example). Both techniques suffer from some well-identified shortcomings when the dimension of the parameter space, and the number of model evaluations alike, increases. Indeed, a PC expansion of total degree p in N variable parameters contains $P = \binom{p+N}{N}$ coefficients. A direct way to compute them is to use a tensor product grid in the parameter space requiring about $q \sim \left(\frac{p}{2}\right)^N$ evaluations of the process. These q runs are very often unaffordable for large parameter spaces and complex configurations, as in CFD for example. Fortunately, the Smolyak algorithm [16] defines sparse grid quadratures involving $q \sim \left(\frac{p}{2}\right)^{\log N}$ points while preserving a satisfactory level of accuracy. In [17], it has been observed that such sparse rules typically become competitive with respect to tensor grids for dimensions $N \geq 4$. Consequently, collocation techniques with sparse quadratures or adaptive regression strategies have been developed in order to circumvent the dimensionality concern [18–25].

In the application presented in this chapter, we adopt the regression approach. We also aim at benefiting from the sparsity of the process outputs themselves to reconstruct their PC representations in a non-adaptive way [26]. Indeed, we rely on the common observation that many cross-interactions between the input parameters are actually smoothened, or even negligible, once that have been propagated to some global quantities of interest processed from complex aerodynamic computations. The corresponding PC expansions should thus involve only low-order polynomials, such that the contribution of the higher-order polynomials is negligible. We can therefore expect to achieve a successful output recovery by the techniques known under the terminology of compressed sensing [27, 28]. In this theory, the reconstruction of a sparse signal on a given, known basis requires only a limited number of evaluations at randomly selected points—at least significantly less than the a priori dimension P of the basis. We thus resort to unstructured sampling sets to recover sparse outputs. Compressed sensing is formulated as a constrained, underdetermined system which may be solved by convex optimization algorithms. The rest of this chapter is organized as follows. The formal framework of non-intrusive UQ is briefly presented in the next section. The gPC expansion method itself is addressed in section "Generalized Polynomial Chaos Expansion," which is more particularly focused on the non-intrusive computation of the PC coefficients by either projection or regression approaches. This framework is illustrated in section "Application to Transonic Turbulent Flow Around a Two-Dimensional RAE 2822 Airfoil" where the application of the foregoing techniques to a two-dimensional transonic turbulent flow around a RAE 2822 airfoil modelized by steady-state Reynolds-averaged Navier–Stokes (RANS) equations is considered. The efficiency of a sparsity-based reconstruction approach is emphasized. Some conclusions are finally drawn in section "Conclusions".

Uncertainty Quantification and Propagation: Model Problem

In CFD applications, numerical models are built to simulate complex fluid flows around rigid or flexible profiles. They are implemented in computer programs which tend to become more and more sophisticated and extensive. These models often exhibit certain features that may be considered as uncertain, or they depend on parameters that may be considered as such. Propagation and quantification of uncertainty aim at establishing a quantitative assessment of and some insight into the impact that these uncertainties have on the predictions given by the models. Statements about the influence of uncertainties may serve to guide the allocation of resources in order to reduce them, but are also essential in the process of validating the models in the presence of uncertain parameters or in the process of optimizing a design, among other objectives [29].

In this chapter, we think of a computational model as a linear or nonlinear mapping F of a set of input parameters $\boldsymbol{\xi}$ in $\Gamma \subseteq \mathbb{R}^N$ into a quantity of interest y; that is,

$$y = F(\boldsymbol{\xi}), \quad F : \mathbb{R}^N \to \mathbb{R}, \tag{1}$$

assuming without loss of generality that the quantity of interest is a scalar. Vector-valued quantities of interest may be considered alike, following the same lines as exposed below. Here, it is also implicitly assumed that the uncertainties affecting the model are embedded in the input variables $\boldsymbol{\xi}$, or a subset of these variables; that is, modeling uncertainties are not accounted for. In other words, modeling errors arising from modeling simplifications and assumptions are not considered here. Output-predictor uncertainties pertaining to the quantity of interest, which may be subject to an additive random noise, are also disregarded in this setting. Therefore, in the so-called parametric approach promoted in this chapter, the uncertain input variables $\boldsymbol{\xi}$ are representative of variable geometrical characteristics, boundary conditions, loadings, physical or mechanical properties, or combinations of them. These input parameters are represented by a $\mathbb{R}^N$-valued random variable $\Xi = (\Xi_1, \Xi_2, \dots, \Xi_N)$, the characterization of which requires the knowledge of its probability distribution $\mathscr{P}_\Xi$. Ξ can be a continuous or a discrete, or even a combination of both, random variable. In the former case, it is understood that the probability distribution thus admits a probability density function (PDF) $\boldsymbol{\xi} \mapsto W_\Xi(\boldsymbol{\xi})$ with values in $\mathbb{R}_+ = [0, +\infty]$ such that $\mathscr{P}_\Xi(\mathscr{B}) = \int_\mathscr{B} W_\Xi(\boldsymbol{\xi}) d\boldsymbol{\xi}$ for any subset $\mathscr{B}$ of $\mathbb{R}^N$.

The mapping of Ξ through the computational model F then provides the characterization of the quantity of interest as a real-valued random variable Y such that $Y = F(\Xi)$. By the causality principle, the probability distribution $\mathscr{P}_Y$ of Y is the probability distribution of Ξ transported by the model F, i.e.:

$$\mathscr{P}_Y(\mathscr{B}) = \mathscr{P}_\Xi(\boldsymbol{\xi} \in \mathbb{R}^N;\, F(\boldsymbol{\xi}) \in \mathscr{B})\,. \tag{2}$$

The subsequent computation of statistical descriptors of the quantity of interest is straightforward. Considering for example the mean μ_Y and variance σ_Y^2, one has:

$$\begin{aligned}
\mu_Y &= \int_\mathbb{R} y \mathscr{P}_Y(dy) = \int_{\mathbb{R}^N} F(\boldsymbol{\xi}) \mathscr{P}_\Xi(d\boldsymbol{\xi})\,, \\
\sigma_Y^2 &= \int_\mathbb{R} (y - \mu_Y)^2 \mathscr{P}_Y(dy) = \int_{\mathbb{R}^N} (F(\boldsymbol{\xi}) - \mu_Y)^2 \mathscr{P}_\Xi(d\boldsymbol{\xi})\,,
\end{aligned} \tag{3}$$

assuming these integrals remain bounded; i.e., Y has finite variance; otherwise, the statistical descriptors do not exist.

The evaluation of the multi-dimensional integrals above requires a thorough knowledge of the computational model F, which of course has no analytical expression for complex configurations of industrial relevance. This can be achieved intrusively, by expliciting all dependences of the underlying physical and computational models with respect to the parameters; see, e.g., [7, 11, 14, 15, 30, 31] and references therein. This approach is seldom considered in CFD because the nonlinearity of the physical model and complexity of the codes makes the implementation tricky. Alternatively, the non-intrusive approach consists in expressing all dependences of the quantity of interest Y with respect to the parameters by carefully chosen sampling procedures or surrogates (also called response surfaces); see, e.g., [17–20, 32–39]. This is the approach retained in the following. More particularly, we address the construction of polynomial surrogates using a family of multivariate orthogonal polynomials with respect to the probability distribution $\mathscr{P}_\Xi$ of the random input parameters.

Generalized Polynomial Chaos Expansion

Polynomial Surrogates

Let $\mathbf{j} = (j_1, j_2, \dots j_N) \in \mathbb{N}^N$ be a so-called multi-index, we denote by $\boldsymbol{\xi}^{\mathbf{j}}$ the multivariate monomial associated to that multi-index as a function from $\mathbb{R}^N$ into $\mathbb{R}$ such that $\boldsymbol{\xi}^{\mathbf{j}} = \xi_1^{j_1} \times \xi_2^{j_2} \times \cdots \times \xi_N^{j_N}$. The modulus of $\mathbf{j}$, namely $|\mathbf{j}| = j_1 + j_2 + \cdots + j_N$, is also the total order of the monomial $\boldsymbol{\xi}^{\mathbf{j}}$. Then a polynomial surrogate model G^p of order p is defined as an N-variate polynomial approximating the computational model F as precisely as possible in the $\mathscr{P}_{\Xi}$-weighted least-squares sense, i.e.:

$$F \simeq G^p = \sum_{|\mathbf{j}|=0}^{p} c_{\mathbf{j}} \boldsymbol{\xi}^{\mathbf{j}}, \quad \mathbf{c} = \arg\min_{\mathbf{d} \in \mathbb{R}^P} \frac{1}{2} \int_{\mathbb{R}^N} \left| F(\boldsymbol{\xi}) - \sum_{|\mathbf{j}|=0}^{p} d_{\mathbf{j}} \boldsymbol{\xi}^{\mathbf{j}} \right|^2 \mathscr{P}_{\Xi}(d\boldsymbol{\xi}), \tag{4}$$

where P is the number of monomials $\boldsymbol{\xi}^{\mathbf{j}}$ such that $0 \leq |\mathbf{j}| \leq p$. Conditions guaranteeing the existence (and possible uniqueness) of a solution to the above problem and convergence of the surrogate G^p as its order p is increased are summarized in [29]. In particular, it should be emphasized that provided the surrogate G^p converges to the computational model in the $\mathscr{P}_{\Xi}$-mean-square sense, that is,

$$\lim_{p \to +\infty} \int_{\mathbb{R}^N} |F(\boldsymbol{\xi}) - G^p(\boldsymbol{\xi})|^2 \, \mathscr{P}_{\Xi}(d\boldsymbol{\xi}) = 0,$$

the probability distribution $\mathscr{P}_{Y^p}$ of the random variable $Y^p = G^p(\Xi)$ converges to the probability distribution $\mathscr{P}_Y$ of Y, Eq. (2), as $p \to +\infty$.

Polynomial Chaos Surrogates

Consider now $\mathscr{B}^p = \{\psi_{\mathbf{j}};\, 0 \leq |\mathbf{j}| \leq p\}$ a set of N-variate polynomials $\psi_{\mathbf{j}}$ spanning the set of all polynomials of total order at most p and orthonormal with respect to $\mathscr{P}_{\Xi}$, that is,

$$\int_{\mathbb{R}^N} \psi_{\mathbf{j}}(\boldsymbol{\xi}) \psi_{\mathbf{k}}(\boldsymbol{\xi}) \mathscr{P}_{\Xi}(d\boldsymbol{\xi}) := \langle \psi_{\mathbf{j}}, \psi_{\mathbf{k}} \rangle = \delta_{\mathbf{jk}}, \quad 0 \leq |\mathbf{j}|, |\mathbf{k}| \leq p, \tag{5}$$

where $\delta_{\mathbf{jk}}$ is the N-dimensional Kronecker symbol such that $\delta_{\mathbf{jk}} = 1$ if $\mathbf{j} = \mathbf{k}$ and $\delta_{\mathbf{jk}} = 0$ otherwise. The existence of this basis is guaranteed provided that all monomials $\boldsymbol{\xi}^{\mathbf{j}}$, $0 \leq |\mathbf{j}| \leq p$, are $\mathscr{P}_{\Xi}$-square integrable and the $\mathscr{P}_{\Xi}$-mean-square norm of all non null polynomials of total order p is nonvanishing. Then, a polynomial surrogate G^p may be constructed equivalently to the $\mathscr{P}_{\Xi}$-weighted least-squares problem (4) by:

$$F \simeq G^p = \sum_{|\mathbf{j}|=0}^{p} G_{\mathbf{j}} \psi_{\mathbf{j}} \,, \quad \mathbf{G} = \arg \min_{\mathbf{g} \in \mathbb{R}^P} \frac{1}{2} \int_{\mathbb{R}^N} \left| F(\boldsymbol{\xi}) - \sum_{|\mathbf{j}|=0}^{p} g_{\mathbf{j}} \psi_{\mathbf{j}}(\boldsymbol{\xi}) \right|^2 \mathscr{P}_{\Xi}(d\boldsymbol{\xi}) \,, \quad (6)$$

where $\mathbf{G} = \{G_{\mathbf{j}};\ 0 \leq |\mathbf{j}| \leq p\}$ and P is again the number of polynomials in the basis $\mathscr{B}^p$. Owing to the orthonormality of the polynomials of that basis, the solution of this problem reads:

$$G_{\mathbf{j}} = \langle F, \psi_{\mathbf{j}} \rangle \,, \quad 0 \leq |\mathbf{j}| \leq p \,. \quad (7)$$

The random variable $Y^p = \sum_{|\mathbf{j}|=0}^{p} G_{\mathbf{j}} \psi_{\mathbf{j}}(\Xi)$ is known as a polynomial chaos expansion of the quantity of interest Y. The basis $\mathscr{B}^p$ is referred to as the polynomial chaos (PC), or homogeneous chaos in the original terminology of Wiener [1]. Here, the random parameters Ξ were Gaussian variables; the extension to non-Gaussian variables in, e.g., [3, 4] has been referred to as generalized polynomial chaos (gPC). Lastly, the vector $\mathbf{G}$ is formed with the so-called polynomial chaos coefficients which completely characterize the surrogate model G^p since the polynomial chaos is known from the probability measure $\mathscr{P}_{\Xi}$ of the random inputs. Several methods have been developed to compute them, which are usually classified into intrusive or non-intrusive approaches; see the introductory chapter "General Introduction to Polynomial Chaos and Collocation Methods".

Computation of Polynomial Chaos Coefficients

The original intrusive approach introduced in [7, 14, 15] is based on a Galerkin-type projection formulation of the model equations (typically partial differential equations, PDEs) onto a prescribed basis of orthonormal polynomials. The procedure results in a so-called spectral problem formulated in terms of the polynomial chaos coefficients of the solution of the model PDEs, which often requires important modifications of the associated computational model. For this reason, intrusive methods are seldom considered in CFD applications as already noticed above. On the other hand, non-intrusive approaches do not require any modification of the computational model. The non-intrusive projection method numerically determines the polynomial surrogate by approximating the integrals in (7) using dedicated structured (typically Gauss integration rules) or unstructured (e.g., Monte Carlo or quasi Monte Carlo methods) sets of samples of the random input parameters Ξ. The interpolatory collocation method determines the polynomial surrogate by interpolating between a set of evaluations of the computational model. It is strongly related to the projection method as outlined in chapter "General Introduction to Polynomial Chaos and Collocation Methods"; see also the discussion in, e.g., [40]. In fact, it is also called the pseudospectral collocation method in [13, 41]. Introducing a quadrature rule of integration $\boldsymbol{\Theta}(N, Q) = \{\boldsymbol{\xi}^l, w^l;\ 1 \leq l \leq Q\}$ formed by Q nodes $\boldsymbol{\xi}^l$ in $\mathbb{R}^N$ and associated weights w^l in $\mathbb{R}$, the integral of any smooth, integrable function $f : \mathbb{R}^N \to \mathbb{R}$ with respect to $\mathscr{P}_{\Xi}$ can be approximated by the weighted sum:

$$\int_{\mathbb{R}^N} f(\boldsymbol{\xi})\mathscr{P}_{\Xi}(d\boldsymbol{\xi}) \simeq \sum_{l=1}^{Q} w^l f(\boldsymbol{\xi}^l)\,.$$

Applying this rule to the computation of the polynomial chaos coefficients of Eq. (7), one arrives at the following surrogate model $G^{p,Q}$:

$$F \simeq G^{p,Q} = \sum_{|\mathbf{j}|=0}^{p} G_{\mathbf{j}}^{Q}\psi_{\mathbf{j}}\,,\quad G_{\mathbf{j}}^{Q} = \sum_{l=1}^{Q} w^l F(\boldsymbol{\xi}^l)\psi_{\mathbf{j}}(\boldsymbol{\xi}^l)\,,\quad 0 \leq |\mathbf{j}| \leq p\,. \tag{8}$$

The families of orthonormal polynomials forming $\mathscr{B}^p$ and associated Gauss quadrature rules are explicitly known and tabulated for some labeled probability distributions $\mathscr{P}_{\Xi}$, such as Gamma distributions (corresponding to Laguerre polynomials), Beta distributions (corresponding to Jacobi polynomials, including Legendre polynomials for uniform distributions), Gaussian distributions (corresponding to Hermite polynomials), Poisson distributions (corresponding to Charlier polynomials), binomial distributions (corresponding to Krawtchouk polynomials); see [4, 11, 13, 30]. Alternatively, they can be generated numerically using the Gramm-Schmidt, Stieltjes or Chebyshev algorithm [42] and the Golub-Welsh algorithm [43] to compute their associated Gauss nodes and weights. This procedure has been applied in, e.g., [44–46]. Another possibility is to consider unstructured quadrature sets, whereby the nodes are sampled randomly or quasi-randomly in the domain of integration $\Gamma \subseteq \mathbb{R}^N$ according to their distribution $\mathscr{P}_{\Xi}$, and the associated weights are typically $w^l = \frac{1}{Q}$: This is the Monte Carlo method and its by-products.

Another approach to determine the polynomial chaos coefficients is linear regression. In this setting, the set $\boldsymbol{\Theta}(N, Q)$ is used to form a linear system in the unknowns $\mathbf{G}$ by simply evaluating the surrogate model G^p at the nodes $\{\boldsymbol{\xi}^l;\ 1 \leq l \leq Q\}$:

$$\boldsymbol{\Phi}\mathbf{G} = \mathbf{y}\,, \tag{9}$$

where $\mathbf{y} = \{y^l = F(\boldsymbol{\xi}^l);\ 1 \leq l \leq Q\}$ is the vector of observations of the computational model F at the sampling nodes, and $[\boldsymbol{\Phi}]_{l\mathbf{j}} = \psi_{\mathbf{j}}(\boldsymbol{\xi}^l)$ is the so-called $Q \times P$ **measurement matrix**. Numerous methods are available to solve this problem whenever $Q \geq P$, for example least-squares minimization [37, 47]. Likewise different strategies exist for the choice of the sampling nodes $\{\boldsymbol{\xi}^l;\ 1 \leq l \leq Q\}$ which are reviewed in, e.g., [29, 48–50]. We do not follow this approach in the subsequent developments though. We are rather interested in the situation whereby $Q \leq P$ and more particularly $Q \ll P$, that is, underdetermined systems. This can be achieved thanks to some recent mathematical results pertaining to the resolution of under-sampled linear systems promoting sparsity of the sought solution, known as compressed sensing or compressive sampling [27, 28]. A recent review of the application of this approach to polynomial surrogates reconstruction is proposed in [51]; see also [17, 26, 52–61]. The compressed sensing approach consists in reformulating Eq. (9) considered as an underdetermined system as a minimization problem with some sparsity constraint, namely:

$$\mathbf{G}^* = \arg\min_{\mathbf{g}\in\mathbb{R}^P} \{ \|\mathbf{g}\|_m ;\ \mathbf{y} = \boldsymbol{\Phi}\mathbf{g} \} , \qquad (P_{m,0})$$

with the ℓ_m-norm $\|\mathbf{g}\|_m = (\sum_{j=0}^{P-1} \left|g_j\right|^m)^{\frac{1}{m}}$, $m > 0$, and $\|\mathbf{g}\|_0 = \#\{j;\ g_j \neq 0\}$ otherwise. **Sparsity** means that only a small fraction of the sought coefficients $\mathbf{G}^*$ are nonzero. Such a constraint is added to ensure well-posedness of (9) when $Q < P$ by seeking a solution with the minimum number of nonzero terms. The case $m = 0$ however yields a non-unique solution in general, which is in addition NP hard to compute as the cost of global search is exponential in P. Further researches in compressed sensing have shown that the convex relaxation of $(P_{0,0})$ by considering the ℓ_1-norm instead yields a unique solution provided that the latter is sufficiently sparse and that the measurement matrix $\boldsymbol{\Phi}$ has some prescribed properties. The problem $(P_{1,0})$ is referred to as basis pursuit [62] in the dedicated literature.

Now, the equality (9) is often too restrictive because the truncated p-th order polynomial chaos basis $\mathscr{B}^p$ is not complete for the exact representation of the observations $\mathbf{y}$. Thus, a truncation error has also to be accounted for in the solution process. This is accommodated by reformulating $(P_{1,0})$ as:

$$\mathbf{G}^* = \arg\min_{\mathbf{g}\in\mathbb{R}^P} \{ \|\mathbf{g}\|_1 ;\ \|\mathbf{y} - \boldsymbol{\Phi}\mathbf{g}\|_2 \leq \varepsilon \} , \qquad (P_{1,\varepsilon})$$

for some tolerance $0 \leq \varepsilon \ll 1$ on the polynomial chaos truncation. The latter problem is known as basis pursuit denoising (BPDN) [62]. The successful recovery of $\mathbf{G}$ of Eq. (9) by solving$(P_{1,\varepsilon})$ is guaranteed by the **restricted isometry property** (RIP) which the measurement matrix $\boldsymbol{\Phi}$ has to satisfy. For each integer $S \in \mathbb{N}^*$, the isometry constant δ_S of $\boldsymbol{\Phi}$ is defined as the smallest number such that:

$$(1 - \delta_S)\|\mathbf{g}_S\|_2^2 \leq \|\boldsymbol{\Phi}\mathbf{g}_S\|_2^2 \leq (1 + \delta_S)\|\mathbf{g}_S\|_2^2$$

for all S–sparse vectors $\mathbf{g}_S \in \{\mathbf{g} \in \mathbb{R}^P;\ \|\mathbf{g}\|_0 \leq S\}$. Then, $\boldsymbol{\Phi}$ is said to satisfy the RIP of order S if, say, δ_S is not too close to 1. This property amounts to saying that all S–column submatrices of $\boldsymbol{\Phi}$ are numerically well-conditioned, or S (or less) columns selected arbitrarily in $\boldsymbol{\Phi}$ are nearly orthogonal. Consequently, they form a near isometry so that $\boldsymbol{\Phi}$ approximately preserves the Euclidean norm of S–sparse vectors. The following theorem by Candès et al. [27, Theorem 1.2], [63, Theorem 3] then states that $(P_{1,\varepsilon})$ can be solved efficiently:

Theorem 1 ([27, 63]) *Assume* $\delta_{2S} < \sqrt{2} - 1$. *Then, the solution* $\mathbf{G}^*$ *to* $(P_{1,\varepsilon})$ *satisfies:*

$$\|\mathbf{G}^* - \mathbf{G}\|_2 \leq C_0 \frac{\|\mathbf{G}_S - \mathbf{G}\|_1}{\sqrt{S}} + C_1 \varepsilon$$

for some $C_0, C_1 > 0$ *depending only on* δ_{2S}.

This result calls for several comments. First, the coefficients $\mathbf{G}$ are actually nearly sparse, rather than strictly sparse, in the sense that only a small fraction of them contribute significantly to the output statistics while the others are not strictly null.

Opportunely, the foregoing theorem deals with all signals and not only the S–sparse ones. Second, it also deals with noiseless recovery if $\varepsilon = 0$. Third, it is deterministic and does not involve any probability for a successful recovery. Lastly, the ℓ_1-minimization strategy is non-adapted because it identifies the sparsity pattern, that is the order (location) of the negligible coefficients in the polynomial chaos basis, and the leading coefficients at the same time. The algorithm can therefore efficiently capture the relevant information of a sparse vector without trying to comprehend that vector [63]. This is clearly a much desirable feature for practical industrial applications.

Application to Transonic Turbulent Flow Around a Two-Dimensional RAE 2822 Airfoil

The foregoing strategy of using a gPC expansion for uncertainty propagation and quantification is applied to an aerodynamic problem taken from [17]. Here, we consider a two-dimensional transonic turbulent flow around a RAE 2822 airfoil modelized by the steady-state Reynolds-averaged Navier Stokes (RANS) equations together with a Spalart-Allmaras turbulence model closure [64]. The nominal flow conditions are the ones described in Cook et al. [65] for the test case #6 together with the wall interference correction formulas derived in [66, pp. 386–387] and their slight modifications suggested in [67, p. 130] (see also the CFD verification and validation Web site of the NPARC Alliance [68]). The nominal free-stream Mach number $\underline{M}_\infty = 0.729$, angle of attack $\underline{\alpha}_\infty = 2.31°$, and Reynolds number $\underline{Re} = 6.50 \cdot 10^6$ (based on the airfoil chord length c, fluid velocity, temperature, and molecular viscosity at infinity) arise from the corrections $\Delta M_\infty = 0.004$ and $\Delta\alpha_\infty = -0.61°$ given in [67, p. 130] for the test case #6 outlined in Cook et al. [65], for which $\underline{M}_\infty = 0.725$, $\underline{\alpha}_\infty = 2.92°$, and $\underline{Re} = 6.50 \cdot 10^6$. At last, the far-field temperature is fixed at $\underline{T}_\infty = 300$ K and the ratio of specific heats of the air at $\gamma = 1.4$.

Definition of the Uncertainties

Our aim is to characterize the influence of variabilities of the free-stream Mach number M_∞ and angle of attack α_∞ (operational parameters), and of the thickness-to-chord ratio $r = h/c$ (geometrical parameter) on some aerodynamic quantities of interest, such as the drag, lift, or pitching moment coefficients C_D, C_L, or C_m, respectively. These variable parameters are assumed to be independent and to follow Beta laws of the first kind β_1. Therefore, their marginal PDFs read:

$$\beta_1(\xi; a, b) = \mathbb{1}_{[X_m, X_M]}(\xi)\frac{\Gamma(a+b)}{\Gamma(a)\Gamma(b)}\frac{(\xi - X_m)^{a-1}(X_M - \xi)^{b-1}}{(X_M - X_m)^{a+b-1}}.$$

Table 1 Symmetric β_{I} laws for the variable geometrical and operational parameters

	$a = b$	X_m	X_M
Ξ_1	4	$0.97 \times \underline{r}$	$1.03 \times \underline{r}$
Ξ_2	4	$0.95 \times \underline{M}_{\infty}$	$1.05 \times \underline{M}_{\infty}$
Ξ_3	4	$0.98 \times \underline{\alpha}_{\infty}$	$1.02 \times \underline{\alpha}_{\infty}$

In the above, $\Gamma(z) = \int_0^{+\infty} t^{z-1}\mathrm{e}^{-t}dt$ is the usual Gamma function, and $[X_m, X_M]$ is the compact support of the random parameter $\Xi \sim \beta_{\mathrm{I}}$. The parameters $a = b$ as well as the bounds X_m, X_M for the three variable parameters $\xi_1 = r$, $\xi_2 = M_\infty$, $\xi_3 = \alpha_\infty$ are gathered in the Table 1. This definition of uncertainties is part of the FP7 UMRIDA Project (http://www.umrida.eu), which gathers a novel data base of industrial challenges with prescribed uncertainties for the validation of UQ techniques against this series of relevant industrial test cases. We note in passing that the β_{I} model is the one arising from Jaynes' maximum entropy principle [69] when constraints on (i) the boundedness of the support $[X_m, X_M]$, and (ii) the values of the averages $\mathbb{E}\{\log(\Xi - X_m)\}$ and $\mathbb{E}\{\log(X_M - \Xi)\}$, are imposed.

Polynomial Basis and Sampling Nodes

From the analysis of section "Generalized Polynomial Chaos Expansion," it is seen that the main ingredients requested for the construction of polynomial surrogates of the quantity of interest $y = C_D$, C_L, or C_m are the truncated polynomial basis $\mathscr{B}^p$ and the quadrature rule $\boldsymbol{\Theta}(N, Q)$, for Q integration nodes and a total number of polynomials $P = \binom{N+p}{N}$. In addition, we have here $N = 3$ for the parameter space dimension. Owing to the choices made for the variable parameters considered for this case (see Table 1), we have $\boldsymbol{\xi} = (\xi_1, \xi_2, \xi_3) \in \Gamma = \prod_{j=1}^{3} [X_m^{(j)}, X_M^{(j)}]$ and $\mathscr{P}_{\Xi}(\boldsymbol{\xi}) = \bigotimes_{j=1}^{3} \beta_{\mathrm{I}}(\xi_j; 4, 4)$. Therefore, the integration nodes should be chosen from a Gauss-Jacobi quadrature rule, and the polynomial basis $\mathscr{B}^p$ should be constituted by the multivariate Jacobi polynomials which are orthogonal with respect to the weight function $\boldsymbol{\xi} \mapsto W_3(\boldsymbol{\xi}) = \prod_{j=1}^{3} W_1(\xi_j)$ on $[-1, 1]^3$ (after a proper renormalization of Γ), where $\xi \mapsto W_1(\xi_j) = (1 - \xi_j^2)^3$. They are computed by:

$$\psi_{\mathbf{j}}(\boldsymbol{\xi}) = \prod_{k=1}^{3} \psi_{j_k}(\xi_k), \quad |\mathbf{j}| \leq p,$$

where $\{\psi_{j_k}; j_k \geq 0\}$ is the family of one-dimensional orthonormal Jacobi polynomials with respect to the weight function W_1. In the present study, the polynomial surrogates G^p constructed for the evaluation of the drag, lift, and pitching moment coefficients are truncated up to the total order $p = 8$. Therefore, $P = \binom{p+3}{3} = 165$ multivariate Jacobi polynomials are ultimately retained in those gPC expansions.

As for the sampling nodes $\{\boldsymbol{\xi}^l;\ 1 \leq l \leq Q\}$, one may choose them as (i) the nodes of a quadrature rule $\boldsymbol{\Theta}(N,Q)$ so that Eq. (8) can be used; or (ii) select them randomly according to their PDF $\mathscr{P}_{\Xi}$ so that $(P_{1,\varepsilon})$ can be used. Indeed, selecting the sampling nodes randomly eases the fulfillment of the RIP and consequently increases the efficiency of ℓ_1-minimization. The use of a structured sampling set (quadrature rule) in the latter procedure is examined further on in [58], though. Quadrature rules for the approximation of multivariate integrals have been introduced in the chapter "General Introduction to Polynomial Chaos and Collocation Methods" of this book. Univariate Gauss-Jacobi-Lobatto (GJL) quadratures $\boldsymbol{\Theta}(1,q)$ are aimed at integrating a smooth function $\xi \mapsto f(\xi)$ defined on $[-1,1]$ by:

$$\int_{-1}^{1} f(\xi)(1-\xi)^{a-1}(1+\xi)^{b-1}d\xi \simeq \sum_{l=1}^{q-2} w^l f(\xi^l) + w^{q-1}f(-1) + w^q f(1)\,, \tag{10}$$

where $a,b > -1$, such that this rule turns to be exact for polynomials up to the order $2q-3$. Here, the bounds ± 1 are explicitly included in the quadrature nodes. The reason why we include them in the rule stems from the fact that the basic engineering practice would typically consider the evaluation of the physical model F at the bounds of the support of the variable parameters. The main advantage of using Gauss-Jacobi quadratures is that they do not add integration nodes for the increased order $a+b-2$ induced by the weight function $(1-\xi)^{a-1}(1+\xi)^{b-1}$. Since in our case, we have chosen a total order $p=8$, $q=10$ GJL modes are needed to recover exactly the orthonormality property (5) for the corresponding univariate Jacobi polynomials. Indeed Q should be defined such that $2q-3 \geq 16$ in this situation.

A multivariate quadrature rule may subsequently be obtained by full or sparse tensorization of the above univariate rule. Firstly, a fully tensorized grid is obtained by the straightforward product rule:

$$\boldsymbol{\Theta}(N,Q) = \bigotimes_{j=1}^{N} \Theta(1,q)\,, \tag{11}$$

which contains $Q = q^N$ nodes in Γ. Secondly, sparse quadrature rules can be derived thank to the Smolyak algorithm [16]. A brief overview is again given in the chapter "General Introduction to Polynomial Chaos and Collocation Methods" of this book. Such sparse rules have been considered in [17, 19, 22–25] among others but will not be used here, though. Note also that a different number of nodes may be used along each dimension for either product or sparse rules, as proposed in some adaptive strategies [18–21, 23–25]. The fully tensorized rule in $N=3$ dimensions based on a 10 nodes univariate GJL rule is displayed on Fig. 1(left). It contains $Q=10^3$ nodes, to be compared with the $Q=80$ nodes sampled randomly for the application of the ℓ_1-minimization algorithm $(P_{1,\varepsilon})$ (right).

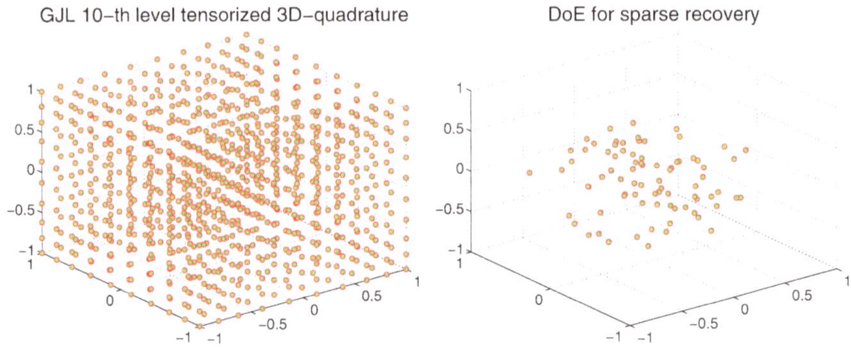

Fig. 1 Left: three-dimensional nodes based on a tensorized univariate GJL rule with $q = 10$ nodes along each dimension and $a = b = 4$ ($Q = 10^3$). Right: randomly sampled nodes following a three-dimensional symmetric β_1 law with $a = b = 4$ ($Q = 80$)

Computational Model

Once the node sets have been defined, either from a fully tensorized quadrature rule or random sampling, it remains to run the computational model for each sampling node $\boldsymbol{\xi}^l$ in those sets to obtain the vector $\mathbf{y}$ of observations of the quantity of interest. The numerical approximation of the aerodynamic problem considered here is computed using the *elsA* software [70]. At first, the RANS equations are discretized in space by a cell-centered finite-volume method. The convective fluxes are computed using the upwind Roe flux and a second-order MUSCL scheme [71] with a van Albada limiter [72]. The diffusive fluxes are computed as the half sum of the normal flux densities in the two adjacent cells sharing the current interface (with due care of the boundary conditions on external interfaces), considering corrected cell-centered gradients of the fluid velocity in these cells. The diffusive fluxes of the transport equation for the turbulent variable in the Spalart-Allmaras model are discretized using a similar approach, whereas first-order Roe fluxes are used for the convective terms. Finally, the source term of the transport equation is computed using the temperature gradients at the center of the cells. Secondly, these semi-discretized RANS equations are solved in time using a backward Euler scheme up to convergence toward a steady-state solution. The linearization of the resulting nonlinear implicit system is performed using the Lower-Upper Symmetric Successive Overrelaxation (LU-SSOR) scheme [73] with four relaxation cycles. The convergence is accelerated by the use of multigrid techniques for steady flows. Uniform flow is considered as the initial conditions for the iterations with respect to the time parameter.

The nominal problem is discretized using a $769c \times 193c$ mesh shown in Fig. 2, where the boundary at infinity was left intensionally far (at about $500c$ from the airfoil). These values proved to be sufficient to avoid spurious reflection with the far-field boundary. Given the large number of simulations to run, the numerical parameters of the steady-state algorithm proved to be essential to insure a fast convergence.

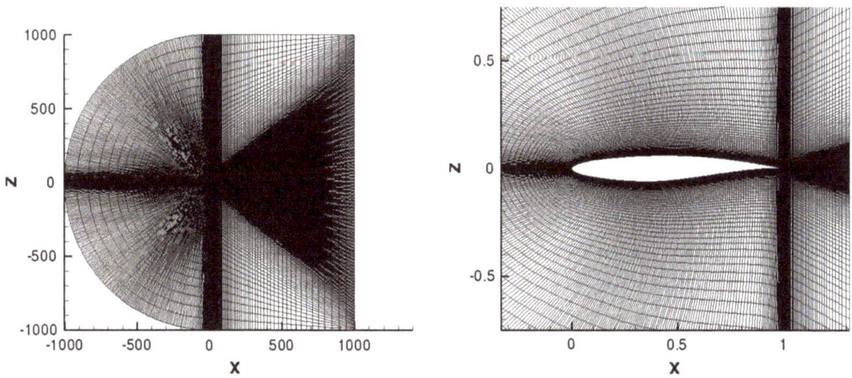

Fig. 2 Computational domain for the baseline configuration: overview of the mesh (left), close view at the airfoil (right)

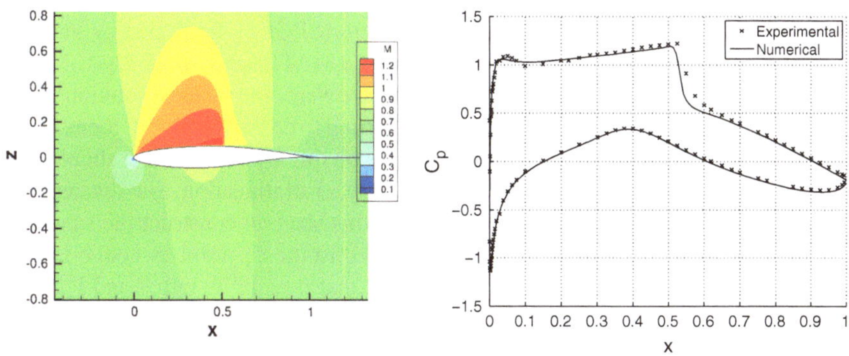

Fig. 3 Magnitude of velocity (left) and static pressure coefficient C_p at the wall (right, solid line) for the baseline configuration $M_\infty = 0.729$, $\alpha = 2.31°$, $Re = 6.50 \cdot 10^6$. The latter is compared with the experiment results gathered on the CFD verification and validation Web site of the NPARC Alliance [68] (crosses)

Once all the foregoing numerical parameters have been fixed, the number of iterations is determined from the evolution of the resulting global forces (not shown here). A number of 2000 iterations appeared to be acceptable, the discrete residuals of all equations and their decrease being checked at every iteration. Hence, this number of iterations has been retained for all subsequent calculations so far. The flow is attached with a weak shockwave on the suction side. The contour plot of the magnitude of the velocity and the static pressure profile at the wall are displayed on Fig. 3 for the baseline configuration $\underline{M}_\infty = 0.729$, $\underline{\alpha}_\infty = 2.31°$, and $\underline{Re} = 6.50 \cdot 10^6$. We see from this latter figure that our numerical results are in good agreement with the experimental results reported in [68].

Results

The mean μ and standard deviation σ of the drag, lift, and pitching moment coefficients C_D, C_L, and C_m, respectively, are gathered in Table 2 for (i) either $Q = 10^3$ calls to the *elsA* computational model F with the 10-th level tensorized GJL rule (see Fig. 1left) to evaluate the polynomial coefficients by projection, Eq. (8); (ii) or $Q = 80$ calls with randomly selected nodes (see Fig. 1right) to evaluate the polynomial coefficients by BPDN $(P_{1,\varepsilon})$. The primary reason why we have chosen this sampling size is for its ease of use with the multithreading setup of our CFD software. However, the sparsity of the polynomial surrogates is observed to be $S \simeq 10$ from our numerical results. A common practical observation is that $Q \geq 4S \simeq 40$ or so is usually enough for a successful recovery. Therefore, a new sampling set of $Q = 40$ randomly selected nodes (not displayed here) has been generated and used to construct other surrogate models by $(P_{1,\varepsilon})$. The corresponding means and standard deviations are also gathered in Table 2. As for solving this ℓ_1-minimization problem $(P_{1,\varepsilon})$, the Spectral Projected Gradient Algorithm (SPGL1) developed by van den Berg and Friedlander [74] and implemented in the MATLAB package `SPGL1` [75], is used. The tolerance was fixed at $\varepsilon = 10^{-5}$ and we were able to find a solution for all surrogates with this a priori choice without resorting to cross-validation, for example [26]. Further investigations should be carried on on this topic, though. It should also be noted that no particular sampling strategy, such as stratification, low-discrepancy series, or preconditioning, has been applied at this stage to construct the sampling sets. Moreover, some weighting could be applied to the ℓ_1-norm as notice in [51] and references therein. We have not considered that possibility either and left it to future works.

The PDFs of the three aerodynamic coefficients considered in this study are displayed on Fig. 4 (in log scale) using the 10-th level tensorized quadrature rule and the 80 samples set. Figure 5 displays the same plots for the 40 samples set. These PDFs were estimated from $N_s = 100{,}000$ evaluations of the gPC surrogates G^p constructed by either approaches and smoothing out the resulting histograms by a normal kernel density function [76]. The horizontal axes are scaled by the mean value of each coefficient computed with the tensorized quadrature rule. A very good agreement is

Table 2 Mean and variance of the aerodynamic coefficients computed by the 10-th level tensorized GJL rule with $Q = 10^3$ and by ℓ_1-minimization with $Q = 80$ and $Q = 40$

	GJL product rule ($Q = 10^3$)		ℓ_1-minimization ($Q = 80$)		ℓ_1-minimization ($Q = 40$)	
	μ	σ	μ	σ	μ	σ
C_D	133.37e–04	34.13e–04	133.34e–04	34.05e–04	133.21e–04	33.78e–04
C_L	72.27e–02	1.67e–02	72.27e–02	1.67e–02	72.29e–02	1.66e–02
C_m	–453.99e–04	32.24e–04	–453.95e–04	32.18e–04	–453.73e–04	31.98e–04

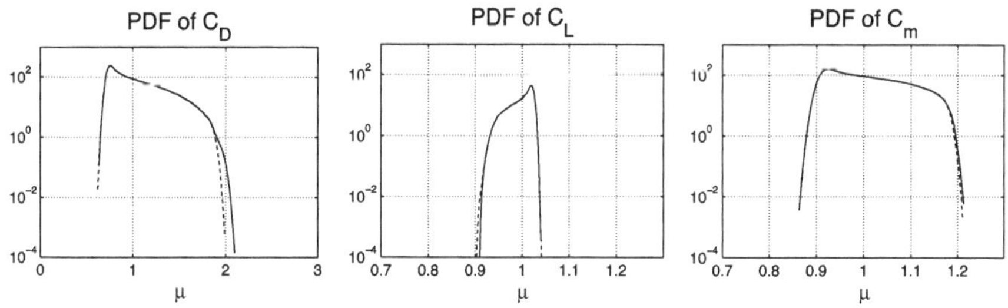

Fig. 4 Comparison of the PDFs of the drag (left), lift (middle) and pitching moment (right) coefficients computed by the 10-th level tensorized GJL rule ($Q = 10^3$, full lines) and ℓ_1-minimization ($Q = 80$, dashed lines)

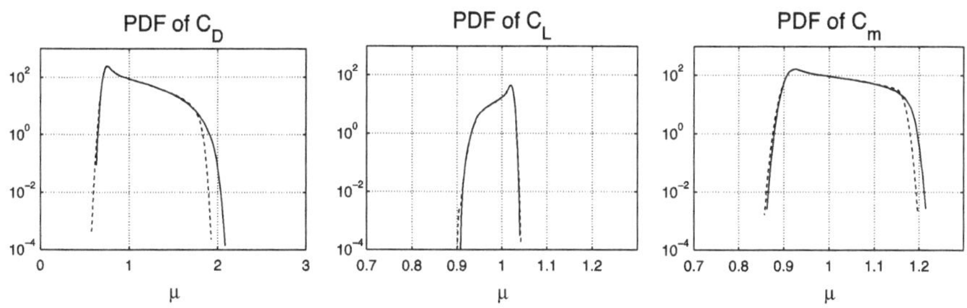

Fig. 5 Comparison of the PDFs of the drag (left), lift (middle) and pitching moment (right) coefficients computed by the 10-th level tensorized GJL rule ($Q = 10^3$, full lines) and ℓ_1-minimization ($Q = 40$, dashed lines)

achieved by both methods. The reduction of the number of samples to the heuristic lower limit of four times the sparsity S for use in the ℓ_1-minimization approach essentially affects the tails of the PDFs.

Conclusions

In this chapter we have addressed various methodologies with relevance to the construction of generalized polynomial chaos expansions (gPC) for parameterized complex processes as encountered in computational fluid dynamics. The presentation has been more particularly focused on the use of adapted sampling sets in the parameter space using either structured or unstructured grids to construct the gPC expansions. These techniques were illustrated with the example of a two-dimensional transonic turbulent flow around a RAE 2822 airfoil considering variable geometrical (the thickness-to-chord ratio) and operational (the free-stream Mach number and angle of attack) parameters. The quantities of interest are the usual drag, lift, and pitching moment coefficients for which polynomial surrogates are sought for using the aforementioned sampling sets as learning sets.

Firstly, multivariate quadrature rules based on univariate Gauss-Jacobi-Lobatto rules have been used for the construction of these polynomials surrogates by projection. Secondly, observing a posteriori that the aerodynamic quantities of interest are sparse in the multivariate polynomial chaos basis associated to the parameters probability density functions, an ℓ_1-minimization approach has been applied in the framework of the theory of compressed sensing. The latter allows to recover the gPC expansion coefficients at a much lower computational cost than the quadrature rules addressed in the first approach. Unstructured sampling nodes are preferred in this process, selecting them randomly in the parameter space. Their number is typically less than the dimension of the polynomial space where the surrogates are sought for, and thus typically much less than the number of nodes of the multivariate quadrature rules that have to be used for a given polynomial order. The ℓ_1-minimization procedure is non-adaptive in the sense that it identifies both the amplitude of the leading expansion coefficients and their ranks. It thus constitutes a promising direction for future developments in practical applications for more complex geometries and flows, where adaptive strategies within the parametric space, weighted minimization, or preconditioned sampling sets may be advantageous.

References

1. Wiener, N.: The homogeneous chaos. Am. J. Math. **60**(4), 897–936 (1938)
2. Cameron, R., Martin, W.: The orthogonal development of nonlinear functionals in series of Fourier-Hermite functionals. Ann. Math. **48**(2), 385–392 (1947)
3. Xiu, D., Karniadakis, G.: The Wiener-Askey polynomial chaos for stochastic differential equations. SIAM J. Sci. Comput. **24**(2), 619–644 (2002)
4. Soize, C., Ghanem, R.G.: Physical systems with random uncertainties: chaos representations with arbitrary probability measure. SIAM J. Sci. Comput. **26**(2), 395–410 (2004)
5. Clouteau, D., Savin, É., Aubry, D.: Stochastic simulations in dynamic soil-structure interaction. Meccanica **36**(4), 379–399 (2001)
6. Debusschere, B.J., Najm, H.N., Pébay, P.P., Knio, O.M., Ghanem, R.G., Le Maître, O.P.: Numerical challenges in the use of polynomial chaos representations for stochastic processes. SIAM J. Sci. Comput. **26**(2), 698–719 (2005)
7. Ghanem, R., Spanos, P.: Stochastic Finite Elements: A Spectral Approach. Springer, New York (1991)
8. Ghanem, R.G.: Ingredients for a general purpose stochastic finite elements implementation. Comput. Methods Appl. Mech. Eng. **168**(1–4), 19–34 (1999)
9. Ghanem, R., Higdon, D., Owhadi, H. (eds.): Handbook of Uncertainty Quantification. Springer International Publishing, Cham (2016)
10. Najm, H.N.: Uncertainty quantification and polynomial chaos techniques in computational fluid dynamics. Ann. Rev. Fluid Mech. **41**, 35–52 (2009)
11. Le Maître, O., Knio, O.: Spectral Methods for Uncertainty Quantification: With Applications to Computational Fluid Dynamics. Springer, Dordrecht (2010)
12. Pettersson, M.P., Iaccarino, G., Nordström, J.: Polynomial Chaos Methods for Hyperbolic Partial Differential Equations. Springer International Publishing, Cham (2015)
13. Xiu, D.: Numerical Methods for Stochastic Computations: A Spectral Method Approach. Princeton University Press, Princeton NJ (2010)
14. Jahedi, A., Ahmadi, G.: Application of Wiener-Hermite expansion to non stationary random vibration of a Duffing oscillator. ASME J. Appl. Mech. **50**(2), 436–442 (1983)

15. Sun, T.C.: A finite element method for random differential equations with random coefficients. SIAM J. Numer. Anal. **16**(6), 1019–1035 (1979)
16. Smolyak, S.: Quadrature and interpolation formulas for tensor products of certain classes of functions. Sov. Math. Dokl. **4**, 240–243 (1963)
17. Savin, É., Resmini, A., Peter, J.: Sparse polynomial surrogates for aerodynamic computations with random inputs. In: AIAA Paper 2016-0433 (2016) 18th AIAA Non-Deterministic Approaches Conference, 4–8 Jan 2016, San Diego, CA
18. Ghiocel, D., Ghanem, R.: Stochastic finite-element analysis of seismic soil-structure interaction. ASCE J. Eng. Mech. **128**(1), 66–77 (2002)
19. Xiu, D., Hesthaven, J.S.: High-order collocation methods for differential equations with random inputs. SIAM J. Sci. Comput. **27**(3), 1118–1139 (2005)
20. Eldred, M.S., Burkardt, J.: Comparison of non-intrusive polynomial chaos and stochastic collocation methods for uncertainty quantification. In: AIAA Paper 2009-0976 (2009) 47th AIAA Aerospace Sciences Meeting, 5–8 Jan 2009, Orlando, FL
21. Blatman, G., Sudret, B.: An adaptive algorithm to build sparse polynomial chaos expansions for stochastic finite element analysis. Probab. Eng. Mech. **25**(2), 183–197 (2010)
22. Nobile, F., Tempone, R., Webster, C.G.: A sparse grid stochastic collocation method for partial differential equations with random input data. SIAM J. Numer. Anal. **46**(5), 2309–2345 (2008)
23. Nobile, F., Tempone, R., Webster, C.G.: An anisotropic sparse grid stochastic collocation method for partial differential equations with random input data. SIAM J. Numer. Anal. **46**(5), 2411–2442 (2008)
24. Resmini, A., Peter, J., Lucor, D.: Sparse grids-based stochastic approximations with applications to aerodynamics sensitivity analysis. Int. J. Numer. Methods Eng. **106**(1), 32–57 (2016)
25. Ma, X., Zabaras, N.: An adaptive hierarchical sparse grid collocation algorithm for the solution of stochastic differential equations. J. Comput. Phys. **228**(8), 3084–3113 (2009)
26. Doostan, A., Owhadi, H.: A non-adapted sparse approximation of PDEs with stochastic inputs. J. Comput. Phys. **230**(8), 3015–3034 (2011)
27. Candès, E.J., Romberg, J.K., Tao, T.: Stable signal recovery from incomplete and inaccurate measurements. Commun. Pure Appl. Math. **59**(8), 1207–1223 (2006)
28. Donoho, D.L.: Compressed sensing. IEEE Trans. Inf. Theory **52**(4), 1289–1306 (2006)
29. Arnst, M., Ponthot, J.P.: An overview of nonintrusive characterization, propagation, and sensitivity analysis of uncertainties in computational mechanics. Int. J. Uncertain. Quantif. **4**(5), 387–421 (2014)
30. Xiu, D., Karniadakis, G.: Modeling uncertainty in flow simulations via generalized polynomial chaos. J. Comput. Phys. **187**(1), 137–167 (2003)
31. Poëtte, G., Després, B., Lucor, D.: Uncertainty quantification for systems of conservation laws. J. Comput. Phys. **228**(7), 2443–2467 (2009)
32. Reagan, M.T., Najm, H.N., Ghanem, R.G., Knio, O.M.: Uncertainty quantification in reacting-flow simulations through non-intrusive spectral projection. Combust. Flame **132**(3), 545–555 (2003)
33. Babuška, I., Nobile, F., Tempone, R.: A stochastic collocation method for elliptic partial differential equations with random input data. SIAM J. Numer. Anal. **45**(3), 1005–1034 (2007)
34. Loeven, G.J.A., Witteveen, J.A.S., Bijl, H.: Probabilistic collocation: an efficient non-intrusive approach for arbitrarily distributed parametric uncertainties. In: AIAA Paper 2007-317 (2007) 45th AIAA Aerospace Sciences Meeting and Exhibit, 8–11 Jan 2007, Reno, NV
35. Chassaing, J.C., Lucor, D.: Stochastic investigation of flows about airfoils at transonic speeds. AIAA J. **48**(5), 938–950 (2010)
36. Hadigol, M., Maute, K., Doostan, A.: On uncertainty quantification of lithium-ion batteries: application to an $LiC_6/LiCoO_2$ cell. J. Power Sources **300**, 507–524 (2015)
37. Hosder, S., Walters, R.W., Balch, M.: Point-collocation nonintrusive polynomial chaos method for stochastic computational fluid dynamics. AIAA J. **48**(12), 2721–2730 (2010)
38. Jones, B.A., Parrish, N., Doostan, A.: Postmaneuver collision probability estimation using sparse polynomial chaos expansions. AIAA J. Guid. Control Dyn. **38**(8), 1425–1437 (2015)

39. Simon, F., Guillen, P., Sagaut, P., Lucor, D.: A gPC-based approach to uncertain transonic aerodynamics. Comput. Methods Appl. Mech. Eng. **199**(17–20), 1091–1099 (2010)
40. Eldred, M.S.: Design under uncertainty employing stochastic expansion methods. Int. J. Uncertain. Quantif. **1**(2), 119–146 (2011)
41. Xiu, D.: Efficient collocational approach for parametric uncertainty analysis. Commun. Comput. Phys. **2**(2), 293–309 (2007)
42. Gautschi, W.: Orthogonal Polynomials: Computation and Approximation. Oxford University Press, Oxford (2004)
43. Golub, G.H., Welsch, J.H.: Calculation of Gauss quadrature rules. Math. Comput. **23**(106), 221–230 (1969)
44. Margheri, L., Meldi, M., Salvetti, M.V., Sagaut, P.: Epistemic uncertainties in RANS model free coefficients. Comput. Fluids **102**, 315–335 (2014)
45. Witteveen, J.A.S., Bijl, H.: Modeling arbitrary uncertainties using Gram-Schmidt polynomial chaos. In: AIAA Paper 2006-896 (2006) 44th AIAA Aerospace Sciences Meeting and Exhibit, 9–12 Jan 2016, Reno, NV
46. Zhang, Z., El-Moselhy, T.A., Elfadel, I.M., Daniel, L.: Calculation of generalized polynomial-chaos basis functions and Gauss quadrature rules in hierarchical uncertainty quantification. IEEE Trans. Comput. Aided Des. Integr. Circuits Syst. **33**(5), 728–740 (2014)
47. Berveiller, M., Sudret, B., Lemaire, M.: Stochastic finite element: a non-intrusive approach by regression. Rev. Eur. Mécanique Numér. **15**(1–3), 81–92 (2006)
48. Gao, Z., Zhou, T.: On the choice of design points for least-square polynomial approximations with application to uncertainty quantification. Commun. Comput. Phys. **16**(2), 365–381 (2014)
49. Hosder, S., Walters, R.W., Balch, M.: Efficient sampling for non-intrusive polynomial chaos applications with multiple uncertain input variables. In: AIAA Paper 2007-1939 (2007) 48th AIAA/ASME/ASCE/AHS/ASC Structures, Structural Dynamics, and Materials Conference, 23–26 April 2007, Honolulu, HI
50. Raisee, M., Kumar, D., Lacor, C.: A non-intrusive model reduction approach for polynomial chaos expansion using proper orthogonal decomposition. Int. J. Numer. Methods Eng. **103**(4), 293–312 (2015)
51. Hampton, J., Doostan, A.: Compressive sampling methods for sparse polynomial chaos expansions. In: Ghanem, R.G., Higdon, D., Owhadi, H. (eds.) Handbook of Uncertainty Quantification, 29 pp. Springer, Cham (2016)
52. Jakeman, J.D., Eldred, M.S., Sargsyan, K.: Enhancing ℓ_1-minimization estimates of polynomial chaos expansion using basis selection. J. Comput. Phys. **289**, 18–34 (2015)
53. Mathelin, L., Gallivan, K.A.: A compressed sensing approach for partial differential equations with random input data. Commun. Comput. Phys. **12**(4), 919–954 (2012)
54. Peng, J., Hampton, J., Doostan, A.: A weighted ℓ_1-minimization approach for sparse polynomial chaos expansions. J. Comput. Phys. **267**, 92–111 (2014)
55. Peng, J., Hampton, J., Doostan, A.: On polynomial chaos expansion via gradient-enhanced ℓ_1-minimization. J. Comput. Phys. **310**, 440–458 (2016)
56. Rauhut, H., Ward, R.: Sparse Legendre expansions via ℓ_1-minimization. J. Approx. Theory **164**(5), 517–533 (2012)
57. Schiavazzi, D., Doostan, A., Iaccarino, G.: Sparse multiresolution regression for uncertainty propagation. Int. J. Uncertain. Quantif. **4**(4), 303–331 (2014)
58. Tang, G., Iaccarino, G.: Subsampled Gauss quadrature nodes for estimating polynomial chaos expansions. SIAM/ASA J. Uncertain. Quantif. **2**(1), 423–443 (2014)
59. West IV, T.K., Brune, A.J., Hosder, S., Johnston, C.O.: Uncertainty analysis of radiative heating predictions for Titan entry. AIAA J. Thermophys. Heat Transf. **30**(2), 438–451 (2016)
60. Yan, L., Guo, L., Xiu, D.: Stochastic collocation algorithms using ℓ_1-minimization. Int. J. Uncertain. Quantif. **2**(3), 279–293 (2012)
61. Yuang, X., Karniadakis, G.E.: Reweighted ℓ_1-minimization method for stochastic elliptic differential equations. J. Comput. Phys. **248**, 87–108 (2013)
62. Chen, S.C., Donoho, D.L., Saunders, M.A.: Atomic decomposition by basis pursuit. SIAM J. Sci. Comput. **20**(1), 33–61 (1998)

63. Candès, E.J., Wakin, M.B.: An introduction to compressive sampling. IEEE Signal Process. Mag. **25**(2), 21–30 (2008)
64. Spalart, P.R., Allmaras, S.R.: A one-equation turbulence model for aerodynamic flows. In: AIAA Paper 1992-0439 (1992) 30th AIAA Aerospace Sciences Meeting and Exhibit, 6–9 Jan 1992, Reno, NV
65. Cook, P.H., McDonald, M.A., Firmin, M.C.P.: Aerofoil RAE 2822—pressure distributions, and boundary layer and wake measurements. In: Experimental Data Base for Computer Program Assessment. AGARD Advisory Report No. 138. NATO (1979) Appendix A6
66. Garner, H.C., Rogers, E.W.E., Acum, W.E.A., Maskell, E.C.: Subsonic wind tunnel wall corrections. In: AGARDograph No. 109. NATO (1966) Chapter 6
67. Haase, W., Bradsma, F., Elsholz, E., Leschziner, M., Schwamborn, D. (eds.): EUROVAL–An European Initiative on Validation of CFD Codes. Notes on Numerical Fluid Mechanics, Vol. 42. Vieweg Verlag, Wiesbaden (1993) Section 5.1
68. http://www.cfd-online.com/Wiki/RAE2822_airfoil
69. Jaynes, E.T.: Information theory and statistical mechanics. Phys. Rev. **106**(4), 620–630 (1957)
70. Cambier, L., Heib, S., Plot, S.: The Onera elsA CFD software: input from research and feedback from industry. Mech. Ind. **14**(3), 159–174 (2013)
71. van Leer, B.: Towards the ultimate conservative difference scheme, V. A second order sequel to Godunov's method. J. Comput. Phys. **32**(1), 101–136 (1979)
72. van Albada, G.D., van Leer, B., Roberts, W.W.: A comparative study of computational methods in cosmic gas dynamics. Astron. Astrophys. **108**(1), 76–84 (1982)
73. Yoon, S.K., Jameson, A.: An LU-SSOR scheme for the Euler and Navier-Stokes equations. In: AIAA Paper 1987-600 (1987) 25th AIAA Aerospace Sciences Meeting, 12–15 Jan 1987, Reno, NV
74. van den Berg, E., Friedlander, M.P.: Probing the Pareto frontier for basis pursuit solutions. SIAM J. Sci. Comput. **31**(2), 890–912 (2008)
75. van den Berg, E., Friedlander, M.P.: `SPGL1`: A solver for large-scale sparse reconstruction (June 2007)
76. Wand, M.P., Jones, M.D.: Kernel Smoothing. Chapman and Hall/CRC, Boca Raton, FL (1995)

Non-intrusive Probabilistic Collocation Method for Operational, Geometrical, and Manufacturing Uncertainties in Engineering Practice

Dirk Wunsch, Rémy Nigro, Grégory Coussement and Charles Hirsch

Introduction

CFD simulations are run today with a unique set of input data such as boundary conditions, physical model, or geometry parameters. In these deterministic simulations, the system response is a single value. The actual operating conditions of industrial products are, however, subject to operational and geometrical uncertainties. In order to include these uncertainties into the CFD-based design process, a new type of simulation is needed: non-deterministic simulations. This type of approach allows to incorporate and propagate uncertainties. The outcome of a non-deterministic simulation is not a single value, but a variation of the system response.

The uncertainty propagation method used in this work is the Non-Intrusive Probabilistic Collocation Method (NIPColM), where the collocation points are selected by means of the Golub–Welsch algorithm [2] for quadrature of general probability density function (PDF) shapes. This approach is combined with a sparse grid technique which allows for a significant reduction in number of deterministic simulations to be run.

D. Wunsch · R. Nigro · C. Hirsch (✉)
Numeca International, Chaussée de la Hulpe 189, Brussels, Belgium
e-mail: charles.hirsch@numeca.be

D. Wunsch
e-mail: dirk.wunsch@numeca.be

R. Nigro
e-mail: remy.nigro@umons.ac.be

R. Nigro · G. Coussement
University of Mons, Mons, Belgium
e-mail: gregory.coussement@umons.ac.be

C. Hirsch et al. (eds.), *Uncertainty Management for Robust Industrial Design in Aeronautics*, Notes on Numerical Fluid Mechanics and Multidisciplinary Design 140, https://doi.org/10.1007/978-3-319-77767-2_9

A particular challenge lies in the combination of operational and geometrical uncertainties in an integrated design and simulation environment. While operational uncertainties are considered by running the CFD solver with the same geometry and the same mesh, but varying boundary conditions, geometrical uncertainties demand in addition a changing geometry and changing mesh. Geometrical uncertainties can be either expressed in parameterized form, also accounting for correlations, or as uncertain fields. Random field and correlated engineering parameter models are suitable for the handling of uncertainties resulting from manufacturing or assembly variability.

This NIPColM is applied to the NASA rotor 37, first, accounting for simultaneous operational and geometrical uncertainties on the inlet total pressure profile, static outlet pressure, the tip clearance, the leading edge angle and leading edge radius [13], and second, as a random field of the blade surface [7].

The importance of accounting for combined operational and geometrical uncertainties during the rotor design process is demonstrated in comparison with the deterministic case. Scaled sensitivity derivatives calculated from the uncertainty propagation method allow to assess objectively the influence of individual uncertainties on the output quantity of interest. Finally, the non-deterministic simulation results are evaluated in form of reconstructed PDFs of scalar output quantities.

Non-intrusive Probabilistic Collocation Method

The uncertainty propagation method used within this work is the Non-Intrusive Probabilistic Collocation Method [4], which can be seen as a generalization of the stochastic collocation method described in [5]. If a generic stochastic partial differential equation is considered such as:

$$L(\xi)u(\vec{x}, t, \xi) = S(\vec{x}, t) \tag{1}$$

- where L being a differential operator containing space and time derivatives;
- S being the source terms;
- ξ being a random input parameter;
- u being the non-deterministic solution.

In the non-intrusive probabilistic collocation method, Lagrange interpolating polynomials are used to construct the following expansion:

$$u(\vec{x}, t, \xi) = \sum_{i=1}^{N_p} u_i(\vec{x}, t) h_i(\xi) \tag{2}$$

- $u_i(\vec{x}, t)$ is the deterministic solution at the collocation point ξ_i
- $h_i(\xi_i)$ is the Lagrange interpolation polynomial corresponding to the collocation point ξ_i.

The Lagrange interpolating polynomial is given by:

$$h_i(\xi) = \prod_{\substack{k=1 \\ k \neq i}}^{N_p} \frac{\xi - \xi_k}{\xi_i - \xi_k} \tag{3}$$

with: $h_i(\xi_j) = \delta_{ij}$ and N_p is the number of collocation points required by the non-intrusive probabilistic collocation method.

The collocation points are selected as the Gauss quadrature points by means of the Golub–Welsch algorithm for general probability density function (PDF) shapes [2]. This has the advantage that any arbitrary PDF type can be used to describe the uncertainties. In order to propagate the input uncertainty modeled by the random variable ξ_i, Eq. (2) is introduced into Eq. (1). This provides a system of N_p uncoupled deterministic simulations:

$$L(\xi_i)u(\vec{x}, t) = S(\vec{x}, t) \tag{4}$$

Once all N_p computations are performed, the first four moments of any output quantity φ are calculated based on the weights w_k from the Gauss quadrature. The mean follows from:

$$\mu_1 = \sum_{k=1}^{N_p} w_k \varphi_k(\vec{x}, t) \tag{5}$$

and the second (variance), third (skewness), and fourth moments (kurtosis) are calculated as:

$$\mu_n = \sum_{k=1}^{N_p} w_k (\varphi_k(\vec{x}, t) - \mu_1)^n \tag{6}$$

Sparse Grids for Multiple Simultaneous Uncertainties

In order to combine several simultaneous uncertainties, the quadrature must be brought to the multi-dimensional case. This multi-dimensional quadrature is achieved by means of tensor products. In the 1D case, the quadrature is given by:

$$Q^{(1)}f = \sum_{i=1}^{N_p} f(\xi_i)\omega_i \tag{7}$$

Multi-dimensional quadrature is then obtained by tensor products:

$$Q^{N_{\text{dim}}}f = (Q^{(1)} \otimes \ldots \otimes Q^{(N_{\text{dim}})})f \tag{8}$$

Each point along one direction is multiplied with each point along all other directions, where the weights are calculated as the products of the 1D-weights.

This approach leads to an exponential increase in the number of points with the number of dimensions. This is the so-called *curse of dimensionality*. Table 1 illustrates how the number of CFD runs scales with the number of dimensions assuming 1D quadrature rules with 3 points. For two simultaneous uncertainties, 9 CFD runs need to be performed, while 10 simultaneous uncertainties would require 59,049 CFD runs, which is not feasible in practice.

Sparse grid quadrature can overcome this curse of dimensionality to a certain extent and make non-intrusive collocation methods accessible for higher stochastic dimensions. The implementation is based on Smolyak's quadrature method [11]. For a review of sparse grid algorithms, the reader may refer to [5] and [13] for more information on the construction of the sparse grid. Sparse grid quadrature works by combining quadrature rules of different orders. To distinguish different orders, a level index is introduced and the number of 1D quadrature points per level grows in function of the chosen quadrature rule. In this work, Gauss quadrature is applied where the number of points on subsequent levels grows with "2 * *level* + 1," with the level 0 quadrature rule being defined as the mid-point rule. To build the sparse grid, difference formulas between two subsequent levels are defined and then summed over all levels; the quadrature rule of level L is thus re-written as a sum of differences over all levels from 0 to L. This allows to change the order of summation and tensor product and truncate the sum to a total level (or total order of the underlying Lagrange polynomials).

Table 2 summarizes the number of quadrature points Np, contained in linear growth sparse grid for levels of 1, 2, and 3 in comparison with a full tensor grid with the same number of 1D quadrature points.

Table 1 Number of CFD runs needed for multi-dimensional uncertainty problems applying tensor product rule, assuming 3 collocation points in 1D

No. of stochastic dimensions	No. of CFD runs
1	3
2	9
3	27
…	…
10	59049
n (m = 3)	m^n

Table 2 Number of quadrature points N_p contained in full tensor grid and sparse grid

Level	1		2		3	
Dimension	Tensor	Sparse	Tensor	Sparse	Tensor	Sparse
1	3	3	5	5	7	7
2	9	5	25	17	49	45
3	27	7	125	31	343	105
4	81	9	625	49	2401	201
5	243	11	3125	71	16807	341
…	…	…	…	…	…	…
10	59049	21	≈9.7e6	241	≈2.8e8	1981

The significant reduction in number of runs makes the simultaneous treatment of many uncertainties in complex 3D CFD problems accessible. This is the basis for the work performed here.

Statistical Output Analysis

The analysis of non-deterministic simulations using the NIPColM relies to a large extent on the computed moments of output quantities, i.e., mean, variance, skewness, and kurtosis. Besides the moments itself, which can be used to plot mean values with bars indicating the variability of the response, PDFs can be reconstructed based on these statistical output moments. Finally, the local surrogate that the Lagrange interpolating polynomial represents, as given in Eq. (2), allows computing scaled sensitivity derivatives.

Statistical Output Moments

Once all simulations are performed, statistical moments of output quantities are automatically calculated, by taking the weights of the individual collocation points into account. The mean, variance, skewness, and kurtosis are calculated following Eqs. (5) and (6). This information is calculated for a selected number of scalar output quantities but can be obtained for any CFD output quantity or output field of interest. It is thus possible to compute mean and variance for a cut through the computational domain, where the values are calculated point-by-point.

Output PDF Reconstruction

Simulations performed with the non-intrusive probabilistic collocation method characterize the output not directly by its PDF, but by the moments of this output PDF, as stated above. The PDF can be reconstructed from its moments, which is important, since the display of non-deterministic results by a mean value with an uncertainty bar corresponding to $\pm\sigma$ implies a symmetric distribution around the mean value, whereas the real system response might be characterized by a skew distribution. To overcome this, the Pearson method [9] is used to reconstruct an approximation of the PDF of a given output quantity from its first four moments. Figure 1 shows a schematic of the standard PDF shapes to be selected in function of the third and fourth statistical output moment by the Pearson method.

Scaled Sensitivity Derivatives

Another important element in the evaluation of non-deterministic solutions is the relative influence of a given input uncertainty on the solution. This influence is predicted by calculation of scaled sensitivity derivatives as done by [12] and applied to the probabilistic collocation method. This increases, on the one hand, the understanding of the system under investigation, and a variation of an uncertain input can be directly linked to a variation in the output quantity of interest. On the other hand, it provides an objective measure of the influence of uncertainties on the output and allows therefore for an efficient reduction in number of uncertainties by identifying uncertainties with little influence on the solution. This is particularly important in systems where many uncertainties are present, and even with an efficient sparse grid quadrature, the benefit of reducing the number of uncertainties can be significant. On the example of a level 2 sparse grid with 10 uncertainties, if the sensitivity analysis would show that 5 of these uncertainties have little

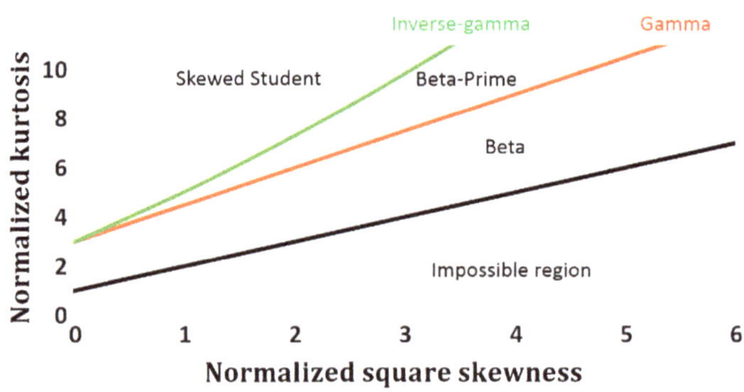

Fig. 1 Schematic of Pearson method

influence, the number of simulations that needs to be performed can be reduced from 241 to 71 as shown in Table 2.

The scaled sensitivity derivative is defined as the partial derivative of the solution $u(\vec{x}, t, \xi)$ with respect to the uncertain input parameter ξ, multiplied by the standard deviation of the uncertain input parameter as:

$$\sigma_{\xi_i} \frac{\partial u(\vec{x}, t, \xi)}{\partial \xi_i} \tag{9}$$

Assuming one single uncertain parameter and by introducing Eqs. (2) and (3) into Eq. (9), the sensitivity derivative $\frac{\partial u}{\partial \xi}$ can be written as:

$$\frac{\partial u}{\partial \xi_i} = \sum_{i=1}^{N_p} u_i \sum_{\substack{j=0 \\ j \neq i}}^{N_p} \frac{1}{\xi_i - \xi_j} \prod_{\substack{k=1 \\ k \neq j \\ k \neq i}}^{N_p} \frac{\xi - \xi_k}{\xi_i - \xi_k} \tag{10}$$

Here ξ_i stands for the value at collocation point i. The evaluation is done around the nominal value. The scaled sensitivity derivative as in Eq. (9) is then obtained by multiplying the partial derivative by the standard deviation of the input uncertainty. This can be calculated for an arbitrary number of input uncertainties, and the individual scaled derivatives can be compared and used to reduce the uncertainty dimension by eliminating the uncertain variable with the lowest absolute values.

It is worth to mention that the here calculated scaled sensitivity derivatives are equivalent to first-order approximations of the standard deviation obtained by perturbation methods [10].

Simultaneous Operational and Geometrical Uncertainties

Another major difficulty on the way to the use of uncertainty quantification (UQ) in a day-to-day engineering practice, after tempering the "curse of dimensionality," is to provide a user-friendly tool for combined operational and geometrical uncertainties in a professional virtual prototyping environment. The principle is by far not limited to CFD but could also be applied, for example, to structural mechanics or aero-acoustics.

The use of operational uncertainties only implies modifying the boundary conditions of a given number of CFD simulations. If a level 2 sparse grid with 10 simultaneous uncertainties is considered, then 241 individual simulations must be initialized with correct modified boundary conditions and weights, respectively. This process alone must be automatized and integrated to be usable in a day-to-day engineering practice. If these 10 uncertainties include geometrical uncertainties, based on the above example of 241 simulations to be set up, not only the boundary conditions must be changed, but also the geometry and the mesh! This is completely unfeasible without an efficient integration into an engineering software suite. An automatic generation of parameterized geometries for turbo-machinery cases, automatic meshing, and initialization of computations in combination with the probabilistic UQ method described above is used in this work.

Operational or geometrical uncertainties are selected in the workflow and attributed with a PDF shape for each uncertainty. The Golub–Welsch algorithm [2] used for the quadrature finds automatically suitable nodes and weights with respect to the input measure.

Based on the sparse grid (or full tensor grid) detailed above, the N_p individual non-deterministic sub-computations are then set up automatically accounting for varying boundary conditions due to operational uncertainties or varying geometry and mesh, which are the result of geometrical uncertainties. The system of equations given by (4) is then solved without any further interference.

Manufacturing and Assembly Uncertainties

Uncertainties resulting from the manufacturing process of components cannot be represented by considering the parameters of a parametric model as independent random variables. The manufacturing variability, which lies within the imposed tolerances, introduces correlations between different points on the surface, i.e., surface points are thus mutually dependent.

One way to account for these correlations is the use of a random field to describe a surface. The NIPColM, however, such as many other UQ propagation methods, cannot use dependent random variables as input in a direct way. To overcome this and to represent the random field with a set of independent variables, principal component analysis (PCA) is applied. In general, a truncation of the transformation is applied to reduce the number of independent variables. The NIPColM is then applied to the reduced orthogonal set of uncorrelated variables.

A second possibility to describe manufacturing uncertainties is by means of correlations between engineering parameters. In analogy to the random fields, the PCA is applied to the set of correlated variables; however, the uncorrelated set obtained by the PCA is not truncated in this case.

Random Fields

A random field is a collection of random variables which includes their mutual dependence, and they are thus suited to represent uncertainties resulting from manufacturing variability. The number of random variables contained in the random field can be either infinite, in the case of a continuous random field, or finite, in the case of a discrete random field. A discretized geometrical surface can be defined

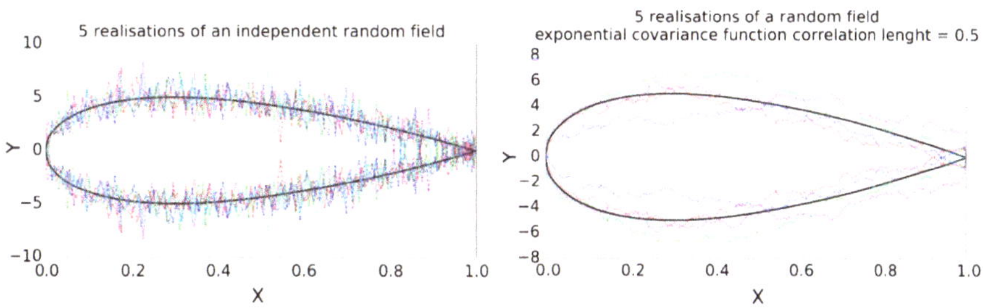

Fig. 2 Realizations of an airfoil discretized with 200 points considering each of these points as independent (left) or as a random field described with a covariance between each of its points (right) (black: nominal value; colors: realizations)

as a discrete random field by considering the collection of the random variables to be defined by the ensemble of the geometry points defining the discrete surface.

The relative influence between two random variables is expressed by the so-called covariance.

Figure 2 shows realization of an airfoil surface considering each surface point as an independent variable, i.e., without correlations (Fig. 2 left) and considering correlations between the surface points (Fig. 2 right) by means of a covariance function. It is clearly seen that correlations introduce preferential deformation shapes on the geometry.

As mentioned above, correlated random variables such as in random fields cannot be used directly in the NIPColM. A transformation of the correlated random variables into a set of uncorrelated orthogonal variables is required. This is achieved by means of a principal component analysis.

Principal Component Analysis

The principal component analysis is a method widely used to represent discrete random fields by a reduced set of uncorrelated random variables; it is also known as proper orthogonal decomposition (POD). The continuous version of a PCA is known as Karhunen–Loeve Expansion (KLE) [3].

The expansion of a random field is expressed in Eq. (11)

$$X_t = \overline{X_t} + \sum_{k=1}^{M} \xi_k^{PCA} * \phi_k \tag{11}$$

Without any loss of generality, Eq. (11) can be modified in order to consider only cases of zero-mean random fields ($\overline{X_t} = 0$) in Eq. (12).

$$X_t = \sum_{k=1}^{M} \xi_k^{PCA} * \phi_k \quad (12)$$

The eigenvectors are found by solving the eigenproblem on the covariance matrix. The covariance matrix determines the covariance between each couple of random variables contained in the random field. Its dimension is thus $M \times M$. As a result, M eigenvectors of dimension $1 \times M$ and M eigenvalues are found. The random variables ξ_k^{PCA} defined by the PCA are uncorrelated. This property allows, under the assumption of a Gaussian random field, to ensure that the random variables ξ_k^{PCA} are independent. Moreover, the assumption of a Gaussian random field leads to the conclusion that each individual random variable ξ_k^{PCA} is Gaussian.

The first two moments of the probability distribution (i.e., mean and variance) of these random variables ξ_k^{PCA} are directly defined. As the random field is centered, the random variables ξ_k^{PCA} are also centered $(\mu = 0)$ and the variance of a Gaussian random variable ξ_k^{PCA} is equal to its corresponding eigenvalue $(\sigma^2 = \lambda_k)$.

In case the imposed random field is non-Gaussian, the use of the "Nataf transformation" [6] allows to transform it into a Gaussian random field. The PCA can then be applied on the transformed Gaussian random field.

The PCA output set can be used to approximate a discrete random field with a number of random variables, which is much smaller than M. This can be achieved by sorting the eigenvectors (i.e., the random variables) in decreasing order of their eigenvalue. After sorting of the eigenvalues, the expansion can be truncated. The number of remaining random variables N is chosen in order to ensure that the error on the random field variation is kept inferior to an imposed value ϵ following Eq. (13).

$$\epsilon > 1 - \frac{\sum_{k=1}^{N} \lambda_k}{\sum_{k=1}^{M} \lambda_k} \, with \, \lambda_1 > \lambda_2 > \cdots > \lambda_M \quad (13)$$

Correlated Parametric Uncertainties

In analogy to the random fields, the PCA is applied to the set of correlated variables, and the correlation is described by a probability distribution and a correlation matrix. The probability distribution of each parameter and their correlation matrix are usually determined by measurements on a set of manufactured geometries.

After application of the PCA, each eigenmode is defined as an uncertainty with a truncated Gaussian probability distribution with a zero mean and a variance equal to the eigenvalue. A zero mean can be assumed without any loss of generality by considering each uncertainty as a perturbation of the geometry. Then the collocation points are determined by the NIPColM on the orthogonal uncorrelated set of variables that result from the PCA. For each of these collocation points, the deformation of each engineering parameter is computed based on the contribution of each of the eigenmodes and the resulting deformation corresponds to a Gaussian random field. Again if the PDF is non-Gaussian, the Nataf transformation is applied. The deformation is then added to the mean value of the parameter to determine the geometry of each collocation point.

Application to Rotor 37

The uncertainty quantification tool chain is applied to the NASA Rotor 37. Detailed description of geometry, experimental setup, and a series of simulations cross-plotting the predictions can be found in [1]. Figure 3 shows the measurement sections and geometry of rotor 37.

Several operation points are run in order to build a performance curve. The static outlet pressure values used to this purpose are given in Table 3.

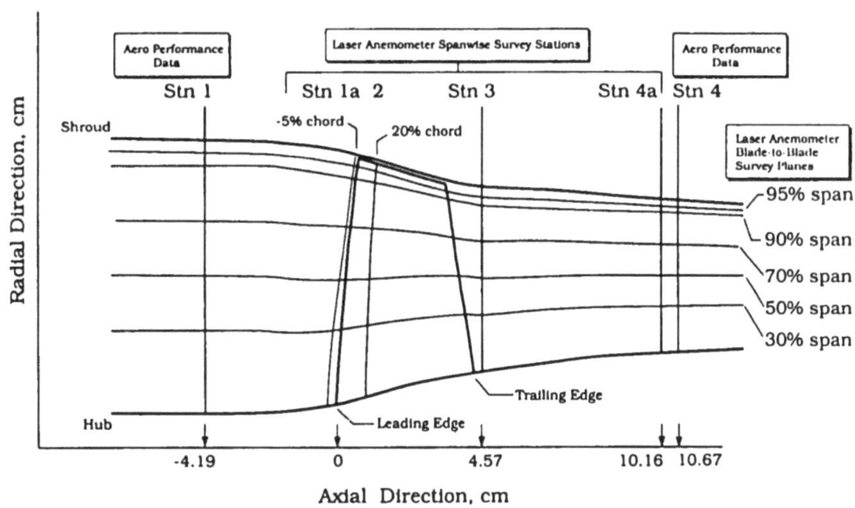

Fig. 3 Rotor 37 measurement sections [1]

Table 3 Operating points for rotor 37 performance map

Running point	Static outlet pressure (Pa)
1	99,215
2	110,000
3	114,074
4	119,035
5	121,033
6	123,008
7	124,027

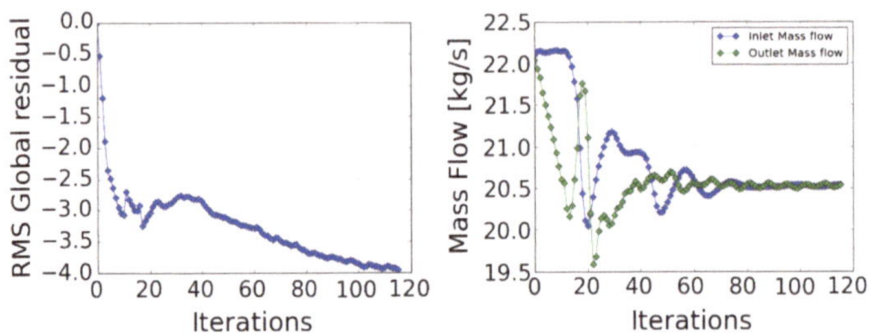

Fig. 4 Convergence for the deterministic computation: (left) root mean square of the global residuals; (right) evolution of the mass flow

Simulations are performed with the flow solver FINE™/Turbo [8], and the uncertainty quantification methodology described above is integrated into FINE™/ Design3D. RANS equations and Spalart–Allmaras turbulence model are solved at a constant rotating hub speed of 17,188 rpm. The convergence of each computation is assessed by the global residuals (sum of all the residuals) and the evolution of the mass flow through the computation domain (cf. Fig. 4). As the residuals decrease of 4 orders of magnitude and the error on the mass flow is below 0.025%, the CFD computation is converged. All the CFD computations used for non-deterministic simulations have the same level of convergence.

The developed tool chain is used to create non-deterministic sub-computations accounting for simultaneous operational and geometrical uncertainties. These simulations are set up in an automatic process accounting for the correct combination of boundary condition modifications due to operational uncertainties and modified geometries due to geometrical uncertainties. For each of these simulations, the correct geometry is built and meshed with Autogrid™, by keeping the number of cells and the global mesh topology constant. Figures 5 and 6 illustrate, on the

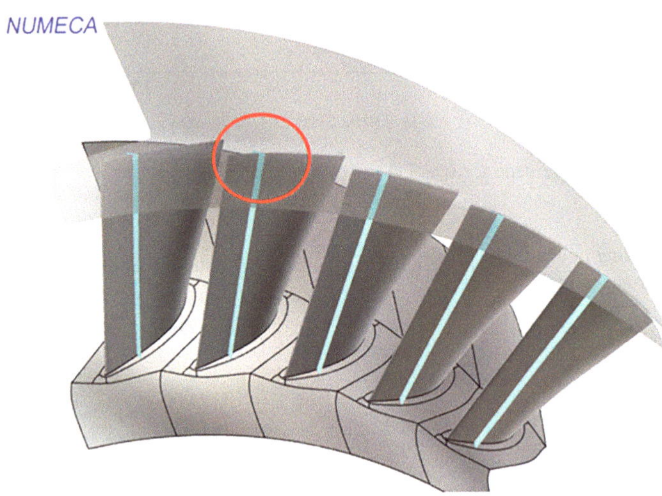

Fig. 5 Global geometry view for tip gap. Circle identifying location of tip gap mesh shown in this figure

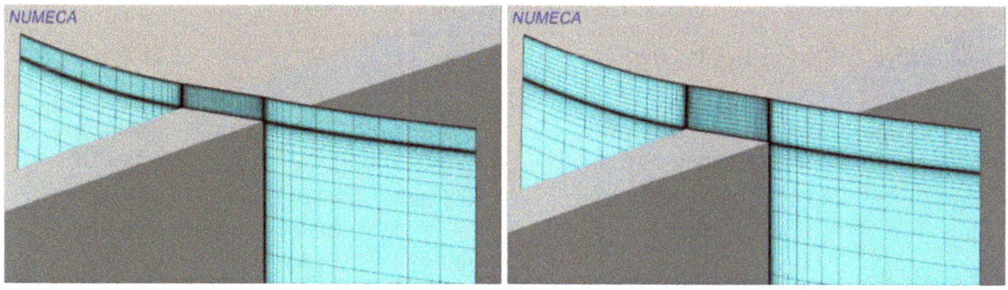

Fig. 6 (left) Geometry and automatically generated mesh with smallest tip gap. (Right) Geometry and automatically generated mesh with largest tip gap

example of a varying tip gap, the automatic mesh generation process. Figure 5 shows a global view of the geometry with the tip gap, while Fig. 6 (left) shows the mesh generated with the smallest tip gap and Fig. 6 (right) the mesh generated with the largest tip gap.

Application of Simultaneous Operational and Geometrical Uncertainties

Industrial configurations are characterized by a multitude of operational and geometrical uncertainties, which can exceed the number of feasible uncertainties even with sparse grid quadrature. The above-developed methodology and especially the use of scaled sensitivities gives a valuable insight into the importance of individual uncertainties on the non-deterministic results. Based on the work in [13], where the trailing edge angle and a span-wise discretization of the blade angles were found to have little influence, the uncertainty model given in Tables 4 and 5 is applied to the rotor 37. Uncertainties on the total inlet pressure, static outlet pressure, tip gap, leading edge angle, and leading edge radius are defined.

This definition of uncertainties on the rotor 37 is also part of the novel database of industrial challenges with prescribed uncertainties, described in chapter "UMRIDA Test Case Database with Prescribed Uncertainties" of the present book.

Table 4 Operational uncertainties imposed on the rotor 37

Uncertainty	Most likely value (m)	Minimum value (%) (m)	Maximum value (%) (m)	PDF type
Inlet total pressure	Table at station 1 [9]	98	102	Symmetric beta-PDF
Static outlet pressure	Table 3 in this document	98	102	Symmetric beta-PDF

Table 5 Geometrical uncertainties imposed on the rotor 37

Uncertainty	Most likely value (m)	Minimum value	Maximum value	PDF type
Leading edge angle	LE_{angle} nominal value from geometry	95% m	105% m	Symmetric beta-PDF
Leading edge radius	LE_{radius} nominal value from geometry	90% m	110% m	Symmetric beta-PDF
Tip clearance	M_{tip} = 0.356 mm	50% M_{tip}	150% M_{tip}	Symmetric beta-PDF

Table 6 Computation time needed for non-deterministic runs in function of sparse grid level, for 5 simultaneous uncertainties

Sparse grid level	No. of simulations for 1 operating point	CPU h-core 1 operating point	CPU h-core performance curve
1	11	29.3	205
2	71	189.3	1325
3	341	909.3	6365

Simulations are performed on a structured full hexahedral mesh, which consists of 2.8 million mesh points, and the CPU-Booster in FINE^{TM}/Turbo is used to accelerate the convergence.

The non-deterministic analysis was performed accounting for the uncertainties as described in Tables 4 and 5, with a level 2 sparse grid (5 collocation points per uncertainty direction), which results in 71 collocation points per UQ run. For comparison and to show the convergence in function of the sparse grid level, additional simulations are performed with level 1 (3 collocation points per uncertainty direction) and level 3 (7 collocation points per uncertainty direction) sparse grids, resulting in 11 and 341 collocation points per run, for levels 1 and 3, respectively.

In order to build a performance curve, the 7 operating points from Table 3 were run. One single simulation was performed in approximately 10 min clock time on 16 parallel cores, which corresponds to *0.95CPUh-core/Million points* per simulation. Table 6 compares the computation times needed for one operating point and for the performance curve with 7 operating points in function of the sparse grid level, accounting for 5 simultaneous uncertainties.

Total pressure ratio and efficiency over mass flow rate.

Accounting for simultaneous operational and geometrical uncertainties as described in Tables 4 and 5, the total pressure ratio and efficiency are plotted over the mass flow rate for the 7 operating points listed in Table 3 and displayed in Figs. 7 and 8, respectively.

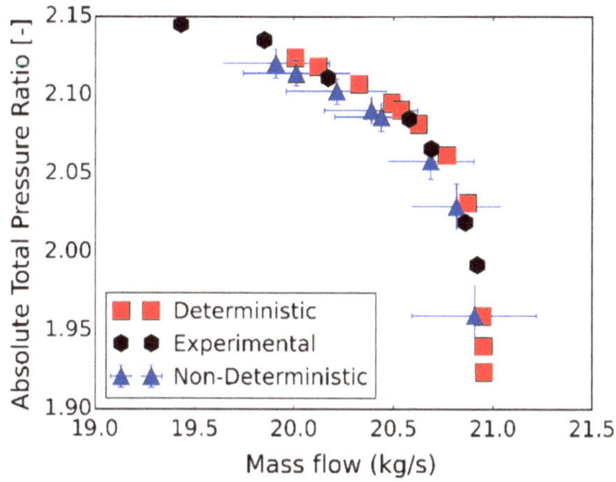

Fig. 7 Absolute total pressure ratio over mass flow for 7 operating points defined in Table 3 and accounting for the 5 operational and geometrical uncertainties listed in Tables 5 and 6. UQ bars are $\pm\sigma$

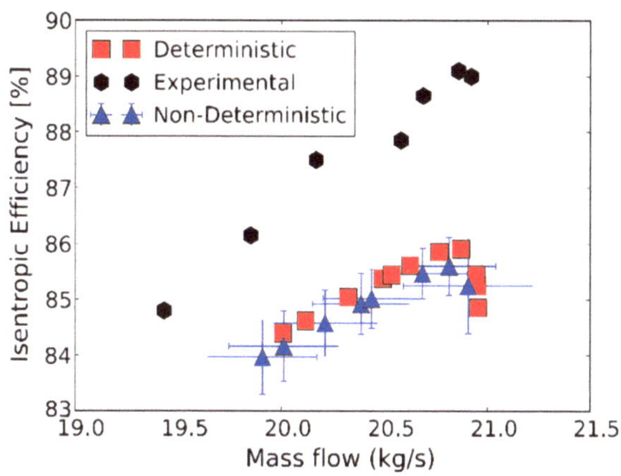

Fig. 8 Isentropic efficiency over mass flow for 7 operating points defined in Table 3 and accounting for the 5 operational and geometrical uncertainties listed in Tables 5 and 6. UQ bars are $\pm\sigma$

As seen in Figs. 7 and 8, the non-deterministic mean is not equal to the result of the deterministic simulation. The deterministic simulation corresponds to a standard CFD simulation where all parameters are set to their nominal values. Figure 8 shows the usual underestimation of the efficiency for this test case in the order of 2%, and the non-deterministic simulation shows slightly lower efficiencies, while the maximum value is shifted to a lower mass flow rate. Displaying these results with mean and standard deviation implies a symmetric distribution of the system response around this mean. Thus assuming a Gaussian distribution, $\pm\sigma$ corresponds the range into which roughly 68% of the results fall. This is the standard approach for displaying these kinds of results, which however is misleading in case the output distribution is not symmetric around the mean.

Indeed, if the first four statistical output moments are calculated and the Pearson approach for PDF reconstruction is applied, it becomes apparent that the results plotted in Figs. 7 and 8 are not symmetrically distributed around the mean. Table 7

Table 7 Higher output moments for mass flow, total pressure ratio, and isentropic efficiency for the operating point at 98% of the choke mass flow

	Mean	Variance	Skewness	Kurtosis
Mass flow (kg/s)	20.444	0.0597	5.6648e−3	9.352e−3
Total pressure ratio (–)	2.087	1.0087e−4	−7.0768e−7	3.1617e−8
Efficiency (–)	0.850	3.1463e−5	−4.3184e−7	1.0205e−8

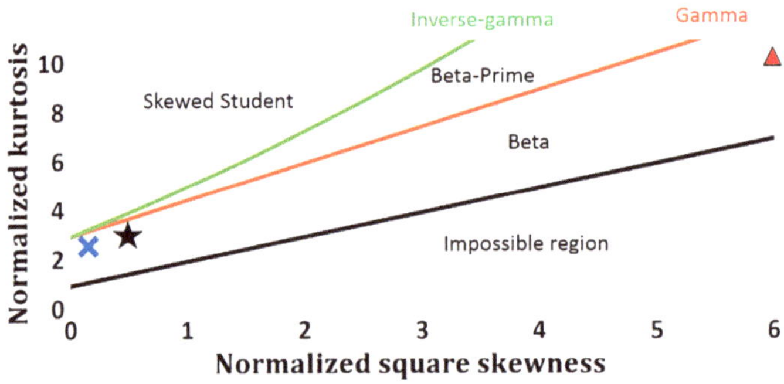

Fig. 9 Pearson diagram, showing the normalized third (squared skewness) and fourth (kurtosis) moment for the operating point at 98% of the choke mass flow. Blue cross: mass flow, black star: total pressure ratio, red triangle: isentropic efficiency

lists the higher output moments for mass flow, total pressure ratio, and isentropic efficiency for the operating point at 98% of the choke mass flow. Figure 9 displays these values in terms of the normalized third (squared skewness) and fourth (kurtosis) moment in a sketch of a Pearson diagram for PDF reconstruction. It must be noted that the higher moments (skewness and kurtosis) are not zero and the distributions are consequently not symmetric, due to nonzero skewness.

Figures 10 and 11 display on the example of the operating point at 98% of the choke mass flow the reconstructed PDFs for total pressure ratio and isentropic efficiency, respectively. The reconstructed PDFs are compared with the mean value and their centered standard deviation to underline the misleading assumption of symmetric distribution of the system response.

Scaled sensitivities in function of operating point.

The evaluation of scaled sensitivities is extremely useful in the analysis of the non-deterministic simulation analysis as it provides an objective criterion for the influence of a given uncertainty on the non-deterministic output. Figures 12 and 13 show the scaled sensitivities on the total pressure ratio and efficiency for the 5 uncertainties included in this study at 3 different operating points: 95.48, 98, and 99.96% of the choke mass flow. It can be seen that for all two quantities the scaled sensitivity in dependence of the leading edge angle changes its sign for the operating point at 95.48% of the choke mass flow. While it is negative for the other two

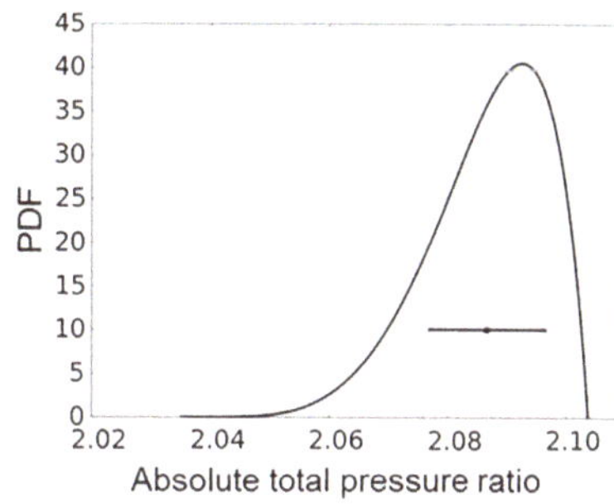

Fig. 10 Reconstructed PDF for total pressure ratio for operating point at 98% of choke mass flow. Reconstructed PDF (solid line), mean with centered $\pm\sigma$ (point with UQ bar)

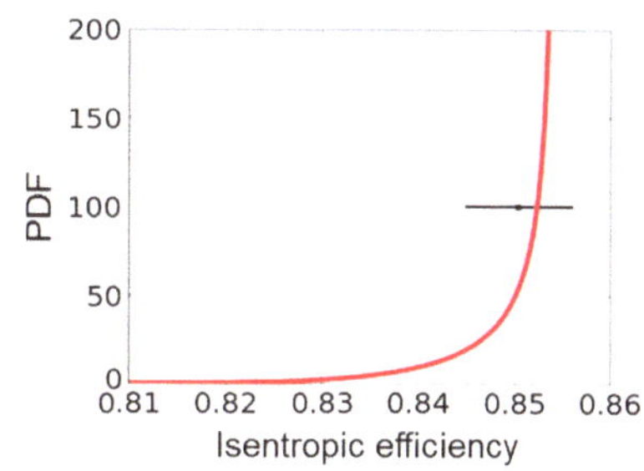

Fig. 11 Reconstructed PDF for efficiency for operating point at 98% of choke mass flow. Reconstructed PDF (solid line), mean with centered $\pm\sigma$ (point with UQ bar)

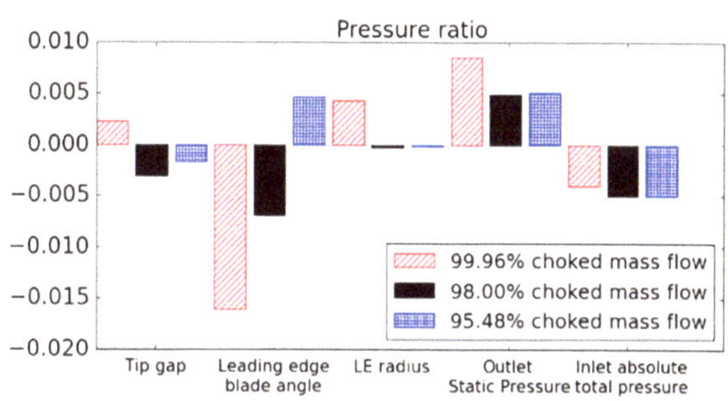

Fig. 12 Scaled sensitivities for total pressure ratio at three different operating points

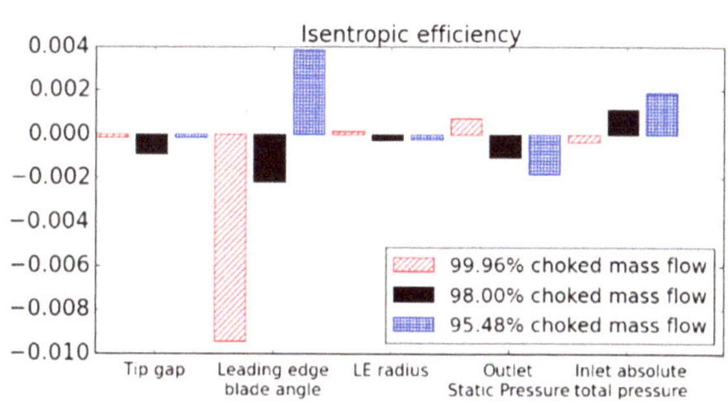

Fig. 13 Scaled sensitivities for efficiency at three different operating points

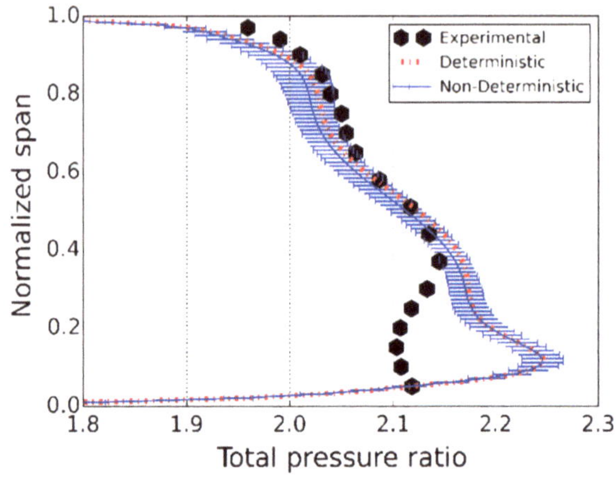

Fig. 14 Non-deterministic absolute total pressure ratio plotted over normalized span at a downstream position of the rotor blade in comparison with deterministic results and experimental data. UQ bars are $\pm\sigma$

operating points, it is positive for this lowest mass flow operating point. This means that increasing the leading edge angle leads to an increase in mass flow, total pressure ratio or efficiency, while increasing the leading edge angle for the other two operating points leads to a decrease in mass flow, total pressure ratio or efficiency.

Pitch-wise averaged quantities at 98% choke mass flow.

In Figs. 14 and 15, the pitch-wise averaged total pressure ratio and efficiency are evaluated at station 4 of the experimental setup of the rotor 37 and compared with the deterministic prediction and experimental data given in [1]. It is seen that the deterministic value differs from the non-deterministic mean. It should be noted that in these plots the mean and standard deviation only are plotted, where the standard

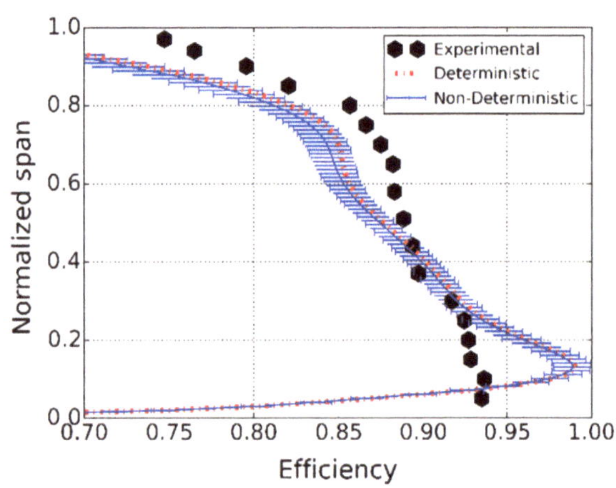

Fig. 15 Non-deterministic isentropic efficiency plotted over normalized span at a downstream position of the rotor blade in comparison with deterministic results and experimental data. UQ bars are $\pm\sigma$

Table 8 Higher output moments for mass flow at the operating point at 98% of the choke mass flow comparing dependence on sparse grid level

Level	Mean	Variance	Skewness	Kurtosis
1	20.437	0.055	0.0113	0.0061
2	20.444	0.060	−0.00567	0.0094
3	20.440	0.062	−0.0055	0.0127
Ratio	Mean	Variance	Skewness	Kurtosis
L2/L1 (%)	100.03	108.91	−50.62	153.96
L3/L2 (%)	99.98	103.30	96.29	135.82

Table 9 Parameters for the covariance models used

Name	Cov. length (m) (=l)	Variance (=σ^2)	Accuracy (=$1-\varepsilon$)	Nb modes (=N)	Nbcolloc points	
					SG lvl 1	SGlvl 2
Model A	0.05	1E−8	0.99	5	11	71
Model B	0.05	1E−6	0.99	5	11	71
Model C	0.025	1E−8	0.99	12	25	337

distribution is centered on the mean value. As seen from the previous section, the bars should be drawn such that they account for asymmetry of the distributions.

Influence of sparse grid level.

Table 8 shows on the example of the operating point at 98% of the choke mass flow that the mean and variances of the mass flow converge with a level 1 sparse grid for this test case. For higher levels, the question of output moment convergences arises. For this operating point, Table 9 lists the higher output moments of the mass flow in function of the sparse grid level and Fig. 16 shows the projection on the Pearson diagram, while Fig. 17 shows the reconstructed PDFs in function of the sparse grid level.

A level 1 sparse grid is sufficient if only mean and variance are of interest. However, it is seen in Fig. 14 that a level 1 sparse grid is not sufficient to reconstruct correctly the output PDFs and higher levels are needed. With a level 3, the higher output moments still vary but the shape of the PDF is already reasonably well represented. A level 2 sparse grid leads to an error in the variance which is three times smaller than the initial approximation error by the parametric model for the variance. Comparing the reconstructed PDFs, a level 2 sparse grid could thus be considered as sufficient for most UQ studies, except when the exact shape of the output PDF must be determined with high precision, such at the tail of the distribution.

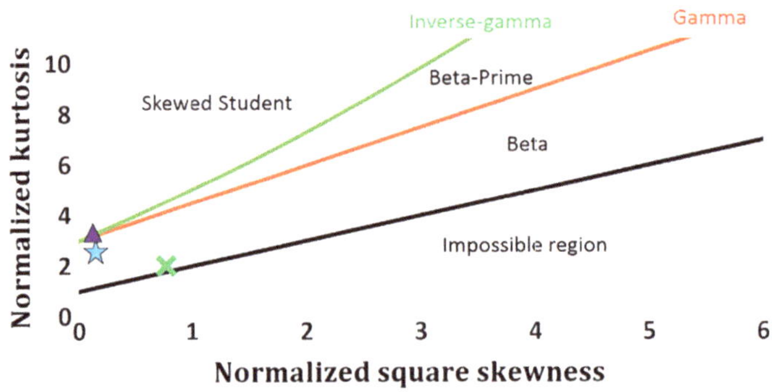

Fig. 16 Pearson diagram, showing the normalized third (squared skewness) and fourth (kurtosis) moment for the operating point at 98% of the choke mass flow. Green cross: level 1, blue star: level 2, purple triangle: level 3

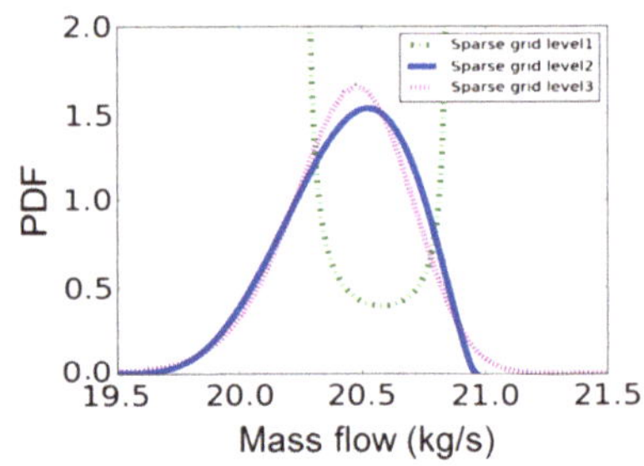

Fig. 17 Reconstructed PDF for mass flow for operating point at 98% of choke mass flow. Reconstructed PDF in function of sparse grid level: level 1 (green dashed-point line), level 2 (solid blue line), level 3 (purple dashed line)

Manufacturing Uncertainties as Random Fields of the Blade Surface

In the following, the compressor blade surface points are considered as a random field. The covariance matrix between these points, which is needed to compute the PCA and therefore propagate the uncertainties with the NIPColM, is determined based on a square exponential covariance function (Eq. (14)).

$$Cov(d) = \sigma^2 * exp\left(- \frac{d^2}{2l^2} \right) \tag{14}$$

- where d is the distance between two points;
- $Cov(d)$ is the covariance function depending on a distance d;
- σ^2 is the variance of the covariance function;

- l is the covariance length.

The two parameters used in this function (σ^2 and l) can be linked to physical properties of the random field: The variance σ^2 is directly proportional to the maximal amplitude of deformation induced by the random field. A high value of this parameter corresponds to a high deformation, and its variation has no influence on the shape of the deformation. The covariance length l represents the distance in-between two geometrical points for which a non-negligible correlation exists. Changing the covariance length changes the shape of the covariance function. If the covariance length is zero, the uncertainties are considered as independent and thus very wavy shapes are possible and if the covariance length is infinite, the resulting uncertainty is represented with 1 mode corresponding to a translation applied on all the geometrical points.

Three different models are used for the comparison and the model parameters are given in Table 9. The influence of the variance and covariance length is shortly shown in the following. The accuracy of the reconstruction, which is used to truncate the expansion of the PCA, is kept at 99%, which was proven to be sufficient [7]. The reader may refer to [7] for more details.

Influence of Variance.

Comparison of models A and B, where model B varies the variance by a factor 100, shows on the example of the first 30 eigenvalues that their cumulative contribution is identical, as seen in Fig. 18. This means that the order of the modes is not influenced by the variance magnitude, but it scales by a factor 100.

Influence of the Covariance Length.

Comparison of models A and C, where model C varies the covariance length by a factor 2 shows at the example of the first 20 eigenvalues that the cumulative value

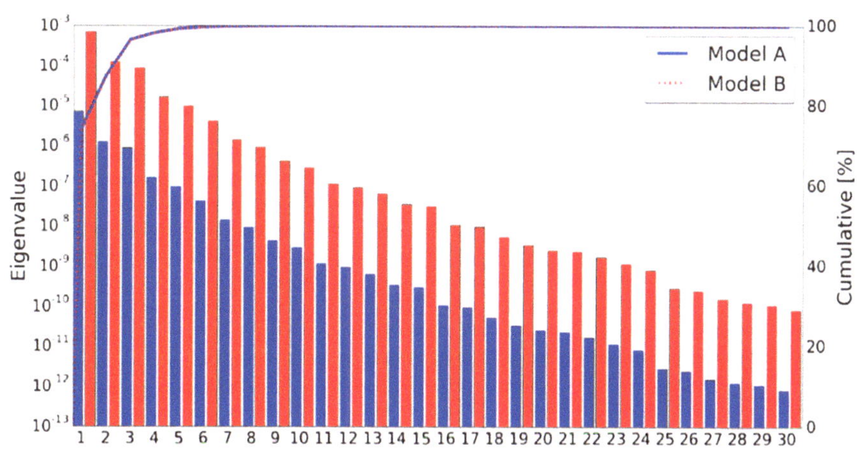

Fig. 18 Order, magnitude, and cumulative value of eigenvalues comparing model A and B

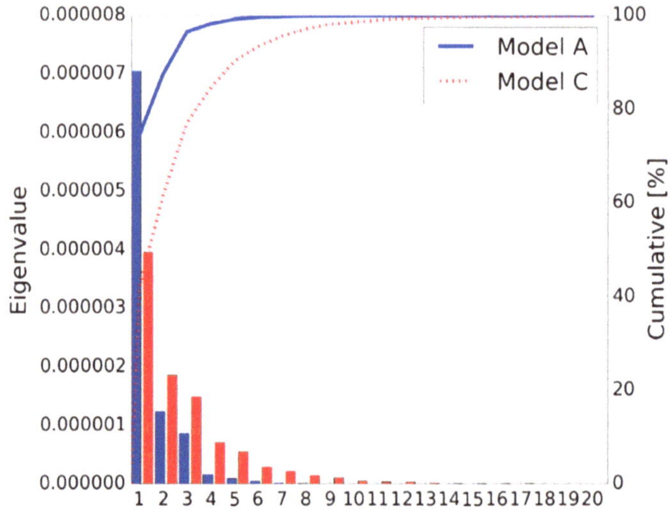

Fig. 19 Order, magnitude, and cumulative value of eigenvalues comparing model A and C

varies with the number of modes, when comparing model A and C as seen in Fig. 19. This is inherently different from the influence of the variance. With an increasing covariance length more eigenvalues are needed to reach the same reconstruction accuracy as in the reference case of model A.

Table 10 shows the first four moments for the pressure ratio and the isentropic efficiency. The mean value for the pressure ratio and the isentropic efficiency does not change by more than 0.03%. The response surface is quasi-linear, given the maximal deformation imposed. For both models, the reduced skewness is close to zero and the reduced kurtosis is close to 3. This indicates that the resulting PDFs are close to Gaussian distributions.

The shape of the PDF can also be deduced from the linearity of the response surface. Indeed, as the input PDF is a Gaussian distribution and that the response surface is linear, it can be concluded that the output PDF is close to a Gaussian distribution. This information is important as it can allow to limit the UQ study to a sparse grid level 1 instead of a sparse grid level 2, which as seen above results in a decrease in the computational cost.

Figure 20 shows the reconstructed PDF for the pressure ratio and the isentropic efficiency for models A and C. The PDF reconstruction shows that the model with a smaller covariance length leads to a wider PDF. This is due to the fact that the cumulative value of the eigenvalues grows slower for model C, i.e., the most influential eigenmodes for model C have higher eigenvalues compared with model A (except for mode 1), and thus larger deformations than the most influential modes for the model A.

Table 10 First four moments for pressure ratio for models A and C

Cov. length (m)	Mean	Variance	Reduced skewness	Reduced kurtosis
0.05	2.0860	1.0180E−05	0.0854	2.8336
0.025	2.0857	1.7566E−05	0.0420	2.9447
Ratios 0.05/0.025 (%)	1.0002	57.953	203.441	96.227

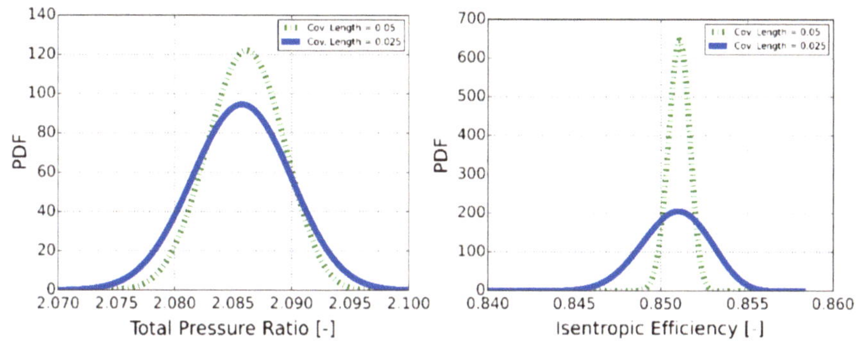

Fig. 20 PDF reconstruction for pressure ratio (left) and isentropic efficiency (right) for models A and C

Influence of the Mode Selection for the UQ Study.

As shown in Fig. 21, the most influential modes on the geometry are not necessarily the most influential modes on different output quantities. This gives particular importance to the percentage of surface reconstruction used, and reconstruction accuracies 95, 99, and 99.9% are compared.

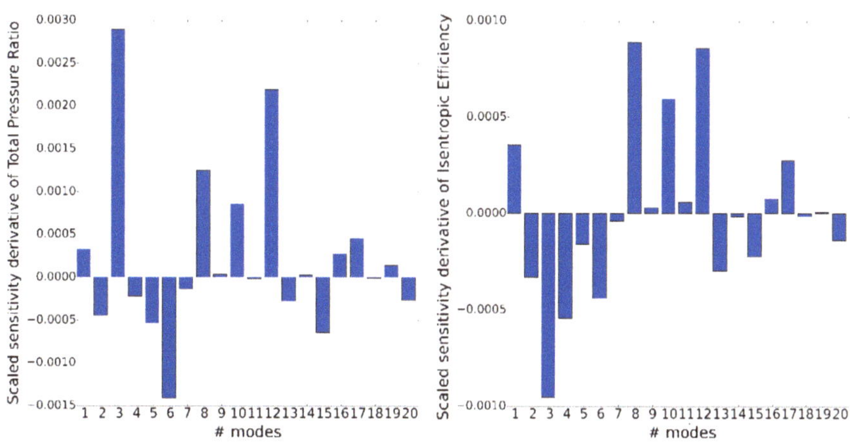

Fig. 21 Scaled sensitivity derivatives for the 20 first modes (corresponds to 99.9% of the geometry variation) for the total pressure ratio (left) and the isentropic efficiency (right)

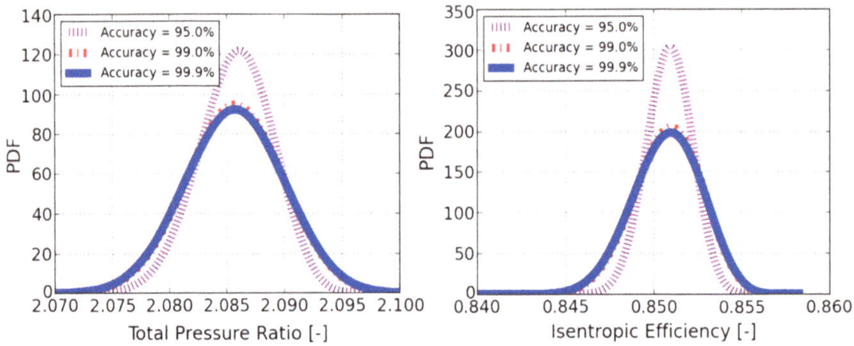

Fig. 22 Reconstructed PDFs 95, 99, and 99.9% of the geometry variation kept for the total pressure ratio (left) and isentropic efficiency (right)

Figure 22 shows the reconstruction of the PDFs accounting for 95, 99, and 99.9% of surface reconstruction accuracy for total pressure ratio and isentropic efficiency, and it is clearly seen that 95% of surface accuracy is not sufficient and 99% must be recommended based on this work.

Conclusions

An industry-ready uncertainty quantification tool chain is developed and successfully applied to both simultaneous operational and geometrical uncertainties and to uncertainties resulting from manufacturing variability, where manufacturing uncertainties are characterized by correlations of the manufactured surfaces. The non-intrusive probabilistic collocation method is combined with a sparse grid approach to drastically reduce the computational cost. If the mean and variance of output quantities are sought, 10 simultaneous uncertainties can be propagated by running 21 runs of a 3D CFD code making use of the sparse grid technique, whereas without sparse grid technique 59049 CFD runs would be needed. The sparse grid technique is thus one of the key features that make UQ in industrial applications feasible. A second required element is the automatization of the entire simulation chain, from uncertainty definition, simulation setup, post-processing and in case of geometrical uncertainties, geometry modification and re-meshing. This process is fully automated including the post-processing of the UQ simulations, which consists of output PDF reconstruction and the calculation of scaled sensitivity derivatives, very useful for the understanding and analysis of the simulations or dimension reduction of the problem.

This tool chain is applied to the rotor 37, demonstrating its capability of handling many simultaneous operational and geometrical uncertainties in turnaround times significantly below the UMRIDA quantitative objective of less than 10 h for 10 simultaneous uncertainties on 100 cores. It is shown that a level 1 sparse grid is sufficient if the mean and variance of output quantities are needed and a level 2 sparse grid is sufficient for the reconstructed PDF shape for most engineering applications. Manufacturing uncertainties are treated in two ways, first, where the manufactured surface is considered as a random field and the covariance matrix of the random field introduces the correlation of the individual surface points and second, as correlated engineering parameters. In both cases, a principal component analysis is used to transform the uncertainties into an uncorrelated set of random variables. Both studies show that the variation of manufacturing uncertainties is rather small due to the imposed manufacturing tolerances and output PDFs are generally close to Gaussian. A level 1 sparse grid can be used for the propagation of manufacturing uncertainties. It is shown that the eigenmodes which have the biggest influence on the surface reconstruction are not necessarily the modes with the biggest influence on the prediction of output quantities and that the most influential modes vary between quantities as shown at the example of isentropic efficiency and total pressure ratio. As a conclusion, a surface reconstruction accuracy of 99% seems necessary for the purpose of UQ studies on manufacturing variability.

References

1. Dunham, J.: CFD Validation For Propulsion System Components, AGARD-AR-355 (1998)
2. Golub, G.H., Welsch, J.H.: Calculation of Gaussian quadrature rules. Math. Comput. **23**, 221–230 (1969)
3. Loève, M.: Probability theory I. In: Graduate Texts in Mathematics, vol. 45, 4th edn. Springer, New York-Heidelberg (1977)
4. Loeven, G.J.A., Witteveen, J.A.S., Bijl, H.: Probabilistic collocation: an efficient non-intrusive approach for arbitrarily distributed parametric uncertainties. In: Proceedings of 45th AIAA Aerospace Sciences Meeting and Exhibit, AIAA paper 2007-317, Reno, United States (2007)
5. Mathelin, L., Hussaini, M.Y.: A Stochastic Collocation Algorithm for Uncertainty Analysis (2003)
6. Nataf, A.: Détermination des distributions de probabilités dont les marges sont données. Comptesrendus de l'académie des sciences **225**, 42–43 (1962)
7. Nigro, R., Wunsch, D., Coussement, G., Hirsch, C.: Chapter 14—Uncertainty quantification in internal flows, STO_TR_AVT_191 (2016)
8. Numeca: User Manual FINE™/Turbo v11.1 (2016)
9. Pearson, K.: Philosophical Transactions of the royal Society of London. 8 laws variant (1895, 1901, 1915)
10. Putko, M.M., Newman, P.A., Taylor III, A.C., Green, L.L.: Approach for uncertainty propagation and robust design in CFD using sensitivity derivatives. Tech. Rep. 2528, AIAA
11. Smolyak, S.: Quadrature and interpolation formulas for tensor products of certain classes of functions. Dokl. Adad. Nauk USSR B, 240–243 (1963)
12. Turgeon, E., Pelletier, D., Borggaard, J.: Applications of continuous sensitivity equations to flows with temperature-dependent properties. Numer. Heat Transf. **44**, 611–624 (2003)
13. Wunsch, D., Nigro, R., Coussement, G., Hirsch C.: Quantification of combined operational and geometrical uncertainties in turbo-machinery design. In: Proceedings of the ASME GT2015, GT2015-43399 (2015)

Non-intrusive Uncertainty Quantification by Combination of Reduced Basis Method and Regression-based Polynomial Chaos Expansion

Mehrdad Raisee, Dinesh Kumar and Chris Lacor

Introduction

Uncertainty in the design and operation of engineering problems may arise from various sources. The uncertainties in physical properties of materials and inevitable randomness in boundary conditions and geometries, as well as physical models uncertainties, are a few examples of such uncertainties that can significantly restrict the reliability of deterministic designs. Gas, steam, wind, and hydraulic turbines are examples of engineering devices that their operational condition and geometry might be uncertain. Design of these turbomachines using deterministic computations may fail in the presence of uncertainties. For a reliable design based on computational fluid dynamics (CFD) predictions, it is necessary to include all sources of uncertainty in the analysis and design process. However, CFD simulation of flows in real-world engineering problems requires a fine 3D computational mesh, small time-step, and high-dimensional stochastic space in the case of a large number of random variables. These dramatically increase the computational cost which is not desirable for design proposes, highlighting the need for employing robust numerical schemes for stochastic analysis of complex industrial flows. While efficient numerical methods for the spatial and temporal discretization of the Navier–Stokes equations are well

M. Raisee (✉)
School of Mechanical Engineering, College of Engineering,
University of Tehran, P. O. Box: 11155-4563, Tehran, Iran
e-mail: mraisee@ut.ac.ir

D. Kumar · C. Lacor
Department of Mechanical Engineering, Vrije Universiteit Brussel, Pleinlaan 2,
1050 Brussels, Belgium
e-mail: dkumar@vub.ac.be

C. Lacor
e-mail: clacor@vub.ac.be

C. Hirsch et al. (eds.), *Uncertainty Management for Robust Industrial Design in Aeronautics*, Notes on Numerical Fluid Mechanics and Multidisciplinary Design 140, https://doi.org/10.1007/978-3-319-77767-2_10

developed, effective numerical schemes for stochastic discretization are still rare (see, e.g., [1, 2]).

In the literature, various techniques have been proposed for uncertainty quantification (UQ). The Monte Carlo (MC) approach [3] is widely used for UQ because of its simplicity and its superior property that the convergence rate does not depend on the number of stochastic dimensions. Unfortunately, the conventional MC methods converge slowly and often require a large number of realizations to achieve reasonable accuracy and thus are impractical for problems with a large number of uncertainties. Over the recent years, a number of other UQ approaches have been developed to represent and propagate uncertainties in engineering problems with a large number of uncertainties. Some examples of commonly developed UQ methods are: the multi-level Monte Carlo [4], the method of moments or the perturbation method [5], and polynomial chaos expansion (PCE) [6, 7]. All these techniques have positive and negative features, and no single technique is optimal for all applications. Following our previous work on UQ [8, 9], we focused on the PCE approach to model uncertainty propagation. PCE methods have been successfully applied to various structural and solid mechanics problems by several researchers [6, 10]. Polynomial chaos (PC) schemes have also been employed to fluid flow and heat transfer problems [7, 8, 11–13]. The polynomial chaos representation can be implemented through either *intrusive* or *non-intrusive* methods. The intrusive approach involves the substitution of all uncertain variables in the governing equations with the polynomial expansions consisting of $P + 1 = (p + n_s)!/p!n_s!$ unknown coefficients, where n_s is the number of stochastic dimensions and p is the polynomial order. Taking the inner product of the equations yields $P + 1$ times the number of original equations that can be solved by the same numerical schemes applied to the original deterministic system. This requires the modification of the CFD codes, and it may be difficult, expensive, and time-consuming for many CFD problems. Moreover, the sources of most commercial codes are not accessible, and thus, it is not feasible to implement the intrusive PC approach in such deterministic codes. For these reasons, here, we focused on non-intrusive polynomial chaos (NIPC) methodology for UQ. The NIPC method performs repeated simulations using deterministic solver on limited number of samples which are chosen properly. Then, the polynomial chaos expansion of output is constructed using deterministic solver evaluations. The two main NIPC approaches used for UQ in CFD are spectral projection (sampling-based and quadrature-based) and regression-based methods. The application of these NIPC schemes to model stochastic problems can be found in [14, 15]. In the present study, the regression-based NIPC scheme, introduced in section "Regression-Based Polynomial Chaos Expansion," is used for the evaluation of PCE coefficients. The main weakness of all NIPC methods is the *curse of dimensionality*. In recent years, some alternative methodologies such as sparse polynomial chaos [16], sparse grid techniques [17], compressive sampling [18], and reduced models [1, 2] have been developed to break the curse of dimensionality. In the framework of the EU FP7 project UMRIDA, this study focuses on the development of an efficient reduced basis model for UQ. In recent years, several model reduction techniques have been proposed for uncertainty quantification. Two examples of such works are [1, 2]. In [2], a generalized spectral decomposition (GSD) was proposed that gives the reduced basis independent of the stochastic discretization scheme. In this method, the solution of the stochastic problem is first approximated as the summation over the product of deterministic functions and random variables. The reduced basis functions then appear as the solutions of a pseudo-eigenvalue problem whose dominant eigenspace is associated with the desired optimal basis. In the final form of GSD, the solution of only a few uncoupled deterministic problems and a few stochastic algebraic equations is required for the computation of deterministic functions and random variables. As shown in [2], the implementation of GSD to a class of stochastic partial differential equations (SPDE) leads to drastic computational saving, although it does not circumvent the curse of dimensionality. In [1], an intrusive model reduction technique was proposed for chaos representation of a SPDE to tackle the curse of dimensionality. They applied it successfully to a 2D solid mechanics problem with randomness in the elastic modulus where for a third-order PC ($p = 3$), they could reduce the number of basis functions to 5 as compared to $P = 165$ in the "standard PCE" using a basis of the classical polynomials of the Askey scheme.

In this study, a regression-based non-intrusive reduced basis technique is developed. The model can be interpreted as a multi-level/multi-fidelity approach, where many low-fidelity model evaluations are combined with few high-fidelity evaluations to ensure accurate results at a lower CPU cost. In the framework of polynomial chaos, such ideas are also explored by Palar et al. [19] and Ng and Eldred [20].

The remaining part of this paper is organized as follows: In section "Regression-Based Polynomial Chaos Expansion," the regression-based polynomial chaos expansion is described. The model reduction methodology is presented in section "Reduced Basis Methodology." In section "Results and Discussion," the numerical results are presented and discussed. Finally, the main findings of the present paper are summarized in section "Conclusions."

Regression-Based Polynomial Chaos Expansion

Let assume $u(\boldsymbol{x};\boldsymbol{\xi})$ is the response of a stochastic system with n_s random variables $\boldsymbol{\xi} = \{\xi_i\}_{i=1}^{n_s}$. In PCE, the uncertain output $u(\boldsymbol{x};\boldsymbol{\xi})$ is decomposed into separable deterministic and stochastic components as:

$$u(\boldsymbol{x};\boldsymbol{\xi}) = \sum_{i=0}^{P} u_i(\boldsymbol{x})\psi_i(\boldsymbol{\xi}), \tag{1}$$

where the total number of output modes, $P + 1 = (p + n_s)!/p!n_s!$, is a function of the order of polynomial chaos (p) and the number of random dimensions (n_s).

The $\psi_i(\boldsymbol{\xi})$'s are orthogonal polynomials with respect to the probability density function (PDF) of input random variables $\boldsymbol{\xi}$:

$$\langle \psi_i \psi_j \rangle = \langle \psi_i^2 \rangle \delta_{ij}. \tag{2}$$

The quadrature-based NIPC scheme may be used for the evaluation of polynomial chaos expansion. However, the application of tensor–product quadrature approach for multi-dimensional problems suffers the curse of dimensionality since the required number of model evaluations grows exponentially with the number of random dimensions n_s (i.e., $(p+1)^{n_s}$). Although sparse quadrature rules are more efficient, still they are impractical for the stochastic problems with high dimensions. A more affordable NIPC scheme to find the response surface of the output is the *regression method*. The regression-based NIPC method starts with Eq. (1). To establish a closed system, $P+1$ sample points ($\boldsymbol{\xi}^s, s = 1, 2, \ldots, P+1$) are generated in the stochastic space for a given PCE with $P+1$ unknown coefficients and the stochastic function, $u(\boldsymbol{x}; \boldsymbol{\xi})$, is evaluated at these sampling points. This yields the following linear system of equations:

$$\underbrace{\begin{pmatrix} \psi_0(\boldsymbol{\xi}^1) & \cdots & \psi_i(\boldsymbol{\xi}^1) & \cdots & \psi_P(\boldsymbol{\xi}^1) \\ \vdots & \vdots & \vdots & \vdots & \vdots \\ \psi_0(\boldsymbol{\xi}^s) & \cdots & \psi_i(\boldsymbol{\xi}^s) & \cdots & \psi_P(\boldsymbol{\xi}^s) \\ \vdots & \vdots & \vdots & \vdots & \vdots \\ \psi_0(\boldsymbol{\xi}^{P+1}) & \cdots & \psi_i(\boldsymbol{\xi}^{P+1}) & \cdots & \psi_P(\boldsymbol{\xi}^{P+1}) \end{pmatrix}}_{\Psi(\boldsymbol{\xi}^s)} \underbrace{\begin{pmatrix} u_0(\boldsymbol{x}) \\ \vdots \\ u_i(\boldsymbol{x}) \\ \vdots \\ u_P(\boldsymbol{x}) \end{pmatrix}}_{U} = \underbrace{\begin{pmatrix} u(\boldsymbol{x}; \boldsymbol{\xi}^1) \\ \vdots \\ u(\boldsymbol{x}; \boldsymbol{\xi}^s) \\ \vdots \\ u(\boldsymbol{x}; \boldsymbol{\xi}^{P+1}) \end{pmatrix}}_{b}, \tag{3}$$

or

$$\Psi U = b. \tag{4}$$

The least squares solution of the linear system (3) is $U = (\Psi^T \Psi)^{-1} \Psi^T b$.

Consistent with the literature (e.g., Hosder et al. [21]), we found that oversampling with $2(P+1)$ model evaluations is necessary to obtain satisfactory results for the PCE. In principle, the sample points can be chosen freely. However, while random sampling is the simplest, its major disadvantage is that the sample points may not be space filling. This will have a repercussion on the accuracy of the results. An alternative to the random sampling technique is the Latin hypercube sampling (LHS) which offers better space filling characteristics. The basic idea is to divide the range of each random variable into n bins of equal probability and then to generate N samples such that, for each random variable, no two values should lie in the same bin. However, LHS suffers from a major difficulty. Indeed, the accuracy of the LHS-based estimates cannot be increased incrementally, i.e., by adding new sample points to the already existing LHS sample set, since the new set will not be a Latin hypercube anymore. An efficient method to build adaptive space filling design is the quasi-random sampling (e.g., Hammersley, Halton, Sobol). In quasi-random sequences, a deterministic sequence of points is generated. The main idea of using a quasi-random sequence is to reduce the discrepancy of the sets of points. In the present work, the coefficients of the PCEs are estimated by the regression-based NIPC, using the Sobol sampling scheme [22].

Due to the orthogonality of the basis, it is straightforward to show that the mean is $\langle u(\boldsymbol{x};\boldsymbol{\xi})\rangle = u_0$, and the variance of the response reads as:

$$\sigma^2 = Var\left(\sum_{i=0}^{P} u_i(\boldsymbol{x})\psi_i(\boldsymbol{\xi})\right) = \sum_{i=1}^{P} u_i^2\langle\psi_i\psi_i\rangle. \tag{5}$$

Reduced Basis Methodology

The above classical PCE (i.e., Eq. (1)) does not represent an optimal PC representation of $u(\boldsymbol{x};\boldsymbol{\xi})$. The optimal chaos expansion is the Karhunen–Loève (KL) expansion (also known as proper orthogonal decomposition (POD)). However, this requires the knowledge of the covariance of the solution, which is unknown. Assuming that the behavior in spatial and random space can be decoupled, the covariance can be obtained via inexpensive calculations on a coarse grid. The size of coarse grid, necessary for the estimation of the optimal basis, can be identified through mesh adaptation in the spatial domain of the problem. Next, the problem can be solved on a fine mesh using the previously defined optimal basis $\{z_i(\boldsymbol{\xi})\}_{i=0}^{m}$ where m is the number of dominant eigenvalues. This indicates that the dimensionality of the KL expansion can be reduced.

The first step in the model reduction scheme is to find an optimal basis using POD, a well-known procedure for extracting basis functions using an ensemble of realizations. To this end, suppose, on a fine grid, expression (6) represents an optimal chaos expansion of the stochastic field $u(\boldsymbol{x};\boldsymbol{\xi})$:

$$u(\boldsymbol{x};\boldsymbol{\xi}) - \langle u(\boldsymbol{x};\boldsymbol{\xi})\rangle = \sum_{i=1}^{m} \hat{u}_i(\boldsymbol{x}) z_i(\boldsymbol{\xi}), \tag{6}$$

where the mean function is the coefficient of the zeroth-order basis (i.e., $\langle u(\boldsymbol{x};\boldsymbol{\xi})\rangle = \hat{u}_0$) and $\{z_i(\boldsymbol{\xi})\}_{i=0}^{m}$ are the $m+1$ dominant modes, forming the optimal basis.

On the coarse grid, the covariance matrix $C(\boldsymbol{x}_i, \boldsymbol{x}_j)$ of the stochastic field can be obtained from:

$$C(\boldsymbol{x}_i, \boldsymbol{x}_j) = \sum_{k=1}^{P} u_k(\boldsymbol{x}_i) u_k(\boldsymbol{x}_j)\langle\psi_k^2\rangle, \tag{7}$$

where u_k's are the classical PCE coefficients obtained using Eq. (3) on the coarse grid.

The corresponding eigenvalues v_i and eigenfunctions $\phi_i(\boldsymbol{x})$ are the solution of the following eigenvalue problem:

$$C\phi_i = v_i\phi_i. \tag{8}$$

For a coarse mesh with n grid nodes the $n \times n$ covariance matrix has the following form:

$$C = \begin{pmatrix} \sum_{k=1}^{P} u_k^2(\mathbf{x}_1)\langle\psi_k^2\rangle & \cdots & \sum_{k=1}^{P} u_k(\mathbf{x}_1)u_k(\mathbf{x}_n)\langle\psi_k^2\rangle \\ \vdots & \vdots & \vdots \\ \sum_{k=1}^{P} u_k(\mathbf{x}_i)u_k(\mathbf{x}_1)\langle\psi_k^2\rangle & \cdots & \sum_{k=1}^{P} u_k(\mathbf{x}_i)u_k(\mathbf{x}_n)\langle\psi_k^2\rangle \\ \vdots & \vdots & \vdots \\ \sum_{k=1}^{P} u_k(\mathbf{x}_n)u_k(\mathbf{x}_1)\langle\psi_k^2\rangle & \cdots & \sum_{k=1}^{P} u_k^2(\mathbf{x}_n)\langle\psi_k^2\rangle \end{pmatrix}. \tag{9}$$

For a large value of $n \gg P$, the solution of the above eigenvalue problem is time-consuming and requires a large amount of memory for the data storage. To overcome this problem, one can notice that the covariance matrix C is symmetric and thus can be decomposed as:

$$C = \underbrace{\begin{pmatrix} u_1(\mathbf{x}_1)\sqrt{\langle\psi_1^2\rangle} & \cdots & u_P(\mathbf{x}_1)\sqrt{\langle\psi_P^2\rangle} \\ \vdots & \vdots & \vdots \\ u_1(\mathbf{x}_i)\sqrt{\langle\psi_1^2\rangle} & \cdots & u_P(\mathbf{x}_i)\sqrt{\langle\psi_P^2\rangle} \\ \vdots & \vdots & \vdots \\ u_1(\mathbf{x}_n)\sqrt{\langle\psi_1^2\rangle} & \cdots & u_P(\mathbf{x}_n)\sqrt{\langle\psi_P^2\rangle} \end{pmatrix}}_{Y(n\times P)}$$

$$\otimes \underbrace{\begin{pmatrix} u_1(\mathbf{x}_1)\sqrt{\langle\psi_1^2\rangle} & \cdots & u_1(\mathbf{x}_i)\sqrt{\langle\psi_1^2\rangle} & \cdots & u_1(\mathbf{x}_n)\sqrt{\langle\psi_1^2\rangle} \\ \vdots & \vdots & \vdots & \vdots & \vdots \\ u_P(\mathbf{x}_1)\sqrt{\langle\psi_P^2\rangle} & \cdots & u_P(\mathbf{x}_i)\sqrt{\langle\psi_P^2\rangle} & \cdots & u_P(\mathbf{x}_n)\sqrt{\langle\psi_P^2\rangle} \end{pmatrix}}_{Y^T(P\times n)}, \tag{10}$$

where the $P \times n$ matrix Y^T is the transpose of the $n \times P$ matrix Y. Substitution of the above decomposition in Eq. (8) and multiplication by Y^T yields:

$$Y^T Y(Y^T \phi_i) = \nu_i(Y^T \phi_i), \tag{11}$$

This indicates that $Y^T Y$ has eigenfunctions $Y^T \phi_i$ and the same eigenvalues as C. However, $Y^T Y$ is only a $P \times P$ matrix, and thus, it is less expensive to find the eigenvalues and corresponding eigenfunctions than from the original covariance matrix C. This makes the size of the eigenvalue problem independent of the coarse grid size. By computing the eigenvalues from Eq. (11), the upper limit m in Eq. (6) can be found by the size of the dominant eigenspace (11) such that $\sum_{i=1}^{m} \nu_i / \sum_i \nu_i$ is sufficiently close to unity. In this work, the upper limit m is chosen to be the minimum integer such that $\sum_{i=1}^{m} \nu_i / \sum_i \nu_i \geq \varepsilon$ for a given ε (for instance $\varepsilon = 0.99$).

Having obtained $u_i(\mathbf{x})$ from the regression-based NIPC (Eq. 3) on the coarse grid and eigenfunctions $\phi_i(\mathbf{x})$ from the solution of the eigenvalue problem (11), the set of optimal basis functions $\{z_i(\boldsymbol{\xi})\}_{i=0}^{m}$ can now be recasted as a linear combination of the set of classical polynomial chaos functions; $\{\psi_i(\boldsymbol{\xi})\}_{i=1}^{P}$ using the following scalar product:

$$z_i(\boldsymbol{\xi}) = [u(\boldsymbol{x};\boldsymbol{\xi}) - \langle u(\boldsymbol{x})\rangle, \phi_i(\boldsymbol{x})] = \sum_{j=1}^{P} \alpha_{ij}\psi_j(\boldsymbol{\xi}), \tag{12}$$

where the coefficients α_{ij} are obtained via the scalar product:

$$\alpha_{ij} = \int_R u_j(\boldsymbol{x})\phi_i(\boldsymbol{x})d\vec{\boldsymbol{x}}. \tag{13}$$

The $m+1$ unknowns $\hat{u}_i$'s in the optimal expansion can be obtained by substitution of $m+1$ random vectors ($\boldsymbol{\xi}^s, s = 1,\dots,m+1$) and the corresponding stochastic outputs $u(\boldsymbol{x};\boldsymbol{\xi}^s)$ in Eq. (6). This yields the following linear system of equations:

$$\underbrace{\begin{pmatrix} z_0(\boldsymbol{\xi}^1) & \cdots & z_i(\boldsymbol{\xi}^1) & \cdots & z_m(\boldsymbol{\xi}^1) \\ \vdots & \vdots & \vdots & \vdots & \vdots \\ z_0(\boldsymbol{\xi}^s) & \cdots & z_i(\boldsymbol{\xi}^s) & \cdots & z_m(\boldsymbol{\xi}^s) \\ \vdots & \vdots & \vdots & \vdots & \vdots \\ z_0(\boldsymbol{\xi}^{m+1}) & \cdots & z_i(\boldsymbol{\xi}^{m+1}) & \cdots & z_m(\boldsymbol{\xi}^{m+1}) \end{pmatrix}}_{Z(\boldsymbol{\xi}^s)} \begin{pmatrix} \hat{u}_0(\boldsymbol{x}) \\ \vdots \\ \hat{u}_i(\boldsymbol{x}) \\ \vdots \\ \hat{u}_m(\boldsymbol{x}) \end{pmatrix} = \begin{pmatrix} u(\boldsymbol{x};\boldsymbol{\xi}^1) \\ \vdots \\ u(\boldsymbol{x};\boldsymbol{\xi}^s) \\ \vdots \\ u(\boldsymbol{x};\boldsymbol{\xi}^{m+1}) \end{pmatrix}. \tag{14}$$

Using Eqs. (12), (14) can be re-expressed as:

$$\underbrace{\begin{pmatrix} z_0(\boldsymbol{\xi}^1) & \cdots & \sum_{j=1}^{P} \alpha_{ij}\psi_j(\boldsymbol{\xi}^1) & \cdots & \sum_{j=1}^{P} \alpha_{mj}\psi_j(\boldsymbol{\xi}^1) \\ \vdots & \vdots & \vdots & \vdots & \vdots \\ z_0(\boldsymbol{\xi}^s) & \cdots & \sum_{j=1}^{P} \alpha_{ij}\psi_j(\boldsymbol{\xi}^s) & \cdots & \sum_{j=1}^{P} \alpha_{mj}\psi_j(\boldsymbol{\xi}^s) \\ \vdots & \vdots & \vdots & \vdots & \vdots \\ z_0(\boldsymbol{\xi}^{m+1}) & \cdots & \sum_{j=1}^{P} \alpha_{ij}\psi_j(\boldsymbol{\xi}^{m+1}) & \cdots & \sum_{j=1}^{P} \alpha_{mj}\psi_j(\boldsymbol{\xi}^{m+1}) \end{pmatrix}}_{Z(\boldsymbol{\xi}^s)} \begin{pmatrix} \hat{u}_0(\boldsymbol{x}) \\ \vdots \\ \hat{u}_i(\boldsymbol{x}) \\ \vdots \\ \hat{u}_m(\boldsymbol{x}) \end{pmatrix} = \begin{pmatrix} u(\boldsymbol{x};\boldsymbol{\xi}^1) \\ \vdots \\ u(\boldsymbol{x};\boldsymbol{\xi}^s) \\ \vdots \\ u(\boldsymbol{x};\boldsymbol{\xi}^{m+1}) \end{pmatrix}. \tag{15}$$

The matrix Z, containing the optimal basis, is already known from Eqs. (3) and (13), and the right-hand side of Eq. (15) can be found from $m+1$ runs of the deterministic solver at $\boldsymbol{\xi}^1,\dots,\boldsymbol{\xi}^s,\dots,\boldsymbol{\xi}^{m+1}$ on the fine mesh. Thus, the expansion coefficients $\hat{u}_i(\boldsymbol{x})$ are obtained by the solution of the above linear system. Here, again oversampling is required. Following the approach used in the regression-based NIPC analysis, $2(m+1)$ sample points were found adequate to give acceptable results. As pointed out, the coefficient of the zeroth-order basis ($z_0(\boldsymbol{\xi})$) is the mean output (i.e., $\langle u(\boldsymbol{x};\boldsymbol{\xi})\rangle = \hat{u}_0$), while the variance is expressed as:

$$\sigma^2 = \sum_{i=1}^{m} \hat{u}_i^2 \langle z_i, z_i \rangle, \tag{16}$$

where $\langle z_i, z_i \rangle = v_i$.

Results and Discussion

In the following subsections, numerical results for three benchmark stochastic problems, namely (I) Ackley function, (II) 2D RAE2822 transonic airfoil, and (III) 3D NASA rotor 37, are presented and discussed.

Highly Irregular Ackley Function

The 2D Ackley function is a challenging test function for the validation of the developed reduced basis methodology due to its complex structural distribution. The stochastic Ackley function is defined as:

$$\begin{aligned} u(\boldsymbol{x}; \boldsymbol{\xi}) = & -20(1 + 0.1\xi_3)\left(\exp\left[-0.2(1 + 0.1\xi_2)\sqrt{0.5(x^2 + y^2)}\right]\right) \\ & - \exp\left(0.5\left[\cos(2\pi(1 + 0.1\xi_1)x) + \cos(2\pi(1 + 0.1\xi_1)y)\right]\right) + 20 + e, \end{aligned} \tag{17}$$

where function coefficients (shown in red in Eq. (17)) are uncertain and the associated random variables $\boldsymbol{\xi} = \{\xi_i\}_{i=1}^3$ are uniformly distributed over $[-1, 1]^3$ with a CoV of 0.0577.

Figure 1 shows the deterministic Ackley function (i.e., $\boldsymbol{\xi} = 0$) on different grids. As expected, the Ackley function is highly irregular in 2D spatial space and is characterized by a nearly flat outer region and a large hole at the center. The mesh refinement from 5×5 to 160×160 reveals more details of the function. It was found that a finer mesh with 400×400 nodes is necessary to reproduce the fine-scale structures of the Ackley function, and thus, such a fine mesh is employed for the fine-scale analysis. Figure 2 shows the distribution of the normalized eigenvalues in the linear and semi-log scales when different grids are used for the coarse-scale analysis. A high polynomial order ($p = 13$) is used for the coarse grid analysis. This is because a regression-based NIPC analysis indicated that such a high polynomial order is necessary to reproduce the details of the mean, variance, and skewness fields. As shown in Fig. 2, the eigenvalues decay rapidly. Thus, only a limited number of modes (or eigenvalues) are needed in the KL expansion. The number of chosen eigenvalues depends on the accuracy of the statistics. For higher accuracy, a larger number of modes should be taken into account. In Fig. 2b, as expected, the normalized eigenvalues distributions decrease slower with the finer grids. Results show that for this

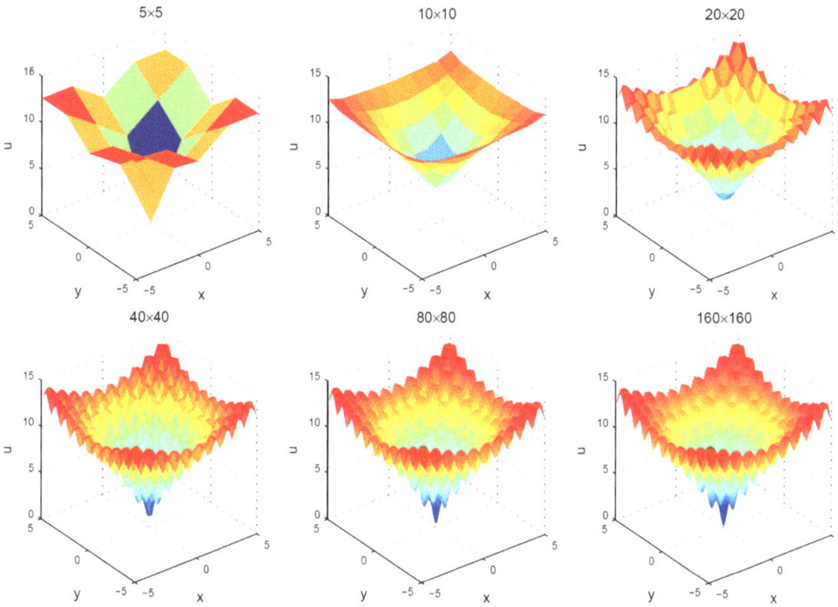

Fig. 1 Representation of Ackley function in different grids

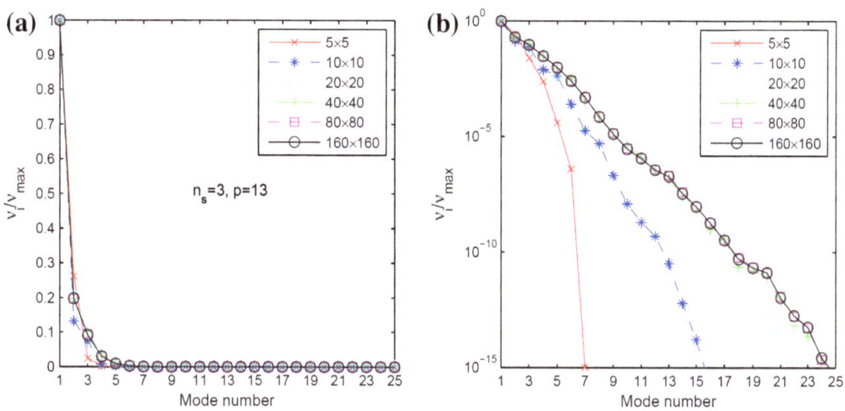

Fig. 2 Normalized eigenvalues using different coarse mesh sizes for the stochastic Ackley function: **a** linear scale; **b** semi-log scale

nonlinear test case, an accurate solution is obtained when a 40×40 mesh is used for the coarse grid analysis. In Fig. 3, the distributions of mean, variance, and skewness fields returned using the reduced basis method are compared with the distributions obtained using regression-based NIPC. It is observed that with a reduced basis of dimension $m + 1 = 15$ (correspond to $\varepsilon = 0.99999999$), the fine-scale results are very close to those of the full NIPC. With reduced basis size $m + 1 = 15$, the

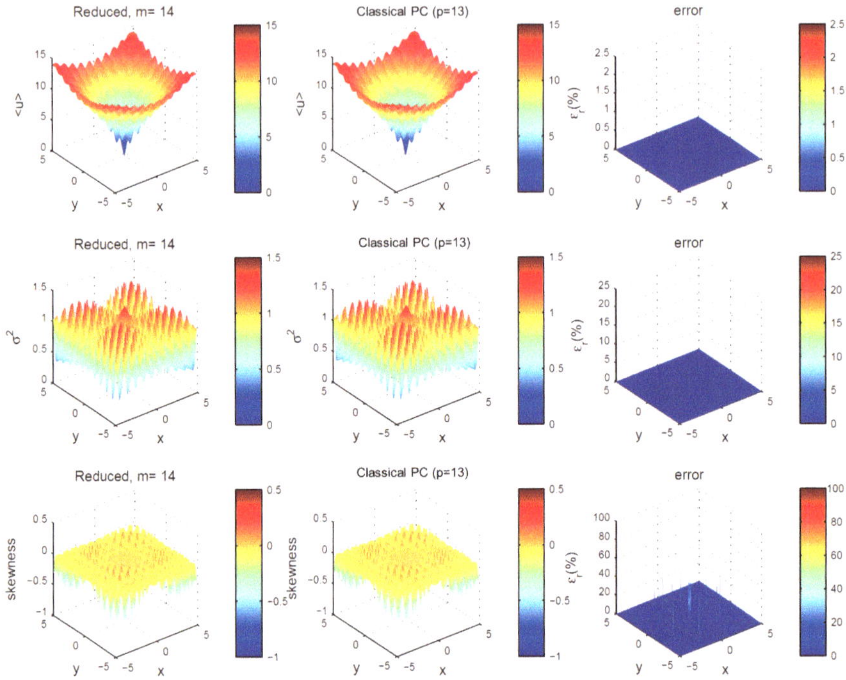

Fig. 3 Comparison of mean, variance, and skewness fields for the Ackley function. First row: mean field; second row: variance field; third row: skewness field

average relative error ($\bar{\varepsilon}_r$) in mean, variance, and skewness is of the order of 10^{-5}, 10^{-3}, and 10^{-2}, respectively. Note, however, that for this case, the full PC analysis needs $2(P+1) = 1120$ expensive function evaluations. Further analysis (not presented here for the sake of brevity) shows the reduced basis methodology is more efficient than the classical PCE by more than one order of magnitude. Further efficiency improvement can be achieved by using a smaller ε (e.g., $\varepsilon = 0.99$) and increasing the allowable relative error in the statistical quantities. More details can be found in [23].

2D Transonic RAE2822 Airfoil

The 2D transonic flow around the RAE2822 airfoil represents a challenging configuration to investigate the performance of the developed reduced-order model due to the shock formation. The nominal flow conditions; free stream Mach number $M = 0.734$, angle of attack $\alpha = 2.79°$, and Reynolds number $Re = 6.5 \times 10^6$ are considered for this test case. For the deterministic solution of the RAE2822 using Ansys/Fluent, the second-order upwind scheme is employed for the approximation of nonlinear convective terms in all transport equations. The Spalart–Allmaras tur-

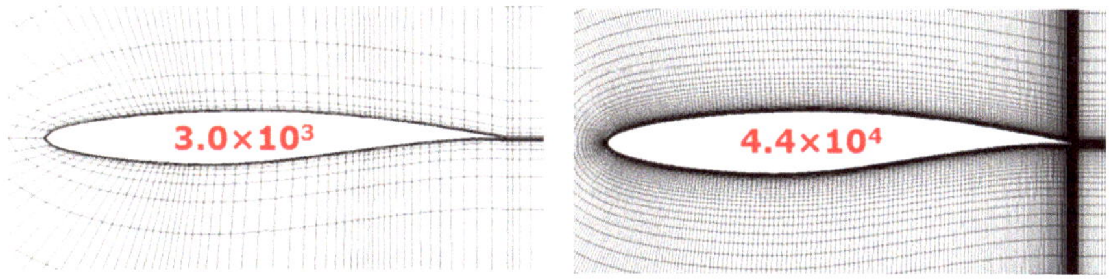

Fig. 4 The coarse and fine C-type meshes with: 3.0×10^3 and 4.4×10^4 grid nodes

bulence model is used for the predictions. To assess the accuracy of the results, a grid study was performed with four different C-type meshes with, respectively, 7.5×10^2, 3.0×10^3, 1.1×10^4, and 4.4×10^4 grid nodes. A coarse mesh with 3.0×10^3 and the finest mesh with 4.4×10^4 grids are shown in Fig. 4. It was found that the predictions on the finest mesh are grid independent and thus are used for the fine-scale analysis. The geometry of the airfoil is assumed to be subject to random deformations, and variations of the airfoil boundary are modeled using the following Gaussian shaped covariance:

$$Cov(s_i, s_j) = \sigma(s_i)\sigma(s_j) \exp\left[-\frac{(s_i - s_j)^2}{2b^2}\right], \tag{18}$$

where s_i and s_j are positions along the airfoil, b is the correlation length, and σ is the variance. For the RAE2822 airfoil, $0 \le s \le 2.032$. Position $s = 0$ corresponds to the leading edge and increases along the upper surface. A constrained standard deviation, $\sigma(s) = \sigma' S(s)$, is considered to freeze the leading and trailing edges of the airfoil. The constraint functions on the upper and lower walls of the airfoil are, respectively, expressed as:

$$S(s) = \begin{cases} \sin\left(\frac{\pi s}{s_u}\right) & 0 \le s < s_u \\ \sin\left[\frac{\pi(s-s_u)}{s_l}\right] & s_u \le s < s_u + s_l \end{cases} \tag{19}$$

where $s_u = \int_{upper} ds$ and $s_l = \int_{lower} ds$.

Using KL expansion, a stochastic process of a given covariance function can always be approximated by a finite sum of products of deterministic spatial functions and uncorrelated random variables. The geometrical uncertainty at the airfoil surface can then be expressed as:

$$\vec{X}(s, \boldsymbol{\xi}) \approx \bar{X}(s) + \sum_{k=1}^{n_s} \sqrt{\lambda_k}\phi_k(s)\xi_k.\hat{n} \tag{20}$$

where $X(s, \boldsymbol{\xi})$ is the airfoil coordinate at sample sample $\boldsymbol{\xi}$, $\bar{X}(s)$ is the airfoil mean coordinate, $\hat{n}$ is a normal vector, and $\phi_k(s)$ and λ_k are eigenvalues and eigenfunctions of the covariance kernel, respectively.

A case with the correlation length $b = 0.05$ and the standard deviation $\sigma' = 0.001$ is considered for the present analysis. The random variables are set to be uniformly distributed over the stochastic space $[-1, 1]^{n_s}$ where n_s is the number of independent random variables. The first ten highest modes of KL expansion are considered as uncertain for the UQ of the RAE2822 test case. The coarse-scale analysis is performed on a mesh with 3.0×10^3 nodes (shown in Fig. 4), a grid size fourteenth times smaller than the finest grid size. A classical PC analysis of third order using regression-based NIPC is performed on the coarse grid to get the covariance in stochastic space of the solution. In this analysis, the covariance matrix is built using all primitive variables (ρ; ρU; ρV; ρE). The criteria of the selection of the coarse grid are based on the analysis presented for the Ackley function. Starting from the POD analysis on a very coarse mesh, the mesh size is gradually increased until sufficient convergence of the POD eigenvalues. This is illustrated in Fig. 5 where the normalized eigenvalues are shown for four different mesh sizes with 7.5×10^2, 3.0×10^3, 1.1×10^4, and 4.4×10^4 grid nodes. It is observed that already on the 3.0×10^3 grid, the eigenvalues have converged. A classical PC analysis of third order using regression is performed on the coarse grid to get the covariance in stochastic space of the solution. In the regression approach, a total of 572 samples are needed as the classical PCE contains 286 polynomials. The Sobol quasi-random sequence is used to generate these sample points. For $\varepsilon = 0.99$, the size of the reduced basis is 22, requiring 44 deterministic CFD calculations on the fine grid. In Fig. 6, the results (pressure coefficient) obtained with the reduced-order model and with the full PC are compared. It is observed that the results of the reduced-order model are in acceptable agreement with the results of the full model. On average, the errors in the mean C_p and its variance are less than 0.2% and 5.0%, respectively. As shown in [24], for

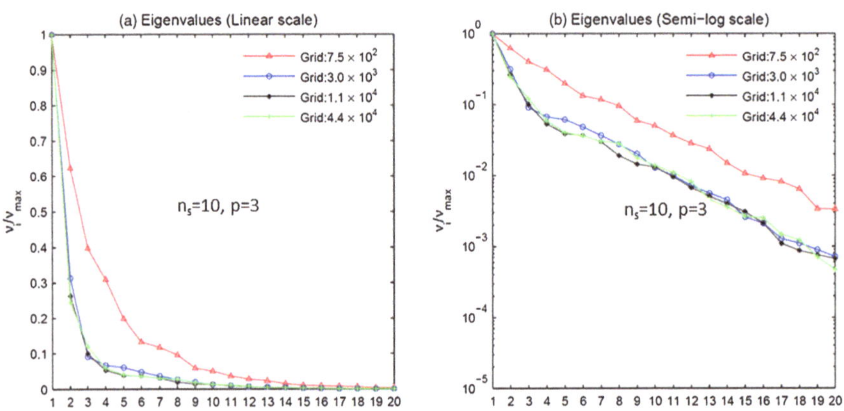

Fig. 5 Normalized eigenvalues from the solution of four different grid sizes analysis

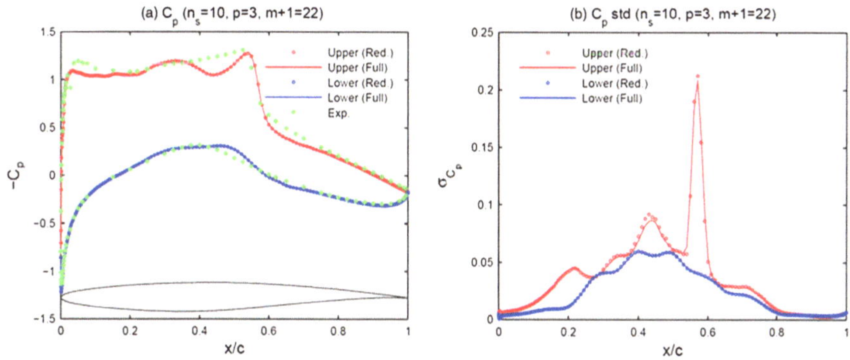

Fig. 6 Comparison of pressure coefficient ($\varepsilon = 0.99$): mean and std deviation using classical PC and model reduction method

the present test case, the reduced basis method (using $\varepsilon = 0.99$) is almost 6–7 times more efficient than the classical PC method. A detailed discussion on the effect of criterion (ε) on the accuracy of the reduced basis method is presented in [24]. A case where the covariance matrix in the reduced basis approach is build using only one primitive variable (e.g., ρU) is also analyzed, and similar results were obtained.

3D Transonic Rotor 37

For the validation of the developed reduced basis approach, uncertainty quantification of the rotor 37, shown in Fig. 7, is further considered. The rotational speed of the rotor is 17188 rpm, and the outlet static pressure is fixed at 110000 Pa. Combination of geometrical and operational uncertainties is considered for this test case. The geometry of the rotor blade is parameterized into sections of 2D airfoils using AutoBlade of NUMECA. The rotor 37 blade is parameterized into three sections of 2D airfoils (at 25, 50 and 75% of the blade height). For each airfoil section, leading and

Fig. 7 Meridional view of the rotor 37 blade with tip gap

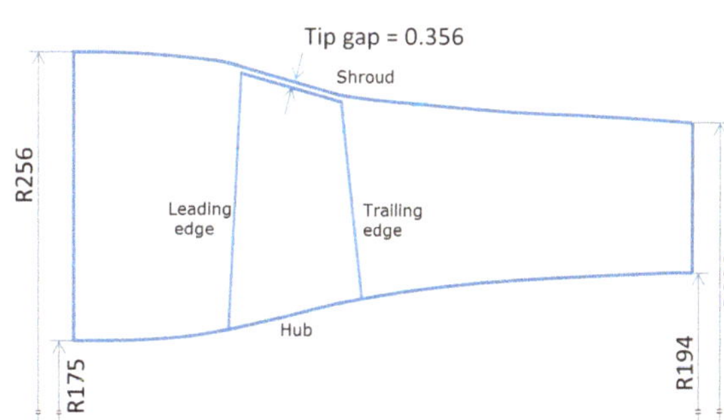

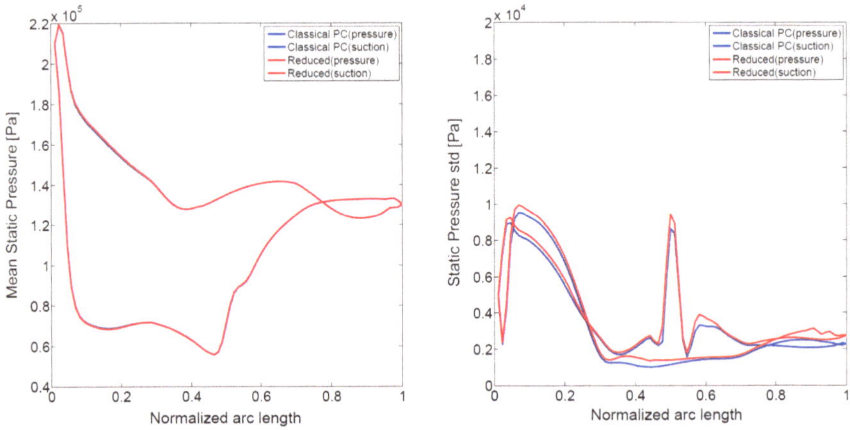

Fig. 8 Mean and std deviation of the pressure distribution around the rotor blade using reduced model for PC order 2

trailing edge angles are considered as uncertain. To model geometrical uncertainty around the blade, the uncertainty is also imposed on four half-thickness parameters (coefficients of half-thickness Bezier curve) of each airfoil section. In addition to these geometrical uncertainties, the tip clearance, the inlet total pressure profile, and the static outlet pressure are also considered uncertain. As a result, a total of 21 uncertain parameters are used for the uncertainty quantification of the NASA rotor 37. Symmetric beta distributions ($\alpha = \beta = 4$) are chosen for all uncertain variables. The details of this test case are given in [24]. Based on experience with previous test cases, a coarse grid with 1.04×10^5 cells is chosen. With a fine grid of 7.66×10^5 cells, the fine-to-coarse grid ratio is almost 7.5. Using a PC order of 2,506 samples on the coarse grid allows to get the covariance matrix. Based on the results from the previous test case, only the static pressure was used to construct the covariance matrix. Similar to the previous test case, a very fast decay was observed in eigenvalues. The ε is set to 0.9999 to capture most of the stochastic information from the coarse grid. The size of the reduced basis is then 21, requiring only 42 deterministic CFD simulations on the fine grid. Figure 8 compares the pressure distribution around the blade at mid-span height with the classical polynomial chaos method for the second-order PC. The mean (left) and the standard deviation (right) of the static pressure are shown. It can be observed that both methods produce similar results. Further analysis of the present test case in [24] indicates that the reduced basis method almost is 5 times more efficient than the regression-based NIPC.

Conclusions

In this paper, an efficient non-intrusive model reduction technique for PCE is presented and discussed. The proposed algorithm relies on the fact that the ideal basis for a stochastic field follows from its POD decomposition. This, however, requires the covariance structure, which in the present approach is obtained from the PCE on a coarse grid, assuming hereby that the stochastic behavior is largely independent from the spatial scales. The size of the ideal basis that results depends on the number of POD modes that are accounted for but is always significantly smaller than the full PCE basis, especially for high stochastic dimensions. The reduced basis approach was successfully applied to: (1) a highly irregular analytical function, (2) the 2D transonic RAE2822 airfoil with ten geometrical uncertainties, and (3) the 3D transonic NASA rotor 37 with 21 geometrical and operational uncertainties. The numerical results show that the reduced basis method is able to produce acceptable results for the statistical quantities. The computation time of the reduced-order model is found to be much lower than that of the classical PCE.

References

1. Doostan, A., Ghanem, R., Red-Horse, J.: Stochastic model reduction for chaos representations. Comput. Methods Appl. Mech. Eng. **196**(37–40), 3951–3966 (2007)
2. Nouy, A.: A generalized spectral decomposition technique to solve a class of linear stochastic partial differential equations. Comput. Methods Appl. Mech. Eng. **196**(45–48), 4521–4537 (2007)
3. Fishman, G.S.: Monte Carlo: Concepts, Algorithms and Applications. Springer, New York (1996)
4. Cliffe, K.A., Giles, M.B., Scheichl, R., Teckentrup, A.L.: Multilevel Monte Carlo methods and applications to elliptic PDEs with random coefficients. J. Vis. Lang. Comput. **14**(1), 3–15 (2011)
5. Taylor, A.C., Green, L.L., Newman, P.A., Putko, M.M.: Some advanced concepts in discrete aerodynamics sensitivity analysis. In: 15th AIAA Computational Fluid Dynamics Conference (AIAA 2001-2529), USA (2001)
6. Ghanem, R., Spanos, P.: Stochastic Finite Elements: A Spectral Approach. Springer, New York (1991)
7. Xiu, D., Karniadakis, G.E.: Modeling uncertainty in flow simulations via generalized polynomial chaos. J. Comput. Phys. **187**(1), 137–167 (2003)
8. Dinescu, C., Smirnov, S., Hirsch, Ch., Lacor, C.: Assessment of intrusive and non-intrusive non-deterministic CFD methodologies based on polynomial chaos expansions. Int. J. Eng. Syst. Model. Simul. **2**(1–2), 87–98 (2010)
9. Wang, X., Hirsch, Ch., Liu, Z., Kang, S., Lacor, C.: Uncertainty-based robust aerodynamic optimization of rotor blades. Int. J. Numer. Methods Eng. **94**(2), 111–127 (2013)
10. Sudret, B., Der Kiureghian, A.: Comparison of finite element reliability methods. Probab. Eng. Mech. **17**(4), 337–348 (2002)
11. Mathelin, L., Hussaini, M.Y., Zang, T.A.: Stochastic approaches to uncertainty quantification in CFD simulations. Numer. Algorithms **38**(1), 209–236 (2005)
12. Xiu, D., Karniadakis, G.E.: A new stochastic approach to transient heat conduction modeling with uncertainty. Int. J. Heat Mass Transf. **46**(24), 4681–4693 (2003)

13. Walters, R.W., Huyse, L.: Uncertainty analysis for fluid mechanics with applications. NASA/CR 2002-211449 (2002)
14. Hosder, S., Walters, R.W., Perez, R.: A non-intrusive polynomial chaos method for uncertainty propagation in CFD simulations. In: 44th AIAA Aerospace Sciences Meeting and Exhibit (AIAA 2006-891), USA (2006)
15. Lucor, D., Enaux, C., Jourdren, H., Sagaut, P.: Stochastic design optimization: application to reacting flows. Comput. Methods Appl. Mech. Eng. **196**(49–52), 5047–5062 (2007)
16. Blatman, G., Sudret, B.: Sparse polynomial chaos expansions and adaptive stochastic finite elements using a regression approach. C. R. Mec. **336**, 2008 (2008)
17. Zabaras, N., Ganapathysubramanian, B.: A scalable framework for the solution of stochastic inverse problems using a sparse grid collocation approach. J. Comput. Phys. **227**(9), 4697–4735 (2008)
18. Hampton, J., Doostan, A.: Compressive sampling of polynomial chaos expansions: convergence analysis and sampling strategies. J. Comput. Phys. **280**, 363–386 (2015)
19. Palar, P.S., Tsuchiya, T., Parks, G.: Decomposition-based evolutionary aerodynamic robust optimization with multi-fidelity point collocation nonintrusive polynomial chaos. In: 17th AIAA Non-Deterministic Approaches Conference, AIAA SciTech. AIAA Paper 2015-1377 (2015)
20. Ng L.W.T., Eldred, M.S.: Multidelity uncertainty quantization using non-intrusive polynomial chaos and stochastic collocation. In: 14th AIAA Non-Deterministic Approaches Conference, Honolulu. AIAA Paper 2012-1852 (2012)
21. Hosder, S., Walters, R.W., Balch, M.: Efficient sampling for non-intrusive polynomial chaos applications with multiple input uncertain variables. In: 9th AIAA Non-Deterministic Approaches Conference, USA. AIAA Paper 2007-1939 (2007)
22. Bratley, P., Fox, B.L.: Algorithm 659: implementing Sobol's quasi-random sequence generator. ACM Trans. Math. Softw. **14**(1), 88–100 (1988)
23. Raisee, M., Kumar, D., Lacor, C.: A non-intrusive model reduction approach for polynomial chaos expansion using proper orthogonal decomposition. Int. J. Numer. Meth. Eng. **103**(4), 293–312 (2015)
24. Kumar, D., Raisee, M., Lacor, C.: An efficient non-intrusive reduced basis model for high dimensional stochastic problems in CFD. Comput. Fluids **138**, 67–82 (2016)

Screening Analysis and Adaptive Sparse Collocation Method

Alberto Clarich and Rosario Russo

Introduction

In many Uncertainty Quantification (UQ) problems, the number of uncertain variables may cause the practical impossibility of building a numerical model which takes into account all the uncertainties of the problem at hand, for a Robust Design Optimization (RDO). The reduction of the number of variables and the reduction of the number of samples to be calculated are the two key issues to apply robust methodologies in real-world design problems.

Screening analysis techniques can help to identify the most important variables and discard the less important, reducing the number of design points to evaluate. Here we will discuss three different methodologies: the smoothing spline ANOVA (SS-ANOVA) algorithm, the principal component analysis (PCA) methodology, and the Morris screening technique.

Different sampling techniques can be used in Robust Design Optimization. The classical are Monte Carlo sampling and Latin hypercube sampling (LHS). Recently, a new technique based on the Polynomial Chaos Expansion (PCE) has proven to be very efficient reaching the same accuracy with less number of samples. Here a variant, the Adaptive Sparse Collocation PCE, is presented, which enables to reduce the number of samples to the only ones needed to have a good estimation of the robust design characteristics.

A. Clarich (✉) · R. Russo
ESTECO SpA Area Science Park, Padriciano 99, 34149 Trieste, Italy
e-mail: engineering@esteco.com; engineering@esteco.it

C. Hirsch et al. (eds.), *Uncertainty Management for Robust Industrial Design in Aeronautics*, Notes on Numerical Fluid Mechanics and Multidisciplinary Design 140, https://doi.org/10.1007/978-3-319-77767-2_11

Screening Analysis Implementing SS-ANOVA

Several tools dedicated to variable screening analysis can be used to determine which uncertainty parameters are really important in a problem and which not.

Correlation coefficient (Pearson or Spearman [1]) or standard ANOVA is useful to understand the degrees of correlation between two series of data, one containing the uncertain parameters and the other containing the output response of a given dataset. The limitation is that these statistical methodologies are fully reliable only when the samples are representative enough, and this occurs, typically, when the database is produced by a kind of Full Factorial Design of Experiments (DOE), whose size increases exponentially with the number of variables.

Generally, databases coming from experimental tests may contain scarce data and could not be produced by a given scheme; therefore, a methodology which is more flexible and applicable also to scarce data is often necessary. An efficient methodology, tested extensively during the UMRIDA project, is called smoothing spline ANOVA (SS-ANOVA) [2]. SS-ANOVA models are a family of smoothing methods suitable for both univariate and multivariate modeling and regression problems characterized by noisy data, given the assumption of Gaussian-type responses.

In particular, SS-ANOVA is a statistical modeling algorithm based on a function decomposition similar to the classical analysis of variance (ANOVA) decomposition and the associated notions of main and interaction. Each term—main effects and interactions—can be used to reveal the percentage of its contribution to the global variance, since in a statistical model the global variance can be explained (decomposed) into single model terms.

The main advantage respect to other statistical methodologies (like standard ANOVA) relies on the fact that the relative significances of each term (therefore main effects and interactions of the variables) are not just computed directly on the available database, with the limitations underlined above, but are contained in the regression model itself defined by the methodology, so they can be computed analytically even from a limited and partially correlated database. In this paragraph, we report the basic formulation of the methodology.

If for the moment we consider a generic univariate regression problem for the function $f(x)$: $[0, 1] \subset \mathbb{R} \rightarrow \mathbb{R}$, we can write the following expression:

$$f_i = f(x_i) + \varepsilon_i, \quad i = 1, \ldots, n \tag{1}$$

where n is the number of designs (sample set size) and $\varepsilon_i \sim N(0, \sigma^2)$ represent independent random errors normally distributed. A suitable solution to the regression problem could be stated as a constrained minimization problem that can be expressed in Lagrangian terms as:

$$\min L(f) + \frac{\lambda}{2} J(f) \tag{2}$$

where *L(f)* is defined as *minus log likelihood* of the model *f(x)* given the data, to be minimized to maximize the data fit, and *J(f)* is defined as a *quadratic roughness functional*, to be subjected by a constraint—$J(f) \leq \rho$—that can be used to control the overfitting (a large roughness guarantees a smoother model, while smaller values imply rougher functions but with better agreement to the data).

The procedure of finding a proper function estimation by minimizing Eq. 2 is called *penalized likelihood method*: In fact, the *J* term represents a penalty on the roughness. This methodology is also referred to as *penalty smoothing* (or smoothing method with roughness penalty), and λ takes the name of *smoothing parameter.*

In particular, with Gaussian-type responses, Eq. 2 can be rewritten as:

$$\min \frac{1}{n} \sum_{i=1}^{n} (f_i - f(x_i))^2 + \lambda \int_0^1 \left[f''(x)\right]^2 dx \tag{3}$$

This means that the likelihood expression of the model takes the usual form of a least square function, while the penalty term (roughness) contains the second derivative of the function f(x)—$f''(x) = d^2f/dx^2$—and takes therefore the aspect of a measure of the curvature of the function over the domain. It can be proved [2] that Eq. (3) corresponds to the definition of *a natural cubic spline* (i.e., a piecewise cubic polynomial, with the knots at all of the distinct values of the sampling set $\{x_i\}$); therefore, the methodology takes the complete name of *Cubic Smoothing Spline* ANOVA.

By extension to a multivariate problem of the type $f(\boldsymbol{x})$: $[0, 1]^N \subset \mathbb{R}^N \rightarrow \mathbb{R}$, Eq. 3 will take the general form of:

$$\min \frac{1}{n} \sum_{i=1}^{n} (f_i - f(x_i))^2 + \sum_{j=1}^{N} \lambda_j \int_0^1 \left[f_j''(x_j)\right]^2 dx \tag{4}$$

If we assume for simplicity that the regression model $f(\boldsymbol{x})$ can be expressed as a sum of *N* independent components $f_j(x_j)$, each one function of a single variable x_j, the regression model would take into account only main effects (the effect of each single variable) and not interaction effects.

A more complete regression model, which has to consider also interaction effects, will include in $f(\boldsymbol{x})$ also the interaction terms $f_{ij}(x_i, x_j)$ keeping unchanged the general formulation of the least square functional (first term of Eq. (4)), while the roughness functional *J(f)*, accordingly to the "*reproducing kernel Hilbert space*" theory, can be expressed as:

$$\lambda J(f) = \lambda \sum_{k=1}^{p} \theta_k^{-1} \langle f, f \rangle_k \tag{5}$$

with <f, f> inner product and λ, θ_k smoothing parameters, to be determined by a proper data-driven procedure, such as the generalized cross-validation (GCV), as described in [3]. The number of decomposition terms p is equal, considering all the interaction terms, to $N(N-1)/2$ with N number of variables, while it would be equal just to N if only main effect terms were considered. Since any term is an additional degree of freedom for the regression model $f(\boldsymbol{x})$, the number p indicates also the minimum number of sampling points needed to solve Eq. (4).

As a conclusion, independently whether only main terms $f_j(x_j)$ or also interaction terms $f_{ij}(x_i, x_j)$ are considered, the regression model $f(\boldsymbol{x})$ will always take the linear combination form of spline basis functions f_k as:

$$f(x) = \sum_{k=1}^{p} f_k \tag{6}$$

We can therefore apply the definition of internal product projecting the $f(\boldsymbol{x})$ to any function f_k obtaining the value of its contribution (or probability), by the (normalized) expression:

$$\pi_k = \frac{\langle f_k, f \rangle}{\|f\|^2} \tag{7}$$

Expression (7) is called *contribution index* π_k and expresses the relative significance of the different terms composing the model, therefore the contribution of each variable main effect or interaction effect. The normalization guarantees the identity $\sum_{k=1}^{p} \pi_k = 1$.

There is still one condition important to be verified in order to guarantee the reliability of the results, which is that the single f_k must not be too much correlated, situation that may occur for instance when the sample size is too low or bad distributed or input variables are actually correlated. The measure of correlation or *collinearity* can be given by the *collinearity indices* K_k defined as:

$$K_k = \sqrt{\frac{\langle f_k, f_k \rangle}{\|f_k\|^2}} \tag{8}$$

These indices have the aspect of the diagonal terms of a normalized correlation matrix; therefore, if the components are almost orthogonal to each other, the collinearity indices have to be close to 1. Conversely, if the collinearity indices are much greater than 1 ($K_k \gg 1$), the values of the relevant π_k indices may become unreliable.

Screening Analysis Using Principal Component Analysis

In many cases, a noisy database may reveal a hidden simplified structure that underlies it, for instance, some variables may be redundant or the uncertainties may be correlated for many reasons. It may be not always possible performing a screening analysis to understand if some uncertainties may be directly excluded by the analysis, those variables being not significant for the problem, but more often (and in particular when experimental data only are available) it might be possible to express the same complexity of the problem by using a different basis of coordinates, characterized by a smaller dimension than the original basis. As an illustration example, we can consider the database of Fig. 1.

The points of the chart represent different samples of the database, which is defined by three variables x, y, z reported in the axis. The three variables may represent three uncertainties of the system, and we assume that the database is obtained by an experimental campaign.

It is possible to note how actually the points lie on a plane, which can be represented by two coordinates only, that we call PC1 and PC2. The definition of the database through the new coordinates system is technically a change of basis, and the advantage of this new formulation is that we can build a numerical model characterized by two uncertainties only instead of three, with a significant reduction of the required sampling size for an accurate Robust Design analysis.

Of course the situation above is very peculiar, because it is rare to find cases for which the uncertainties may be so highly correlated, but what is important is to find a general methodology able to compute the most meaningful basis to re-express a noisy and/or correlated dataset.

The primary need is to find a new basis of coordinates for which the system is equivalent to the original one in terms of variance, since we want to express the same uncertainty distribution of the original problem with the uncertainty distribution of the new system.

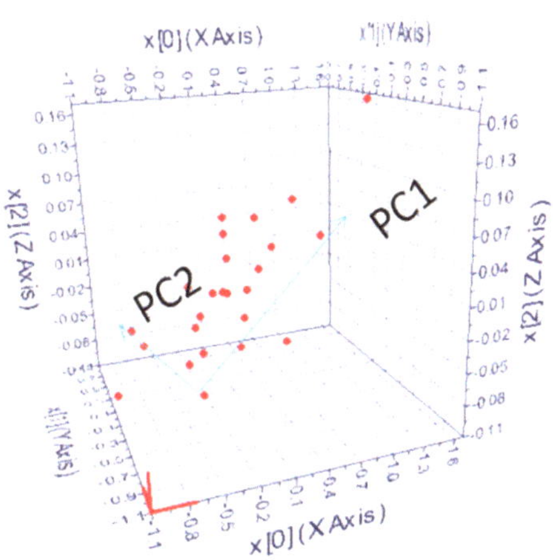

Fig. 1. Example of change of basis in a database

In addition, we want to find a basis whose coordinates are linearly independent (orthogonal basis) so that it would not be possible to express any of them by a linear combination of the others.

This means that we need to express the covariance matrix of the system in function of the original coordinates and diagonalize the matrix in order to find its eigenvalues and eigenvectors: A basis of coordinates coincident with the eigenvectors (principal components) will in fact be characterized by the largest values of variance (eigenvalues) arranged in decreasing order up to the lowest and by null covariance (orthogonal basis).

The eigenvectors characterized by small eigenvalues will therefore express a coordinate of the system for which the variance of the database is negligible. Therefore, by ranking the eigenvectors in order of decreasing eigenvalues, from the highest (corresponding to the first principal component) to the smallest eigenvalue, it is possible to keep only the main principal components neglecting the last ones: The main principal components will in fact retain the highest fraction of the system global variance, and, expressing the system in a new basis made by them only, will also allow to express almost the same variance of the original system database with a reduced number of coordinates. The methodology which aims to follow this purpose is called principal component analysis (PCA).

The first step considered by PCA is the definition of the covariance matrix of a database [4].

Let us consider by hypothesis a database made by m variables (the original basis of coordinates) and n samples. If we call two of the coordinates as A and B, we will have two vectors of measurements defined as:

$$\boldsymbol{a} = [a_1, a_2, \ldots, a_n], \quad \boldsymbol{b} = [b_1, b_2, \ldots, b_n] \tag{9}$$

The *variance* of each vector can be expressed by the internal product notations as:

$$\sigma_A^2 = \langle a_i a_i \rangle \tag{10}$$

while the *covariance* between two variables, and therefore the degree of correlation, can be expressed using the same notations as follows:

$$\sigma_{AB}^2 = \langle a_i b_i \rangle \tag{11}$$

Considering that Eq. (11) is a particular case obtained when A = B, we may build a matrix whose elements are expressed as:

$$\sigma_{ab}^2 = \frac{1}{n-1} \boldsymbol{a}\boldsymbol{b}^T \tag{12}$$

In this way, Eq. (12) is equivalent to Eq. (10), besides the introduction of a constant for normalization. If we generalize the definition of the original basis of coordinates calling the single components as $\mathbf{x_i}$ (where for instance $\mathbf{x_1}= \mathbf{a}$ and $\mathbf{x_2} = \mathbf{b}$), the complete database can be defined in the original basis as a matrix **X** made by $m \times n$ elements, and Eq. (12) becomes:

$$C_x = \frac{1}{n-1} XX^T \tag{13}$$

Matrix of Eq. (13) is defined as *Covariance Matrix.*

Note that covariance matrix is defined by a x suffix, because its elements are expressed in the original basis **X**, so on the diagonal we will find the variance of the system database for each original variable $\mathbf{x_i}$ while on the off-diagonal, we will find the covariance between the variables $\mathbf{x_i}$ and $\mathbf{x_j}$.

At this point, the next step of the PCA is to diagonalize the covariance matrix. This procedure in linear algebra corresponds to the search of the *eigenvalues* of the matrix, i.e., into expressing the matrix by a new basis of (orthogonal) coordinates (the *eigenvectors*), for which it becomes diagonal: The fact that the off-diagonal terms are equal to 0 means that the covariance of the new coordinates is null (orthogonal or not correlated components), and in addition it can be proved that the first eigenvectors retain the highest values of variance.

In other words, we have to find a basis of m vectors $\mathbf{P_i}$ for which the original variables **X** become expressed as **Y** and the covariance matrix $\mathbf{C_X}$ becomes $\mathbf{C_y}$ with:

$$\begin{cases} \sigma_{Y_k}^2 = \frac{1}{n-1}\lambda_k \\ \sigma_{Y_iY_j}^2 = 0 \end{cases} \tag{14}$$

Expressing the problem with the new base **P**, the original database **X** will become defined as (considering that the change of basis is a linear operation):

$$Y = PX \tag{15}$$

At this point, we can express the covariance matrix $\mathbf{C_X}$ in the new base, obtaining:

$$C_Y = \frac{1}{n-1} YY^T = \frac{1}{n-1} (PX)(PX)^T = \frac{1}{n-1} PXX^T P \tag{16}$$

A symmetric matrix like $XX^T = A$ can in general (if not *degenerate*) be diagonalized by an orthogonal matrix of its eigenvectors, E, obtaining $A = EDE^T$ where D is the diagonal matrix (it follows in fact that $A = EDE^{-} \rightarrow AE = ED$ which means $A\mathbf{e_i} = \lambda_i \mathbf{e_i}$ for every i, which is the definition of the eigenvalue equation).

At this point, if we assume as we want to prove that the needed basis P is coincident with the basis of the eigenvectors $P = E^T$, we can get from Eq. 16:

$$C_Y = \frac{1}{n-1} P\left(EDE^T\right)P^T = \frac{1}{n-1} P\left(P^T DP\right)P^T \frac{1}{n-1} D \tag{17}$$

The last passage is obtained considering that eigenvectors are linearly independent, so $\mathbf{P}^\mathrm{T} = \mathbf{P}^{-1}$. Finding the eigenvectors of the covariance matrix or the XX^T matrix can therefore satisfy Eq. (6).

There are several methodologies to compute the eigenvalues of the matrix C, but we have preferred to use the method based on the transformation to a *tridiagonal* matrix by *Householder* orthogonal transformations, then extracting its eigenvalues and eigenvectors by the tridiagonal QL algorithm. Further details of the methodology can be found in [5].

Screening Analysis Using the MORRIS Methodology

The last UQ methodology we propose can be used as a general method to identify the most significant variables in any process.

Several different screening methods have been proposed in the literature (see, e.g., [6]); most of them are really efficient, but they work well only in specific cases: For example, the *Sequential Bifurcation* method, proposed by Bettonvil and Kleijnen in [7], requires that main and interaction effects must be nonnegative; others, as the *Full Factorial* or the *Fractional Factorial* methods, are computationally expensive.

The method presented by Morris in [8], with some improvements introduced by Campolongo et al. in [9], has two important characteristics: It can be applied to all problems, and it is computationally cheaper than other methods. For this reason, we propose this methodology as the last UQ methodology; for simplicity, from here we call the methodology [9] as Morris screening analysis.

Conversely from the first two methods presented here, this approach requires the definition of a series of analysis under a pre-defined scheme; therefore, we cannot apply it directly on an available experimental database, but rather we can apply it on a numerical simulation model (for instance, a CFD model) where the uncertainties have already been quantified with their distributions, in order to understand if it is possible to represent the same response uncertainties of the problem by a reduced number of parameters, and therefore by a reduced sampling size for any design proposed in a Robust Design Optimization.

Consider the function $f(x_1, \ldots, x_n)$ of n variables (the uncertain parameters) and let $\{\tilde{\mathbf{x}}_1, \ldots, \tilde{\mathbf{x}}_N\}$ be a set of N points in the variable space (the database). For each of these points, we define n new points which define a *trajectory*, in such a way that:

$$\begin{aligned} \tilde{\mathbf{x}}_{io} &= \tilde{\mathbf{x}}_i \\ \tilde{\mathbf{x}}_{ik} &= \tilde{\mathbf{x}}_{ik-1} + \Delta_k e_k \end{aligned} \tag{18}$$

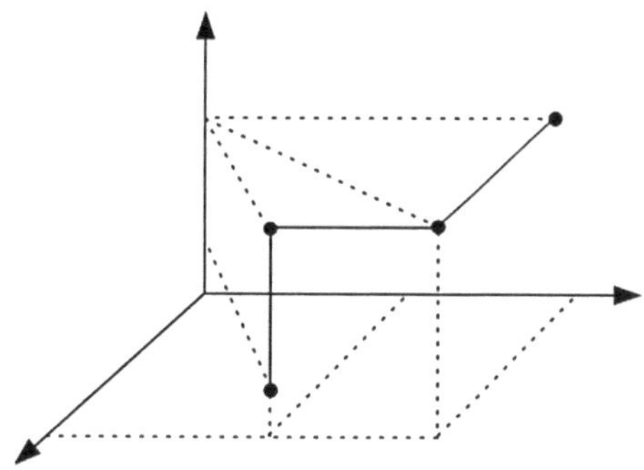

Fig. 2. Example of i-trajectory made by $n + 1$ points accordingly to Eq. (20)

with $i = 1, \ldots, N$ and $k = 1, \ldots, n$. $\{e_1, \ldots, e_n\}$ is the canonical base of R^n and Δ_k is a proper percentage of the variable range (see Fig. 2 for an example of a trajectory in a 3-D variable space). The purpose of defining the trajectories is the one to measure how the function f changes when only one of the parameters is perturbed, and in this way it is possible to compute the *elementary effects* d_k^i as:

$$d_k^i = \frac{f(\tilde{\mathbf{x}}_{ik}) - f(\tilde{\mathbf{x}}_{ik-1})}{\Delta_k} \tag{19}$$

Since the elementary effect of any variable k is given by N different factors (one for each i), the definitive measures of sensitivity can be obtained by mean and standard deviation of d_k^i, obtaining in Eq. (20) respectively the *mean deviation* μ_k and *the standard deviation* σ_k as:

$$\begin{gathered} \mu_k = \frac{1}{N} \sum_{i=1}^{N} d_k^i \\ \sigma_k = \sqrt{\frac{1}{N-1} \sum_{i=1}^{N} \left(d_k^i - \mu_k\right)^2} \end{gathered} \tag{20}$$

The first measure of Eq. (20) represents the importance of the kth input factor over the output (i.e., its main effect), while the second measure by the variance of the elementary effects estimates the interactions of the kth input factor with the other ones. Thanks to these two measures, we can sort the input variables by order of importance. The procedure is generic for any problem, but we can apply it specifically for UQ: The input variables $x_1, \ldots, x_n$ are the uncertain parameters of the process, and the database $\{\tilde{\mathbf{x}}_1, \ldots, \tilde{\mathbf{x}}_N\}$ is given by a sampling of their uncertain distributions. By the evaluation of the elementary effects and the consequent mean and standard deviation factors, the relative effect of each parameter can be estimated, and the parameters with marginal effects can be excluded by the process.

The method proposed until this point is the original Morris algorithm [8], which basically has two problems:

(i) If the output sign changes, some information could be lost;
(ii) If n is too large, the procedure could become too expensive, indeed it requires $N \cdot (n + 1)$ evaluations.

Campolongo et al. in [9] proposed the following sensitivity measure in order to overcome the first problem:

$$\mu_k^* = \frac{1}{N} \sum_{i=1}^{N} \left| d_k^i \right| \tag{21}$$

Moreover, as a possible solution for the second problem, they proposed to consider the factors of the problem $x_1, \ldots, x_n$, and to divide the input variables into M sets: $G_1, \ldots, G_M$ with $M \ll n$. Let $I_1, \ldots, I_M$ be the sets of indices of the *variables* within every group.

For each point $\tilde{\mathbf{x}}_i$, $i = 1, \ldots, N$ we build a trajectory with $m = 1, \ldots, M$ new points:

$$\begin{aligned} \tilde{\mathbf{x}}_{io} &= \tilde{\mathbf{x}}_i \\ \tilde{\mathbf{x}}_{im} &= \tilde{\mathbf{x}}_{im-1} + \boldsymbol{\Delta}_m \cdot \boldsymbol{d}_m \end{aligned} \tag{22}$$

where $\boldsymbol{d}_m = (d_{m1}, \ldots, d_{mn})$ is a vector associated with the mth group, such that:

$$\begin{aligned} d_{mi} &= 1 \quad i \in I_m \\ d_{mi} &= 0 \quad \textit{otherwise} \end{aligned} \tag{23}$$

In this way, it is possible to compute the mean and the standard deviation for every group, selecting the groups of variables which have the highest values of the effects, eventually iterating the process P times by grouping the variables in different ways, selecting then the factors which occur more times in the most important groups.

In any way, the grouping methodology is convenient when $M\,P \ll n$, because the overall number of evaluations will be equivalent to $N \cdot P \cdot (M + 1)$.

Sampling Methodologies for Robust Design Optimization

The first aspect to be taken in account when defining a robust design optimization is the definition of an accurate sampling methodology.

The samples are defined in the variable space neighborhood of the nominal design, following the statistical distribution of each uncertain parameter, and from the evaluation of its performance, the response distribution can be evaluated.

This is a crucial point of analysis under uncertainties and consequently for a RDO problem, since the procedure is to be repeated for any design proposed by the optimization procedure.

The sample points must be defined in such a way that the method for the uncertainty quantification (UQ) of the performances could be accurate enough, but at the same time, for the RDO purpose, the number of sample points should be low enough to allow the application to cases of industrial relevance. The choice of the sampling points is therefore strongly correlated to the method used for the uncertainty quantification.

Monte Carlo and Latin Hypercube Sampling (LHS) are classical sampling methodologies which can be used to compute directly statistical momentums in function of sampling points database: As alternative, they can be also used to populate the training database for methodologies like [10] based on DACE or RSM (Response Surface) meta-models, by which momentum functions can be computed on a large number of extrapolated samples.

As reported above, the simplest approaches for the UQ quantification are based on the discretization of the momentum equations, in the sense that mean and standard deviations may be computed directly following the expressions below:

$$\bar{C}_D(x) = \frac{\sum_{i=1}^{n} C_{Di}(x)}{n} \tag{24}$$

$$\sigma_{CD}(x) = \sqrt{\frac{\sum_{i=1}^{n} \left(C_{Di}(x) - \bar{C}_D(x)\right)^2}{(n-1)}} \tag{25}$$

In these expressions the statistical momentums (here related to a drag coefficient performance, C_D) are computed directly in function of the N samples of the database. As already noted, this statistical approximation may be improved following methods DACE or other meta-models, to let the analysis be more significant from the statistical point of view, just simply raising the number of sampling points using a meta-model.

In any case, the approach followed by these types of UQ methods requires the definition of a sampling set to be as much representative as possible of the uncertain parameters distributions. For this purpose, two methodologies of Design of Experiments (DOE) are available: Monte Carlo and Latin hypercube.

Monte Carlo is the simplest sampling methodology and is just based on a random definition of sample points. More in particular, from Fig. 3, it is illustrated how the Monte Carlo algorithm defines a sampling set of points: From a random sequence of points defined by a uniform probability (ordinate), the sample points are obtained by the projection of the former points onto a Gaussian Cumulative Distribution (defined as Uncertainty Distribution of the input variable). The process is then extended to a case with a generic number of variables. It is possible to prove that the statistical moments, and as a result the mean and the standard deviation, of a random sample converges to the exact moments of the full distribution as the inverse of squared root of N, where N is the sample size.

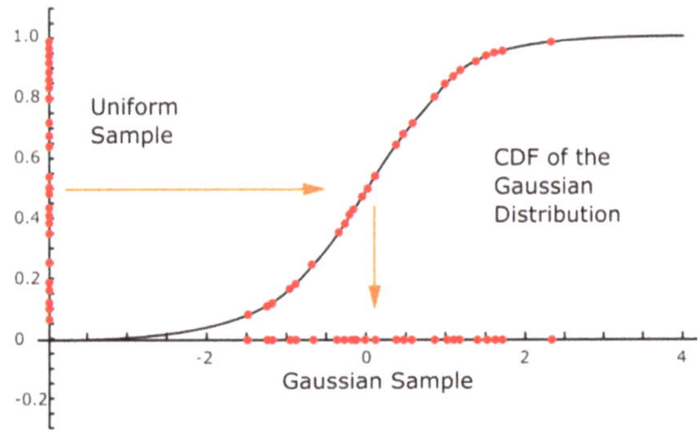

Fig. 3. Example of Monte Carlo sampling

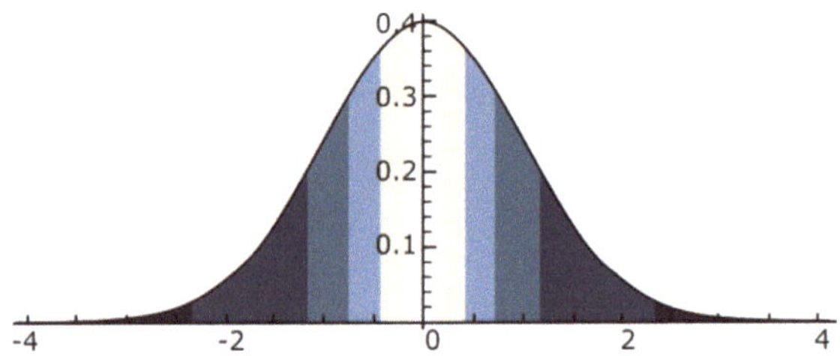

Fig. 4. Example of Latin hypercube sampling

A more efficient sampling methodology, in the sense that the convergence of the random sample to the exact full distribution is more rapid, and in particular proportional to the inverse of N, is Latin hypercube sampling (LHS).

In this case (Fig. 4), if a LHS is composed of N points, and every variable is divided in N strata with equal probability (the same portion of area under the Gaussian distribution), every single stratum will be occupied by exactly one point.

Latin hypercube sampling has been designed specifically to produce better accuracy than Monte Carlo in uncertainty quantification.

Non-intrusive Polynomial Chaos Expansion

In order to improve the efficiency for UQ of the sampling methodologies introduced in previous paragraph, an important methodology that has been proposed is the Polynomial Chaos Expansion [11].

This methodology consists essentially in expanding the uncertain variable response in a suitable series and then determining analytically (and thus exactly) the statistical moments of the truncated expansion. The expansion itself is referred to as the "Chaos," while the maximum degree of the expansion is called the "Chaos order."

As a result, it can be proved that the estimate of the statistical moments converges to true values at exponential rate, i.e., the error in the estimates scales as $exp^{(-N)}$, where N is the sample size.

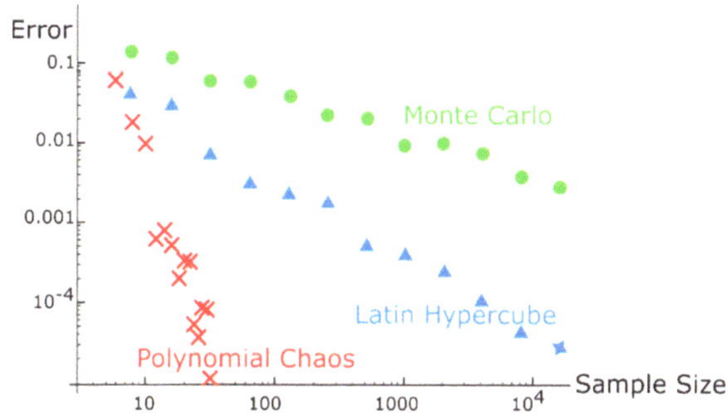

Fig. 5. Example of UQ sampling convergence rate

Figure 5 above reports a comparison of the statistical moment relative errors, relative to a mathematical test function [12], obtained using Monte Carlo or Latin hypercube sampling and Polynomial Chaos Expansion. It emerges clearly the large advantage of this methodology with respect to the former ones, since for a RDO problem it is extremely important to obtain a good accuracy of UQ for each design with the lowest possible number of samples.

There exist basically two types of Polynomial Chaos methodologies, the intrusive and the non-intrusive one. In the following part of this chapter, we describe only the second methodology, because it is more versatile for every industrial application and any simulation software, not requiring the modification of the solver equations, but working directly on the results computed by the samples.

The basic idea of Polynomial Chaos methodology for UQ is that, under specific conditions [11], a stochastic process can be expressed as a spectral expansion based on suitable orthogonal polynomials, with weights associated with a particular probability density function.

The idea is to project the variables of the problem onto a stochastic space spanned by a set of complete orthogonal polynomials ψ that are functions of random variables $\xi(\theta)$, where θ is a random event [13]. For example, the variable ϕ has the following spectral infinite dimensional representation:

$$\Phi(\boldsymbol{x},\, t,\, \theta) = \sum_{i=0}^{\infty} \Phi_i(\boldsymbol{x},\; t)\Psi_i(\xi(\theta)) \tag{26}$$

Equation (26) divides the random variable $\varphi(\mathbf{x},\, t,\, \theta)$ into a deterministic part, i.e., the coefficient $\phi_i(\mathbf{x}, t)$, and a stochastic part, i.e., the Polynomial Chaos $\psi_i(\xi(\theta))$. The set of polynomials $\{\psi_i\}$ forms a complete orthogonal basis in the Hilbert space determined by their support. The orthogonality relation takes the form of:

$$\langle \psi_i \psi_j \rangle = \langle \psi_i^2 \rangle \delta_{ij} \tag{27}$$

where δ_{ij} is the Kronecker delta and <.,.> denotes the ensemble average, i.e., the inner product in the Hilbert space of the variables ξ, which reads:

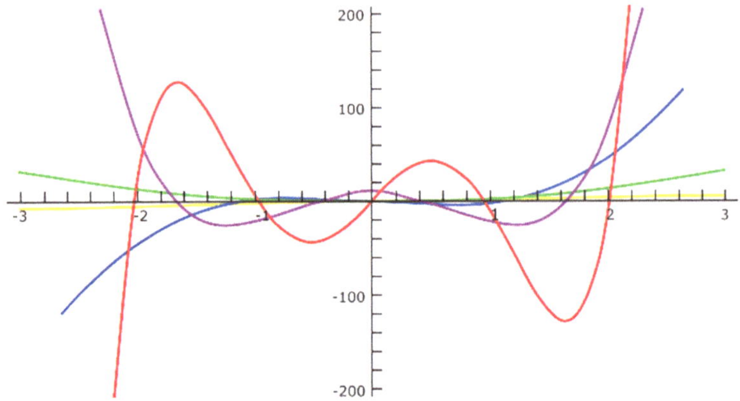

Fig. 6. Example of Polynomial Chaos Expansion (Hermite polynomials)

$$\langle f(\xi), g(\xi)\rangle = \int f(\xi)g(\xi)w(\xi)d(\xi) \tag{28}$$

with $w(\xi)$ a weighting function. According to the Askey-chaos theory [14], a specific weighting functions $w(\xi)$ is to be selected for each probability density function. For example, the weight for Gaussian distribution corresponds to Hermite polynomials (Fig. 6), Exponential and Gamma distributions to Laguerre and generalized Laguerre polynomials respectively, Beta distribution to Jacobi polynomials, and Uniform distribution to Legendre ones (Wiener–Askey scheme).

Thus, in the case of random inputs with Gaussian distribution, we represent the variable $\varphi(\mathbf{x}, t, \theta)$ in terms of an *Hermite* spectral representation (or *Wiener* chaos) for which the weighting function reads

$$w(\xi) = e^{-\frac{1}{2}\xi^T\xi} \tag{29}$$

apart from normalization factors.

In numerical applications, the series in Eq. (26) has to be truncated to a finite number of terms (here denoted with *N*) or, equivalently, to a finite order. Hence, for a Wiener chaos, Eq. (26) becomes::

$$\begin{aligned}\phi(\boldsymbol{x}, t, \theta) &= \sum_{i=0}^{N} \phi_i(\boldsymbol{x}, t)\Psi_i(\xi) \\ &= \sum_{p_1=0}^{P_1}\sum_{p_2=0}^{P_2}\cdots\sum_{p_n=0}^{P_n} \Phi_{p_1p_2p\ldots p_n}(\boldsymbol{x}, t)H_{p_1}(\xi_1)H_{p_2}(\xi_2)\ldots H_{p_n}(\xi_n)\end{aligned} \tag{30}$$

where $H_{p_k}(\xi_k)$ is the *Hermite* polynomial of order p_k in terms of the *k*th random variable ξ_k with Gaussian distribution *N*(0, 1). The number of total terms of the series in Eq. (31), where a tensorial-expanded representation has been adopted, is determined by:

$$N + 1 = \frac{(n+p)!}{n!p!} \tag{31}$$

with n the dimension of the uncertain variables ξ and p the highest order of polynomials $\{\Psi_i\}$.

By applying the orthogonality condition to the truncated spectral expansion, the expectation value and variance of $\phi(\mathbf{x}, t, \theta)$ are straightforwardly found to be, respectively, given by

$$E_{PC}(\Phi) = \mu_\Phi = \Phi_0(\boldsymbol{x}, t) \tag{32}$$

$$Var_{Pc}(\Phi) = \sigma_\Phi^2 = \sum_{I=1}^{N} \left[\Phi_i^2(\boldsymbol{x}, t)\langle\psi_i^2\rangle\right] \tag{33}$$

with $\langle\psi_i^2\rangle$ the polynomial normalization.

Thus, the problem of uncertainty quantification is shifted to the determination of the polynomial expansion coefficients $\phi_i(\mathbf{x}, t)$. Various techniques have been developed for the solution of this problem (see Ref. [15]), and normally the sampling points are evaluated at the roots of the polynomials. We follow the *regression* or *collocation method* based on the least square minimization of the discrepancy between $\phi(\mathbf{x}, t, \theta)$ and its truncated expansion:

$$\min \sum_{j=1}^{N_S} \left| \Phi(X, t, \theta) - \sum_{J=0}^{N} \Phi_j(X, t)\, \psi_j(\xi) \right|^2 \tag{34}$$

with N_S the number of sampling points used to evaluate the discrepancy.

This approach has in fact the advantage that the samples can be arbitrarily chosen, except for their number which has to be equal to or greater than the number of points reported in Eq. (31). For the UQ tool implemented in modeFRONTIER [11] optimization platform software from ESTECO, the sample can be generated by means of a Latin hypercube sampling, and the solver of the least square problem of Eq. 34 is the standard Levenberg–Marquardt method [16].

Adaptive Sparse Collocation

An efficient method to apply UQ with large number of variables is based on the application of a regression analysis directly on the Polynomial Chaos Expansion (PCE) expression, in other words the PCE will keep only those terms which actually affect the output, discarding the others.

The methodology consists first in ranking the terms using a least angle regression (LAR) technique [17] and then in assessing how many PCE terms should be kept.

The order of selection of the PCE terms will reflect a ranking based on how much each term affects the output. Once the ranking is done, it is necessary to establish a way to choose how many PCE terms should be kept. The criterion for this is based on the mean leave-one-out error (Err_{LOO}).

$$Err_{LOO} = \frac{1}{N} \sum_{i=1}^{N} \Delta_i^2 \tag{35}$$

where N is the number of samples and $\Delta_i = \text{output}(\mathbf{x}_i) - \hat{M}(\mathbf{x}_i)$ is the difference between the output corresponding to the ith sample and the output computed from the PCE obtained excluding from the training samples the ith. The criterion to select the number of terms consists in monitoring the quantity:

$$R_{LOO} \equiv 1 - \frac{Err_{LOO}}{var(\mathbf{output})(\mathbf{\#samples} - \mathbf{\#terms})} \tag{36}$$

R_{LOO} is function of the number of PCE terms: It tends initially to increase as the number of terms increases, but from a certain number of terms on, it starts showing a decreasing trend. This index is in fact sensitive to overfitting problems, so that optimal number of terms is generally much lower than the maximum possible number equal to the original degree of freedom of the complete polynomial.

The described approach gives the important benefit of reducing the global number of unknown coefficients for the PCE, and therefore giving the possibility as well of reducing the number of sampling points, needed for the PCE training.

References

1. Spearman, C.: The proof and measurement of association between two things. Am. J. Psychol. **15**, 72–101 (1904)
2. Gu, C.: Smoothing Spline ANOVA Models, Springer, New York (2002)
3. Craven, P., Wahba, G.: Smoothing noisy data with spline functions: estimating the correct degree of smoothing by the method of generalized cross-validation. Numer. Math. **31**, 377–403 (1979)
4. Shlens, J.: A Tutorial on Principal Component Analysis, Dec 10, 2005, Version 2
5. Er-xiong, J.: QL method for symmetric tridiagonal matrices. J. Shanghai Univ. (English Edition) **8**(4), 369–377 (2004)
6. Saltelli, A., Chan, K., Scott, E.M. (eds.): Sensitivity Analysis. Wiley, Chichester (2000)
7. Bettonvil, B., Kleijnen, J.P.C.: Searching for important factors in simulation models with many factors: sequential bifurcation. Eur. J. Oper. Res. **96**(1), 180–194 (1997)
8. Morris, M.D.: Factorial sampling plans for preliminary computational experiments. Technometrics **33**, 161–174 (1991)
9. Campolongo, F., Cariboni, J., Saltelli, A.: An effective screening design for sensitivity analysis of large models. Environ. Model Softw. **22**(10), 1509–1518 (2007)
10. Sacks, J., Welch, W.J., Mitchell, T.J., Wynn, H.P.: Design and analysis of computer experiment. Statist. Sci. **4**(4), 409–453 (1989)

11. Xiu, D., Karniadakis, G.E.: The Wiener-Askey polynomial chaos for stochastic differential equations. SIAM J. Sci. Comput. **24**(2), 619–644 (2002)
12. Clarich, A., Pediroda, V.: Robust design applications with modeFRONTIER, applying NODESIM-CFD tools, NODESIM-CFD Workshop on Quantification of CFD Uncertainties, Bruxelles, 29–20 Oct 2009
13. Parussini, L., Pediroda, V., Poloni, C.: Prediction of geometric uncertainty effects on fluid dynamics by polynomial chaos and fictitious domain method. Comput. Fluids **39**(2010), 137–151 (2009)
14. Askey, A., Wilson, J.A.: Some basic hypergeometric orthogonal polynomials that generalize Jacobi polynomials. Mem. Am. Math. Soc. 319
15. Loeven, G., Witteveen, J., Bijl, H.: Probabilistic collocation: an efficient non-intrusive approach for arbitrarily distributed parametric uncertainties. In: 45th AIAA Aerospace Science Meeting and Exhibit. Reno, Nevada (2007)
16. Marquardt, D.: An algorithm for least-squares estimation of nonlinear parameters. SIAM J. Appl. Math. **11**(2), 431–441 (1963)
17. Efron, B., Hastie, T., Johnstone, I., TIbshirani, R.: Least angle regression. Ann. Statist. **32**(2) 407–499 (2004)

General Introduction to Surrogate Model-Based Approaches to UQ

Daigo Maruyama, Stefan Görtz and Dishi Liu

Introduction

The quantification of aerodynamic uncertainties with CFD is computationally challenging due to two reasons: a large number of variables and high cost of CFD models, which calls for efficient numerical methods. Table 1 shows a categorization of some typical uncertainty quantification (UQ) methods to obtain some target statistics of aerodynamic performance. They are basically of two types: direct integration and surrogate-based integration. The former obtains the statistics by directly integrating the CFD samples, while the latter does so by integrating on a much cheaper surrogate built on the basis of the samples. These methods adopt either scattered or regular grid sampling scheme; the former admits samples anywhere in the variable domain, while the latter only samples on a prescribed grid (either tensorial or sparse).

In UQ problems with a large number of variables, like those with geometrical uncertainties, the methods based on regular grid are less favored due to two reasons. The first is their vulnerability to "curse of dimensionality", even with sparse grid. The second is their lack of tolerance to sample failures (which are not rare in CFD computations). With methods based on scattered sampling schemes, one has much more freedom in choosing sample number N and robustness against sample failures; this group includes mainly Monte Carlo (MC) quadrature and quasi-Monte Carlo

D. Maruyama (✉) · S. Görtz · D. Liu
German Aerospace Center (DLR), Institute of Aerodynamics and Flow Technology, Braunschweig, Germany
e-mail: Daigo.Maruyama@dlr.de

S. Görtz
e-mail: Stefan.Goertz@dlr.de

D. Liu
e-mail: Dishi.Liu@dlr.de

C. Hirsch et al. (eds.), *Uncertainty Management for Robust Industrial Design in Aeronautics*, Notes on Numerical Fluid Mechanics and Multidisciplinary Design 140, https://doi.org/10.1007/978-3-319-77767-2_12

Table 1 Some typical uncertainty quantification methods

	Direct integration	Surrogate-based integration
Scattered sampling	Monte Carlo Quasi-Monte Carlo	Non-intrusive polynomial chaos Kriging Radial basis functions
Regular grid sampling	Sparse grid quadrature	Stochastic collocation

(QMC) quadrature. The MC quadrature have dimension-independent error convergence rate $O(N^{-1/2})$, while QMC, which does not belong to the MC-type quadrature methods, has an upper-bound rate of $O(\log^d(N)N^{-1})$ with d the number of variables. Due to higher degree of sample uniformity, the latter can be more accurate even with a large d.

In the other dimension in Table 1, surrogate-based methods are gaining more use recently, e.g., radial basis functions, Kriging, and polynomial chaos expansion. Surrogate-based methods have an advantage when the gradients can be sampled at a relatively lower cost (e.g., by using an adjoint solver), because the direct integrations cannot effectively utilize the gradient information (augmenting samples generated by Taylor's expansion are not statistically independent of the original samples and hence bring no benefit to the accuracy of the statistics). Due to the cheaper cost of gradients (in case that number of response quantities is smaller than number of variables), surrogate-based methods utilize more information with the same computational cost, and this advantage is expected to increase with number of variables.

Surrogate models are frequently being used in the context of efficient quantification of uncertainties. A surrogate model of the quantity of interest is much cheaper to evaluate than the full-order model and can be used to obtain the target statistics through sampling on the surrogate models. There are several types of surrogate modeling methods that are being used, including response surface methods using polynomial models, Kriging, radial basis functions (RBF), support vector regression, and stochastic collocation. In this chapter, a general introduction to Kriging and RBF models is presented for UQ applications. Both models are constructed based on scattered data obtained by sampling the full-order model. Different sampling methods are in use, such as Latin hypercube sampling, Monte Carlo, and quasi-Monte Carlo sampling, and are typically referred to as design of experiment methods (DoE). A discussion of the different DoE methods is, however, beyond the scope of this chapter.

Kriging

Kriging is a statistical interpolation method suggested by Krige [1]. Kriging has received growing popularity due to its good capability of predicting multidimensional and highly nonlinear responses based on sampled data while providing a useful error estimation (indicating the uncertainty of the prediction). Here, an approximation of a function of interest $f(\mathbf{x})$ with d variables (uncertain parameters in UQ applications) $\mathbf{x} = [x_1, x_2, \ldots, x_d]^T$ is sought after. The approximation of $f(\mathbf{x})$ at an arbitrary $\mathbf{x}$ by the Kriging model is generally represented as follows when n samples of the function $f(\mathbf{x})$ are given as $f(\mathbf{x}^{(1)}), f(\mathbf{x}^{(2)}), \ldots, f(\mathbf{x}^{(n)})$:

$$\hat{f}(\mathbf{x}) = \beta + \varepsilon(\mathbf{x}) \tag{1}$$

where $\hat{f}(\mathbf{x})$ is the predicted function value at an arbitrary point $\mathbf{x}$, β is a global constant, and $\varepsilon(\mathbf{x})$ is a stationary random process. Different types of Kriging exist. The most commonly used types of Kriging are ordinary Kriging and universal Kriging. When β is expressed by the mean of all of the given samples as $\beta = E[f(\mathbf{x})]$, the model is called ordinary Kriging. On the other hand, when β is expressed by regression such as expressed by a general linear or polynomial trend model, the model is called universal Kriging. Some more detailed classification of the different types of Kriging model can be found in [2].

The stationary random process $\varepsilon(\mathbf{x})$ indicates deviation from the constant value β, having a Gaussian type PDF as $N(0, \sigma)$, where σ is the process variance. The stationary random process $\varepsilon(\mathbf{x})$ is a covariance of arbitrary two points and satisfies the following correlation matrix:

$$Corr\left[\varepsilon\left(\mathbf{x}^{(i)}\right), \varepsilon\left(\mathbf{x}^{(j)}\right)\right] = \sigma^2 \mathbf{R}\left(\mathbf{x}^{(i)}, \mathbf{x}^{(j)}\right) \tag{2}$$

where $\mathbf{R}$ is a spatial correlation function depending on $\mathbf{x}^{(i)}$ and $\mathbf{x}^{(j)}$. There are several forms of this correlation function such as Gaussian or cubic spline. Here Gaussian is given as an example. In this case, the correlation function is expressed as:

$$\mathbf{R}\left(\mathbf{x}^{(i)}, \mathbf{x}^{(j)}\right) = \prod_{k=1}^{d} \exp\left(-\theta_k \left|x_k^{(i)} - x_k^{(j)}\right|^2\right) \tag{3}$$

where $\boldsymbol{\theta} = [\theta_1, \theta_2, \ldots, \theta_d]^T$ are the distance weights with respect to the variables $\mathbf{x} = [x_1, x_2, \ldots, x_d]^T$. Here in this stochastic model (Kriging model), one can define the likelihood function L, which is a function denoted by β, σ, and $\boldsymbol{\theta}$ as:

$$L(\beta, \sigma, \boldsymbol{\theta}) = \frac{1}{\sqrt{(2\pi\sigma^2)^n |\mathbf{R}|}} \exp\left(-\frac{(\mathbf{f} - \beta\mathbf{F})^T \mathbf{R}^{-1} (\mathbf{f} - \beta\mathbf{F})}{2\sigma^2}\right) \tag{4}$$

where $\mathbf{f} = \left[f(\mathbf{x}^{(1)}), f(\mathbf{x}^{(2)}), \ldots, f(\mathbf{x}^{(n)})\right]^T$, and $\mathbf{F}$ is a regression matrix in the case of universal Kriging. When ordinary Kriging is adopted, $\mathbf{F} = \mathbf{1}$, which is the n-dimensional unit vector. The likelihood function $L(\beta, \sigma, \boldsymbol{\theta})$ is to be maximized to realize stochastically the best fitting to the data, which is called maximum likelihood estimation (MLE). The optimum β and σ can explicitly be determined when $L(\beta, \sigma, \boldsymbol{\theta})$ is maximized:

$$\beta(\boldsymbol{\theta}) = \left(\mathbf{F}^T\mathbf{R}^{-1}\mathbf{F}\right)^{-1}\mathbf{F}^T\mathbf{R}^{-1}\mathbf{f}$$
$$\sigma^2(\boldsymbol{\theta}) = \frac{1}{n}(\mathbf{f} - \beta\mathbf{F})^T\mathbf{R}^{-1}(\mathbf{f} - \beta\mathbf{F}) \tag{5}$$

The optimum values of the hyperparameters $\boldsymbol{\theta}$ are usually found by maximizing the logarithm of likelihood function:

$$\ln[L(\boldsymbol{\theta})] = -n\ \ln\left(\sigma^2(\boldsymbol{\theta})\right) - \ln|\mathbf{R}(\boldsymbol{\theta})| \tag{6}$$

Finally after the MLE, the predicted function value $\hat{f}(\mathbf{x})$ at an arbitrary unsampled location $\mathbf{x}^*$ can be represented as follows:

$$\hat{f}(\mathbf{x}^*) = \beta + \mathbf{r}^T\mathbf{R}^{-1}(\mathbf{f} - \beta\mathbf{F}) \tag{7}$$

where $\mathbf{r} = \left[r^{(1)}, r^{(2)}, \ldots, r^{(n)}\right]^T$ is the correlation expressed by Eq. (2) between an arbitrary unsampled location $\mathbf{x}^*$ and the sample points; e.g., at the ith sample point, it is expressed as:

$$r^i = Corr\left[\varepsilon(\mathbf{x}^*), \varepsilon\left(\mathbf{x}^{(i)}\right)\right] \tag{8}$$

This leads us to estimate stochastic values such as mean, standard deviation, maximum or minimum value, and probability density function (pdf) when the prediction points $\mathbf{x}$ are considered as uncertain input parameters and the function $f(\mathbf{x})$ is the quantity of interest (QoI).

Co-Kriging and Gradient-Enhanced Kriging

The correlation described in Eq. (2) is defined on the function values $f(\mathbf{x})$ at the sample points $\mathbf{x}$. Usually, the more the number of samples is increased, the better the quality of the Kriging model becomes. Instead of increasing the number of samples, secondary or auxiliary information available at the existing sample points or at other locations, where no primary information is available, could be incorporated into the Kriging model. This idea is the so-called co-Kriging. Gradient-enhanced Kriging is obtained if the partial gradients of QoI with respect to the variables are available at the primary sample points in addition to the function values and used as secondary information [3–6]. For instance, if the gradient information is calculated efficiently by an adjoint CFD solver, the accuracy of the Kriging model could be significantly improved. Gradient-enhanced Kriging

(GEK) is introduced in brief in the following because of its importance in the context of efficient quantification of aerodynamic uncertainties.

The correlation of Eq. (2) is modified as follows in GEK:

$$\begin{aligned}
Corr\left[\varepsilon\left(\mathbf{x}^{(i)}\right),\ \varepsilon\left(\mathbf{x}^{(j)}\right)\right] &= \sigma^2\mathbf{R}\left(\mathbf{x}^{(i)},\ \mathbf{x}^{(j)}\right) \\
Corr\left[\varepsilon(\mathbf{u}_i),\ \frac{\partial\varepsilon\left(\mathbf{u}_j\right)}{\partial u_k}\right] &= \sigma^2\frac{\partial\mathbf{R}\left(\mathbf{x}^{(i)},\ \mathbf{x}^{(j)}\right)}{\partial x_k^{(j)}} \\
Corr\left[\frac{\partial\varepsilon(\mathbf{u}_i)}{\partial u_k},\ \varepsilon\left(\mathbf{u}_j\right)\right] &= \sigma^2\frac{\partial\mathbf{R}\left(\mathbf{x}^{(i)},\ \mathbf{x}^{(j)}\right)}{\partial x_k^{(i)}} \\
Corr\left[\frac{\partial\varepsilon(\mathbf{u}_i)}{\partial u_l},\ \frac{\partial\varepsilon\left(\mathbf{u}_j\right)}{\partial u_k}\right] &= \sigma^2\frac{\partial^2\mathbf{R}\left(\mathbf{x}^{(i)},\ \mathbf{x}^{(j)}\right)}{\partial x_l^{(i)}\partial x_k^{(j)}}
\end{aligned} \tag{9}$$

Then, the size of the correlation matrix becomes $n(1+d)$, $\mathbf{f}$ and $\mathbf{F}$ in Eqs. (4), (5), and (7) are:

$$\mathbf{f} = \left[f\left(\mathbf{x}^{(1)}\right),\ \ldots,f\left(\mathbf{x}^{(n)}\right),\ \frac{\partial f\left(\mathbf{x}^{(1)}\right)}{\partial\mathbf{x}_1},\ \frac{\partial f\left(\mathbf{x}^{(1)}\right)}{\partial\mathbf{x}_2},\ \ldots,\ \frac{\partial f\left(\mathbf{x}^{(n)}\right)}{\partial\mathbf{x}_d}\right]^T,$$

$$\mathbf{F} = [1,\ \ldots,\ 1,\ 0,\ 0,\ \ldots,\ 0]^T,$$

respectively. Further enhancement by using the Hessian information has also proposed, and some applications to uncertainty quantification (UQ) and aerodynamic shape optimization were shown in [7]. This is a powerful surrogate modeling method when the Hessian is evaluated efficiently with the help of automatic differentiation (AD) tools [8].

Statistical Indicators in Kriging

Since the Kriging model is based on Bayesian statistics, it contains not only the predicted values but also information of uncertainty or, in other words, the approximation error at unsampled locations. For example, the mean square error (MSE) $s(\mathbf{x})$ at arbitrary points $\mathbf{x}$ is given by

$$s(\mathbf{x}) = \sigma^2\left[1-\mathbf{r}^T\mathbf{R}^{-1}\mathbf{r}+\left(\mathbf{r}^T\mathbf{R}^{-1}\mathbf{F}\right)\left(\mathbf{F}^T\mathbf{R}^{-1}\mathbf{F}\right)^{-1}\left(\mathbf{r}^T\mathbf{R}^{-1}\mathbf{F}\right)^T\right] \tag{10}$$

The MSE is considered as one of the major indicators frequently used for error estimation and adaptive sampling techniques [5, 9]. One other popular indicator where the MSE in Eq. (10) is made use of is in the so-called expected improvement

(EI) function [10]. This function indicates the probability that the predicted function $\hat{f}(\mathbf{x})$ is improved in terms of the current maximum or minimum sampled value of $f(\mathbf{x})$ if a new sample is added at some location, or, in other words, this metric helps to select the point which is statistically most advantageous to sample next. The EI is represented as follows:

$$EI(\mathbf{x}) = \begin{cases} (f_{\min} - \hat{f})\Phi\left(\frac{f_{\min} - \hat{f}}{s(\mathbf{x})}\right) + s(\mathbf{x})\phi\left(\frac{f_{\min} - \hat{f}}{s(\mathbf{x})}\right), & \mathit{if}\ s(\mathbf{x}) > 0 \\ 0, & \mathit{if}\ s(\mathbf{x}) = 0 \end{cases} \tag{11}$$

where $f_{\min}$ is the current best sampled point (minimum function value). Φ and ϕ denote the cumulative density function (CDF) and the probability density function (PDF), respectively. EI for the maximum can be calculated by replacing $f_{\min} - \hat{f}$ with $\hat{f} - f_{\max}$, where $f_{\max}$ is the current maximum sample value. A new sample is filled in at the location where the EI function is maximal. EI is often used in surrogate-based optimization to infill additional samples in order to enhance the accuracy of the Kriging model around the maximum/minimum values. This optimization strategy is called efficient global optimization (EGO).

An extension to multiobjective optimization problems, the so-called expected hypervolume improvement (EHVI), has also been proposed [11, 12]. The relative expected improvement (REI) function has been also studied for the purpose of evaluating the mean value more efficiently in the context of UQ [13]. These different measures help to infill additional samples efficiently to refine and improve the surrogate model. This possibility is being made use of in the following chapters estimating stochastic quantities more accurately.

Radial Basis Functions (RBF)

The radial basis function (RBF) method is an approximation of an unknown function by linear combination of radial basis functions expressed as follows:

$$\begin{aligned} \hat{f}(\mathbf{x}^*) &= \boldsymbol{\beta}^T \mathbf{g}(\mathbf{x}^*, \mathbf{x}) = \sum_{i=1}^{n} \beta_i g_i \\ \mathit{where} \quad g_i &= \phi\left(r_i \left\| \mathbf{x}^* - \mathbf{x}^{(i)} \right\|\right) \end{aligned} \tag{12}$$

where $\boldsymbol{\beta}^T$ is a vector of weighted parameters which is uniquely determined by a set of the given samples and corresponding function values by solving the concurrent linear system in Eq. (12). n is the number of sample points, and ϕ is a radial basis function, determined by the Euclidean distance r between an arbitrary unsampled point $\mathbf{x}^*$ and the sample point $\mathbf{x}$. Then, the predicted function $\hat{f}(\mathbf{x}^*)$ at an arbitrary unsampled locations $\mathbf{x}^*$ is calculated by Eq. (12).

Table 2 Radial basis functions with global support

Name	$\phi(r)$
Thin plate spline	$r^2 \log(r)$
Gaussian	e^{-r^2}
Cubic	$(a+r)^3$
Multiquadric biharmonics	$\sqrt{a^2+r^2}$
Inverse multiquadric biharmonics	$\sqrt{\frac{1}{a^2+r^2}}$
Quadric biharmonics	$1+r^2$
Inverse quadric biharmonics	$\frac{1}{1+r^2}$

There are several classes of radial basis functions. A general classification is to distinguish between compact support and global support. The radial basis functions with compact support can be applied in mesh deformation [14]. In applications to surrogate models, those with global support are often used. Table 2 shows seven radial basis functions which are frequently used. The parameter a in Table 2 is usually tuned for fitting to the samples.

This method can also utilize gradient information of the QoI, e.g., the gradient-assisted RBF method proposed in [15]. Also, the gradients of the predicted function $\hat{f}(\mathbf{x}^*)$ can be obtained analytically by the following equation:

$$\begin{aligned} \frac{\partial \hat{f}(\mathbf{x}^*)}{\partial \mathbf{x}^*} &= \boldsymbol{\beta}^T \frac{\partial \mathbf{g}}{\partial \mathbf{x}^*} = \sum_{i=1}^{n} \beta_i \phi' \left(r_i \left\| \mathbf{x}^* - \mathbf{x}^{(i)} \right\| \right) \frac{\partial r_i}{\partial \mathbf{x}^*} \\ \text{where} \quad \frac{\partial r_i}{\partial \mathbf{x}^*} &= \frac{\left(\mathbf{x}^* - \mathbf{x}^{(i)} \right)^T}{r_i \left\| \mathbf{x}^* - \mathbf{x}^{(i)} \right\|} \end{aligned} \tag{13}$$

where $\phi'(r_i) = \partial\phi/\partial r_i$. A more detailed discussion on gradients and Hessians of surrogated models based on RBF can be found in [16].

Summary

Radial basis function models and different types of Kriging models are often used to approximate unknown functions based on scattered data. One of the major differences between these two methods is that Kriging provides the mean squared error of the prediction. This statistical information can be used to improve the surrogate model by adaptively sampling with respect to the targeted statistics of the quantity of interest (QoI). This is of particular interest in UQ and robust design in aerodynamics, where CFD computations are very costly.

It is reported by one of the authors [17] that GEK and Kriging are more efficient methods to propagate uncertainties through a CFD model than many others, especially when the sample number is not large, e.g., N < 200 in a 9D nonlinear model. The RBF method is suitable for scattered samples and high-dimensional models. However, a large sample number could result in a large condition number in the linear system which reduces the accuracy of RBF approximations [18], and the gradient-assisted RBF method is not as widely used as gradient-enhanced Kriging method.

References

1. Krige, D.G.: A statistical approach to some basic mine valuations problems on the Witwatersrand. J. Chem. Metall. Min. Eng. Soc. S. Afr. **52**(6), 119–139 (1951)
2. Hengl, T.: A practical guide to geostatistical mapping of environmental variables. European Comission, JRC Scientific and Technical Reports. EUR 22904 EN (2007)
3. Chung, H.S., Alonso, J.J.: Using gradients to construct cokriging approximation models for high-dimensional design optimization problems. AIAA Paper 2002-0317 (2002)
4. Han, Z.H., Görtz, S., Zimmermann, R.: Improving variable-fidelity surrogate modeling via gradient-enhanced kriging and a generalized hybrid bridge function. J. Aerosp. Sci. Technol. **25**(1) (2013)
5. Dwight, R., Han, Z.H.: Efficient uncertainty quantification using gradient-enhanced Kriging. AIAA Paper 2009–2276 (2009)
6. Lockwood, B.A., Anitescu, M.: Gradient-enhanced universal kriging for uncertainty propagation. Preprint ANL/MCS-P1808-1110 (2010)
7. Yamazaki, W., Rumpfkeil, M.P., Mavriplis, D.J.: Design optimization utilizing Gradient/ Hessian enhanced surrogate model. In: 28th AIAA Applied Aerodynamics Conference. AIAA paper 2010-4363 (2010)
8. Rumpfkeil, M.P., Mavriplis, D.J.: Efficient Hessian calculations using automatic differentiation and the adjoint method. AIAA-2010-1268 (2010)
9. Shimoyama, K., Kawai, S., Alonso, J.J.: Dynamic adaptive sampling based on kriging surrogate models for efficient uncertainty quantification. In: 54th AIAA/ASME/ASCE/AHS/ ASC Structures, Structural Dynamics, and Materials Conference. AIAA paper 2013-1470 (2013)
10. Jones, D.R., Schonlau, M., Welch, W.J.: Efficient global optimization of expensive black-box functions. J. Glob. Optim. **13**(4), 455–492 (1998)
11. Emmerich, M.: Single- and multi-objective evolutionary design optimization using gaussian random field metamodels. Ph.D. Dissertation, TU Dortmund (2005)
12. Shimoyama, K., Sato, K., Jeong, S., Obayashi, S.: Updating Kriging surrogate models based on the hyper volume indicator in multi-objective optimization. J. Mech. Des. **135**(9). Article ID 094503 (2013)
13. Łaniewski-Wołłk, L.: Relative expected improvement in kriging based optimization. In: EUROGEN 2009 Kwakow Conference (2009)
14. de Boer, A., van der Schoot, M.S., Bijl, H.: Mesh deformation based on radial basis function interpolation. Comput. Struct. **85**(11–14), 784–795 (2007)
15. Giannakoglou, K.C., Papadimitriou, D.I., Kampolis, I.C.: Aerodynamic shape design using evolutionary algorithms and new gradient-assisted metamodels. Comput. Methods Appl. Mech. Eng. **195**(44–47), 6312–6329 (2006)
16. McDonald, D.B., Grantham, W.J., Tabor, W.L., Murphy, M.J.: Global and local optimization using radial basis function response surface models. Appl. Math. Model. **31**(10) (2007)

17. Liu, D.: A best practice guide: effcient quantification of aerodynamic uncertainties. Report of the Institute of Aerodynamics and Flow Technology, DLR, DLR-IB 124-2012/2. ISSN 1614-7790 (2012)
18. Lazzaro, D., Montefusco, L.B.: Radial basis functions for the multivariate interpolation of large scattered data sets. J. Comput. Appl. Math. **140**, 521–536 (2002)

Comparing Surrogates for Estimating Aerodynamic Uncertainties of Airfoils

Daigo Maruyama, Dishi Liu and Stefan Görtz

Introduction

In aerodynamic analysis and design of airfoils, the quantities of interest (QoI) are typically the drag coefficient (C_d), the lift coefficient (C_l) (or the lift-to-drag ratio L/D), the pitching moment coefficient (Cm), and the pressure distributions (C_p). In the context of uncertainty quantification and robust design, the QoI are statistics of the aerodynamic coefficients. Uncertain input parameters may be operational parameters, such as the angle of attack (AoA), the Mach number (M), or the Reynolds number (Re), and an inherently large number of geometric parameters.

There are two issues in terms of the computational cost and the accuracy when constructing a surrogate model in applications to uncertainty quantification (UQ) and robust design of airfoils. One is that the computational cost associated with the construction of the surrogate model may not be negligible any more when the space of input parameters is high-dimensional which is often of the case when geometrical uncertainties are considered. In this chapter, we consider a test case where every surface grid point of the CFD grid is considered uncertain, but we assume a correlation between the points and make use of a truncated Karhunen–Loève expansion (KLE) to reduce the number of parameters [1, 2].

The other issue is how the accuracy depends on the number of samples used to build the different surrogate models. Therefore, we first present a general study in

D. Maruyama (✉) · D. Liu · S. Görtz
German Aerospace Center (DLR), Institute of Aerodynamics and Flow Technology, Braunschweig, Germany
e-mail: Daigo.Maruyama@dlr.de

D. Liu
e-mail: Dishi.Liu@dlr.de

S. Görtz
e-mail: Stefan.Goertz@dlr.de

C. Hirsch et al. (eds.), *Uncertainty Management for Robust Industrial Design in Aeronautics*, Notes on Numerical Fluid Mechanics and Multidisciplinary Design 140, https://doi.org/10.1007/978-3-319-77767-2_13

terms of accuracy of statistics of aerodynamic performance of an airfoil using different surrogate models that are conducted. A larger number of sample points are expected to enhance the quality of all of the surrogate models, but this will inevitably lead to larger computational cost.

Suitable sampling techniques are required for different target statistics of the aerodynamic performance statistics, such as mean of C_d, maximum value of C_d, or minimum value of C_l. This is the second topic here and is of interest in the context robust design as an application of UQ to aerodynamic design of airfoils. Therefore, in this chapter, the following two topics are discussed:

(a) UQ using various surrogate models and different number of samples and comparison to reference statistics;
(b) Sampling techniques for specific stochastic quantities of interest, including adaptive sampling.

For the surrogate modeling part, we made DLR's surrogate and reduced-order modeling toolbox called SMARTy [3–5].

Comparison of Different Surrogate Models

As for the first topic, to investigate the efficiency of various surrogate models in the context of UQ, we made two comparisons in which the methods are compared in their efficiency of quantifying aerodynamic performance uncertainties of airfoils caused by operational and/or geometric uncertainties. First, we compare Kriging and gradient-enhanced Kriging (GEK) with direct integration for a viscous test case. Second, we compare direct integration and four UQ methods based on surrogate-based integration; i.e., we compare quasi-Monte Carlo (QMC) quadrature with GEK, polynomial chaos (PC) combined with a sparse Gauss–Hermite (SGH) quadrature, gradient-enhanced radial basis functions (GERBF), and a gradient-enhanced polynomial chaos (GEPC) method for an inviscid test case (an introduction to these methods can be found in [1]).

Unless otherwise stated, the computational cost is measured in terms of "compensated evaluation number" N_c. For the gradient-employing methods, $N_c = mN$ with N denoting the number of sample Points and m denoting that of response quantity, since the cost of computing the partial gradients of one response quantity with respect to all uncertain parameters using an adjoint CFD solver equals approximately to the cost of one nonlinear CFD evaluation and m denotes the number of response quantities. For Kriging, QMC, and PC-SGH methods, $N_c = N$.

The reference statistics (against which the error of estimated statistics is judged) in the first comparison is obtained by a direct integration of ten thousand QMC samples, while that for the second comparison is acquired by an integration of four million QMC samples. The accuracy of the latter reference statistics is estimated by using the multipartition method [6].

Viscous Test Case: Comparison of Kriging and GEK

This comparison is based on a CFD model of the viscous flow around the RAE2822 airfoil. We opt for DLR's unstructured RANS solver TAU [7–9], the Spalart-Allmaras turbulence model, a central flux discretization scheme, matrix dissipation, and a "3v" multigrid cycle. The domain is discretized by a hybrid unstructured grid in which the airfoil has 380 surface nodes, as shown in Fig. 1. The uncertainties come from a random Mach number and angle of attack, together with a random perturbation to the original airfoil geometry at every surface grid point. The two operational variables are assumed to be beta-distributed around $M = 0.729$ and $\alpha = 2.31°$, respectively. The perturbations in Mach number and angle of attack are with a support within ±2% of the nominal values. The geometry perturbation is modeled by a random field parameterized into 24 independent Gaussian variables through a truncated KLE. We kept all eigenvalues larger than 10^{-7} in the truncated KLE approximation [1, 2]. The upper and lower surfaces of the airfoil were treated as two separate random fields as their correlation in geometric variations is assumed weak in this test case.

Quasi-Monte Carlo quadrature and two surrogate-based UQ methods, gradient-enhanced Kriging and plain Kriging, are applied to the test case, and their efficiency is compared in estimating two statistics (mean and standard deviation) of the coefficient of lift (C_l). The accuracy of the estimates is judged by comparison with reference statistics which are based on 10,000 QMC samples. The computational cost is measured in terms of "compensated evaluation number" N_c. Note that for the gradient-employing method (GEK) N_c is set to 2N in this particular comparison since here we only handle one system response quantity (SRQ), i.e., C_l.

Figure 2 shows the results of the comparison in terms of two different statistics of C_l as a function of N_c. GEK is seen to converge faster than Kriging in estimating

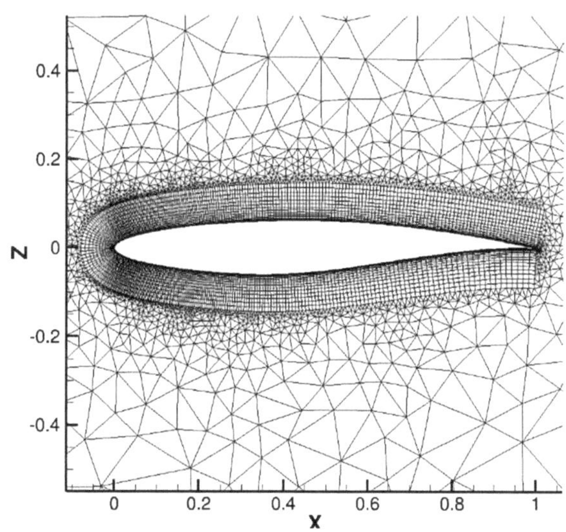

Fig. 1 Grid of RAE2822 airfoil in viscous test case 1: zoom around the airfoil

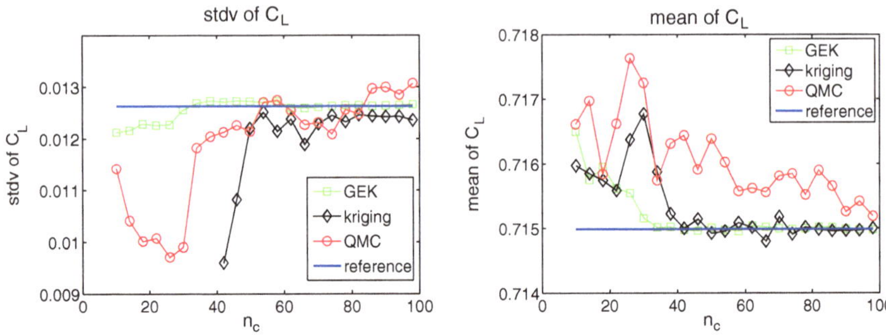

Fig. 2 Convergence of estimate C_l statistics to the reference statistics by various UQ methods (note that $N_c = 2N$ in case of GEK because only the gradient of C_l was considered, while $N_c = N$ for Kriging and QMC) [2]

all two statistics. This can obviously be attributed to the more information utilized by the former with the help of the adjoint TAU solver. It can also be observed that if the sample number is small, Kriging may perform worse than QMC. Figure 3 shows a comparison of the probability density functions (pdf) obtained by GEK and QMC for the same computational cost ($N_c = 40$). It can be confirmed that the pdf by GEK has much better agreement with that of the reference.

Inviscid Test Case: Comparison of Four Surrogate-Based Integration UQ Methods and Direct Integration

This comparison is based on a CFD model of the inviscid flow around the RAE2822 airfoil at a Mach number of 0.73 and an angle of attack 2.0°. We use the TAU flow solver, opting for a central flux discretization, scalar dissipation, a backward Euler solver, and a “4w” multigrid cycle. The domain is discretized by a 193-by-33 structured grid in which the airfoil has 128 surface nodes, as shown in Fig. 4.

The source of uncertainty is a random perturbation to the original airfoil geometry, which is modeled by a random field parameterized into nine independent Gaussian variables through a KLE. Figure 5 shows three examples of the perturbed geometry of the airfoil. With this parameterization of the uncertain geometry, the five UQ methods mentioned above are applied to the test case and compared in terms of their efficiency in estimating statistics of C_l and C_d, as well as the probability distribution functions (pdf), although here we only show results in terms of C_d. The full set of the results can be referred to in [2].

The following statistics of C_d are estimated as:

1. Mean of C_d, μ_l, and μ_d,
2. Standard deviation of C_d, σ_d,
3. Exceedance probabilities $P_{D,\kappa} = \Pr\{C_D \geq \mu_D - \kappa \cdot \sigma_D\}$ with $\kappa = 2, 3$.

The reference values of these statistics are obtained from a relatively large number ($N = 4 \times 10^6$) of quasi-Monte Carlo (QMC) samples of the CFD model.

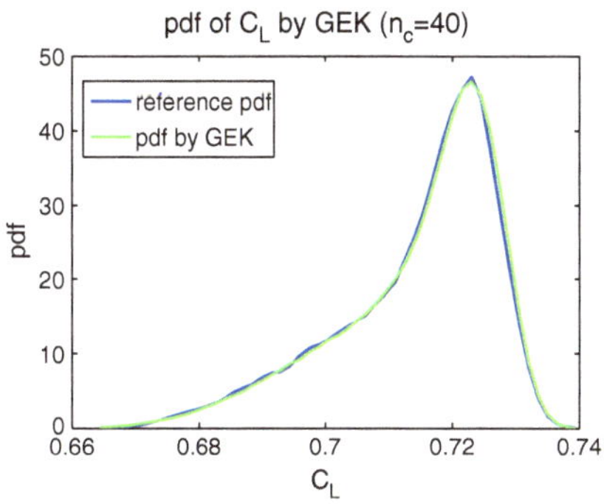

Fig. 3 Comparison of probability density functions (pdf) of lift coefficient obtained by GEK and QMC for the same computational cost ($N_c = 40$) (note that $N_c = 2N$ in case of GEK because only the gradient of C_l was considered, while $N_c = N$ for Kriging and QMC)

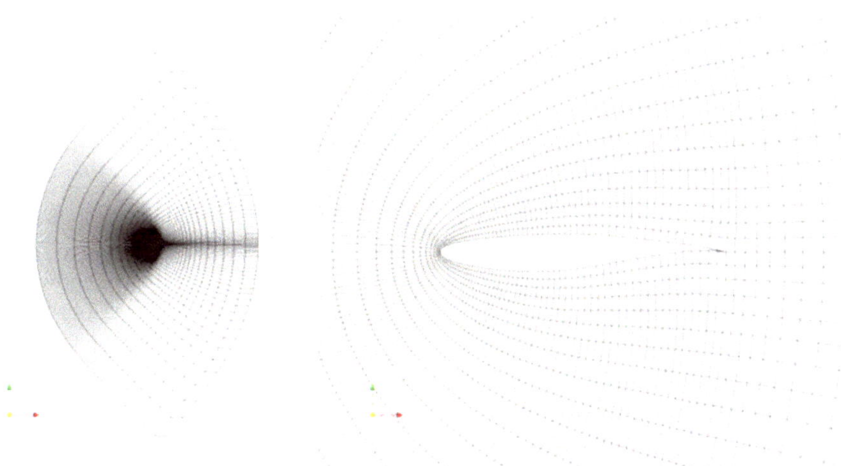

Fig. 4 Grid of RAE2822 airfoil in inviscid test case 2: the total grid (left) and zoom around the airfoil (right)

Since the theoretical error bound of QMC integration is not a practical accuracy indicator, the standard deviations (ς) of these reference values are estimated by using Snyder's multipartition method [6] and tabulated in Table 1.

In the efficiency comparison of the UQ methods below the smallest measured errors (measured against these reference values) in mean and standard deviation (stdv) are at least 10 times larger than 3ς, which means, by taking the assumption that these reference values are Gaussian distributed around the true values of the statistics, the measured errors have a 99.73% confidence interval of at widest ±10%. For exceedance probabilities, this confidence interval is also valid except for a few

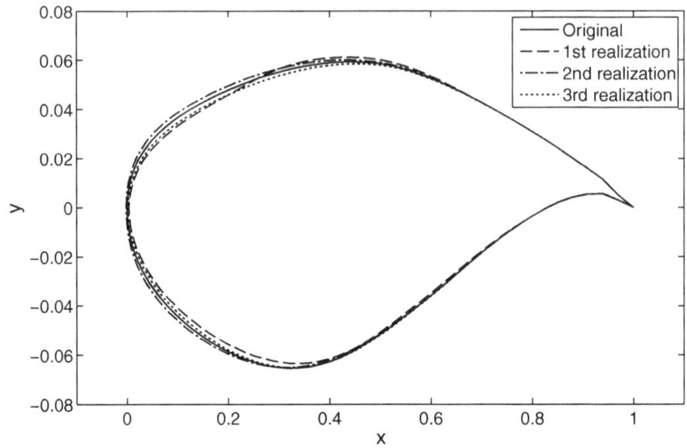

Fig. 5 Three realizations of perturbed geometry

Table 1 Estimated standard deviation of reference statistics for C_d [2]

$\varsigma(\mu_D)$	$\varsigma(\sigma_D)$	$\varsigma(P_{D,2})$	$\varsigma(P_{D,3})$
3.0e−9	1.4e−8	1.2e−5	6.8e−6

measured errors, as shown in Figs. 6 and 7, where the values of the corresponding 3ς are depicted by dashed lines without symbols.

The results of this comparison are shown in Figs. 6 and 7. The figures show the errors of the five methods in estimating the target statistics of C_d. It is observed there that generally the gradient-employing surrogate methods perform better than direct integration methods. This can be ascribed to the fact that the former utilize more information at the same computational cost. This advantage comes from the cheaper cost of the gradients computed by an adjoint solver in the case that the number of response quantities of interest is smaller than the number of variables (in our case, 2 vs. 9), and the advantage would increase for a larger number of variables or fewer response quantities. The PC-SGH method has only two data points due to the very limited choice of sample numbers. It is hard to evaluate its error convergence property based on only two data points.

This shows a fundamental shortcoming of sparse grid quadratures for a relative high number of variables (9 in this test case). The GEPC method shows constant convergence in error, but is not as efficient as the other two gradient-employing methods. The reason could be that the increase of admissible order of polynomial along increasing N_c is slow in this 9-variate problem and that the polynomial surrogate tends to "overshoot" in the outskirts of the domain. But the PC methods have a merit that they do not need a parameter optimization procedure. GEK and GERBF are the most efficient methods as far as seen from these results. Beside the cheaper gradients, this could be attributed to properties of the kernel functions they use and the effort of tuning model parameters. Convergence rate of inverse

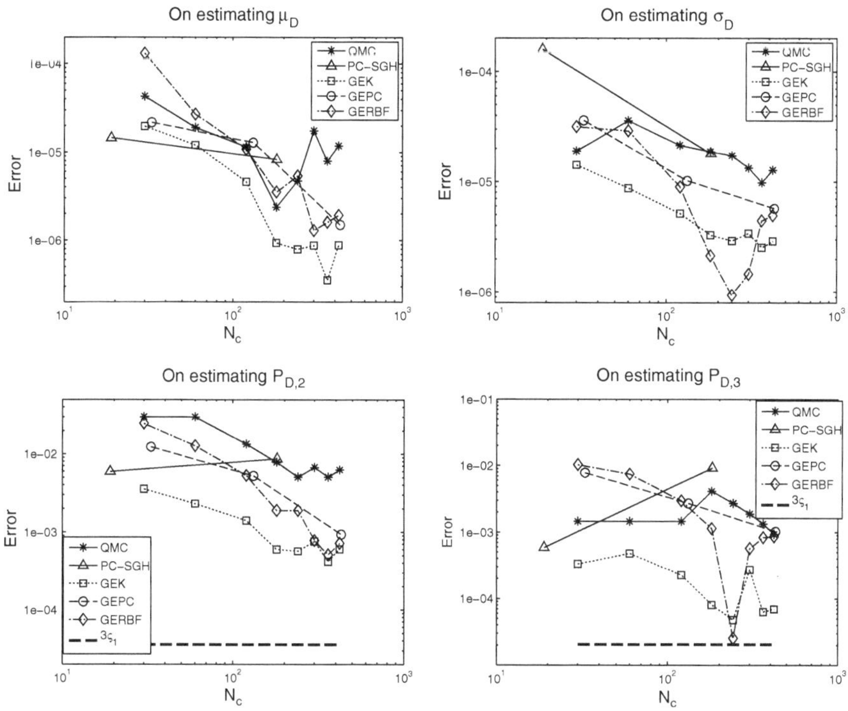

Fig. 6 Error in estimating mean, standard deviation (upper row), and exceedance probabilities (lower row) of Cd (note that $N_c = 3N$ in case of gradient-employing methods because the gradients of C_l and C_d were considered, while $N_c = N$ for PC-SGH and QMC) [2]

multiquadric RBF was estimated $O(e^{-c/h})$ with h the "fill distance" and c a constant, which translates to a rate in N at $O(e^{-cN^{1/d}})$ [10]. We assume the error in the statistics is proportional to that in point-wise approximations, and find in this 9-variate test case the observed convergence rate is much better than $O(e^{-cN^{1/9}})$—this hints that the effective fill distance h reduces faster than $O(N^{-1/d})$ due to that some variables are less important than the others (typical for a KLE parameterization). GEK seems slightly better than GERBF in these results, especially at smaller N_c values. But this cannot be generalized to other applications. In this particular case, the advantage of GEK could come from the possible advantage of the cubic spline kernel used over inverse multiquadric.

Different ways of utilizing gradient information by GEK (involving first- and second-order derivatives and generating a symmetric kernel matrix) and GERBF (involving only first-order derivatives and generating non-symmetric matrix) could also contribute to the difference in their performance. At larger N_c values, a "rebound" of error can be observed in GERBF and GEK results, more obvious in the former. This is caused by the stabilizing treatment of these methods. Though an uncertain relation (or trade-off principle) exists stating that accuracy and stability

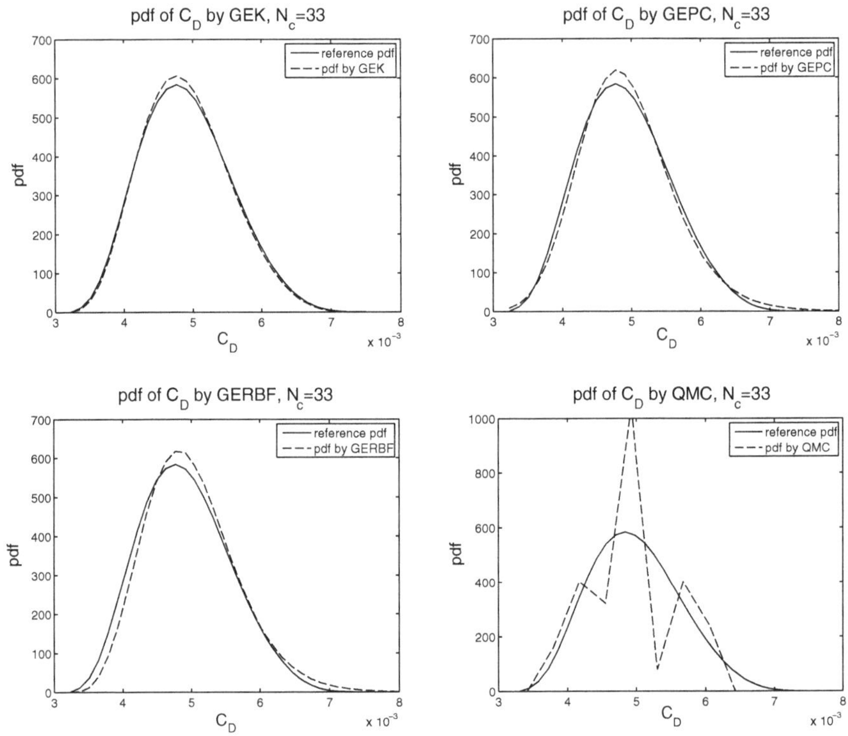

Fig. 7 Estimated pdf (in dash line) of C_d by QMC, GEK, GEPC, and GERBF at $N_c = 33$ [2]

cannot be both good, more advanced stabilizing techniques, e.g., pivoted Cholesky decomposition [11], could improve the convergence as sample number is large. Figure 7 shows the pdf of C_d estimated by QMC and the three gradient-employing surrogate methods with $N_c = 33$ (the smallest possible N_c value for GEPC method), in comparison with the reference pdf (computed by four million QMC samples). There one observes that for the same computational cost, the surrogate methods yield much more accurate pdfs. This is consistent with their relative performance in estimating the statistics. One question naturally arising here is whether the advantage of gradient-employing surrogate methods observed in this test case can be generalized to other cases. Our answer is "yes" for at least high-dimensional and almost-everywhere-differentiable problems. This is based on the following arguments. The particular configuration of the test case, e.g., discretization size and excluding/including viscous effect, influences the precision at which the CFD model approximates the true physics, while our study is on the precision at which numerical integral approximates CFD statistics. These two precisions can be studied separately. Yet this particular configuration could affect the CFD behavior upon the geometric uncertainty, but it would not change our conclusion because with cheaper gradients the former methods utilize more information than do the latter at the same cost. This advantage is independent of the configuration. And we have been careful to make only a characteristic conclusion rather than a quantitative one which could depend on the test case and its configuration.

One should notice that the advantage of the gradient-employing surrogate methods should not be taken as universal. For example, on problems with few variables or problems that are mostly non-differentiable the advantage might not exist. If the problem has a very oscillatory topography, the advantage could only be manifested with relatively more samples.

Sampling Techniques for Specific Stochastic Quantities of Interest

Based on the results in the previous section, efficient sampling techniques suitable for stochastic quantities of interest are discussed. In this section, we only consider GEK and a fixed number of 30 samples. GEK with a Gaussian kernel (correlation function) was adopted, and the hyperparameters were optimized by a global optimizer (a differential evolution algorithm was used) by maximum likelihood estimation (MLE). Note that setting up the entire GEK model took around 15 s. On one core on our HPC cluster.

Here, we focus on different sampling techniques to evaluate the statistics of quantities of interest (QoI) accurately for a fixed number of samples. The statistics of QoI are, for instance, mean and standard deviation of the drag coefficient, which are of interest in robust design applications with different measures of robustness, for example, the "expectation measure" and the "mean-risk approach." The maximum or minimum value of the QoI is necessary when the so-called "worst-case risk measure" is used. It would be expensive in terms of the number of samples to construct a versatile yet accurate surrogate model suitable to quantify any kind of statistical value. Here, we focus on how to efficiently and accurately quantify the statistical values which are directly necessary to evaluate a given objective function. The statistical values to be considered in this section are mean, standard deviation, maximum or minimum values of the drag coefficient (C_d), which is QoI.

The test case used in this section is the UMRIDA test case BC-02, which is based on the RAE2822 airfoil. The flow solver and the grid used (see Fig. 1) are the same as those described in the previous section for the viscous test case except that the negative Spalart-Allmaras turbulence model [12] was used. The considered operational and geometrical uncertainties are basically identical as well except that the mean values of angle of attack (α) and Mach number (M) are 2.79° and 0.734, respectively, to match the UMRIDA test case description. All of the 380 surface grid nodes of the airfoil are considered uncertain and are parameterized by 10 independent variables ξ furnished by a truncated KLE [1, 2] to keep the relative information content (the ratio between the principal eigenvalues and the total sum of all eigenvalues) larger than 99% [13]. The uncertain input parameters are assumed to have normal distributions as $\alpha \sim N(2.79°, 0.1°)$, $M \sim N(0.734, 0.005)$, and $\boldsymbol{\xi} \sim N(0, 1)$.

When UQ methods are applied to robust design, one approach is to evaluate the aerodynamic performance statistics for each nominal design, i.e., for different combinations of design parameters. This means that a surrogate model needs to be constructed at every iteration of the optimization process. Realizing high reproducibility and accuracy with a limited number of samples and suitable surrogate model construction for focusing on statistics of interest are required. Therefore, the issues here in terms of surrogate model accuracy are dependent on the statistics of interest on the following two issues:

(1) Initial set of points;
(2) Sample distributions and adaptive sampling.

Ideally, the statistics that are computed based on the surrogate model are independent of the set of samples used to construct the surrogate model. Different sets of points for a given number of samples can be obtained by extracting arbitrary continuous rows from a Sobol sequence [14–16] used as QMC sampling, which can maintain "uniformity" (also called "low discrepancy") of the samples. This is related to the first issue. Details of this strategy can be found in [17] and [18]. Discrepancy, which is a mathematical definition to measure the uniformity of samples, of these different sets of 30 samples is constant.

These original sample sets form uniform distributions. Here, it is assumed that the probability density functions (pdf) of the uncertain input parameters are given by normal distributions. The uniformly distributed samples can be transformed into normal distributions by using their cumulative density functions (cdf). Additional samples can be infilled based on statistical information in Kriging, which is a so-called adaptive sampling technique. Details are described later. These topics are related to the second issue mentioned in the above.

The input parameters are the uncertain operational and geometrical parameters: $\mathbf{u} = (\mathbf{a}, \boldsymbol{\xi})$, where $\mathbf{a} = (\alpha, M)$, and their normal distributions are truncated to $\mu_{\mathbf{u}} \pm 3\sigma_{\mathbf{u}}$. The statistics of interest are the following two kinds of statistical values f as the output uncertainties:

$$f \equiv \mu_{C_d} + \sigma_{C_d} \tag{1}$$

$$f \equiv \max_{\mathbf{u}}(C_d(\mathbf{u})) \tag{2}$$

These statistical values are named as "robust" and "reliability," respectively. They are used in the context of "expectation measure" and "mean-risk approach" as robust design by Eq. (1) and "worst-case risk measure" as reliability-based design by Eq. (2), respectively, to formulate objective functions. The evaluation of "reliability" as expressed by Eq. (2) can also be regarded as a surrogate-based optimization problem considering **u** as "design variables". Maximizing C_d on a surrogate model leads us to the idea of applying adaptive sampling techniques to

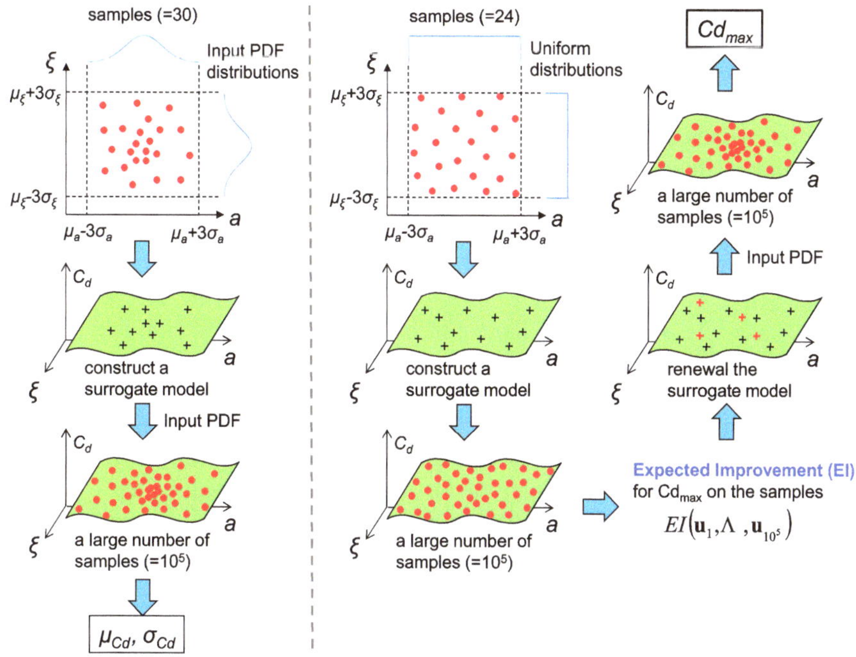

Fig. 8 Schematic diagram of two different procedures used to compute different statistics of interest based on sampling techniques and surrogate modeling

construct more accurate surrogate models for the statistics of interest. If Kriging is used, then the expected improvement (EI) function is a suitable indicator to yield more accurate surrogate models. The procedure of evaluating the stochastic value f is as follows (see also Fig. 8):

Step 1. Initial set of points are generated in the uncertain input parameter space $\mathbf{u} = (\mathbf{a}, \boldsymbol{\xi})$ in truncated ranges of $\mu_{\mathbf{u}} \pm 3\sigma_{\mathbf{u}}$ by using QMC sampling.
Step 2. The CFD and adjoint calculations are performed for the sample points to obtain C_d and its partial gradients with respect to the uncertain parameters $\mathbf{u} = (\mathbf{a}, \boldsymbol{\xi})$.
Step 3. A GEK surrogate model of C_d is constructed based on the initial samples.
Step 4. If necessary, adaptive sampling is applied to refine the surrogate model with additional samples, and go back to Step 2. Otherwise, the construction of a surrogate model finishes and any statistics is calculated on the surrogate model.

The time-consuming part of this procedure is Step 2, unless many samples are added in Step 4. At Step 1, we used a Sobol sequence for QMC sampling as mentioned above. The distributions of the generated samples are either transformed into the input pdf or a uniform is kept. At Step 4, stochastic values can be almost uniquely obtained by using Monte Carlo (MC) or QMC on the surrogate model with a sufficiently large number of samples (>10^5). During the steps from 1 to 4, the surrogate model is reconstructed when additional samples are filled in. Note that in any case, we chose to restrict the total number of samples to 30 to maintain the same computational costs throughout, even in the case of adaptive sampling.

In the case of evaluation of "robust" as expressed by Eq. (1), "uniformity" of the samples fitted to the input pdf could be an important factor to evaluate μ_{C_d} and σ_{C_d}. On the contrary, in the case of evaluating "reliability," i.e., evaluating $\max_{\mathbf{u}}(C_d(\mathbf{u}))$, any statistical information in the domains where the accuracy is needed cannot be driven by the input pdf. In this case, statistical indicators such as the mean-square error (MSE) of the prediction and the EI extracted from the Kriging model can be useful. The statistical indicators are first uniformly generated on the input uncertainties domain. More details about the adaptive sampling strategy are explained here by following the procedure described above. To start with, 24 initial samples with a uniform distribution are generated in Step 1. At Step 4, the expected improvement (EI) function is evaluated on the sufficiently large number of samples with the uniform distributions by using QMC on the surrogate model. EI is an indicator to find the sample that will most probably result in the largest improvement of the surrogate if computed with CFD. Here, we aim to maximize the C_d (see Eq. (2)). The formulation of EI in mono-objective maximization is represented by the following equation:

$$EI(\mathbf{u}) = \left(\hat{C}_d - C_{d\max}\right)\Phi\left(\frac{\hat{C}_d - C_{d\max}}{s(\mathbf{u})}\right) + s(\mathbf{u})\phi\left(\frac{\hat{C}_d - C_{d\max}}{s(\mathbf{u})}\right) \quad (3)$$

where $s(\mathbf{u})$ is the MSE of the surrogate model, $\mathbf{u}$ is the uncertain parameters, $\hat{C}_d$ is the value of the drag coefficient on the surrogate, $C_{d\max}$ is the current maximum value of all the initial samples evaluated with CFD, and Φ and ϕ denote the cumulative density function (cdf) and the probability density function (pdf), respectively. The EI value itself gives the probability that a sample point would become the global optimum (maximum in this case). Here, the six highest EI values are selected to determine six additional sample locations. Then, going back to Step 2, the CFD including adjoint calculations are executed on the six new sample points. The surrogate model is reconstructed at Step 3 by using the all 30 samples. Finally, $\max_{\mathbf{u}}(C_d(\mathbf{u}))$ is obtained on the reconstructed surrogate model at Step 4. In practice, it is estimated based on the surrogate model by using a random sampling method using MC or QMC with a sufficiently large number of samples (>10^5) instead of using an optimization algorithm. It is evaluated by the quantity at 99‰ of the cdf of C_d expressed as $\max_{\mathbf{u}}(C_d(\mathbf{u})) \approx Cd_{99\%\max}$ rather than taking $\max_{\mathbf{u}}(C_d(\mathbf{u}))$.

Now, we discuss how accurately the stochastic values f as represented by Eqs. (1) and (2) are evaluated in terms of the issues (1) and (2). Figures 9 and 10

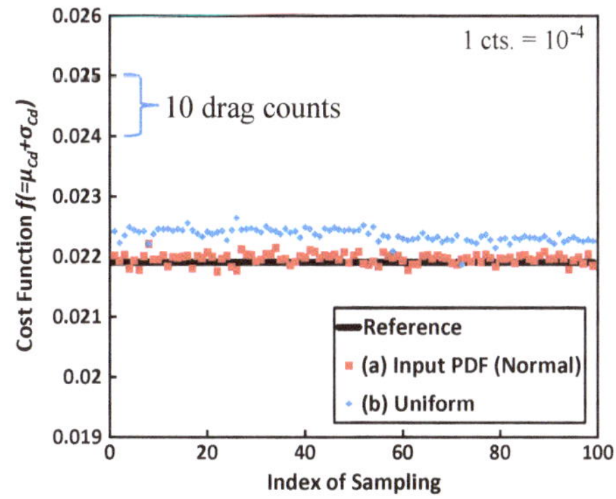

Fig. 9 Mean and standard deviation of C_d for different sample distributions as a function of the sampling sequence (for a given number of 30 samples used to construct a surrogate model)

show how the accuracy of f depends on the sample distribution used for constructing a surrogate model for each stochastic value. Both a uniform distribution and a normal distribution of the samples used to construct the surrogate model were considered. In addition, adaptive sampling was combined with a uniform initial sample distribution, but only for the case of evaluating the "reliability" $(Cd_{99\%\mathrm{max}})$. The horizontal axis is the index of the different set of points. Each index corresponds to a different set of 30 samples which were continuously extracted from the Sobol sequence as explained in the above. The reference values to evaluate the accuracy were obtained by using 10^5 direct CFD calculations on the Sobol sequence-based QMC sampling.

The results indicate that different sampling strategies are required for different stochastic values. This means that we should focus on the accuracy of the surrogate model in specific domains of the (uncertain input) parameter space. In the case of evaluating the "robustness" shown in Fig. 9, the samples are distributed according to the input pdf (normal distributions in this case, red square symbols), leading to higher accuracy in terms of the evaluated f than a uniform distribution (blue diamond symbols). As for "reliability," it can be observed in Fig. 10 that the combination of a uniform sample distribution and adaptive sampling (green square

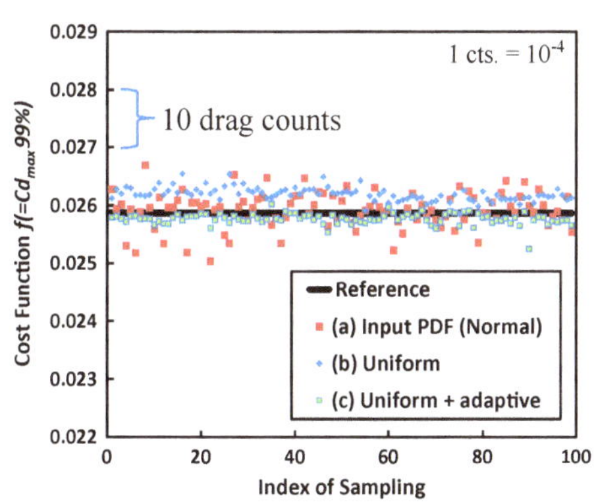

Fig. 10 Value of C_d at the 99‰ of cdf for different sample distributions as a function of the sampling sequence (for a given number of 30 samples used to construct a surrogate model)

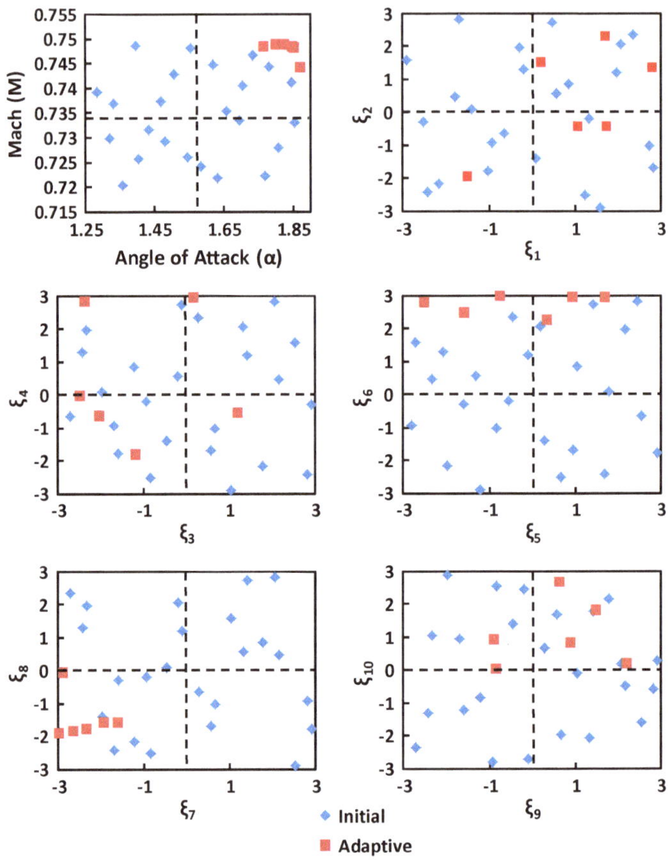

Fig. 11 An example of the initial and adaptive sampling distributions in the input uncertainty parameter space $\mathbf{u} = (\mathbf{a}, \boldsymbol{\xi})$ for evaluation of the 99‰ of cdf of C_d

symbols) effectively reduces the noise in the cost function f and can improve the accuracy in terms of the absolute value of f compared to a uniform sample distribution without adaptive sampling (diamond symbols). In this case, the accuracy of the surrogate model in terms of $Cd_{99\%\max}$ was improved by adaptive sampling, which plays an important role to determine the statistics of interest.

Figure 11 shows one example of the initial and adaptive sample distributions for the worst-case risk measure, $Cd_{99\%\max}$. The adaptively added sample points (red square symbols) are mostly located close to the borders of the parameter space where samples used in the case of the "mean risk" are usually sparse due to their normal distribution. This shows that there is a trade-off in when trying to evaluate different stochastic values with a single surrogate model. To obtain $Cd_{99\%\max}$ accurately, a uniform initial sample distribution is preferable to avoid strong extrapolation toward the borders of the parameter space where the maximum C_d

may be located. If adaptive sampling is applied, a uniform initial sample distribution is preferable to avoid a lack of accuracy of the EI function over the parameter space which is required to determine the samples to be added. On the contrary, to obtain μ_{C_d} and σ_{C_d}, applying the input pdf of the uncertain parameters to the sample distribution is recommended. Further adaptive sampling techniques, e.g., a combination of MSE with the input pdf and EI, are of interest that account for this problem.

Summary

In the first section, we evaluated statistics of aerodynamic performance (lift and drag coefficients) such as mean, standard deviation, skewness, kurtosis, and probability density function, on a variety of surrogate models with gradient-assisted or not, direct integration, and polynomial chaos with gradient assist or supported by sparse Gauss–Hermite. The test case used here was the RAE2822 transonic airfoil. The results show that globally to say the gradient-enhanced Kriging (GEK) is the most accurate in calculating any kinds of statistics, especially for the probability density function (pdf), which stands out compared with other methods.

Based on the results in the first section, the second section showed accuracy results using adaptive sampling techniques to obtain specific stochastic quantities which are mean and standard deviation (mean risk), and maximum (the worst-case risk) of drag coefficient. The surrogate model, the number of the input uncertainties, and the number of samples were fixed at GEK, 12 and 30, respectively. No specific sampling strategy is necessary for evaluating mean and standard deviation as far as "uniformity" of the samples, which is achieved by the Sobol sequence, is maintained and the samples distributions are fitted to the input probability density function (pdf). On the contrary, adaptive sampling techniques using the expected improvement (EI) in the uniformly distributed sample can efficiently assist to evaluate the maximum value.

References

1. Liu, D., Görtz, S.: Efficient quantification of aerodynamic uncertainty due to random geometry perturbations. In: New Results in Numerical and Experimental Fluid Mechanics IX, vol. 124 of the series Notes on Numerical Fluid Mechanics and Multidisciplinary Design, pp. 65–73. Springer (2014). ISBN 978-3-319-03157-6. ISSN 1612-2909
2. Liu, D., Litvinenko, A., Schillings, C., Schulz, V.: Quantification of airfoil geometry-induced aerodynamic uncertainties—comparison of approaches. SIAM/ASA J. Uncertain. Quantif. **5**(1), (2016)
3. Han, Z.H., Görtz, S., Zimmermann, R.: Improving variable-fidelity surrogate modeling via gradient-enhanced Kriging and a generalized hybrid bridge function. J. Aerosp. Sci. Technol. **25**(1) (2013)

4. Liu, D.: Efficient quantification of aerodynamic uncertainties using gradient-employing surrogate methods. In: Management and Minimisation of Uncertainties and Errors in Numerical Aerodynamics, vol. 122 of the series Notes on Numerical Fluid Mechanics and Multidisciplinary Design, pp. 283–296. Springer. ISBN 978-3-642-36184-5. ISSN 1612-2909 (2013)
5. Han, Z.H., Görtz, S., Zimmermann, R.: Improving variable-fidelity surrogate modeling via gradient-enhanced Kriging and a generalized hybrid bridge function. J. Aerosp. Sci. Technol. **25**(1), 177–189 (2013)
6. Snyder, W.C.: Accuracy estimation for quasi-Monte Carlo simulations. Math. Comput. Simul. **54**(1–3), 131–143 (2000)
7. Galle, M., Gerhold, T., Evans, J.: Parallel computation of turbulent flows around complex geometries on hybrid grids with the DLR-TAU code. In: Ecer, A., Emerson, D.R. (eds.) Proceedings of 11th Parallel CFD Conference, Williamsburg, VA, North-Holland, 23–26 May 1999
8. Gerhold, T., Hannemann, V., Schwamborn, D.: On the validation of the DLR-TAU code. In: Nitsche, W., Heinemann, H.J., Hilbig, R. (eds.) New Results in Numerical and Experimental Fluid Mechanics. Notes on Numerical Fluid Mechanics, Vieweg, vol. 72, pp. 426–433 (1999). ISBN 3-528-03122-0
9. Schwamborn, D., Gerhold, T., Heinrich, R.: The DLR TAU-code: recent applications in research and industry, invited lecture. In: Wesseling, P., Oate, E., Priaux, J. (eds.) Proceedings of the European Conference on Computational Fluid Dynamics (ECCOMAS CFD 2006), The Netherlands (2006)
10. Wendland, H.: Scattered Data Approximation. Cambridge University Press (2005)
11. Liu, D., Matthies, H.G.: Speeding up Kriging by using pivoted Cholesky decomposition and low-rank structures. In: SIAM Annual Meeting 2016, Boston, USA, 11–15 Jul 2016
12. Allmaras, S.R., Johnson, F.T., Spalart, P.R.: Modifications and clarifications for the implementation of the Spalart-Allmaras turbulence model. In: Seventh International Conference on Computational Fluid Dynamics (ICCFD7), ICCFD7-1902, Hawaii, July 2012
13. Laloy, E., Rogiers, B., Vrugt, J.A., Mallants, D., Jacques, D.: Efficient posterior exploration of a high-dimensional groundwater model from two-stage Markov chain Monte Carlo simulation and polynomial chaos expansion. Water Resour. Res. **49**, 1–19 (2013)
14. Sobol, I.M.: Distribution of points in a cube and approximate evaluation of integrals. Zh. Vychisl. Mat. Mat. Fiz **7**(4), 784–802 (1967)
15. Joe, S., Kuo, F.Y.: Remark on algorithm 659: implementing Sobol's quasirandom sequence generator. ACM Trans. Math. Softw. **29**, 49–57 (2003)
16. Joe, S., Kuo, F.Y.: Constructing Sobol sequences with better two-dimensional projections. SIAM J. Sci. Comput. **30**, 2635–2654 (2008)
17. Maruyama, D., Liu, D., Görtz, S.: An efficient aerodynamic shape optimization framework for roubst design of airfoils using surrogate models. In: Proceedings of the European Congress on Computational Methods in Applied Sciences and Engineering (ECCOMAS 2016), Crete, Greece (2016)
18. Maruyama, D., Liu, D., Görtz, S.: Surrogate model based approaches to UQ and their range of applicability. In: Uncertainty Management for Robust Industrial Design in Aeronautics. Chapter No. 43

Ordinary Kriging Surrogates in Aerodynamics

Antoine Dumont, Jean-Luc Hantrais-Gervois, Pierre-Yves Passaggia, Jacques Peter, Itham Salah el Din and Éric Savin

Introduction

Surrogate models become increasingly popular in performing various optimization or uncertainty quantification (UQ) analyses. The principle of a surrogate model relies on an efficient approximation method which estimates a scalar or vector-valued input/output functional using a data set constituted by observations of this functional, often computationally expensive [1]. For example, the computational resources necessary to evaluate stochastic integrals are known to grow exponentially with the number of dimensions due to the so-called curse of dimensionality. This increase becomes even more critical when the integrands need to be evaluated by intensive computations. Surrogate models are a non-intrusive alternative for such

A. Dumont · J.-L. Hantrais-Gervois · I. Salah el Din
Applied Aerodynamics Department, Onera–The French Aerospace Lab,
8 rue des Vertugadins, 92190 Meudon, France
e-mail: antoine.dumont@onera.fr

J.-L. Hantrais-Gervois
e-mail: jean-luc.hantrais-gervois@onera.fr

I. Salah el Din
e-mail: itham.salah_el_din@onera.fr

P.-Y. Passaggia
Marine Sciences Department, University of North Carolina at Chapel Hill,
3311 Venable Hall, Chapel Hill, NC 27599, USA
e-mail: passaggia@unc.edu

P.-Y. Passaggia · J. Peter · É. Savin (✉)
Computational Fluid Dynamics Department, Onera–The French Aerospace Lab,
29 avenue de la Division Leclerc, 92322 Châtillon, France
e-mail: eric.savin@onera.fr

J. Peter
e-mail: jacques.peter@onera.fr

C. Hirsch et al. (eds.), *Uncertainty Management for Robust Industrial Design in Aeronautics*, Notes on Numerical Fluid Mechanics and Multidisciplinary Design 140, https://doi.org/10.1007/978-3-319-77767-2_14

computations which can outcome the limitation emphasized before by providing cheap representations of the integrands in the parameters space. These representations are typically computed using an interpolation or a regression procedure. However, when considering a moderate to large parameters space, efficient algorithms are mandatory to derive accurate representations of parameterized integrands.

The Kriging procedure [2, 3] is a good candidate because of its robustness, accuracy, and ability to provide a description of the error done by replacing a complex function by a Kriging surrogate. This approach is interpolating and assumes that the function to approximate is the realization of a second-order Gaussian process. The construction of this interpolation using a sampled data set is based on a covariance function, often chosen as Gaussian, of which inner parameters have to be tuned up in order to provide a reliable surrogate. Here, we apply this procedure to the simulation of airflow around a wing profile with the consideration of variable shape parameters of the profile. The present study is more particularly dedicated to a two-dimensional RAE 2822 airfoil simulated at transonic speed [4, 5]. It is a popular, well-documented test case in the literature and has received much attention over the past decades; see [6, 7] and references therein for an application to the development of algorithms for shape optimization. Our objective is to obtain an accurate surrogate for evaluating the aerodynamic performance of the airfoil when its shape is altered, considering a dense data set with a moderately high dimension of the parameters space (four in the present investigation). This accuracy is required for a subsequent robust optimization of the airfoil. We note at this stage that the Kriging procedure becomes computationally intractable for high-dimensional parameters space and large data sets, and alternative techniques need be advocated in these situations as emphasized in [8]. They include, for example, sparse grid-based polynomial projection or regression techniques using either structured or unstructured nodes, as detailed in the previous chapters "General Introduction to Polynomial Chaos and Collocation Methods" and "Generalized Polynomial Chaos for Non-intrusive Uncertainty Quantification in Computational Fluid Dynamics" and references therein.

The rest of this chapter is organized as follow. The next section "Test Problem" introduces the parameterized test problem considered in this work and the numerical tools used to construct the data set. Details about the Kriging method are presented in section "Kriging Surrogate Model." The inputs and results of a robust UQ study based on a Kriging surrogate are described in section "Robust Optimization Based on the Surrogate Model." Finally, some conclusions are drawn in section "Conclusions."

Test Problem

Actual cruise flight conditions are mostly transonic, meaning that the flow can become locally supersonic due to the geometry of the wing profiles. Its acceleration on the upper surface of the profiles induces a depression which is responsible for the lift force. If this depression is too sharp, a discontinuity, or shock wave, is created to balance the pressure gradient at the trailing edge between the upper and lower surfaces. The position and strength of the shock wave are responsible for a part of the drag force, and hence, these features are of major importance in optimization, for instance to minimize the drag force at constant lift force. Modifications of the shape of the profiles can alleviate such issues by smoothing out the discontinuity, hence increasing the lift force while decreasing the drag force. We thus consider a test problem where the objective is to develop a numerical strategy to quantify the influence of variable geometrical parameters of an airfoil on its aerodynamic performances such as the lift and/or drag force.

More particularly, the present test case is a transonic turbulent flow around a RAE 2822 airfoil. The flow is modeled by the steady-state Reynolds-averaged Navier–Stokes (RANS) equations together with a Spalart–Allmaras turbulence model closure [9], which are now routinely used for the quantification of design issues involving compressible aerodynamics. The baseline conditions of the flow are described in [4] for the test case #6 together with the wall interference correction formulas derived in [10, pp. 386–387] and their slight modifications suggested in [11, p. 130]. The operational parameters considered here are thus $M_\infty = 0.729$ for the free-stream Mach number, $\alpha_\infty = 2.31°$ for the angle of attack, and $Re = 6.50 \cdot 10^6$ for the Reynolds number based on the chord length c, fluid velocity, temperature, and molecular viscosity at infinity. They arise from the corrections $\Delta M_\infty = 0.004$ and $\Delta\alpha_\infty = -0.61°$ given in [11, p. 130] for the test case #6 outlined in [4], for which $M_\infty = 0.725$, $\alpha_\infty = 2.92°$, and $Re = 6.50 \cdot 10^6$.

Discretization and Numerical Parameters

In this section, we describe how the numerical simulations of the foregoing example are performed. The flow is computed using the cell-centered finite volume CFD software *elsA* [12]. The RANS equations are discretized using a $641c \times 129c$ mesh, shown in Fig. 1, where the boundary at infinity was left intentionally far at $450c$ from the airfoil. These values proved to be sufficient to avoid spurious reflections with the far-field boundary. The discretized numerical solution is obtained based on:

- A Roe flux using a second-order MUSCL scheme [13] (based on van Albada limiter [14]) for the convective term of the RANS system;
- First-order Roe fluxes for the advection term of the turbulent variable;
- Corrected second-order diffusive terms based on the corrected mean of the cell-centered gradients of the two adjacent cells;
- Source terms for the turbulent transport computed using the temperature gradients at the center of the cells.

The flow is attached with a weak shock wave on the suction side. The static pressure at the wall as well as the iso-Mach number levels are presented in Fig. 2. Given the large number of simulations to run, the parameters of the steady-state algorithm proved to be essential to insure a fast convergence. This was achieved using the following set of parameters:

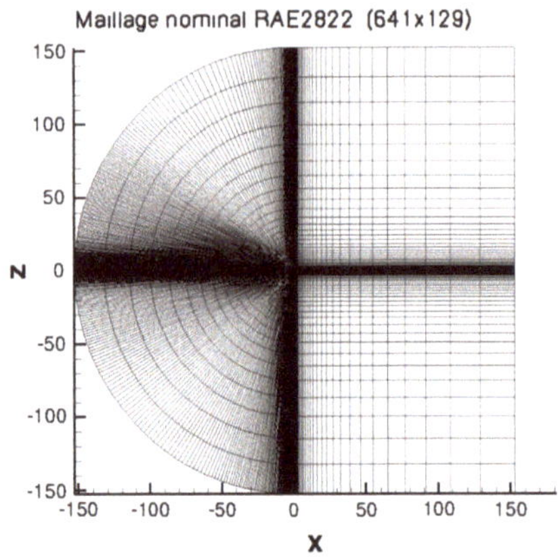

Fig. 1 Computational domain (left) and close view (right) of the mesh for the baseline RAE 2822 configuration

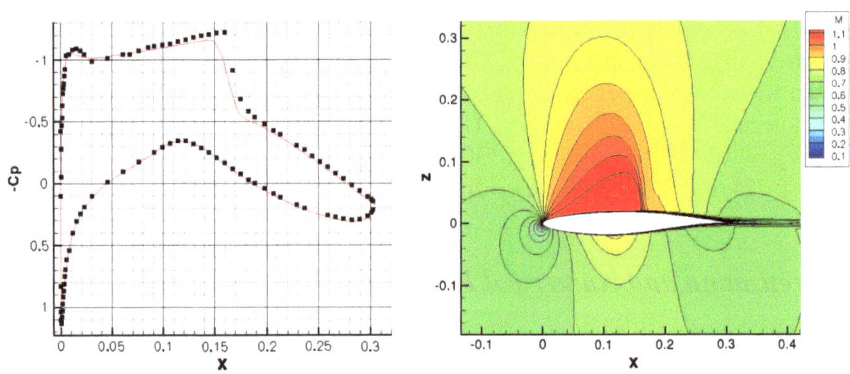

Fig. 2 Static pressure coefficient $-C_p$ at the wall (left) and iso-Mach number levels (right) for the baseline RAE 2822 configuration at $M_\infty = 0.729$, $\alpha_\infty = 2.31°$, $Re = 6.50 \cdot 10^6$ compared with the experimental data reported in [5]

- An implicit Lower-Upper Symmetric Successive Overrelaxation (LU-SSOR) scheme [15] using four relaxation cycles, increasing the CFL conditions after 100 iterations to CFL = 50;
- A multi-grid approach for the Navier–Stokes system over two grid levels with two iterations on the coarse grid;
- A single fine level iteration for the turbulent equation alternating with a multi-grid iteration for the RANS system.

The corresponding decrease of the explicit residual for the mass flow is shown in Fig. 3 (left). Once the numerical parameters have been fixed, the number of iterations is determined from the evolution of the global forces, through a 30,000 iterations calculation (decreasing discrete residuals of all equations being checked at every iteration, the final reduction of the mass conservation residual, shown in Fig. 3, is

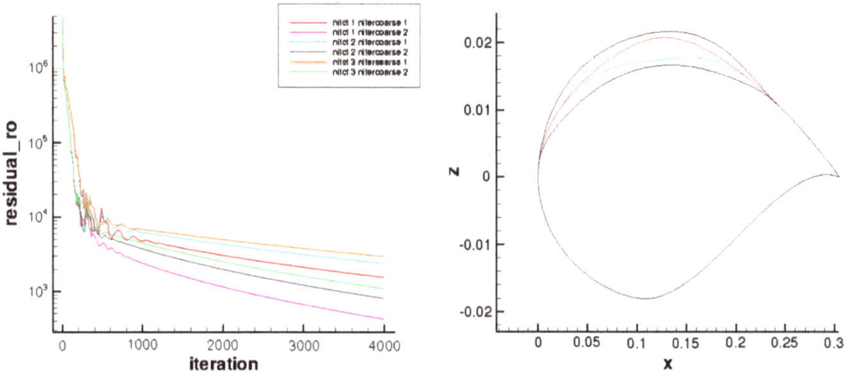

Fig. 3 Convergence of the multi-grid simulations (one level of coarse grid) for the baseline mesh (left) and maximum and minimum displacement (black lines) of the shape of the RAE 2822 profile using the four parameters of the present study (left). Maximum (red) and minimum (blue) displacement of the third bump alone

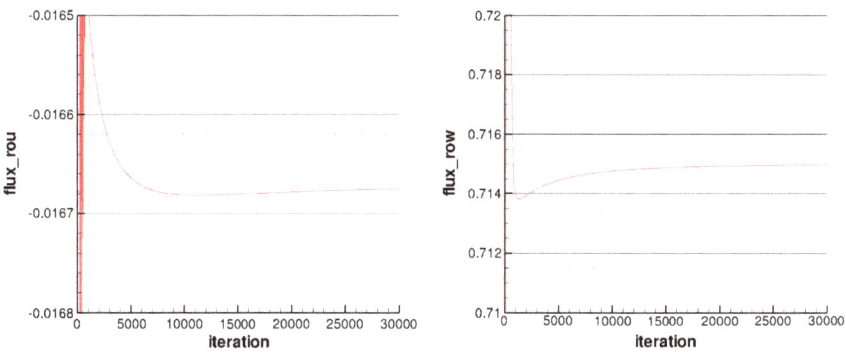

Fig. 4 Total forces along the x horizontal axis (left) and z vertical axis (right) as a function of the number of iterations

$2.5 \cdot 10^4$). After 6000 multi-grid cycles, the force values appeared to remain within the range $[-2/10{,}000, +2/10{,}000]$ which is acceptable, given the convergence error; see Fig. 4. Hence, this number of iterations has been retained for all subsequent calculations.

Shape Parameters

The shape of the baseline RAE 2822 profile is subsequently altered using four B-spline functions, located on the suction side of the airfoil. These alterations are responsible for triggering the pressure gradient induced by the geometry of the profile. The pressure gradient is significantly altered even for small variations of the airfoil shape, which leads to nonlinear variations of the lift and drag forces. The shape parameters used to alter the baseline configuration are:

- Four control points located at 5, 20, 40, and 60% of the chord c;
- The deformation is described by B-spline functions denoted by $S_n(s)$ such that:

$$S_n(s) = \sum_{j=0}^{m-n-1} \alpha_j b_n^{(j)}(s)\,, \tag{1}$$

  where the α_j's are the values of the deformed shape at the control points and s is the curvilinear abscissa of the airfoil. The $m-n$ basis functions $b_n^{(j)}$ of degree n are defined by the recurrence relation:

$$b_0^{(j)}(s) := \begin{cases} 1 & \text{if } s_j \leq s < s_{j+1}\,, \\ 0 & \text{otherwise,} \end{cases}$$

  and:

$$b_n^{(j)}(s) := \frac{s - s_j}{s_{j+n} - s_j} b_{n-1}^{(j)}(s) + \frac{s_{j+n+1} - s}{s_{j+n+1} - s_{j+1}} b_{n-1}^{(j+1)}(s)\,.$$

  Here, $m = 4$ and $n = 2$ have been chosen;
- The amplitude ξ_k, $k = 1, 2, 3$, or 4, of each B-spline varies in the range $\pm 0.0025c$ in the direction of the outward normal vector to the nominal (unaltered) profile;
- The leading edge and the trailing edge are considered as fixed.

The shape of the airfoil thus evolves continuously as a function of the shape parameter vector $\boldsymbol{\xi} := (\xi_1, \xi_2, \xi_3, \xi_4)$ which varies in the design domain $D = [-0.0025c, 0.0025c]^4$. The support of each spline independently of the others is depicted in Figs. 5 and 6. The effect of the maximum and minimum displacements is shown in Fig. 3 (right).

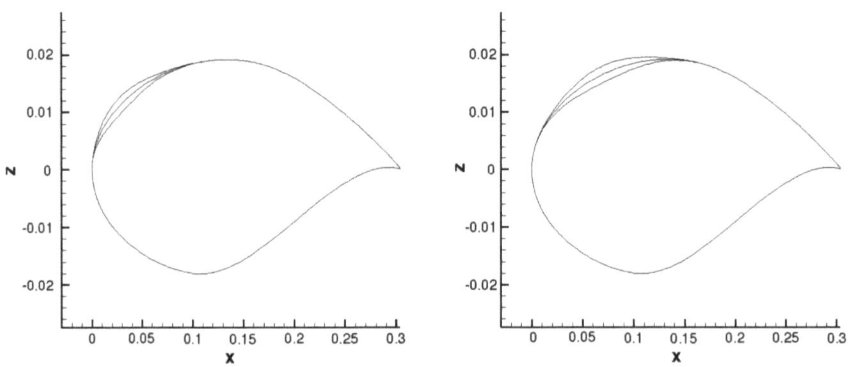

Fig. 5 Baseline configuration and deformation of each spline with an amplitude in the range $[-0.0025c, 0.0025c]$ at $0.05c$ (left) and $0.2c$ (right)

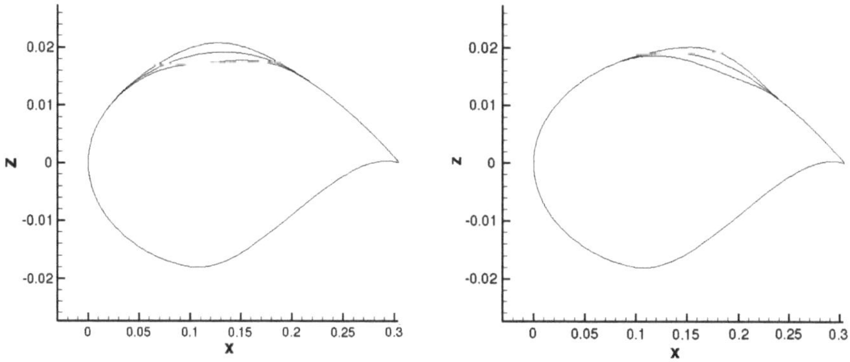

Fig. 6 Baseline configuration and deformation of each spline with an amplitude in the range $[-0.0025c, 0.0025c]$ at $0.4c$ (left) and $0.6c$ (right)

Quantity of Interest

The quantity of interest is the lift-to-drag ratio Y where the lift coefficient is denoted by C_L and the drag coefficient by C_D, such that:

$$Y(\boldsymbol{\xi}) = \frac{C_L(\boldsymbol{\xi})}{C_D(\boldsymbol{\xi})}. \tag{2}$$

Note that in the present work the computation of the drag coefficient has been performed using the far-field approach detailed in [16]. Our aim is thus to provide an optimal approximation (in a sense clarified in the subsequent section) for the function $\boldsymbol{\xi} \mapsto Y(\boldsymbol{\xi})$. This is achieved on the basis of $Q > 0$ repeated runs of the CFD software for the altered configurations at selected parameter values $\boldsymbol{\xi}^l = (\xi_1^l, \xi_2^l, \xi_3^l, \xi_4^l)$, $1 \leq l \leq Q$, in the normalized parameter domain $\hat{D} = [0, 1]^4$. The data set containing all the simulations is denoted by $S_Q = \{Y(\boldsymbol{\xi}^l);\ 1 \leq l \leq Q\}$ in the following, with $Q = \#S_Q$.

Kriging Surrogate Model

This section outlines the numerical procedure used to compute an optimal interpolant, or surrogate model, of the quantity of interest Y by the Kriging method [1–3, 17, 18]. It also describes the strategy retained to obtain optimized inner parameters for the Kriging interpolant. For that purpose, we basically follow the presentation in [8]. As already stated above, our aim is to use this surrogate to characterize the influence of alterations of the baseline profile of the RAE 2822 airfoil described in section "Shape Parameters" on its aerodynamic performance, as expressed here by the ratio between the lift and drag coefficients (2).

The Kriging Method

The Kriging method is based on the assumption that the quantity of interest $Y(\boldsymbol{\xi})$ is a random variable which can be decomposed into [1]:

$$Y(\boldsymbol{\xi}) = \mu(\boldsymbol{\xi}) + \varepsilon(\boldsymbol{\xi}), \tag{3}$$

where $\mu(\boldsymbol{\xi})$ and $\varepsilon(\boldsymbol{\xi})$ are sought as a deterministic contribution and a random fluctuation, respectively. The mean μ is typically sought as a constant or a low-order polynomial—hence the following terminology [3] of: (i) simple Kriging if μ is constant and known a priori; (ii) ordinary Kriging if μ is constant but unknown; or (iii) universal Kriging if μ is an unknown polynomial of known order (a polynomial chaos expansion, for example, as in [19]). The random process ε indexed by the parameters $\boldsymbol{\xi}$ is assumed to be a second-order, centered, mean-square stationary process, or stationary covariance process in the terminology of [18], such that its mean $\mu_\varepsilon = \mathbb{E}\{\varepsilon(\boldsymbol{\xi})\}$ and variance $\sigma_\varepsilon^2 = \mathbb{E}\{\varepsilon(\boldsymbol{\xi})^2\}$ are, respectively, null and constant. In these definitions, $\mathbb{E}\{X\}$ stands for the average of the random variable X. Thus, its covariance function depends only on the difference $\boldsymbol{\xi} - \boldsymbol{\xi}'$, and what is more the lag distance $\|\boldsymbol{\xi} - \boldsymbol{\xi}'\|$, that is to say $\mathrm{Cov}\{\varepsilon(\boldsymbol{\xi}), \varepsilon(\boldsymbol{\xi}')\} = \mathbb{E}\{\varepsilon(\boldsymbol{\xi})\varepsilon(\boldsymbol{\xi}')\} = \phi(\boldsymbol{\xi}, \boldsymbol{\xi}') = \varphi(\|\boldsymbol{\xi} - \boldsymbol{\xi}'\|)$. Here, the covariance of two random variables X and Y is defined by $\mathrm{Cov}\{X, Y\} := \mathbb{E}\{(X - \mathbb{E}\{X\})(Y - \mathbb{E}\{Y\})\}$ and the variance of a random variable X by $\mathrm{Var}\{X\} := \mathrm{Cov}\{X, X\}$. In this framework, the approximation (or linear predictor) of $Y(\boldsymbol{\xi}^*)$ at the unobserved coordinates $\boldsymbol{\xi}^\star$, denoted by $\mathscr{I}_Q Y(\boldsymbol{\xi}^\star)$, is a linear combination of the samples $Y(\boldsymbol{\xi}^l) \in S_Q$ such that:

$$\mathscr{I}_Q Y(\boldsymbol{\xi}^\star) = \sum_{l=1}^{Q} \lambda^l(\boldsymbol{\xi}^\star) Y(\boldsymbol{\xi}^l), \tag{4}$$

where $\lambda^l(\boldsymbol{\xi}^\star)$ are weights. The aim of the Kriging method is then to derive the best linear unbiased predictor in the sense of a mean-square error. It can be sought as the solution of a constrained minimization problem where the objective function is the variance $\sigma^2(\boldsymbol{\xi}^\star) = \mathrm{Var}\{\mathscr{I}_Q Y(\boldsymbol{\xi}^*) - Y(\boldsymbol{\xi}^\star)\}$ with the unbiased constraint $\mathbb{E}\{\mathscr{I}_Q Y(\boldsymbol{\xi}^\star)\} = \mathbb{E}\{Y(\boldsymbol{\xi}^\star)\}$, that is:

$$\{\lambda^l(\boldsymbol{\xi}^\star)\}_{1\leq l\leq Q} = \arg \min_{\{\Lambda^l\}_{1\leq l\leq Q}} \left\{ \sigma^2(\boldsymbol{\xi}^\star);\ \sum_{l=1}^{Q} \Lambda^l \mathbb{E}\{Y(\boldsymbol{\xi}^l)\} = \mathbb{E}\{Y(\boldsymbol{\xi}^\star)\} \right\}, \tag{5}$$

with:

$$\begin{aligned} \sigma^2(\boldsymbol{\xi}^\star) &= \mathrm{Var}\{Y(\boldsymbol{\xi}^\star)\} + \mathrm{Var}\{\mathscr{I}_Q Y(\boldsymbol{\xi}^\star)\} - 2\,\mathrm{Cov}\{Y(\boldsymbol{\xi}^\star), \mathscr{I}_Q Y(\boldsymbol{\xi}^\star)\} \\ &= \phi(\boldsymbol{\xi}^\star, \boldsymbol{\xi}^\star) + \sum_{l=1}^{Q} \sum_{m=1}^{Q} \Lambda^l \Lambda^m \phi(\boldsymbol{\xi}^l, \boldsymbol{\xi}^m) - 2 \sum_{l=1}^{Q} \Lambda^l \phi(\boldsymbol{\xi}^l, \boldsymbol{\xi}^\star). \end{aligned} \tag{6}$$

The covariance between Y and $\mathscr{I}_Q Y$ in Eq. (6) endows the set S_Q with an inner product and avoids the explicit knowledge of $Y(\boldsymbol{\xi})$. This technique is also known as the kernel trick and avoids explicit mappings where only the foregoing inner product between samples in S_Q is used in the evaluation of the covariance [20]. In addition, the kernel function ϕ has to be positive definite; see [18] for a comprehensive review. In the present work, the product of k one-dimensional Gaussian functions satisfying the stationary covariance process assumption has been considered. However, other kernels could be envisaged as in, e.g., [1, 18]. The covariance kernel used in the present investigation is thus given by:

$$\mathrm{Cov}\{Y(\boldsymbol{\xi}), Y(\boldsymbol{\xi}')\} = \phi(\boldsymbol{\xi}, \boldsymbol{\xi}') = \exp\left[-\frac{1}{2}\sum_{k=1}^{N}\left(\frac{\xi_k - \xi_k'}{\zeta_k}\right)^2\right], \tag{7}$$

where the kernel parameters $\boldsymbol{\zeta} = (\zeta_1, \zeta_2, \ldots, \zeta_N)$ have to be computed in order to minimize the quadratic error between $\mathscr{I}_Q Y(\boldsymbol{\xi}^\star)$ and $Y(\boldsymbol{\xi}^\star)$ in Eq. (5). Here, N is the dimension of the parameters space, which is $N = 4$ in the present work.

Optimization of the Inner Parameters of the Covariance Kernel

The computation of the inner parameters of the covariance kernel is a crucial step for the accuracy of the interpolation and requires to solve a nonlinear problem in N dimensions. The choice was made to compute the kernel parameters $\boldsymbol{\zeta}$ using a cross-validation procedure [17]. The so-called leave-one-out (LOO) method consists in removing each point in the data set S_Q one by one and compute the mean-square difference E between the Kriging $\mathscr{I}_Q Y$ and its LOO counterpart $\mathscr{I}_{Q-1}^l Y$ computed from $S_Q \setminus \{Y(\boldsymbol{\xi}^l)\}$, at the location $\boldsymbol{\xi}^l$, such that:

$$E(Y)_{\mathrm{LOO}} = \frac{1}{Q}\sum_{l=1}^{Q}\left(\mathscr{I}_{Q-1}^l Y(\boldsymbol{\xi}^l) - Y(\boldsymbol{\xi}^l)\right)^2. \tag{8}$$

The computation of the Q new predictors $\mathscr{I}_{Q-1}^l Y$ can be done efficiently using Rippa's method [21]. A differential evolution algorithm has been subsequently used to solve the optimization problem for the kernel parameters $\boldsymbol{\zeta}$ associated with Eq. (8). Details about the algorithm can be found in [22].

Direct and Dual Estimate of Ordinary Kriging

Ordinary Kriging [3] is by far the most popular approach and has been considered in the present investigations. It starts with the assumption that the mean of $Y - \mathscr{I}_Q Y$ vanishes although μ is unknown. The unbiased constraint implies that:

$$\sum_{l=1}^{Q} \lambda^l(\boldsymbol{\xi}^\star)\mu - \mu = 0\,, \tag{9}$$

and thus $\sum_{l=1}^{Q} \lambda^l(\boldsymbol{\xi}^\star) = 1$. Introducing the Lagrange multiplier χ and canceling the gradient with respect to Λ^l in Eq. (6), the system to be solved, together with the condition (9), is given by:

$$\begin{aligned} 2\sum_{l=1}^{Q} \lambda^l(\boldsymbol{\xi}^\star)\phi(\boldsymbol{\xi}^l,\boldsymbol{\xi}^m) - 2\phi(\boldsymbol{\xi}^\star,\boldsymbol{\xi}^m) + 2\chi = 0\,, \quad \forall\, m = 1,2,\ldots,Q\,,\\ \sum_{l=1}^{Q} \lambda^l(\boldsymbol{\xi}^\star) = 1\,, \end{aligned} \tag{10}$$

where the weights $\{\lambda^l(\boldsymbol{\xi}^\star)\}_{1\leq l\leq Q}$ are the coefficients of the linear interpolation (4) defining the Kriging surrogate. They are thus computed as the solution of the linear system:

$$\begin{bmatrix} \phi(\boldsymbol{\xi}^1,\boldsymbol{\xi}^1) & \cdots & \phi(\boldsymbol{\xi}^1,\boldsymbol{\xi}^Q) & 1 \\ \vdots & \ddots & \vdots & \vdots \\ \phi(\boldsymbol{\xi}^Q,\boldsymbol{\xi}^1) & \cdots & \phi(\boldsymbol{\xi}^Q,\boldsymbol{\xi}^Q) & 1 \\ 1 & \cdots & 1 & 0 \end{bmatrix} \begin{pmatrix} \lambda^1 \\ \vdots \\ \lambda^Q \\ \chi \end{pmatrix} = \begin{pmatrix} \phi(\boldsymbol{\xi}^1,\boldsymbol{\xi}^\star) \\ \vdots \\ \phi(\boldsymbol{\xi}^Q,\boldsymbol{\xi}^\star) \\ 1 \end{pmatrix}. \tag{11}$$

However, this method becomes very expensive when the data set S_Q is large, and thus, numerous output values $Y(\boldsymbol{\xi}^\star)$ have to be computed. This difficulty can be overcome by observing that the linear predictor $\mathscr{I}_Q Y(\boldsymbol{\xi}^\star)$ also reads:

$$\mathscr{I}_Q Y(\boldsymbol{\xi}^\star) = \sum_{l=1}^{Q} \pi^l \phi(\boldsymbol{\xi}^l,\boldsymbol{\xi}^\star) + \pi^{Q+1}\,, \tag{12}$$

where the weights $\{\pi^l\}_{1\leq l\leq Q}$ are obtained as the solution of the system:

$$\begin{bmatrix} \phi(\boldsymbol{\xi}^1,\boldsymbol{\xi}^1) & \cdots & \phi(\boldsymbol{\xi}^1,\boldsymbol{\xi}^Q) & 1 \\ \vdots & \ddots & \vdots & 1 \\ \phi(\boldsymbol{\xi}^Q,\boldsymbol{\xi}^1) & \cdots & \phi(\boldsymbol{\xi}^Q,\boldsymbol{\xi}^Q) & 1 \\ 1 & \cdots & 1 & 0 \end{bmatrix} \begin{pmatrix} \pi^1 \\ \vdots \\ \pi^Q \\ \pi^{Q+1} \end{pmatrix} = \begin{pmatrix} Y(\boldsymbol{\xi}^1) \\ \vdots \\ Y(\boldsymbol{\xi}^Q) \\ 0 \end{pmatrix}. \tag{13}$$

This approach is more convenient than (4) and (11) because the weights are now independent of the unobserved shape parameters $\boldsymbol{\xi}^\star$.

Robust Optimization Based on the Surrogate Model

A tensorized grid of nine equidistributed abscissas for each of the four shape parameters of section "Shape Parameters" is used to obtain the data set S_Q; therefore, $Q = 9^4 = 6561$. The values of the lift-to-drag ratios $Y(\boldsymbol{\xi}^l)$ in S_Q as computed by the *elsA* CFD software vary in the range $[36.788, 75.198]$. The Kriging surrogate model for the present parametric problem is computed using Onera's in-house software `META_NUMF` [17]. It implements the various algorithms described in section "Kriging Surrogate Model". The inner parameters selected by the cross-validation procedure are $\boldsymbol{\zeta} \simeq (0.1020, 0.0472, 0.0957, 0.1434)$. Note that the second parameter ζ_2 (which can be interpreted as a correlation length) is rather small compared to the others. This may result in small amplitude oscillations in this direction. Also, note that the computation for the surrogate model has been carried out on eight Intel® Xeon® processors E5540 (8M cache, 2.53 GHz) for 5 days for the inner parameters optimization. The surface response obtained using the Kriging surrogate is shown in the two-parameters plan (ξ_1, ξ_3) at $\xi_2 = -0.0025c$, $\xi_4 = 0$ and $\xi_2 = 0.0025c$, $\xi_4 = 0$ in Fig. 7.

The response surface has been further characterized using a particle swarm optimization algorithm [23] to localize its extrema. The surrogate model is characterized by 28 local maxima $\{Y_m\}_{1 \leq m \leq 28}$, where the global maximum is $Y_{\max} \simeq 76.6724$ located at:

$$\boldsymbol{\xi}_{\max} \simeq (0.002143, -0.001873, 0.001125, 0.001162)\,,$$

such that $Y_{\max} = \mathscr{I}_Q Y(\boldsymbol{\xi}_{\max})$. The 28 local maxima extracted from the above procedure are listed below in Table 1 in 16 digits precision for illustration purposes. They proved to verify $\|\boldsymbol{\nabla}_{\boldsymbol{\xi}} Y(\boldsymbol{\xi})\| < 10^{-2}$, where the gradient was computed using a second-order centered finite difference scheme with a step $\Delta\xi_k = 0.5 \cdot 10^{-3} \times 0.0025c$ in each direction $k = 1, 2, 3$, and 4 in the design domain D. These maxima have subsequently been considered for an UQ study where the mean and the variance of each maximum have been computed conditionally upon a prior probability distribution of the shape parameters chosen as a Beta law of the first kind $\beta_{\rm I}$ for each of them.

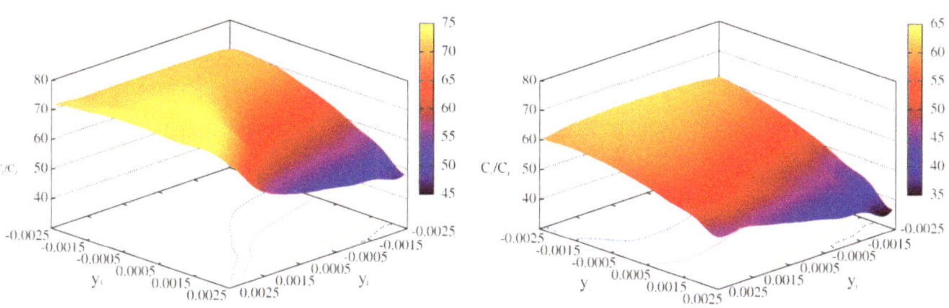

Fig. 7 Surface response in the (ξ_1, ξ_3) plan for $\xi_2 = -0.0025$ (left) and $\xi_2 = 0.0025$ (right) with $\xi_4 = 0$ in both cases

Table 1 List of local maxima $\{Y_m\}_{1\leq m\leq 28}$ identified on $\mathscr{I}_Q Y(\boldsymbol{\xi})$

m	ξ_1	ξ_2	ξ_3	ξ_4	$Y(\boldsymbol{\xi})$
1	0.0009247281603877587	-0.002477328481440417	-0.0020522618807618394	-0.001006637007272001	72.96062013330138
2	-0.0019877340165556532	-0.0012501286098216477	-0.0022230288250307834	1.637485303221777e-05	72.9053124038243
3	-0.0021080331586727672	-0.0024760806714095025	-0.0022790922661521593	-0.0009996692634806229	72.8083039561 6871
4	0.0009227995269777299	-0.0012491170657324849	-0.0020015373479271 56	0.0003111090101579929	73.11460103259566
5	0.0014184152803302856	-0.00248199225007820 6	0.0009674096676959162	-0.0013436457334901153	75.13385342547113
6	0.0021005751944998536	-0.0024838327139435935	0.0016378831262294602	-0.0007521259615468634	76.19703358554321
7	0.00226082471620925 1	-0.00187536060913814 4	-0.000147906043850845	0.0011438578569142157	75.33061972327575
8	0.001495913190949574	-0.0018758598395562961	0.0015100035322869972	-0.0014295665683347527	74.01965083298452
9	0.0007217459064240732	5.673767819445487e-07	-0.002037392586569452	0.0007255817623195416	73.37227470479218
10	-0.0020511434400430797	-0.0012442591298048122	-0.0012727785963404998	0.000999942602579444	72.48039437600411
11	-0.00046863745004457153	-7.173866486613438e-07	-0.002172688079996433	0.0003997664586228222	73.63518562977276
12	0.00011892753917269912	-0.002479381700009219	-0.0007452975881961537	-0.002277292293674497	74.32636561284727
13	0.0021112462457524967	-0.0012465346002510994	0.001580251301382974	0.0016964375010017608	76.1863078631464
14	0.002144283961154753	-0.0018726431452980397	0.00112667245068374	0.0011645115201617874	76.67235940824781
15	0.0009176228156363136	-0.0012482712347895557	0.0004398328525607037	0.0006259803524756003	75.76076285356561
16	0.002252700336863187	4.600109478019969e-07	-0.0007018136569205303	0.0022622430964318965	75.04867663515955
17	-4.1945374307096866e-05	0.0006262230074350205	-0.0021470974826768028	0.001330451966658052	73.65690479624587
18	0.0008007997438743406	-0.002481211158065615	0.0003250360152117076 6	-0.002393635032754219	74.99121577439612
19	0.0002768385784311889 7	-0.0012451399796126398	-8.168292353796704e-05	0.0004131959512205995 7	75.42284453664654
20	0.0010060672108371622	-0.0006244922023885663	-0.0007618898135975365	0.0011409171442569492	74.47172633897203
21	0.002129228704884845	-0.0012506867133831355	0.000426988917664922	0.0017446918994712614	75.34062327397848
22	0.0010563407826690357	0.0006259797249405174	-0.0010642802734270236	0.002184313671369558 7	74.19290076614386
23	-0.0003057993918150065	-0.001247433430195173	-0.0007151826244908299	-4.83642984228606e-05	74.68696951011597
24	0.0009907930610100206	0.001875515814836005	-0.0022052419551779243	0.0022553625771607956	74.0466960120915
25	0.0013825430485137993	-0.0018710126427371548	0.0005284653042229091	0.0004806569969429608	75.84927930947487
26	-0.0010037161479930269	-0.0012503314535020984	-0.002172701898798983	-0.0008923292318644253	73.695550780483
27	0.002240997891465996	0.0018755852596172282	-0.0021457313740432917	0.0023055559834000336	74.34508504051362
28	0.002293961593751379	-0.0006239914756562181	-0.00018841428563772227	0.0021561109756890805	74.97687454582714

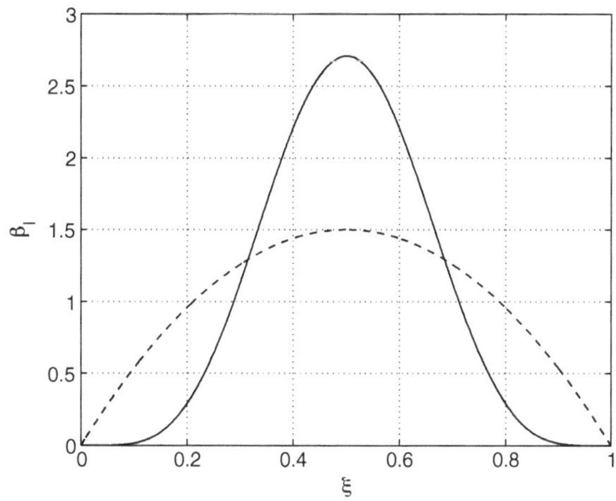

Fig. 8 $\beta_{\rm I}$ probability distribution functions for $\alpha = \beta = 6$ (full line) and $\alpha = \beta = 2$ (dashed line) used for modeling the uncertainty on the design of the shape parameters

This distribution function appears as a suitable choice when considering manufacturing defects since it has a compact support. In addition, it is the law arising from Jaynes' maximum entropy principle [24] if the averages $\mathbb{E}\{\log \xi\}$ and $\mathbb{E}\{\log(1-\xi)\}$ are given. The convention for the definition (on the normalized parameter domain $\hat{D}$) of the Beta distribution $\beta_{\rm I}$ used in the present study is:

$$\beta_{\rm I}(\xi;\alpha,\beta) = \mathbb{1}_{[0,1]}(\xi)\frac{\Gamma(\alpha+\beta)}{\Gamma(\alpha)\Gamma(\beta)}\xi^{\alpha-1}(1-\xi)^{\beta-1}, \tag{14}$$

where $z \mapsto \Gamma(z)$ is the usual Gamma function defined by $\Gamma(z) = \int_0^{+\infty} t^{z-1}e^{-t}dt$. The probability density functions are taken symmetric with $\alpha = \beta = 6$ for the bump located near the leading edge at 5% of the chord c (parameter ξ_1) and $\alpha = \beta = 2$ for the three other bumps (parameters ξ_2, ξ_3 and ξ_4); see Fig. 8. The amplitude of the uncertainty was taken in the range $\pm 5 \cdot 10^{-4}c$ which is consistent with observations from the industry [25]. The present mean and variance are computed by a Monte Carlo procedure using $n_s = 10^4$ samples. Given the convergence of the Monte Carlo method as $O(1/\sqrt{n_s})$, the statistics have been rounded to two significant digits in Table 2.

The solution of the UQ is shown in Table 2 for the maxima identified in Table 1. The mean of the maxima appears to be lower than the maximum found during the optimization process with variances in the range [0.1, 1.2]. We define the robust maximum $Y_{\rm R}$ such that:

$$Y_{\rm R}(\boldsymbol{\xi}) = \mathbb{E}\{Y(\boldsymbol{\xi})\} - 3\sqrt{\mathrm{Var}\{Y(\boldsymbol{\xi})\}} \tag{15}$$

has the largest value. Based on that definition, the maximum identified during the optimization procedure (labeled $m = 14$ in Table 1 and highlighted in gray) is not the robust one. Indeed, for the global deterministic maximum, $Y_{\max} \simeq 76.67$ that decreases to $\mathbb{E}\{Y_{14}\} \simeq 75.5$ with $\sqrt{\mathrm{Var}\{Y_{14}\}} \simeq 0.74$ which gives $Y_{{\rm R},14}(\boldsymbol{\xi}) \simeq 73.28$.

Table 2 List of mean maxima of $\mathscr{I}_Q Y(\boldsymbol{\xi})$ and their variance for the uncertainty propagation procedure described in section "Robust Optimization Based on the Surrogate Model"

m	ξ_1	ξ_2	ξ_3	ξ_4	$\mathbb{E}\{Y(\boldsymbol{\xi})\}$	$\sqrt{\text{Var}(Y(\boldsymbol{\xi}))}$
1	0.00092472816038	-0.00247732848144	-0.00205226188076	-0.00100663700727	72.22	0.51
2	-0.0019877340165	-0.00125012860982	-0.00222302882503	1.63748530322e-05	72.24	0.27
3	-0.0021080331586	-0.00247608067141	-0.00227909226615	-0.00099966926348	71.99	0.49
4	0.00092279952697	-0.00124911706573	-0.00200153734793	0.000311109010158	72.54	0.26
5	0.00141841528033	-0.00248199225008	0.000967409667696	-0.00134364573349	74.01	0.87
6	0.00210057519450	-0.00248383271394	0.001637883126230	-0.00075212596154	74.81	1.11
7	0.00226082471621	-0.00187536060914	-0.00014790604385	0.001143857856910	74.59	0.39
8	0.00149591319095	-0.00187585983956	0.001510003532290	-0.00142956656833	73.10	0.44
9	0.00072174590642	5.67376781945e-07	-0.00203739258657	0.000725581762320	72.78	0.27
10	-0.0020511434400	-0.00124425912980	-0.00127277859634	0.000999942602579	71.66	0.35
11	-0.0004686374500	-7.1738664866e-07	-0.00217268808000	0.000399766458623	72.98	0.30
12	0.00011892753917	-0.00247938170001	-0.00074529758819	-0.00227729229367	73.38	0.70
13	0.00211124624575	-0.00124653460025	0.001580251301380	0.001696437501000	74.87	0.83
14	0.00214428396115	-0.00187264314530	0.001126672450680	0.001164511520160	75.50	0.74
15	0.00091762281563	-0.00124827123479	0.000439832852561	0.000625980352476	74.77	0.57
16	0.00225270033686	4.60010947802e-07	-0.00070181365692	0.002262243096430	74.29	0.39
17	-4.194537430e-05	0.000626223007435	-0.00214709748268	0.001330451966660	73.00	0.30
18	0.00080079974387	-0.00248121115807	0.000325036015212	-0.00239363503275	73.85	0.86
19	0.00027683857843	-0.00124513997961	-8.1682923538e-05	0.000413195951221	74.47	0.54
20	0.00100606721084	-0.00062449220238	-0.00076188981359	0.001140917144260	73.81	0.35
21	0.00212922870488	-0.00125068671338	0.000426988917665	0.001744691899470	74.50	0.46
22	0.00105634078267	0.000625979724941	-0.00106428027343	0.002184313671370	73.52	0.34
23	-0.0003057993918	-0.00124743343020	-0.00071518262449	-4.8364298229e-05	73.91	0.42
24	0.00099079306101	0.001875515814840	-0.00220524195518	0.002255362577160	73.33	0.31
25	0.00138254304851	-0.00187101264274	0.000528465304223	0.000480656996943	74.84	0.57
26	-0.0010037161479	-0.00125033145350	-0.00217270189880	-0.00089232923186	73.04	0.29
27	0.00224099789147	0.001875585259620	-0.00214573137404	0.002305555983402	73.58	0.32
28	0.00229396159375	-0.00062399147565	-0.00018841428563	0.002156110975691	74.23	0.37

On the other hand, the local robust maximum is the point labeled $m = 7$ in Table 2 and highlighted in light gray, for which $Y_7 \sim 75.33$, $\mathbb{E}\{Y_7\} \simeq 74.59$ with a variance of $\sqrt{\mathrm{Var}\{Y_7\}} \approx 0.39$ and $Y_{\mathrm{R},7}(\boldsymbol{\xi}) \simeq 73.42$. This surrogate model appears therefore as an interesting test case for robust optimization as the global maximum, when subjected to uncertainty, happens to be less robust than a local maximum which maximizes the mean of the quantity of interest and minimizes the standard deviation associated with a stochastic process.

Conclusions

In this chapter, we have addressed an uncertainty quantification and shape optimization problem for a RAE 2822 airfoil at transonic speeds using a Kriging-based surrogate model of the lift-to-drag ratio. More specifically, the shape of the baseline profile has been altered by four localized bumps of variable amplitudes on its suction side. Their locations have been selected so that they modify the position of the shock which primarily drives the aerodynamic performance of the profile, such as the lift and drag forces. The transonic flow about the airfoil has been simulated using the Onera software *elsA* for each instance of the bump amplitudes in a tensorized grid with nine equidistant sampling values for each amplitude parameter, resulting in a dense data set of 9^4 samples. Each simulation has been performed for carefully chosen numerical parameters optimized on the basis of the baseline configuration compared with the available experimental results. The high-quality, dense Kriging surrogate model interpolating the data set uses Gaussian covariance kernels and exhibits multiple local maxima. They have been identified by a particle swarm algorithm, although its results might not be exhaustive and more maxima might be found using an alternative algorithm.

An uncertainty quantification analysis has subsequently been carried out about each local maxima, assigning a variation range and a prior probability density function to the shape parameters to mimic the effect of design tolerances and aging. This analysis has revealed that the global maximum, for example, is not necessarily the most robust one in terms of its standard deviation or of a robust design criterion such as the mean minus three times the standard deviation. It is thus believed that the data set and results obtained in the present work constitutes an interesting test case for assessing uncertainty quantification and/or robust optimization strategies in future investigations.

References

1. Sacks, J., Welch, W.J., Mitchell, T.J., Wynn, H.P.: Design and analysis of computer experiments. Stat. Sci. **4**(4), 409–423 (1989)
2. Krige, D.G.: A statistical approach to some basic mine valuation problems on the Witwatersrand. J. Chem. Metall. Min. Soc. South Afr. **52**(6), 119–139 (1951)
3. Matheron, G.: Principles of geostatistics. Econ. Geol. **58**(8), 1246–1266 (1963)
4. Cook, P.H., McDonald, M.A., Firmin, M.C.P.: Aerofoil RAE 2822—pressure distributions, and boundary layer and wake measurements. In: Experimental Data Base for Computer Program Assessment. AGARD Advisory Report No. 138. NATO (1979) Appendix A6
5. http://www.cfd-online.com/Wiki/RAE2822_airfoil
6. Karakasis, M.K., Koubogiannis, D.G., Kyriakos, C.G.: Hierarchical distributed metamodel-assisted evolutionary algorithms in shape optimization. Int. J. Numer. Methods Fluids **53**(3), 455–469 (2007)
7. Laurenceau, J., Sagaut, P.: Building efficient response surfaces of aerodynamic functions with Kriging and Cokriging. AIAA J. **46**(2), 498–507 (2008)
8. Chkifa, A., Cohen, A., Passaggia, P.Y., Peter, J.: A comparative study between Kriging and adaptive sparse tensor-product methods for multi-dimensional approximation problems in aerodynamics design. ESAIM: Proc. Surv. **48**, 248–261 (2015)
9. Spalart, P.R., Allmaras, S.R.: A one-equation turbulence model for aerodynamic flows. In: 30th AIAA Aerospace Sciences Meeting and Exhibit, 6–9 January 1992, Reno NV AIAA paper 1992–0439 (1992)
10. Garner, H.C., Rogers, E.W.E., Acum, W.E.A., Maskell, E.C.: Subsonic wind tunnel wall corrections. In: AGARDograph No. 109. Chapter 6 NATO (1966)
11. Haase, W., Bradsma, F., Elsholz, E., Leschziner, M., Schwamborn, D., (eds.): EUROVAL—An European Initiative on Validation of CFD Codes. Notes on Numerical Fluid Mechanics, vol. 42, Section 5.1. Vieweg Verlag, Wiesbaden (1993)
12. Cambier, L., Heib, S., Plot, S.: The Onera elsA CFD software: input from research and feedback from industry. Mech. Ind. **14**(3), 159–174 (2013)
13. van Leer, B.: Towards the ultimate conservative difference scheme, V. A second order sequel to Godunov's method. J. Comput. Phys. **32**(1), 101–136 (1979)
14. van Albada, G.D., van Leer, B., Roberts, W.W.: A comparative study of computational methods in cosmic gas dynamics. Astron. Astrophys. **108**(1), 76–84 (1982)
15. Yoon, S.K., Jameson, A.: An LU-SSOR scheme for the Euler and Navier-Stokes equations. In: 25th AIAA Aerospace Sciences Meeting, 12–15 January 1987, Reno NV, AIAA paper 1987-600 (1987)
16. Destarac, D., van der Vooren, J.: Drag/thrust analysis of jet-propelled transonic transport aircraft; Definition of physical drag components. Aerosp. Sci. Technol. **8**(6), 545–556 (2004)
17. Bompard, M.: Modèle de substitution pour l'optimisation globale et de forme en aérodynamique et méthode locale sans paramétrisation. Ph.D. thesis, Université Nice-Sophia Antipolis (2011)
18. Kleijnen, J.: Kriging metamodeling in simulation: a review. Eur. J. Oper. Res. **192**(3), 707–716 (2009)
19. Schöbi, R., Sudret, B., Wiart, J.: Polynomial-chaos-based Kriging. Int. J. Uncertain. Quantif. **5**(2), 171–193 (2015)
20. Aizerman, M.A., Braverman, E.A., Rozonoer, L.I.: Theoretical foundations of the potential function method in pattern recognition learning. Autom. Remote Control **25**, 821–837 (1964)
21. Rippa, S.: An algorithm for selecting a good value for the parameter c in radial basis function interpolation. Adv. Comput. Math. **11**(2), 193–210 (1999)
22. Storm, R., Price, K.: Differential evolution—A simple and efficient heuristic for global optimization over continuous spaces. J Glob. Optim. **11**(4), 341–359 (1997)
23. Kennedy, J., Eberhart, R.: Particle swarm optimization. In: Proceedings of IEEE International Conference on Neural Networks, vol. 4, pp. 1942–1948 (1995)

24. Jaynes, E.T.: Information theory and statistical mechanics. Phys. Rev. **106**(4), 620–630 (1957)
25. Resmini, A., Peter, J., Lucor, D.: Sparse grids-based stochastic approximations with applications to aerodynamics sensitivity analysis. Int. J. Numer. Methods Eng. **106**(1), 32–57 (2016)

Surrogates for Combustion Instabilities in Annular Combustors

M. Bauerheim, A. Ndiaye and F. Nicoud

Introduction

Large power densities encountered in gas turbines and engines can lead to large oscillations of pressure and heat release, known as combustion instabilities [1–3]. First described by Lord Rayleigh in the late nineteenth century, these thermo-acoustic instabilities aroused the interest of industries in the 60's when they became a major issue in the design of the F-1 Apollo engine [2]. Nowadays, gas turbines for electricity generation and aircraft engines also suffer from combustion instabilities. Indeed, lean conditions which weaken the flame stabilizing mechanisms as well as the annular shape of modern combustors (i.e., along e_θ, Fig. 1) promote low-frequency azimuthal modes. Typical geometries feature an annular combustion chamber connected to several burners (typically 10–20), which therefore increases the number of uncertain parameters [4] and complexifies their statistical relationship (for instance "Does neighboring flames are statistically correlated?") [5, 6]. Because of this complexity, experiments are still rare [7, 8] and numerical simulations are CPU demanding [9–11]. Since Uncertainty Quantification (UQ) requires the evaluation of several numerical and experimental realizations, new low-cost flexible tools have to be developed to evaluate the risk of instability of such a system.

To investigate azimuthal modes, 3D acoustic tools based on the non-homogeneous Helmholtz equation [12] have been adapted to annular configurations [13, 14].

M. Bauerheim (✉)
DAEP, ISAE-SUPAERO, 10 avenue Edouard Belin, 31400 Toulouse, France
e-mail: michael.bauerheim@isae-supaero.fr

A. Ndiaye
CERFACS, 42 avenue Gaspard Coriolis, 31052 Toulouse, France
e-mail: ndiaye@cerfacs.fr

F. Nicoud
Université Montpellier 2, Place Bataillon, 34095 Montpellier, France
e-mail: franck.nicoud@umontpellier.fr

C. Hirsch et al. (eds.), *Uncertainty Management for Robust Industrial Design in Aeronautics*, Notes on Numerical Fluid Mechanics and Multidisciplinary Design 140, https://doi.org/10.1007/978-3-319-77767-2_15

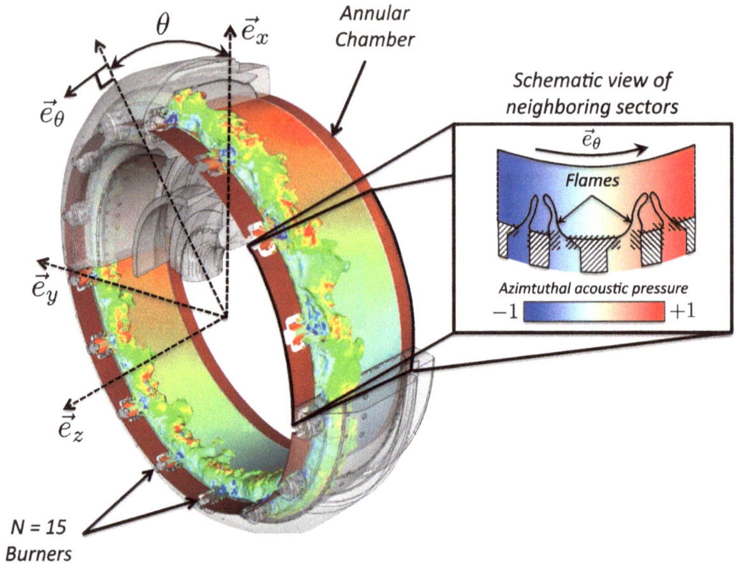

Fig. 1 3D view (left) of an azimuthal combustion instability (pressure fluctuations along the azimuthal direction e_θ). Schematic view (right) of two neighboring burners

Recently, such a tool has been used for the UQ analysis of longitudinal modes in a swirled combustor containing a few uncertain parameters [15], requiring thousands of acoustic simulations to reach the statistical convergence. For annular cases containing many uncertain parameters, more efficient low-order tools should be employed. The principle is usually based on a network model of multiple connected blocks [8, 14, 16, 17]. Galerkin projections, state-space models, or simple 1D tubes can be used for blocks, and their connections are made using jump conditions [18, 19] or scattering matrices [8, 17]. In section "Network Model for Annular Combustors," a network model called Analytical Tool to Analyze and Control Azimuthal Modes in Annular Combustors (ATACAMAC) is derived for an annular combustion chamber fed by an arbitrary number of burners connected to an annular plenum [20–22]. This methodology allows the reduction of the system's complexity leading to an explicit analytical dispersion relation. This equation is implicit and nonlinear in the complex angular frequency ω. Solving this dispersion relation, either analytically or numerically, gives access to ω and therefore the growth rate $\Im(\omega)$ which determines the stability of the mode: If $\Im(\omega) > 0$, the mode is unstable, whereas $\Im(\omega) < 0$ implies a stable mode. Several cases will be analyzed in section "Analytical and Numerical Resolutions," revealing key parameters controlling the stability such as the coupling between the chamber and the burners and symmetry breaking. These parameters are partly unknown, which calls for UQ analysis. To tackle this problem, the ATACAMAC framework is combined with adjoint [23, 24] and Active Subspace [6, 25] techniques in section "Advanced Techniques for UQ

Analysis," leading to an efficient approach to perform UQ analysis at low costs on non-trivial annular configurations [26].

Network Model for Annular Combustors

An analytical model for combustion instabilities can be constructed using inter-connected 1D tubes. In each tube, representing a burner or an annular sector, the upstream ($w^+ = p' + \rho_0 c_0 u'$) and downstream ($w^- = p' - \rho_0 c_0 u'$) acoustic waves propagate from the curvilinear coordinate s_0 to $s_0 + \Delta s$ at the sound speed c^0 (assuming a null Mach number for simplicity [27]):

$$w^{\pm}(s_0 + \Delta s) = w^{\pm}(s_0)e^{\pm j\omega \Delta s/c^0} \tag{1}$$

where the value of c^0 depends on the location (c_u^0 in the burners and plenum, but c_b^0 in the chamber, Fig. 2). Thus, using Eq. (1), the azimuthal propagation in the ith sector of the annular plenum and chamber can be combined to form a propagation matrix $R_i(\omega)$ such that:

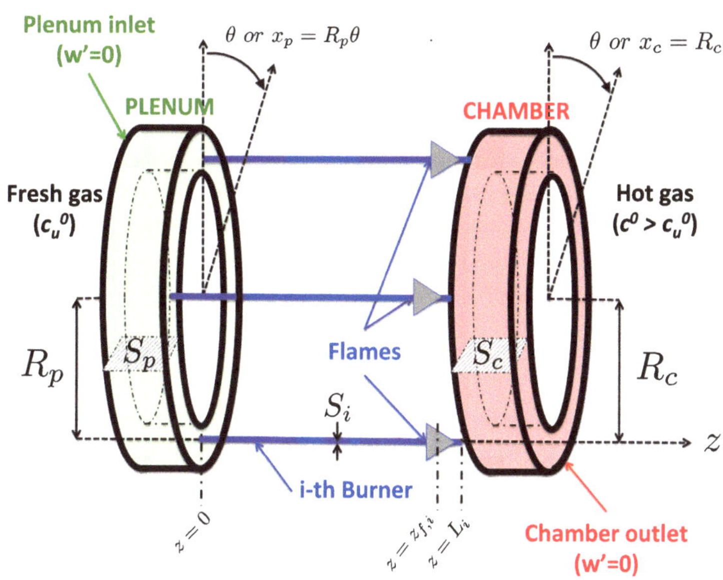

Fig. 2 Sketch of an annular combustion chamber fed by burners connected to a common annular plenum. Because of the flame, the annular plenum and burners contain a fresh mixture characterized by a density ρ_u^0 and sound speed c_u^0, whereas hot products with ρ_b^0 and c_b^0 are located in the combustion chamber

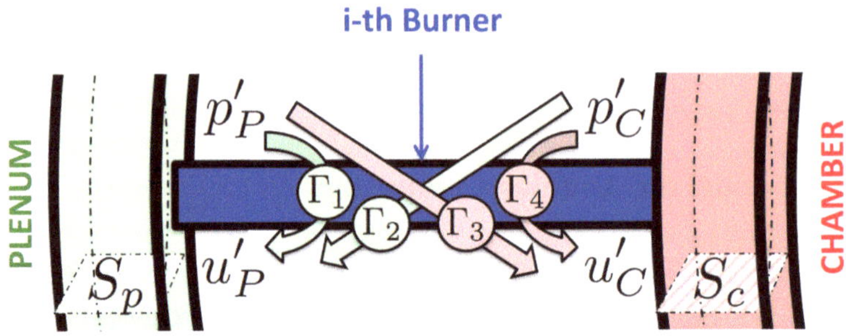

Fig. 3 H-junction connecting the annular plenum to the combustion chamber through the *i*th *burner*. The analytical derivation by [20] leads to four coupling parameters $\Gamma_{i=1,\dots,4}$

$$\begin{bmatrix} w_p^+ \\ w_p^- \\ w_c^+ \\ w_c^- \end{bmatrix}(s_{i+1}) = \begin{bmatrix} e^{jk_u 2L_p/N} & 0 & 0 & 0 \\ 0 & e^{-jk_u 2L_p/N} & 0 & 0 \\ 0 & 0 & e^{jk_b 2L_c/N} & 0 \\ 0 & 0 & 0 & e^{-jk_b 2L_c/N} \end{bmatrix} \begin{bmatrix} w_p^+ \\ w_p^- \\ w_c^+ \\ w_c^- \end{bmatrix}(s_i) = [R_i] \begin{bmatrix} w_p^+ \\ w_p^- \\ w_c^+ \\ w_c^- \end{bmatrix}(s_i) \tag{2}$$

where $k_u = \omega/c_u^0$ and $k_b = \omega/c_b^0$ are the wave numbers in the fresh and hot gases, and L_p and L_c are the half-perimeter of the annular plenum and chamber. N is the number of burners; consequently, $2L_p/N$ and $2L_c/N$ are the length of one sector delimited by the coordinates s_i and s_{i+1}.

Each of the N sectors of the plenum is connected to the chamber through a burner, creating a H-junction as shown in Fig. 3. Thus, the acoustic pressure p' and velocity u' in the plenum are related to the pressure and velocity in the chamber. Using jump conditions [18, 19], the acoustic propagation in the burner described by Eq. (1), and a $n-\tau$ model for the unsteady heat release Q' produced by the flame ($Q' = \mathscr{F} u' = n e^{j\omega\tau} u'$, where n and τ are the gain and the time-delay of the Flame Transfer Function denoted $\mathscr{F}$), an interaction matrix $[T_i]$ is derived by [20]. It relates acoustic quantities before the ith junction (coordinate s_i^-) to the ones after the junction (s_i^+):

$$\begin{bmatrix} w_p^+ \\ w_p^- \\ w_c^+ \\ w_c^- \end{bmatrix}(s_i^+) = [P]^{-1} \begin{bmatrix} 1 & 0 & 0 & 0 \\ \Gamma_{i,1} & 1 & \Gamma_{i,2} & 0 \\ 0 & 0 & 1 & 0 \\ \Gamma_{i,3} & 0 & \Gamma_{i,4} & 1 \end{bmatrix} [P] \begin{bmatrix} w_p^+ \\ w_p^- \\ w_c^+ \\ w_c^- \end{bmatrix}(s_i^-) = [T_i] \begin{bmatrix} w_p^+ \\ w_p^- \\ w_c^+ \\ w_c^- \end{bmatrix}(s_i^-) \tag{3}$$

where $[P]$ is the matrix relating the Riemann invariants $w^\pm$ to the acoustic pressure and velocity, and $\Gamma_{i,k=1,\dots,4}$ are the coupling parameters derived by [20]:

$$\begin{cases} \Gamma_{i,1} = -\frac{S_i}{2S_p}\mathrm{cotan}(k_u L_i) \\ \Gamma_{i,2} = \frac{S_i}{2S_p}\frac{1}{\sin(k_u L_i)} \\ \Gamma_{i,3} = \frac{S_i}{2S_c}\frac{\rho_b^0 c_b^0}{\rho_u^0 c_u^0}\frac{1+ne^{j\omega\tau}}{\sin(k_u L_i)} \\ \Gamma_{i,4} = -\frac{S_i}{2S_c}\frac{\rho_b^0 c_b^0}{\rho_u^0 c_u^0}(1+ne^{j\omega\tau})\mathrm{cotan}(k_u L_i) \end{cases} \tag{4}$$

where L_i is the i-th burner length and S_i its cross section.[1] These coupling parameters are also obtained in longitudinal configurations [28] and characterize how cavities are coupled and interact (Fig. 3). Decoupling can be achieved using a large section change at the burner junction, but it can be also affected by the flame itself (i.e., by n and τ). Note that if $\Gamma_{i,1} = \Gamma_{i,2} = 0$ for all junctions $i = 1, \dots, N$, then the annular plenum is disconnected from the rest of the system.

Using the propagation and interaction matrices to connect the annular sectors, the annular periodicity leads to the equation governing the acoustic modes behavior in the annular plenum and chamber:

$$\left(\prod_{i=1}^{N}[R_i][T_i]\right)\begin{bmatrix} w_p^+ \\ w_p^- \\ w_c^+ \\ w_c^- \end{bmatrix} = \begin{bmatrix} w_p^+ \\ w_p^- \\ w_c^+ \\ w_c^- \end{bmatrix} \tag{5}$$

Equation (5) has non-trivial solutions if and only if the determinant is null, which yields the dispersion relation to be solved:

$$\det\left(\prod_{i=1}^{N}[R_i][T_i] - I_d\right) = 0 \tag{6}$$

where I_d is the 4-by-4 identity matrix.

Analytical and Numerical Resolutions

The dispersion relation (6) is nonlinear in ω, and thus, no general solution exists. Numerical solvers can efficiently solve this equation, but explicit expressions are still useful to understand key mechanisms controlling combustion instabilities.

[1]These expressions assume a flame located exactly at the burner/chamber junction, i.e., $z_{f,i} = L_i$. This location plays a crucial role for plenum modes.

Axisymmetric Configurations

In many practical applications, annular combustors are axisymmetric, meaning that all sectors and flames are identical. It implies that in the analytical model, all matrices $[R_i]$ and $[T_i]$ are similar (the subscript i can be omitted), which simplifies the dispersion relation: $\det(\{[R][T]\}^N - I_d) = 0$. This equation can be recast as

$$\prod_{p=1}^{N} \det([R][T] - e^{j2p\pi/N}) = 0 \quad \Leftrightarrow \quad \det([R][T] - e^{j2p\pi/N} I_d) = 0 \ \text{ for } p = 1, \ldots, N \tag{7}$$

This simplification highlights that in axisymmetric configurations, each sector has the same acoustic behavior: The stability of the system can be deduced using one sector (matrix $[R][T]$) which necessarily acts as a pure phase-lag $2p\pi/N$, where p corresponds physically to the azimuthal order. In the UQ context, it suggests that investigating one azimuthal order in an axisymmetric annular chamber is not more expensive than studying a longitudinal configuration. This result can be applied to 3D acoustic tools where the Bloch theory allows the computations of the annular combustor using only one sector [29], and therefore, it can drastically reduce the computational effort to evaluate the risk of instability.

Despite this apparent simplicity, annular configurations containing a chamber and a plenum can exhibit complex lock-in and veering phenomena, for which the active flames are a key ingredient.

- Considering first the simple case where all coupling parameters are null, i.e., $\Gamma_1 = \Gamma_2 = \Gamma_3 = \Gamma_4 = 0$, Bauerheim et al. [20] showed that the plenum and the chamber are fully decoupled from the burners and flames. It results that eigenfrequencies are $f_p^0 = pc_u^0/2L_p$ (pure azimuthal decoupled mode in the plenum) or $f_c^0 = pc_b^0/2L_c$ (pure azimuthal decoupled mode in the chamber). Since the cold mixture and hot gases have a different temperature, and the half-perimeter of the plenum and chamber is different, eigenmodes in the plenum and chamber are distinct.
- When coupling factors are not null but satisfy $|\Gamma_{k=1,\ldots,4}| \ll 1$, solutions are close to the fully decoupled case. Consequently, they can be searched as $f_c = f_c^0 + \delta f$ and $f_p = f_p^0 + \delta f$. A Taylor expansion of the dispersion relation yields the solutions in the case where the two annular cavities are not naturally coupled, i.e., when f_p^0 and f_c^0 are not multiple of each other:

$$f_c = \frac{pc_b^0}{2L_c} - \frac{c^0 N \Gamma_4^0}{4\pi L_c} \quad \text{and} \quad f_p = \frac{pc_u^0}{2L_p} - \frac{c_b^0 N \Gamma_1^0}{4\pi L_p} \tag{8}$$

where Γ_1^0 (respectively Γ_4^0) is the value of the coupling parameter Γ_1 (respectively Γ_4) at the frequency $f = f_p^0$ (respectively $f = f_c^0$): These modes are called "weakly coupled." This analytical expression is validated against a semi-analytical method

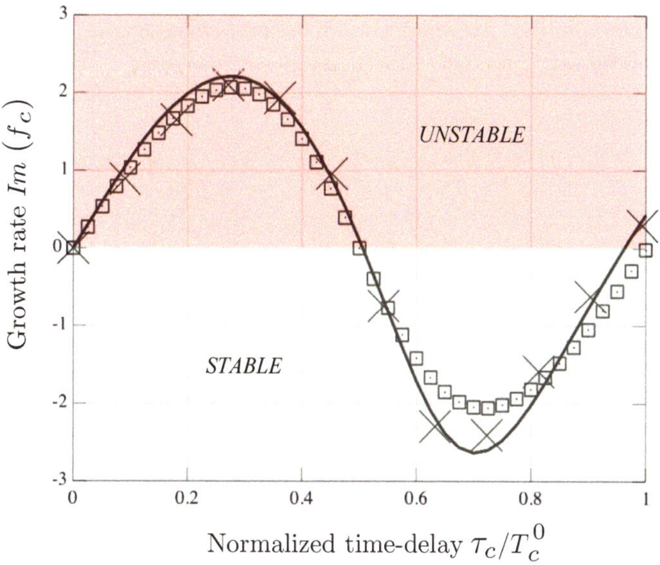

Fig. 4 Growth rate $\Im(f_c)$ of the first azimuthal chamber mode ($p = 1$) in a combustor containing an annular chamber and plenum with $N = 4$ burners. The FTF corresponds to $n = 1.57$ and the time-delay τ is varied from 0 to T_c^0, where $T_c^0 = 1/f_c^0$ is the period of the mode. The growth rate is estimated using Eq. (8) (□), a semi-analytical solution (−), and by a 3D Helmholtz solver (×)

(the analytical dispersion relation in Eq. (6) is solved numerically using a Newton–Raphson technique) as well as a 3D Helmholtz solution on an annular case with $N = 4$ burners in Fig. 4. Note that the growth rate of these modes scales with the number of burners N.

- When the weakly coupled assumption is not satisfied, the two annular cavities can couple and oscillate at the same frequency, even if f_p^0 and f_c^0 do not match: The burners and flames tune one of the two cavities so that they can both resonate. In this case, the acoustic mode cannot be identified as "plenum" or "chamber" modes because the whole combustor is resonating: A bifurcation occurs as shown by the stability map in Fig. 5: These modes are called "strongly coupled." Compared with WC modes where at least one time-delay stabilizes the configuration, one SC mode is always unstable (the other one being always stable) whatever the time-delay. It suggests that stabilizing the configuration requires first to decouple the system before acting on the flames. In a UQ context, capturing the steep bifurcation between WC/SC modes is a challenging task, especially since no explicit solutions to guide the analysis are available for this case (Eq. (8) holds only for weakly coupled modes).

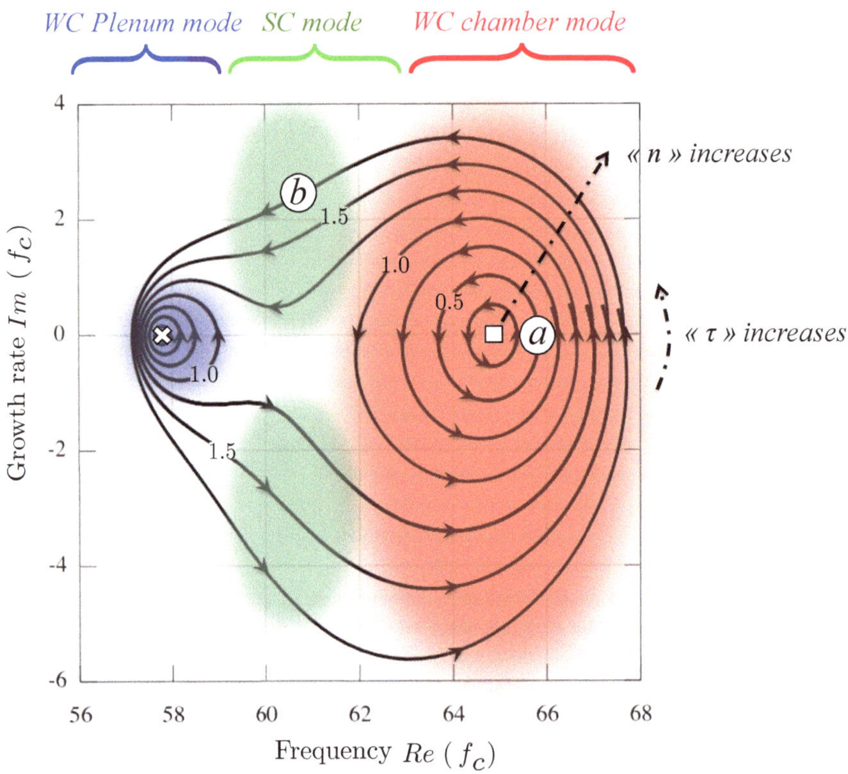

Fig. 5 Stability map $\{\Re(f_c), \Im(f_c)\}$ of a configuration with $N = 4$ burners and the flame parameters are varied: n from 0.25 to 1.75 and τ/T_c^0 from 0 to 1. It highlights the bifurcation between the weakly coupled (WC) modes, close to the fully decoupled solutions (× and □), and the strongly coupled (SC) modes for which no distinction plenum/chamber can be observed

Non-symmetric Configurations

The ATACAMAC tool applied to axisymmetric cases has revealed in section "Axisymmetric Configurations" that the coupling parameters $\Gamma_{k=1,\ldots,4}$ governed the stability of the system as well as the bifurcation between weakly and strongly coupled modes. For UQ analysis, this result is of prime importance since the problem contains only two uncertain parameters: The geometric parameters and the sound speed are usually well known, whereas the most uncertain parameters are due to the flame response to acoustics, i.e., the gain n and the time-delay τ. While uncertainties on the fuel or global equivalence ratio can impact all the N flame responses and associated FTFs, turbulence or manufacturing tolerances yield uncertainties with low flame–flame correlations. In other words, flames and burners are not necessarily identical, and thus, the framework developed in section "Axisymmetric Configurations" does not hold anymore. Such a case also appears when passive control devices are introduced in the annular chamber to stabilize the system, e.g., Helmholtz resonators [30, 31] or baffles [32]. To account for this symmetry breaking effect, the N flames are now considered different, i.e., with different gains n_i and time-delays τ_i. To ease the formulation, no plenum is considered here, i.e., $\Gamma_{i,1} = \Gamma_{i,2} = 0$. The 2-by-2 restrictions of the matrices $[R_i]$ and $[T_i]$ to the chamber modes make that only $\Gamma_{i=1,\ldots,N,4}$ are relevant. Nevertheless, N is usually of the order of 10–20. In a combustor containing 19 burners for example, it yields 38 uncertain parameters (the N statistically independent gains and time-delays), which leads already to a large UQ problem when solved with 3D acoustic tools.

Following a similar approach than the one considered for symmetric cases, Bauerheim et al. [4] showed that for weakly coupled modes, the complex frequencies of the azimuthal mode of order p are:

$$f_c^{\pm} = \frac{pc_b^0}{2L_c} - \frac{c_b^0}{4\pi L_c}\left(\Sigma_0 \pm \mathscr{S}_0\right) \tag{9}$$

where Σ_0 is the "coupling strength" defined as

$$\Sigma_0 = \sum_{i=1}^{N} \Gamma_i^0 \tag{10}$$

This parameter is the sum of all the coupling parameters of the system and is independent of the pattern used to distribute the burner uncertainties along the annular chamber. It corresponds to a symmetric effect.

The parameter $\pm\mathscr{S}_0$ is the "splitting strength" which "splits" the two azimuthal mode frequencies f_c^+ and f_c^-. A convenient form of this parameter is obtained by using the spatial Fourier transform of the coupling parameter distribution γ:

$$\mathscr{S}_0 = \sqrt{\gamma(2p)\gamma(-2p)} \text{ where } \gamma(k) = \sum_{i=1}^{N} \Gamma_i^0 e^{-j2k\pi i/N} \tag{11}$$

Note that the "coupling strength" can be also recast in this form, i.e., $\Sigma_0 = \gamma(0)$. It shows that only few specific patterns can affect the azimuthal mode stability. They correspond to the 0th and the $\pm 2p^{th}$ Fourier coefficients γ of the coupling parameter or heat release distribution [33]. Unlike the coupling strength Σ_0, the splitting parameter $\mathscr{S}_0$ can be changed by modifying the pattern of the burner types along the annular chamber. Such a modification can be intended as when controlling devices are introduced, or unintended when turbulence or uncertainties affect randomly the flame response to acoustics. In a UQ perspective, the explicit solution of Eq. (11) allows the CPU cost to be drastically reduced since only patterns associated with $\gamma(0)$ and $\gamma(\pm 2p)$ can be retained [4]. Recently, Ghirardo et al. [34] also show that nonlinearities of the flame response itself can produce a splitting effect [22, 34].

The azimuthal mean flow induced by swirlers or modern effusive plates can also promote such a splitting [27].

Advanced Techniques for UQ Analysis

Compared with LES or 3D acoustic tools, the ATACAMAC tool allows brute-force UQ techniques, like the Monte Carlo method, to be performed. The objective is to estimate the uncertainty on the growth rate $\Im(f_c)$ from the uncertainties of the N_q inputs. Indeed, while most studies intend to predict if the system is stable or unstable, the benefit of a UQ analysis is to assess the risk of instability, in other words the probability of the mode to be unstable. This can be measured by the risk factor introduced by Bauerheim et al. [5, 6]:

$$\mathrm{RF}\ (\%) = 100 \int_0^{\infty} \mathrm{PDF}(\omega_i) d\omega_i \tag{12}$$

where $\omega_i = 2\pi\Im(f_c)$ and $\mathrm{PDF}(\omega_i)$ is the probability density function of the growth rate ω_i. If the risk factor is $\mathrm{RF} = 0\%$, it implies that the mode is stable, whereas $\mathrm{RF} = 100\%$ corresponds to an unstable mode. Between these two extreme cases associated with the classical binary stability prediction, the risk factor quantifies how a mode is stable regarding the uncertainties of the inputs. Such an example is displayed in Fig. 6 where a Monte Carlo method is used to estimate the risk factor of the first azimuthal mode of a combustor containing $N = 19$ burners. Each point in the stability map $\{\Re(f_c), \Im(f_c)\}$ (left) corresponds to an ATACAMAC simulation. Here, 10,000 simulations were required to reach the convergence and estimate the

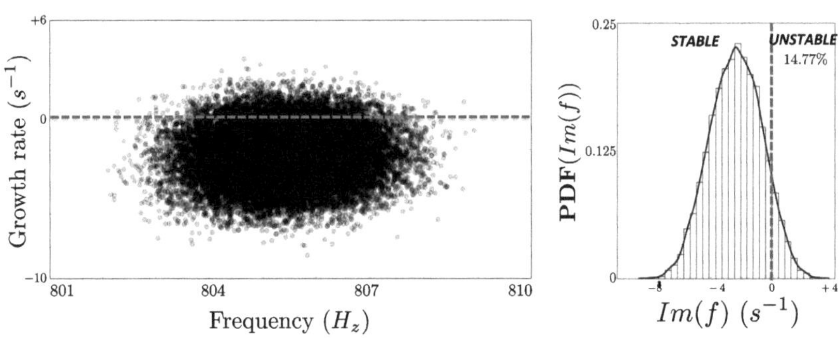

Fig. 6 Monte Carlo analysis using 10,000 ATACAMAC simulations for the first weakly coupled azimuthal chamber mode of a combustor containing $N = 19$ burners. Each point on the stability map (left) corresponds to one ATACAMAC simulation. This method allows a robust estimation of the growth rate PDF and its associated risk factor

growth rate PDF (right). The integration of this PDF for positive growth rate gives an approximation of the risk factor, here RF = 14%. In this case, it suggests that the mode is probably stable, but further investigations should be done to ensure a safe stable mode, i.e., RF = 0%. Note that a risk factor close to 50% is a useless prediction since the mode can be either stable or unstable. This might be due to a lack of knowledge of the key parameter uncertainties, or because the system is highly sensitive to small errors.

Adjoint Analysis

Situations where the systems are highly sensitive to uncertainties suggest that a sensitivity analysis could be used before performing an expensive Monte Carlo analysis. The ATACAMAC framework allows a low-cost method for sensitivity computations based on the adjoint analysis. The objective is not to compute the surface response $[\Im(f_c)](q_k)$ in the whole parameter space $\{q_{k=1,\dots,N_q}\}$, as in a UQ analysis, but only its gradients $[\Im(f_c)](q_k)/dq_k$. This knowledge allows the determination of (1) the sensitivity of the system and (2) the critical parameters which strongly affect the system's ouput. Whereas a sensitivity analysis can be CPU demanding if estimated using a finite difference, the adjoint method allows the computation of the gradients with only one additional simulation whatever the number of input parameters N_q.

The equation governing the system (Eq. (5)) can be recast as $[M](\omega, q)X = X$, where $[M] = \prod_{i=1}^{N}[R_i][T_i]$ is the matrix of the system depending on the angular frequency ω and a set of parameters q, and X is the state vector corresponding to the acoustic pressure and velocity at one specific annular coordinate. Its adjoint form can be written as

$$\left([M]^H(\omega_0^*, q_0) - I_d\right) X_0^\dagger = 0 \tag{13}$$

where $[M]^H$ is the conjugate transpose of the matrix system $[M]$, $X_0^\dagger$ is the adjoint eigenvector, and ω_0^* is the conjugate of the complex angular frequency ω_0 obtained when solving the direct problem.

The knowledge of the adjoint eigenvector $X_0^\dagger$ allows the derivation of an approximate solution of the direct problem around q_0: In other words, $X_0^\dagger$ provides a first-order approximation $\omega = \omega_0 + (\epsilon q_1)\omega_1$ of the problem for a set of parameters $q_0 + \epsilon q_1$ close enough to q_0. Compared with finite differences where each direction of the parameter space should be probed by changing one direction at a time, the adjoint analysis provides this information for all the N_q directions using only one additional computation. Magri et al. [24] have shown that this first-order eigenvalue drift reads

$$\omega_1 = -\frac{\langle X_0^\dagger, \delta M(\omega_0, \epsilon q_1)X_0\rangle}{\langle X_0^\dagger, \frac{\partial M(\omega, q_0)}{\partial \omega}X_0\rangle} \tag{14}$$

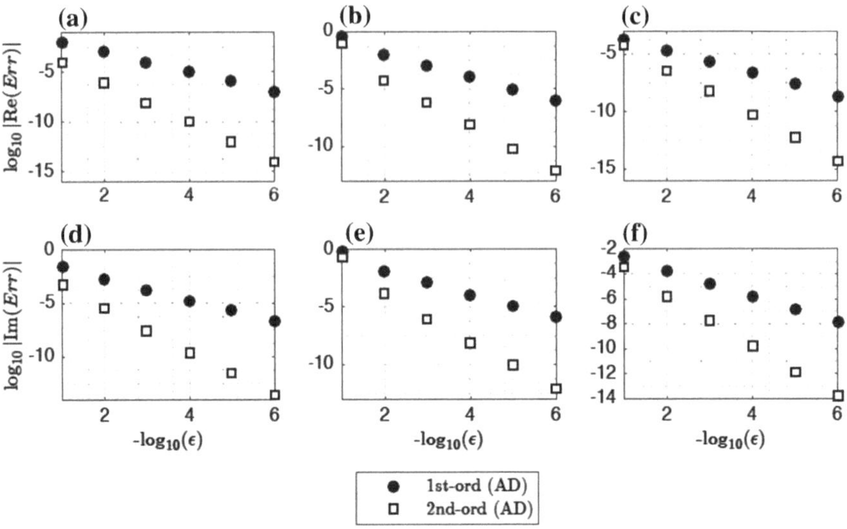

Fig. 7 Relative error between finite difference and the adjoint-based first- and second-order sensitivities on three annular configurations: a weakly coupled mode in a symmetric case (a and d); a strongly coupled mode in a symmetric combustor (b and e); a strongly coupled mode with symmetry breaking

where $\delta M(\omega_0, \epsilon q_1) = M(\omega_0, q_0 + \epsilon q_1) - M(\omega_0, q_0)$ is the difference between the matrix system at the frequency ω_0 for the two different yet close parameter sets q_0 and $q_0 + \epsilon q_1$, and $\langle . \rangle$ is the inner product. Note that Magri et al. also derived a second-order approximation, not detailed here for the sake of simplicity [24]. These first- and second-order adjoint-based approximations are validated against the finite-difference technique in Fig. 7 on three different cases:

- a weakly coupled mode in a symmetric case (a and d);
- a strongly coupled mode in a symmetric combustor (b and e);
- a strongly coupled mode with symmetry breaking (c and f).

The sensitivity analysis performed in Fig. 7 is associated with small deterministic changes in the parameter space. The same approach can be employed in a UQ context where the parameter change is chosen randomly, assuming this change is small. Using two computations only (one direct and one adjoint computations), the eigenvalue drifts can be computed for any random perturbation ϵq_1 at a low cost (no additional computation is needed, and only Eq. (14) has to be evaluated numerically for each sample). Such an adjoint-based method is validated against a classical and CPU demanding Monte Carlo technique (a and b, red dashed lines) on a strongly coupled mode in a non-symmetric annular combustor in Fig. 8. Results show that only small errors are obtained using a first-order approximation (c and d), whereas the growth rate PDF is well approximated by a second-order adjoint-based method (e and f). This result reveals how low-order methods which allow sophisticated

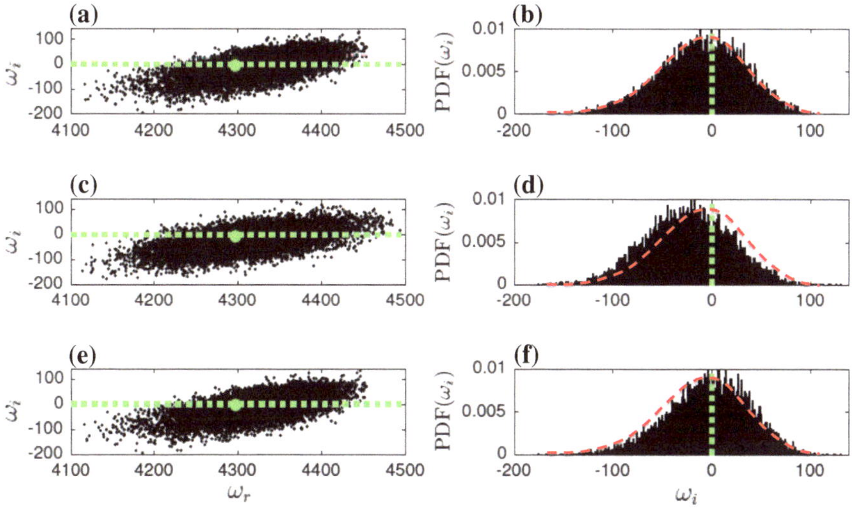

Fig. 8 Stability map (left) and PDF of the growth rate (right) of a strongly coupled case in a non-symmetric annular combustor (10% of uncertainties on the flame parameters) estimated using a finite difference (a and b and the red dashed lines), a first-order adjoint approximation (c and d), and a second-order adjoint approximation (e and f). The dotted lines highlight the stability limit $\Im(\omega) = 0$, and the green dot corresponds to the deterministic computation

techniques (adjoint analysis is much more difficult when using reactive 3D Navier–Stokes solvers for instance) can provide efficient methods for UQ analysis of industrial combustion chambers.

Active Subspace

The adjoint analysis presented in section "Adjoint Analysis" provides an efficient approach to tackle UQ problems involving a large number of uncertain parameters. Nevertheless, such a method has some limitations:

- The adjoint method is well suited for smooth functions where a second-order approximation is sufficient to capture the major fluctuations of the output.
- No indication is provided to quantify which and how parameters affect the surface response.
- The adjoint method allows an efficient computation of the surface response using the low-order model, but cannot be transferable to other tools. Note that applying adjoint to more complex solvers (3D acoustic tools, LES solvers etc.) is a tedious task [29, 35].

To overcome these constraints, the Active Subspace method developed by Constantine et al. [25] was adapted to thermo-acoustics by Bauerheim et al. [6]. It can be used as a stand-alone approach or combined with the adjoint-based technique proposed in section "Adjoint Analysis." The objective is to reduce the number of the uncertain parameters from N_q to just a few before employing other UQ technique efficient in low dimensions. It is based on the uncentered covariance matrix of the output gradient, which can be evaluated using finite difference or the adjoint:

$$C = \mathbb{E}[\nabla_q \omega (\nabla_q \omega)^T] \approx \frac{1}{M} \sum_{k=1}^{M} [\nabla_q \omega(q^{(k)}) (\nabla_q \omega(q^{(k)}))^T] \tag{15}$$

where the column vector $\nabla_q \omega = [\partial\omega/\partial q_1, \ldots, \partial\omega/\partial q_{N_q}]^T$ is the eigenvalue's sensitivity with respect to the N_q uncertain parameters, $\mathbb{E}(.)$ is the expectation operator, and $q^{(k)}$ corresponds to the kth random sample of parameters among the M samples available for the Active Subspace determination. Because the matrix C is symmetric, it admits a real eigenvalue decomposition

$$C = W \Lambda W^T \tag{16}$$

where Λ is the diagonal matrix $[\lambda_k]$ and W_k is the associated eigenvector. Based on the relative importance in the eigenvalues λ_k (i.e., gaps between eigenvalues), this method selects new variables, called the "active variables," corresponding to directions (i.e., linear combinations of the initial uncertain parameters) of significant change in the output response. Other directions can be disregarded because of not affecting the output. An example of spectra is given in Fig. 9 for a weakly (left) and strongly (right) coupled modes in an annular combustion chamber containing

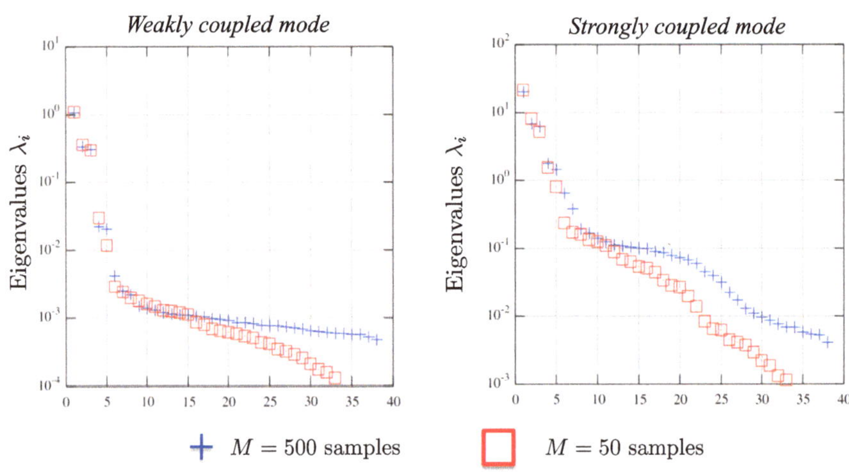

Fig. 9 Active Subspace spectra for a weakly (left) and strongly (right) coupled modes in an annular combustion chamber containing 19 burners and 38 uncertain parameters (each FTF gains and time-delays)

19 burners and 38 uncertain parameters corresponding to the FTF gains and time-delays. Figure 9 shows that only five active variables are significant. Note that the first relevant eigenvalues converge rapidly (50 finite-difference samples for 38 uncertain parameters). These spectra can be evaluated using the adjoint-based method to reduce the cost:

- If the response surface is smooth, all random perturbations can be computed using Eq. (14) based on the unperturbed case (i.e., no uncertainties), thus requiring only two computations (one direct and one adjoint) whatever the number of parameters N_q and the number of samples M.
- If the response surface is not smooth enough, the perturbed case $\omega(q^{(k)})$ cannot be obtained through Eq. (14): A direct computation is mandatory. However, in the Active Subspace context, the adjoint can still be applied to extract the gradients $\nabla_q \omega$ around $\omega(q^{(k)})$ at a low cost. Consequently, the method needs a direct and adjoint computations for each sample $q^{(k)}$, in contrast with finite differences which require $N_q + 1$ direct computations for each sample: The adjoint-based approach now scales linearly with the number of sample, i.e., $2 \times M$, but is still independent of the number of inputs N_q, whereas the finite differences scale like $2 \times M \times (1 + N_q)$. It shows that an Active Subspace method combined with adjoint computations allows efficient uncertainty quantifications at low cost, even for large UQ problems (i.e., large N_q).

The benefit of such a method is that the "active variables" obtained by ATACAMAC in this configuration can be reused on the same case but using more complex acoustic tools, for instance a 3D Helmholtz solver. Moreover, Bauerheim et al. [6] have shown that these active variables are not a mathematical artifact of the method, but correspond to physical parameters. Using a spatial Fourier transform, they showed that the problem can be reduced to only three active variables corresponding to $\gamma(0)$ and $\gamma(\pm 2p)$, as predicted by Eq. (11) for weakly coupled modes. The Active Subspace approach reveals that these three parameters are also governing the strongly coupled modes for which no explicit solutions are available yet. In a UQ perspective, this result highlights that a UQ analysis involving only three uncertain parameters can be applied to study weakly or strongly coupled modes in annular combustors. This dimension reduction is crucial since most of the UQ techniques are effective only in low dimensions. As a conclusion, the Active Subspace approach combined with an adjoint method constitutes an efficient method to perform a systematic UQ analysis for combustion instabilities in annular combustors at the design stage.

Conclusion

While large UQ analysis for combustion instabilities is still out of reach when using high-fidelity simulations or 3D acoustic tools, low-order models for thermo-acoustics can be developed for annular combustors at a reduced cost. The

ATACAMAC tool (Analytical Tool to Analyze and Control Azimuthal Modes in Annular Combustors) is described to tackle this problem in complex geometries where the combustor contains several burners connected to an annular plenum and a combustion chamber. In some cases, called "weakly coupled modes," fully explicit analytical solutions can be derived for the complex frequencies of the system, for which the imaginary part gives the stability of the system. It reveals key parameters governing the stability of the system: (1) the "coupling strength" associated with a symmetric effect and (2) the "splitting strength" due to symmetry breaking. This framework allows the development of sophisticated techniques such as adjoint-based sensitivity analysis. This approach gives the gradients of the growth rate using only two computations whatever the number of parameters involved in the system. In a UQ context, it provides at low cost a first- and second-order approximations of the surface response. Results revealed that the PDF of the growth rate can be captured efficiently for both weakly and strongly coupled modes. Finally, this method can be combined with the Active Subspace approach to assess the risk of instability in complex annular combustors even at the design stage.

Acknowledgements Part of this work has been realized in the European LEMCOTEC project (PhD financed by Snecma) as well as during the Summer Program of CTR at Stanford University with the collaborations of P. Constantine, L. Magri and M. Juniper.

References

1. Lieuwen, T., Yang, V.: Combustion instabilities in gas turbine engines. Operational experience, fundamental mechanisms and modeling. In: Progress in Astronautics and Aeronautics, AIAA, vol. 210 (2005)
2. Culick, F.E.C., Kuentzmann, P.: Unsteady motions in combustion chambers for propulsion systems. In: NATO Research and Technology Organization (2006)
3. Poinsot, T., Veynante, D.: Theoretical and numerical combustion, 3rd edn. www.cerfacs.fr/elearning
4. Bauerheim, M., Salas, P., Nicoud, F., Poinsot, T.: J. Fluid Mech. **760**, 431–465 (2014)
5. Bauerheim, M., Ndiaye, A., Constantine, P., Iaccarino, G., Moreau, S., Nicoud, F.: In: 2014 Proceedings of the Summer Program, Center for Turbulence Research, Stanford University, pp. 209–218 (2014)
6. Bauerheim, M., Ndiaye, A., Constantine, P., Moreau, S., Nicoud, F.: J. Fluid Mech. **789**, 534–566 (2016)
7. Worth, N., Dawson, J.: Proc. Combust. Inst. **34**, 3127–3134 (2013)
8. Bourgouin, J.-F.: Dynamique de flamme dans les foyeres annulaires comportant des injecteurs multiples, Ph.D. thesis, Ecole Centrale de Paris (EM2C), (2014)
9. Staffelbach, G., Gicquel, L., Boudier, G., Poinsot, T.: Proc. Combust. Inst. **32**, 2909–2916 (2009)
10. Wolf, P., Staffelbach, G., Gicquel, L., Muller, J.: Combust. Flame **159**, 3398–3413 (2012)
11. Bauerheim, M.: Theoretical and numerical study of symmetry breaking effects on azimuthal thermoacoustic modes in annular combustors, Ph.D. thesis, INP Toulouse & Cerfacs, (2014)
12. Nicoud, F., Benoit, L., Sensiau, C., Poinsot, T.: AIAA J. **45**, 426–441 (2007)
13. Sensiau, C., Nicoud, F., Poinsot, T.: Int. J. Aeroacoust. **8**, 57–68 (2009)
14. Bauerheim, M., Staffelbach, G., Worth, N., Dawson, J., Gicquel, L., Poinsot, T.: Proc. Combust. Inst. **35**, 3355–3363 (2014)

15. Ndiaye, A., Bauerheim, M., Moreau, S., Nicoud, F.: ASME Paper 2015-GT-44133 (2015)
16. Stow, S.R., Dowling, A.P.: In: ASME Paper 2001-GT-0037, New Orleans, Louisiana
17. Schuermans, B., Bellucci, V., Paschereit, C.: In: International Gas Turbine and Aeroengine Congress & Exposition, ASME Paper, vol. 2003-GT-38688
18. Dowling, A.P.: J. Sound Vib. **180**, 557–581 (1995)
19. Bauerheim, M., Nicoud, F., Poinsot, T.: Combust. Flame **162**, 60–67 (2014)
20. Bauerheim, M., Parmentier, J., Salas, P., Nicoud, F., Poinsot, T.: Combust. Flame **161**, 1374–1389 (2014)
21. Bauerheim, M., Poinsot, T.: Annual Research Briefs. CTR (2015)
22. Bauerheim, M., Nicoud, F., Poinsot, T.: Phys. Fluids **28**, 021303 (2016)
23. Magri, L., Juniper, M.: Int. J. Spray Combust. Dyn. **6**, 225–246 (2014)
24. Magri, L., Bauerheim, M., Juniper, M.: J. Comput. Phys. **325**, 395–410 (2016)
25. Constantine, P.G., Dow, E., Wang, Q.: SIAM J. Sci. Comput. **36**, 1500–1524 (2014)
26. Magri, L., Bauerheim, M., Nicoud, F., Juniper, M.: J. Comput. Phys. (2016)
27. Bauerheim, M., Cazalens, M., Poinsot, T.: Proc. Combust. Inst. **35**, 3219–3227 (2014)
28. Schuller, T., Durox, D., Palies, P., Candel, S.: Combust. Flame **159**, 1921–1931 (2012)
29. Mensah, G., Moeck, J.: ASME paper GT2015-43476 (2015)
30. Stow, S.R., Dowling, A.P.: In: ASME Paper 2003-GT-38168, Atlanta, Georgia, USA
31. Campa, G., Camporeale, S.: ASME Paper 2014-GT-25228 (2014)
32. Dawson, J., Worth, N.: Proc. Combust. Inst. **35**, 3283–3290 (2014)
33. Noiray, N., Bothien, M., Schuermans, B.: Combust. Theory Model. **15**, 585–606 (2011)
34. Ghirardo, G., Juniper, M., Moeck, J.: ASME paper GT2015-43127 (2015)
35. Silva, C., Magri, L., Runte, T., Polifke, W.: J. Eng. Gas Turbines Power **139** GTP-16-1224 (2016)

General Introduction to Monte Carlo and Multi-level Monte Carlo Methods

Robin Schmidt, Matthias Voigt, Michele Pisaroni, Fabio Nobile, Penelope Leyland, Jordi Pons-Prats and Gabriel Bugeda

Introduction

With Monte Carlo (MC) methods, we identify a broad class of approaches that rely on the idea of approximating statistics of the response of a system by simulation through sampling. Because of its simplicity, robustness, and dimension independent convergence rate, MC methods can be used to characterize, in principle, any system that has a probabilistic interpretation. MC methods are often the easiest way, and sometimes the only feasible one, to solve a wide range high-dimensional problems.

Hereafter, we will denote random variables with capital letters and realizations of those with lower case letters. Vectors and matrices are shown in bold.

R. Schmidt (✉) · M. Voigt
Institute of Fluid Mechanics, Technische Universität Dresden, 01062 Dresden, Germany
e-mail: robin.schmidt@tu-dresden.de

M. Voigt
e-mail: matthias.voigt@tu-dresden.de

M. Pisaroni · F. Nobile · P. Leyland
Scientific Computing and Uncertainty Quantification, Ecole Polytechnique Fédérale de Lausanne, 1015 Lausanne, Switzerland
e-mail: michele.pisaroni@epfl.ch

F. Nobile
e-mail: fabio.nobile@epfl.ch

P. Leyland
e-mail: penelope.leyland@epfl.ch

J. Pons-Prats · G. Bugeda
CIMNE, Aeronautical Group, 08034 Barcelona, Spain
e-mail: jpons@cimne.upc.edu

G. Bugeda
Universitat Politecnica de Catalunya - BarcelonaTech, 08034 Barcelona, Spain
e-mail: bugeda@cimne.upc.edu

C. Hirsch et al. (eds.), *Uncertainty Management for Robust Industrial Design in Aeronautics*, Notes on Numerical Fluid Mechanics and Multidisciplinary Design 140, https://doi.org/10.1007/978-3-319-77767-2_16

Suppose we are interested in computing the expected value $\mathbb{E}[Q]$ of a quantity of interest (QoI) $Q = Q(\mathbf{X})$ of a problem having some random elements $\mathbf{X}$. Assume we can generate N independent and identically distributed (i.i.d.) realizations $X^{(i)}$, $i = 1, ..., N$ and for each of them evaluate the corresponding QOI $Q^{(i)} = Q(X^{(i)})$. Then the MC estimator for the expectation $\mathbb{E}[Q]$ of Q is defined as:

$$\mathrm{E}^{\mathrm{MC}}[Q] := \frac{1}{N}\sum_{i=1}^{N} Q(X^{(i)}). \tag{1}$$

The simulation procedure that makes use of i.i.d. samples and the MC estimator introduced in Eq. (1) to approximate $\mathbb{E}[Q]$ is generally called Crude Monte Carlo (CMC).

Thanks to the Strong Law of Large Numbers, the approximation of $\mathrm{E}^{\mathrm{MC}}[Q]$ converges with probability one (converges almost surely) to $\mathbb{E}[Q]$ as $N \to \infty$ as long as Q is integrable.

Moreover, $\mathrm{E}^{\mathrm{MC}}[Q]$ is an unbiased estimator:

$$\mathbb{E}[\mathrm{E}^{\mathrm{MC}}[Q]] = \mathbb{E}[Q] \tag{2}$$

meaning that the expected value of the MC estimator equals $\mathbb{E}[Q]$ for any N.

The rate of convergence of MC method can be described by the Central Limit Theorem (CLT) using the concept of convergence in distribution (weak convergence, size of the error with some probability). If the variance of Q, hereafter denoted with $\mathbb{V}ar[Q]$, is finite then the CLT asserts that

$$\sqrt{N}\left(\mathrm{E}^{\mathrm{MC}}[Q] - \mathbb{E}[Q]\right) \implies \sqrt{\mathbb{V}ar[Q]}\mathcal{N}(0, 1) \tag{3}$$

as $N \to \infty$, where $\mathcal{N}(0, 1)$ is a normal random variable with mean zero and unit variance and $\implies$ means convergence in distribution. From Eq. 3, for N large enough, we can derive confidence intervals for the estimator $\mathrm{E}^{\mathrm{MC}}[Q]$ of $\mathbb{E}[Q]$:

$$\left|\mathrm{E}^{\mathrm{MC}}[Q] - \mathbb{E}[Q]\right| \leq C_\alpha \frac{\sqrt{\mathbb{V}ar[Q]}}{\sqrt{N}} \tag{4}$$

with probability $1 - \alpha$, where C_α is a confidence parameter such that the cumulative distribution function Φ of a standard normal random variable $\Phi(C_\alpha) = 1 - \frac{\alpha}{2}$.

From Eq. (4), we can draw three crucial conclusions:

- the rate of convergence of MC is $O(N^{-1/2})$,
- for large N the error is normally distributed,
- the complexity of the computation depends solely on $\mathbb{V}ar[Q]$.

If an exact representation of the QoI is not accessible and we rely on a numerical approximation (e.g., a finite volume (FV) or finite element (FE) approximation in fluid dynamics computations) with a discretization parameter M (number of spatial degrees of freedom), then Q is approximated by Q_M. The accuracy in estimating $\mathbb{E}[Q]$ by $\mathrm{E}^{\mathrm{MC}}[Q_M]$ can be quantified by considering the mean square error (MSE) of the estimator:

$$e(\mathrm{E}^{\mathrm{MC}}[Q_M])^2 := \mathbb{E}[(\mathrm{E}^{\mathrm{MC}}[Q_M] - \mathbb{E}[Q])^2] = \underbrace{(\mathbb{E}[Q_M - Q])^2}_{(\text{B-E}^{\text{MC}})} + \underbrace{\frac{\mathbb{V}\mathrm{ar}[Q_M]}{N}}_{(\text{SE-E}^{\text{MC}})}. \quad (5)$$

On the right-hand side, we can isolate two distinct contributions. The first term, the discretization error or bias (B-E^{MC}), is the square error in mean between Q_M and Q and depends solely on the space discretization parameter M. The second term, the statistical error (SE-E^{MC}), represents the variance of the estimator and decays inversely with the number of samples N.

The Crude Monte Carlo (CMC) approach is summarized in the algorithm below.

Algorithm 1: Crude Monte Carlo Algorithm

CMC(N, M)
 for $i = 0 : N$ **do**
  Generate random samples: $Q_M(X^{(i)})$
 compute $\mathrm{E}^{\mathrm{MC}}[Q_M] := \frac{1}{N}\sum_{i=1}^{N} Q_M(X^{(i)})$

CMC is a very elegant approach and has been proven to be robust and accurate for non-smooth problems, nevertheless its very slow convergence rate $O(N^{-1/2})$ prevents to achieve reasonably estimations in acceptable time for large-scale aerodynamic applications that require the solution of computational expensive CFD simulations.

Different strategies have been investigated in the last decades to accelerate MC methods. They are all based on the idea of reducing the ratio on the right-hand side of Eq. (4) $\frac{\sqrt{\mathbb{V}ar[Q]}}{\sqrt{N}}$. The two most prominent categories of approaches are:

- *Alternative Sampling techniques*: increase the denominator term to converge more rapidly by using deterministic (low-discrepancy) sequences, stratified sampling, or Latin Hypercube Sampling rather than pseudorandom numbers.
- *Variance Reduction techniques*: reduce the numerator term $\mathbb{V}ar[Q]$ by suitably modifying the quantity Q in a consistent way (i.e., without changing the expectation) as in the Multi-Level Monte Carlo approach.

These methodologies will be presented in the following sections and chapters.

Choice of Sampling Sequences

The generation of the $x^{(i)}$ samples with predefined probability distribution is a pivotal procedure in MC methods. In this section, we review different approaches used to generate pseudorandom and quasi-random numbers and methodologies required to prescribe appropriate correlations to random variables.

Pseudorandom Numbers

The simplest procedure is random sampling. However, true random numbers are the result of physical phenomenon as, for example, radioactive decay processes. Practical applications utilize pseudorandom numbers. Those results from pseudorandom number generators (PRNGs), also referred to as deterministic random number generators, are based on some reproducible mathematical formulation. Starting from a certain seed, the goal is to produce a sequence of uniform pseudorandom numbers in the interval (0, 1) with statistical properties that are in very good agreement with those of a true sequence of i.i.d. random variables. The period length of the PRNG describes the number of random numbers until the sequence repeats itself. In general, a small period seems bad; however, a larger period is not necessarily better. A good PRNG has good performance in different criteria. A variety of theoretical and empirical tests, see, e.g., [1], can be conducted to decide whether a PRNG can be considered a good one.

The most common PRNG is based on recursive arithmetic, as, for example, linear congruential generators. Popular PRNG is the Mersenne Twister [2] or the combined multiple recursive generator according to [3].

Quasi-random Numbers

Quasi-random numbers are the result of low-discrepancy sequences. The resulting realizations are uniformly distributed in the interval [0,1). They exhibit much more uniformity compared to random or pseudorandom numbers. Therefore, they increase the convergence rate if used within MC methods. In order to specify the application of low-discrepancy sequences in MC methods, the term Quasi-Monte Carlo (QMC) is used. The convergence rate of QMC is usually close to $O(N^{-1})$, which is higher compared to CMC, see Eq. (4).

Uniformity is measured by utilizing discrepancy which is defined as follows. Let B be a rectangle in the d-dimensional unit hypercube J with sides parallel to the coordinate axes and $m(B)$ its volume. The discrepancy of a set of N points in $[0, 1)^d$ is defined as

$$D_N = \sup_{B \in J} \left| \frac{\text{Number of points in } B}{N} - m(B) \right| . \tag{6}$$

The most common low-discrepancy sequences are *Halton* and *Sobol* sequences. Both are based on the *van der Corput* sequence which is constructed by reversing the base-b representation of the sequence of natural numbers. For more details, concerning the construction of low-discrepancy sequences the interested reader is referred to [1, 4, 5].

Although low-discrepancy sequences possess high uniformity in low dimensions d (and large N), they can exhibit poor space-filling behavior for small N and large d. The d-dimensional *Halton* sequence, e.g., is constructed by pairing d one-dimensional sequences based on d different prime numbers (usually the first d primes). In the case of high dimensions, the base b must be large. The corresponding *van der Corput* sequences with large bases produce long linearly growing segments. If these are paired with each other, a strongly linear space filling of the unit square is obtained. Different techniques designated *leap* [6] and *scramble* [7] were created in order to overcome these problems.

Pseudorandom Variables with Non Uniform Distribution

In order to generate a random variable X from an arbitrary distribution the following two steps are involved.

1. Generation of uniform random numbers $U_1, ..., U_N$ with the PRNG.
2. Transformation of U_i according to its respective probability density function $f(X)$ or joint probability density function $f(\mathbf{X})$.

In the previous sections, we briefly presented how uniform pseudorandom numbers or uniform quasi-random numbers can be created. Here, we will describe two transformation methods in order to get a random variable X from such a uniform distributed random variable. The most notable transformation methods are the inverse transform method and the acceptance–rejection method.

In the inverse transform method, the random variable is calculated with the inverse of the CDF $F(X)$, see Algorithm 2.

Algorithm 2: Inverse-Transform Method

Generate U from U(0, 1)
Return $X = F^{-1}(U)$

The acceptance–rejection method directly works with the PDF $f(X)$ of the considered random variable X. Moreover a further PDF $g(X)$ is needed, such that for some $c \geq 1$, $c\,g(X) \geq f(X)$ for all x. It is assumed that random numbers can be easily generated from $g(X)$. The resulting method is described in Algorithm 3.

Algorithm 3: Acceptance-Rejection Method

Generate X from $g(X)$
Generate U from $\mathrm{U}(0, 1)$
If $U \leq f(X)/(c\,g(X))$ use X, otherwise reject X and repeat the process.

Stratification

If it is possible to divide a heterogeneous population into subpopulations each of which is homogeneous, a precise estimate of, e.g., the subpopulations mean can be obtained from a small sample. A combination of such estimates can deliver a precise estimate of the whole population with smaller number of realizations compared to CMC. This line of thought leads to *stratified sampling*. The idea behind stratified sampling is to divide the population of N units into m non-overlapping subpopulations, called strata. Each strata has N_i units with $i = 1, ..., m$ and $\sum_i N_i = N$. A sample of size n_i with $i = 1, ..., m$ and $\sum_i n_i = n$ is selected by some design within each stratum. In case of a random sample in each stratum, the term *stratified random sampling* is used. How to chose the strata depends on the particular problem. The population mean per unit $\mathrm{E}^{\mathrm{St}}[Q]$ can be estimated with

$$\mathrm{E}^{\mathrm{St}}[Q] = \frac{\sum_{i=1}^{m} N_i\, \mathrm{E_i}[Q]}{N} = \sum_{i=1}^{m} W_i\, \mathrm{E_i}[Q]\,, \tag{7}$$

where W_i denotes the stratum weight. Only when the sampling fraction is the same in all strata which means e.g.

$$\frac{n_i}{n} = \frac{N_i}{N}\,, \tag{8}$$

the population mean corresponds to the sample mean. Such a stratified sampling is called *proportional*. If a predefined cost function is available, an optimal allocation of sample size can be achieved, e.g., in order to minimize the variance for $\mathrm{E}^{\mathrm{St}}[Q]$. A simple cost function can be a linear one where the cost is proportional to the size of the sample but varies from stratum to stratum.

The variance of an estimated mean of random sampling is denoted V_{ran}, of stratified sampling with proportional sample allocation V_{prop} and with optimum allocation for fixed n it is V_{opt}. It is shown in [8] that the following relation holds.

$$V_{opt} \leq V_{prop} \leq V_{ran} \tag{9}$$

Therefore, it can be argued that stratified sampling is always better compared to random sampling when enough information is available for its appropriate implementation. However, enough information is represented, e.g., by the frequency distribution of the result quantity, which is often only estimated prior to a probabilistic investigation. Therefore, the necessity of defining the strata is a major problem in stratified sampling. In case of one result quantity and if it is known a priori, for example, due to a reasonable number of measurements, a procedure to calculate the strata and number of strata is given in [8]. The determination of the strata becomes further complicated when many result quantities should be considered. The strata definition for one result quantity may be inappropriate for other quantities.

Correlation and Discrepancy Control

So far, only the marginal distributions of single variables were taken into account when creating random vectors. An $N \times d$ sample vector can be obtained by repeating d times the generation of one-dimensional random variables with N realizations. This naive approach can lead to undesired dependencies between the variables which must be avoided. On the other hand, specific interrelationships between the input variables might be explicitly desired for a variety of probabilistic simulations, for example, when treating measurements of a real system or in the context of sensitivity analyses, where correlation is of great importance and must be considered.

Relations between input variables can be represented by correlation, for example, using Pearson correlation coefficients (generally denoted with ρ) or Spearman rank correlation coefficient (denoted with r). According to Pearson, the correlation coefficient for two random variables X_i and X_j is defined as:

$$\rho_{ij} = \frac{\sum_{k=1}^{N}(x_{ki} - \overline{x}_i)(x_{kj} - \overline{x}_j)}{\sqrt{\sum_{k=1}^{N}(x_{ki} - \overline{x}_i)^2\sqrt{\sum_{k=1}^{N}(x_{kj} - \overline{x}_j)^2}}}. \tag{10}$$

The rank correlation coefficient is calculated with the ranks of the data.

A correlation matrix **C** of size $d \times d$ is obtained from a sample of the size $N \times d$. In the case of the correlation coefficient according to Pearson $C_{ij} = \rho_{ij}$. Furthermore, it is assumed that the desired correlation structure is known and predefined by a target correlation matrix **T**.

There are two main groups of methodologies used to generate correlated random vectors with arbitrary given marginal distribution and correlations:

1. Methods that transform a correlated standard normal random variable into a target non-normal variable
2. Methods that optimize the rank correlation structure of a sample.

The popular Nataf model [9] belongs to the first group. A standard normal random vector $\mathbf{Z}$ with a correlation matrix $\mathbf{T}'$ is transformed component-wise into the desired vector $\mathbf{X}$ with a correlation matrix $\mathbf{T}$. The marginal transformation is obtained by:

$$X_i = F_i^{-1}(\Phi(Z_i)), \tag{11}$$

where Φ is the standard normal CDF and $F_i(X_i)$ the CDF of X_i. The Nataf model approach assumes that Z is jointly normal and uses the Pearson correlation coefficient (invariant under nonlinear strictly increasing transformations) as in Eq. (11). Thus, the relation $\mathbf{T}' \neq \mathbf{T}$ holds. In order to get the unknown matrix $\mathbf{T}'$, each element ρ'_{ij} must be computed by solving:

$$\rho_{ij} = \int_{-\infty}^{\infty} \int_{-\infty}^{\infty} \frac{x_i - \mu_i}{\sigma_i} \frac{x_j - \mu_j}{\sigma_j} \varphi_2(z_i, z_j, \rho'_{ij}) \mathrm{d}z_i \mathrm{d}z_j , \tag{12}$$

where $\varphi_2(z_i, z_j, \rho'_{ij})$ designates the PDF of the bivariate standard normal distribution.

In order to avoid the elaborate solution of Eq. (12), empirical equations have been developed such that $\rho'_{ij} = \mathrm{f}(\rho_{ij})$ can be computed, see e.g. [10].

If the matrix $\mathbf{T}'$ is available, uncorrelated standard normal distributed random vectors can be transformed into correlated ones by means of Cholesky transformation. The Cholesky decomposition $\mathbf{T}' = \mathbf{L}\mathbf{L}^T$ provides the lower triangular matrix $\mathbf{L}$. The correlated random vectors are then obtained by applying $\mathbf{X}\mathbf{L}^T$.

The idea of converting uncorrelated random variables into correlated ones by orthogonal transformation is also the basis of methods belonging to the second group. One of these was developed by Iman and Conover [11] and is known as *Restricted Pairing*. The random vectors of individual random variables are generated according to their respective probability distribution without taking into account correlations. The restricted pairing technique uses the rank correlation coefficient. Compared to the Pearson correlation coefficient, the latter has the advantage of being invariant under monotonic transformations of the marginals. Algorithm 4 describes the procedure of Restricted Pairing.

Algorithm 4: Restricted Pairing

Calculate $\mathbf{C}$ from $\mathbf{X}$
Calculate lower triangular matrix $\mathbf{Q}$ from $\mathbf{Q}\mathbf{Q}^T = \mathbf{C}$
$\mathbf{P}$ from $\mathbf{P}\mathbf{P}^T = \mathbf{T}$
$\mathbf{S} = \mathbf{P}\mathbf{Q}^{-1}$
$\mathbf{R} = \mathbf{X}\mathbf{S}^T$
Change the order in $\mathbf{X}$ in order to achieve the same ranks as in $\mathbf{R}$

The method proceeds from uncorrelated random variables. Practically, this is only possible to a limited extent. Therefore, the correlation of the input sample is taken into account by incorporating the rank correlation matrix **C** which results from the available sample. In case of perfect correlation $r_{ij} = 1$ with $i \neq j$ in the input sample, the rows in each column of **X** can be randomly shuffled.

An iteratively improved implementation of the Restricted Pairing technique has been presented in [12].

Besides the two aforementioned groups of methodologies, other approaches exist, as, e.g., the usage of Copulas to construct a multivariate random vector of dependent components.

A desired order within the sample can also be set up by solving a combinatorial optimization problem. The optimization is based on a scalar quantity which measures the deviation $\mathbf{E} = \mathbf{T} - \mathbf{A}$ between target correlation matrix and the actual correlation matrix **A**. Vořechovský and Novák [13] described the deviation by *root mean square correlation* r_{rms} and minimized it by using Simulated Annealing and interchanging a pair of two realizations x_{ik} and x_{jk}.

$$r_{rms} = \sqrt{\frac{2\sum_{i=1}^{d-1}\sum_{j=i+1}^{d}(E_{ij})^2}{d(d-1)}} \tag{13}$$

A suitable matrix norm can also be used to measure the maximum absolute correlation error:

$$r_{max} = \max_{1 \leq i \leq j \leq d} |E_{ij}| \tag{14}$$

If the correlation adjustment can be formulated as an optimization problem, a discrepancy improvement can be obtained with the same approach only by exchanging the objective. As an example, Liefvendahl and Stocki [14] used a genetic algorithm to solve the optimization problem.

The description of the space-filling proprieties of samples by means of a scalar quantity is possible with a multitude of criteria. An overview and an evaluation of existing criteria can be found in [15]. Beside all, the *centered L^2 discrepancy* [16] shows good performances for projections in 2D subspaces.

$$\begin{aligned} C^2 = {} & \left(\frac{13}{12}\right)^d - \frac{2}{N}\sum_{i=1}^{N}\prod_{k=1}^{d}\left(1 + \frac{1}{2}|x_k^{(i)} - 0.5| - \frac{1}{2}|x_k^{(i)} - 0.5|^2\right) \\ & + \frac{1}{N^2}\sum_{i,j=1}^{N}\prod_{k=1}^{d}\left(1 + \frac{1}{2}|x_k^{(i)} - 0.5| - \frac{1}{2}|x_k^{(j)} - 0.5| - \frac{1}{2}|x_k^{(i)} - x_k^{(j)}|\right) \end{aligned} \tag{15}$$

Latin Hypercube Sampling Method

Latin Hypercube Sampling (LHS) was first published by McKay et al. [17] and further developed by Iman and Conover [18]. The method can reduce the variance of an estimator compared to random sampling, which results in a reduction of the sample size while maintaining the statistical significance.

A mathematical proof of the variance reduction compared to CMC was given by McKay et al. [17] under the condition that the system behavior is monotonic in each of its inputs. Iman and Conover [18] show for an additive model with uniform inputs that the variance of an estimated mean converges with a factor of N^{-2} faster compared to CMC. Stein [19] demonstrated that the amount of variance reduction increases with the degree of additivity in the model response. An experimental comparison of LHS against CMC was carried out by Manteufel [20]. LHS estimates an unbiased mean value as well as the distribution function. The bias in the estimation of the variance is low and associated with a significantly lower sampling variability.

The idea behind LHS relates to stratified sampling. In LHS, only the marginal distributions are stratified in such a way that each random variable X is divided into N contiguous intervals of equal probability with respect to the corresponding CDF $F(X)$. For that purpose, the unit probability is divided into N intervals of identical probability 1/N. These probability intervals are bounded by a lower ϕ_{k-1} and upper bound ϕ_k.

$$\phi_k = \frac{k}{N} \quad \text{with} \quad k = 1, ..., N \tag{16}$$

The calculation of the corresponding interval bounds ξ_k over the values of the random variable X can be performed by utilizing the inverse of the CDF $F(X)$

$$\xi_k = F^{-1}(\phi_k)\,. \tag{17}$$

In each probability interval one realization x_k must be selected. Therefore, $x_k \in (\xi_{k-1}, \xi_k)$ holds. Besides *random* LHS where each realization x_k is uniformly distributed in its respective interval, *mean* and *median* LHS exist. For those methods different ways of selecting the sample values from the probability intervals are applied. In case of *median* LHS each probability interval is selected by taking the following set of sampling probabilities.

$$\mathbf{p} = (p_1, p_2, ..., p_k, ..., p_N) \quad \text{with} \quad p_k = \frac{k - 0.5}{N} \tag{18}$$

The samples are selected using the inverse transformation of the probabilities in $\mathbf{p}$.

$$x_k = F^{-1}(p_k) \tag{19}$$

The mean in each interval is selected for *mean* LHS. It makes a numerical integration of the PDF $f(X)$ necessary. The samples are selected using

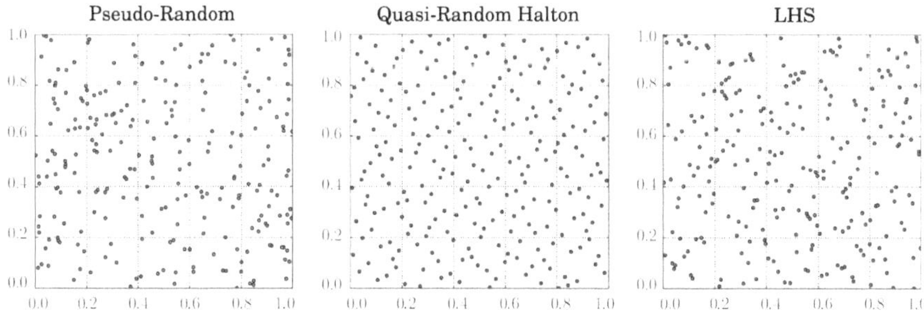

Fig. 1 Visual comparison of 250 pseudorandom, Halton quasi-random and *random* Latin Hypercube samples

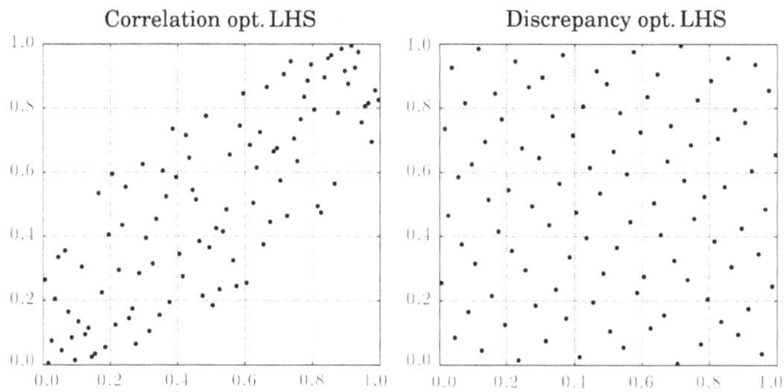

Fig. 2 Visual comparison of correlation optimized and discrepancy optimized *median* LHS; $N = 100$

$$x_k = \frac{\int_{\xi_{k-1}}^{\xi_k} x f(X) \mathrm{d}x}{\int_{\xi_{k-1}}^{\xi_k} f(X) \mathrm{d}x}. \tag{20}$$

Figure 1 shows a visual comparison of different sampling techniques for $N = 250$ realizations. Figure 2 extends the visual comparison and shows correlation and discrepancy control for a *median* LHS with $N = 100$.

Multi-level Monte Carlo

As previously stated, Crude Monte Carlo (CMC) sampling has a dimension independent convergence rate which is not affected by the presence of possible discontinuities in the parameter space. However, the CMC approach converges very slowly and is impractical in complex applications that require accurate solutions. The Multi-Level Monte Carlo (MLMC) method has been introduced by Heinrich [21, 22] in the context of parametric integration and extended by Giles [23] to approximate stochastic differential equations (SDEs) in financial mathematics as a way to improve the efficiency of MC simulations. Applications to PDE models with random parameters can be found in [24–28].

Algorithm 5: Multi-Level Monte Carlo Algorithm

MLMC(L, $\{N_l\}_{l=0}^{L}$)
  for $l = 0 : L$ **do**
    for $i = 0 : N_l$ **do**
      Generate random input: $\omega^{(i,l)}$
      $Q_{M_l}^{(i)} \leftarrow \mathsf{PROBLEM}_l\,(\omega^{(i,l)})$
      $Q_{M_{l-1}}^{(i)} \leftarrow \mathsf{PROBLEM}_{l-1}\,(\omega^{(i,l)})$
      $Y_l^{(i)} = Q_{M_l} - Q_{M_{l-1}}$
    compute $\mathrm{E}^{\mathrm{MC}}[Y_l]$
  return $\mathrm{E}^{\mathrm{MLMC}}[Q_M] = \sum_{l=0}^{L} \mathrm{E}^{\mathrm{MC}}[Y_l]$

The key idea of MLMC is to simultaneously draw MC samples on several approximations Q_{M_l} of Q built on a hierarchy of nested computational grids (with discretization parameters $M_0 < M_1 < \cdots < M_L = M$) thanks to the linearity propriety of the expectation operator. Indeed the expectation of a QoI computed on the finest level can be written as a telescopic sum of the expectation of the QoI on the coarsest level plus a sum of correction terms adding the difference in expectation between evaluations on consecutive levels:

$$\mathbb{E}[Q_{M_L}] = \mathbb{E}[Q_{M_0}] + \sum_{l=1}^{L} \mathbb{E}[Q_{M_l} - Q_{M_{l-1}}] = \sum_{l=0}^{L} \mathbb{E}[Y_l] \tag{21}$$

with $Y_l = Q_{M_l} - Q_{M_{l-1}}$ and $Y_0 = Q_{M_0}$.

The MLMC estimator for $\mathbb{E}[Q]$ can be written as:

$$\mathrm{E}^{\mathrm{MLMC}}[Q_M] := \sum_{l=0}^{L} \frac{1}{N_l} \sum_{i=1}^{N_l} Y_l(\omega^{(i,l)}) = \sum_{l=0}^{L} \mathrm{E}^{\mathrm{MC}}[Q_{M_l} - Q_{M_{l-1}}], \tag{22}$$

where the same realization $\omega^{(i,l)}$ is used to compute the correction $Y_l(\omega^{(i,l)}) = Q_{M_l}(\omega^{(i,l)}) - Q_{M_{l-1}}(\omega^{(i,l)})$ on both levels whereas corrections on different levels should be sampled independently.

The accuracy in estimating $\mathbb{E}[Q]$ by $\mathrm{E}^{\mathrm{MLMC}}[Q_M]$ can be quantified by considering the mean square error (MSE) of the estimator:

$$e(\mathrm{E}^{\mathrm{MLMC}}[Q_M])^2 := \mathbb{E}[(\mathrm{E}^{\mathrm{MLMC}}[Q_M] - \mathbb{E}[Q])^2] = \underbrace{(\mathbb{E}[Q_M - Q])^2}_{(\mathrm{B\text{-}E}^{\mathrm{MLMC}})} + \underbrace{\sum_{l=0}^{L} \frac{\mathbb{V}\mathrm{ar}[Y_l]}{N_l}}_{(\mathrm{SE\text{-}E}^{\mathrm{MLMC}})}. \tag{23}$$

The standard MLMC algorithm is summarized in Algorithm 5. The notation $\mathsf{PROBLEM}_l$ denotes a general 'black-box' CFD solver that computes the QoI of the problem under investigation given a set of input values at the grid discretization level l. The description of the treatment of specific geometric or operating input random parameters will be provided in the following chapters.

References

1. Kroese, D.P., Taimre, T., Botev, Z.I.: Handbook of Monte Carlo Methods. Wiley, Inc. (2011)
2. Matsumoto, M., Nishimura, T.: Mersenne twister: a 623-dimensionally equidistributed uniform pseudo-random number generator. ACM Trans. Model. Comput. Simul. (TOMACS) **8**(1), 3–30 (1998)
3. L'Ecuyer, P.: Good parameters and implementations for combined multiple recursive random number generators. Oper. Res. **47**(1), 159–164 (1999)
4. Niederreiter, H.: Random number generation and quasi-Monte Carlo methods. In: Society for Industrial and Applied Mathematics, CBMS-NSF Regional Conference Series in Applied Mathematics, vol. 63, Pa (1992)
5. Morokoff, W.J., Caflisch, R.E.: Quasi-Monte Carlo integration. J. Comput. Phys. **122**(2), 218–230 (1995)
6. Kocis, L., Whiten, W.J.: Computational investigations of low-discrepancy sequences. ACM Trans. Math. Softw. (TOMS) **23**(2), 266–294 (1997)
7. Hong, H.S., Hickernell, F.J.: Algorithm 823: implementing scrambled digital sequences. ACM Trans. Math. Softw. (TOMS) **29**(2), 95–109 (2003)
8. Cochran, W.G.: Sampling Techniques, 3rd edn. Wiley Series in Probability and Mathematical Statistics. Wiley, New York (1977)
9. Nataf, A.: Dtermination des distributions de probabilits dont les marges sont donns. Comptes Rendus de l'Acadmie des Sciences **225**, 42–43 (1962)
10. Ditlevsen, O., Madsen, H.O.: Structural Reliability Methods. Wiley, Chichester, New York (1996)
11. Iman, R.L., Conover, W.J.: A distribution-free approach to inducing rank correlation among input variables. Commun. Stat. Simul. Comput. **11**(3), 311–334 (1982)
12. Dandekar, R.A., Cohen, M., Kirkendall, N.: Sensitive micro data protection using Latin hypercube sampling technique. In: Inference Control in Statistical Databases, Springer, pp. 117–125 (2002)
13. Vořechovský, M., Novák, D.: Correlation control in small-sample Monte Carlo type simulations I: a simulated annealing approach. Probab. Eng. Mech. **24**(3), 452–462 (2009)
14. Liefvendahl, M., Stocki, R.: A study on algorithms for optimization of Latin hypercubes. J. Stat. Plan. Inference **136**(9), 3231–3247 (2006)
15. Damblin, G., Couplet, M., Iooss, B.: Numerical studies of space-filling designs: optimization of Latin hypercube samples and subprojection properties. J. Simul. **7**(4), 276–289 (2013)
16. Hickernell, F.: A generalized discrepancy and quadrature error bound. Math. Comput. Am. Math. Soc. **67**(221), 299–322 (1998)
17. McKay, M.D., Beckman, R.J., Conover, W.J.: A comparison of three methods for selecting values of input variables in the analysis of output from a computer code. Technometrics **42**(1), 55–61 (1979)

18. Iman, R.L., Conover, W.: Small sample sensitivity analysis techniques for computer models with an application to risk assessment. Commun. Stat. Theory Methods **9**(17), 1749–1842 (1980)
19. Stein, M.: Large sample properties of simulations using Latin hypercube sampling. Technometrics **29**(2), 143–151 (1987)
20. Manteufel, R.D.: Evaluating the convergence of Latin hypercube sampling. In: 41st AIAA/ASME/ASCE/AHS/ASC Structures, Structural Dynamics and Materials Conference (2000)
21. Heinrich, S.: Monte Carlo complexity of global solution of integral equations. J. Complex. **14**(2), 151–175 (1998)
22. Heinrich, S., Sindambiwe, E.: Monte Carlo complexity of parametric integration. J. Complex. **15**(3), 317–341 (1999)
23. Giles, M.B.: Multilevel Monte Carlo path simulation. Oper. Res. **56**(3), 607–617 (2008)
24. Barth, A., Lang, A., Schwab, C.: Multilevel Monte Carlo method for parabolic stochastic partial differential equations. BIT Numer. Math. **53**(1), 3–27 (2013)
25. Barth, A., Schwab, C., Zollinger, N.: Multi-level Monte Carlo finite element method for elliptic PDEs with stochastic coefficients. Numer. Math. **119**(1), 123–161 (2011)
26. Charrier, J., Scheichl, R., Teckentrup, A.L.: Finite element error analysis of elliptic PDEs with random coefficients and its application to multi level Monte Carlo methods. SIAM J. Numer. Anal. **51**(1), 322–352 (2013)
27. Cliffe, K., Giles, M., Scheichl, R., Teckentrup, A.L.: Multilevel Monte Carlo methods and applications to elliptic PDEs with random coefficients. Comput. Vis. Sci. **14**(1), 3–15 (2011)
28. Teckentrup, A., Scheichl, R., Giles, M., Ullmann, E.: Further analysis of multilevel Monte Carlo methods for elliptic PDEs with random coefficients. Numer. Math. **125**(3), 569–600 (2013)

Latin Hypercube Sampling-Based Monte Carlo Simulation: Extension of the Sample Size and Correlation Control

Robin Schmidt, Matthias Voigt and Ronald Mailach

Introduction

In order to calculate an MC estimator for the expectation $\mathbb{E}[Q]$ of Q

$$\mathrm{E}^{\mathrm{MC}}[Q] := \frac{1}{N}\sum_{i=1}^{N} Q(X^{(i)}), \tag{1}$$

where Q is a quantity of interest with $Q = \mathcal{M}(X)$ with the deterministic model $\mathcal{M}$. The sample $\mathbf{X}$ of some random variables X_i, $i = 1, ..., d$ can be generated with different sampling approaches as already described in chapter "General Introduction to Monte Carlo and Multi-level Monte Carlo Methods". One of such methods is Latin hypercube sampling (LHS). The method can reduce the variance of an estimator compared to random sampling, which results in a reduction of the sample size while maintaining the statistical significance. A drawback of the LHS compared to random sampling is the procedure for the generation of the realizations and the associated limitations in the extension of the sample. Depending on the selection of the sample values, an extension of the sample size can be achieved by doubling or adding an even multiple of its sample size.

The results from a Monte Carlo simulation in terms of sensitivities or robustness measures mostly are in the form of point estimators available. The statistical qual-

R. Schmidt (✉) · M. Voigt · R. Mailach
Institute of Fluid Mechanics, Technische Universität Dresden,
01062 Dresden, Germany
e-mail: robin.schmidt@tu-dresden.de

M. Voigt
e-mail: matthias.voigt@tu-dresden.de

R. Mailach
e-mail: ronald.mailach@tu-dresden.de

C. Hirsch et al. (eds.), *Uncertainty Management for Robust Industrial Design in Aeronautics*, Notes on Numerical Fluid Mechanics and Multidisciplinary Design 140, https://doi.org/10.1007/978-3-319-77767-2_17

ity of these estimators can be determined only after completion of a Monte Carlo simulation by using, e.g., confidence intervals. If the quality is not sufficient, the sample size needs to be increased. Here, it may occur that deterministic calculations must be discarded and a new LHS must be performed. This approach is especially doubtful in conjunction with time-consuming and therefore very "expensive" deterministic calculations. In the literature, various methods for extending a Latin Hypercube sampling are described.

In [1] Pleming and Manteufel, use a level group approach, where at each level a certain number of groups are added. Once the initial LHS design is created, an equal number of realizations in the form of a group are added. Thereby the initial bins are considered. A certain number of groups are necessary to maintain the LHS design in each level. In the extension approach presented by Pleming and Manteufel, the realizations are set at fixed bin fractions and not uniformly distributed in the interval. This approach can be combined with an algorithm for correlation or discrepancy control. A extension algorithm for stratified sampling was proposed by Tong [2]. In case of LHS, the algorithm is able to extend the sample to an integral multiple of its size. However, used in a hierarchical manner the sample size in each step is extended at least twice the size of the predecessor matrix. In [3], Vořechovský explored an approach that consists of an exponent and a base value. The added realizations follow from this approach. Two methods with the names LLHS and HSLHS are presented. They differ with respect to the calculation rule and the base value. The realizations are set at the median of the corresponding probability interval. A disadvantage of the LLHS is the deviation from the LHS design, which results in a grouping around the mean. If all existing realizations are considered the HSLHS yields to a LHS design, however each extension by itself is not a LHS design. Recently, further improvements were proposed by Vorechovskỳ [4] with an approach named HLHS. The HLHS concept reminds of the HSLHS approach, however, it has been generalized and combined with a correlation control and optimization algorithm based on Simulated Annealing; see also [5]. Sallaberry et al. [6] presented an approach to extend a LHS by doubling the sample while considering an existing correlation structure. No subsequent correlation control procedure is necessary.

A method for extending a LHS should ideally have the following features:

- The concept of the LHS design should be maintained or at least it should be possible to reach a LHS design at a certain stage.
- The extension algorithm should be flexible.
- Correlation or discrepancy adjustment should be possible within the method or with an external method.

The first two criteria usually contradict each other. Another reasonable feature can be.

- Each extension is a separate LHS design in the same parameter space.

Latin Hypercube Sampling

The procedure to create a *random* LHS is described in chapter "General Introductionto Monte Carlo and Multi-level Monte Carlo Methods". However, minor changes are carried out in order to increase the representation of the random variables X marginal PDF, if the distribution is an asymptotically expiring one, as, e.g., the normal distribution. Statistical measures as, e.g., $\text{Var}(X)$, which are a result of the sample, are very sensitive to the position of the realizations x_1 and x_N in the outermost probability intervals. In such intervals, the mean is used instead of an uniformly distributed value. Therefore, Eq. (2) also presented in chapter "General Introduction to MonteCarlo and Multi-level Monte Carlo Methods" is used to calculate x_1 and x_N. For further reading see also [7].

$$x_k = \frac{\int_{\xi_{k-1}}^{\xi_k} x f(X) \mathrm{d}x}{\int_{\xi_{k-1}}^{\xi_k} f(X) \mathrm{d}x} \quad \text{with} \quad k = 1, N \tag{2}$$

As already emphasized, the usage is limited to realizations in the tails of asymptotically expiring distributions.

Extended Latin Hypercube Sampling

The extension algorithm *extended Latin Hypercube sampling* (eLHS) is based on the choice of a sample group size denoted with N_g. The algorithm allows to extend an existing sample of N_g realizations with any positive multiple of the same size. The positive integer l, named *level*, denotes the different extension steps.

If a probabilistic simulation is already present, the existing number of samples corresponds to the group size and is referred to as level $l = 1$. If no sample is already available, level $l = 1$ results from an LHS with a sample size of N_g. The basic idea is to select a small group size, e.g., $N_g = d + 5$ with d as the number of input variables, and reach the desired total number of realizations $N^{(l)}$ by extending l times.

$$N^{(l)} = l \cdot N_g \tag{3}$$

The superscript (l) denotes the respective level. The flowchart in Fig. 1 describes the extension procedure. Each random variable is handled individually. In each extension step, only N_g realizations are added to the sample matrix. That means, the division of each random variable X into $N^{(l)}$ contiguous intervals of equal probability $1/N^{(l)}$ with respect to the corresponding CDF $F(X)$ is not maintained in each level of the eLHS. Besides the level l, we introduce a control variable λ. The relation between λ and the level l states whether a recalculation of the probability interval bounds is necessary. If $l = 2^{(\lambda-1)} - 1$, the probability interval bounds need to be recalculated.

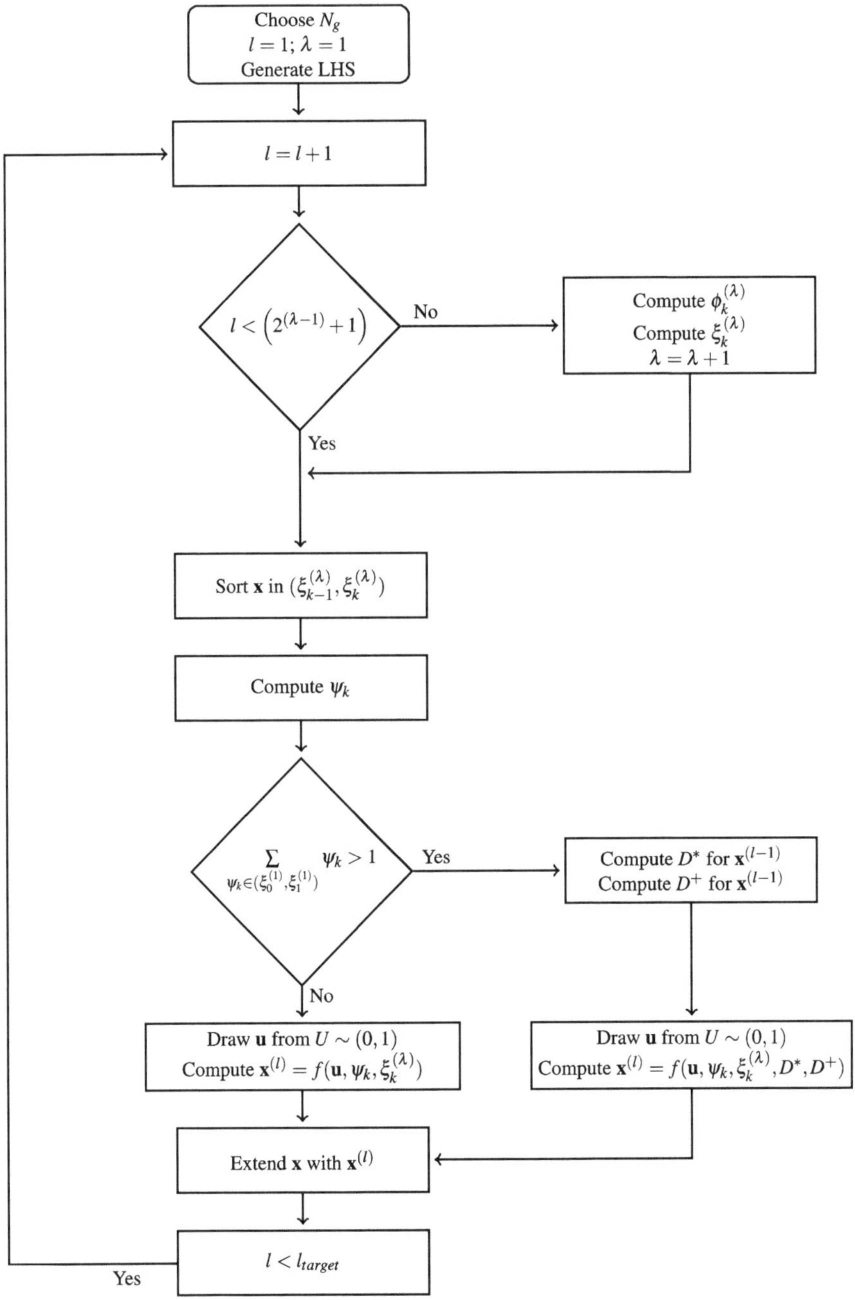

Fig. 1 Flowchart of eLHS

If $l < 2^{(\lambda-1)} - 1$, vacant intervals are still available. The probability interval bounds are defined as

$$\phi_k^{(\lambda)} = \frac{k}{2^{\lambda} N_g} \quad \text{with} \quad k = 1, ..., 2^{\lambda} N_g \,. \tag{4}$$

The probability intervals are denoted with $[\mathfrak{I}_k^{(\lambda)}]$. The corresponding interval bounds over the values of the random variable X are denoted with ξ_k. An integration of the PDF over the interval $(\xi_{k-1}^{(\lambda)}, \xi_k^{(\lambda)})$ would result in the probability interval $1/(2^{\lambda} N_g)$.

$$\int_{\xi_{k-1}^{(\lambda)}}^{\xi_k^{(\lambda)}} f(X) \mathrm{d}x = \frac{1}{2^{\lambda} N_g} \tag{5}$$

Equation (5) is solved numerically in order to compute the interval bounds $\xi_k^{(\lambda)}$. The corresponding intervals are denoted with $[I_k^{(\lambda)}]$. The interval bounds of the first level and therefore with $\lambda = 1$ are designated as *root* bounds $\phi_k^{(1)}$ and $\xi_k^{(1)}$. The corresponding intervals are designated as *root* intervals. The existing realizations $\mathbf{x}$ are assigned to the intervals $[I_k^{(\lambda)}]$. Subsequently, an indicator value ψ_k is determined for each interval. For an empty interval, $\psi_k = 1$, for an occupied one $\psi_k = 0$. Based on the indicator values, the number of free intervals per root interval can be determined. The realizations $\mathbf{x}^{(l)}$ are assigned uniformly distributed to the empty intervals bounded by $\xi_k^{(\lambda)}$. However, per root interval bounded by $\xi_k^{(1)}$ only one realization is added. This approach ensures that exactly N_g samples are added to the sample matrix and, furthermore, that each extension $\mathbf{x}^{(l)}$ represents a separate LHS. However, it introduces a deviation from the LHS design in such levels.

As a result more than one empty interval per root interval is in certain levels available, e.g., in level $l = 3$. In order to reduce the error compared to the LHS design, the selection of the interval in such levels is based on the largest negative distance between continuous and discrete distribution function of the present realizations $\mathbf{x}$. For this purpose, the lower distance between continuous and discrete distribution function is defined as

$$D^{*}(x_k) = \left(\frac{k-1}{N^{(l-1)}} - F(x_k) \right) \text{ with } k = 1, ..., N^{(l-1)} \,. \tag{6}$$

In Eq. (6), $F(x_k)$ denotes the continuous CDF of X. The minimum value of D^* is a measure of the lack of realizations in the PDF. Furthermore, the value D^+, describing the upper distance, is calculated with the present realizations.

$$D^{+}(x_k) = \left(F(x_k) - \frac{k}{N^{(l-1)}} \right) \text{ with } k = 1, ..., N^{(l-1)} \tag{7}$$

For each root interval $[I_j^{(1)}]$ with $j = 1, ..., N_g$, the newly occupied interval is located below the interval of the realization with

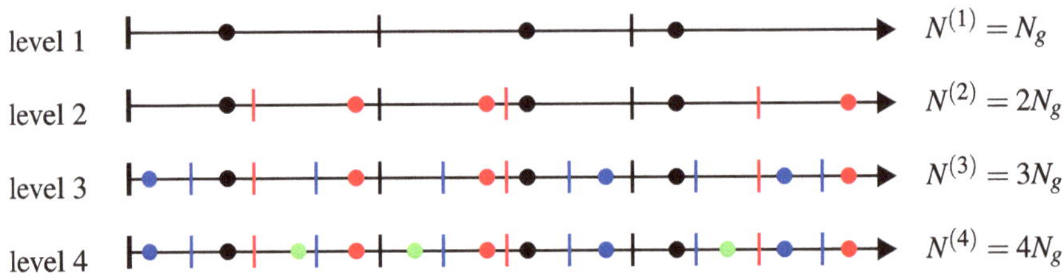

Fig. 2 Schematic example with group size $N_g = 3$ and 4 levels

$$d^* = \min_{x_k \in (\xi_{j-1}^{(1)}, \xi_j^{(1)})} (D^*(x_k)) . \tag{8}$$

Furthermore, the constraint, $|d^*| > |D^+(x_j)|$ must be satisfied where x_j designates the highest random number in each root interval. Otherwise, the interval above x_j is selected.

Following from the described extension approach, we can summarize that the combined extended sample of the size $N^{(l)}$ is only for $l = 2^{(\lambda-1)}$ a true fully occupied LHS; otherwise, certain probability intervals $[\mathfrak{I}_k^{(\lambda)}]$ stay empty.

Example: In Fig. 2 a, one-dimensional extension is shown for a uniform distribution $U(0, 1)$ and a group size $N_g = 3$. The realizations are illustrated on a number ray schematically. Level one contains $N^{(1)} = N_g = 3$ realizations. In the second level, the root sample is duplicated and the new realizations (in red) are assigned to the empty intervals. In the third level, the interval bounds $\phi_k^{(2)}$ and $\xi_k^{(2)}$ are recalculated again. However, two new empty intervals are per root interval available now. The intervals to be occupied are selected by using d^* and D^+. In the fourth level, no recalculation of the interval bounds in necessary. The present empty intervals are occupied uniformly distributed (in green).

Selected Performance Results of eLHS

In order to evaluate the eLHS, two standard normal distributed random variables $X \sim N(0, 1)$ are generated according to the experimental matrix shown in Table 1. The rank correlation values of $r = [0, 0.25, 0.5, 0.75, 0.99]^T$ are examined for each test point.

The following samples were generated with LHS. The same correlation values as for the eLHS are used. Each test point of the two experimental matrices was repeated 100 times.

$$\begin{aligned} N =& [15, 20, 30, 40, 50, 60, 65, 80, 100, 120, \\ & 130, 160, 200, 240, 260, 320]^T \end{aligned} \tag{9}$$

Table 1 Experimental matrix eLHS

l	$N^{(l)}$		
	$N_g = 5$	$N_g = 10$	$N_g = 20$
3	15	30	60
4	20	40	80
6	30	60	120
8	40	80	160
10	50	100	200
12	60	120	240
13	65	130	260
16	80	160	320

Table 2 Realization in interval $[I_1]$ for $N_g = 10$

l	x^{LHS}	x^{eLHS}	$\xi_0^{(l)}, \xi_1^{(l)}$
1	−1.7550	−1.7550	−10, −1.2816
2	−2.0627	−1.7550	−10, −1.6449
3	−2.2270	−2.3378	−10, −1.9600
4	−2.3378	−2.3378	−10, −1.9600
5	−2.4209	−2.3378	−10, −2.2414
…	…	…	…
8	−2.5887	−2.3378	−10, −2.2414
9	−2.6293	−2.8207	−10, −2.4977

Due to brevity of this chapter, the eLHS will be evaluated against the LHS concerning the reproduction of the standard deviation $s(X)$. For further results, the reader is referred to [8]. The comparison is performed with the mean value of 100 repetitions. Error bars describe the standard deviation $s(s(X))$ due to the repetition. The target value of the estimated standard deviation s equals one.

The error of the LHS with respect to the target value decreases asymptotically with increasing sample size N. The eLHS results deviate from the LHS due to the empty intervals $[I_k]$ at the levels $\{3, 6, 10, 12, 13\}$. Also, the scatter of the standard deviation increases. In level $\{4, 8, 16\}$, all intervals are occupied and the standard deviation of the eLHS samples should correspond to those of the LHS. However, this can only be approximately achieved at level 16. This behavior is not visible for the reproduction of the distribution function or mean value (see [8]). The reason behind the deviations in $\{4, 8\}$ is the allocation of the outermost intervals $[I_1]$ and $[I_N]$ with the mean x-coordinate (see Eq. 2). Table 2 presents the realization of eLHS and LHS in the interval $[I_1]$ as well as the corresponding bounds using the example of $N_g = 10$. The value of $\xi_0 = -10$ is selected for presentation purposes.

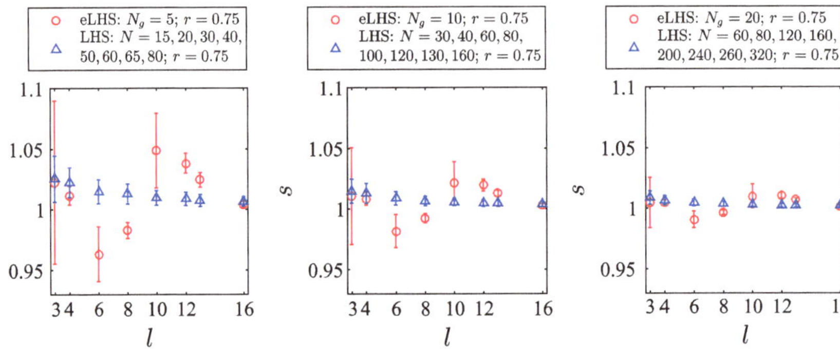

Fig. 3 Standard deviation over the number of level

At level $l = 1$ eLHS and LHS posses the same value for the realization in the interval $[I_1]$. At level $l = 2$, the outermost interval is still occupied. Moreover, the corresponding realization of eLHS is higher compared to the one of the LHS. The LHS value is the required value for $N = 20$ in order to achieve a good representation of σ. Therefore, the estimated standard deviation is reduced. The described behavior also applies to the levels $l = 5...8$ which is shown in Fig. 3.

The opposite case occurs in the levels $l = 3, 9...15$. The realization is smaller compared to the required value for the appropriate number of realizations $N^{(l)}$. Therefore, the estimated standard deviation of eLHS is increased compared to LHS. The influence of the realizations in the intervals next to the outermost intervals is shown at level $l = 4$. These intervals are occupied by mean instead of randomly distributed values which is a result of lower levels. In the case of $l = 4$, the intervals $[I_2]$ and $[I_{N-1}]$ are occupied with the mean value from level $l = 1$. The result is a reduction of the scatter of the standard deviation and a reduction of the deviation to the target value.

Note, the described problem due to the usage of mean values is limited to asymptotically expiring distributions. Avoidance is possible by random assignment of the outermost intervals $[I_1]$ and $[I_N]$. From Fig. 3, we can see that in general the deviations to the LHS decrease with increasing group size.

Correlation Control and Maintenance

As stated in chapter "General Introduction to Monte Carlo and Multi-level MonteCarlo Methods", the univariate random variables have to be assembled to the sample matrix $\mathbf{X}$ of the size $N \times d$ according to their dependence structure. For this purpose, an iterative Restricted Pairing (see chapter "General Introduction to Monte Carloand Multi-level Monte Carlo Methods") inspired by the work of Dandekar et al. [9] is coupled with a simple random search procedure. The correlation adjustment procedure denoted Random Search Restricted Pairing (RSRP) is shown in Algorithm 1.

Algorithm 1: Correlation adjustment procedure

Initialize $\mathbf{X}$ and $\mathbf{X}^*$
while $\Omega(\mathbf{X}^*) > \Omega_{term}$ **do**
  Generate $\mathbf{X}^*$
  Iterative Restricted Pairing with $\mathbf{X}^*$

The sample matrix $\mathbf{X}$ is initialized by linking the univariate random vectors. Each column represents a random variable. The matrix $\mathbf{X}^*$ is generated with each column being a random permutation of the same column from $\mathbf{X}$.

The use of Restricted Pairing means the use of the Cholesky transformation. For this purpose, it is necessary that the target and sample correlation matrices are symmetrical and positive definite. This property is granted in applications with physical background. The requirement is violated as soon as the number of realizations of the sample N is less than the number of random variables d. Therefore, the present method can only be applied for $N > d$.

A comparison with other methods for correlation control is shown in Table 3. The example deals with $d = 15$ independent uniform distributed variables $U \sim (0, 1)$ and a sample size of $N = 16$. The optimization criterion

$$E = \sum_{i=1}^{d-1} \sum_{j=i+1}^{d} (E_{ij})^2 , \tag{10}$$

was used where E_{ij} denotes the elements of the error matrix $\mathbf{E}$ (see chapter "GeneralIntroduction to Monte Carlo and Multi-level Monte Carlo Methods"). The tests were repeated 10 times, and the mean value for maximum absolute correlation deviation r_{max} and root mean square correlation deviation r_{rms} is shown. All three methods provide comparable results, with the genetic algorithm [10] performing best.

The rank correlation of the added sample matrix $\mathbf{X}^{(l)}$ is set with the above presented RSRP after each marginal variable is extended. The flowchart of the approach is shown in Fig. 4. Sallaberry et al. [6] showed that two samples with the same correlation structure in turn lead to a similar correlation structure when they are joined

Table 3 Performance of correlation adjustment for $d = 15$ and $N = 16$

	RSRP	Simulated annealing [5]	Genetic algorithm [10]
r_{max}	0.079	0.087	0.065
r_{rms}	0.049	0.044	0.023

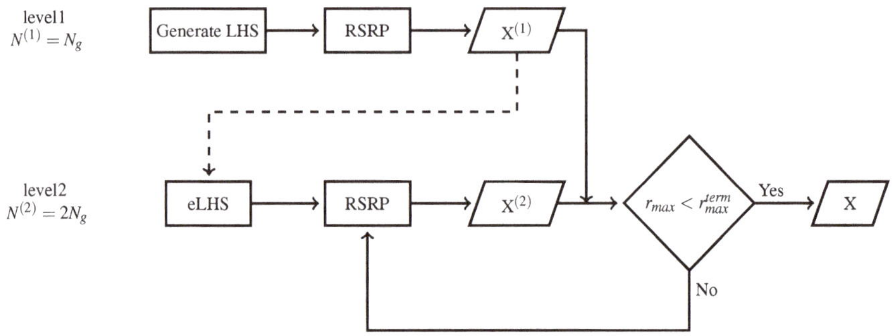

Fig. 4 Correlation control and eLHS for level $l = 2$

together. In the presented procedure, the whole sample matrix **X** is tested concerning its maximum absolute rank correlation error, after adjusting the correlations of the new realizations and joining them with the existing sample. The correlation adjustment in the new realizations $\mathbf{X}^{(l)}$ is repeated until $r_{max}(\mathbf{X})$ is less than a predefined criterion, as, e.g., in [11].

Performance of RSRP and eLHS

The evaluation of RSRP in the extension procedure is done with the example described in section "Selected Performance Results of eLHS." Figure 5 shows the maximum absolute correlation deviation over the extension level for the three group sizes. A distinction is made within the five computed rank correlation values. Generally, it can be observed that the RSRP leads to very low maximum absolute correlation deviations of at most 10^{-2}. The samples with low group size, that also means low group size variable ratio N_g/d, and high correlation coefficient result in the highest

Fig. 5 Maximum absolute correlation deviation r_{max} over number of levels

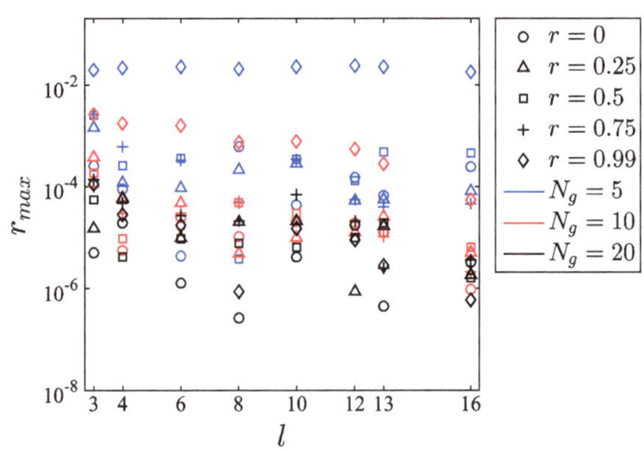

deviations. Comparing the samples with a group size of $N_g = 10$, the largest absolute correlation deviations also arise from high correlations of $r = 0.99$. At a group size of $N_g = 20$, which corresponds to 10 times the number of variables, this behavior is no longer present. With increasing group size and thus variability of each level, the correlation deviation decreases. Also, an increase in the number of levels results in a decrease of the deviation.

Pleming et al. [1] propose to choose the size of the sample four times the number of the random variables when applying Restricted Pairing with LHS. In the context of eLHS and RSRP, we can deduce that for low up to medium correlations even a factor of 2, that means $N_g = 2d$, can be enough to achieve low correlation deviations. In case of very high correlations, we propose a factor >5.

References

1. Pleming, J.B., Manteufel, R.D.: Replicated Latin hypercube sampling. In: 46th AIAA/ASME/ASCE/AHS/ASC Structures, Structural Dynamics and Materials Conference, University of Texas at San Antonio (2005)
2. Tong, C.: Refinement strategies for stratified sampling methods. Reliab. Eng. Syst. Saf. **91**(10–11), 1257–1265 (2006)
3. Vořechovský, M.: Extension of sample size in Latin hypercube sampling with correlated variables. In: 4th International Workshop on Reliable Engineering Computing (2010)
4. Vořechovský, M.: Hierarchical refinement of Latin hypercube samples. Comput. Aided Civ. Infrastruct. Eng. **30**(5), 394–411 (2015)
5. Vořechovský, M., Novák, D.: Correlation control in small-sample Monte Carlo type simulations I: a simulated annealing approach. Probab. Eng. Mech. **24**(3), 452–462 (2009)
6. Sallaberry, C., Helton, J., Hora, S.: Extension of Latin hypercube samples with correlated variables. Reliab. Eng. Syst. Saf. **93**(7), 1047–1059 (2008)
7. Huntington, D.E., Lyrintzis, C.S.: Improvements to and limitations of Latin hypercube sampling. Probab. Eng. Mech. **13**(4), 245–253 (1998)
8. Schmidt, R., Voigt, M., Vogeler, K.: Extension of Latin hypercube samples while maintaining the correlation structure. In: Proceedings of the 12th International Probabilistic Workshop, Weimar (2014)
9. Dandekar, R.A., Cohen, M., Kirkendall, N.: Sensitive micro data protection using Latin hypercube sampling technique. In: Inference Control in Statistical Databases, Springer, pp. 117–125 (2002)
10. Liefvendahl, M., Stocki, R.: A study on algorithms for optimization of Latin hypercubes. J. Stat. Plan. Inference **136**(9), 3231–3247 (2006)
11. Voigt, M.: Probabilistische simulation des strukturmechanischen verhaltens von turbinenschaufeln. Ph.D. Thesis, TU Dresden (2010)

Multi-level Monte Carlo Method

Jordi Pons-Prats and G. Bugeda

Introduction

Uncertainty quantification has gained interest during the recent years. Two clear examples are NODESIM-CFD and, the just finished, UMRIDA projects.

NODESIM-CFD project analysed the efficiency of Monte Carlo methods in front of other uncertainty quantification techniques: Polynomial Chaos Expansion, Probabilistic Collocation Methods, etc. It was concluded that Monte Carlo method is not suitable for industrial application in CFD due to its demanding computational cost, although this cost remains constant, or almost constant, with respect to the number of uncertain parameters. The curse of dimensionality was a key issue to keep the interest in Monte Carlo method.

UMRIDA project has focused its attention on a new and promising method related to Monte Carlo, named Multi-Level Monte Carlo. The available literature demonstrates how different levels of successive discretization levels of the analysis lead to a better performance compared to standard Monte Carlo method. The big number of evaluations remains on the coarse levels, while the refinements help to converge to the mean and variance estimators more quickly and efficiently.

This present chapter describes the implementation tasks of the Multi-Level Monte Carlo method for its application to the efficient handling of a large number of uncertainties. It is foreseen that the resulting tool can provide a stand-alone

J. Pons-Prats (✉) · G. Bugeda
CIMNE, c/Jordi Girona s/n, Campus Nord UPC, 08034 Barcelona, Spain
e-mail: jpons@cimne.upc.edu

G. Bugeda
e-mail: bugeda@cimne.upc.edu

G. Bugeda
UPC – Barcelonatech, Campus Nord UPC, 08034 Barcelona, Spain

C. Hirsch et al. (eds.), *Uncertainty Management for Robust Industrial Design in Aeronautics*, Notes on Numerical Fluid Mechanics and Multidisciplinary Design 140, https://doi.org/10.1007/978-3-319-77767-2_18

methodology for uncertainty quantification, or it can be combined with optimization tools to get a robust design tool.

To define a reference point, the Monte Carlo method is briefly introduced so a clear comparison will be done, highlighting the differences between MC and MLMC.

Description of Monte Carlo Method

Monte Carlo method is a well-established technique for sampling and producing statistical analysis. It is well known that its brute force approach leads to a high computational cost compared to other methodologies, but it is true that the computational cost is independent of the number of uncertainties and their typology. The accuracy getting the statistical moments of the results and as well as the easy implementation of any sampling probabilistic density function (PDF) are advantages of the Monte Carlo method.

This method formulates the mean value of a functional as:

$$E_M\left[U_\tau^n\right] = \frac{1}{M}\sum_{k=1}^{M} U_\tau^{k,n}$$

being

- E the mean value expectation,
- M the number of samples,
- U the functional investigated,
- τ the discretization of the functional,
- n the last time step of the numerical simulation (just to ensure the numerical convergence of the simulation).

The other three statistical moments, namely the variance, skewness and kurtosis, can be easily obtained in a similar way from the set of samples.

The whole Monte Carlo procedure can be summarized as:

1. Define the desired number of samples.
2. Calculate the stochastic points derived from the PDF of the input parameters.
3. Calculate the value of the functional for each stochastic point.
4. Calculate the statistical moments of the results.

It is clear from the formulation that the cost is directly related to the cost of evaluating each sample. The total cost is the sum of each individual evaluation. Due to the fact that a large number of samples are required to get accurate estimators of the statistical moments, the final cost can be unaffordable.

As shown in Figs. 2 and 3, the statistical estimators converge to their final values after an initial range with large oscillations. Figure 3 shows the detail of the

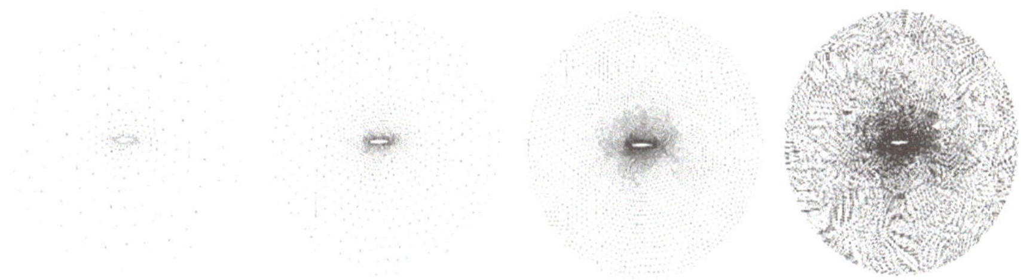

Fig. 1 Coarse-to-fine mesh refinement

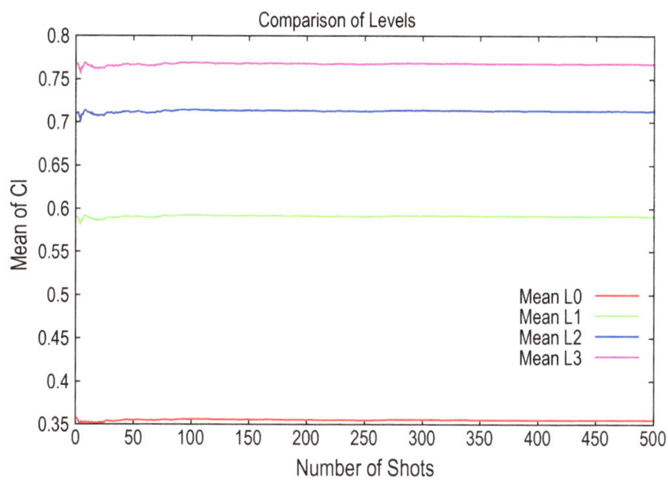

Fig. 2 Mean value of QoI versus shots

finest mesh. The plot shows that in the analysed case about 500 samples were required to get an almost stabilized value for the mean, although larger amounts are accepted as necessary to fully stabilize these oscillations. This statement is reinforced by the data shown in Fig. 4, where the variance is plot regarding different levels of accuracy and number of samples. To get an almost converged variance value, more than 500 samples are required to really stabilize the oscillations. The oscillations will be a problem-related issue, but it is clear that they will be related to the number of uncertainties, the range of their variance and the influence of the uncertain input parameter to the output variable. Then, it is easy to understand that the number of samples to ensure variance converge can be large. These figures show how increasing the level of accuracy and refining the mesh, the mean, and the variance of the QoI also are refined. The meshes used are shown in Fig. 1, which define the levels of accuracy named as L1, L2, L3, L4.

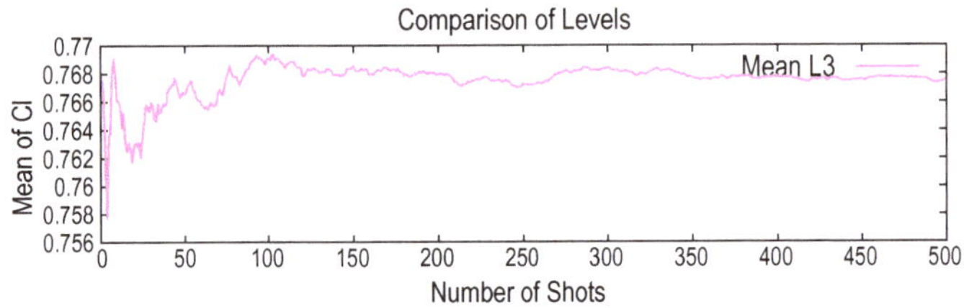

Fig. 3 Evolution of the mean value w.r.t. number of samples

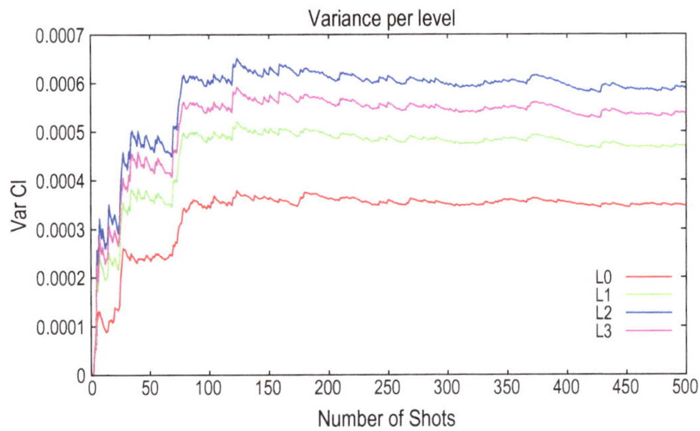

Fig. 4 Evolution of the variance value w.r.t. number of samples. Comparison of different levels of accuracy

Description of the Multi-level Monte Carlo Method

The Multi-Level Monte Carlo method is based on similar criteria as standard Monte Carlo. It also uses a set of samples to estimate the statistical moments, but in MLMC case the samples are organized in successive refinement levels. Thanks to this refinement, the number of samples to be used is expected to reduce each step while speeding up the convergence of the statistical moments. The mentioned refinement can be applied to the mesh size or to other parameters leading to an improved accuracy of the output. The improvement in the performance of the MLMC method, compared with the classical MC one, is based on the combination of a big number of cheap samples with low accuracy providing a good convergence of the statistical moments with a small number of expensive samples with high accuracy providing good quality results for the estimation of the output variables.

As mentioned, the MLMC method is based on a combination of a large number of evaluations with a low level of accuracy with a small number of evaluations with a high level of accuracy. Let us assume a sequence $P_0, P_1, \ldots, P_L$ which approximates a quantity of interest (QoI) not only with increasing accuracy, but also with increasing cost. Due to the linear property of the mean value operator, we have (see Ref. [1])

$$E[P_L] = E[P_0] + \sum_{l=1}^{L} E[P_l - P_{l-1}]$$

and therefore we can use the following unbiased estimator for $E[P_L]$,

$$E[P_L] \approx N_0^{-1} \sum_{n=1} P_0^{(0,n)} + \sum_{l=1}^{L} \left\{ N_l^{-1} \sum_{n=1}^{N_l} \left(P_l^{(l,n)} - P_{l-1}^{(l,n)} \right) \right\}$$

In order to calculate the standard deviation,

$$V[P_L] = \sum_{l=0}^{L} N_l^{-1} V_l[P_l^n]$$

$$V_l[P_l^n] = \frac{\sum_{n=1}^{N_l} \left(P_l^n - E[P_l^n] \right)^2}{n-1}$$

being:

- N_l the number of samples for the l level of accuracy, and n the number of the sample calculated at each step,
- $P_l^{(l,n)}$ the result of the evaluation of the n sample of the l level of accuracy (notice that the sampling points used for each level of accuracy are independent),
- n the number of evaluated samples.

If we define C_l as the computational cost of the analysis of one sampling point belonging to the first level of accuracy, then the overall cost of the multi-level estimator is

$$C_L = \sum_{l=0}^{L} N_l C_l$$

For a fixed computational cost, and choosing the appropriate Lagrangian multiplier λ, the standard deviation is minimized by choosing

$$N_l = \lambda \sqrt{\frac{V_l}{C_l}}$$

A typical computer implementation of the MLMC method is as follows:

1. Start L = 0,
2. Estimate standard deviation V_l defining an initial number of samples N_l.

Calculate optimal N_l using $N_l = 2\epsilon^{-2}\sqrt{V_l h_l}\left(\sum_{l=0}^{L}\sqrt{V_l/h_l}\right)$, being h_l the computational cost associated with the analysis of each sampling point and ϵ a user-defined tolerance. The values that ϵ can usually take are from 10^{-1} to 10^{-4}. These typical values help to the convergence of the method, although it is greatly dependant on the behaviour of the QoI.
3. Evaluate extra samples if optimal N_l is larger than the initial estimation
4. If $L \geq 2$, test convergence using:

$$\max\{M^{-1}|Y_{L-1}|, |Y_L|\} < \frac{1}{\sqrt{2}}(M-1)\epsilon$$

or

$$|Y_L - M^{-1}Y_{L-1}| < \frac{1}{\sqrt{2}}\left(M^2 - 1\right)\epsilon$$

5. If $L < 2$, or not converged, $L = L + 1$
6. Go to (2), i.e. estimate standard deviation V_l defining an initial number of samples N_l

To avoid misunderstandings, it should be clarified that what the Multi-Level Monte Carlo method defines as "Level" means the difference of two consecutive levels of discretization. See more details in the next section.

Look at Refs. [2–6] for more details.

On the following section, some analyses will be described using Multi-Level Monte Carlo for two different refinement criteria.

Refinement Levels

As described, the MLMC method is based on successive refinement applied to the solver. When dealing with FEM solvers, the easiest way to achieve this refinement is using the mesh size. But it is not limited to, while other types of refinements are also possible. The following sections present the one based on the mesh size and another one based on the number of time steps of the simulation.

In order to clearly describe the refinement strategy, together with the application of the MLMC method, two main analyses are presented. The first one is an approximation of the BC-02 test case. It uses the same RAE2822 geometry, but slightly different values for the AoA and M. Table 1 describes the related values.

The solver used in any case is based on Euler equations. The quantities of interest are the coefficients of lift, drag and momentum, or a selection of one of them.

Table 1 Definition of uncertainties for the first analysis

Parameter	Mean	Deviation
AoA	2.79	0.1
M	0.734	0.005

Mesh Refinement

The initial analysis is based on the definition of four different mesh refinements. Each mesh is used to calculate the behaviour of the statistical moments of the quantity of interest. The objective is to compare the results from a single refinement to those obtained combining the different levels of refinement, as the Multi-Level Monte Carlo does. Each mesh has been calculated over the same set of stochastic samples in order to reduce the variability associated with the sampling. The Multi-Level Monte Carlo only determines that the two sublevels must use the same set of samples, which is not the case in this preliminary analysis. Five hundred samples are defined in each case. The defined meshes are the following:

- R0; coarse mesh; 318 nodes, 571 triangular elements
- R1; refinement 1; 1207 nodes, 2284 triangular elements
- R2; refinement 2; 4698 nodes, 9136 triangular elements
- R3; refinement 3; 18,532 nodes, 36,544 triangular elements

The procedure of the MLMC analysis is the following:

- Each level uses two consecutive refinements, except the initial level, which only uses the coarser refinement.
- The statistical data of the quantity of interest is calculated in a standard way at each level as described in previous equation. Each level contributes to the final value of the estimators of the statistical moments of the QoI.

The refinement on the mesh is defined as described in Table 2.

The number of initial samples per levels has been chosen as 10 or 50, in regard to define subcases enabling a comparison of the results.

The calculation of these test cases has been initiated from scratch. No previous information has been used, so no estimation of the required number of levels and samples is known a priori.

Mesh refinement discretization with AoA, M and thickness-to-chord ratio stochastically defined.

With a preliminary analysis of the results, one can conclude that the mean value is converging quite fast, which is important to reduce the computational cost. On the other hand, the variability that the three uncertain parameters introduces can be easily identified on the value of the variance. Figure 5 shows the evolution of the mean value of the QoI along the multi-level analysis. The steps on the plot identify the contribution of each new level. The convergence shown in this figure is typical for such a problem definition, where the contribution of each level produces a new step increment, while it tends to converge to the mean of the level. Defining different numbers of uncertain input parameters will not lead to a different behaviour. Step size could be different, the local convergence could be faster, and of course, the final converged value could be also different, but the overall behaviour one should expect looks like in Fig. 5.

Table 2 Refinement of the mesh

Parameter	Level refinement	Factor S
Mesh size	$h(l) = s^{\wedge} - 1 * h(l - 1)$	2

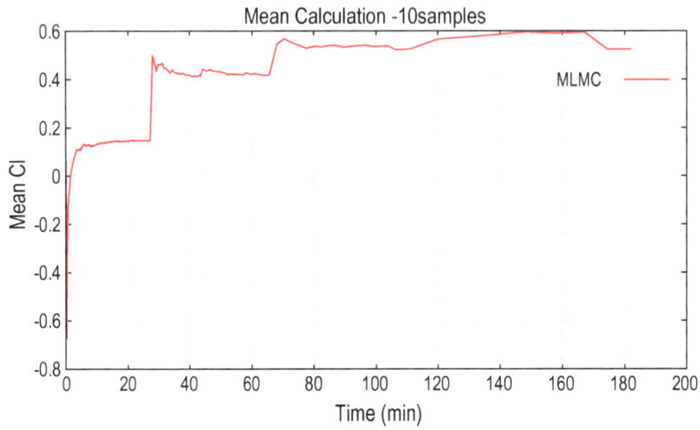

Fig. 5 Mean convergence

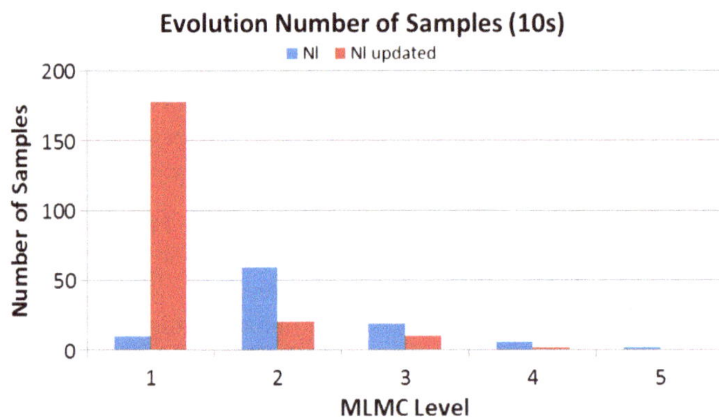

Fig. 6 Samples per level

Figure 6 shows the behaviour of the analysis from the point of view of the method. It shows the number of samples defined and updated to fulfil the procedure criteria of convergence. According to the tolerance the user defines, the updated number of samples can be lower or higher. Typically, the lower the number of the initial samples, the more probable to need additional samples is. For the successive levels of refinement, the tolerance value λ also affects the number of samples each level requires. In addition to this tolerance, two other values, also referred as tolerance, are used to ensure the accuracy of the mean and variance during the analysis. By default, the number of samples of the next level is directly related to the final number of samples used in the previous level, as a way to manage the final cost and keep it as reduced as possible.

Results obtained with MLMC analysis with levels of discretization are based on the number of time steps.

Now, the different levels of accuracy correspond with the different levels of discretization, which correspond to the number of time steps of the finite element analysis.

The selection of the number of time-steps, as the discretization criteria, is based on the need to get a lower computational cost and an easy implementation of such a test case.

Four tests are presented to assess the effect of tolerance of the mean and the variance on the stopping criteria, as well as to assess the effect of defining the λ value which is used to calculate the updated number of samples in each level. Table 3 describes the values defined in each test case. It is clear that the three parameters will influence the convergence ratio and will determine the final amount of samples in each level as well as the number of levels to be calculated. It is expected that the effect of the two tolerances will mainly determine the number of levels, while the λ value will determine the number of samples.

The number of initial samples is not relevant to determine the convergence of the method, from the point of view of reducing the number of samples of the successive levels. Figures 7 and 8 show the evolution of the DMx and BKr test cases. Both of them were using λ equal to 10^{-2}. For low values of epsilon, the updated number of samples (N_l) is usually lower than the calculated number of samples, which means that no additional samples should be calculated. Figures 9 and 10 show the evolution of required samples when λ is equal to 10^{-3} for 10 and 50 initial samples. In these cases, the larger value of ε leads to the need to calculate more samples than those initially defined, increasing the final total cost of the analysis. If the value of λ is larger (10^{-4} in some cases), the number of samples presents a complete lack of convergence to a minimum. The additional samples to calculate rapidly increase to an absolutely unaffordable number of samples (millions of samples for instance), which means the analysis should be discarded and stopped.

Let us focus on the convergence of the mean value and the value of the variance of the quantity of interest (QoI), which in this case is the mean and the variance of the Cl coefficient. One can easily realize that a large total number of samples are obtained when a larger number of initial samples are defined. It is true that the final convergence is also influenced by the stopping criteria the user defines. On the other hand, the accuracy of the method, compared to a standard Monte Carlo analysis, is greater with a reduced number of samples.

Table 3 Set-up of the MLMC tests

Test name	Initial samples	Tolerance of the mean	Tolerance of the variance	Tolerance ε
DMx	10	10^{-6}	10^{-9}	10^{-2}
BKr	50	10^{-6}	10^{-6}	10^{-2}
DBe	10	10^{-6}	10^{-9}	10^{-3}
D1Kr	50	10^{-6}	10^{-9}	10^{-3}

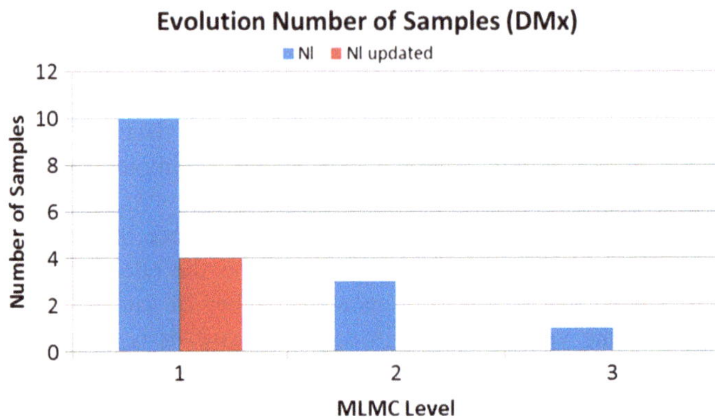

Fig. 7 Number of samples defining 10 initial samples

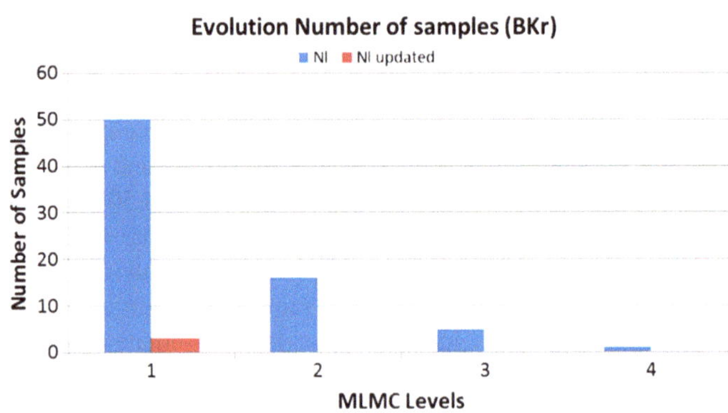

Fig. 8 Number of samples defining 50 initial samples

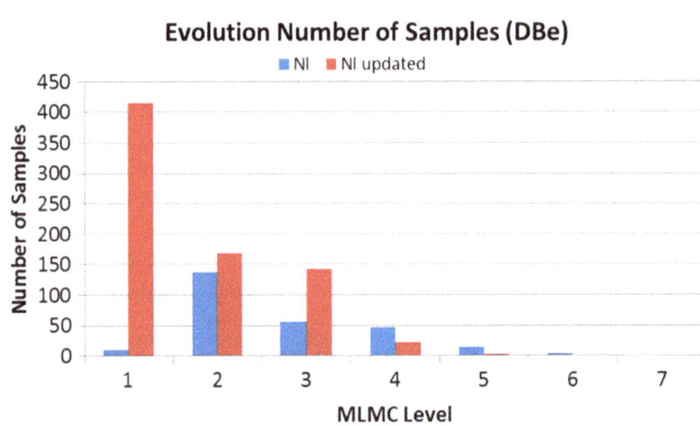

Fig. 9 Number of samples defining 10 initial samples

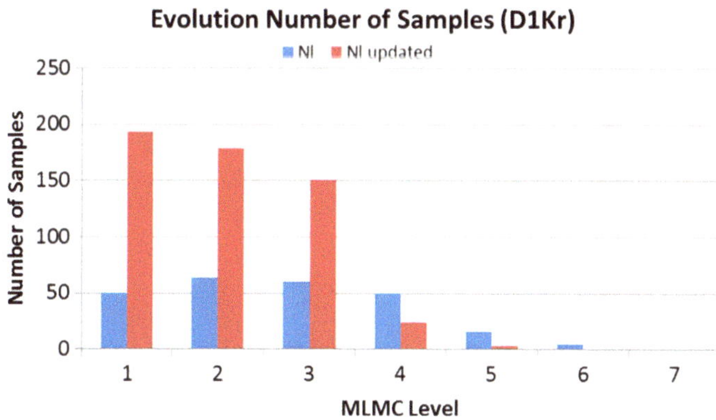

Fig. 10 Number of samples defining 50 initial samples

Figures 11 and 12 show the evolution of the mean value over the calculation time. The total number of samples (not the initial one) is used, as reference, to compare the evolution of the convergence of the analysis. This evolution is compared with the equivalent Monte Carlo analysis (with equal number of samples). The first point to highlight is the reduction in cost we get using the Multi-Level Monte Carlo. Even with the same number of samples the final cost is reduced thanks to the fact that the individual cost of each sample is also reduced. On the other hand, the accuracy in both cases is similar. To get a MLMC final value closer to the MC, one greatly depends on the intrinsic random character of the method. One could conclude that the gain using MLMC is OK, but could still prefer using MC.

Then, if we analyse the variance, a greater difference between the two methods can be identified.

Figures 13 and 14 show the evolution of the variance along the analysis, the first over the time and the second over the number of total samples. The benefit

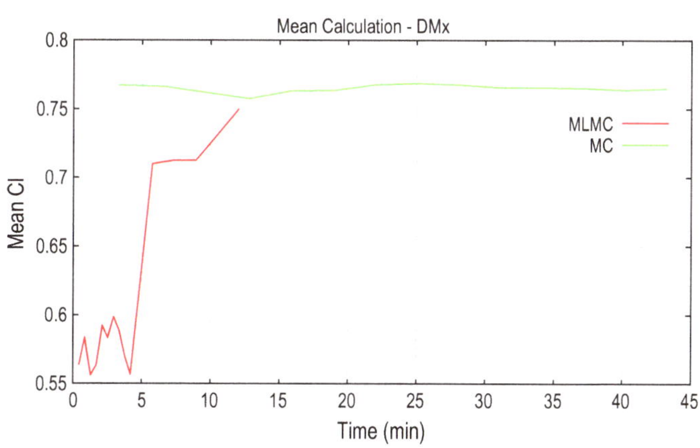

Fig. 11 Evolution of mean value defining 10 initial samples

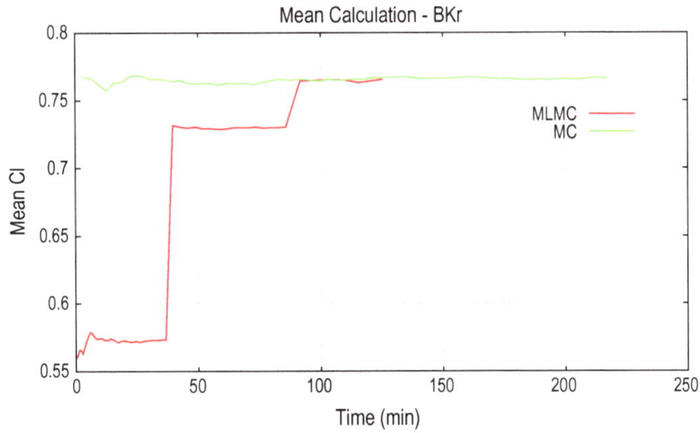

Fig. 12 Evolution of mean value defining 50 initial samples

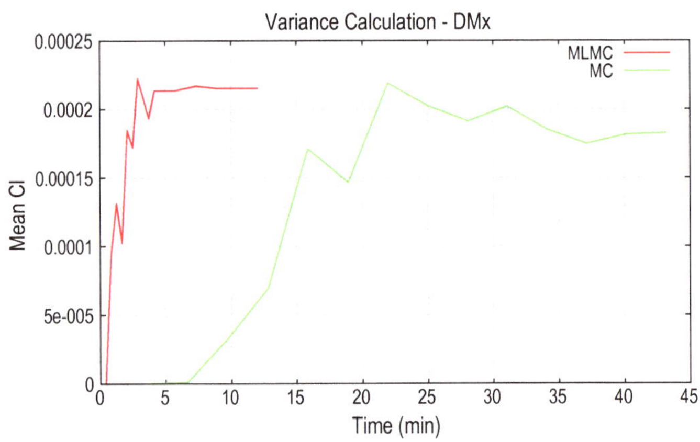

Fig. 13 Evolution of variance defining 10 initial samples

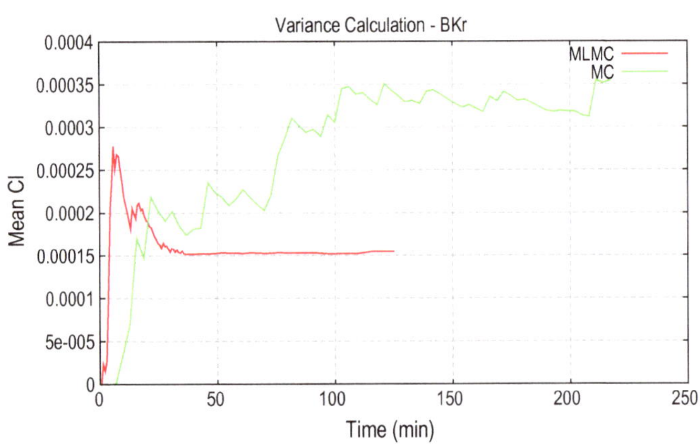

Fig. 14 Evolution of variance defining 50 initial samples

of MLMC is now clear; the convergence of the variance is obtained easily with the MLMC method for the same number of calculated samples. As the figures show, MLMC variance takes less time to better converge than in the MC analysis.

Discussion

MLMC was initially used in the financial analysis. As far as the author knows, no engineering application was developed. The MLMC method was described as an upgrade of the standard Monte Carlo method, and the results justifying this statement were confirming it. It is clear that the computational cost associated with the financial analysis is usually lower than the cost of a CFD analysis. In the literature, examples calculating thousands of samples are quite typical, but the reader should consider that in CFD problems, launching thousands of computations is not affordable and less practical. If, in addition, the user is considering industrial CFD applications, the limitations are even more strict.

MC methods, including MLMC, are interesting when dealing with large number of uncertainties. The curse of dimensionality is not a real problem in such methods, as happened with other UQ methods also described in this book. The results described in the previous chapters already included geometrical and non-geometrical uncertainties without leading to a major problem.

The introduction of geometrical uncertainties did not represent a drawback for the method. From the point of view of the uncertainty management, it means that a geometrical uncertainty is treated as any other uncertainty.

Anyway, it is true that an issue can arise when dealing with geometrical uncertainties. It is related to the geometry management and associated meshing process. On the process of sampling and evaluating each sample, a geometrical uncertainty will require to update the geometry and the mesh in order to evaluate each sample. Some basic points should be considered to avoid problems:

- Range of the uncertain parameter: if it is not well defined, it could lead to a wrong geometry. It could happen that either the geometry or the mesh was impossible to be created/updated. A clear and simple example could be if the uncertainty on one 2D airfoil parameter leads to the situation where the lower profile crosses the upper one.
- Robustness of the geometry updating process: more generally speaking, the previous point can be extended to ensure that the update of the geometry is robust for any combination of the values of each uncertainty.
- Automatic update of the geometry: in addition to the robustness, the repeatability and automation of the update process must be ensured. It is an important point not only regarding the re-meshing process, but regarding the repeatability of the whole process of evaluating the sample. The automation should be complete in order to avoid user interaction which can halt the analysis unnecessary.

- Computational cost: an important issue is the computational cost of updating the geometry and, mainly, of the finite element mesh, which can be considerable and should be taken into account when defining the uncertainty analysis.

The present chapter has presented the MLMC in comparison with MC method. This book describes other UQ methods that demonstrate impressive and interesting performance when dealing with industrial applications. Anyway, Monte Carlo methods, especially Multi-Level Monte Carlo, are always a safe bet for the accuracy one can get, for the easy of the implementation, and for the robustness of the method regarding the curse of dimensionality, mainly.

References

1. Peter, J., Marcelet, M., Burguburu, S., Pediroda, V.: Comparison of surrogate models for the actual optimization of a 2D turbomachinery flow. In: Proceedings of the 7th WSEAS International Conference on Simulation, Modelling and Optimization (SMO '07), p. 46. ISSN: 1790-5117. ISBN: 978-960-6766-05-3
2. Giles, M.B.: Multilevel Monte Carlo path simulation. Oper. Res. **256**, 981–986 (2008)
3. Giles, M.B., Reisinger, C.: Stochastic finite differences and multilevel Monte Carlo for a class of SPDEs in finance. SIFIN **1**(3), 575–592 (2012)
4. Mishra, S., Schwab, Ch., Sukys, J.: Multi-level Monte Carlo finite volume methods for nonlinear systems of conservation laws in multi-dimensions. J. Comp. Phys. **231**(8), 3365–3388 (2011)
5. Mishra, S., Schwab, Ch., Sukys, J.: Multilevel Monte Carlo finite volume methods for shallow water equations with uncertain topography in multi-dimensions. SIAM J. Sci. Comput. **34**(6), B761–B784 (2012). https://doi.org/10.1137/110857295
6. Mishra, S., Schwab, Ch., Sukys, J.: Monte Carlo and multi-level Monte Carlo finite volume methods for uncertainty quantification in nonlinear systems of balance laws. Seminar für Angewandte Mathematik ETH Zurich. Research Report No. 2012-08, April 2012

Continuation Multi-level Monte Carlo

Michele Pisaroni, Fabio Nobile and Penelope Leyland

Multi-level Monte Carlo

As stated in the introductory chapter [1], the key idea of MLMC is to simultaneously draw MC samples on several approximations Q_{M_l} of a quantity of interest (QoI) Q built on a hierarchy of computational grids (with discretization parameters $M_0 < M_1 < \cdots < M_L = M$). For the sake of explanation, we recall the MLMC estimator for $\mathbb{E}[Q]$:

$$\mathrm{E}^{\mathrm{MLMC}}[Q_M] := \sum_{l=0}^{L} \frac{1}{N_l} \sum_{i=1}^{N_l} Y_l(\omega^{(i,l)}) = \sum_{l=0}^{L} \mathrm{E}^{\mathrm{MC}}[Q_{M_l} - Q_{M_{l-1}}], \tag{1}$$

The accuracy in estimating $\mathbb{E}[Q]$ by $\mathrm{E}^{\mathrm{MLMC}}[Q_M]$ can be quantified by considering the mean squared error (MSE) of the estimator:

$$e(\mathrm{E}^{\mathrm{MLMC}}[Q_M])^2 := \mathbb{E}[(\mathrm{E}^{\mathrm{MLMC}}[Q_M] - \mathbb{E}[Q])^2] = \underbrace{\left(\mathbb{E}[Q_M - Q]\right)^2}_{(\text{B-E}^{\text{MLMC}})} + \underbrace{\sum_{l=0}^{L} \frac{\mathbb{V}\mathrm{ar}[Y_l]}{N_l}}_{(\text{SE-E}^{\text{MLMC}})}. \tag{2}$$

M. Pisaroni (✉) · F. Nobile · P. Leyland
Scientific Computing and Uncertainty Quantification,
Ecole Polytechnique Fédérale de Lausanne, 1015 Lausanne, Switzerland
e-mail: michele.pisaroni@epfl.ch; michele.pisaroni@gmail.com

F. Nobile
e-mail: fabio.nobile@epfl.ch

P. Leyland
e-mail: penelope.leyland@epfl.ch

C. Hirsch et al. (eds.), *Uncertainty Management for Robust Industrial Design in Aeronautics*, Notes on Numerical Fluid Mechanics and Multidisciplinary Design 140, https://doi.org/10.1007/978-3-319-77767-2_19

We now make the following assumptions:

a1. The cost to compute one sample Q_{M_l} at level l is:

$$Cost(Q_{M_l}(\omega^{(i)})) \leq c_\gamma M_l^\gamma, \tag{3}$$

a2. $\mathbb{E}[Q_{M_l}]$ converges to $\mathbb{E}[Q]$ with rate α w.r.t. M_l, i.e.,

$$\left|\mathbb{E}[Q_{M_l} - Q]\right| \leq c_\alpha M_l^{-\alpha} \tag{4}$$

for some $c_\alpha, \alpha > 0$

a3. $\mathbb{V}\mathrm{ar}[Y_l]$ decays with rate β w.r.t. M_l, i.e.,

$$\mathbb{V}\mathrm{ar}[Y_l] \leq c_\beta M_l^{-\beta}, \tag{5}$$

for some $c_\beta, \beta > 0$ and $\alpha \geq \min(\beta, \gamma)$.

For standard problems and discretization techniques, such assumptions are verified for some rates α, β, γ and constants c_α, c_β, c_γ. The rates α, β however will heavily depend on the specific problem and discretization chosen and can vary considerably.

It can be shown [2, 3] that these assumptions **a1**–**a3**, for any $\varepsilon > 0$, we can estimate an optimal number of levels $L = L(\varepsilon)$ and an optimal sample size $\left\{N_l\right\}_{l=0}^{L}$ on each level such that

$$e(\mathrm{E}^{\mathtt{MLMC}}[Q_M])^2 < \varepsilon^2 \tag{6}$$

and the cost to achieve a RMSE of ε is:

$$\mathbf{Cost}(\mathrm{E}^{\mathtt{MLMC}}[Q_M]) = \sum_{l=0}^{L} N_l \cdot Cost(Q_{M_l}(\omega^{(i)})) \lesssim \begin{cases} \varepsilon^{-2} & \beta > \gamma, \\ \varepsilon^{-2}(\log \varepsilon)^2 & \beta = \gamma, \\ \varepsilon^{-2-(\gamma-\beta)/\alpha} & \beta < \gamma. \end{cases} \tag{7}$$

where $\lesssim$ means that the bound holds up to a multiplicative constant of ε.

For $\beta > \gamma$, the computation effort is primarily on the coarsest levels, whereas for $\beta < \gamma$ the primary cost is on the finest levels. For $\beta = \gamma$, the efforts are spread across all levels.

The complexity bound (7) improves in all cases the complexity of a Crude MC (CMC) approach, which can be estimated as:

$$\mathbf{Cost}(\mathrm{E}^{\mathtt{MC}}[Q_M]) = \sum_{i=1}^{N} Cost(Q_M(\omega^{(i)})) \lesssim \varepsilon^{-2-\gamma/\alpha} \tag{8}$$

In particular, the rate β, that defines the decay of the variance of consecutive differences, plays an important role in the complexity gain of MLMC over CMC.

The result (7) is a theoretical bound on the best complexity achievable by a MLMC method, but it brings also practical recipes to select the maximum level L and the sample sizes $\{N_l\}_{l=0}^{L}$ to achieve the optimal complexity.

We introduce a splitting parameter $\theta \in (0, 1)$ (usually taken as $\theta = \frac{1}{2}$) and require in our simulations that:

$$\mathit{Bias}: \ \mathbf{B} := \left|\mathbb{E}[Q] - \mathbb{E}[Q_M]\right| \leq (1-\theta)\varepsilon, \tag{9a}$$

$$\mathit{Statistical\,Error}: \ \mathbf{SE} := \mathbb{V}ar[\mathrm{E}^{\mathrm{MLMC}}[Q_M]] = \sum_{l=0}^{L} \frac{\mathbb{V}ar[Y_l]}{N_l} \leq \theta(2-\theta)\varepsilon^2 \tag{9b}$$

so that the MSE:

$$e(\mathrm{E}^{\mathrm{MLMC}}[Q_M]) = \mathbf{B}^2 + \mathbf{SE} \leq \varepsilon^2 \tag{10}$$

From (4), the bias constraint (9a) is satisfied by choosing:

$$L = L(\varepsilon) : \ M_{L(\varepsilon)} \geq \left(\frac{(1-\theta)\varepsilon}{c_\alpha}\right)^{-\frac{1}{\alpha}} \tag{11}$$

On the other hand, following the optimization argument in [2], the statistical error constraint (9b) is satisfied by choosing:

$$N_l(\varepsilon) = \left\lceil \left(\frac{1}{\theta(2-\theta)\varepsilon^2}\right) \sqrt{\frac{\mathbb{V}ar[Y_l]}{C_l}} \sum_{k=0}^{L} \sqrt{C_k \mathbb{V}ar[Y_k]} \right\rceil \tag{12}$$

In practical aerodynamic applications, we are usually required to compute quantity of interest $\mathcal{Q}(x, \omega)$ that are scalar fields defined on a certain domain D (e.g., pressure coefficient around an airfoil). In this case, we enforce the MSE to be smaller than ε^2, where in the definition of the MSE, we measure for convenience the spatial error in the L^2 norm (mean-square sense) [4].

$$\begin{aligned} e(\mathrm{E}^{\mathrm{MLMC}}[\mathcal{Q}_M])^2 &:= \mathbb{E}[\|\mathrm{E}^{\mathrm{MLMC}}[\mathcal{Q}_M] - \mathbb{E}[\mathcal{Q}]\|^2_{L^2(D)}] \\ &= \underbrace{\|\mathbb{E}[\mathcal{Q}_M - \mathcal{Q}]\|^2_{L^2(D)}}_{(\text{B-E}^{\text{MLMC}})} + \underbrace{\sum_{l=0}^{L} \frac{1}{N_l} \|\mathbb{V}\mathrm{ar}[\mathcal{Y}_l]\|_{L^1(D)}}_{(\text{SE-E}^{\text{MLMC}})}. \end{aligned} \tag{13}$$

where $\mathcal{Y}_l = \mathcal{Q}_{M_l} - \mathcal{Q}_{M_{l-1}}$.

Following the same arguments presented above, we enforce:

$$Bias: \ \mathbf{B} := \|\mathbb{E}[\mathcal{Q}_M - \mathcal{Q}]\|_{L^2(D)} \leq (1-\theta)\varepsilon, \tag{14a}$$

$$Statistical\,Error: \ \mathbf{SE} := \mathbb{V}ar[\mathrm{E}^{\mathtt{MLMC}}[\mathcal{Q}_M]] = \sum_{l=0}^{L} \frac{\|\mathbb{V}\mathrm{ar}[\mathcal{Y}_l]\|_{L^1(D)}}{N_l} \leq \theta(2-\theta)\varepsilon^2 \tag{14b}$$

From (4), the bias constraint (14a) is satisfied by choosing L as in (11) and the statistical error constraint (14b) is satisfied by choosing:

$$N_l = \left\lceil \left(\frac{1}{\theta(2-\theta)\varepsilon^2}\right) \sqrt{\frac{\|\mathbb{V}\mathrm{ar}[\mathcal{Y}_l]\|_{L^1(D)}}{C_l}} \sum_{k=0}^{L} \sqrt{C_k \|\mathbb{V}\mathrm{ar}[\mathcal{Y}_l]\|_{L^1(D)}} \right\rceil \tag{15}$$

Given a hierarchy of discretizations with $M_0 < M_1 < \cdots$ from the practical point of view the standard MLMC algorithm is generally composed of four steps:

1. Theoretical or computational estimation of the problem-dependent rates and constants ($\mathcal{P} = \{c_\alpha, \alpha, c_\beta, \beta, c_\gamma, \gamma\}$).
2. Estimation of $\mathbb{V}ar[Y_l]$ (or $\|\mathbb{V}\mathrm{ar}[\mathcal{Y}_l]\|_{L^1(D)}$ in case of scalar field).
3. Estimation of the optimal number of levels L from (11) and sample sizes N_l from (12) (or (15) in case of scalar field).
4. Run the hierarchy $\{0, \ldots, L\}$ with an optimal $\{N_l\}_{l=0}^{L}$.

Theoretical estimates for the rates α and β exist for certain classes of PDEs with random parameters [5–8] and depend on the smoothness of the data of the problem as well as the smoothing proprieties of the differential operator. On the other hand, the parameter γ depends on the number of spatial dimensions of the deterministic problem and the efficiency of the deterministic solver.

The total cost of MLMC strongly depends also on the problem-dependent constants $c_\alpha, c_\beta, c_\gamma$ as they enter in the choice of the optimal parameters L, $\{N_l\}_{l=0}^{L}$, and these have to be estimated numerically as accurately as possible. The common practice is to compute the rates and the constants by performing an initial *screening* over the first few levels $\{0, \ldots, \overline{L}\}$ with a predefined sample size $\overline{N}$ and fit the rates and constants via a least squares procedure.

Algorithm 1: MLMC Screening

SCREENING$(\overline{N}, \overline{L})$
  for $l = 0 : \overline{L}$ **do**
    for $i = 0 : \overline{N}$ **do**
      Generate random samples: $\mathbf{O}(\omega^{(i,l)})$, $\mathbf{G}(\omega^{(i,l)})$
      $Q^{(i)}_{M_l} \leftarrow \mathsf{PROBLEM}_l\ (\mathbf{O}(\omega^{(i,l)}), \mathbf{G}(\omega^{(i,l)}))$
      $Q^{(i)}_{M_{l-1}} \leftarrow \mathsf{PROBLEM}_{l-1}\ (\mathbf{O}(\omega^{(i,l)}), \mathbf{G}(\omega^{(i,l)}))$
      $Y^{(i)}_l = Q_{M_l} - Q_{M_{l-1}}$
  estimate $\{C_l\}, \left\{\left|\mathbb{E}[Q_{M_l} - Q]\right|\right\}, \{\mathbb{V}\mathrm{ar}[Y_l]\}$
  compute $\mathscr{P} = \{c_\alpha, c_\beta, c_\gamma, \alpha, \beta, \gamma\}$ using least squares fit
  compute L using (11) and N_l using (12) for scalar QoI or (15) for scalar field QoI
  return L, $\{N_l\}_{l=0}^{L}$

The main drawback of this procedure is that for computationally expensive problems, this screening phase, whose cost is usually not accounted for in the literature, can be quite time-consuming. In particular, if $\overline{L}$ and $\overline{N}$ are chosen too large the screening phase might turn out to be more expensive than the overall MLMC simulation on the optimal hierarchy $\{0, \ldots, L\}$. On the other hand, if $\overline{L}$ and $\overline{N}$ are chosen too small, the extrapolation of the convergence rates α and β on finer levels might be quite unreliable. The screening procedure for standard MLMC is summarized in Algorithm 1.

Continuation Multi-level Monte Carlo

To overcome the above-mentioned limitations of the standard MLMC algorithm concerning the screening phase, we consider here the Continuation Multi-Level Monte Carlo (C-MLMC) algorithm proposed in [1]. The key idea of C-MLMC is to solve for the QoI with a sequence of decreasing tolerances and progressively improve the estimation of the problem-dependent parameters $\mathscr{P}$ that, as presented before, directly control the number of levels and sample sizes. To achieve a certain RMSE of ε, we set a slightly smaller tolerance $\frac{\varepsilon}{r_2}$ with $r_2 > 1$ and define a sequence of decreasing tolerances $\varepsilon_0 > \varepsilon_1 > \cdots > \varepsilon_i > \cdots > \varepsilon_k = \frac{\varepsilon}{r_2}$ with $\varepsilon_i = r_1 \varepsilon_{i+1}$ where, for a given $\varepsilon_0 > \varepsilon$, the number k of iterations is given by:

$$k = \left\lfloor \frac{-log(\frac{\varepsilon}{r_2}) + log(\varepsilon_0)}{log(r_1)} \right\rfloor. \tag{16}$$

Eventually, we might still run the algorithm for few more iterations with tolerances $\varepsilon_{k+j} = \frac{\varepsilon_k}{r_2 j}$ until the actual estimated RMSE is below ε.

At the ith iteration of the C-MLMC algorithm with prescribed tolerance ε_i, we compute the optimal number of levels by solving the following discrete optimization problem and by exhaustive search:

$$(L_i, \theta_i) = \underset{\substack{L \in [L_{i-1}, \ldots, L_{MAX}], \theta \in (0,1) \\ s.t.\ c_\alpha M_L^{-\alpha} = (1-\theta)\varepsilon_i}}{\arg\min} \mathbf{Cost}(\varepsilon_i, \theta, L) \tag{17}$$

using the cost model:

$$\mathbf{Cost}(\varepsilon_i, \theta, L) = \left(\frac{1}{\theta(2-\theta)\varepsilon_i^2}\right)\left(\sum_{l=0}^{L} \sqrt{C_l \mathbb{V}ar[Y_l]}\right)^2 \tag{18}$$

obtained with an optimal choice of N_l (computed with 12).

The essential feature of the C-MLMC with respect to standard MLMC algorithm is that the parameter set $\mathscr{P}$ is computed on the fly and updated at each iteration of the algorithm. The estimation of the parameters that describe the cost (c_γ, γ) and the bias (c_α, α) is relatively straightforward since these quantities can be estimated also with just few realizations per level. The estimation of the variances $\mathbb{V}ar[Y_l]$, on the other hand, can be quite inaccurate with a small sample size. In a standard MLMC, such variances are usually computed using a sample variance estimator:

$$\mathbb{V}ar[Y_l] \approx \mathrm{V}^{\mathrm{MC}}[Y_l] = \frac{1}{N_l - 1} \sum_{n=1}^{N_l} \left(Y_l(\omega^{(n,l)}) - \mathrm{E}^{\mathrm{MC}}[Y_l]\right)^2 \tag{19}$$

At the deepest levels, usually we do not have enough realizations to accurately compute $\mathrm{V}^{\mathrm{MC}}[Y_l]$ (asymptotically accurate only as $N_l \to \infty$) and estimate the sample sizes N_l for the next iteration, as well as the parameters (c_β, β) needed to extrapolate $\mathbb{V}ar[Y_l]$ hence N_l on new levels that are added at the next iteration.

Collier et al. [1] presented an intuitive methodology based on Bayesian updates that use samples generated on all levels to locally improve the estimation of $\mathbb{V}ar[Y_l]$. Using the bias model $\mathbb{E}[Y_l] \approx \widehat{\mu}_l := c_\alpha M_l^{-\alpha}$ and variance model $\mathbb{V}ar[Y_l] \approx \widehat{\lambda}_l^{-1} := c_\beta M_l^{-\beta}$ with $c_\alpha, \alpha, c_\beta, \beta$ estimated from the previous iteration of the C-MLMC algorithm. The idea is to describe Y_l as a Gaussian random variable $\mathscr{N}(\mu_l, \lambda_l^{-1})$ and perform a Bayesian update of μ_l and λ_l^{-1} based on the collected values $Y_l(\omega^{(n,l)})$ and a Normal-Gamma prior distribution with maximum at $\widehat{\mu}_l$ and $\widehat{\lambda}_l$. The posterior is also a Normal-Gamma, with maximum at

$$\mu_l^{MAP} = \frac{N_l \mathrm{E}^{\mathrm{MC}}[Y_l] + k_0 \widehat{\mu}_l}{k_0 + N_l} \quad \text{and} \quad \lambda_l^{MAP} = \frac{\Xi_{1,l} - \frac{1}{2}}{\Xi_{2,l}} \tag{20}$$

with:

$$\Xi_{1,l} = \frac{1}{2} + k_1 \widehat{\lambda}_l + \frac{N_l}{2}, \tag{21a}$$

$$\Xi_{2,l} = k_1 + \frac{N_l - 1}{2} \mathrm{V}^{\mathrm{MC}}[Y_l] + \frac{k_0 N_l (\mathrm{E}^{\mathrm{MC}}[Y_l] - \widehat{\mu}_l)^2}{2(k_0 + N_l)}. \tag{21b}$$

The parameters k_0 and k_1 represent our "certainty" on $\widehat{\mu}_l$ and $\widehat{\lambda}_l^{-1}$. The resulting update formula for $\mathbb{V}ar[Y_l] \approx \lambda_l^{-1}$ is then:

$$\mathrm{V}^{\mathrm{C}}[Y_l] := \frac{\Xi_{2,l}}{\Xi_{1,l} - \frac{1}{2}} \quad l > 0 \tag{22}$$

We can easily show that:

$$N_l \to \infty \quad \Longrightarrow \quad \mathrm{V}^{\mathrm{C}}[Y_l] \to \frac{1}{N_l - 1} \sum_{n=1}^{N_l} \left(Y_l(\omega^{(n,l)}) - \mathrm{E}^{\mathrm{MC}}[Y_l] \right)^2 \tag{23}$$

thus recovering the sample variance estimator, whereas

$$N_l = 0 \quad \Longrightarrow \quad \mathrm{V}^{\mathrm{C}}[Y_l] = \frac{1}{\widehat{\lambda}_l} = c_\beta M_l^{-\beta} \tag{24}$$

thus using just the prior model with fitted parameters (c_β, β).

Finally, following the above arguments, we approximate the variance of the MLMC estimator as:

$$\mathbb{V}ar[\mathrm{E}^{\mathrm{MLMC}}[Q_M]] = \sum_{l=0}^{L} \frac{\mathbb{V}ar[Y_l]}{N_l} \approx \sum_{l=0}^{L} \frac{\mathrm{V}^{\mathrm{C}}[Y_l]}{N_l} \tag{25}$$

and the total MSE as:

$$e(\mathrm{E}^{\mathrm{MLMC}}[Q_M]) \approx \left| \mathrm{E}^{\mathrm{MC}}[Y_L] \right|^2 + \sum_{l=0}^{L} \frac{\mathrm{V}^{\mathrm{C}}[Y_l]}{N_l}. \tag{26}$$

The resulting algorithm is described in Algorithm 2.

Algorithm 2: Continuation Multi-Level Monte Carlo.

```
CMLMC(N̄, L̄, L_MAX, k_0, k_1, r_1, r_2, ε_M, ε_0)
  for l = 0 : L̄ do
    for i = 0 : N̄ do
      Generate random samples: O(ω^(i,l)), G(ω^(i,l))
      Q^(i)_{M_l}     ← PROBLEM_l (O(ω^(i,l)), G(ω^(i,l)))
      Q^(i)_{M_{l-1}} ← PROBLEM_{l-1} (O(ω^(i,l)), G(ω^(i,l)))
      Y^(i)_l = Q^(i)_{M_l} − Q^(i)_{M_{l-1}}
  compute {C_l}, {|E^MC[Y_l]|}, {V^MC[Y_l]}
  compute 𝒫 by least squares fit
  compute V^C[Y_l] using (22)
  compute i_E using (16)
  while i < i_E or TErr > ε_M do
    compute ε_i
    compute L_i and θ_i using (17)
    for l = 0 : L_i do
      compute N_l using (12)
      for i = 0 : N_l do
        Generate random samples: O(ω^(i,l)), G(ω^(i,l))
        Q^(i)_{M_l}     ← PROBLEM_l (O(ω^(i,l)), G(ω^(i,l)))
        Q^(i)_{M_{l-1}} ← PROBLEM_{l-1} (O(ω^(i,l)), G(ω^(i,l)))
        Y^(i)_l = Q^(i)_{M_l} − Q^(i)_{M_{l-1}}
    update {C_l}_{l=0}^{L_i}, {|E^MC[Y_l]|}_{l=0}^{L_i}, {V^MC[Y_l]}_{l=0}^{L_i}
    update E^MLMC[Q_M] = Σ_{l=0}^{L} E^MC[Y_l]
    compute (c_α, α) ← {|E^MC[Y_l]|}_{l=0}^{L_i} using least squares fit
    compute (c_γ, γ) ← {C_l}_{l=0}^{L_i} using least squares fit
    compute (c_β, β) ← {V^MC[Y_l]}_{l=0}^{L_i} using least squares fit
    update 𝒫 = {c_α, c_β, c_γ, α, β, γ}
    update V^C[Y_l],
    estimate B ≈ E^MC[Y_{L_i}]
    estimate 𝕍ar[E^MLMC[Q_{M_i}]] using (25)
    compute TErr = B + sqrt(𝕍ar[E^MLMC[Q_{M_i}]])
    i = i+1
  return E^MLMC[Q_M]
```

Model Problems

We consider turbulent compressible flows modeled by the Favre-averaged (density-weighted average $\tilde{f} = \overline{\rho f}/\overline{\rho}$) Navier–Stokes equations [9, 10]:

$$\frac{\partial \overline{\rho}}{\partial t} + \frac{\partial}{\partial x_i}\left(\overline{\rho}\tilde{u}_i\right) = 0, \tag{27a}$$

$$\frac{\partial}{\partial t}\left(\overline{\rho}\tilde{u}_i\right) + \frac{\partial}{\partial x_j}\left(\overline{\rho}\tilde{u}_i\tilde{u}_j\right) = -\frac{\partial p}{\partial x_i} + \frac{\partial \overline{\sigma}_{ij}}{\partial x_j} + \frac{\partial \tau_{ij}}{\partial x_j} \tag{27b}$$

$$\frac{\partial}{\partial t}\left(\overline{\rho}\tilde{E}\right) + \frac{\partial}{\partial x_j}\left(\overline{\rho}\tilde{u}_j\tilde{H}\right) = \frac{\partial}{\partial x_j}\left[\tilde{u}_i\overline{\sigma}_{ij} + \overline{\sigma_{ij}u_i''}\right] \tag{27c}$$

$$+ - \frac{\partial}{\partial x_j}\left[-q_j + c_p\overline{\rho u_j''T''} - \tilde{u}_i\tau_{ij} + \frac{1}{2}\overline{\rho u_i''u_i''u_j''}\right] \tag{27d}$$

where we denote with $\tau_{ij} = -\overline{\rho u_i''u_j''}$ the Reynolds stress term and $\overline{\sigma}_{ij}$ the viscous stress tensor. The former is approximated by solving the Spalart–Allmaras turbulence model with quadratic constitutive relation [11]. We approximate the turbulent heat flux $c_p\overline{\rho u_j''T''} \approx -\frac{c_p\tilde{\mu}_t}{Pr_t}\frac{\partial \tilde{T}}{\partial x_j}$ following a Reynolds analogy, and we use a constant turbulent Prandtl number $Pr_t = 0.9$. Sutherland's Law is employed to compute the dynamic viscosity from the temperature of the ideal gas ($Pr = 0.72$).

The above-mentioned equations are discretized on structured grids (finite volume method) and advanced in time using a fully implicit time-stepping scheme. Local time-stepping and algebraic multigrid (AMG) are used for convergence acceleration to the steady-state solution.

Model Problems: NASA ROTOR-37

The first problem we consider is the well-established turbomachinery test case NASA ROTOR-37 (UMRIDA BC-01), a transonic axial flow compressor. The rotor has 36 blades and an aspect ratio of 1.19, rotates at 17188.7 (rpm) [1800 (rad/s)], leading to a tip speed of 454 (m/s). A detailed description of the geometry, the original experimental setup, and a series of simulations can be found in [12, 13].

The design parameters of the rotor are summarized in the following Table 1.

Determinsitic results The computational model (Fig. 1b) consists of one blade with periodic boundary conditions. The rotation is imposed to the hub and the blade, while the shroud is kept fixed. Total pressure and total temperature profiles derived from

Table 1 Design values for the NASA ROTOR-37 problem

Quantity	Symbol	Design value
Rotor total pressure ratio	P_2/P_1	2.106
Rotor total temperature ratio	T_2/T_1	1.270
Rotor adiabatic efficiency	η_{ad}	0.877
Mass flow (kg/s)	$\dot{m}$	20.188

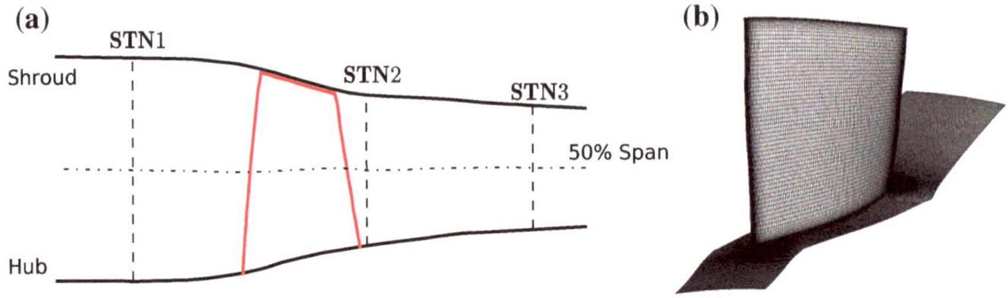

Fig. 1 **a** NASA rotor 37 and **b** computational model

Table 2 MLMC 4-levels grid hierarchy for the ROTOR-37 problem. *CTime*[*s*] is the real time in seconds required to compute one deterministic simulation on the prescribed number of cpus

Level	Spanwise nodes	Cells	y+	*CTime*[*s*] (*n.cpu*)
*L*0	33	156,769	1–2	110 (80)
*L*1	53	536,669	1–2	225 (128)
*L*2	73	1,244,133	1–2	435 (192)
*L*3	93	2,241,801	1–2	837 (224)
*L*4	113	4,253,889	1–2	1588 (256)

experiments [12] are imposed at the inlet boundary, and the static pressure is varied at the outlet to change the mass flow.

The proprieties of the 4-levels grid hierarchy used in the C-MLMC, generated using NUMERCA Autogrid, are presented in the following Table 2 along with the average computational time required to compute one deterministic simulation using CFD++ software environment.

We ensure an appropriate refinement near the small tip clearance [0.356 (mm)] and that near the boundaries the $y+$ is between 1 and 2, for all the grid levels, to accommodate the requirements of Spalart–Allmaras turbulence model employed in the CFD simulations. In Fig. 2, we observe a good agreement between the computational results obtained with the finest grid level (L4) and experimental measurements of Reid and Moore [13]. The significant differences between numerical results and measurements are in the rotor stall region. For this reason, we will only consider operating points before stall conditions [$\dot{m} > 20.5$ (kg/s)].

Figure 3 presents the flow features on the suction and pressure side of the blade and at 50% of the span for the maximum adiabatic efficiency conditions ($\eta_{ad} = 0.876$). We distinguish the bow shock at the leading edge of the blade and a classical λ–shock region (Fig. 3g) on the suction side where the shock impacts the boundary layer. In downstream of the shock–boundary layer interaction, we identify a flow separation region. Such separation can be inferred also by looking at the skin friction (Fig. 3c), the boundary layer transition and at the turbulence index (Fig. 3e) at the wall. Additionally, the boundary layer transition induces a sudden increase of eddy viscosity (Fig. 3h).

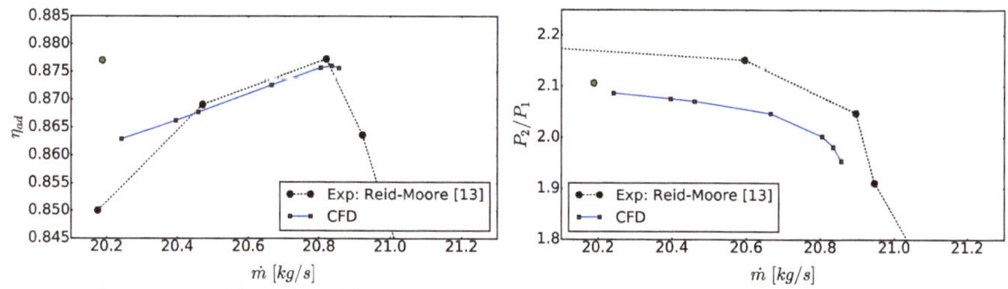

(a) Rotor adiabatic efficiency (b) Rotor total pressure ratio

Fig. 2 Experimental and computational compressor maps of the ROTOR-37. The green circles indicate the design parameters presented in the previous table

Table 3 Operating uncertainties for the ROTOR-37 stochastic analysis

	Quantity	Reference (r)	Uncertainty $\mathscr{TN}(\mu, \sigma, X_{LO}, X_{UP})$
INLET	P_{tot}	18 pt. profile (see Fig. 4)	$\mathscr{TN}(r, 1\%, -2\%, +2\%)$
	T_{tot}	18 pt. profile (see Fig. 4)	$\mathscr{TN}(r, 1\%, -2\%, +2\%)$
OUTLET	p_o	$C1 = 92500.0$ (Pa)	$\mathscr{TN}(r, 1\%, -2\%, +2\%)$
		$C2 = 99215.0$ (Pa)	$\mathscr{TN}(r, 1\%, -2\%, +2\%)$
		$C3 = 110000.0$ (Pa)	$\mathscr{TN}(r, 1\%, -2\%, +2\%)$

Stochastic Results using C-MLMC After assessing the validity of the CFD model, we now propagate uncertainties to study their effects on the performances of the rotor using the C-MLMC approach presented in section "Continuation Multi-level Monte Carlo." We consider operating uncertainties in the inlet total pressure and total temperature profile and the outlet static pressure. The uncertainties on the parameters are modeled as truncated Gaussian random variables where we use the notation $y \sim \mathscr{TN}(\mu, \sigma^2, a, b)$ to denote a r.v. with density function

$$p(y) = \begin{cases} 0 & y < a \\ \frac{1}{z}\frac{1}{\sqrt{2\pi}\sigma}e^{-\frac{(y-\mu)^2}{2\sigma^2}} & a \leq y \leq b \\ 0 & y > b. \end{cases} \quad \text{and} \quad z = \int_a^b \frac{1}{\sqrt{2\pi}\sigma}e^{-\frac{(y-\mu)^2}{2\sigma^2}}dy \tag{28}$$

The following Table 3 summarizes the reference operating parameters and the uncertainties considered for the following simulations.

Figure 4 depicts the inlet uncertain total pressure and total temperature profiles. The same random perturbation from the reference profile of the total pressure and temperature is applied to every point on the inlet (fully correlated perturbation).

In Fig. 5, we present the stochastic results for the adiabatic efficiency, rotor total pressure ratio, stage total pressure ratio, and mass flow for the ROTOR-37 affected by operating uncertainties (3 uncertain parameters). For the three analyzed cases ($C1$, $C2$, $C3$ in the mean outlet pressure p_0), we plot the compressor map with mean $\pm$ standard deviation for the four quantities of interest. We notice that the mean values

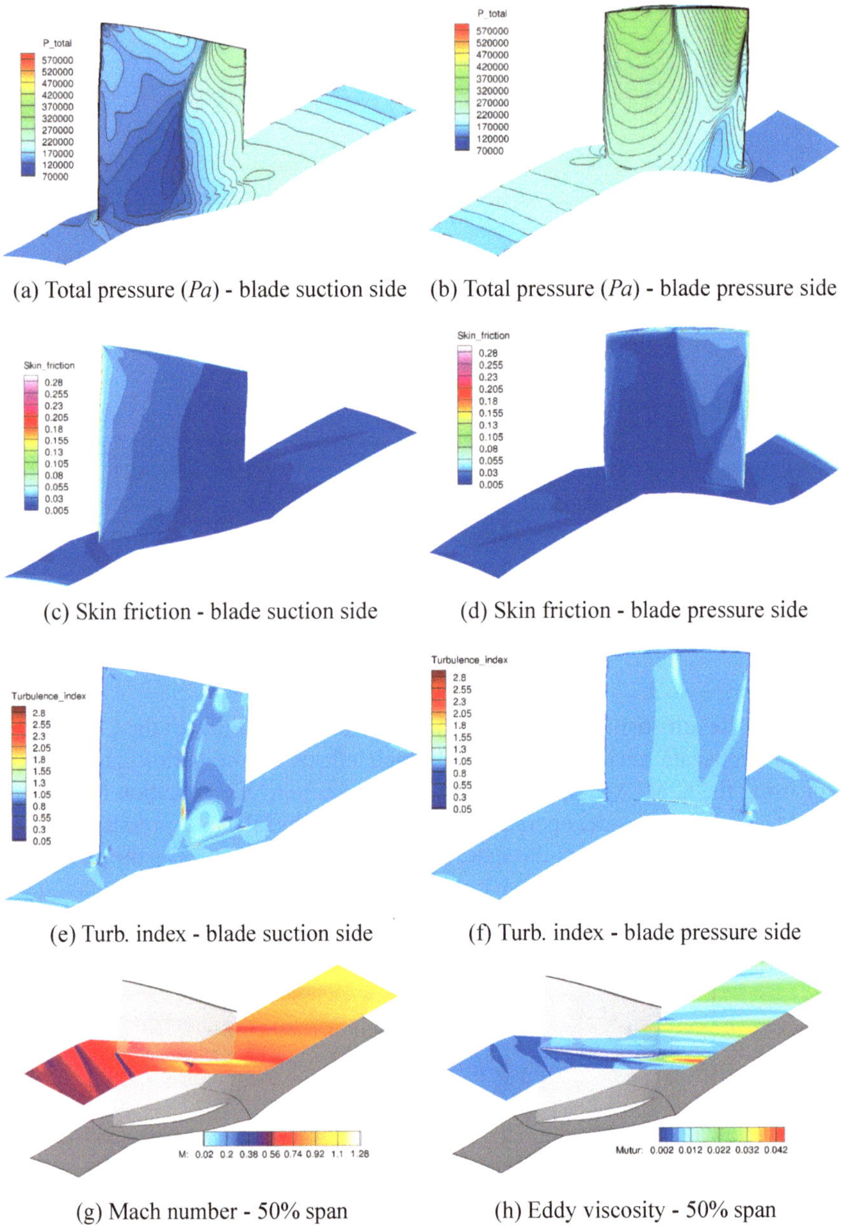

(a) Total pressure (*Pa*) - blade suction side (b) Total pressure (*Pa*) - blade pressure side

(c) Skin friction - blade suction side (d) Skin friction - blade pressure side

(e) Turb. index - blade suction side (f) Turb. index - blade pressure side

(g) Mach number - 50% span (h) Eddy viscosity - 50% span

Fig. 3 Deterministic results for the ROTOR-37. Left: suction side; right: pressure side

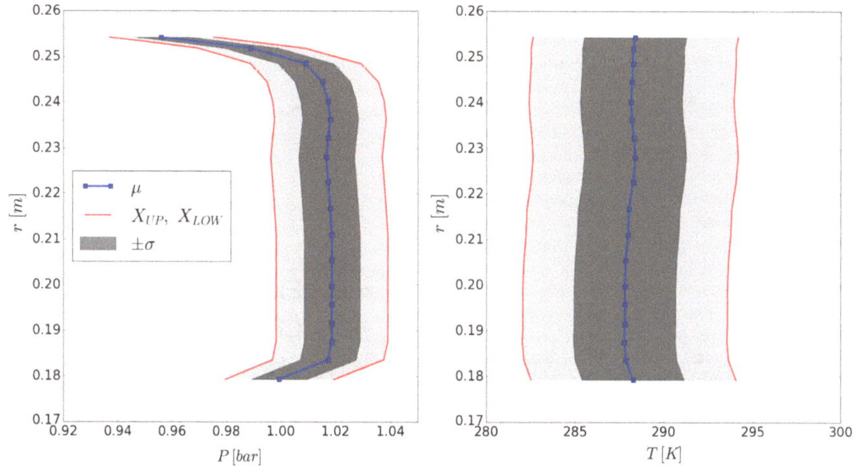

Fig. 4 Uncertain total pressure and total temperature inlet profiles. The blue line represents the mean profile (μ), the shaded gray area is one standard deviation ($\pm\sigma$), and the red lines are the upper and lower boundaries of the uncertain range (X_{LOW}, X_{UP})

of $\dot{m}$, η_{ad}, P_2/P_1 and P_3/P_1 in the stochastic case are comparable with the deterministic ones, as observed by [14, 15]. Additionally, we also observe that the mass flow $\dot{m}$ is the most sensitive quantity to variations in the operating parameters as quantified in Table 4. The variability of $\dot{m}$, η_{ad}, P_2/P_1 and P_3/P_1 seems to increase as we approach the stall conditions.

Model Problem: RAE2822

The second problem we consider is the 2D RAE2822 (UMRIDA BC-02), a supercritical airfoil which has become a standard test case for transonic flows. A detailed description of the airfoil geometry, the original experimental setup and a series of simulations can be found in [16, 17]. For this specific problem, we consider as scalar field QoI the pressure coefficient C_p of the RAE 2822 affected by operating and geometric uncertainties due to fluctuations in the surrounding flow and manufacturing tolerances. The nominal geometry of the RAE2822 airfoil is defined with a set of PARSEC parameters [18]. The following table summarizes these parameters and the operating conditions considered hereafter (corrected flow conditions for case 6 in [17]).

Figure 6 illustrates the nominal geometry of the RAE 2822 and the meaning of the parameters in Table 5.

Determinsitic results The proprieties of the 4-levels structured C-grid hierarchy used in the C-MLMC simulations are presented in the following Table 6 along with

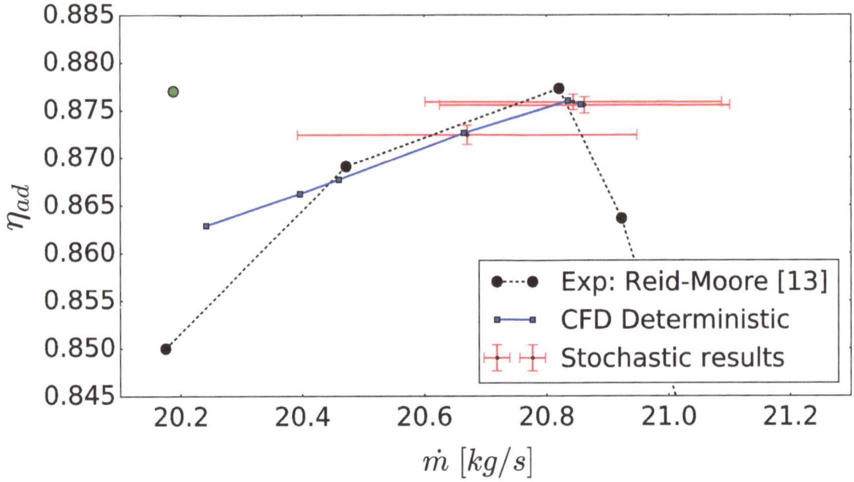

(a) Rotor adiabatic efficiency

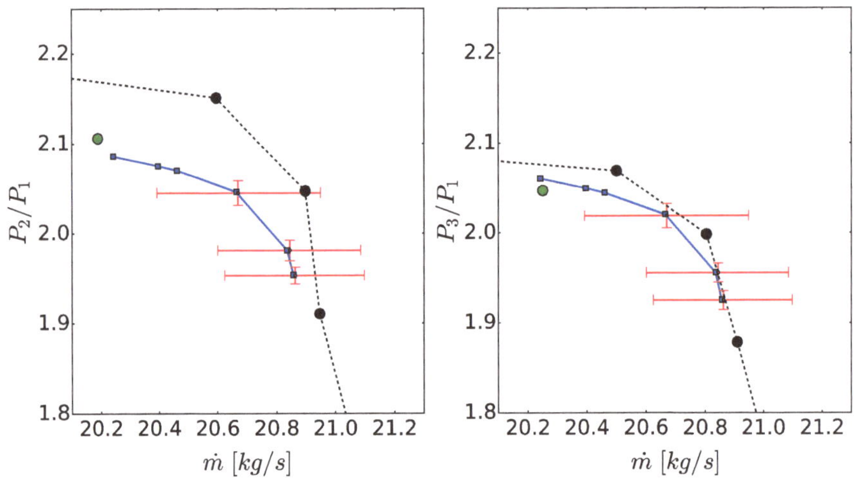

(b) Rotor total pressure ratio (P_2/P_1) and Stage total pressure ratio (P_3/P_1)

Fig. 5 Experimental, deterministic and stochastic results for the compressor map of the ROTOR-37. Each red interval correspond to mean ± standard deviation

the average computational time required to compute one deterministic simulation using CFD++ software environment. A close-up view of the structured grid in the proximity of the leading edge for level 0 and level 1 is presented in Fig. 7.

As for the ROTOR-37 previous problem, we ensure that near the boundaries the $y+$ is between 1 and 2 for all the grid levels to fulfill the requirements of the Spalart–Allmaras turbulence model. In Fig. 8, we compare the computational results obtained

Table 4 Deterministic and stochastic results for the ROTOR-37

Case-p_o	Deterministic	Stochastic (% σ)
$C1 = 92500.0$ (Pa)	$\dot{m} = 20.8564$ (kg/s)	$\dot{m} = 20.8621 \pm 0.2371$ (kg/s) (1.13%)
	$\eta_{ad} = 0.8756$	$\eta_{ad} = 0.8755 \pm 0.0009$ (0.10%)
	$P_2/P_1 = 1.9540$	$P_2/P_1 = 1.9534 \pm 0.0093$ (0.47%)
	$P_3/P_1 = 1.9255$	$P_3/P_1 = 1.9252 \pm 0.0105$ (0.54%)
$C2 = 99215.0$ (Pa)	$\dot{m} = 20.8564$ (kg/s)	$\dot{m} = 20.8440 \pm 0.2424$ (kg/s) (1.16%)
	$\eta_{ad} = 0.8760$	$\eta_{ad} = 0.8758 \pm 0.0008$ (0.09%)
	$P_2/P_1 = 1.9813$	$P_2/P_1 = 1.9812 \pm 0.0113$ (0.57%)
	$P_3/P_1 = 1.9559$	$P_3/P_1 = 1.9558 \pm 0.0106$ (0.54%)
$C3 = 110000.0$ (Pa)	$\dot{m} = 20.6653$ (kg/s)	$\dot{m} = 20.6706 \pm 0.2777$ (kg/s) (1.34%)
	$\eta_{ad} = 0.8726$	$\eta_{ad} = 0.8724 \pm 0.0010$ (0.11%)
	$P_2/P_1 = 2.0464$	$P_2/P_1 = 2.0451 \pm 0.0137$ (0.67%)
	$P_3/P_1 = 2.0204$	$P_3/P_1 = 2.0190 \pm 0.0135$ (0.67%)

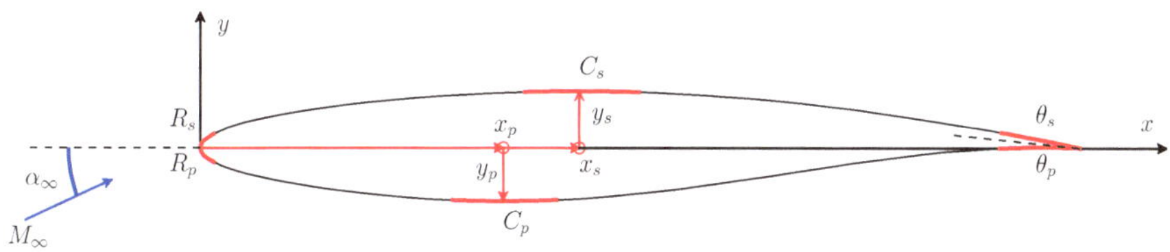

Fig. 6 Geometry of the RAE 2822 transonic airfoil and PARSEC parameters that define the geometry of the airfoil

with the finest grid level (L4) and experimental measurements [17] and we observe a good agreement.

Stochastic Results using C-MLMC We now propagate geometric and operating uncertainties in the model to study their effects on the C_p profile of the airfoil using the C-MLMC approach. We consider operating uncertainties in the far-field Mach number and angle of attach and geometric uncertainties in the PARSEC coefficients that define the shapes of the airfoil. In case of geometric uncertainties that affect the shape of the airfoil, for each random geometry (set of PARSEC coefficients) we deform the existing grid levels by solving a linear elasticity problem on the volume grid to accommodate the new boundary definition (Fig. 9).

Table 5 Geometric and operating reference parameters for the RAE2822 problem

	Symbol	Reference value
Operating	α_∞	2.31
	M_∞	0.729
	Re_c	$6.5 \cdot 10^6$
	p_∞ (Pa)	101,325
	T_∞ (K)	288.5
	Symbol	Design value
Geometric	R_s	0.00839
	R_p	0.00853
	x_s	0.431
	x_p	0.346
	y_s	0.063
	y_p	−0.058
	C_s	−0.432
	C_p	0.699
	θ_s	−11.607
	θ_p	−2.227

Table 6 MLMC 4-levels grid hierarchy for the RAE2822 problem. *CTime*[*s*] is the real time in seconds required to compute one deterministic simulation on the prescribed number of cpus

Level	Airfoil nodes	Cells	$y+$	*CTime*[*s*] (*n.cpu*)
*L*0	160	7722	1–2	13.9 (16)
*L*1	320	31,442	1–2	49.7 (24)
*L*2	640	126,882	1–2	336.9 (32)
*L*3	1280	509,762	1–2	2145.5 (40)
*L*4	2560	2,043,522	1–2	6854.3 (48)

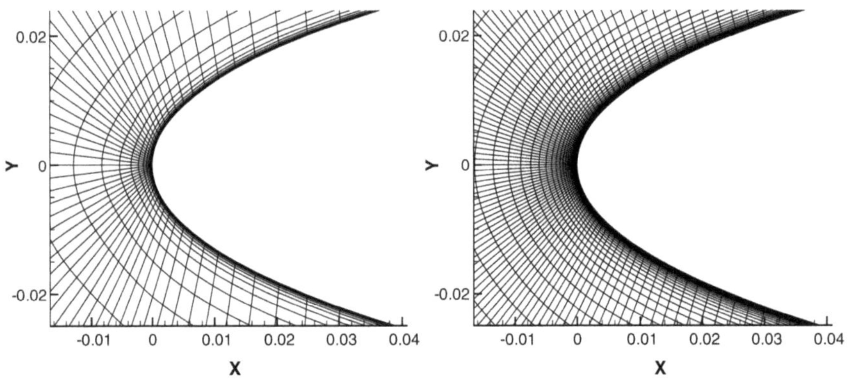

(a) *L*0 (b) *L*1

Fig. 7 Leading edge close-up view of level 0 and 1 grids for the RAE2822 problem

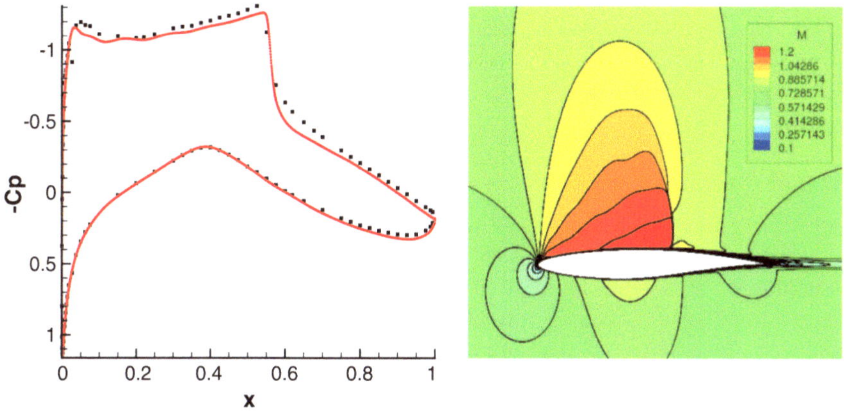

(a) Experimental (black squares) and computational (red line) C_p coefficient around the airfoil

(b) Mach number

Fig. 8 Deterministic results for the RAE2822 airfoil

Fig. 9 Grid deformation to accommodate the geometric uncertainty

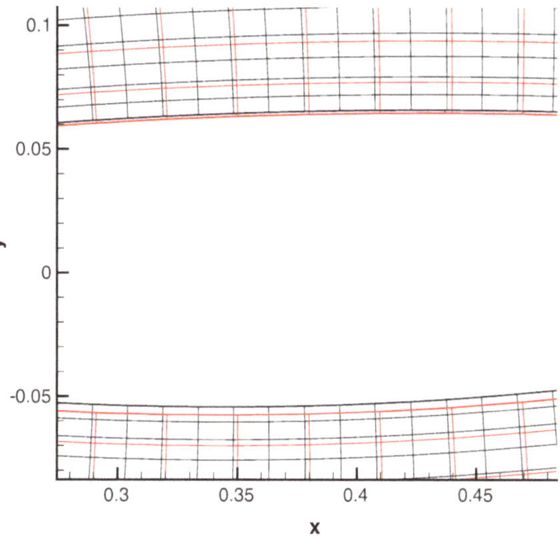

The following Table 7 and Fig. 10 summarize the operating and geometric parameters and their uncertainties modeled as truncated Gaussian random variables (see (28)).

In Fig. 11, we present the stochastic results for the pressure coefficient profile C_p around the airfoil under operating uncertainties (2 uncertain parameters hereafter denoted as **OPER(2)**), geometric uncertainties (8 uncertainties denoted as **GEOM(8)**), and operating and geometric uncertainties at the same time (10 uncertainties denoted as **OPER(2) + GEOM(8)**) presented in Table 7.

Table 7 Operating and geometric uncertainties for the RAE2822 stochastic analysis

	Quantity	Reference (r)	Uncertainty $\mathcal{T}\mathcal{N}(\mu, \sigma, X_{LO}, X_{UP})$
Operating	α_∞	2.31	$\mathcal{T}\mathcal{N}(r, 2\%, -2\%, +2\%)$
	M_∞	0.729	$\mathcal{T}\mathcal{N}(r, 5\%, -5\%, +5\%)$
	Re_c	$6.5 \cdot 10^6$	–
	p_∞ (Pa)	101,325	–
	T_∞ (K)	288.5	–
Geometric	R_s	0.00839	$\mathcal{T}\mathcal{N}(r, 0.25\%, -1\%, +1\%)$
	R_p	0.00853	$\mathcal{T}\mathcal{N}(r, 0.25\%, -1\%, +1\%)$
	x_s	0.431	$\mathcal{T}\mathcal{N}(r, 0.5\%, -1\%, +1\%)$
	x_p	0.346	$\mathcal{T}\mathcal{N}(r, 0.5\%, -1\%, +1\%)$
	y_s	0.063	$\mathcal{T}\mathcal{N}(r, 0.5\%, -3\%, +3\%)$
	y_p	−0.058	$\mathcal{T}\mathcal{N}(r, 0.5\%, -3\%, +3\%)$
	C_s	−0.432	$\mathcal{T}\mathcal{N}(r, 0.5\%, -1\%, +1\%)$
	C_p	0.699	$\mathcal{T}\mathcal{N}(r, 0.5\%, -1\%, +1\%)$
	θ_s	−11.607	–
	θ_p	−2.227	–

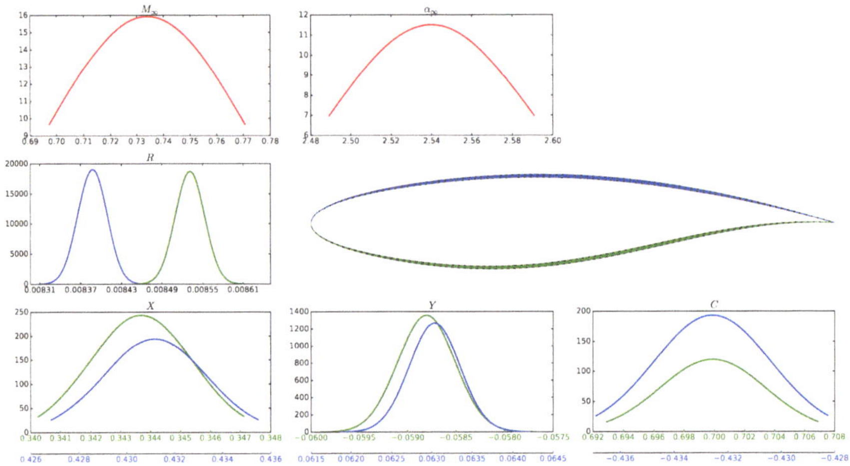

Fig. 10 Probability density functions of the operating (red) and geometric (blue suction side and green pressure side) parameters for the RAE2822 stochastic analysis

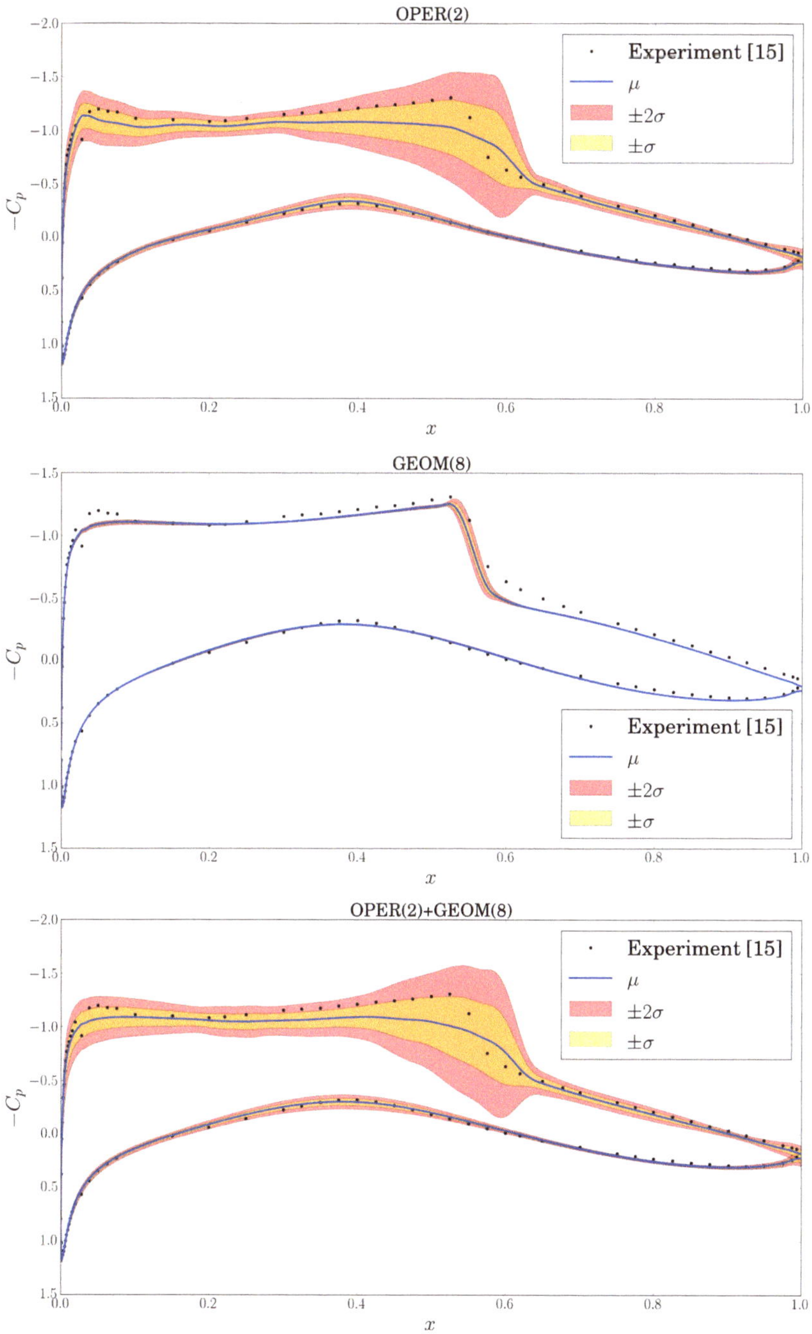

Fig. 11 UQ analysis results for the RAE2822 presenting the mean pressure coefficient profile around the airfoil and its standard deviation. Experimental data from [17]

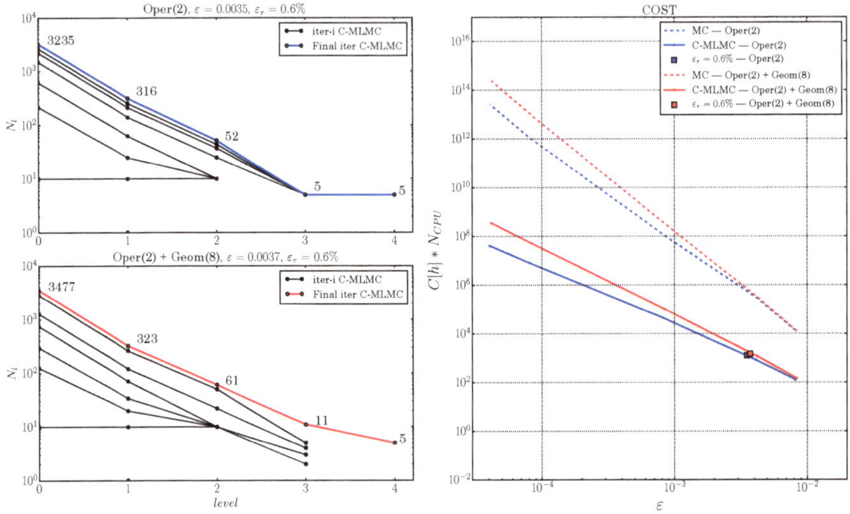

Fig. 12 C-MLMC hierarchies for two different sets of uncertain parameters (left) and aggregate computational cost compared with MC (right). The solid lines in the cost plot are an extrapolated model based on the rates and constants $(\alpha . c_\alpha,\ \beta, c_\beta, \gamma, c_\gamma)$ fitted in C-MLMC. The red and blue squares are the actual computed cost and error in the C-MLMC simulations

Lastly, we present in Fig. 12 the level sample sizes at each iteration of the C-MLMC algorithm to achieve a relative error $\varepsilon_r = 0.6\%$ on the L2 norm of the pressure coefficient for the **OPER(2)** and **OPER(2) + GEOM(8)** cases. Additionally, we compare the aggregate cost (total CPU time) required by our implementation of C-MLMC with the MC method to achieve a RMSE of ε. Notice how the performance of the C-MLMC is only mildly affected by the number of uncertain parameters. Moreover for the target relative tolerance $\varepsilon_r = 0.6\%$, the gain in computational cost of C-MLMC over MC is about two orders of magnitude and is expected to increase even further if smaller tolerances are prescribed. The results match nicely the theoretical estimates.

References

1. Collier, N., Haji-Ali, A.L., Nobile, F., von Schwerin, E., Tempone, R.: A continuation multilevel Monte Carlo algorithm. BIT Numer. Math. 1–34 (2014)
2. Giles, M.B.: Multilevel Monte Carlo path simulation. Oper. Res. **56**(3), 607–617 (2008)
3. Teckentrup, A., Scheichl, R., Giles, M., Ullmann, E.: Further analysis of multilevel Monte Carlo methods for elliptic PDEs with random coefficients. Numer. Math. **125**(3), 569–600 (2013)
4. Pisaroni, M., Leyland, P., Nobile, F.: A Multi Level Monte Carlo algorithm for the treatment of geometrical and operational uncertainties in internal and external aerodynamics. In: 46th AIAA Fluid Dynamics Conference, 4398 (2016)

5. Barth, A., Schwab, C., Zollinger, N.: Multi-level Monte Carlo finite element method for elliptic PDEs with stochastic coefficients. Numer. Math. **119**(1), 123–161 (2011)
6. Cliffe, K., Giles, M., Scheichl, R., Teckentrup, A.L.: Multilevel Monte Carlo methods and applications to elliptic PDEs with random coefficients. Comput. Vis. Sci. **14**(1), 3–15 (2011)
7. Barth, A., Lang, A., Schwab, C.: Multilevel Monte Carlo method for parabolic stochastic partial differential equations. BIT Numer. Math. **53**(1), 3–27 (2013)
8. Charrier, J., Scheichl, R., Teckentrup, A.L.: Finite element error analysis of elliptic PDEs with random coefficients and its application to multilevel Monte Carlo methods. SIAM J. Numer. Anal. **51**(1), 322–352 (2013)
9. Gatski, T.B., Bonnet, J.P.: Compressibility, Turbulence and High Speed Flow. Academic Press (2013)
10. Hirsch, C.: Numerical Computation of Internal and External Flows: The Fundamentals of Computational Fluid Dynamics. Butterworth-Heinemann (2007)
11. Mani, M., Babcock, D., Winkler, C., Spalart, P.: Predictions of a supersonic turbulent flow in a square duct. AIAA Paper **860** (2013)
12. Dunham, J.: CFD validation for propulsion system components (la validation CFD des organes des propulseurs). Technical report, DTIC Document (1998)
13. Reid, L., Moore, R.D.: Design and overall performance of four highly loaded, high speed inlet stages for an advanced high-pressure-ratio core compressor. NASA Technical Report (1978)
14. Loeven, G., Bijl, H.: The application of the probabilistic collocation method to a transonic axial flow compressor. In: American Institute of Aeronautics and Astronautics (AIAA) (2010)
15. Gopinathrao, N.P., Bagshaw, D., Mabilat, C., Alizadeh, S.: Non-deterministic CFD simulation of a transonic compressor rotor. In: ASME Turbo Expo 2009: Power for Land, Sea, and Air, American Society of Mechanical Engineers, pp. 1125–1134 (2009)
16. Haase, W., Brandsma, F., Elsholz, E., Leschziner, M., Schwamborn, D.: EUROVAL An European Initiative on Validation of CFD Codes: Results of the EC/BRITE-EURAM Project EUROVAL, 1990–1992, Vol. 42. Springer (2013)
17. Schmitt, V., Charpin, F.: Experimental data base for computer program assessment. Report of the Fluid Dynamics Panel Working Group. AGARD-AR-138 (1979)
18. Sobieczky, H.: Parametric airfoils and wings. In: Fujii, K., Dulikravich, G.S. (eds.) Notes on Numerical Fluid Mechanics, vol. 68, pp. 71–88 (1998)

Introduction to Intrusive Perturbation Methods

Alain Dervieux

Introduction

Perturbation methods generally refer to methods in which the evaluation of several values of some function is made easier by introducing derivatives of it. Starting from a given black box simulation software, perturbation methods involve to get some derivative of the software, which cannot be performed without an *intrusion* into it, that is, some effort in the application-dependant software transformation and further development.

We can regroup in "Intrusive Perturbation Methods" two of such intrusions, in relation to two different problematics.

- (**a**) In uncertainty propagation by perturbation or moments, the probabilistic behavior of the physical process output is evaluated by a two-step scheme: (1) derive a surrogate black box by derivation of the initial black box, and (2) identify the moments of the surrogate black box or use it in a sampling-based PDF evaluator such as the Monte Carlo method.
- (**b**) In robust optimization, the uncertainty propagation in the state system can be applied in a non-intrusive manner, but the research of an optimum may then be addressed with an extended gradient method, which relies on an intrusive adjoint-based functional derivation. The robust/probabilistic optimization then can use, for example, the following three steps ([4, 7] and "Second Order Derivatives for Geometrical Uncertainties" in this book): (1) build probabilistic state and functional with, for example, an NIPC, (2) compute the gradients of the mean and of the variance of the functional, and (3) update the design variable with a combination of the two gradients.

A. Dervieux (✉)
Ecuador, INRIA, 2002 Route des Lucioles, 06902 Sophia-Antipolis, France
e-mail: alain.dervieux@inria.fr

C. Hirsch et al. (eds.), *Uncertainty Management for Robust Industrial Design in Aeronautics*, Notes on Numerical Fluid Mechanics and Multidisciplinary Design 140, https://doi.org/10.1007/978-3-319-77767-2_20

As already mentioned problematic (b) is addressed in chapter "Second Order Derivatives for Geometrical Uncertainties".

The rest of this section is devoted to problematic (a).

Some Perturbation Methods for Uncertainty Management

Considering an unknown random variable, one way to try to know its PDF is to approximate it. In some cases, this can be done from a collection of known PDFs as in PC methods. Otherwise, the straightforward method is the full nonlinear Monte Carlo (MC) technique or more efficiently Multi-level Monte Carlo as described in papers included in chapter "Multi-level Monte Carlo Method". A full nonlinear Monte Carlo method gives us complete and exact information about uncertainty propagation in the form of its PDF, but with a prohibitively expensive cost in terms of CPU time. Then, it can be useful to use surrogate models which are designed in order to be less CPU consuming. A particular class of surrogate model is built by a Taylor expansion of the initial functional. An example of derivatives computation with Automatic Differentiation (AD) is given in chapters "Introduction to Intrusive Perturbation Methods and Their Range of Applicability" and "Use of Automatic Differentiation Tools at the Example of TAPENADE". Derivatives can also be used through the classical method of moments or through MC/Moments hybrid methods like the inexpensive Monte Carlo. The two next paragraphs explain these approaches.

Method of Moments

To reduce the computational cost, we may think to use only some (derivate) quantities characterizing the distribution of the input variables instead of an entire sample drawn from a population with a given PDF. The idea behind the Method of Moments is based on the Taylor series expansion of the original nonlinear functional around the *mean value* $\mu_\gamma = E[\gamma]$ of the input control γ, and then computing some statistical moments of the output (usually mean and variance). In this way, we are assuming that the input control γ can be decomposed as sum of a fully deterministic quantity μ_γ with a stochastic perturbation $\delta\gamma_u$ with the property $E[\delta\gamma_u] = 0$. With these definitions, the Taylor three-term expansion of the functional $j(\gamma)$ around the mean value μ_γ is

$$j(\gamma) = j(\mu_\gamma + \delta\gamma_u) = j(\mu_\gamma) + G\delta\gamma_u + \frac{1}{2}\delta\gamma_u^* H\delta\gamma_u + O(\|\delta\gamma_u\|^3) \qquad (1)$$

where $G = \left.\frac{\partial j}{\partial \gamma_u}\right|_{\mu_\gamma}$ is the gradient of the functional with respect to the uncertain variables and $H = \left.\frac{\partial^2 j}{\partial {\gamma_u}^2}\right|_{\mu_\gamma}$ is the Hessian matrix, both evaluated at the mean of the input variables μ_γ.

By considering various orders of the Taylor expansion (1) and taking the first and the second statistical moment, we can approximate the mean μ_j and the variance σ_j^2 of the functional $j(\gamma)$ in terms of its derivatives evaluated at μ_γ and in terms of statistical moments of the control γ.

First-order moment methods:

$$\begin{cases} \mu_j = j(\mu_\gamma) + O\left(E\left[\delta\gamma_u^2\right]\right) \\ \sigma_j^2 = E\left[\left(G\delta\gamma_u\right)^2\right] + O\left(E\left[\delta\gamma_u^3\right]\right) \end{cases} \tag{2}$$

Second-order moment methods:

$$\begin{cases} \mu_j = j(\mu_\gamma) + \frac{1}{2}E\left[\delta\gamma_u^* H \delta\gamma_u\right] + O\left(E\left[\delta\gamma_u^3\right]\right) \\ \sigma_j^2 = E\left[\left(G\delta\gamma_u\right)^2\right] + E\left[\left(G\delta\gamma_u\right)\left(\delta\gamma_u^* H\delta\gamma_u\right)\right] - \frac{1}{4}E\left[\delta\gamma_u^* H\delta\gamma_u\right]^2 + \\ \quad + \frac{1}{4}E\left[\left(\delta\gamma_u^* H\delta\gamma_u\right)^2\right] + O\left(E\left[\delta\gamma_u^4\right]\right) \end{cases} \tag{3}$$

With this method, it is clear that we are using only some partial information about the input uncertainties; in fact, we are using only some statistical moments of the control variable instead of full information available with its PDF, and we will not have anymore the PDF of the propagated uncertainty, but only its approximate mean and variance. Another important point is that the Method of Moments is applicable only for *small* uncertainties, due to the local nature of Taylor expansion approximation.

Two things should be noted here: the first one is that for applying the Method of Moments, *we need the derivatives* of the functional with respect to the control variables affected by uncertainties: in particular, we need the gradient for the first-order method, and gradient and Hessian for the second-order method.

We are interested by an equation-constrained functional of the form $j(\gamma) = J(\gamma, W)$, where $W = W(\gamma)$ is solution of the state equation

$$W = W(\gamma) \quad \Leftrightarrow \quad \Psi(\gamma, W) = 0,$$

we have for the derivative

$$\frac{\partial j}{\partial \gamma_u} = \frac{\partial J}{\partial \gamma_u} + \frac{\partial J}{\partial W}\frac{\partial W}{\partial \gamma_u}.$$

We know the solution $W(\gamma)$ by its numerical values as results of a program assembling an appropriate system and solving it with an appropriate solution algorithm, e.g., a fixed point method. Its differentiation can be considered as a from-scratch software development. It can be much more safe and efficient to use an Automatic Differentiation tool (like TAPENADE, [3]) in order to obtain the needed derivatives. The same remarks apply to the computation of the Hessian matrix. In particular, we note that the derivatives are computed at the mean value of the control μ_γ, so they are fully deterministic and can be picked out from the expectations in Eqs. (2) or (3). In other words, we can write

$$\begin{aligned} E\left[\left(G\delta\gamma_u\right)^2\right] &= \sum_{i,k} G_i G_k E\left[\delta\gamma_u^{(i)}\delta\gamma_u^{(k)}\right] = \sum_{i,k} G_i G_k C_{ik} \\ E\left[\delta\gamma_u^* H\delta\gamma_u\right] &= \sum_{i,k} H_{ik} E\left[\delta\gamma_u^{(i)}\delta\gamma_u^{(k)}\right] = \sum_{i,k} H_{ik} C_{ik} \\ E\left[\left(G\delta\gamma_u\right)\left(\delta\gamma_u^* H\delta\gamma_u\right)\right] &= \sum_{i,k,l} G_l H_{ik} E\left[\delta\gamma_u^{(i)}\delta\gamma_u^{(k)}\delta\gamma_u^{(l)}\right] \\ E\left[\left(\delta\gamma_u^* H\delta\gamma_u\right)^2\right] &= \sum_{i,k,l,m} H_{ik} H_{lm} E\left[\delta\gamma_u^{(i)}\delta\gamma_u^{(k)}\delta\gamma_u^{(l)}\delta\gamma_u^{(m)}\right] \end{aligned} \tag{4}$$

where $G_i = \left.\frac{\partial j}{\partial\gamma_u^{(i)}}\right|_{\mu_\gamma}$ are the elements of the gradient, $H_{ik} = \left.\frac{\partial^2 j}{\partial\gamma_u^{(i)}\partial\gamma_u^{(k)}}\right|_{\mu_\gamma}$ are the elements of the Hessian matrix, and $C_{ik} = E\left[\delta\gamma_u^{(i)}\delta\gamma_u^{(k)}\right] = \mathrm{cov}(\gamma_u^{(i)}, \gamma_u^{(k)})$ are the elements of the *covariance matrix*. Every expectation term $E[\dots]$ in Eq. (4), is defined by the statistical model of the uncertainties and could be computed in a preprocessing phase.

For example, for the important case where the uncertainties are normally distributed, we have:

$$\begin{aligned} E\left[\delta\gamma_u^{(i)}\delta\gamma_u^{(k)}\delta\gamma_u^{(l)}\right] &= 0 \\ E\left[\delta\gamma_u^{(i)}\delta\gamma_u^{(k)}\delta\gamma_u^{(l)}\delta\gamma_u^{(m)}\right] &= C_{ik}C_{lm} + C_{il}C_{km} + C_{im}C_{kl} \end{aligned}$$

and if these (normal) uncertainties are independent, then holds the relation $C_{ik} = \sigma_i^2\delta_{ij}$ where $\sigma_i^2 = E\left[\delta\gamma_u^{(i)}\delta\gamma_u^{(i)}\right]$ and Eq. (4) becomes

$$\begin{aligned} E\left[\left(G\delta\gamma_u\right)^2\right] &= \sum_i G_i^2\sigma_i^2 \\ E\left[\delta\gamma_u^* H\delta\gamma_u\right] &= \sum_i H_{ii}\sigma_i^2 \\ E\left[\left(G\delta\gamma_u\right)\left(\delta\gamma_u^* H\delta\gamma_u\right)\right] &= 0 \\ E\left[\left(\delta\gamma_u^* H\delta\gamma_u\right)^2\right] &= \sum_{i,k}\left(H_{ii}H_{kk} + 2H_{ik}^2\right)\sigma_i^2\sigma_k^2. \end{aligned} \tag{5}$$

The second comment to do for the Method of Moments concerns the equation of the variance for second-order moment (3): although we have taken into account the term $E\left[\left(\delta\gamma_u^* H \delta\gamma_u\right)^2\right]/4$, the error is still of the order of $E\left[\delta\gamma_u^4\right]$. This is because the other terms of the same kind require the knowledge of order of derivatives higher than the second. In fact, it can be shown that the fourth-order term in $O(E\left[\delta\gamma_u^4\right])$ depends on the third derivative of the functional.

From the previous discussion, it is clear that in order to apply the Method of Moments, we need to solve only one (expensive) nonlinear system with derivatives (at the mean μ_γ) and then apply (inexpensive) Eqs. (2) or (3) where, for the fully nonlinear Monte Carlo approach of the previous subsection, we need to solve $N \gg 1$ nonlinear state systems.

An application of this method for aerospace applications can be found in [6].

Inexpensive Monte Carlo

This method, developed by [2], is based on the idea of adjoint error correction as proposed by [5], and it could be viewed at midpoint of Monte Carlo and Method of Moments. The discrete adjoint equation corresponding to state equation and functional is

$$\left(\frac{\partial \Psi}{\partial W}\right)^* \Pi = \left(\frac{\partial J}{\partial W}\right)^* \tag{6}$$

where Π is the adjoint solution. The key here is to perform a Taylor series expansion for the functional *and* for the state equation with respect to the state variables W, and then using the adjoint Eq. (6).

For the functional, the first-order Taylor expansion with respect to the generic state W_0 is

$$J(\gamma, W) = J(\gamma, W_0) + \left.\frac{\partial J}{\partial W}\right|_{(\gamma, W_0)} (W - W_0) + O\left(\|W - W_0\|^2\right).$$

Using the adjoint Eq. (6), we have

$$J(\gamma, W) = J(\gamma, W_0) + \left.\left(\Pi^* \frac{\partial \Psi}{\partial W}\right)\right|_{(\gamma, W_0)} (W - W_0) + O\left(\|W - W_0\|^2\right). \tag{7}$$

A first-order Taylor expansion of the state equation gives us

$$0 = \Psi(\gamma, W) = \Psi(\gamma, W_0) + \left.\frac{\partial \Psi}{\partial W}\right|_{(\gamma, W_0)} (W - W_0) + O\left(\|W - W_0\|^2\right).$$

Replacing the first-order derivative in the (7) with the analogous term in the last equation, we obtain

$$J(\gamma, W) = J(\gamma, W_0) - \left(\Pi^* \Psi\right)\big|_{(\gamma, W_0)} + O\left(\|W - W_0\|^2\right) \tag{8}$$

where the state equation Ψ and the adjoint solution Π^* are both evaluated at the point (γ, W_0). Finally, if we use an approximate adjoint solution $\Pi^* = \widetilde{\Pi}^* + O(\|\Pi - \widetilde{\Pi}\|)$ instead of the exact one, we have

$$J(\gamma, W) = J(\gamma, W_0) - \widetilde{\Pi}^* \Psi(\gamma, W_0) + O\left(\|W - W_0\|^2, \|W - W_0\| \; \|\Pi - \widetilde{\Pi}\|\right). \tag{9}$$

From Eq. (6), we can see the adjoint solution Π as a function of the variables γ and W, and then, remembering that the state variables depend on the control ($W = W(\gamma)$) through the state equation, we have

$$\Pi(\gamma, W) = \Pi(\gamma, W(\gamma)) = \pi(\gamma).$$

Thus, if we expand at the first-order the latter equation around the mean value of the control μ_γ (as we did in the previous subsection), we obtain

$$\Pi(\gamma, W) = \pi(\mu_\gamma) + O(\|\delta\gamma_u\|) = \Pi(\mu_\gamma, W(\mu_\gamma)) + O(\|\delta\gamma_u\|) \tag{10}$$

where $\pi(\mu_\gamma) = \Pi(\mu_\gamma, W(\mu_\gamma))$ is the solution of the adjoint system (6) with the derivatives computed at the point $(\mu_\gamma, W(\mu_\gamma))$. It is clear that considering Π as a function on the only variable γ and identifying the approximate adjoint solution $\widetilde{\Pi}$ with $\pi = \Pi(\mu_\gamma, W(\mu_\gamma))$ we have $O(\|\Pi - \widetilde{\Pi}\|) = O(\|\delta\gamma_u\|)$, and Eq. (9) becomes

$$J(\gamma, W) = J(\gamma, W_0) - \pi^* \Psi(\gamma, W_0) + O\left(\|W - W_0\|^2, \|W - W_0\| \; \|\delta\gamma_u\|\right). \tag{11}$$

Now it only remains to decide how to choose the approximation W_0. As usual, we perform a first-order Taylor expansion of $W(\gamma)$ around μ_γ

$$W(\gamma) = W(\mu_\gamma) + \frac{dW}{d\gamma_u}\bigg|_{\mu_\gamma} \delta\gamma_u + O(\|\delta\gamma_u\|^2)$$

then we have two options

IMC1 $W_0 = W(\mu_\gamma)$ and then

$$J(\gamma, W) = J(\gamma, W(\mu_\gamma)) - \pi^* \Psi(\gamma, W(\mu_\gamma)) + O\left(\|\delta\gamma_u\|^2\right) \tag{12}$$

IMC2 $W_0 = W(\mu_\gamma) + \frac{dW}{d\gamma_u}\Big|_{\mu_\gamma} \delta\gamma_u$ and then

$$J(\gamma, W) = J(\gamma, W_0) - \pi^* \Psi(\gamma, W_0) + O\left(\|\delta\gamma_u\|^3\right) \tag{13}$$

where the first approach has an overall leading error of second order, while the second approach has an overall leading error of third order.

In contrast to the fully nonlinear Monte Carlo, where has to be solved one nonlinear system $\Psi(\gamma, W) = 0$ for each sample point $\gamma_i = \mu_\gamma + \delta\gamma_{u,i}$, in the inexpensive Monte Carlo we evaluate the residual $\Psi(\gamma_i, W_0)$ for each sampling point, resulting in a computationally less expensive method.

To be more clear, the algorithm is summarized as follows:

- choose the mean control μ_γ and solve the nonlinear system $\Psi(\mu_\gamma, W) = 0$
- solve the adjoint linear system $\left(\frac{\partial \Psi}{\partial W}\Big|_{(\mu_\gamma, W(\mu_\gamma))} \right)^* \pi = \left(\frac{\partial J}{\partial W}\Big|_{(\mu_\gamma, W(\mu_\gamma))} \right)^*$
- (if we use the IMC2) compute the matrix $\frac{\partial W}{\partial \gamma_u}\Big|_{\mu_\gamma}$ (this is could be done solving m linear system, where $\gamma_u \in \mathbb{R}^m$)
- construct the N sampling point $\gamma_i = \mu_\gamma + \delta\gamma_{u,i}$ with $i = 1, \ldots, N$
- (if we use the IMC2) compute the N extrapolations of the state variables $W_{0,i} = W(\mu_\gamma) + \frac{\partial W}{\partial \gamma_u}\Big|_{\mu_\gamma} \delta\gamma_{u,i}$
- compute the N values $j(\gamma_i)$ using Eq. (12) for IMC1 or Eq. (13) for IMC2

To see the relations with the results obtained with the Method of Moments (Eqs. 2–3), we can note that in the equation there are two terms (J and Ψ) as function of γ that could be expanded up to the first order without increasing the error order, i.e.,

$$\begin{aligned} J(\gamma, W) &= J(\mu_\gamma, W(\mu_\gamma)) + \frac{\partial J}{\partial \gamma_u}\Big|_{(\mu_\gamma, W(\mu_\gamma))} \delta\gamma_u - \pi^* \Psi(\gamma, W(\mu_\gamma)) + O\left(\|\delta\gamma_u\|^2\right) \\ &= j(\mu_\gamma) + \frac{\partial J}{\partial \gamma_u}\Big|_{(\mu_\gamma, W(\mu_\gamma))} \delta\gamma_u - \pi^* \frac{\partial \Psi}{\partial \gamma_u}\Big|_{(\mu_\gamma, W(\mu_\gamma))} \delta\gamma_u + O\left(\|\delta\gamma_u\|^2\right) \end{aligned} \tag{14}$$

where we used the property $\Psi(\mu_\gamma, W(\mu_\gamma)) = 0$. Taking the expectation (and remembering that $E[\delta\gamma_u] = 0$), we have

$$\mu_j = j(\mu_\gamma) + O\left(E\left[\delta\gamma_u^2\right]\right) \tag{15}$$

i.e., the same equation for the mean of the functional obtained with the first-order moment method (2).

Orientation

Independently of which of the above formulations are chosen, the main effort is devoted in the computation of derivatives. The chapters “Algorithmic Differentiation for Second Derivatives” and “Second Order Derivatives for Geometrical Uncertainties” are devoted to this issue.

Acknowledgements For the chapters "Introduction to Intrusive Perturbation Methods", "Algorithmic Differentiation for Second Derivatives", "Introduction to Intrusive Perturbation Methods and Their Range of Applicability" and "Use of Automatic Differentiation Tools at the Example of TAPENADE", the author used a lot of material from the publication [1] and the reports delivered in NODESIM project.

References

1. Belme, A., Martinelli, M., Hascoët, L., Pascual, V., Dervieux, A.: AD-based perturbation methods for uncertainties and errors. Int. J. Eng. Syst. Model. Simul. **2**(1–2) (2010)
2. Ghate, D., Giles, M.B.: Inexpensive Monte Carlo Uncertainty Analysis. Recent Trends in Aerospace Design and Optimization, pp. 203–210. Tata McGraw-Hill, New Delhi (2006)
3. Hascoët, L., Pascual, V.: TAPENADE 2.1 user's guide. Technical Report No. 0300, INRIA, September 2004
4. Miranda, J., Kumar, D., Lacor, C.: Adjoint-based robust optimization using polynomial chaos expansions. In: Papadrakis, M., Papadopoulos, V., Stefanou, G., Plevris, V. (eds.) European Conference on Computational Fluid Dynamics ECCOMAS, Crete Island, Greece, 5–10 June 2016
5. Pierce, N.A., Giles, M.B.: Adjoint recovery of superconvergent functionals from PDE approximations. SIAM Rev. **42**(2), 247–264 (2000)
6. Putko, M.M., Newman, P.A., Taylor, A.C., Green, L.L.: Approach for uncertainty propagation and robust design in CFD using sensitivity derivatives. Technical Report No. 2528, AIAA (2001)
7. Shankaran, S., Jameson, A.: Robust optimal control using polynomial chaos and adjoints for systems with uncertain output. Technical Report No. 3069, AIAA. In: 20th AIAA Computational Fluid Dynamics Conference, Honolulu, Hawaii, 27–30 June 2011

Algorithmic Differentiation for Second Derivatives

Alain Dervieux

Introduction

In this section, we study two methods for obtaining second derivatives of a functional subject to a state equation. We introduce analytic formulas which can then be used in combination with a non-global application of an Automatic Differentiation (AD) tool like TAPENADE [7]. This second step is not addressed here but will be presented in details in chapter "Use of Automatic Differentiation Tools at the Example of TAPENADE".

More precisely, we recall that we are interested by obtaining the first and second derivatives of a functional j depending of $\gamma \in \mathbb{R}^n$ and expressed in terms of a state $W \in \mathbb{R}^N$ as follows:

$$\begin{cases} \psi(\gamma) = \Psi(\gamma, W(\gamma)) = 0 \\ j(\gamma) = J(\gamma, W(\gamma)). \end{cases} \quad (1)$$

We observe that $\gamma \to W(\gamma)$ is a function *implicitly* defined through the state equation $\Psi(\gamma, W(\gamma)) = 0$ and the functional $j(\gamma) = J(\gamma, W(\gamma))$ is evaluated from the solution of the state equation corresponding to γ. Then, we distinguish two different points of view:

- *Implicit differentiation:* it consists to differentiate directly the *implicit* function j as a function of the control variable γ. This means that the entire program involving the solution algorithm for state equation is differentiated in one way.
- *Differentiation of explicit parts:* the second point of view is to apply differentiation only to routines which compute *explicit* functions, that is functions Ψ and J and then use one of the available composite derivative formula in order to get $dj/d\gamma$.

A. Dervieux (✉)
Ecuador, INRIA, 2002 Route des Lucioles, 06902 Sophia-Antipolis, France
e-mail: alain.dervieux@inria.fr

C. Hirsch et al. (eds.), *Uncertainty Management for Robust Industrial Design in Aeronautics*, Notes on Numerical Fluid Mechanics and Multidisciplinary Design 140, https://doi.org/10.1007/978-3-319-77767-2_21

Also, two different formulas of composite derivatives can be applied:

- *The Tangent mode* applied to a program computing Φ produces another routine computing, from u and from an arbitrary direction $\dot{u}$ (of same dimension as u), the derivative in direction $\dot{u}$:

$$u, \dot{u} \mapsto \frac{\partial \Phi}{\partial u}(u)\dot{u} = \phi_p'(w_{p-1})\phi_{p-1}'(w_{p-2}) \cdots \phi_1'(w_0)\dot{u}. \tag{2}$$

- *The Reverse mode* when applied to the previous program computing Φ produces a routine which computes, from u and from an arbitrary direction $\bar{v}$ (of same dimension as v), the following product of same dimension as u:

$$u, \bar{v} \mapsto \left(\frac{\partial \Phi}{\partial u}(u)\right)^* \bar{v} = \phi_1'^*(w_0)\phi_{p-1}'^*(w_{p-2}) \cdots \phi_p'^*(w_{p-1})\bar{v}. \tag{3}$$

If we choose an implicit differentiation, differentiating the entire routine j can be performed with either Tangent or Reverse mode of the AD tool; see chapter "Surrogate Model-Based Approaches to UQ and Their Range of Applicability". It permits to obtain directly a differentiated program, in a black box manner, with the risk that this program has a rather good reliability but not sufficiently good performances.

To analyze this issue, we observe that since the routine computing the functional j contains the iterative solver method for the state equation, the differentiated routines will contain this solver in differentiated form. We assume that we need n_{iter} iterations in order to obtain the nonlinear solution, and that for each iteration, we have an unitary cost.

Tangent mode produces a program that we need to apply n time for computing the entire gradient. The cost is $n(n_{\text{iter}}\alpha_T)$ where α_T is the overhead associated with the differentiated code with respect to the original one. One has usually $1 < \alpha_T < 4$; see for example [5]. Further, the memory requirements will be about the same as for the undifferentiated code.

With Reverse mode, we are able to obtain the entire gradient with a single evaluation of the differentiated routine. But the usual Reverse mode produces a code which involves two successive parts:

- a *forward sweep* close to original code, and
- a (*backward sweep*) performed in the opposite order of the original code.

The problem is that the *backward sweep* needs data computed in the *forward sweep*, but in the opposite order, that is from last data computed in the *forward sweep* to first ones. In one way to solve this, the Store-All (SA) strategy, these data are stored during the *forward sweep*.

The total cost (in terms of CPU time and memory) strongly depends on the strategy applied by the AD tool to solve the problem of inverse order differentiation for the original routine. For the case of the SA strategy, the CPU cost will be $(n_{\text{iter}}\alpha_R)$ with $1 < \alpha_R < 5$, i.e., α_R times the undifferentiated code, but the required memory will be n times greater. For a Recompute-All (RA) strategy, the CPU cost will be

$(n_{\text{iter}}^2 \alpha_R)$, i.e., $(n_{\text{iter}} \alpha_R)$ the nonlinear solution, but the memory will be the same of the undifferentiated routine. For real large programs, neither SA nor RA strategy can work, we need a special storage/recomputation trade-off in order to be efficient using *checkpoints* (see [7]). Obviously, with checkpointing, the CPU cost will be greater than the cost of SA strategy and it can be shown [6] that the cost for the differentiated code can be of the order of $\sqrt[s]{n_{\text{iter}}}$ (where s is the number of snapshots available).

In many cases, these strategies (SA, RA, or SA/RA) are compulsory. This is the case for unsteady nonlinear systems, where main iteration is time advancing, and for which we do not know a strategy working without the intermediate-time state variable values. *Checkpointing* can be applied quasi-automatically by the Automatic Differentiator or applied by hand-coding.

When the iterative algorithm is a fixed point one, with enough convergence to the fixed point, only the final state variable is necessary for *backward sweep*. Some intervention of the programmer is useful for avoiding the storage of unnecessary intermediate values of state variable.

Lastly, the behavior (efficiency, stability) of differentiated *solution algorithms* is questionable, while a natural and conservative tendancy is to re-use the algorithm used for state—preconditioner, iterator—for solving adjoint systems.

From the previous arguments, in case of simple explicit solvers, the implicit differentiation is a good option, but for more sophisticated solvers (GMRES, ...) we recommand the differentiation of explicit parts/routines, followed by an assembly of these by the programmer.

We present now in more details this latter strategy.

Computation of First Derivative

Using the chain rule, the gradient of the functional $j(\gamma) = J(\gamma, W(\gamma))$ is given by

$$\frac{dj}{d\gamma} = \frac{\partial J}{\partial \gamma} + \frac{\partial J}{\partial W}\frac{dW}{d\gamma}$$

where the derivatives of the state variables $W(\gamma)$ are obtained solving the linear system

$$\frac{d\psi}{d\gamma} = \frac{\partial \Psi}{\partial \gamma} + \frac{\partial \Psi}{\partial W}\frac{dW}{d\gamma} = 0.$$

Two strategies can be applied.

Direct Differentiation (Tangent Mode)

It consists in computing the Gâteaux or directional derivatives with respect to each component direction ($e_i = (0, \dots 0, 1, 0, \dots, 0)^T$, where 1 is at the i-th component):

$$\frac{dj}{d\gamma_i} = \frac{dj}{d\gamma} e_i = \frac{\partial J}{\partial \gamma_i} + \frac{\partial J}{\partial W} \frac{dW}{d\gamma_i} \tag{4}$$

with:

$$\frac{\partial \Psi}{\partial W} \frac{dW}{d\gamma_i} = -\frac{\partial \Psi}{\partial \gamma} e_i \tag{5}$$

This has to be applied to each component of γ, i.e., n times and the cost is n linearised N-dimensional systems to solve. If we choose to solve the single system (5) with an iterative matrix-free method, and the solution is obtained after $n_{\mathrm{iter},T}$ step, the total cost will be of the order of $\alpha_T n_{\mathrm{iter},T}$, i.e., $n_{\mathrm{iter},T}$ evaluation of the matrix-by-vector operation $\left(\frac{\partial \Psi}{\partial W}\right)x$, where each evaluation costs α_T times the evaluation of the state residual $\Psi(\gamma, W)$ (and the cost of the state residual is taken as reference equal to 1). Therefore, the cost of the full gradient will be $n\alpha_T n_{\mathrm{iter},T}$.

Inverse Differentiation (Reverse Mode)

The complete gradient is given by the equation

$$\left(\frac{dj}{d\gamma}\right)^* = \left(\frac{\partial J}{\partial \gamma}\right)^* - \left(\frac{\partial \Psi}{\partial \gamma}\right)^* \Pi_0 \tag{6}$$

where Π_0 is the solution of the linear system

$$\left(\frac{\partial \Psi}{\partial W}\right)^* \Pi = \left(\frac{\partial J}{\partial W}\right)^* . \tag{7}$$

This computation needs only one extra linearised N-dimensional system, the adjoint system (some methods for calculation of the adjoint solutions are described in [4]). If we choose to solve the adjoint system (7) with an iterative matrix-free method, we can apply the same estimate done as in the case of the Tangent mode differentiation, but this time the overhead associated with the evaluation of the matrix-by-vector operation $\left(\frac{\partial \Psi}{\partial W}\right)^* x$ with respect to the state residual evaluation will be α_R (and usually $\alpha_R > \alpha_T$) and the number of iteration $n_{\mathrm{iter},R}$ for the convergence of the solution could be different from $n_{\mathrm{iter},T}$ of the previous case (but the asymptotical rate of convergence will be the same of the original linear system $\left(\frac{\partial \Psi}{\partial W}\right)x = b$, see [4]). Therefore, the cost for the gradient will be $\alpha_R n_{\mathrm{iter},R}$, and the Reverse mode differentiation for the gradient computation is cheaper than the Tangent mode if $n \gg 1$.

For second derivatives, we have different possibilities.

Second Derivative by Tangent on Tangent

These methods were initially investigated by [12] along with various other algorithms, but the publication does not go into the implementation details for a generic fluid dynamic code. Here, we present the mathematical background behind the idea and the efficient AD implementation of Ghate and Giles [3] but with a different analysis of the computational cost.

Main Property

Starting from the derivative (4), we perform another differentiation with respect to the variable γ_k obtaining

$$\frac{d^2j}{d\gamma_i d\gamma_k} = D^2_{i,k}J + \frac{\partial J}{\partial W}\frac{d^2W}{d\gamma_i d\gamma_k} \tag{8}$$

where

$$D^2_{i,k}J = \frac{\partial}{\partial \gamma}\left(\frac{\partial J}{\partial \gamma}e_i\right)e_k + \frac{\partial}{\partial W}\left(\frac{\partial J}{\partial \gamma}e_i\right)\frac{dW}{d\gamma_k} + \frac{\partial}{\partial W}\left(\frac{\partial J}{\partial \gamma}e_k\right)\frac{dW}{d\gamma_i} + \frac{\partial}{\partial W}\left(\frac{\partial J}{\partial W}\frac{dW}{d\gamma_i}\right)\frac{dW}{d\gamma_k}.$$

Differentiating Eq. (5), we get

$$D^2_{i,k}\Psi + \frac{\partial \Psi}{\partial W}\frac{d^2W}{d\gamma_i d\gamma_k} = 0 \tag{9}$$

where

$$D^2_{i,k}\Psi = \frac{\partial}{\partial \gamma}\left(\frac{\partial \Psi}{\partial \gamma}e_i\right)e_k + \frac{\partial}{\partial W}\left(\frac{\partial \Psi}{\partial \gamma}e_i\right)\frac{dW}{d\gamma_k} + \frac{\partial}{\partial W}\left(\frac{\partial \Psi}{\partial \gamma}e_k\right)\frac{dW}{d\gamma_i} + \frac{\partial}{\partial W}\left(\frac{\partial \Psi}{\partial W}\frac{dW}{d\gamma_i}\right)\frac{dW}{d\gamma_k}.$$

Substituting the second derivatives of the state with respect to the control variables $\frac{d^2W}{d\gamma_i d\gamma_k}$ in Eq. (8) from Eq. (9), we get

$$\begin{aligned}\frac{d^2j}{d\gamma_i d\gamma_k} &= D^2_{i,k}J - \frac{\partial J}{\partial W}\left(\frac{\partial \Psi}{\partial W}\right)^{-1} D^2_{i,k}\Psi \\ &= D^2_{i,k}J - \Pi_0^* D^2_{i,k}\Psi\end{aligned} \tag{10}$$

where Π_0 is the solution of the adjoint system (7) evaluated at the point $(\gamma, W(\gamma))$ solution of the state equation $\Psi(\gamma, W) = 0$.

The n derivatives $\frac{dW}{d\gamma_i}$ should be computed (and stored) using tangent mode differentiation of the nonlinear solver algorithm, and each derivative costs $n_{\text{iter},T}\alpha_T$. If we need the full Hessian matrix, we have to evaluate the quantity (10) $n(n+1)/2$ times; i.e., we have to evaluate the terms $D^2_{i,k}\Psi$ and $D^2_{i,k}J$ for $i = 1, \ldots, n$ and $j = i, \ldots, n$ due to the symmetry of the Hessian, and each evaluation of $D^2_{i,k}\Psi$ costs α_T^2 (the evaluation of $D^2_{i,k}J$ is negligible with respect to $D^2_{i,k}\Psi$). Therefore, the full Hessian costs $n\alpha_T[n_{\text{iter},T} + (n+1)\alpha_T/2]$. With similar arguments, if we want only the diagonal part of the Hessian, the cost is $n\alpha_T[n_{\text{iter},T} + \alpha_T]$.

Global ToT Algorithm

Solve

$$\begin{array}{c|c} \Psi(\gamma, W) = 0 & n_{\text{iter}}\Psi \\ \left(\frac{\partial \Psi}{\partial W}\right)^* \Pi_0 = \left(\frac{\partial J}{\partial W}\right)^* & n_{\text{iter}}\alpha_R\Psi + \alpha_R J \end{array}$$

For $i = 1, \ldots, n$ solve

$$\frac{\partial \Psi}{\partial W}\frac{dW}{d\gamma_i} = -\frac{\partial \Psi}{\partial \gamma}e_i \;\Big|\; (n_{\text{iter}} + 1)\alpha_T\Psi$$

For $i = 1, \ldots, n$ and $j = 1, \ldots, i$

$$\begin{array}{l|l} \frac{\partial}{\partial \gamma}\left(\frac{\partial J}{\partial \gamma}e_i\right)e_j & \alpha_T^2 J \\ \frac{\partial}{\partial \gamma}\left(\frac{\partial J}{\partial W}\frac{dW}{d\gamma_j}\right)e_i & \alpha_T^2 J \\ \frac{\partial}{\partial W}\left(\frac{\partial J}{\partial W}\frac{dW}{d\gamma_j}\right)\frac{dW}{d\gamma_i} & \alpha_T^2 J \\ \hline \frac{\partial}{\partial \gamma}\left(\frac{\partial \Psi}{\partial \gamma}e_i\right)e_j & \alpha_T^2 \Psi \\ \frac{\partial}{\partial \gamma}\left(\frac{\partial \Psi}{\partial W}\frac{dW}{d\gamma_j}\right)e_i & \alpha_T^2 \Psi \\ \frac{\partial}{\partial W}\left(\frac{\partial \Psi}{\partial W}\frac{dW}{d\gamma_j}\right)\frac{dW}{d\gamma_i} & \alpha_T^2 \Psi \end{array}$$

Second Derivative by Tangent on Reverse

This consists in the direct derivation in any direction $e_i, i = 1, n$ of the (non-scalar) function:

$$\left(\frac{\partial j}{\partial \gamma}\right)^*(\gamma, W(\gamma)) = \left(\frac{\partial J}{\partial \gamma}\right)^*(\gamma, W(\gamma)) - \left(\frac{\partial \Psi}{\partial \gamma}\right)^* \Pi(\gamma, W(\gamma))$$

where $W(\gamma)$ and $\Pi(\gamma, W(\gamma))$ are solutions of the above two state systems.

Main Property

The mains computations to apply are summed up in the following lemma:

Lemma *the second derivatives can be computed as:*

$$\boxed{\begin{aligned}\frac{\partial}{\partial \gamma_i}\left(\frac{\partial j}{\partial \gamma}\right)^* = \left(\frac{\partial^2 j}{\partial \gamma^2}\right) e_i = \frac{\partial}{\partial \gamma}\left(\frac{\partial J}{\partial \gamma}\right)^* e_i + \frac{\partial}{\partial W}\left(\frac{\partial J}{\partial \gamma}\right)^* \frac{dW}{d\gamma_i} + \\ - \frac{\partial}{\partial \gamma}\left[\left(\frac{\partial \Psi}{\partial \gamma}\right)^* \Pi_0\right] e_i - \frac{\partial}{\partial W}\left[\left(\frac{\partial \Psi}{\partial \gamma}\right)^* \Pi_0\right] \frac{dW}{d\gamma_i} - \left(\frac{\partial \Psi}{\partial \gamma}\right)^* \lambda_i\end{aligned}} \tag{11}$$

For each $i = 1, \ldots, n$, Eq. (11) needs Π_0, the solution of the adjoint system

$$\left(\frac{\partial \Psi}{\partial W}\right)^* \Pi = \left(\frac{\partial J}{\partial W}\right)^* \tag{12}$$

and two perturbed N-dimensional linear systems:

$$\begin{cases} \dfrac{\partial \Psi}{\partial W}\dfrac{dW}{d\gamma_i} = -\dfrac{\partial \Psi}{\partial \gamma} e_i \\ \left(\dfrac{\partial \Psi}{\partial W}\right)^* \lambda_i = \dfrac{\partial}{\partial \gamma}\left(\dfrac{\partial J}{\partial W}\right)^* e_i + \dfrac{\partial}{\partial W}\left(\dfrac{\partial J}{\partial W}\right)^* \dfrac{dW}{d\gamma_i} + \\ \qquad - \dfrac{\partial}{\partial \gamma}\left[\left(\dfrac{\partial \Psi}{\partial W}\right)^* \Pi_0\right] e_i - \dfrac{\partial}{\partial W}\left[\left(\dfrac{\partial \Psi}{\partial W}\right)^* \Pi_0\right] \dfrac{dW}{d\gamma_i} \end{cases} \tag{13}$$

where all the functions in Eqs. (11)–(13) are evalued at the final state (in order to verify $\Psi(\gamma, W(\gamma)) = 0$). In addition, the second linear system in (13) is of the same type of the adjoint system (12) but with a different right-hand side, so we can use the same algorithm (but with different initial data) for both equations.

It is useful to note that Eq. (11) gives us an entire column (or row, by symmetry) of the Hessian matrix, where the Tangent-on-Tangent approach (10) gives us a single element. Another interesting point is that the Tangent-on-Reverse method does not requires to know in advance all the derivatives of the state variables respect to the control; i.e., we can compute the derivative $\frac{dW}{d\gamma_i}$ for a given index i, then use it in Eqs. (11) and (13) to obtain the i-th column (or row) of the Hessian and then we can avoid to store it for the next index $i+1$.

The cost associated to the full Hessian evaluated with the Tangent-on-Reverse approach is

- $n(\alpha_T n_{\text{iter},T})$ for computing the derivatives of the state variables respect to the control $\frac{dW}{d\gamma_i}$ $(i=1,\ldots,n)$
- $n\alpha_R\alpha_T$ for evaluating the quantities on the right-hand side of Eqs. (11) and (13)
- $n(\alpha_R n_{\text{iter},R})$ for solving the n adjoint systems in the (13)

Therefore, the full Hessian costs $n[\alpha_T n_{\text{iter},T}+\alpha_R\alpha_T+\alpha_R n_{\text{iter},R}]$.

Details of Calculations

First of all, we note that the tangent derivative along the direction δ of a (n-dimensional) function $f(\gamma)=F(\gamma,W)$ is

$$\begin{aligned}\frac{df}{d\gamma}\delta &= \frac{\partial F}{\partial\gamma}\delta+\frac{\partial F}{\partial W}\frac{dW}{d\gamma}\delta\\ &= \frac{\partial F}{\partial\gamma}\delta-\frac{\partial F}{\partial W}\left(\frac{\partial\Psi}{\partial W}\right)^{-1}\frac{\partial\Psi}{\partial\gamma}\delta\\ &= \frac{\partial F}{\partial\gamma}\delta+\frac{\partial F}{\partial W}\theta\end{aligned}$$

where θ is the solution of the linear system

$$\frac{\partial\Psi}{\partial W}\theta=-\frac{\partial\Psi}{\partial\gamma}\delta.$$

Now, we can perform the tangent derivative (along the direction δ) of $\left(\frac{dj}{d\gamma}\right)^*$

$$\left(\frac{d^2j}{d\gamma^2}\right)\delta=\frac{d}{d\gamma}\left(\frac{dj}{d\gamma}\right)^*\delta=\frac{d}{d\gamma}\left(\frac{\partial J}{\partial\gamma}\right)^*\delta-\frac{d}{d\gamma}\left[\left(\frac{\partial\Psi}{\partial\gamma}\right)^*\Pi\right]\delta.$$

The first term is

$$\frac{d}{d\gamma}\left(\frac{\partial J}{\partial\gamma}\right)^*\delta=\frac{\partial}{\partial\gamma}\left(\frac{\partial J}{\partial\gamma}\right)^*\delta+\frac{\partial}{\partial W}\left(\frac{\partial J}{\partial\gamma}\right)^*\theta.$$

The second one is a little more complex

$$\frac{d}{d\gamma}\left[\left(\frac{\partial\Psi}{\partial\gamma}\right)^*\Pi\right]\delta = \left[\frac{d}{d\gamma}\left(\frac{\partial\Psi}{\partial\gamma}\right)^*\Pi\right]\delta + \left(\frac{\partial\Psi}{\partial\gamma}\right)^*\frac{d\Pi}{d\gamma}\delta$$
$$= \frac{d}{d\gamma}\left[\left(\frac{\partial\Psi}{\partial\gamma}\right)^*\Pi_0\right]\delta + \left(\frac{\partial\Psi}{\partial\gamma}\right)^*\lambda$$
$$= \frac{\partial}{\partial\gamma}\left[\left(\frac{\partial\Psi}{\partial\gamma}\right)^*\Pi_0\right]\delta + \frac{\partial}{\partial W}\left[\left(\frac{\partial\Psi}{\partial\gamma}\right)^*\Pi_0\right]\theta + \left(\frac{\partial\Psi}{\partial\gamma}\right)^*\lambda$$

where Π_0 is the solution of the linear system (solved at the final state)

$$\left(\frac{\partial\Psi}{\partial W}\right)^*\Pi = \left(\frac{\partial J}{\partial W}\right)^* \tag{14}$$

and $\lambda = \frac{d\Pi}{d\gamma}\delta$. Now, we perform a tangent derivative of the adjoint Eq. (14)

$$\frac{d}{d\gamma}\left[\left(\frac{\partial\Psi}{\partial W}\right)^*\Pi\right]\delta = \frac{d}{d\gamma}\left[\left(\frac{\partial\Psi}{\partial W}\right)^*\Pi_0\right]\delta + \left(\frac{\partial\Psi}{\partial W}\right)^*\frac{d\Pi}{d\gamma}\delta$$
$$= \frac{\partial}{\partial\gamma}\left[\left(\frac{\partial\Psi}{\partial W}\right)^*\Pi_0\right]\delta + \frac{\partial}{\partial W}\left[\left(\frac{\partial\Psi}{\partial W}\right)^*\Pi_0\right]\theta + \left(\frac{\partial\Psi}{\partial W}\right)^*\lambda$$
$$= \frac{\partial}{\partial\gamma}\left(\frac{\partial J}{\partial W}\right)^*\delta + \frac{\partial}{\partial W}\left(\frac{\partial J}{\partial W}\right)^*\theta.$$

So λ is the solution of the linear system

$$\left(\frac{\partial\Psi}{\partial W}\right)^*\lambda = \frac{\partial}{\partial\gamma}\left(\frac{\partial J}{\partial W}\right)^*\delta + \frac{\partial}{\partial W}\left(\frac{\partial J}{\partial W}\right)^*\theta +$$
$$- \frac{\partial}{\partial\gamma}\left[\left(\frac{\partial\Psi}{\partial W}\right)^*\Pi_0\right]\delta - \frac{\partial}{\partial W}\left[\left(\frac{\partial\Psi}{\partial W}\right)^*\Pi_0\right]\theta.$$

Reassembling all the terms, we have that the projection of the Hessian in a direction δ is given by

$$\boxed{\begin{aligned}\left(\frac{d^2j}{d\gamma^2}\right)\delta = \frac{\partial}{\partial\gamma}\left(\frac{\partial J}{\partial\gamma}\right)^*\delta + \frac{\partial}{\partial W}\left(\frac{\partial J}{\partial\gamma}\right)^*\theta +\\ - \frac{\partial}{\partial\gamma}\left[\left(\frac{\partial\Psi}{\partial\gamma}\right)^*\Pi_0\right]\delta - \frac{\partial}{\partial W}\left[\left(\frac{\partial\Psi}{\partial\gamma}\right)^*\Pi_0\right]\theta - \left(\frac{\partial\Psi}{\partial\gamma}\right)^*\lambda\end{aligned}}$$

where

$$\begin{cases} \left(\frac{\partial \Psi}{\partial W}\right)^* \Pi_0 = \left(\frac{\partial J}{\partial W}\right)^* \\ \qquad \frac{\partial \Psi}{\partial W}\theta = -\frac{\partial \Psi}{\partial \gamma}\delta \\ \left(\frac{\partial \Psi}{\partial W}\right)^* \lambda = \frac{\partial}{\partial \gamma}\left(\frac{\partial J}{\partial W}\right)^* \delta + \frac{\partial}{\partial W}\left(\frac{\partial J}{\partial W}\right)^* \theta + \\ \qquad\qquad - \frac{\partial}{\partial \gamma}\left[\left(\frac{\partial \Psi}{\partial W}\right)^* \Pi_0\right]\delta - \frac{\partial}{\partial W}\left[\left(\frac{\partial \Psi}{\partial W}\right)^* \Pi_0\right]\theta \,. \end{cases}$$

Global ToR Algorithm

Solve

$$\begin{array}{l|c} \Psi(\gamma, W) = 0 & n_{\text{iter}}\Psi \\ \left(\frac{\partial \Psi}{\partial W}\right)^* \Pi_0 = \left(\frac{\partial J}{\partial W}\right)^* & n_{\text{iter}}\alpha_R\Psi + \alpha_R J \end{array}$$

For $i = 1, \ldots, n$

$$\text{solve} \quad \frac{\partial \Psi}{\partial W}\frac{dW}{d\gamma_i} = -\frac{\partial \Psi}{\partial \gamma}e_i \,\Big|\, (n_{\text{iter}} + 1)\alpha_T\Psi$$

$$\begin{array}{l|l} \dot{\bar{J}}_W & \alpha_T\alpha_R J \\ \dot{\bar{J}}_\gamma & \alpha_T\alpha_R J \\ \dot{\bar{\Psi}}_W & \alpha_T\alpha_R\Psi \\ \dot{\bar{\Psi}}_\gamma & \alpha_T\alpha_R\Psi \end{array}$$

$$\text{solve} \quad \left(\frac{\partial \Psi}{\partial W}\right)^* \lambda_i = \dot{\bar{J}}_W - \dot{\bar{\Psi}}_W \,\Big|\, n_{\text{iter}}\alpha_R\Psi$$

$$\dot{\bar{J}}_\gamma - \dot{\bar{\Psi}}_\gamma - \left(\frac{\partial \Psi}{\partial \gamma}\right)^* \lambda_i \,\Big|\, \alpha_R\Psi$$

Comparison Between ToT and ToR

At this point, the natural question arising from the previous analysis is about the choice of the method that is less expensive for a given problem. We have seen that the algorithms for ToT and ToR approaches have some common parts (the solution of the adjoint system and the evaluation of the state derivatives), so the comparison of the two approach should be done in cost of the part that characterize the two algorithms:

- $\frac{n(n+1)}{2}\alpha_T^2$ for ToT
- $n\alpha_R[\alpha_T + n_{\text{iter},R}]$ for ToR

For a very large number of variables, hundreds in practical cases, the ToR can be more efficient since its complexity is not quadratic. In practice, the ToT is more performant.

Remark If we are interested in a (scalar!) functional depending on the gradient, then it can be interesting to apply a second inverse differentiation. We do not consider here this direction. □

Concluding Remarks

This section has presented the main lines of the second differentiation of a functional subject to a state equation.

The next step in the obtention of a second-order Taylor surogate software model is the effective implementation of these second derivatives. That step is presented in details in chapter "Use of Automatic Differentiation Tools at the Example of TAPENADE".

Many other informations concerning the method and concerning application to robuste optimization can be found in [1, 2, 8–11].

References

1. Belme, A., Martinelli, M., Hascoët, L., Pascual, V., Dervieux, A.: AD-based perturbation methods for uncertainties and errors. Int. J. Eng. Syst. Model. Simul. 2(1–2), (2010)
2. Duvigneau, R., Martinelli, M., Chandrashekarappa, P.: Uncertainty quantification for robust design. In: Breitkopf, P., Coelho, R.F. (eds.) Multidisciplinary Design Optimization in Computational Mechanics. Wiley, ISTE (2010)
3. Ghate, D., Giles, M.B.: Efficient hessian calculation using automatic differentiation. In: 25th Applied Aerodynamics Conference AIAA, Miami (florida), pp. 4059 (2007)
4. Giles, M.B., Duta, M.C., Muller, J.D., Pierce, N.A.: Algorithm developments for discrete adjoint methods. AIAA J. **41**(2), (2003)

5. Griewank, A.: Evaluating derivatives principles and techniques of algorithmic differentiation, SIAM. Front. Appl. Math. **19**, (2000)
6. Griewank, A.: Achieving logarithmic growth of temporal and spatial complexity in reverse automatic differentiation. Optim. Methods Softw. **1**, 35–54 (1992)
7. Hascoët, L., Pascual, V.: TAPENADE 2.1 user's guide. Technical Report 0300, INRIA, September 2004
8. Martinelli M., Duvigneau, R.: On the use of second-order derivative and metamodel-based Monte-Carlo for uncertainty estimation in aerodynamics. Comput Fluids **37**(6), (2010)
9. Martinelli, M., Duvigneau, R.: Comparison of second-order derivatives and metamodel-based Monte-Carlo approaches to estimate statistics for robust design of a transonic wing. In: 10th AIAA Non-Deterministic Approaches Conference, April 2008, pp. 2008–2071. Schaumburg, USA, AIAA (2008)
10. Martinelli, M.: Sensitivity evaluation in aerodynamic optimal design. Ph.D thesis, Scuola Normale di Pisa, Italy (2007)
11. Martinelli, M., Hascoët, L.: Tangent-on-tangent vs. tangent-on-reverse for second differentiation of constrained functionals. Lect. Notes Comput. Sci. Eng. **64**, 151–161 (2008)
12. Sherman, L.L., Taylor III, A.C., Green, L.L., Newman, P.A.: First and second-order aerodynamic sensitivity derivatives via automatic differentiation with incremental iterative methods. J. Comput. Phys. **129**, 307–331 (1996)

Second-Order Derivatives for Geometrical Uncertainties

Marcin Wyrozębski, Łukasz Łaniewski-Wołłk and Jacek Rokicki

Introduction

Today, computational fluid dynamics (CFD) methods form a very important part of a modern design process. CFD solvers have reached high level of maturity and they are able to simulate even very complex flows. As the computing power constantly increases, the interest in the shape optimization methods starts to grow rapidly. The introduction of adjoint method and their application for the computation of the gradient and for the sensitivity analysis allows for an important reduction in time required for the optimization process [1–3]. Further development is possible through application of second-order sensitivity analysis [4–6]. This new method takes advantage of using the Hessian matrix within the optimization process.

In the classical approach, the simulated models are treated as fully deterministic, while the real, manufactured products might differ from the designed version. Moreover, computations are often carried out only for a specified operational conditions, while they might be variable in real-life scenarios. Apart from the variability of operational conditions, variability of the geometry itself has a significant impact on the system performance. Therefore, this variability should be included in the design process.

The geometrical uncertainty can be regarded as a continuous field, that has to be somehow discretized. In our approach, we propose to use for this purpose the existing CFD surface mesh. This leads to a large number of correlated uncertainties corre-

M. Wyrozębski (✉) · Ł. Łaniewski-Wołłk · J. Rokicki
Department of Aerodynamics, Institute of Aeronautics and Applied Mechanics,
ul. Nowowiejska 24, 00-665 Warsaw, Poland
e-mail: mwyrozebski@meil.pw.edu.pl

Ł. Łaniewski-Wołłk
e-mail: llaniewski@meil.pw.edu.pl

J. Rokicki
e-mail: jack@meil.pw.edu.pl

C. Hirsch et al. (eds.), *Uncertainty Management for Robust Industrial Design in Aeronautics*, Notes on Numerical Fluid Mechanics and Multidisciplinary Design 140, https://doi.org/10.1007/978-3-319-77767-2_22

sponding to the normal shift of the surface nodes. The resulting set of uncertainties is too large to be analyzed in a reasonable time with such methods as Polynomial Chaos [7] or Meta-Modeling [8].

Handling of such uncertainties is frequently done with the help of some reduction techniques. Most of these techniques, e.g., Karhunen-Loeve (KL) decomposition, are based on the correlation structure [9]. Some, such as Active Subspace [10, 11], gain from the use of the gradient information, requiring however the Monte Carlo sampling which increases computational complexity. All these methods provide a new basis in a linear space, which is optimal for discretization and quantification of the given uncertainty field. The presented approach has the same goal, but takes into account the information both from the correlation structure as well as from the local behavior of the objective near a selected design point.

Still different approach that is presented in this paper is called a Method of Moments (MoM) and allows to compute some stochastic parameters of the objective function basing on its local behavior. In order to do that, it uses Taylor series expansion, and therefore, computation of the derivatives of the objective is required. In CFD applications, the adjoint method is used in order to obtain sensitivity information. However, in most cases MoM methods require not only the gradient of the objective but also its Hessian. Martinelli and Duvigneau [12] present an application of the second-order Method of Moments to an uncertainty quantification test case, considering free-stream Mach number and angle of attack as random variables. In this paper, we are using similar approach, but for a very large set of geometrical uncertainties.

Methodology

In this Section, we present a novel approach for geometrical uncertainties quantification based on the information on the second derivatives. First, the details of the method are presented, and subsequently, the equations for the required Hessian matrix computations are derived. Finally, we provide a method to reduce the set of nodal uncertainties using for this purpose the Radial Basis Functions.

Proposed Method

Let us now assume that the discretised random field ζ has a zero mean value $\mathrm{E}[\zeta] = 0$ and the covariance matrix $C_{ij} = \mathrm{E}[\zeta_i \zeta_j]$. Let us also assume that the analyzed objective function f depends on the set of deterministic parameters x describing the geometry for which ζ forms an additive uncertainty defining geometry variability. Our main goal is the quantification of the random variable $f(x + \zeta)$. The expected value of the objective with respect to the uncertainties can be written, using Taylor series, as:

$$\begin{aligned}\mathrm{E}[f(x+h\zeta)] &= \mathrm{E}\left[f(x) + h\zeta_i\frac{\partial f}{\partial x_i} + \frac{1}{2}h^2\zeta_i\zeta_j\frac{\partial^2 f}{\partial x_i \partial x_j} + o(h^3)\right] \\ &= f(x) + h\frac{\partial f}{\partial x_i}\mathrm{E}\left[\zeta_i\right] + \frac{1}{2}h^2\frac{\partial^2 f}{\partial x_i \partial x_j}\mathrm{E}\left[\zeta_i\zeta_j\right] + o(h^3)\end{aligned}$$

The substitution of the mean and the covariance of ζ ($C_{ij} = \mathrm{E}[\zeta_i\zeta_j]$) gives:

$$\mathrm{E}[f(x+h\zeta)] = f(x) + \frac{1}{2}h^2 C_{ij}\frac{\partial^2 f}{\partial x_i \partial x_j} + o(h^3) \tag{1}$$

Repeating the same operation for the variance gives:

$$\mathrm{Var}^2 f(x+h\zeta) = h^2\frac{\partial f}{\partial x_i}\frac{\partial f}{\partial x_j}C_{ij} + o(h^3)$$

In computational practice, we can find the deterministic value of the objective $f(x)$ by converging the discrete CFD solution. The covariance matrix C_{ij} is (by assumption) known, e.g., from sampling measurements. The remaining element is the evaluation of the second derivative Hessian matrix $\frac{\partial^2 f}{\partial x_i \partial x_j}$. Unfortunately, the dimension of the full Hessian matrix is extremely high being proportional to the number of surface nodes of the mesh.

Therefore, the proposed approach is based on the idea to reduce the number of analyzed uncertainties. This reduction process has to approximate the expected value as accurately as possible. Considering the second-order term in the formula (1), we notice that the product of the two matrices can be intepreted as a trace of their product:

$$C_{ij}\frac{\partial^2 f}{\partial x_i \partial x_j} = \mathrm{Tr}(\mathrm{CH}) = \sum_{\mathrm{p}} \lambda_{\mathrm{p}}$$

where λ_p are the eigenvalues of the generalized eigenvalue problem:

$$\frac{\partial^2 f}{\partial x_i \partial x_j}v_p = \lambda_p C^{-1} v_p \tag{2}$$

This in turn gives a straightforward way to approximate the expected value (1) by calculating the eigenvalues starting from the largest one (highest absolute value).

As the multiplication $\frac{\partial^2 f}{\partial x_i \partial x_j}v$ has a cost of the same level as a standard adjoint calculation, the cost of solving the eigenvalue problem (2) can be greatly reduced by using iterative methods such as the Power Method or the Inverse Power Method. The total cost of the proposed algorithm is proportional to the number of analyzed eigenvalues; therefore, the trade-off between accuracy and the computational cost has to be considered. Additionally, the eigenvectors of (2) can be used as candidates for the reduction of the set of geometrical uncertainties for other UQ methods such as Monte Carlo.

As eigenvectors of (2) are orthogonal to both the Hessian and the covariance matrix, their use can be very beneficial. This second-order cross-derivatives expressed in such basis are zero. Thus, the number of coefficients that has to be found in the polynomial approximation is reduced by an order of magnitude and this can be useful for the Polynomial Chaos and the Meta-Modeling methods. The orthogonality with respect to the covariance matrix means that the uncertainties are independent, which is a significant assumption for many UQ methods.

Hessian Matrix Computation

The ability to efficiently calculate the directional second-order derivative of an objective function is very important for the presented approach. By the directional derivative, we understand $\frac{\partial^2 J}{\partial x_i \partial x_j}\beta_j$, where β is an arbitrary vector. The applied method is called "Tangent-on-Reverse" and is presented in detail in chapter "Introduction to Intrusive Perturbation Methods." In this Section, we briefly present the resulting algorithm.

Assume we have a state vector $u(\alpha)$ depending on a vector of parameters α. Let the objective function be defined as (e.g., lift force) $J(u, \alpha)$ and the governing equation (discretisation's residual of the, e.g., Navier–Stokes equation) as $R(u) = 0$.

To efficiently calculate the second-order derivative, one can apply the following steps:

1. Find the flow solution u by solving the primal equation

$$R(u) = 0 \tag{3}$$

2. Find the adjoint solution v by solving the adjoint equation

$$\frac{\partial R_i}{\partial u_j} v_i = -\frac{\partial J}{\partial u_j} \tag{4}$$

3. Compute the gradient of the objective function

$$\frac{d}{d\alpha_k} J = \frac{\partial J}{\partial \alpha_k} + v_i \frac{\partial R_i}{\partial \alpha_k}$$

4. For each direction β

    (a) Find b by solving the tangent equation

$$\frac{\partial R_i}{\partial u_q} b_q = -\beta_r \frac{\partial R_i}{\partial \alpha_r} \tag{5}$$

(b) Find a by solving the adjoint equation

$$\frac{\partial R_i}{\partial u_j} a_i = -\left[\left(b_q \frac{\partial}{\partial u_q} + \beta_r \frac{\partial}{\partial \alpha_r}\right) \frac{\partial}{\partial u_j}(J + v_s R_s)\right] \tag{6}$$

(c) Calculate the second-order derivative of the objective function (in the direction of β) using

$$\beta_p \frac{d^2}{d\alpha_k d\alpha_p}(J) = a_i \frac{\partial R_i}{\partial \alpha_k} + \left(b_q \frac{\partial}{\partial u_q} + \beta_r \frac{\partial}{\partial \alpha_r}\right) \frac{\partial}{\partial \alpha_k}(J + v_s R_s)$$

As it can be seen from the above, the evaluation of the second-order derivative in a specific direction is divided into two parts. Firstly, we solve the adjoint equation for the gradient of the objective function (4). Subsequently, we solve two additional equations: one tangent and one adjoint. By solving these equations, one obtains second derivative of the objective J w.r.t. design parameters (α_k, α_p) computed in the direction β.

Uncertain Deformations

In this section, a method to reduce the number of uncertainties is presented. For this purpose, Radial Basis Function (RBF) interpolation is used. This can be interpreted as a method of Low-Rank Approximation [13] of the covariance matrix.

As a result of discretization of the continuous geometrical uncertainty field, one obtains a large number of correlated uncertainties with one random variable describing the uncertainty of position along a specific direction (in this paper, it is always z-axis). Due to the fact that the number of nodes of the surface mesh can be very high, especially for complex cases, the resulting set of random variables is too large for further analysis.

One possible approach is called a Low-Rank Approximation and attempts to approximate the covariance matrix C with a product VV^T where V has only few nonzero columns.

Let us assume that the set of nodal positions is denoted by x_j while the corresponding uncertainties are ζ_i $(i = 1, \dots, N)$. Then, our uncertain geometry q_j can be written as

$$q_i = x_i + \zeta_i$$

Set of nodal uncertainties spans the whole $\mathbb{R}^N$ space. Recalling formula (1), we can express the expected value of the objective function under the full basis of uncertainties as:

$$\mathrm{E}[f(x+h\zeta)] = f(x) + \frac{1}{2}h^2 C_{ij}\frac{\partial^2 f}{\partial x_i \partial x_j} + o(h^3)$$

Computation of the full Hessian matrix (with dimension $N \times N$) is too expensive; therefore, we have to apply the reduction. Let us assume that the vectors $\phi_1 \ldots \phi_M \in \mathbb{R}^N$, $(M \ll N)$ represent some arbitrary chosen nodal displacements (value ϕ_{ij} denotes the displacement of a i-th node generated by ϕ_j) and the corresponding random variables $\alpha_1, \ldots, \alpha_M$ that will create our new, reduced set of uncertainties. Using our new basis, we can approximate full variability of the geometry as:

$$\zeta_i \approx \sum_{j=1}^{M} \Phi_{ij}\alpha_j, \quad i = 1, \ldots, N$$

$$q_i \approx \hat{q}_i = x_i + \sum_{j=1}^{M} \Phi_{ij}\alpha_j, \quad i = 1, \ldots, N$$

where $\Phi = (\phi_{ij})$ is a rectangular matrix $N \times M$. Using this reduced basis, we can express our formula for the mean value of the objective as:

$$\mathrm{E}[f(x+h\zeta)] \approx f(x) + \frac{1}{2}h^2 \hat{C}_{ij}\hat{H}_{ij} + o(h^3)$$

where the Hessian matrix $\hat{H}_{ij}$ and the covariance matrix $\hat{C}_{ij}$ are defined as:

$$\hat{H}_{ij} \overset{\text{def}}{=} \frac{\partial^2 f}{\partial \hat{q}_i \partial \hat{q}_j} = \sum_{k,l=1}^{M} \Phi_{ik}\Phi_{jl}\frac{\partial^2 f}{\partial \alpha_k \partial \alpha_l}, \quad i,j = 1, \ldots, N$$

$$\hat{C}_{ij} \overset{\text{def}}{=} \mathrm{Cov}(\hat{q}_i, \hat{q}_j) = \sum_{k,l=1}^{M} \Phi_{ik}\Phi_{jl}\mathrm{Cov}(\alpha_k, \alpha_l), \quad i,j = 1, \ldots, N$$

In order to provide the best approximation accuracy, one can specify covariance matrix of the new random variables α by solving the Least Square Problem:

$$\left\| \mathrm{Cov}(\hat{q}_i, \hat{q}_j) - \mathrm{Cov}(q_i, q_j) \right\|_2 = \min!$$

Low-Rank Approximation allows for the dimension reduction; however, this method has limited accuracy that depends on M—the number of the considered directions ϕ (higher M increases the required computational cost, smaller M lowers the approximation accuracy).

The selection of the reduced basis $\boldsymbol{\Phi}$ can be arbitrary. In the present case, the Gaussian Radial Basis Functions (RBF) is of the form

$$\phi(\xi) = \exp\left(-\frac{(\xi - \xi_0)^2}{2\mu^2}\right) \tag{7}$$

where ξ denotes an independent variable (nodal position), ξ_0 points at the center of RBF while μ is the RBF radius.

It is important to understand that the Low-Rank Approximation is reducing the number of uncertainties to a level which is computationally (and memory-wise) achievable. For a high radius of correlation of the uncertainty field, the covariance matrix is dense and it is impossible to store it all for a very high number of nodal uncertainties (hundreds of thousands). With the use of LRA, we can reduce this number to a feasible value such as hundreds or thousands, without loosing the quality of the covariance structure. This procedure is effective because the full covariance matrix has a low rank in the first place.

On the other hand, the eigenvalue problem presented earlier is aimed at reducing the number of uncertainties even further using not only the covariance information, but also the properties of the objective. These two methods are coupled to obtain the maximum reduction of the computational cost.

Results

In this section, we present the demonstration of the proposed methodology. Firstly, however, details regarding solver implementation are provided. Then, the test case is briefly described including the geometry, the boundary conditions, the used meshes and the assumed uncertain deformations. Following this, the spectrum of the eigenvalues of the generalized eigenproblem (2) is presented and the relation between the required number of modes and the radius of RBF is discussed. The objective mean value for the defined uncertainties and its comparison with those obtained by the other methods is shown at the end.

The results presented in the present paper were obtained by the in-house solver **Flow2/RED**. The solver is based on the Residual Distribution (RD) method [14, 15] has been developed at the Warsaw University of Technology (WUT) [16, 17]. RD is an alternative method to Finite Volume and Finite Element and has few important advantages. It allows to apply multidimensional upwind scheme and has a compact nearest-neighbor stencil. Moreover, it can capture discontinuities occurring in the flow in a monotone and parameter-free manner.

This solver is capable to simulate steady and compressible Euler, Navier–Stokes, and RANS equations in 2D and 3D. The common turbulence models, such as Spalart-Allmaras and k-ω, are implemented. **Flow2/RED** solver can be executed across multiple nodes with MPI communication. The solver has been validated in the earlier projects ADIGMA [18] and IDIHOM [16] and proved to be a highly efficient tool capable of addressing industrial-scale cases.

The presented UQ methodology was implemented in **Flow2/RED**. The authors extended its functionality by creating a new module which is responsible for computation of the gradient of the objective function using an adjoint method (at present is is a leading method in aerodynamic optimization). The most important parts of the adjoint module are (i) the assembling algorithm for the Jacobian matrix and (ii) the solver of the linear system. In the Residual Distribution Method, the global residual vector and the Jacobian matrix (for a full mesh) is assembled from the local residuals and the matrices computed for each cell. The important property of the RD method is the locality of computations. For a particular element, the computations do not depend on other cells, and therefore, the algorithm can be easily parallelized. The second part of the module is responsible for computation of the Hessian-by-vector multiplication and mainly consists of the assembling and solving of two linear system—the tangent (5) and the adjoint (6). As the Jacobian matrix is non-symmetric, the Generalized Minimal Residual Method (GMRES) is used.

The proposed method requires computation of the derivatives of residual R and of the objective J. In order to compute them numerically, one can use Finite Difference Method. This method works fast and is easy to implement; nevertheless, it allows only for approximate calculation of the derivatives. This problem can be alleviated by the use of Automatic Differentiation (AD) tools, which are capable of reproducing derivatives exactly (up to the floating point precision), but usually are difficult to apply properly in a mature code, as they have specific requirements regarding the way the initial code is written. One can identify two approaches used in the AD tools: the operator overloading and the source transformation [19]. The first one has higher memory requirements, because all operations that have to be differentiated must be stored at run-time in memory on a so-called tape. In contrast, the source transformation approach analyzes directly the source code and produces the extended code, that can calculate the required derivatives. In the present work, we use the TAPENADE [20], source transformation tool developed at INRIA. The previously mentioned property of the RD method regarding the cell locality of residuum is very beneficial, as it is sufficient to differentiate only part of the code responsible for residuum calculation at the generic cell. This significantly simplifies the source transformation process.

BC-03 Test Case

The test consisted in simulations of the 3D, transonic, inviscid flow (see UMRIDA database, case BC-03 [21]). The geometry is a DLR F6 fuselage, wing, pylon, and nacelle configuration [22]. As an objective, we have considered the lift force for the free-stream Mach number $M = 0.75$ and AoA corresponding to the lift coefficient $C_L = 0.5$.

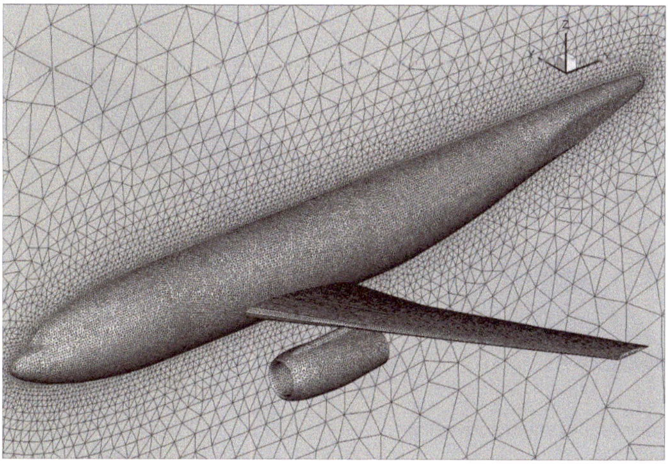

Fig. 1 Inviscid mesh ($2.6 \cdot 10^6$ tetrahedral elements)

Fig. 2 Contour plot of the non-dimensional density ρ/ρ_0 (final solution)

For the sake of simplicity, the flow was treated as inviscid. We have used four meshes with approximately $0.3 \cdot 10^6$, $1.1 \cdot 10^6$, $1.7 \cdot 10^6$ and $2.6 \cdot 10^6$ tetrahedral elements. The presented results were obtained with the finest mesh shown in Fig. 1. The contours of the density for the converged flow solution are shown in Fig. 2.

For this test case, we have assumed the covariance matrix structure using the Gaussian function:

$$C(x_i, x_j) = \psi(x_i)\psi(x_j)\exp\left(-\frac{|x_i - x_j|^2}{2\sigma^2}\right)$$

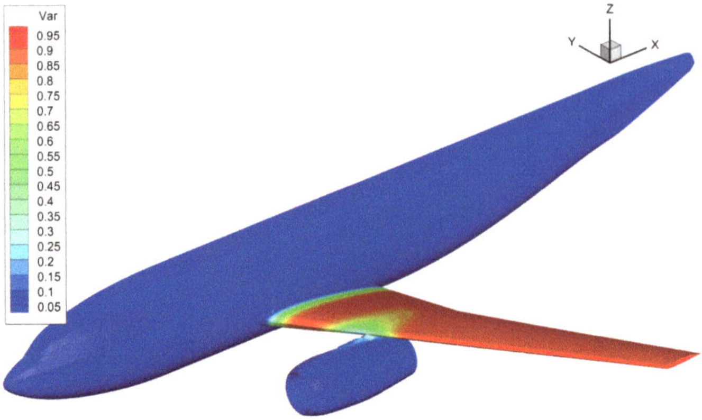

Fig. 3 Contour plot of a nodal variance

where x_i are the points on the surface, $|x - y|$ denotes the euclidean distance, σ is a radius of correlation, and $\psi(x)$ is the "fixing" function. This function is not zero only there where the geometry is affected by uncertainties (in this case on the wing). The function is smooth to preserve the continuity of the geometry and is defined as:

$$\psi(\xi) = 1 - \exp\left(-\frac{d_{wall}(\xi)}{2\sigma^2}\right)$$

where $d_{wall}(\xi)$ defines the distance of a node ξ from the nearest wall.

For the analyzed case, we have the fixed fuselage and the nacelle, leaving wing, and pylon surfaces variable. Surface plot of nodal variance is shown in Fig. 3. Such approach can lead to a limitation of geometry variability, but, as it can be seen, its range is limited to a region close to a fixed zone.

For the representation of geometry variability, we have assumed 40 uncertain deformations represented by the Gaussian Radial Basis Functions. Centers of RBFs are located in such way that the best approximation of covariance structure for a full set of nodes ζ is achieved. The centers are distributed sequentially, one by one, using a maximum variance condition proposed in [23]. The collection of RBFs is initialized with the node for which the uncertainty variance $\mathrm{Var}(x) = C(x, x)$ is maximal. For a given collection of RBFs, one can decompose the uncertainty into two parts: the part which is fully represented by a linear combination of these RBFs and the residual part. The variance of the residual part can be efficiently calculated and it is used for the selection of the next RBF. This procedure is continued until a maximum value of the residual variance function is lower than a given threshold or until M reaches the specified size.

In order to verify the implementation of the code, we have arbitrarily selected 3 RBFs and compared the Hessian matrix calculated by the Kriging method based on the full factorial computation. Comparison of these matrices is presented in the

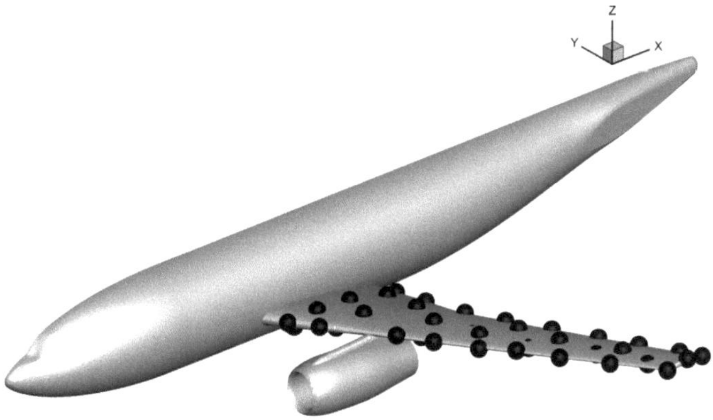

Fig. 4 Radial Basis Functions distributed on the surface

Table 1 Comparison of the Hessian matrices obtained with the **Flow2/RED** code and the Kriging method

Flow		
−12.016278	3.071211	0.452490
3.071211	−36.655157	1.389806
0.452489	1.389807	−21.116542
Kriging		
−11.845916	2.986513	0.453560
2.986513	−37.395125	1.386518
0.453560	1.386518	−21.049571
Relative error		
−1.44%	−2.84%	0.24%
−2.84%	1.98%	−0.24%
0.24%	−0.24%	−0.32%

Table 1. Results obtained with these two approaches are in a good agreement, with a maximum relative error of 2.84%.

In order to verify the proposed methodology, we have computed the mean value with the present method using the in-house solver and we have compared it with the results of Monte Carlo (2000 samples) and the Kriging methods (full factorial of $343 = 7 \times 7 \times 7$ samples). Comparison of the objective correction due to geometrical uncertainties $\Delta f = \mathrm{E}[f(x + h\zeta)] - f(x)$ can be found in the Table 2. It can be seen that the Monte Carlo and Kriging are in close agreement while the analyzed method predicts the correction different by 13%. However, if we compare the influence only up to second-order in the Kriging method, we can see that both results are similar, which means that the major part of the error in the mean value prediction comes from the Taylor series truncation.

Table 2 Comparison of the objective correction for different methods

Method	Δf
Monte Carlo	−0.4026461
Kriging	−0.4025724
Our method	−0.3514531
Kriging (second order)	−0.3489399

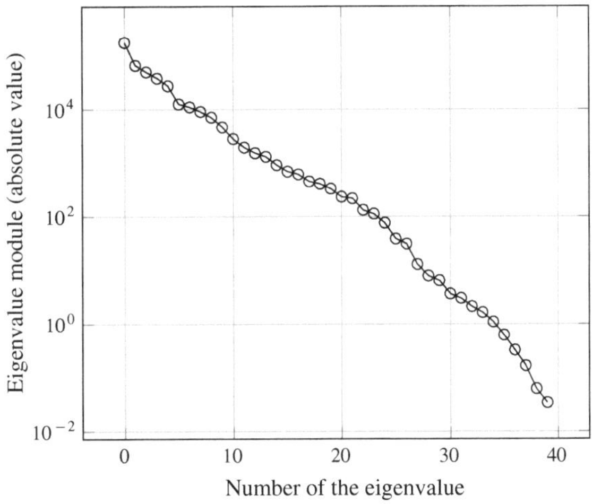

Fig. 5 Spectrum of eigenvalues module

Once we have successfully verified the correction of the objective function caused by geometrical uncertainties Δf, we can proceed to the analysis of the spectrum of the generalized eigenvalue problem (2).

The proposed Hessian reduction method is based on the assumption that one can represent the most of the second-order contribution with only few eigenvalues and the corresponding eigenvectors (so-called modes). In order to verify this assumption, we are showing the spectrum of eigenvalues moduli (Fig. 5). It can be seen that the eigenvalues decrease almost exponentially. Therefore, it seems possible to keep sufficient level of accuracy using only a limited number of largest eigenvalues.

The further investigation is focused on finding the number of required modes which are needed to represent the assumed level (in this case 99%) of second-order contribution in the Taylor expansion (1) as a function of correlation radius of our assumed uncertainty deformations ϕ. We have solved the eigenproblem in three different ways and compared the representation based on the eigenvectors of the Hessian (A) and the covariance matrix (B) treated separately with the present method based on the generalized eigenproblem solution (C). The results can be found in Fig. 6.

One can see that for the case (A), the number of modes remains the same for all cases. It is worth mentioning that only 6 out of 40 eigenvectors can be neglected as they provide only 1% of information. Considering case (B), one can notice that the

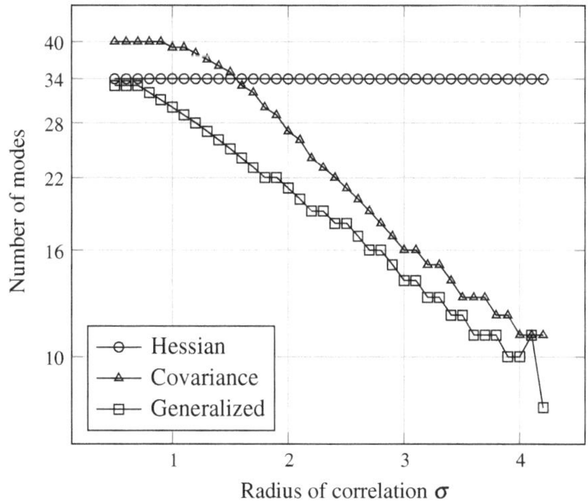

Fig. 6 Number of modes required to represent 99% of second-order contribution

number of modes is decreasing as the correlation radius increases. Compared with (A), such behavior is beneficial and can reduce computational time when correlation radius is large. However for smaller radii, the situation is opposite and the approach (B) might even require the full spectrum of eigenvalues. Considering the proposed method (C), one can notice that the number of required modes is never larger than it is for the two other approaches. For the full range of correlation radius, our method requires always smaller number of modes than in the approach (B), and for a small radius, we are at least as good as (A).

Summary

The presented second-order sensitivity method for handling geometrical uncertainties was proven to give good approximation of the objective function mean value and effectively reduce the number of uncertainties.

This method is based on the assumption that the full Hessian matrix can be replaced by the reduced form obtained using eigen decomposition and selection of only a small number of largest eigenvalues without an important loss in the second-order contribution. Numerical results shown in this paper confirm that for the analyzed test case, this assumption is verified. Moreover, the analyzed method always gives a larger reduction of the uncertainty space than the techniques that use the Hessian and the covariance matrices separately. The number of the required eigenvalues (and the corresponding eigenvectors) is smaller for a full range of correlation radius (of our uncertain deformations ϕ).

Another very important property of the proposed method is that the resulting eigenvectors of the generalized eigenvalue problem can be used for further UQ analysis. The covariance matrix expressed in such basis is diagonal, which means that the uncertainties are independent. This significantly simplifies further UQ analysis. On the other hand, the Hessian matrix in this basis is also diagonal, which means that there are less coefficients to be determined in the polynomial approximation.

The method needs further development; in particular, the mean value is not the only stochastic parameter of interest (e.g., one might be interested also in the variance). Similarly, the current implementation of the method requires explicit assemble of the full Hessian matrix. It would be beneficial in terms of computational time to use a method in which only the largest eigenvalues are found (e.g., power method, Arnoldi/Lanczos), without assembling of the Hessian.

As results shown in this paper cover only inviscid Euler flow, the method has to be verified also for the Navier-Stokes and the RANS models.

Acknowledgements The majority of the present research was done within the FP7 project UMRIDA—Uncertainty Management for Robust Industrial Design in Aeronautics (Grant agreement: ACP3-GA-2013-605036). It was also partly supported by the European Union in the framework of European Social Fund through the Didactic Development Program of the Faculty of Power and Aeronautical Engineering of the Warsaw University of Technology.

References

1. Giles, M.B., Duta, M.C.: Müller, J.D., Pierce, N.A.: Algorithm developments for discrete adjoint methods. AIAA J. **41**(2), 198–205 (2003)
2. Jameson, A.: Aerodynamic design via control theory. J. Sci. Comput. **3**(3), 233–260 (1988)
3. Jameson, A.: Aerodynamic shape optimization using the adjoint method (2003)
4. Papadimitriou, D.I., Giannakoglou, K.C.: Aerodynamic shape optimization using first and second order adjoint and direct approaches. Arch. Comput. Methods Eng. **15**(4), 447–488 (2008)
5. Ghate, D., Giles, M.: Efficient Hessian calculation using automatic differentiation. In: 25th AIAA Applied Aerodynamics Conference. American Institute of Aeronautics and Astronautics
6. Wang, Z., Navon, I.M., Dimet, F.X.L., Zou, X.: The second order adjoint analysis: theory and applications. Meteorol. Atmos. Phys. **50**(1–3), 3–20
7. Smith, R.C.: Uncertainty quantification: theory, implementation, and applications. SIAM-Soc. Ind. Appl. Math. (2013)
8. Jin, R., Du, X., Chen, W.: The use of metamodeling techniques for optimization under uncertainty. Struct. Multidiscip. Optim. **25**(2), 99–116 (2003)
9. Ghanem, R.G., Spanos, P.D.: Stochastic Finite Element: A Spectral Approach, vol. 224. ResearchGate, Springer, New York (2003)
10. Bauerheim, M., Ndiaye, A., Constantine, P., Iaccarino, G., Moreau, S., Nicoud, F.: Uncertainty quantification of thermo-acoustic instabilities in annular combustors. In: Proceedings of the Summer Program, pp. 209–218 (2014)
11. Constantine, P.G., Dow, E., Wang, Q.: Active subspace methods in theory and practice: applications to kriging surfaces. SIAM J. Sci. Comput. **36**(4), A1500–A1524 (2014). arXiv: 1304.2070
12. Martinelli, M., Duvigneau, R.: On the use of second-order derivatives and metamodel-based Monte-Carlo for uncertainty estimation in aerodynamics. Comput. Fluids **39**(6), 953–964 (2010)
13. Markovsky, I.: Low rank approximation: algorithms, implementation, applications. Springer Science & Business Media (2011)

14. Ricchiuto, M.: Construction and analysis of compact residual discretizations for conservation laws on unstructured meshes. Ph.D. Thesis, Universite Libre de Bruxelles and von Karman Institute for Fluid Dynamics (2005)
15. Abgrall, R., Mezine, M.: Construction of second-order accurate monotone and stable residual distribution schemes for steady problems. J. Comput. Phys. **195**(2), 474–507 (2004)
16. Majewski, J., Szałtys, P., Rokicki, J.: Anisotropic adaptation for simulation of high-reynolds number flows past complex 3d geometries. In: Kroll, N., Hirsch, C., Bassi, F., Johston, C., Hillewaer, K. (eds.) IDIHOM: Industrialization of High-Order Methods—A Top-Down Approach. Notes on Numerical Fluid Mechanics and Multidisciplinary Design, vol. 128, pp. 101–124. Springer, Berlin, Heidelberg (2015)
17. Majewski, J., Szałtys, P., Wyrozębski, M.: Residual distribution method for high Reynolds number simulations on complex geometries. Comput Fluids **166**(0045–7930), 104–116 (2018). https://doi.org/10.1016/j.compfluid.2018.01.028
18. Majewski, J., Rokicki, J.: Anisotropic mesh adaptation in the presence of complex boundaries. In: Kroll, N., Bieler, H., Deconinck, H., Couaillier, V., Ven, H., Sørensen, K. (eds.) ADIGMA—A European Initiative on the Development of Adaptive Higher-Order Variational Methods for Aerospace Applications. Notes on Numerical Fluid Mechanics and Multidisciplinary Design, vol. 113, pp. 441–453. Springer, Berlin, Heidelberg (2010)
19. Naumann, U.: The Art of Differentiating Computer Programs. An Introduction to Algorithmic Differentiation (2012)
20. Hascoët, L., Pascual, V.: The tapenade automatic differentiation tool: principles, model, and specification. ACM Trans. Math. Softw. **39**(3) (2013)
21. UMRIDA Consortium: Test case description innovative database for UQ and RDM. Technical Report (2014)
22. NASA: 3rd AIAA cfd drag prediction workshop. https://aiaa-dpw.larc.nasa.gov/Workshop3/workshop3.html. Accessed 17 Nov 2016
23. Łaniewski Wołłk, L.: Automatic parametrization and mesh deformation for CFD optimization. arxiv.org/abs/1311.6190 [math] November 2013

Part III
Application of Uncertainty Quantification to Industrial Challenges

Application of UQ for Turbine Blade CHT Computations

K. Vinogradov and G. Kretinin

Introduction

Thermal state of the parts in the gas turbine engine hot section defines the structural integrity, as well as the service time and the engine lifecycle cost.

A HPT blade is one of the most crucial and loaded parts. A great number of efforts are applied in order to provide the required thermal state and the necessary level of turbine blade cooling efficiency. However, the most part of numerical studies of phenomena and processes, as well as efforts to optimize and improve structural characteristics (enormous number of studies) are put into practice on model problems (flat planes, model blades) [1–6].

The HPT blades characteristics are affected by a great number of factors and uncertainties such as geometrical variations (e.g., turbine blade tip clearance), operational deviations (temperature and pressure of hot gas before the HPT blade, as well as pressure of cooling air under the HPT blade).

Consideration of uncertainty influence on a HPT blade is particularly urgent in the course of actual operation. Operational environment of a HPT blade is significantly affected by operational variations, as well as by efficiency decreasing over engine life. Stochastic fluctuations of the HPT blade operational parameters require the implementation of probabilistic optimization criteria in order to obtain robust solution of the optimization task. In this connection, the authors set the main task of the research: to define how the operational and geometrical uncertainties influence the thermal state of one of the key blade areas (leading edge and blade tip) with convection–film cooling and aerodynamic efficiency of a HPT blade by means of a computational study. The second task was the possibility of finding a steady robust solution for film-cooling holes configuration on the blade leading edge and robust

K. Vinogradov (✉) · G. Kretinin
NPO Saturn, 152903 Lenin Avenue, 163, Rybinsk, Yaroslavl Region, Russia
e-mail: keryisdex@mail.ru

C. Hirsch et al. (eds.), *Uncertainty Management for Robust Industrial Design in Aeronautics*, Notes on Numerical Fluid Mechanics and Multidisciplinary Design 140, https://doi.org/10.1007/978-3-319-77767-2_23

optimum tip gap values for blade tip cooling and kinetic energy losses. So, the authors make an effort to optimize the thermal state of the most thermally loaded blade components providing a compromise with aerodynamic efficiency.

Geometry and CFD Domain

The 3D-model of a cooled HPT blade in Siemens NX v8.5 was used to generate the computational area. The 3D-model is presented in Fig. 1.

The cooling system of the blade is of the convection–film type. The internal serpentine passages are used as a convection cooling system. The ribbing of internal passages is applied to intensify the convective heat transfer. Film cooling is made by of two rows of film-cooling holes on the leading edge. The cooled air is supplied into serpentine channels and discharged through the holes in trailing edge of the blade.

In this study, two regions with maximal thermal loads are considered as the blade areas. There are the leading edge and the blade tip.

It should be noted that radial form of film-cooling holes conforms to the hot gas leakage line on the blade (film-cooling rows in the initial variant are placed along it). Their form is presented here as parametric in order to automate the row placement change. Both hole rows on the leading edge are parameterized as follows: α is the hole axis angle calculated in relation to stagnation point, L is the distance to the stagnation point in the targeted section. The parameterization layout and the parameters applied are presented in Fig. 2.

In our study, the arrangement of both rows changes simultaneously and the distance between them remains invariable. The developed method of parameterization allows changing the film-cooling hole row arrangement with minimal parameter number (at that the radial distribution does not change) providing approximately steady value of convective heat transfer intensity.

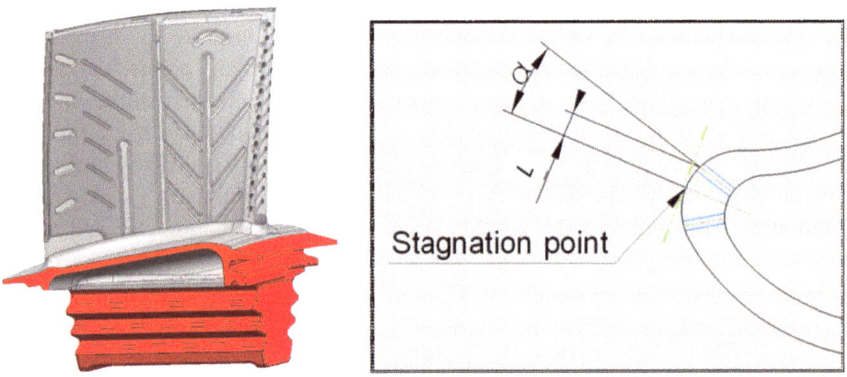

Fig. 1 HPT blade geometry and film-cooling holes rows parameterization

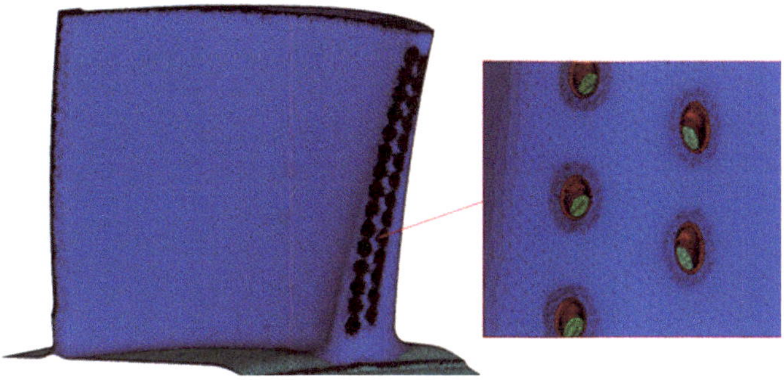

Fig. 2 Blade body mesh

The computational model includes the blade solid domain, the internal cooling passages, and the hot gas path. The blade profile surface, internal passages, and hub end-wall surface are considered as the conjugate heat transfer interfaces. The blade foot is modeled by means of heat boundary conditions (surface temperature, flow temperature, and heat transfer coefficient).

A non-structured tetrahedral grid was used for mesh generation of the blade channel and the internal blade cavity with seven prismatic layers (growth ratio 1.5 and the first cell size 1–6 m). The non-structured tetrahedral grid was applied to mesh generation of the blade body. The mesh model of the blade body is shown in Fig. 2.

ANSYS ICEM CFD was used to generate this type of mesh. The unstructured tetrahedral mesh was developed to carry out heat transfer calculations and to determine temperature distribution in the solid domain. The node-to-node technology was used to couple the internal cooling passages domain and the solid domain. The boundary layer mesh of hot gas path domain was refined by dimensionless wall distance. $y+$ values around 1 were achieved to accommodate the full solution through the viscous sublayer. The height of the first element next to the wall was 1–6 m.

A preliminary grid dependency test as in [1] was performed in order to check the grid density influence on the results and to determine the optimal number of grid nodes. Figure 4 presents the efficiency for a blade leading edge at different grids. The number of nodes in the considered grids is as follows: from about 1.5 million nodes to about 8 million nodes. As it was demonstrated, the results from about 6.5 million nodes showed a good fit with the results from about 8 million nodes. Therefore, the former number (6.5 million) was selected as the optimal number of nodes. The number of nodes for internal cooling passages domain is about 2.2 million, for the hot gas path domain is about 2.5 million, for solid domain it amounts to approximately 1.7 million. Views of the chosen grid for leading edge film-cooling holes are presented in Fig. 3. The computations were equated for one blade with periodic boundary conditions.

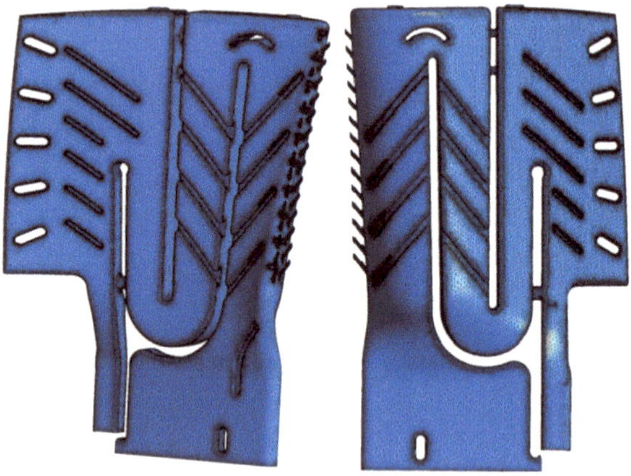

Fig. 3 Internal cooling passages mesh

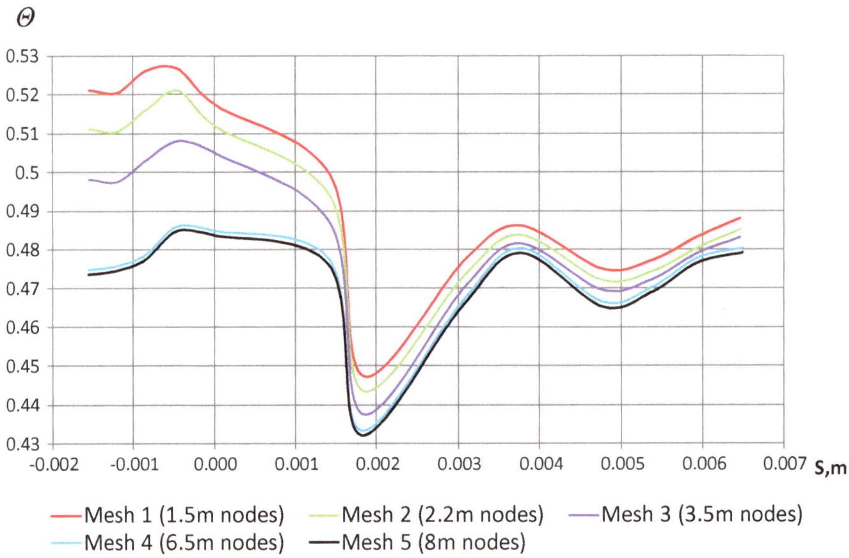

Fig. 4 Grid dependency test

Three-dimensional Reynolds-averaged Navier–Stokes analyses of fluid flow and film cooling were performed with the help of ANSYS CFX 14.5 [7] which employs an unstructured grid system. The solutions were obtained using the finite volume method of discretization of the compressible RANS equations. The shear stress transport (SST) turbulence model [8] is used as a turbulence viscosity equation solving and closure system of equations. The SST model is put into action by solving a turbulence frequency-based model (k-ω) near the wall and a (k-ε) model in the main flow. The blending function developed by Menter [9] ensures a smooth transition between the two models. Bardina et al. [10] showed that the SST model captures a flow separation more effectively than other eddy viscosity models, and, thus, predicts the near-wall turbulence that is essential in prediction of turbulent heat transfer.

SST turbulence model is used in a significant amount of the works [11] in the field of CFD modeling and analysis of film-cooling process. Thus, Didenko et al. in the paper [12] accomplished verification of this approach to simulation of conjugate heat transfer on HPT blades.

The chemically ideal air–gas mixture from ANSYS CFX was used as a working medium. Air and gas had only differences for their thermodynamic state. Air entered the blade cooling passages and went out into the air–gas channel through the holes on the leading edge and the exit slit. Gas entered the inlet of the blade passage. As the material for the blade, heat-resistant alloy was accepted. Main boundary conditions are shown in Fig. 5.

The boundary conditions at the inlet of the computational model (radial distribution of temperature, pressure, and flow angle) were accepted according to the design gas-dynamic calculations for the gas turbine under consideration. Temperature and pressure radial distributions are presented in Fig. 6. Initial design Reynolds number is based on blade chord length (b), and outlet velocity at midspan—Reb = 1.15 * 10^6. Isentropic Mach number at midspan—Ma = 0.715. The results of CHT computations of cooling efficiency for the blade areas under consideration were area averaged on the blade surfaces shown in Fig. 7. The two most important

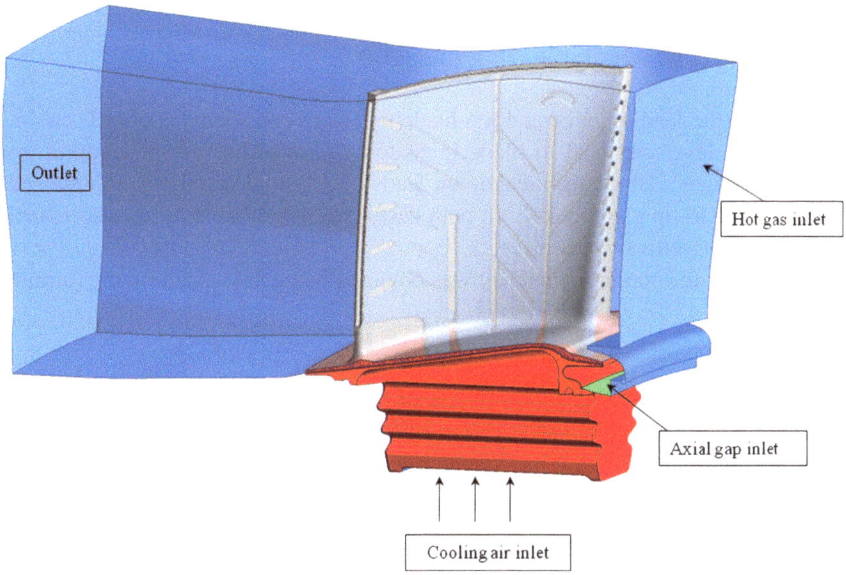

Fig. 5 Boundary conditions

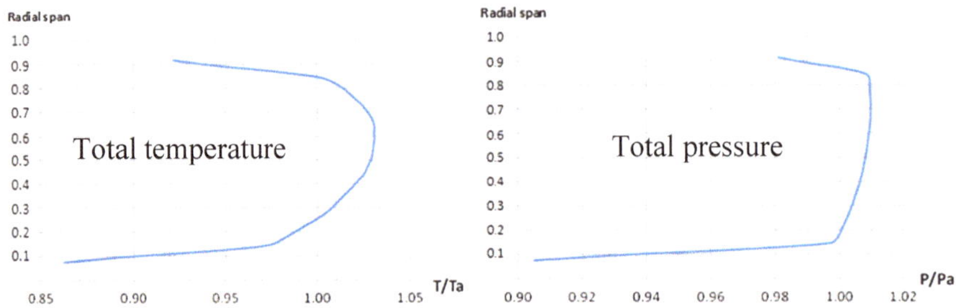

Fig. 6 Inlet hot gas parameters radial distributions

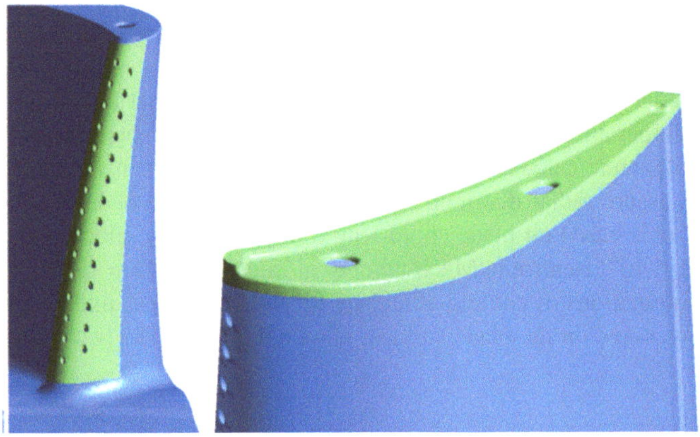

Fig. 7 Surfaces of averaging for leading edge and tip

blade areas (the leading edge and the blade tip) were considered. Low LE cooling effectiveness level can lead to blade breakdown (due to hot gas penetration into cooling passages). Blade tip breakdown leads to essential aerodynamic efficiency reduction due to tip gap increase. In both cases, we will be able to have acceptable average cooling effectiveness.

As a result, the cooling efficiency was obtained being calculated by the formula:

$$\Theta = \frac{\boldsymbol{T}_{rec,m} - \boldsymbol{T}_{cw}}{\boldsymbol{T}_{rec,m} - \boldsymbol{T}_{tc}} \tag{1}$$

To assess aerodynamic performance kinetic energy loss value ζ was applied. The loss value was calculated by the Sieverding formula [13]:

$$\zeta = 1 - \frac{1 - \left(P_1/P_1^*\right)^{(k-1)/k}}{1 - \left(P_1/P_0^*\right)^{(k-1)/k}} \tag{2}$$

This parameter can be used to access aerodynamic characteristics of the blade. In this case, changes in flow rate of coolant are not taken into account. However, hot gas and cooling air pressure ratios were set as constants (hot gas path total pressure P* and cooling air total pressure P^*_{rel} were considered as correlated parameters). Therefore, the total coolant flow rate does not have essential changes.

UQ and RDO Problem Statement

In the test case proposed, the HPT blade tip gap is the only considered geometrical uncertainty resulting from HPT manufacture tolerance and assembling deviations. The authors chose the theoretical Gaussian distribution for tip gap pdf function, due to several experimental investigations on prototype engines. The nominal dimensions' assembly draft value of the tip gap was obtained as 0.5 mm and set as the mean value. The standard deviation was 0.1 mm.

In this test case, the following operational deviations are examined: hot gas total pressure and total temperature mean values T^*, P^* in the channel inlet are the first and the second operational uncertainties. Deviations of these parameters result from uncertainties of manufacture and assembling deviations of engines parts (compressor, combustion chamber, turbine, etc.) and from different engines operating time and conditions.

Cooling air total pressure under the blade in relative coordinate frame P^*_{rel} is the third operational uncertainty. This operational uncertainty has the same pdf distribution as the hot gas pressure because both parameters depend on compressor pressure ratio. These parameters were considered as correlated.

Hot gas temperature and pressure inlet values in the channel were obtained from experimental investigations of 90 industrial gas turbine engines. These engines are similar to the engine optimized in the paper, in terms of dimension, power, and efficiency. During the abovementioned engine tests, the exhaust gas temperature (EGT) and the compressor pressure ratio were measured. The EGT and the compressor pressure ratio for these engines are presented in Fig. 8.

Fig. 8 EGT and compressor pressure ratio for considered engines

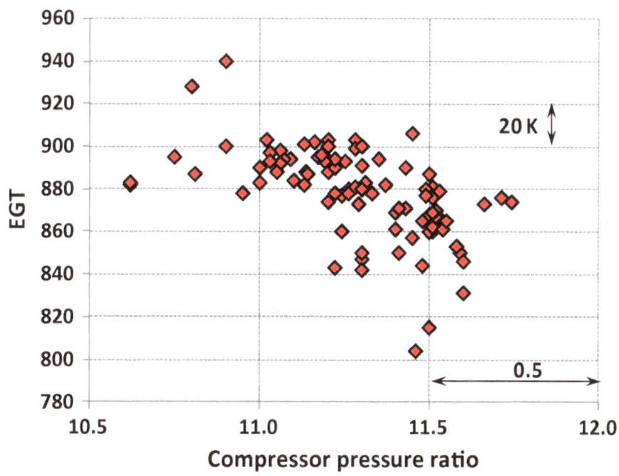

Hot gas parameters in the channel inlet and cooling air parameters are calculated from EGT values (experimental data is showed in Fig. 8) by means of verified thermodynamic calculations methodology used at NPO Saturn. Correlation and dependence between EGT and hot gas parameters were obtained on several test engines and implemented into a thermodynamic model. The radial distribution of total temperature after the combustion chambers has almost the same character. The data was acquired from the combustion chambers separate tests of the reviewed engines. Therefore, we can assume the constant shape of the radial temperature distribution. In spite of that, the mean values of the exhaust temperature of the combustion chamber and pressure in real engines change according to variation of EGT from Fig. 8. Meanwhile, deviations of the mean temperature and pressure in the hot gas path inlet ΔT^*, ΔP^* were set as operational deviations. ΔP^*_{rel} was considered as a parameter correlated to ΔP^*. In our research, the epistemic uncertainties were not taken into consideration due to dimensionality problem. More than 60 calculations for the six parameters considered were executed.

The general statement of the RDO and UQ procedure is described in chapter "Uncertainties Identification and Quantification." In our research, we consider a tip gap being the value of the blade tip clearance, α the angle of the film-cooling hole arrangement axis, and L the distance to the leakage point as variable parameters (vector x). Thus, variables α and L are regarded as the deterministic ones, and tip gap is considered as the stochastic value distributed according to the normal law with mean square deviation 0.1 mm. As an external conditions' vector (e), we take into consideration corrections for input pressure and temperature, as well as those for cooling air pressure ΔP^*, ΔT^*, ΔP^*_{rel}. These parameters are the stochastic values, which distribution laws were obtained from the experimental data and are shown in Fig. 9. Standard deviation for ΔT^* was 21.42 K and 0.236 bar for ΔP^*.

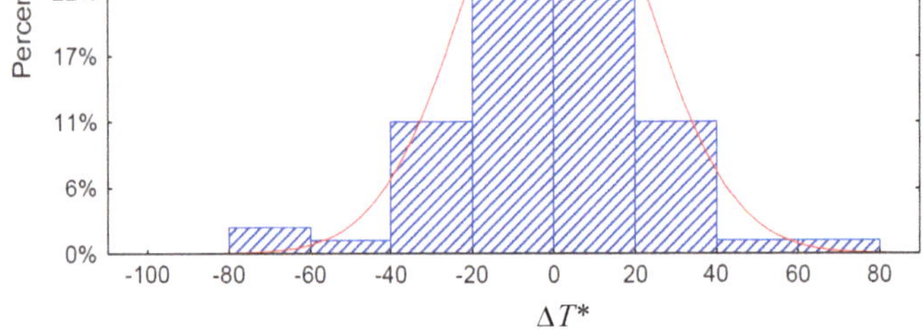

Fig. 9 External conditions vector components pdf (for hot gas temperature deviations)

It is necessary to note that in robust optimization task ΔP^* and ΔP^*_{rel} were considered as correlated parameters with correlation coefficient equal to 1.

When solving robust design optimization, efficiency values $y = f(x, e)$ are the random ones. In this case, it is necessary to use probabilistic optimization criteria $\tilde{y}(\bar{x})$. In our study, we regard the cooling efficiency values (for the leading edge and the tip) and those of the kinetic energy loss provided with 90% probability as probabilistic criteria $P_c\{f(x, e) \leq \tilde{y}(\bar{x})\} \geq P_p$ the efficiency value ensured with probability no less than the one given [14]. It is the most general criterion, used for most robust optimization tasks. Sometimes we need to solve few robust optimization tasks at different given probability levels. A probabilistic constraint tip gap value ≥ 0.2 mm provided with 90% probability was considered in our work.

The main problem occurring while solving robust design optimization problem is that of determining probabilistic criteria values. The simplest and the most universal method of evaluation of probabilistic criteria is the Monte Carlo method. The main advantage of this method, as applied to RDO problems, is the absence of necessity to make any priori assumptions about the goal function peculiarities (smoothness, monotony, continuity, differentiability, and so on). However, the efficiency of the Monte Carlo method applied while solving real-life problems largely depends on the required accuracy of definition of probabilistic criteria. Particularly, applying the gradient methods of optimization, requiring high accuracy of definition of probabilistic criteria, results in high computational expense (required number of tests at each iteration of extremum search makes up $\approx 10^6$ to 10^9).

When solving the task under consideration, we used the Monte Carlo method along with the method of multi-criterion optimization IOSO [15] as well as the procedure of multi-level optimization involving surrogate models [16–18].

In the present work, APPROX software was used for surrogate model construction. This software was developed by Sigma technology and NPO Saturn in EU FP6 program NODESIM-CFD [19]. The response surfaces constructed based on the artificial neural network modeling (radial basis function) [20] in combination with modified full square regression as the low-fidelity model.

HPT Blade RDO Problem Solution

To construct surrogate model iterative procedure was implemented to CHT computations in ANSYS CFX. Within the applied procedure of multi-level optimization, four iterations were carried out. Initial Design of Experiments (DoE) had 32 calculations with additional 10 calculations at every multi-level optimization iteration. Therefore, total number of calculations in database to construct surrogate model was ≈ 70. At the first step, the initial DoE based on 32 CHT computations was constructed. The Sobol uniform sequence algorithm was applied to generate initial combinations of variable parameters. In Fig. 10, two flow patterns and

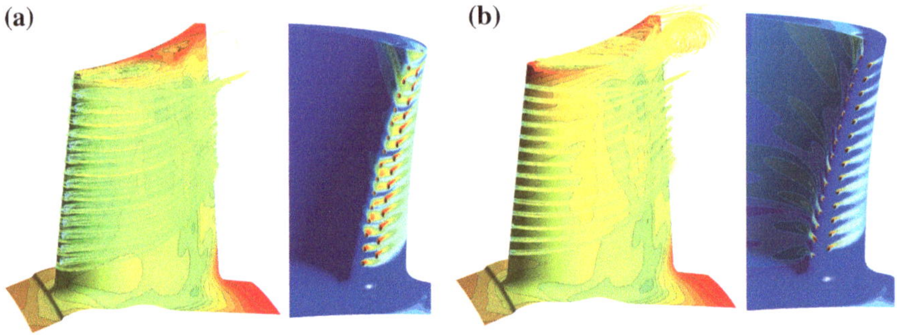

Fig. 10 Distribution of metal temperature and cooling flow concentration on blade surface for points 33 (**a**) and 40 (**b**) from surrogate model database

distributions of the cooling flow concentration on the blade surface for points from 1 multi-level iteration were presented.

Point 33 (a) is characterized by a high level of leading edge cooling effectiveness due to hot gas and cooling flow interaction (mixing) in this region. Point 40 (a) has a high level of blade tip cooling effectiveness and low kinetic energy losses ζ. However, LE cooling effectiveness is extremely low for this point. Therefore, Fig. 15 shows that different combination of uncertainty parameters provides different interaction between hot gas and cooling air with a wide range of considered areas cooling effectiveness and aerodynamic efficiency.

Surrogate model database modifications (criteria and variable parameters) via multi-level iterations for leading edge and blade tip cooling effectiveness are presented in Fig. 11. Variations of accuracy for low-fidelity models via iterations for the first criterion (leading edge cooling efficiency) are shown in Fig. 12.

It should be noted that accuracy of approximation of the first criterion (leading edge cooling efficiency) has the minimal value less than 2%.

The tendency is noticeable toward increase of low-fidelity model accuracy under the growth of the number of iterations for all the considered criteria. It should be noted that accuracy of approximation of the second criterion (blade tip cooling efficiency) has the minimal value about 1.5%.

We can stop multi-level optimization iterations process when prediction accuracy is compared with database average accuracy. We have attained this necessary accuracy level at the 4th multi-level optimization iteration.

Therefore, the obtained surrogate model can provide the necessary level of prediction accuracy to carry out investigations of the different robust optimization task statement. It is possible to increase the number of iterations or to apply another response surfaces technique to improve the quality of approximation (if it is necessary).

Using the obtained surrogate model, we have solved different optimization tasks, based on deterministic principles and different robust criteria.

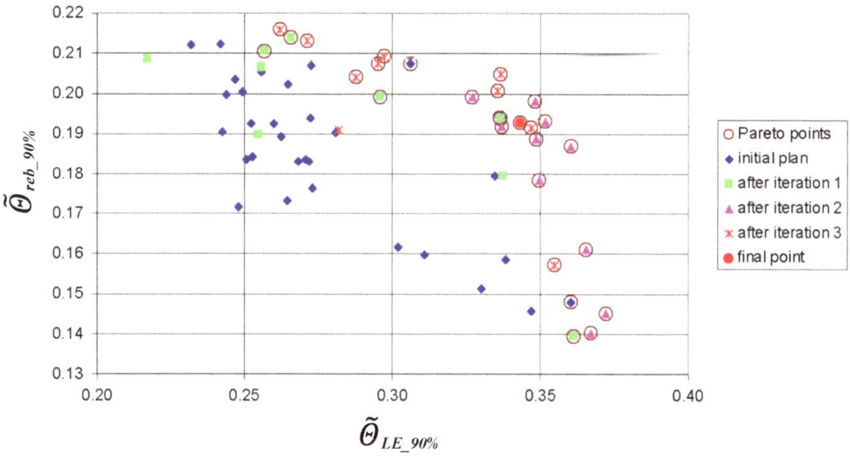

Fig. 11 Surrogate model database modifications (criteria) via iterations for leading edge and blade tip cooling effectiveness

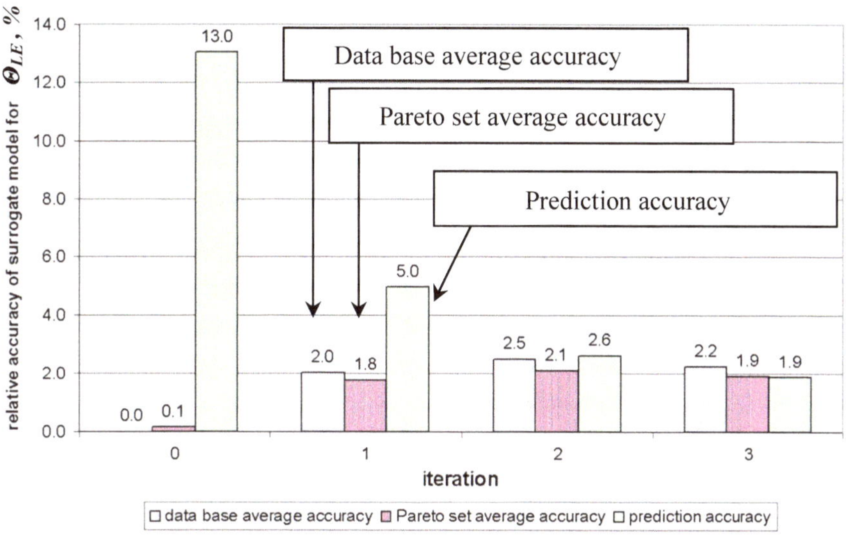

Fig. 12 Variation of surrogate models accuracy for leading edge cooling effectiveness

Using probabilistic criteria, the robust optimal solutions for the mentioned RDO task were obtained. Pareto set and the chosen point are presented in Fig. 13. We choose an "angle point" (red dot in Fig. 13) from Pareto set. Obviously, the desire to reduce the kinetic energy loss to minimal values can result in sufficient decrease

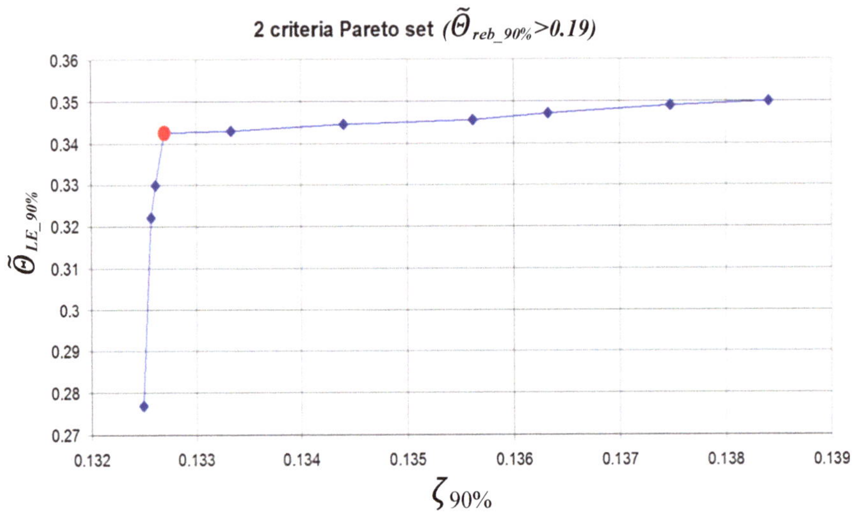

Fig. 13 Results of the two-criteria RDO task ($\tilde{\Theta}_{reb\_90\%} > 0.19$-constraint)

in cooling efficiency both as to the tip and as the leading edge (left part of the graph in Fig. 13). At the same time, the trade-offs make it possible to get reasonable values for both the cooling efficiency and the loss. The point regarded as a sought for solution of the task is indicated in red. In such a case, choosing of the angle point is the most evident optimization result.

It should be mentioned that we obtained necessary levels of robust criteria characterizing the leading edge cooling effectiveness, the blade tip cooling effectiveness, and kinetic energy losses at the chosen point from the Pareto set (angle point).

- $\Theta_{LE} > 0.343$ with 90% probability;
- $\Theta_{reb} > 0.193$ with 90% probability;
- $\zeta < 0.133$ with 90% probability;

Comparison of the obtained robust solution with a deterministic optimum point is presented in Fig. 14.

The deterministic criterion (Θ_{LE}) is 4.5% more for Point 3 (the deterministic optimum point). It should be noted that the robust criterion (with 90% probability) is 1.8% less than for the RDO procedure result.

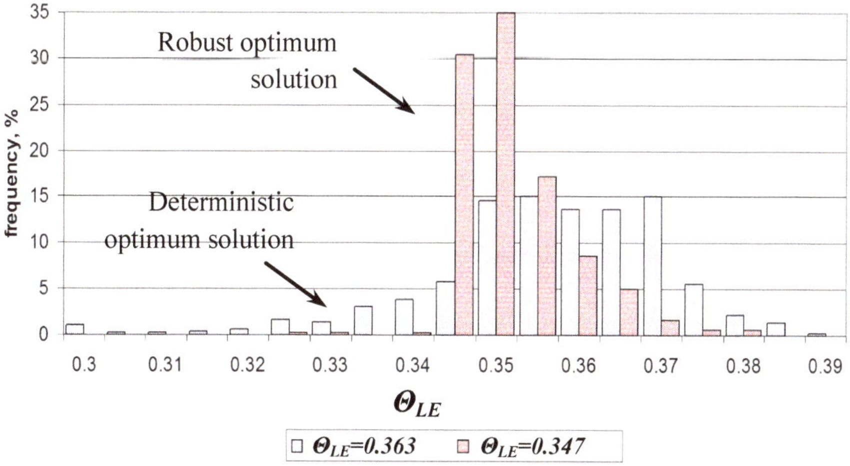

Fig. 14 Probability density functions for RDO solution and deterministic optimum point

UQ Based on RDO Results via Surrogate Model Refinement

On the basis of the analysis of the robust optimization results, three points were chosen from the obtained set of the Pareto-optimal solution for the detailed investigation and verification of the execution of the uncertainty quantification during the robust optimization process on a single surrogate model.

These points were chosen because of the necessity to investigate optimization space subdomains with the maximal values of individual optimization criteria.

This approach was made possible by the particular features of the algorithms used (Monte Carlo+RSM+IOSO). Additional generation of the Design of Experiments, refinement of the approximating functions (surrogate model), and evaluation of the probabilistic criteria were conducted in the vicinity of the chosen points. Areas of interests chosen for the refinement of a surrogate model by generating additional points for the Design of Experiments are shown in Fig. 15.

For the refinement of the surrogate model, a special procedure was developed. Generally, it consists of seven steps.

1. Generation of 16 uniformly distributed points in the neighboring areas of each of the chosen points using Sobol generator; the areas are shown in Fig. 15 (red rectangles);
2. Selection of 30 points from the resulting set by a specific criterion. Maximization of minimal distance from existing in the DoE points in the parameter space; $B_{min} \rightarrow max$;
3. Computation of the obtained 30 points on a high-level 3D CFD model (CHT of HPT blade);

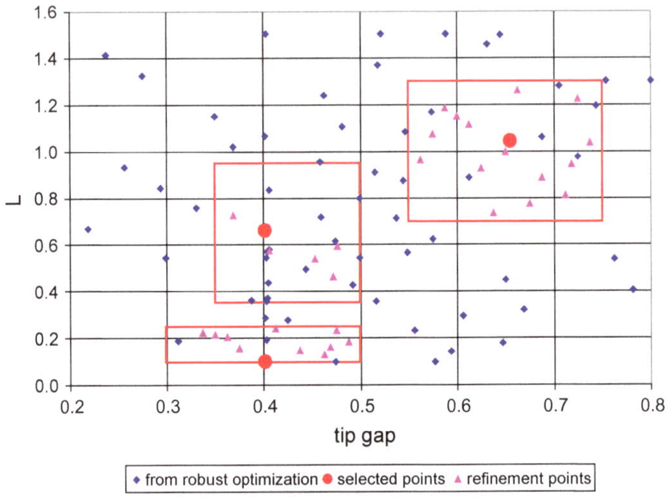

Fig. 15 Areas of interests chosen for the generation of additional points

4. Extending DoE after processing result of the CFD computations;
5. Modification of the approximating functions on the basis of CFD results;
6. Evaluation of the probabilistic criteria and distribution laws on the refined surrogate model;
7. Comparison of the robust optimization results and UQ results.

Employed procedure allows to carry out UQ during solving process of robust optimization problem on the various level of surrogate model refinement. At the same time, it allows to evaluate probabilistic criteria for every point in the original DoE.

Comparison with evaluation results according to robust optimization model for the point 1 is shown in Fig. 16 (comparison of the pdf for leading edge cooling).

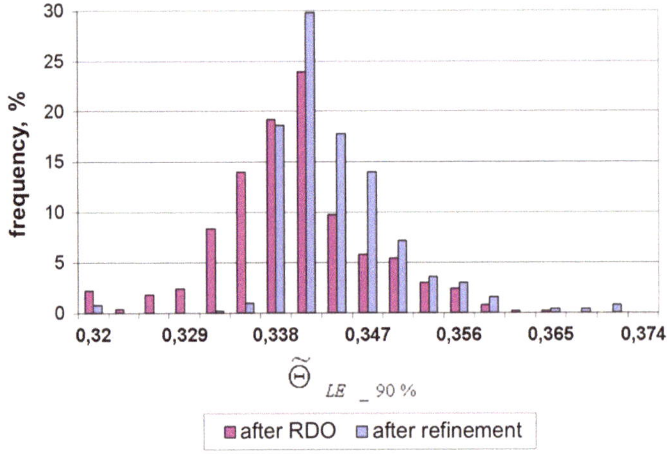

Fig. 16 Refinement of selected point's parameters evaluation. LE cooling effectiveness

Table 1 Accuracy comparison between RDO surrogate model and UQ model for point №1

	$\tilde{\theta}_{LE\_90\%}$	$\tilde{\theta}_{reb\_90\%}$	$\tilde{\zeta}_{reb\_90\%}$
RDO	0.337	0.192	0.132
UQ_refinement	0.331	0.192	0.133
Δ (%)	1.9	−0.4	−0.1

Comparison of the integral accuracy of the surrogate models for the point 1 before and after refinement (robust optimization and uncertainty quantification) is shown in Table 1.

Table 1 shows that maximal deviation between probabilistic criteria for the two surrogate models is observed for the $\tilde{\theta}_{LE\_90\%}$ values and is less than 2% (0.006 by absolute cooling efficiency), which indicates similarity between results of both models and possibility, in principle, of using robust optimization surrogate model for uncertainty quantification.

Results of computations and evaluations of probabilistic criteria on the refined surrogate model in comparison with evaluation results according to robust optimization model for the point 3 are shown of Fig. 17 (comparison of the probability densities for all criteria under consideration).

Comparison of the integral accuracy of the surrogate models for the point 2 before and after refinement (robust optimization and uncertainty quantification) is shown in Table 2.

Table 2 shows that maximal deviation between probabilistic criteria for the two surrogate models is observed for the $\theta_{reb\_90\%}$ values and is equal to 2.7% (0.005 by absolute cooling efficiency). This discrepancy was caused by using this criterion as a constraint for the robust optimization. The other two parameters are equal

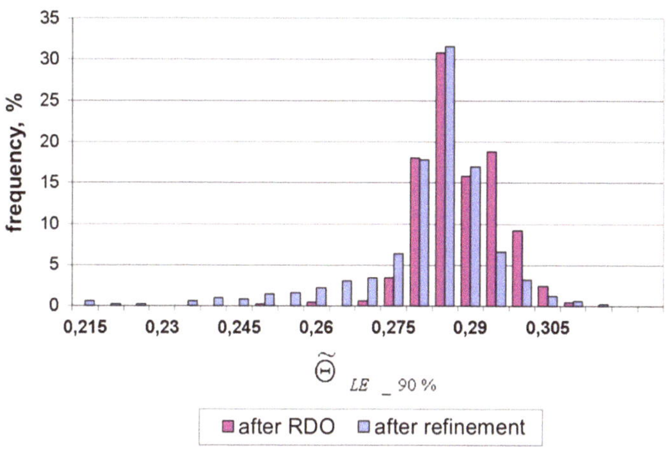

Fig. 17 Refinement of selected point's parameters evaluation. LE cooling effectiveness

Table 2 Accuracy comparison between RDO surrogate model and UQ model for point №3

	$\tilde{\theta}_{LE\_90\%}$	$\tilde{\theta}_{reb\_90\%}$	$\tilde{\zeta}_{reb\_90\%}$
RDO	0.262	0.216	0.143
UQ_refinement	0.267	0.222	0.142
Δ (%)	−1.8	−2.7	0.3

between models with even higher precision, which indicates similarity between results of both models and possibility, in principle, of using robust optimization surrogate model for uncertainty quantification.

Conclusions

1. One of the most promising techniques to solve UQ and RDO problems in coupling is, for example, usage of approximate assessments of probabilistic criteria under Monte Carlo combined with direct optimization techniques IOSO.
2. Application of the multi-level optimization procedure offers a significant reduction of the computation cost and time expenditures for the solution of complex real-life problems while maximizing the probability of manufacturing.
3. Uncertainty quantification using combined method (Monte Carlo+RSM+IOSO) was performed. Primary characteristics and probabilistic criteria of the thermal state and aerodynamic characteristics of the HPT blade were obtained.
4. Conducted investigation has shown that surrogate model constructed during robust optimization provides acceptable accuracy of the predictions of the numerical characteristics of the model and required probabilistic criteria (mean, standard deviation, etc.). Initial surrogate model (RDO) has shown acceptable accuracy in comparison with extended and refined models on all three examined regions. On average, the difference between results is below 2.5–3%.
5. The largest difference between surrogate models (about 6%) was observed in the case of probabilistic criteria of the blade tip thermal state, which was used as a constraint. It was caused by the low number of computational points in the subregion used in solving robust optimization problem.
6. Results of solving robust optimization problem let us estimate probabilistic criteria with acceptable accuracy in the required regions of the space of parameters and optimization criteria. This proves that employed approach to solving problems of UQ and RDO simultaneously is highly efficient.

References

1. Lee, K.-D., Kim, S.-M., Kim, K.-Y.: Optimization of a fan-shaped hole for film cooling using a surrogate model. In: ASME Paper GT2009-59520 (2009)
2. Johnson, J.J., King, P.I., Clark, J.P., Ooten, M.K.: Genetic algorithm optimization of an HPT vane pressure side film cooling array. In: ASME Paper GT2012-68049 (2012)
3. Fadlun, E.A., Michelizzi, I., De Iaco, M.: Measurement error influence on gas turbine operability for condition-based maintenance and reliability/availability improvement. In: ASME Turbo Expo 2008, Berlin, Germany, GT2008-50749 (2008)
4. Bunker, R.S.: The effect of manufacturing tolerances on gas turbine cooling. J. Turbomach. **131**, 41018-41018-11 (2009)
5. Carnevale, M., Montomoli, F., D'Ammaro, A., Salvadori, S.: Uncertainty quantification: a stochastic method for heat transfer prediction using LES. In: ASME Paper GT2012-68142 (2012)
6. Montomoli, F., D'Ammaro, A., Uchida, S.: Uncertainty quantification and conjugate heat transfer: a stochastic analysis. In: ASME Paper GT2012-68203 (2012)
7. ANSYS CFX Theory, vol. 14. Ansys Inc. (2011)
8. Menter, F.R.: Two-equation eddy-viscosity turbulence models for engineering applications. AIAA J. **32**(8), 1598–1605 (1994)
9. Menter, F.R.: A comparison of some recent eddy-viscosity turbulence models. ASME Trans. J. Fluid Eng. **118**(3), 514–519 (1996)
10. Bardina, J.E., Huang, P.G., Coakley, T.: Turbulence modeling validation. In: 28th Fluid Dynamics Conference, AIAA Paper 1997-2121 (1997)
11. El Ayoubi, C., Ghaly, W., Hassan, I.: Optimization of film cooling holes on the suction surface of a high pressure turbine blade. In: ASME Paper GT2012-69773 (2012)
12. Didenko, R.A., Karelin, D.V., Ievlev, D.G., Shmotin, Y.N., Nagoga, G.P.: Pre-swirl cooling air delivery system performance study. In: ASME Paper GT2012-68342 (2012)
13. Sieverding C.H.:Recent progress in the understanding of basic aspects of secondary flows in turbine blade passages A.S.M.E. J. Eng. Gas Turbines Power, **107**, 248–257 (1984)
14. Egorov, N., Kretinin, G.V., Leshchenko, I.A.: How to execute robust design optimization. In: Proceedings of 8th AIAA/USAF/NASA/ISSMO Symposium on Multidisciplinary Analysis and Optimization, Atlanta, GA, USA, AIAA Paper 2002-4328 (2002)
15. Egorov, I.N.: Indirect optimization method on the basis of self-organization. In: International Conference on Optimization Techniques and Applications (ICOTA'98), vol. 2, pp. 683–691. Curtin University of Technology, Perth, Australia (1998)
16. Egorov, I.N., Kretinin, G.V., Leshchenko, I.A., Babiy, Y.I.: Optimization of complex engineering systems using variable-fidelity models. In: Proceedings of the 1st ASMO UK/ISSMO Conference on Engineering Design Optimization, pp. 143–149. MCB University Press (1999). ISBN: 0-86176-650-4
17. Egorov, I.N.: Optimization of a multistage axial compressor. Stochastic approach. In: ASME Paper 92-GT-163 (1992)
18. Egorov, I.N., Kretinin, G.V.: Optimization of gas turbine engine elements by probability criteria. In: ASME Paper 93-GT-191 (1993)
19. EU FP6 program NODESIM-CFD. http://www.nodesim.eu/
20. Yokoyama, R., Ito, K.: Robust optimal design of a gas turbine cogeneration plant based on minimax regret criterion. In: ASME Paper 99-GT-128 (1999)
21. Wu, Y.-T.: Methods for efficient probabilistic analysis of systems with large numbers of random variables. In: Proceeding of 7th AIAA/USAF/ NASA/ISSMO Symposium on Multidisciplinary Analysis and Optimization, St. Louis, USA, AIAA-98-4908 (1998)

Application of Uncertainty Quantification Methodologies to Falcon

G. Rogé and X. Loyatho

Geometry, Design Point, Modelization

Geometry:

The geometry of the generic Falcon Jet has been delivered to workshop participants. The CAD model was available under several formats: .CATPart, Catia V5R20 ServicePackage 6 HotFix74; igs; .stp.

Access rights have been granted upon request for the own work of the only Parties (Alenia Aermacchi SPA, Institut National de Recherche en Informatique et en Automatique, NUMECA, Vrije Universiteit Brussel and Technische Universiteit Delft) involved in IC-03 test for the purposes and the duration of UMRIDA project.

Any publication of data resulting from the use of this "Géométrie Falcon Générique" will be subject to the prior authorization of DASSAULT AVIATION in accordance with article 8.3.

All the data were written on a CD ROM (confidential and the intellectual property of DASSAULT AVIATION. It may not be used, reproduced, modified, or disclosed without its authorization) (Fig. 1).

Design Point:

The general description of the Falcon Jet test case is the following: Mach number = 0.8, Angle of Attack = 2°, Altitude = 40,000 ft = 12,192 m, Reynolds number = 14.512 Million (based on mean aerodynamic chord), Reference density = 0.30132 kg/m^3, Reference pressure = 18820.15 Pa, Reference temperature = 216.65 K, Reference velocity = 236.57 m/s, Reference length = Mean

G. Rogé (✉) · X. Loyatho
Dassault Aviation, 78 quai Marcel Dassault, 92552 Saint-Cloud, France
e-mail: gilbert.roge@dassault-aviation.com

X. Loyatho
e-mail: ximun.loyatho@dassault-aviation.com

C. Hirsch et al. (eds.), *Uncertainty Management for Robust Industrial Design in Aeronautics*, Notes on Numerical Fluid Mechanics and Multidisciplinary Design 140, https://doi.org/10.1007/978-3-319-77767-2_24

Fig. 1 Generic Falcon Jet (IC-03)

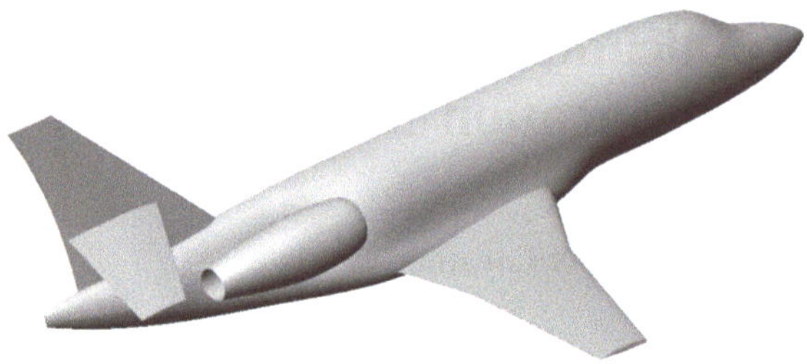

aerodynamic chord = 2.888 m, Reference area = 49 m^2 (full aircraft), Moment reference point: (x = 9.0355 m, y = 0 m, z = 0 m).

Modelization:

The modelization is the following: Reynolds-Averaged Navier–Stokes (RANS) 3D, Wall treatment: adiabatic no-slip wall, Flow regime: fully turbulent. Turbulence modeling: Spalart–Allmaras or k-ε SST two layers or k-ω SST Menter, Volume mesh: y+ ~ 1.

Uncertainties

Aerodynamic parameters:

Two aerodynamic parameters have been considered: the angle of attack and the free stream Mach number. Stochastic modelization follows bounded, asymmetric beta PDF distributions (Holland approach) with the following parameters: angle of attack min = 1.97°, angle of attack max = 2.1°, most likely value for angle of attack = 2°, free stream Mach number min = 0.795°, free stream Mach number max = 0.807°, most likely value for free stream Mach number = 0.8 (Figs. 2 and 3).

Fig. 2 Beta law for AoA

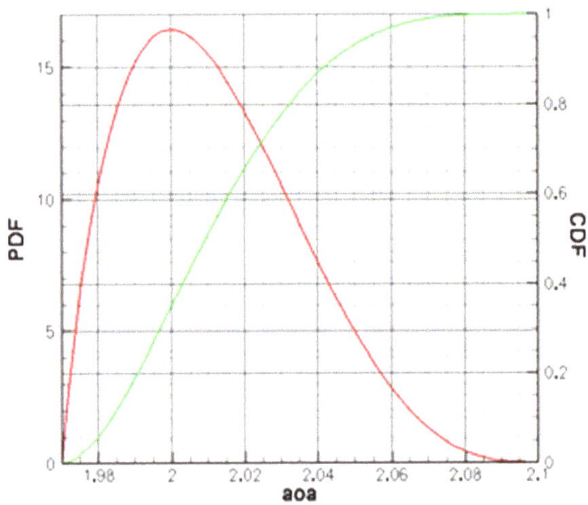

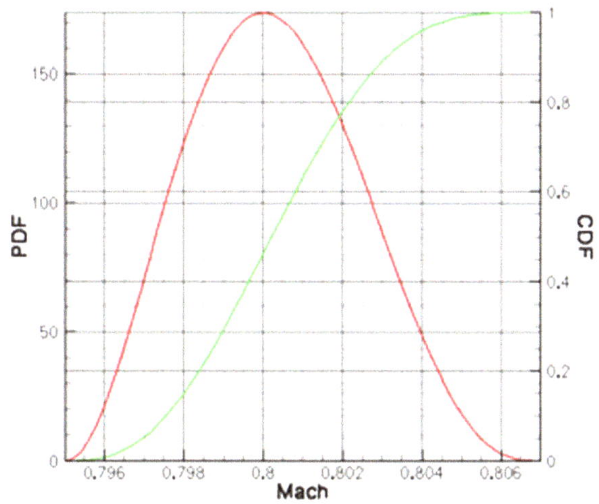

Fig. 3 Beta law for free stream Mach number

Geometric parameters:

Eight geometric parameters have been considered: These parameters control wing spanwise twist angle distribution (x/c = 0.25; rotation/control section plane). The eight control sections (S2, …, S9) are the following: 0.999 y + 0.008 z = 1240.614; 2205.710; 3131.413; 3848.613; 5293.181; 6628.500; 8074.377; 9244.835 mm.

Stochastic modelization follows *Bounded, Asymmetric Beta PDF distributions (Holland approach)* with the following parameters: *Delta twist angle min = −0.5°, Delta twist angle max = 0.2°, Most likely value for Delta twist angle = 0.01°* (Figs. 4 and 5).

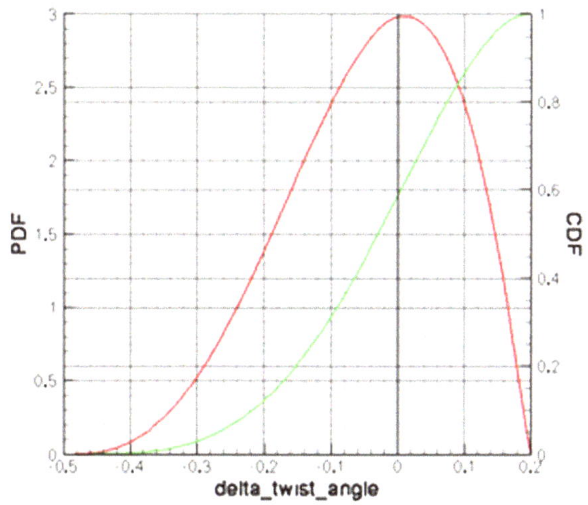

Fig. 4 Beta law for twist angle

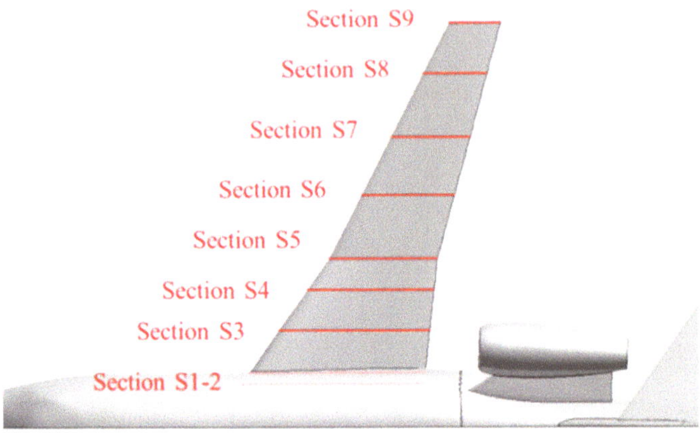

Fig. 5 Control sections

Modeling uncertainties (epistemic):

Modelization uncertainties have been considered both using different RANS Turbulence models: Spalart–Allmaras, k-ε two layers, k-ω SST Menter, and different set of coefficients for these models.

Numerical error:

Numerical error has been considered through mesh: volume mesh: Y+ = 1, Y+ < 1, Y+ > 1; skin mesh density: 2; other effects.

Expected Output

For total drag coefficient (CD), lift coefficient (Cl), pitch moment coefficient (Cm), wing spanwise cCl(y) distribution, Cp distribution (cutter y = 3848 mm), the goal is to compute both 4 first statistical moments (mean value, standard deviation, skewness, kurtosis) and PDF/CDF.

One of the additional objectives is to perform the following cross comparisons between partners: RANS exploitation, sensitivity study (first and second derivatives), uncertainty propagation study (Method of Moment, Monte Carlo family, Polynomial Chaos family, other methods), study of cross effects (e.g., turbulence modeling—mesh), global UQ on the database incorporating modelization (turbulence), numerical error (mesh), operational (AoA, free stream Mach number), and geometrical (wing spanwise twist angle distribution) uncertainties.

Results Dassault Aviation

CFD framework:

Dassault Aviation in-house Navier–Stokes solver AETHER is based on Streamline Upwind Petrov Galerkin (SUPG: more precisely, GLS, Galerkin Least Square) stabilized finite element approach. Navier–Stokes equations are symmetrized using entropy variables. Finite element spatial discretization is based on unstructured tetrahedral elements. A fully implicit iterative time-marching procedure based on GMRES (BSOR preconditioning) algorithm allows us to an efficient resolution. Domain decomposition and parallelized (using MPI) led us to a scalable and out-performing code.

Source code differentiated with Tapenade (Automatic Differentiation product of INRIA) led us to Direct and Adjoint approach.

Baseline calculation:

In order to build a reference point for uncertainties study, our choice is the following: Spalart–Allmaras and an unstructured tetrahedral mesh (7.17 Million vertices, 42.6 Million tetrahedral elements) (Figs. 6 and 7).

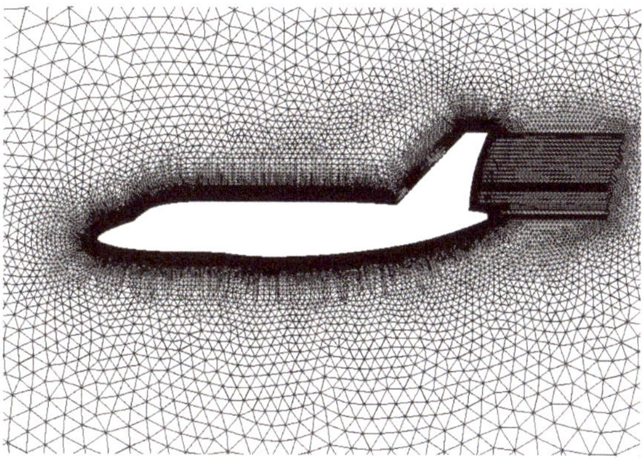

Fig. 6 Mesh symmetry plane, y = 0. Zoom

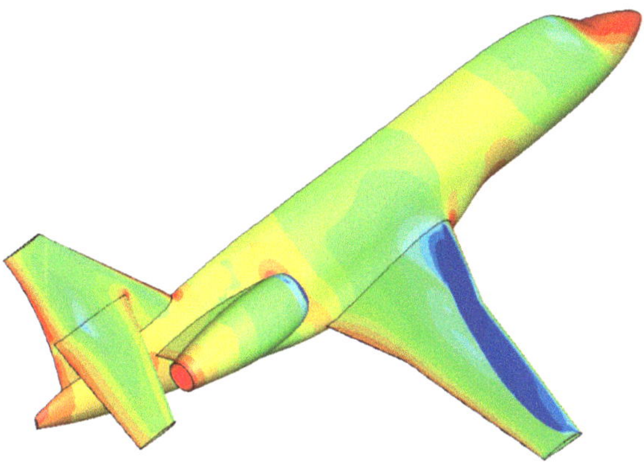

Fig. 7 Cp distribution

Mesh effect:

Four unstructured tetrahedral meshes have been built starting from the same surface mesh: MESH 1, 7.17 Million vertices; MESH 2, 8.60 Million vertices; MESH 3, 8.91 Million vertices; MESH 4, 9.66 Million vertices (Fig. 8).

Turbulence modeling effect:

Three modelizations have been compared: Spalart–Allmaras; k-omega SST MENTER; k-epsilon 2 layers (Fig. 9).

Fig. 8 Mesh effect (Spalart–Allmaras)

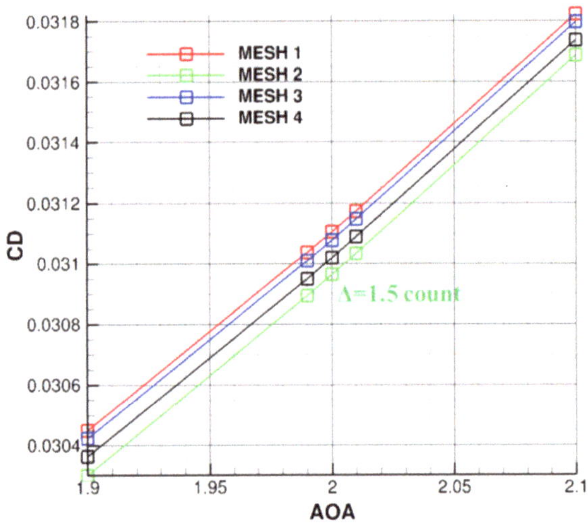

Fig. 9 Model effect (Mesh 1)

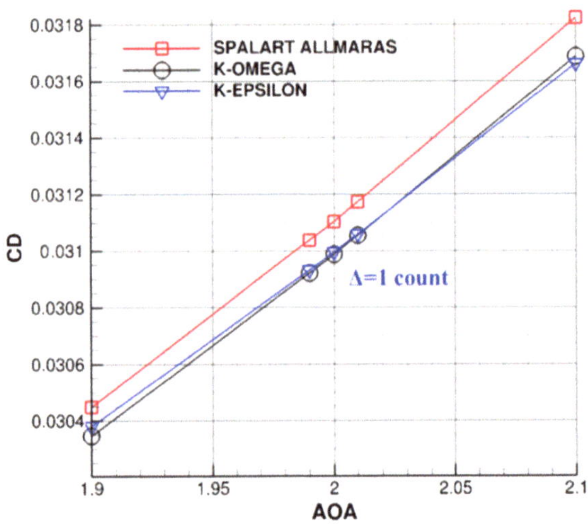

Aerodynamic parameters:

For uncertainty propagation approach, Method of Moments, order 1 has been used (Figs. 10, 11, 12 and 13).

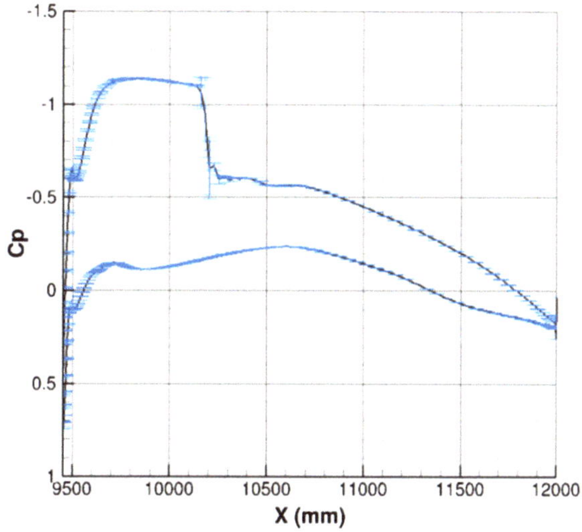

Fig. 10 Mach number effect (Spalart–Allmaras, Mesh 1). Cp. (± standard deviation) Cutter y = 3848 mm

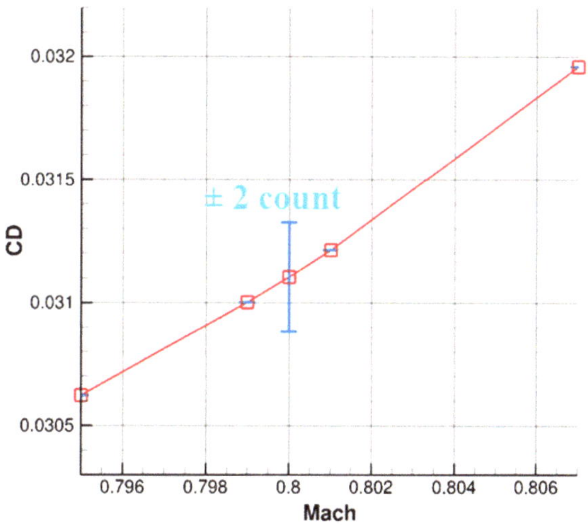

Fig. 11 Mach number effect (Spalart–Allmaras, Mesh 1). Drag (± standard deviation) versus Mach

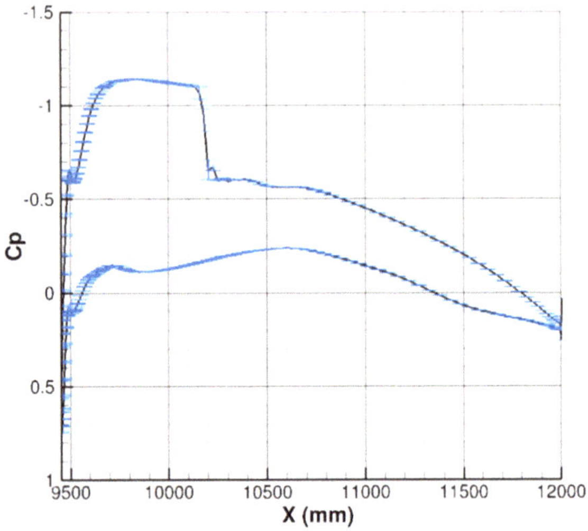

Fig. 12 AoA effect (Spalart–Allmaras, Mesh 1). Cp. (± standard deviation) Cutter y = 3848 mm

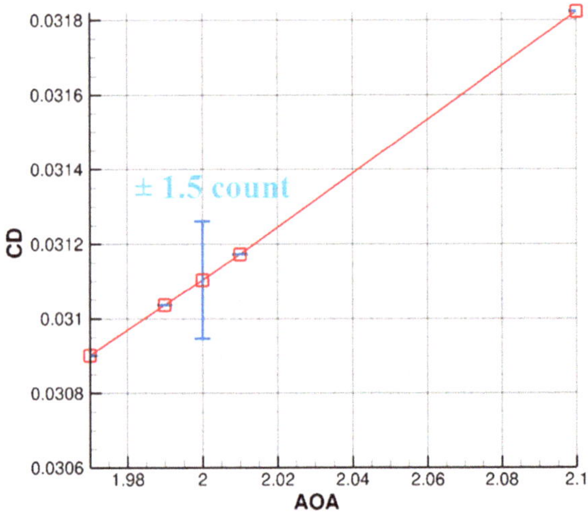

Fig. 13 AoA effect (Spalart–Allmaras, Mesh 1). Drag (± standard deviation) versus AoA

Geometric parameters:

For uncertainty propagation approach, Method of Moments, order 1, has been used (Figs. 14, 15 and 16).

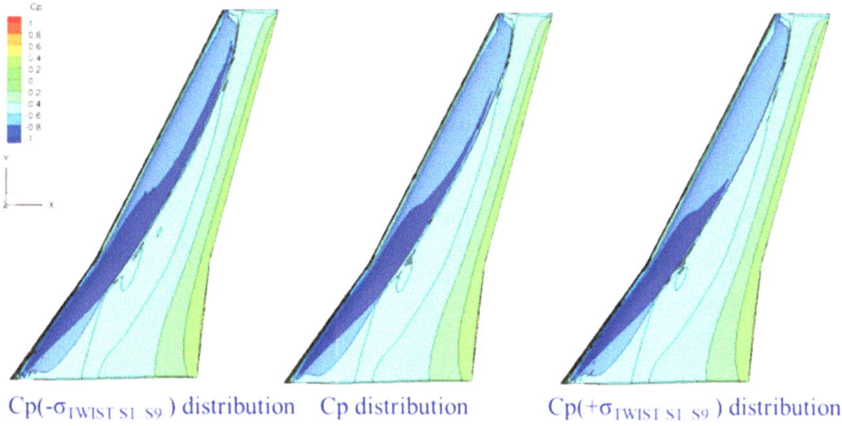

Fig. 14 Eight twist angles. Cp ± standard deviation distribution

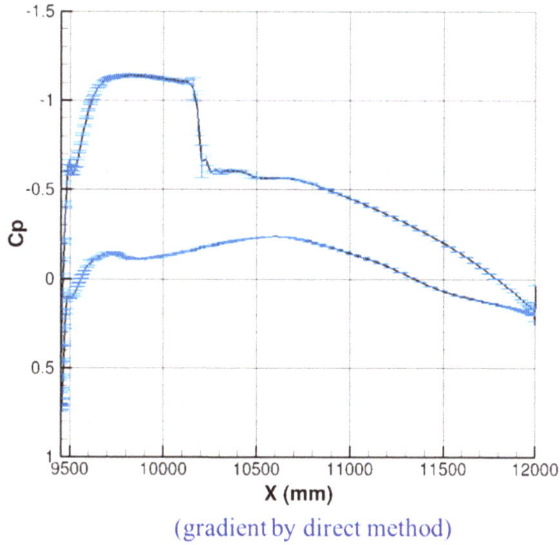

Fig. 15 Dassault Aviation. Eight twist angles. Cp ± standard deviation. Cutter y = 3848 mm

Partner's Results

NUMECA:

CFD framework:

NUMECAFINE TM/OPEN Solver Navier–Stokes is based on a second-order central difference scheme. Convergence criterion is to reach 7 orders of magnitude on relative residual. For a simulation starting from scratch, CPU is: 16 h on 1 core.

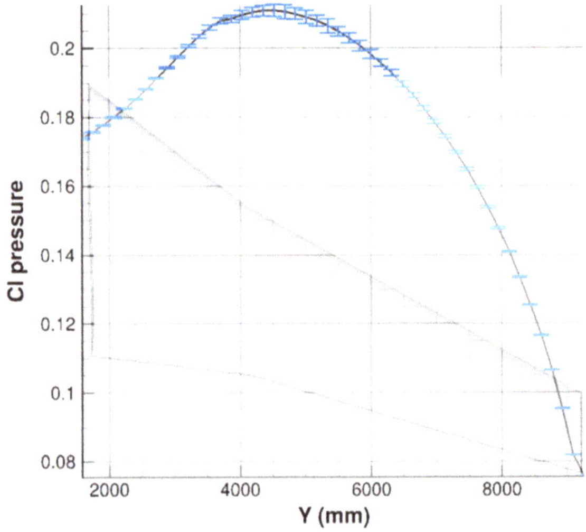

Fig. 16 Dassault Aviation. Eight twist angles. cCl ± standard deviation distribution

Baseline calculation:

In order to build a reference point for uncertainty study, NUMECAS's choice is the following: Spalart–Allmaras and a coarse mesh (no viscous layer) with around 3 Million of unstructured hexahedral cells, 3.5 Million of vertices and refinement close to: LE, TE, and wing/nacelles wake (Figs. 17 and 18).

For uncertainty propagation approach, Sparse Grid Level 1 NIPC, 31 collocation points (Smolyak quadrature) have been used.

CPU is 5 h on 100 cores (10 uncertainty variables: 8 twist angles, AoA, Mach) (results are given on Figs. 19, 20 and 21).

Fig. 17 Surface Mesh (Hexpress/Hybrid mesh generator)

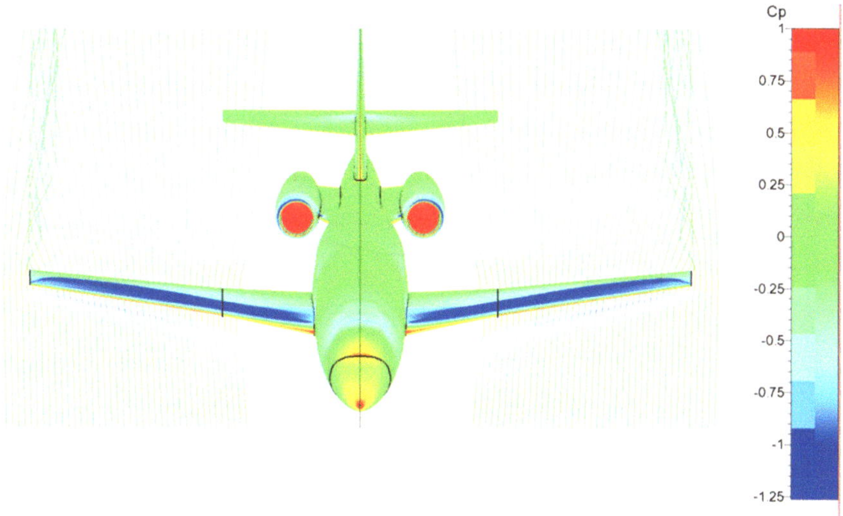

Fig. 18 Pressure coefficient contours and streamlines on undeformed geometry at nominal operating conditions

Fig. 19 Cutter y = 3848 mm. Mean value of Cp. Red color: RANS Dassault Aviation; Green color: NUMECA (coarse mesh without viscous layer → too backward shock, as with Euler approximation)

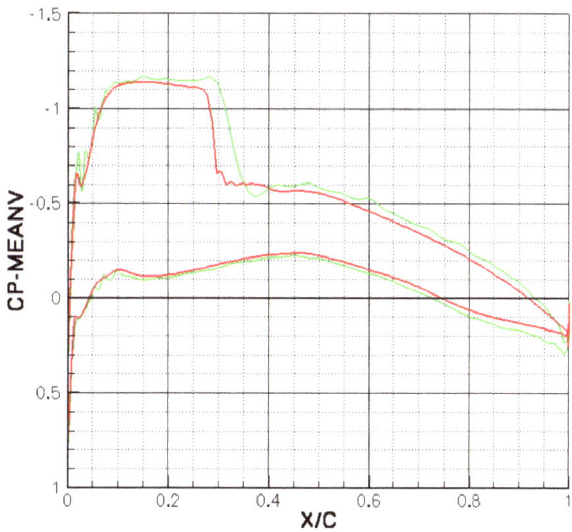

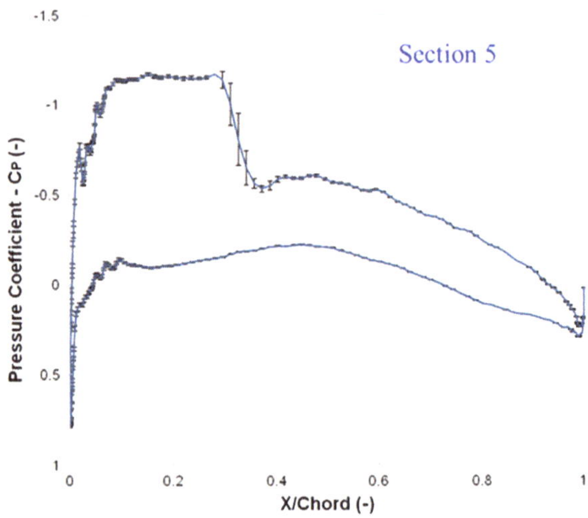

Fig. 20 NUMECA. Eight twist angles. Cp ± standard deviation. Cutter y = 3848 mm

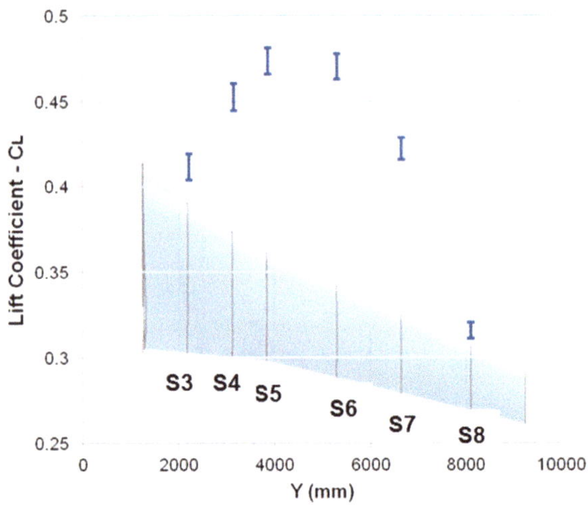

Fig. 21 NUMECA. Eight twist angles. cCl ± standard deviation distribution

Alenia Aermacchi (Leonardo):

Leonardo performs only Euler calculations. Of course, Euler instead of RANS approach let Leonardo a too backward shock.

For uncertainty propagation approach, Method of Moment Order 2 has been used (Figs. 22 and 23).

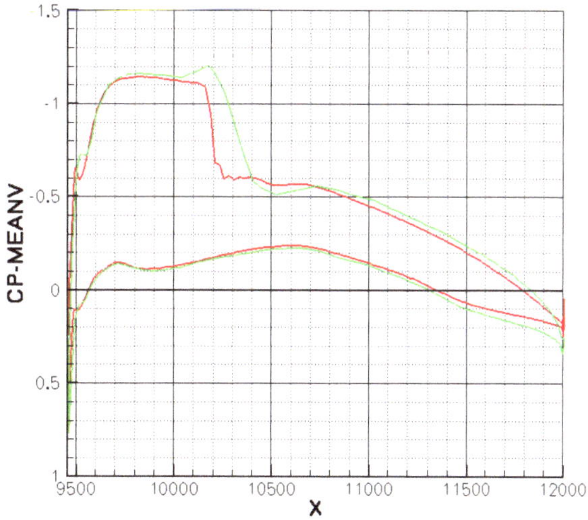

Fig. 22 Cutter y = 3848 mm. Mean value of Cp. Red color: RANS Dassault Aviation; Green color: Leonardo

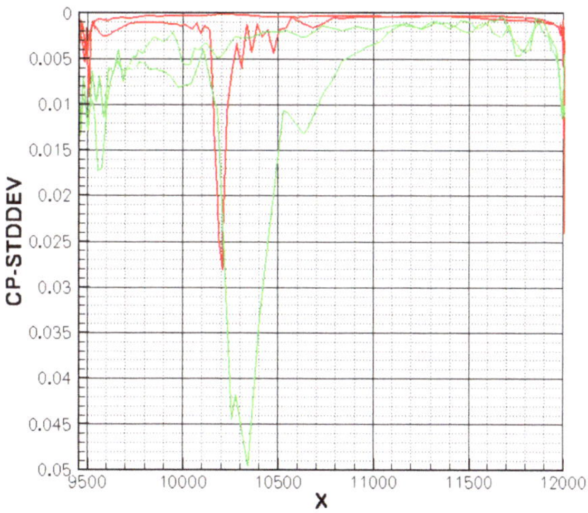

Fig. 23 Cutter y = 3848 mm. Standard deviation of Cp. Red color: RANS Dassault Aviation; Green color: Leonardo

Global coefficients comparison:

	Dassault Aviation	NUMECA	LEONARDO
CD	0.03110	0.00620	
Cl	0.37291	0.39300	
Cm	−0.07537	−0.17500	
Std CD	0.00030	0.00027	0.0201
Std Cl	0.00411	0.00640	0.0101
Std Cm	0.00076	0.00320	0.0107

Collaboration TU DELFT—Dassault Aviation

TU Delfts UQ approach is based on Bayesian Model Scenario Averaging (BMSA).

For the Spalart–Allmaras, 14 sets of 7 coefficients (KAPPA, CW2, CW3, CV1, CB1, CB2, SIGMA) have been built by TU Delft.

RANS calculations, using Dassault Aviation RANS code Aether has been performed (Figs. 24 and 25).

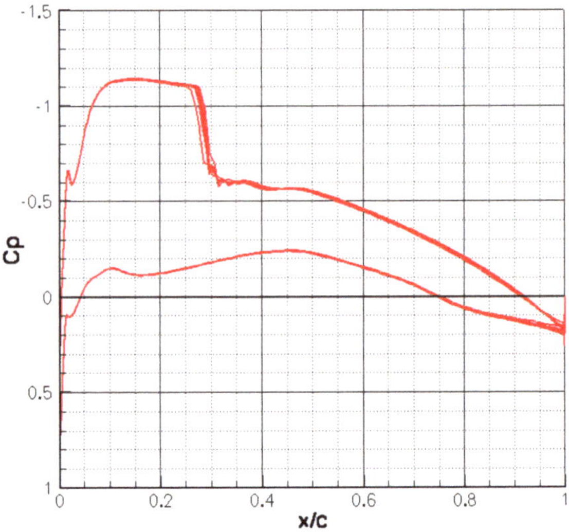

Fig. 24 Uncertainty on Spalart's coefficients. Cp for the 14 sets of coefficients. Cutter y = 3848 mm

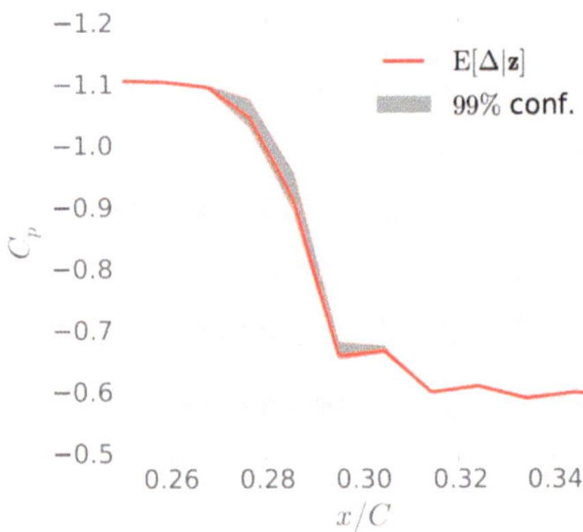

Fig. 25 UQ on Spalart's coefficients. Cp ± Std. Cutter y = 3848 mm. Zoom close to shock

Conclusion/Lessons Learned

A *UQ* Database incorporating modelization (turbulence), numerical error (mesh), operational (AoA) and geometrical (wing spanwise twist distribution) uncertainties have been established for IC-03 Falcon Test Case.

Recommendations:

1. Error

- Machine zero solution residual convergence (implicit scheme)
- Results approach constant CD, Cl, Cm, Cp, Cf with grid size adaption (consistent mesh family)

2. Modelization

- Uncertainty on model and model's coefficients (BMSA approach) has to be checked at least on the finer mesh

3. Shape uncertainty

- A large number of uncertain variables (with associated PDF) has to be taken into account. Variables could be dependent or not.
- Despite shock arising in transonic domain, first and second derivatives with respect to shape design parameters (like twist angle, relative thickness) are pertinent. MoM order 1 is very efficient. MoM order 2 is needed in case where first-order derivative is close to zero.
- NIPC/Sparse Grid is an efficient approach up to ten uncertain variables.

References

1. Umrida Project: Test Case Description Innovative Database For UQ and RDM: IC-03
2. UMRIDA Delft Workshop on UQ: IC-03 Falcon Test Case Presentation

Application of UQ to Combustor Design

S. Richard, J. Lamouroux, A. Ndiaye and F. Nicoud

Introduction

Combustion instabilities refer to the sustained pressure fluctuations of acoustic phenomena in combustors where unsteady combustion takes place [5]. Historically, the handling of combustion instabilities has been achieved through hardware design changes, and these modifications have included changes in fuel delivery system, changes in fuel injection distribution pattern and modifications to the combustor or combustor liner geometry. Design modifications have been made in the aim to considerably reduce the oscillations behavior of the system so as to avoid the coupled oscillations which lead to combustion instabilities. Because of the inherent nonlinearities associated with the turbulent flow or chemical reactions, the coupled interactions of acoustic waves and heat release fluctuations, combustion instabilities mechanisms are very complex to study. Besides, the numerous parameters involved may be huge. It is necessary to control them by taking care of uncertainties on input parameters and underlying models. Typically, the flame models obtained experimentally or numerically are known to be uncertain and these uncertainties can lead to strong variations of the system. In the present case, the flame model parameters

S. Richard (✉) · J. Lamouroux
Safran Helicopter Engines, Avenue Joseph Szydlowski,
64511 Bordes, France
e-mail: stephane.richard@turbomeca.fr

J. Lamouroux
e-mail: franck.lamouroux@safran.fr

A. Ndiaye · F. Nicoud
CERFACS, Avenue Gaspard Coriolis, 31000 Toulouse, France
e-mail: ndiaye@cerfacs.fr

F. Nicoud
e-mail: franck.nicoud@umontpellier.fr

C. Hirsch et al. (eds.), *Uncertainty Management for Robust Industrial Design in Aeronautics*, Notes on Numerical Fluid Mechanics and Multidisciplinary Design 140, https://doi.org/10.1007/978-3-319-77767-2_25

Fig. 1 Full helicopter engine (right) and detailed view of the annular combustor (left)

are extracted from self-excited LES computation of a whole annular helicopter combustor (Fig. 1). Uncertainties on such parameters are associated to grid refinement, combustion and turbulence model quality, numerics and boundary conditions which may affect the flame response to acoustic perturbations.

This configuration has been studied using large eddy simulation techniques by the mean of the AVBP code [2, 4] that are powerful for studying the dynamics of turbulent flames and their interactions with the acoustic waves [6].

However, these simulations are very CPU demanding and faster tools such as low-order tools are required in the design process of new burners. Therefore, a sensitivity analysis of thermo-acoustic results with respect to the input parameters is thus a necessary step toward reliable predictions of unstable modes in such complex systems.

In Fig. 2, a typical result of a thermo-acoustic analysis is showed, i.e., a set of modes, each with its own frequency and growth rate. When no uncertainty is present, each mode corresponds to a single point (black symbols) in the frequency plane. Here, modes 1, 4, and 5 should be controlled or avoided since the growth rate ω_i is positive. If uncertainties are present, each mode belongs to an admissible region of the frequency plane. Mode 2 (and maybe 6) should also be considered. It suggests that taking into account uncertainties is required for reliable predictions of combustion instabilities. The study of thermos-acoustic instabilities can be achieved

Fig. 2 Location of the first six thermo-acoustic modes in a typical combustor

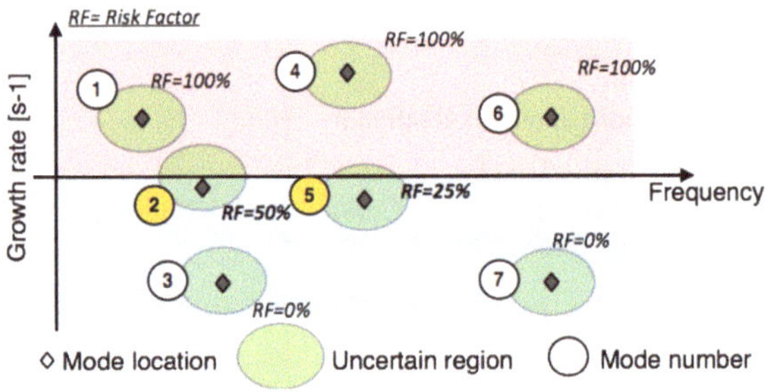

from several approaches: theoretical models, full-scale large eddy simulations (LES) or by low-order network methods. The problem using theoretical models remains the large number of hypothesis required simplifying the problem, and hence, only academic or highly simplified configurations can be studied. Using large eddy simulation (LES) to predict the dynamics of turbulent flames and their interactions with the acoustic waves is a good way to reproduce and describe the behavior of these instabilities. However, even when LES simulations confirm that a combustor is unstable, they neither suggest how to control the instability, nor give a measure of the results uncertainty. The CPU cost when using LES techniques is also very high, and faster tools are required to simplify the design process of the system. For example, performing parametric studies where hundreds of simulations are required is not tractable with LES, even on today's computers. A linear wave equation for small pressure perturbations $p_1(x; t)$ can be derived from the reactive Navier–Stokes equations by neglecting turbulence and viscous effects and assuming a frozen baseline flow:

$$\nabla \cdot \left(\frac{1}{\rho_0}\nabla p_1\right) - \frac{1}{\gamma p_0}\frac{\partial^2 p_1}{\partial t^2} = -\frac{\gamma - 1}{\gamma p_0}\frac{\partial q_1}{\partial t} \tag{1}$$

where q_1 is the heat release fluctuation. Assuming harmonic pressure $p_1 = \hat{p}(\mathbf{x})e^{i\omega t}$ and heat release fluctuations $q_1 = \hat{q}(\mathbf{x})e^{i\omega t}$, this equation becomes:

$$\nabla \cdot \left(\frac{1}{\rho_0}\nabla \hat{p}\right) + \frac{\omega^2}{\gamma p_0}\hat{p} = i\omega\frac{\gamma - 1}{\gamma p_0}\hat{q}(\mathbf{x}) \tag{2}$$

where $\hat{p}$ and $\hat{q}$ are the complex amplitude of the pressure and heat release disturbance, $\omega = 2\pi f$ is the complex angular frequency of the thermos-acoustic mode. The density ρ_0 and the specific heat ratio γ can depend on space $\mathbf{x}$ and are known quantities related to the baseline flow; the thermodynamic pressure p_0 is constant under the zero Mach number assumption. In this equation, the right-hand side term is closed by a flame transfer function (FTF) using the n-τ model. It is a global complex valued quantity that links the heat release fluctuations integrated over the flame zone to the velocity fluctuation of the acoustic wave at some reference location $\mathbf{x}_{ref}$ which is generally placed at the burner inlet:

$$\begin{aligned} \hat{Q}(\omega) &= \mathcal{F}(\omega)\hat{u}(\mathbf{x}_{ref}) \cdot \mathbf{n}_{ref}, \\ \hat{Q} &= \int_{\Omega} \hat{q}\, d\Omega \end{aligned} \tag{3}$$

The FTF is also associated to a unit vector $\mathbf{n}_{ref}$. The flame response can be deduced either from:

- Theoretical model for simple flames
- Large eddy simulation. In the present study, the FTF parameters have been deduced from a self-excited LES calculation of a Safran Helicopter Engines combustor

- Experimental data

The global expression of the FTF reads:

$$\mathcal{F}(\omega) = n(\omega)e^{i\omega\tau} \tag{4}$$

where n is the amplitude of the flame response and corresponds to the time delay of the latter. To be exploited numerically, the FTF information must be recast in order to give access to the local heat release $\hat{q}$ instead of only the volume integrated $\widehat{Q}$. This is done by introducing the local amplitude and time delay in such a way that: $\hat{q}(\mathbf{x}) = n_{local}(\mathbf{x})e^{i\omega\tau_{local}(\mathbf{x})}$. The preceding equations then provide the following constraint:

$$ne^{i\omega\tau} = \int_{\Omega} n_{local}\, e^{i\omega\tau_{local}} d\tau \tag{5}$$

Further assuming that n_{local} has a constant value within the flame region and is zero elsewhere allows relating the local FTF to the global one, experimentally measured. The time delay local can then be taken everywhere equal to the global value given from the experiment. They are known to control combustion instabilities although being uncertain. One objective of this report is to assess the sensitivity of the computed complex angular frequency to these uncertain parameters.

The test case targeted corresponds to a full annular helicopter combustor fed with N injectors and flames schematically represented in Fig. 3. Each of these injectors has 2 uncertain input parameters, n the FTF amplitude and τ the FTF time delay. It is therefore a high-dimensional UQ problem since 2 × N uncertain parameters are involved. Also, these types of annular system favor azimuthal modes that consist on a radial propagation of acoustic waves into the system. To deal with the curse of dimensionality, two strategies have been adopted (a) a brute force Monte Carlo analysis is performed directly on the 2 × N uncertain parameters and (b) the dimension of the system is reduced using a UQ methodology called active

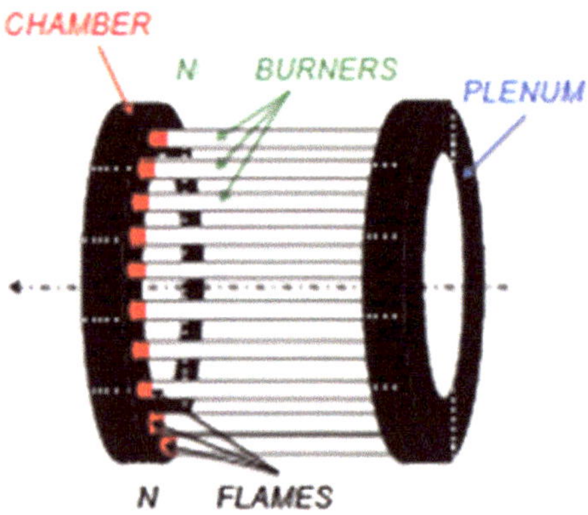

Fig. 3 The annular configuration studied with one plenum connecting N burners and an annular chamber

subspace. As Monte Carlo methods are known to be expensive, the goal is to understand the system variations by using only the strongest dimensions provided by the active subspace approach.

The present chapter is organized as follows. The second section presents the full annular combustor studied and the LES setup used to simulate this system. The third one presents the Helmholtz solvers used to study thermos-acoustic activity of the combustor. In section "Uncertainty Quantification of the Annular Helicopter Combustor," a brute force Monte Carlo methodology is applied to evaluate the uncertain region associated to the first acoustic mode of the system. The active subspace methodology proposed by CERFACS in chapter "Surrogates for Combustion Instabilities in Annular Combustors" of this book is used to reduce the system dimension. Then, the relevant dimensions of the system are used to fit a linear surrogate model. The purpose is to compute uncertain region of the first acoustic mode of the system with an affordable number of simulations using the model. Finally, the prediction and the accuracy of the low-order model are validated against the initial full Monte Carlo analysis.

LES of a Safran Helicopter Engines Annular Helicopter Combustor

The industrial validation case targeted in this study is an annular helicopter combustion chamber equipped with N burners. Each burner is composed of swirlers in which fuel is injected to efficiently mix kerosene with air prior to combustion. In this study, LES is first used to retrieve FTF parameters from a detailed compressible CFD unsteady computation with the AVBP solver (Fig. 4). Such a procedure avoids the realization of expansive tests based on pressure and heat release records. To reduce uncertainties on boundary conditions, the chamber casing is also simulated. The computational domain starts after the inlet diffuser and ends between the high-pressure stator and rotor. In this section, the flow is choked, allowing an accurate acoustic representation of the outlet. The air flowing in the casing feeds the combustion chamber through the swirlers, cooling films, and dilution holes, all of those being explicitly meshed and resolved. Multi-perforated walls used to cool the liners are taken into account by a homogeneous boundary condition. Such a condition is not suited to account for acoustic damping at the combustor wall, resulting in a zero dissipation of acoustic waves at the combustor liner, thus often leading to an overestimation of the acoustic activity in the combustion chamber.

In the present LES, strong acoustic oscillations are found at a frequency close to 500 Hz (Fig. 5), allowing to directly retrieve FTF parameters of the burners. These parameters are the following: $n = 7612$ J/m and $\tau = 1.77$ ms.

These parameters will be used for acoustic computations in the following. At this stage, it is not possible to fully understand the origin of the acoustic instability, even if a longitudinal mode is suspected from the analysis of the pressure field. In this

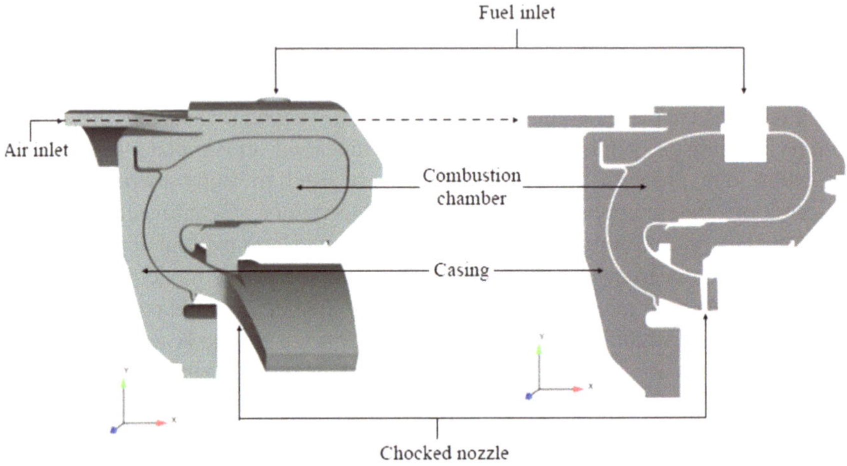

Fig. 4 Full annular Safran Helicopter Engines helicopter combustor (validation case)

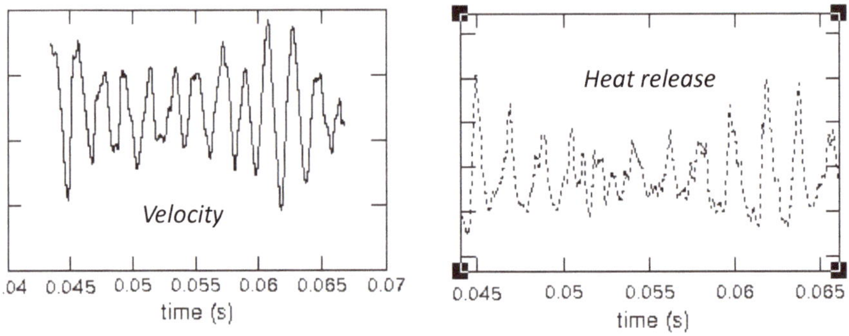

Fig. 5 Velocity and heat release fluctuations from Safran Helicopter Engines' LES computation

report, the risk factor associated to the different acoustic modes of the system—including both azimuthal and longitudinal modes, i.e., the probability of the mode to become unstable is targeted.

Acoustic Computations

The 3D Helmholtz Solver AVSP

To provide an acoustic analysis of the system studied, the Helmholtz solver developed at CERFACS and called AVSP is initially used to solve Eq. (1) in the full annular combustor. The aim of this tool is to provide acoustic analysis in three-dimensional configurations by taking into account the flame acoustics interactions. For solving the problem given by Eq. (1), it is necessary to give to this code the mean pressure p_0, the mean density ρ_0, and the mean field of sound speed $c_0^2 = \gamma \frac{p_0}{\rho_0}$ obtained from the time-average LES results. The outputs of the solver are the pressure field $\hat{p}$ and the complex angular frequency ω of the acoustic mode: its real part ω_r corresponds to the frequency of oscillation and ω_i is the growth rate of the acoustic disturbances. AVSP is used to study initially thermos-acoustic activity of the system. As the system is complex, one Helmholtz computation using AVSP takes 1–10 h running on 32 cores for the full annular combustor.

Single-Sector Computation with AVSP

In order to focus on the longitudinal mode observed in the LES, single-sector computations are performed with AVSP. Such computations do not allow retrieving azimuthal modes since rigid walls are considered as boundary conditions between two different sectors. First, passive flame simulations (i.e., without introducing acoustic–flame interactions) are performed to identify eigenacoustic modes of the combustor (Table 1). Several frequencies are found, the first one being around 500 Hz, i.e., close to the one observed in the LES. Active flame simulations are then conducted, in which the 500 Hz frequency is targeted. These simulations exhibit the presence of an unstable longitudinal mode at a frequency of 490 Hz (Table 2 and Fig. 6). This result allows to state that combustion instabilities appearing in the LES may be related to a longitudinal mode of the combustor.

Full Annular Combustor Computation with AVSP

Single-sector computations can only give access to longitudinal or transverse modes of the combustor. However, azimuthal modes could also appear, needing for full combustor (360°) acoustic computations. Passive flame simulations lead to the results presented in Table 3. The longitudinal mode suspected in the LES analysis and confirmed in the single-sector calculations is retrieved, while possible azimuthal modes are also identified at higher frequencies. At this stage, it is however

Table 1 Combustor eigenfrequencies from single-sector passive flame simulations

Passive flame			
Mode index	Mode structure	Frequency (Hz)	Growth rate (s^{-1})
1	–	495.55	0.0
2	–	1005.99	0.0
3	–	1591.93	0.0

Table 2 Combustor eigenfrequencies and associated growth rates from single-sector active flame simulations

Active flame			
Mode index	Mode structure	Frequency (Hz)	Growth rate (s^{-1})
1	Longitudinal	490.14	1.0e−1
2	Longitudinal	1004.22	9.22e−2
3	Longitudinal	1591.48	1.1e−1

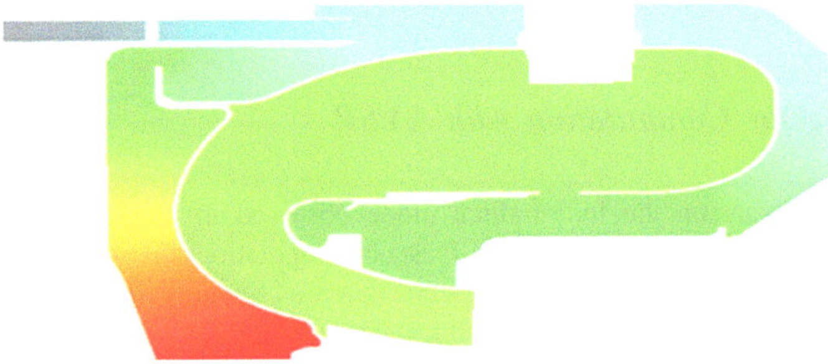

Fig. 6 Pressure perturbation from active flame simulations

Table 3 Combustor eigenfrequencies from full combustor passive flame simulations

Passive flame			
Mode index	Mode structure	Frequency (Hz)	Growth rate (s^{-1})
1	Longitudinal	495.55	0.0
2	Azimuthal	683.20	0.0
3	Azimuthal	683.228	0.0
4	Azimuthal	950.46	0.0

not possible to know if these modes could be unstable or not. This is why a new set of Helmholtz simulations is performed including an active flame model based on the FTF parameters to represent the flame/acoustic interactions. Each simulation aims at converging one target mode extracted from the passive flame computation. Results obtained from the new simulations are presented in Table 4 and Fig. 7. It is shown that both longitudinal and azimuthal modes may be unstable, as their growth rate is found slightly positive. It should also be underlined that the quite low values of the growth rate bring uncertainties to this statement, underlying the interest of performing a UQ analysis of this configuration.

In this UQ analysis, $2 \times N$ uncertain parameters are involved thus, it is worth investigating on a way to speed up and get at lower cost thermos-acoustic modes of the system. A new analytical tool developed at CERFACS and called ATACAMAC is therefore used as a surrogate model for AVSP and is more suitable for UQ analysis in the following.

Table 4 Combustor eigenfrequencies from full combustor active flame simulations

Active flame			
Mode index	Mode structure	Frequency (Hz)	Growth rate (s^{-1})
1	Longitudinal	490.14	1.22e−1
2	Azimuthal	680.17	6.05e−2
3	Azimuthal	680.24	6.27e−2
4	Azimuthal	950.42	−6.16e−2

Fig. 7 Pressure fluctuation modulus field. Left—first longitudinal mode (490 Hz); right—first azimuthal (standing) mode (680 Hz)

Analytical Tool to Analyze and Control Azimuthal Mode in Annular Chambers

The Analytical Tool to Analyze and Control Azimuthal Mode in Annular Chambers (ATACAMAC) has been developed to handle more complex geometries with an arbitrary number of burners [1] or configurations with two annular cavities which can couple an annular plenum and an annular chamber [1]. This study focuses on such a configuration with a 1D annular plenum which connects N burners and a 1D annular combustion chamber, representative of a real industrial gas turbine. The ATACAMAC tool is based on the annular network reduction, which first splits the full annular combustor into N sectors containing one sector of the annular chamber and plenum and an H connector between the annular cavities and the burners.

Each sector is split into two parts:

- Propagation in annular cavities (plenum or chamber): the plenum and the chamber can be treated separately. Pressure (p') and velocity (u') fluctuations in the ith sector of the plenum and chamber at the azimuthal position θ, $X_\theta = [p'_{p(\theta)}; u'_{p(\theta)}; p'_{c(\theta)}; u'_{c(\theta)}]^T$ are linked to acoustics quantities $X_{\theta+\Delta\theta}$ the location $\theta+\Delta\theta$ by a 4-by-4 rotation matrix $R_i(\Delta\theta)$.
- Interactions which couple the annular plenum and the annular chamber via the burner and the flame: acoustic pressure and velocity before the junction (θ^-) are linked to the acoustic quantities after the junction (θ^+) by a 4-by-4 matrix T_i

which depends on four coupling parameters [1]: $\Gamma_{i,1}$ corresponds to the plenum/burner interaction while $\Gamma_{i,4}$ is associated with the chamber/burner coupling. $\Gamma_{i,2}$ and $\Gamma_{i,3}$ are cross-interactions between the plenum and the chamber. The matrix T_i is obtained from jump conditions at null Mach number. The 4-by-4 matrix associated with the complete ith sector (interaction+propagation, which corresponds to $\Delta\theta = 2\pi/N$) is then equal to:

$$X_i(\theta_0 + 2\pi/N) = R_i X_i(\theta_0^+) = R_i T_i X_i(\theta_0^-) \tag{6}$$

Since the quantities between neighboring sectors are equivalent $X_i(\theta_0^-) = X_{i-1}(\theta_0 + 2\pi/N)$, using the periodicity condition $X_{N+1}(\theta_0) = X_1(\theta_0)$ yields:

$$X_1(\theta_0) = \left(\prod_{k=N}^{1} R_i T_i\right) X_1(\theta_0) \tag{7}$$

The eigenvector problem of Eq. (7) has non-trivial solutions if, and only if, its determinant is null leading to the dispersion relation:

$$\det\left(\prod_{k=N}^{1} R_i T_i - I_d\right) = 0 \tag{8}$$

The annular chamber is studied with identical burners and flames so that all coupling parameters Γ_i and matrices R_i or T_i are the same (the subscript i can be omitted in this section).

Since ATACAMAC relies on a simplified description of the combustor geometry, an adjustment of some geometrical parameters has first to be performed to fit 3D results from AVSP. Such an adjustment is based on the objective of reproducing both the real and imaginary part of the targeted eigenmode for a number of imposed FTF delays τ, which is a key parameter for flame instabilities prediction. In practice, this is mostly done slightly varying the burner length, since the burner (or injector) geometry is complex and its acoustic length is not easy to extract from a CAD. Results from this fitting procedure are presented in Fig. 8 and Table 5.

Uncertainty Quantification of the Annular Helicopter Combustor

Principle

ATACAMAC is used to study how the uncertainties on n and τ propagate into uncertainties on the growth rate ω_i and to determine the risk factor of the acoustic mode, i.e., the probability for a mode to be unstable ($\omega_i > 0$):

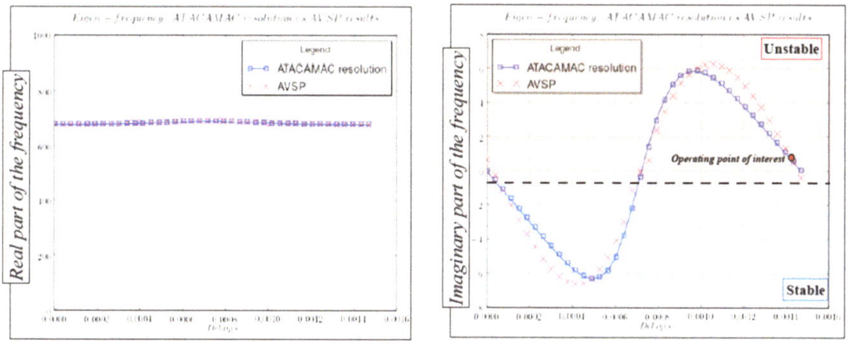

Fig. 8 Real (left) and imaginary (right) parts of a targeted combustor eigenmode for several FTF delays—comparisons between AVSP and ATACAMAC after adjustment of the burner length parameter

Table 5 Comparison of the frequency and growth rate of a targeted unstable mode obtained from AVSP and ATACAMAC computations after adjustment of the burner length parameter. Results are obtained with the FTF delay extracted from the LES

Active flame: AVSP versus ATACAMAC			
Mode index	Mode structure	Frequency (Hz)	Growth rate (s^{-1})
AVSP	1st azimuthal	680.17	6.05e−2
ATACAMAC	1st azimuthal	681.25	6.024e−2

$$Risk\ Factor\,(\%) = 100 \int_0^{\infty} PDF(\omega_i)\, d\omega_i \qquad (9)$$

Parameters of the FTF (n and τ) are varied to produce stability maps of the configuration. The literature does not confer a clear accurate analysis on the uncertainty range of the FTF parameters n and τ. The latter may have a non-negligible impact on the stability of the system. We thus retained the same values as those proposed by CERFACS on the first industrial test case of the project (see chapter "Surrogates for Combustion Instabilities in Annular Combustors"): $\sigma_n/n = 10\%$ and $\sigma_\tau/\tau = 5\%$, where $n_{crocco} = 3.916$ and $\tau_{crocco} = 7.097e{-}4$ correspond to the nominal LES values.

Brute Force Monte Carlo Analysis

Uncertainty quantification analysis is adopted to determine the risk factor of the first azimuthal mode of the system using the ATACAMAC tool ($f_0 = \omega_0/2\pi = 680$ Hz), i.e., the probability for this mode to be unstable. To do so, a brute force Monte

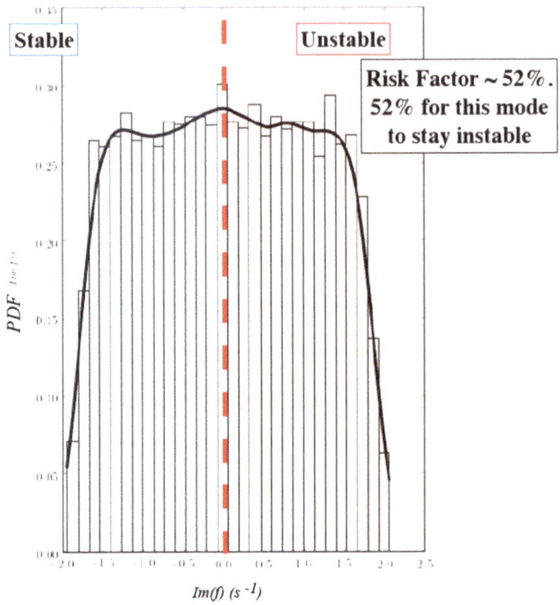

Fig. 9 Monte Carlo results using 8,000 samples: PDF of the mode to be unstable. Such a computation takes around 10 min versus days with the AVSP code

Carlo is first performed on the 2 × N uncertain input parameters. Convergence tests have proved that 8,000 samples are sufficient to evaluate the factor as well as mean and standard deviation of the growth rate (Fig. 9).

This study was done by taking care of the fact that the uncertainty on τ may lead to strong variations of the growth rate. The risk factor defined as the probability that the mode is unstable is approximately 30%. Monte Carlo methods are expensive to implement, that is the reason why a new UQ approach called active subspace (Constantine et al. 2014) is adopted to reduce the system dimensions.

Active Subspace Methodology

Active subspace (Constantine et al. 2014) is used to reduce the dimension of the parameter space from 2 × N dimensions to just a few. To find active variables of the system, the method requires gradient evaluation to detect which directions in the parameter space lead to strong variations of the growth rate. Other directions leading to a flat response surface are not useful for describing the combustor stability and hence can be disregarded. The gradient of the growth rate $g(p_k)$ with respect to the input parameters $p_k = (n_i, \tau_i)_{i=1,N}$ of the kth sample denoted $\nabla g(p_k) = (\nabla_p^k g)$ was thus computed by finite differences.

The uncentered covariance matrix C of the gradient vector can then be expressed as the following:

$$C = \mathbb{E}[(\nabla_p^k g)(\nabla_p^k g)^T] \simeq \frac{1}{M}\sum_{k=1}^{M} (\nabla_p^k g)(\nabla_p^k g)^T \tag{10}$$

where $\mathbb{E}$ is the expectation operator and M is the number of samples. Since this matrix is symmetric, positive, and semidefinite, it admits a real eigenvalue decomposition:

$$C = W\Lambda W^T, \Lambda = diag(\lambda_1, \ldots, \lambda_m), \lambda_1 \geq \cdots \geq \lambda_m \geq 0 \tag{11}$$

where W is the eigenvector corresponding to the coefficients of a linear combination of input parameters (W_p^T) and are the eigenvalues which quantify the effect of the active variable W_p^T on the growth rate response $g(p)$: the higher λ_i is, the more significant the active variable $W_i^T p$ is on the average output response. Figure 10 displays the eigenvalues spectrum obtained by the active subspace method using N = 500 samples.

It shows that only the first 3–4 active subspaces are relevant and control the system variations. The first active variable is a constant corresponding to an equi-weighted linear combination, i.e., associated with the mean flame transfer function over the N burners. Furthermore, a change of variables could be used to ease the physical interpretation of active variables as well as improve the accuracy of the singular value decomposition:

$$(n_i, \tau_i) \triangleright \Re(\hat{F}(\omega^0)), \Im(\hat{F}(\omega^0)) \tag{12}$$

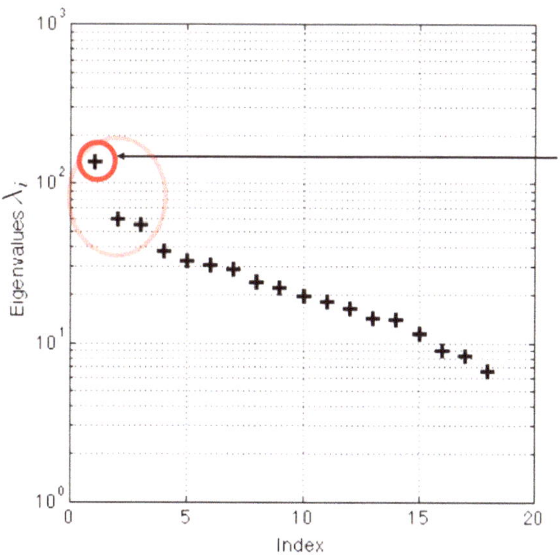

Fig. 10 Eigenvalues spectrum obtained by the active subspace method using N = 500 samples

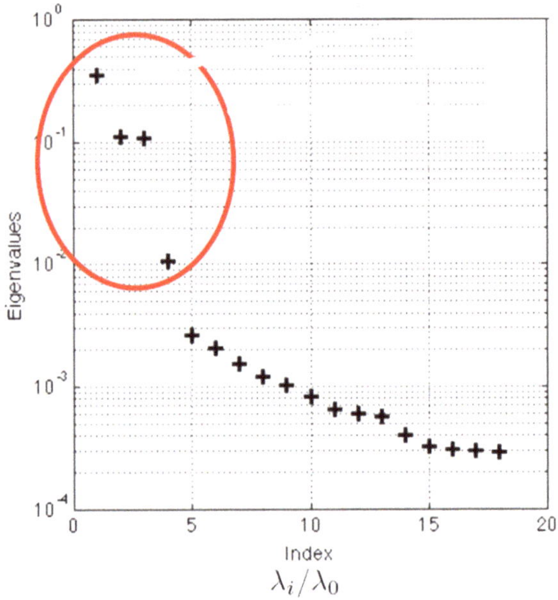

Fig. 11 Eigenvalues spectrum obtained by the active subspace method using N = 500 samples and $\Re(\hat{F}(\omega^0)), \Im(\hat{F}(\omega^0))$ formalism

where $F = [F_1, \ldots, F_N]^T$ is the array of FTF of all the N burners viewed as a periodic complex signal, $\widehat{F}$ is the he Fourier transform of the collection of FTF, $F_i(\omega^0)$ is the FTF of the ith flame depending on the parameter n_i, τ_i and the pulsation of the chamber in the absence of burners $\omega^0 = \pi c^0/L_c$.

Figure 11 confirms that only 4D active subspace are relevant. Other coefficients play only a minor role. The idea is using the 3D active subspace to t a surrogate linear model. This algebraic model is used to perform a Monte Carlo analysis at reduce cost to determine the risk factor of the first azimuthal mode.

Surrogate Linear Model Regression

The active subspace approach applied in the previous section has provided successful results by reducing the 2 × ND parameter space into a 3D space involving physical quantities associated with the Fourier transform of the FTF. A two-step method is therefore proposed to estimate at low cost the risk factor of the first azimuthal mode:

- few samples are used to fit a low-order model

$$\hat{\omega}_i = \alpha_0 + \alpha_1 Av1 + \alpha_2 Av2 + \alpha_3 Av3 + \alpha_4 Av4 \tag{13}$$

where α's coefficient are tuned using only few samples of the initial Monte Carlo database.

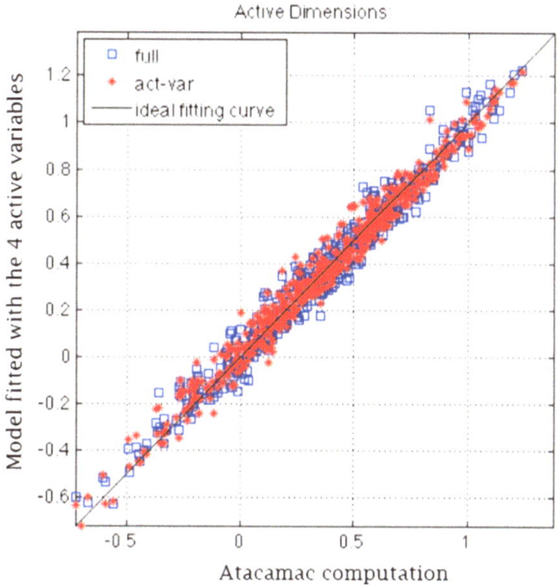

Fig. 12 ATACAMAC growth rates (100 samples) are compared with estimations obtained with the low-order model

- a Monte Carlo is applied on this low-order model.

Results of the model fitting are displayed in Fig. 12 where ATACAMAC growth rates (using 100 samples) are compared with estimations obtained with the low-order model. The low-order model has been replayed 100,000 to assess the growth rate as well as the risk factor of the first azimuthal mode at low cost. Therefore, the risk factor is approximately equal to 49.87%. The latter is close to the one assessed previously using Monte Carlo approach (52%).

Conclusion

In this chapter, a new methodology for UQ proposed by CERFACS in this project and called active subspace was used. This approach [3] detects which directions (or linear combination of directions) in the parameter space lead to strong variations of the growth rate. Other directions leading to fit response surface are not useful for describing the combustor stability and are disregarded, thus reducing the dimension of the surface response. In this study, computing the active variables requires a singular value analysis of the gradient of the surface response, which is to be performed with respect to all the 2 × N dimensions. For the helicopter combustor, gradient computations were achieved by finite differences to build the active variables $W = (W_1, W_2, W_3)$ and tune the linear regression model.

Being able to assess the gradient of the growth rate $\omega_i(p)$ at much lower cost than performing finite differences is, thus, necessary to achieve an efficient UQ analysis.

This can be achieved with the adjoint methods and would provide several benefits as sensitivity analysis using a 3D nonlinear Helmholtz solver as AVSP code. It would directly take into account all geometry features.

References

1. Bauerheim, M., Parmentier, J.F., Salas, P., Nicoud, F., Poinsot, T.: An analytical model for azimuthal thermos-acoustic modes in an annular chamber fed by an annular plenum. Combust. Flame **161**(5), 1374–1389 (2014)
2. Boileau, M., Staffelbach, G., Cuenot, B., Poinsot, T., Berat, C.: LES of an ignition sequence in a gas turbine engine. Combust. Flame **154**(1), 2–22 (2008)
3. Constantine, P., Dow, E., Wang, Q.: Active subspace methods in theory and practice: applications to kriging surfaces. SIAM J. Sci. Comput. **36**(4), A1500–A1524 (2014)
4. Moureau, V., Lartigue, G., Sommerer, Y., Angelberger, C., Colin, O., Poinsot, T.: Numerical methods for unsteady compressible multi-component reacting flows on fixed and moving grids. J. Comput. Phys. **202**(2), 710–736 (2005)
5. Ndiaye, A., Bauerheim, M., Moreau, S., Nicoud, F.: Uncertainty quantification of thermo-acoustic instabilities in a swirled stabilized combustor. In: ASME Turbo Expo (2015)
6. Wolf, P.: Ph.D. thesis, INP Toulouse (2010)

Manufacturing Uncertainties for Acoustic Liners

N. Magnino

Introduction

Reduction of aeronautical noise pollution is a design challenge with social, commercial, and technical implications because it eases the acceptance of new airports to match the always increasing demand of air transportations and reduces taxes and fee from government agencies and local administrations imposed to the airliner operating a noisy fleet, but further reduction of noise levels can be achieved only with introduction of innovative concepts of the design and the manufacturing of almost all the aircraft subsystems especially if they want newly produced aircrafts to respect the certification requirements imposed by authorities that are envisaged to be more and more stringent in the future. In modern turbofan engines, the fan contribution is one of the most important sources as the predominant portion of the thrust derives from the fan with the noise propagating both upstream irradiating from the inlet and downstream from the exhaust. As Leonardo Finmeccanica is one of the world's top suppliers of engine nacelles systems for big airliners, regional and business jets, we will focus only on the inlet noise which can be attenuated using sound-absorbing acoustic panels, aka acoustic liners installed just before the fan. In general, an acoustic liner consists of a porous external layer (facing sheet), which permits the passage of sound waves, one or more layers of honeycombs that act as Helmholtz resonators (eventually separated by a porous septum in the case of DDOF or multiple degree of freedom as they are called) and, finally, by a rigid backing skin. As acoustic liners have to prove effective at least in all the certification conditions prescribed by customer and authorities (i.e., Approach, Sideline, and Flyover), the design process (especially in the preliminary phase where a lot of scenarios are to be considered) is by nature a multi-objective optimization between eventually

N. Magnino (✉)
Leonardo S.p.A., Via P. Foresio 1, 21040 Venegono Superiore, VA, Italy
e-mail: nicola.magnino@leonardocompany.com

C. Hirsch et al. (eds.), *Uncertainty Management for Robust Industrial Design in Aeronautics*, Notes on Numerical Fluid Mechanics and Multidisciplinary Design 140, https://doi.org/10.1007/978-3-319-77767-2_26

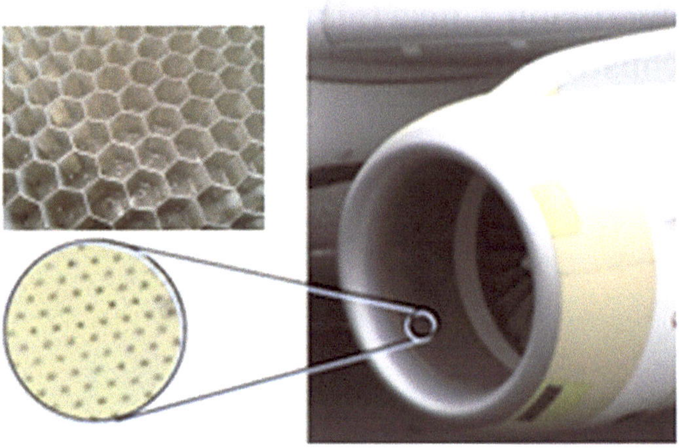

Fig. 1 Full engine nacelle (right) and detailed views of the acoustic panel interior (upper left) and perforated facing skin (lower left)

conflicting requirements that can be done manually with a traditional trial and error method as in the past or automatically [2] with various algorithms but with the risk to get local or just ideal optima whose real performance falls short in the field due to uncertainties (e.g., manufacturing tolerances) not being accounted, that is the rationale behind the participation of Leonardo Finmeccanica in UMRIDA (Fig. 1).

Acoustic Model of a Leonardo Finmeccanica Engine Nacelle Liner

The industrial validation case targeted in this study is a typical regional jet engine nacelle integrating an acoustic liner consisting of N Helmholtz resonating cavities. Each one is essentially composed of a honeycomb cell of a specific height targeted to attenuate the design frequency of the impinging noise being closed at the bottom by a rigid backing skin and on top by a porous face sheet whose holes dissipate energy as the acoustic waves travel back and forth. A consolidated (all metal SDOF) acoustic liner technology has been selected (Fig. 2) from vast Leonardo Finmeccanica production of acoustic liners, based on usefulness, readiness, and cost. The name of the program and the customer as well as engine data and liner impedance together with some technological process details will not be disclosed, as they are strictly company confidential. A quick note about the acronym used to identify the type of panel, single degree of freedom, means there is no division in the honeycomb cell as opposed to DDOF where a septum is inserted into it. The main geometrical uncertain parameters of typical SDOF with their driving factors and the four uncertainties considered in UMRIDA marked in solid black while others are grayed out:

- *Face sheet open area* (PoA_{eff}); f(Machinery Tool);
- *Honeycomb height* (h); f(Material);
- *Holes diameter* (d); f(Nacelle Geom., Machinery Tool);
- *Holes shape*; f(Manufacturing process);
- *Holes pitch* (p); f(Nacelle Geom., Manufacturing process);
- *Perforation angle* (α); f(Machinery Tool);
- *Face sheet thickness* (t_{fs}); f(Material, Manufacturing process);
- *Backing skin configuration*; f(Material, Manufacturing process).

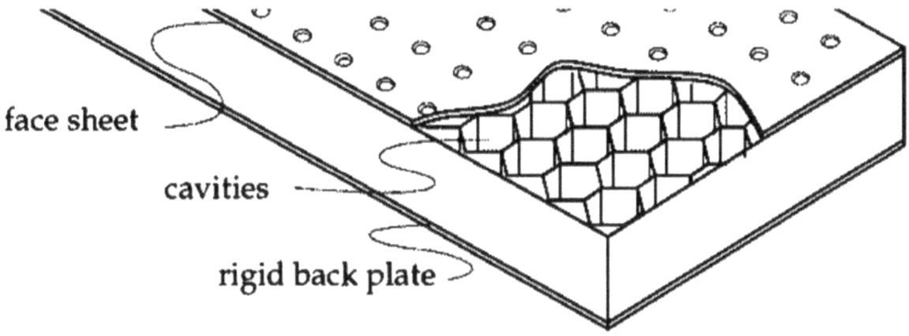

Fig. 2 Nacelle SDOF liner acoustic panel

Acoustic requirements are prescribed by customers and aviation agencies for three typical flight conditions: Approach; Sideline (or take-off); and Flyover (or cut-back) illustrated in Fig. 3.

As these requirements can be in contrast to each other and an acoustic liner providing the best attenuation in one flight condition not necessarily performs well in the other conditions, hence liners shall be designed through a multi-objective optimization procedure [2, 3]. Objective function represented by noise attenuation computed at certification flight conditions in the far-field surrounding the nacelle is considered. FNM has established an experimental database relevant for the investigation of manufacturing tolerances of acoustic panels. Specifically, a number of samples have been manufactured and tested to detect the steady (or DC) flow resistance for a set of given impinging air velocities by means of the FNM DC flow rig. Section "Results of Experimental Database Distribution Fitting" will illustrate how the experimental database has been used to quantify the manufacturing uncertainties. At the same time, FNM has prepared a numerical model representative of the nacelle with the acoustic liner using the commercial aero-acoustic software MSC Actran. In this finite element model, the liner is represented by a boundary condition of acoustic admittance whose values are taken from the in-house [4] impedance model via a spreadsheet updated by a macro as a function

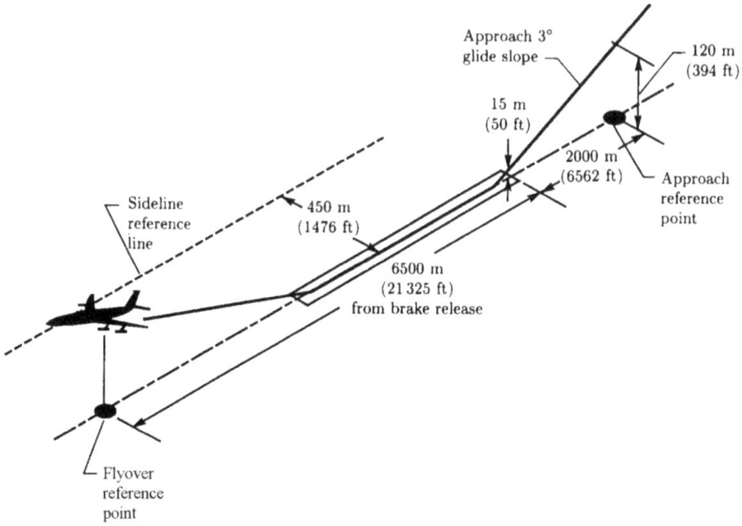

Fig. 3 Noise certification reference points

of geometrical and operational parameters. More details about Actran model are reported in next paragraph, while in section "Integration of Aeroacoustic Model Within modeFRONTIER Framework" we will describe how modeFRONTIER software from ESTECO has been used to automate the execution of a series of acoustic simulations for the purpose of an accurate UQ of the system response.

Aero-acoustic Model of the Liner

For the sake of UQ, the CAA simulations have been restricted to only one condition (the most computational demanding one, i.e., take-off), with the purpose to consider more in the further tasks of the UMRIDA project, after having identified the most efficient strategy. The geometrical parameters of the acoustic panel described in the previous chapter are entered into a proprietary impedance model [4] together with values depending on flight condition to get the admittance table in the format accepted by the boundary condition in the computational acoustic finite element model. Simulations have been performed with MSC Actran software developed by Free Field Technology, on an axisymmetric 2D model comprising of about 30,000 quadratic elements. The level of refinement of the mesh must be enough to resolve the convected waves associated with the frequencies and flow speed considered (highest for take-off) and could be eventually refined to gain accuracy despite an increase in computational costs. The simulation time to complete one configuration design was about 2 min running in parallel on eight cores of one of the computational nodes in the FNM's HPC. As cost function ideally it should be used the effective Perceived Noise Level (EPNL) as prescribed by the certification authorities, but in order to calculate it it's necessary to have the whole airplane's noise modeled, not only the fan and also all the 1/3 octave band center frequencies should be considered leading to computational costs not compatible with UMRIDA. In this work, we will use a simpler metric that can be computed more easily: the overall sound pressure level (OASPL) defined in:

$$OASPL = 10Log \sum_{j} 10^{\frac{SPL_j}{10}} \tag{1}$$

$$SPL(f) = 10Log\left(\frac{\sum_{40^\circ}^{90^\circ} \sum_{m,n} p_{rms}{}^{2}{}_{(\vartheta,f,m,n)}}{p_{ref}^{2}}\right) \tag{2}$$

The OASPL uniformly sums over the directivity angles and over frequencies the Sound Pressure Level (SPL) which is a function of the frequency, the root mean square of the acoustic pressure function of radial and azimuthal modes (that propagate out of the nacelle) and directivity angle at a distance of 150 ft according to the scheme in Fig. 4.

All the Actran simulations were performed at two frequencies—the BPF and its first harmonic 2BPF—and the range of directivity used in this work is 50°–80°, considered the most critical for the take-off condition (the only one considered for UQ). In fact, noise radiating from engine inlet is tonal and therefore at the Blade Passage Frequency (BPF) and its harmonics higher sound levels are propagated. In this view, a liner should be optimized to attenuate the noise mainly at these frequencies.

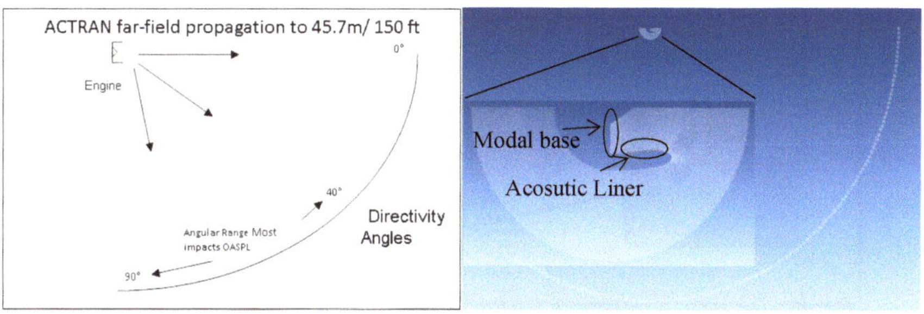

Fig. 4 Schematic of far-field propagation and ACTRAN mesh with acoustic panel circled

UQ of Experimental Database Parameters

The first step is the UQ of the geometrical uncertainties of the acoustic liner. A campaign of experiments has been conducted by FNM, which has performed the geometrical measurements of a series of acoustic panels. The database could therefore be imported in modeFRONTIER software from ESTECO and analyzed with distribution fitting tool in order to find the statistical distributions which better fit the experimental data. In order to perform the UQ of the acoustic performances through a numerical model (sections "Integration of Aeroacoustic Model Within modeFRONTIER Framework" and "UQ of Acoustic Response"), it is in fact necessary to first define analytically a distribution which can be considered when defining the numerical sampling, in order to be accurately representative of the real database. Next sections will describe the basics of the distribution tool, while after the theoretical background exposition we will report the results of the analysis applied to the experimental data.

Distribution Fitting Tool

The distribution fitting tool available in modeFRONTIER software is useful to find, from a database of 11 statistical distributions, the one that best fits an existing experimental data distribution.

As illustrated in Fig. 5, the input is an ASCII file containing the database relative to uncertainties or fluctuations of an existing system. The tool elaborates as output

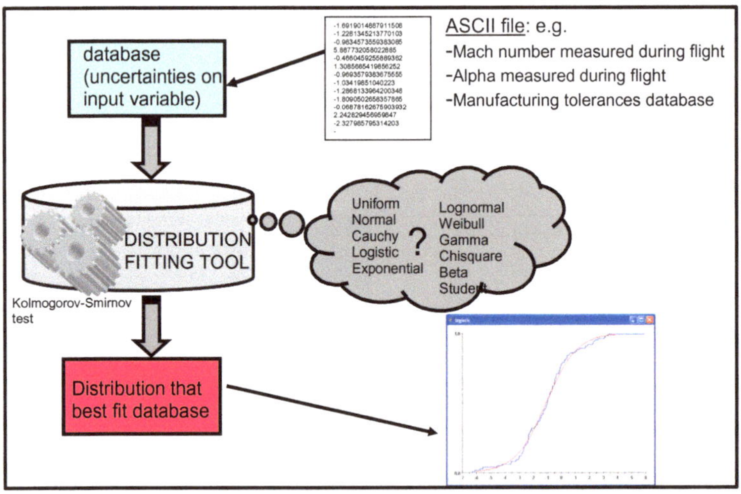

Fig. 5 ESTECO's distribution fitting tool

the statistical distribution that best fits these data, giving as output also the most relevant parameters that identifies it. The tool uses two indexes to quantify how much the statistical distribution fits the experimental one, the Kolmogorov–Smirnov (K–S) test and the maximum likelihood function, explained in sections "Kolmogorov–Smirnov Test" and "Maximum Likelihood Function". An internal optimization is performed, the input parameters being the distribution type and the relative parameters (i.e., mean and standard deviation for Normal), and the objective function being the maximization of the K–S coefficient. In particular, a genetic algorithm is used (i.e., NSGAII [5] from modeFRONTIER) as global search, and then a gradient-based algorithm (i.e., BFGS [6]) is used for local refinement. As output, the 11 different statistical distributions are sorted from the one with lowest value of KS test coefficient (worst fitting) to the one with highest value of KS test (best fitting). As discussed above, for all the distributions the relative parameters are reported. In addition, for the best fitting distribution, a plot that compares the cumulative statistical distribution with the database one is given.

Kolmogorov–Smirnov Test

With respect to Fig. 6, consider a database cumulative distribution FN(x) and a generic statistical distribution F(x). The distance D is computed as the maximum absolute value difference between the database and statistical distribution:

$$D = \max_{-\infty < x < +\infty} |F_N(x) - F(x)| \tag{3}$$

Fig. 6 Definition of K–S test

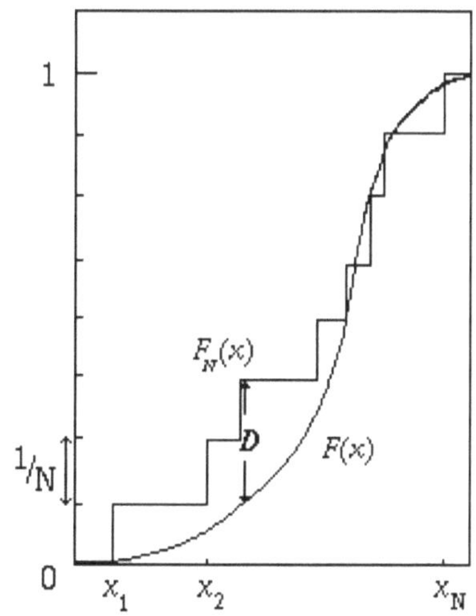

The Kolmogorov–Smirnov function is then defined as 1D:

$$KS = 1 - D \tag{4}$$

In the optimization run by the distribution fitting tool, this quantity KS is to be maximized, and the maximum value will have a superior limit of 1.

Maximum Likelihood Function

An alternative to the K–S parameter to quantify the quality of the statistical distribution fitting is the maximum likelihood function [7]. Actually, the internal optimization of the algorithm is performed through the KS test, but the user can read out for each statistical distribution also this function:

$$\begin{aligned} L &= \prod_{i=1}^{n} P(X_i, \alpha) \\ -NLL &= -\frac{1}{n}\ln(L) \end{aligned} \tag{5}$$

In the expression above, α represents the statistical distribution parameter, and P(xi) represents the probability of the statistical distribution corresponding to the point x_i. The function L is thus computed by the product of the probability corresponding to each point of the database that results *particularly* high only if the statistical probability is high in correspondence of the database points, that happens only if the database distribution is similar to the statistical one. The normalized likelihood function (–NLL) is computed as described in Eq. 5, and thus it is a function to be minimized.

Results of Experimental Database Distribution Fitting

The different geometrical measurements obtained by the experimental campaign on the acoustic panel, relatively to: cell depth, face sheet thickness, holes diameter, and percentage of open area, are then analyzed. Figs. 7, 8, and 9 report the outcome of the distribution fitting tool of modeFRONTIER, applied to each parameter's measurements. The first two charts in every figure are relative to the PDF and the CDF (probability and cumulative density function) of the experimental data. Selecting any of the available statistical distribution, the latter one can be over-imposed on the chart to compare the fitting. Another chart on the left-bottom corner reports Lilliefors diagram: Basically, a good fitting is obtained when all the database points (the points of the chart) are contained inside the red lines, which represent the limits of the selected analytical distribution. Moreover, the parameter

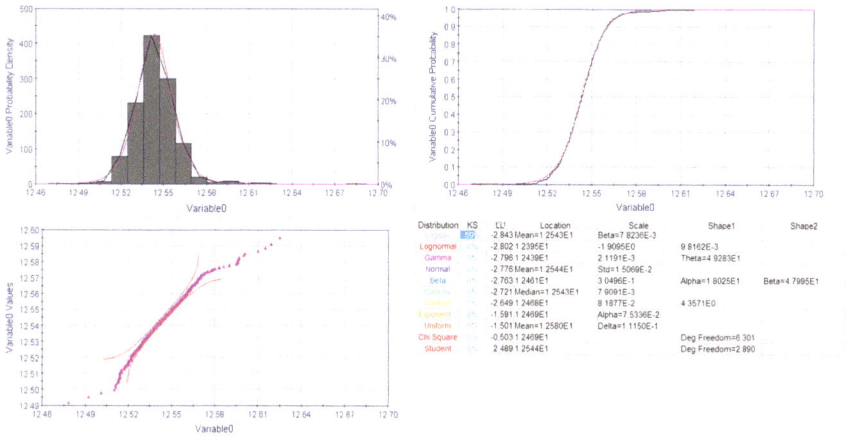

Fig. 7 Distribution fitting tool results of cell depth *h*

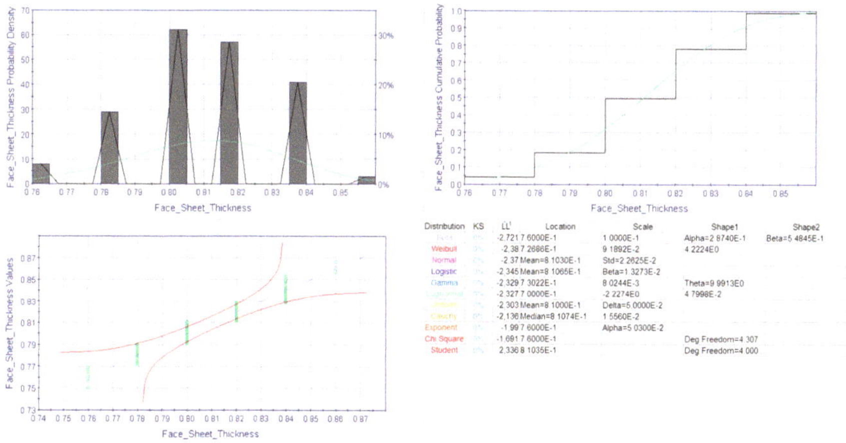

Fig. 8 Distribution fitting tool results of face sheet thickness *t*

to quantify correctly the fitting, as noted in section "Maximum Likelihood Function", is the maximum log likelihood (LL) function. If we order the LL values from the lowest to the highest, we get the distribution which better fits the experimental data. The chart in the right-bottom corner therefore indicates for each distribution the optimal parameters to fit the database distribution and which is the best one. These results will then be used by the numerical model (section "Integration of Aeroacoustic Model Within modeFRONTIER Framework") to evaluate a series of sampling points needed to quantify the performance uncertainties.

The optimal distribution regarding cell depth parameter is the logistic, defined by a location and a scale parameter. The fitting in this case is pretty good (59% of KS

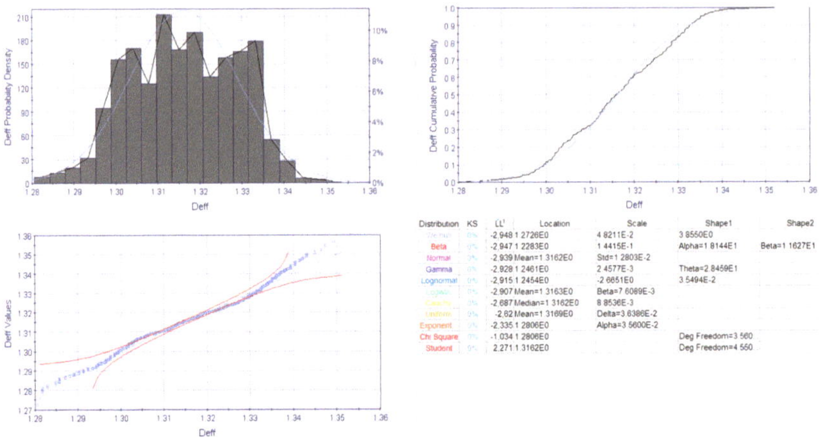

Fig. 9 Distribution fitting tool results of hole diameter d_{eff}

test), but the likelihood is still high considering a Normal distribution (−2.77 vs. −2.84 of the Logistic), indicating that a Normal distribution could be used as well to represent this database.

The optimal distribution regarding face sheet thickness parameter could be the Normal or the Weibull, defined by a location and a scale parameter in the second case, and by other two shape parameters in the first. The fitting in this case is not pretty good accordingly to KS test, but just because the distribution is strongly characterized by a coarse discretization (due to measurements limitation). Nevertheless, the LL parameter indicates that the likelihood with these distributions is high (also Lilliefors test confirms it); therefore, it is right to assume one of these ones as the theoretical distribution for this parameter (that is not limited in reality to discrete values).

The optimal distribution for the face sheet hole diameter parameter is the Normal or the Weibull, defined by a location and a scale parameter. The fitting is less good than in the other cases, but the LL is still the highest considering these distributions (−2.94 for Normal and −2.95 for Weibull) (Fig. 10).

For the PoA_{eff} distribution, few data are available (eight measurements), but they are nevertheless enough to give high values of KS factor for many distributions, confirming that the uncertainty can be modeled correctly (previous series have even too many samples that may give overfitting an outliers problem). In particular, Normal distribution has a LL equal to −5.4. For the sake of simplicity, it is therefore been decided to consider a Normal distribution for all the parameters of the database, since this type of distribution matches correctly all the variables. The distributions parameters are summarized in Table 1.

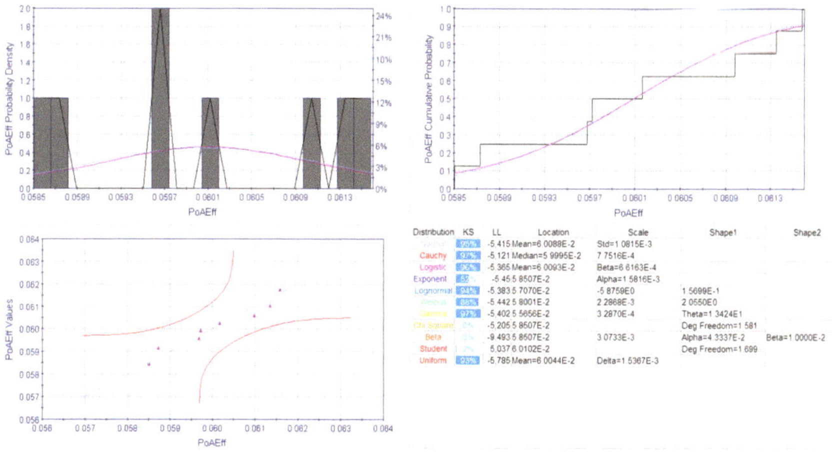

Fig. 10 Distribution fitting tool results of PoA_{eff}

Table 1 Summary of uncertain parameters UQ

Uncertainty	Distribution type	Mean	Standard deviation
Cell depth *h*	Normal	12.544	1.5069E−2
Sheet thickness *t*	Normal	0.8103	2.2625E−2
Hole diameter *d*	Normal	1.3162	1.2803E−2
Percentage of open area *PoA*	Normal	0.060088	1.0815E−3

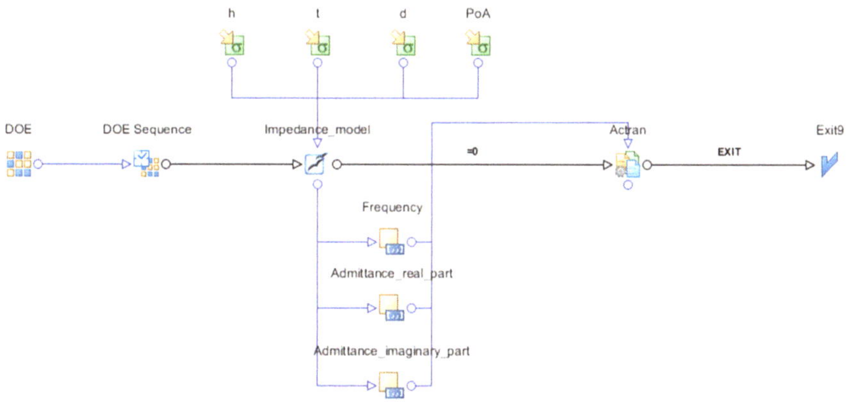

Fig. 11 modeFRONTIER workflow for aero-acoustic model process automation

Integration of Aeroacoustic Model Within ModeFRONTIER Framework

Figure 11 represents the workflow created in modeFRONTIER, (the optimization platform software developed by ESTECO), utilized for the automatic execution of the aero-acoustic simulations.

The horizontal flow is related to the application process which has to be executed for each design simulation, and a dedicated node module is placed in the loop connected to the other ones. The first node is the direct interface with OpenOffice, linking to a spreadsheet version of the proprietary semi-empirical model of the acoustic impedance, developed by Leonardo Finmeccanica [4] to predict the impedance of the acoustic liner as a function of the given geometrical and flow parameters associated with flight condition considered. The input parameter nodes contain all the info related to the distributions, as defined in section "UQ of Experimental Database Parameters". Below the OpenOffice interface, the data flow continues with the extraction of the response data; in this case, three vectors are defined by three buffer nodes: a set of 24 1/3 octave bands central frequencies in the 50–10,000 Hz range, the real and the imaginary part of the admittance. The buffer nodes are therefore linked to the following application of the process flow, which is the direct interface to MSC Actran software developed by Free Field Technology. This node contains all the information necessary for the aero-acoustic simulation: (i) An input template file is used to indicate in which position the impedance data have to be written, (ii) a driver panel contains the batch command to execute the software by giving in input the model file, and (iii) an output template is used to indicate which output files have to be extracted by the analysis. The Actran node contains also a call to MSC PLTViewer with a Python script to calculate out of all the solution files the OASPL, the final response of the workflow. Apart from the process automation, the first part of the workflow contains the DOE and the scheduler nodes. The first one is used to define the design of experiments to be evaluated; in this case, this is just the nominal or mean value of the distributions, the values being defined in Table 1. The other node contains the algorithm selected for the execution of the designs, in this case just a DOE sequence but the same architecture serves also for optimization not shown here. The RDO properties are defined in the dedicated panel: To analyze the proper sampling strategy, we could change the number of sampling points, the order of the polynomial chaos, and the application of the adaptive sparse methodology that will be described in next section. In the next chapter, we will then report the results obtained for the UQ of the response following different methodologies, and finally, it will be selected as the methodology giving the most accurate results with the least possible number of sampling simulations.

UQ of Acoustic Response

The numerical simulations can be performed automatically accordingly to the specified scheduler algorithm. In order to find the most efficient UQ methodology, we first performed a large DOE series, to be used to validate the different methodologies applied. For this purpose, initially we have run a series of 500 simulations, driven by Latin Hypercube DOE from modeFRONTIER, and then have obtained the UQ of the acoustic performances by the application of a polynomial chaos expansion of the second order.

Adaptive Sparse Polynomial Chaos Methodology for UQ

Since in this UQ problem the number of geometrical uncertainties is not large, the application of a SS-ANOVA screening tool in order to reduce the number of important uncertainties is not applicable. We will therefore apply a modified version of regression analysis, called adaptive sparse polynomial chaos methodology [8], which aims not to reduce the number of uncertainties to apply polynomial chaos expansion, but rather aims to reduce the number of terms of the same polynomial chaos expansion. The intention is the same (i.e., to perform an accurate UQ by a reduced number of sampling points, thus making the subsequent robust design optimization task feasible), but in this way it can be achieved without necessarily discarding completely any of the uncertain parameters. The methodology is basically defined by a stepwise regression analysis [1], where the polynomial chaos expansion terms are iteratively added or removed, according to the way they improve or worsen the regression model, which is quantified by a R-square value, computed on the complete database. Once the optimal set of polynomial terms (i.e., the set giving the highest R-square) is found, the UQ of the response can be obtained by the reduced polynomial chaos expansion as sketched in Fig. 12.

As an example from [8], Fig. 13 shows how the R-square changes with the number of polynomial terms, and how the optimal value is normally reached by an

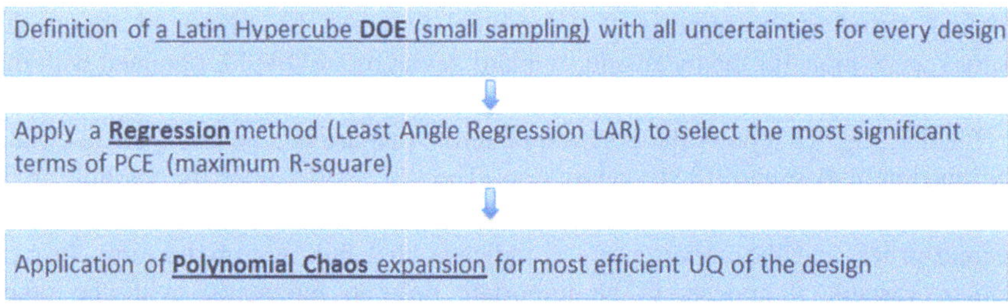

Fig. 12 Adaptive sparse polynomial chaos methodology

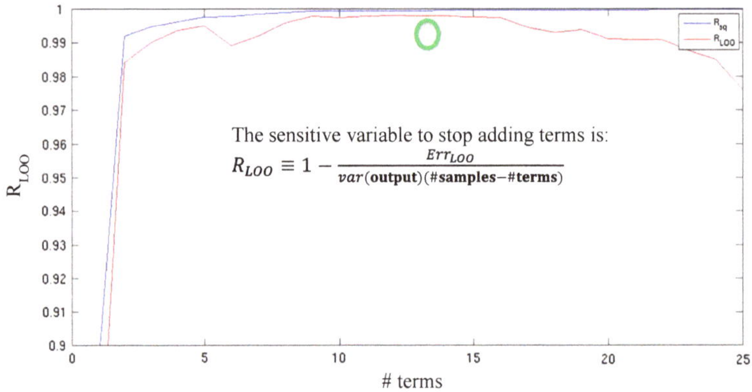

Fig. 13 Example of R-square versus number of polynomial terms for UQ

intermediate number of polynomial terms (too few are not accurate, and too many gives overfitting).

Another important advantage of the proposed approach, beside the one of improving the quality of the UQ with the available database, is the possibility to reduce progressively the number of sampling points, until a satisfactory threshold of accuracy is achieved.

UQ Results

In Table 2 are reported the results obtained with a progressively lower number of sampling points together with the optimal number of polynomial chaos terms (full or sparse).

It can be noted that the mean value of the OASPL response of the system can be obtained with high accuracy even by a low number of samples, and also the standard deviation is very accurate (less than 0.05% of error) with just 30 samples and accurate enough even with 15 samples (2% of error), which is the minimum number required for a second order. To improve further the quality of the approximation with 15 samples, the sparse polynomial chaos methodology has been applied, reducing the number of polynomial terms to 10. Figure 14 shows the convergence plots for the mean and standard deviation of OASPL obtained with the different methodologies applied.

Finally, Fig. 15 reports the polynomial terms that are retained in order to obtain the maximum R-square (0.99) regression: The "^1" suffix after the variable indicates that this term is of the first order, the "^2" indicates that this term is of the second order, and the underscore "_" indicates that this is a mixed term. In particular, variable *h* appears to be important only in interaction with the other parameters, while variables *PoA*, and subsequently, *t,* and *d*, are the most important

Table 2 Comparison of UQ of acoustic performances by different methodologies

Sampling	μ(OASPL)	σ(OASPL) (error)
500 samples—full PCE	133.29711	1.4613E−5 (0%)
100 samples—full PCE	133.29711	1.4612E−5 (0.0068%)
30 samples—full PCE	133.29711	1.4606E−5 (0.048%)
15 samples—full PCE	133.29712	1.4319E−5 (2.0%)
15 samples—*sparse* PCE	133.29711	1.4612E−5 (0.0068%)

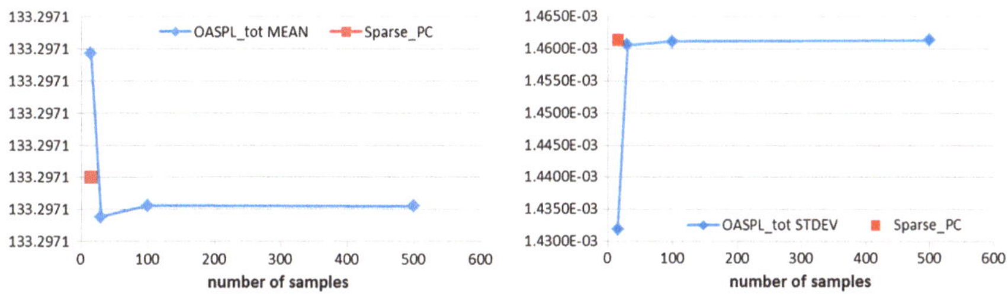

Fig. 14 Convergence of OASPL moments versus samples number for different methodologies

Constant term: 123.08068569420057
PoA^1: 0.06357629613405774
t^1: 0.03167859637193685
d^1: 0.013300302380818661
PoA^1 * d^1: 3.247963573688598E-4
PoA^1 * h^1: -6.270497206753558E-5
t^2: 4.2307676081786903E-4
PoA^2: 0.001357632026433811
h^1: 0.001258112198596739
PoA^1 * t^1: -9.558315593705808E-4
h^1 * t^1: 2.8109939595050817E-5
d^2: 9.556623581704204E-5

R-squared: 0.9999999843434352
R-squared LOO: 0.999996606178817
Adjusted R-squared LOO: 0.9999841621678126
AIC: -327.21239031597696
BIC: -318.71578790275044

Fig. 15 Sparse polynomial chaos expansion terms for optimal OASPL regression

ones. R-squared value (close to 1.0) and other regression indicators are also explicitly readable in Fig. 15, confirming the good fitting of the PCE expansion. The error on standard deviation is therefore reduced (from 2% to less than 0.1% as reported in Table 2) by avoiding the overfitting on those terms having no effect in the system.

Conclusion

In this chapter, a new methodology for UQ based on regression analysis, developed by ESTECO within the UMRIDA project [8] and called adaptive sparse polynomial chaos expansion [1], was used. This approach aims not to reduce the number of uncertainties but the number of the terms of the polynomial chaos expansion without necessarily discarding any parameter. This methodology was selected as perfectly fitting the industrial case of the engine nacelle acoustic liner, where the number of uncertain geometrical parameters subjected to manufacturing tolerances considered here is as low as four. The adaptive sparse polynomial chaos expansion methodology proved very successful not only allowing to improve the quality of the UQ but also to reduce the number of samples to 15 and the terms of the polynomial expansion to only 10, avoiding overfitting and thus achieving an error for the standard deviation of the OASPL of less than 0.1% much lower than the 2% error produced by the 15 samples full PCE.

References

1. Blatman, G., Sudret, B.: Adaptive Sparse Polynomial Chaos Expansion Based on Least Angle Regession. Elsevier Science (2010)
2. Copiello, D., Ferrante, P., Iemma U., Maci A.: Preliminary design of aero-engine intake acoustic liners by means of the multi-objective approach. In: AIAA 2010-3828
3. Copiello, D., Ferrante, P.: Multi-objective optimization of inlet true zero-splice liners. In: 15th AIAA/CEAS Aeroacoustics Conference, Miami, AIAA 2009-3107 (2009)
4. Murray, P.B., Ferrante, P., Scofano A.: Manufacturing process and boundary layer influences on perforate liner impedances. In: 11th AIAA/CEAS Aeroacoustics Conference, Monterey, California (2005)
5. Deb, K., Pratap, A., Agarwal, S., Meyarivan, T.: A fast and elitist multi-objective genetic algorithm: NSGA-II. IEEE Trans. Evol. Comput. J. **6** (2002)
6. Broyden, C.G.: J. Inst. Math. Appl. **6**, 76–90 (1970)
7. Aldrich, J.: R.A. Fisher and the making of maximum likelihood 1912–1922. Stat. Sci. **12** (1997)
8. Clarich, A., Russo, R.: Efficient UQ of large number of uncertainties: adaptive sparse Polynomial Chaos expansion with modeFRONTIER. In: Proceedings of UMRIDA Workshop on Uncertainty Quantification, Delft, 15 April 2015

Manufacturing Uncertainties in High-Pressure Compressors

Rémy Nigro, Dirk Wunsch, Grégory Coussement and Charles Hirsch

Introduction

Uncertainty quantification (UQ) methods are now a major topic in academic research. However, these methods are not yet widely used in daily engineering design practice. In this chapter, we demonstrate the industrial applicability of the methodology described in chapter "Non-intrusive Probabilistic Collocation Method for Operational, Geometrical and Manufacturing Uncertainties in Engineering Practice".

The problems classically used for the validation of UQ methodologies contain a limited number of uncorrelated uncertainties. Thus, two major research topics must be tackled to be able to industrialize the UQ methods: the treatment of a high number of uncertainties and the use of correlations.

The first problem comes from the dimensionality of the problem: the cost of several methods increases exponentially with the number of uncertainties, leading to the so-called curse of dimensionality. As industrial test cases may contain several simultaneous uncertainties, some methods, such as the Smolyak sparse grid,

R. Nigro · D. Wunsch · C. Hirsch (✉)
Numeca International, Brussels, Belgium
e-mail: charles.hirsch@numeca.be

R. Nigro
e-mail: remy.nigro@umons.ac.be

D. Wunsch
e-mail: dirk.wunsch@numeca.be

R. Nigro · G. Coussement
University of Mons, Mons, Belgium
e-mail: gregory.coussement@umons.ac.be

C. Hirsch et al. (eds.), *Uncertainty Management for Robust Industrial Design in Aeronautics*, Notes on Numerical Fluid Mechanics and Multidisciplinary Design 140, https://doi.org/10.1007/978-3-319-77767-2_27

which is used in this work, are developed to temper this curse (more details on this technique in chapter "Non-intrusive Probabilistic Collocation Method for Operational, Geometrical and Manufacturing Uncertainties in Engineering Practice").

The second problem refers to the correlation between geometrical parameters which may modify deeply the deformations caused by the uncertainties.

Figure 1 shows the differences in the realizations of two different sets of uncertainties: one without any correlations (left) and one with correlations between the uncertainties (random field) (right). This figure clearly shows that the deformations imposed are highly dependent on the correlations. Therefore, as correlations exist in industrial problems, the UQ method must be able to tackle them. In this work, the principal component analysis (PCA) is used to tackle this problem, and for the detailed description of the method, the reader may refer to chapter "Non-intrusive Probabilistic Collocation Method for Operational, Geometrical and Manufacturing Uncertainties in Engineering Practice".

The non-intrusive probabilistic collocation method (NIPColM) is used to tackle an industrial test case from the UMRIDA database. A total of 15 correlated uncertainties are defined on the geometry of a rotor blade, which is computed with its two adjacent stator rows (1.5-stage high-pressure compressor in chapter "UMRIDA Test Case Database with Prescribed Uncertainties"). The results are discussed in terms of computational cost for a given accuracy and in terms of sensitivity of some quantities of interest with respect to the input correlated uncertainties.

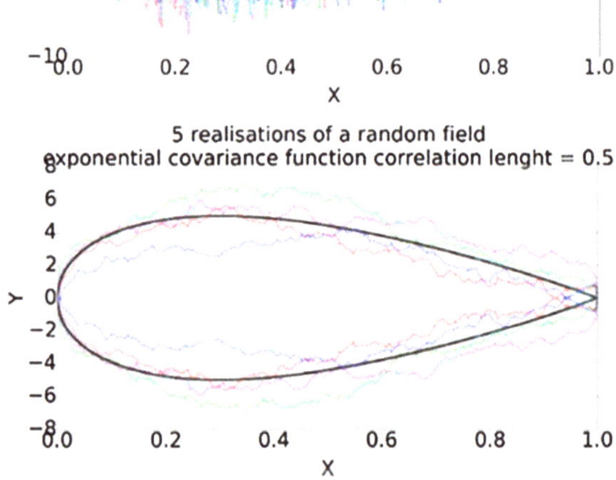

Fig. 1 Realizations of an airfoil discretized with 200 points considering each of these points as independent (top) or as a random field described with a covariance between each of its points (bottom) (black: nominal value; colours: realizations)

Description of the IC-09: 1.5-Stage High-Pressure Compressor

The uncertainties are defined by optical measurements of industrially manufactured blades. A parametric model, which has been built to reflect the geometry variation of this system, is used to determine a set of correlated uncertainties (they are the parameters of the model) [1] (Fig. 2).

The uncertainty quantification is performed using the NIPColM method coupled with the PCA as described in chapter "Non-intrusive Probabilistic Collocation Method for Operational, Geometrical and Manufacturing Uncertainties in Engineering Practice". The geometry variation is only considered on the rotor stage (no uncertainties considered on the stator stages).

The NIPColM is applied by solving a set of deterministic CFD runs (at each collocation points defined by the method). The CFD problem is defined with the RANS equations and Spalart–Allmaras turbulence model. A mixing plane technique is used at the interface between each row and one blade passage is meshed, as shown in Fig. 3. The rotor stage has a constant rotating hub speed. Each stator has a shroud fillet and a hub gap, while the rotor has a hub fillet and a shroud gap. The inlet boundary conditions are defined by azimuthal-averaged radial profiles of total conditions (pressure and temperature) and flow direction (angles), while the outlet is defined by an azimuthal-averaged static pressure profile. The mesh consists of 3.6 million cells and has been selected after a mesh convergence study.

The CFD simulations are performed with the flow solver FINE™/Turbo [2], and the uncertainty quantification methodology described in the reference documents is integrated in FINE™/Design3D. The convergence of each computation is assessed by the analysis of the global residuals (sum of all the residuals) and the evolution of the mass flow through the computation domain (cf. Fig. 4). As the residuals decrease of five orders of magnitude and the error on the mass flow is below 0.01%,

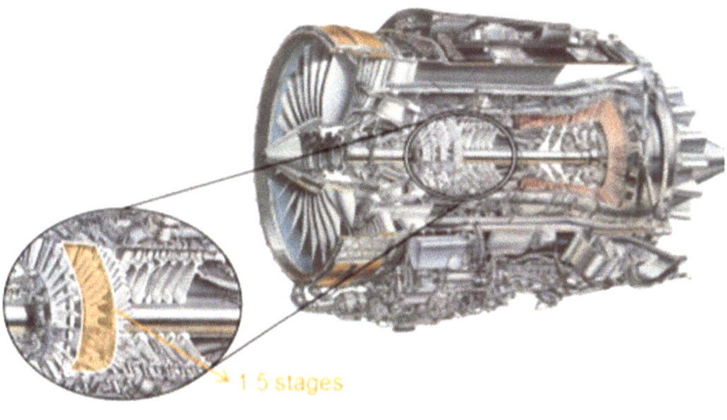

Fig. 2 One-and-a-half axial high-pressure compressor stages of an aero-engine as physical system to be investigated under real manufacturing uncertainties (courtesy of Rolls-Royce Deutschland)

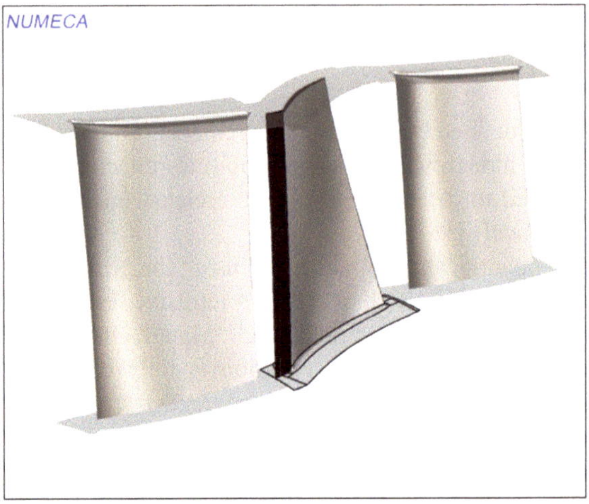

Fig. 3 Computational domain for the 1.5-stage high-pressure compressor

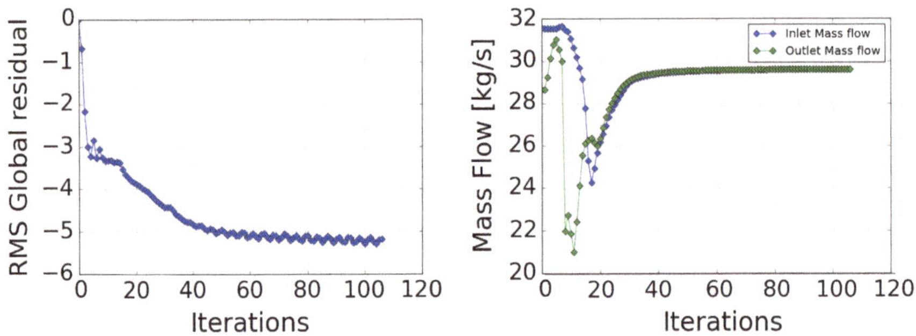

Fig. 4 Convergence for the deterministic computation: (left) root mean square of the global residuals; (right) evolution of the mass flow

the CFD computation is converged. All the CFD computations have the same level of convergence. The deterministic flow of this 1.5-stage compressor presents secondary flows through the gaps on each blade and a shock on the rotor blade (cf. Fig. 5).

Parametric Model

The uncertainties are defined on a parametric model which has been constructed to represent the blade deformations with a set of parameters which have a physical meaning. The parametric model developed by Lange et al. [3] consists in a

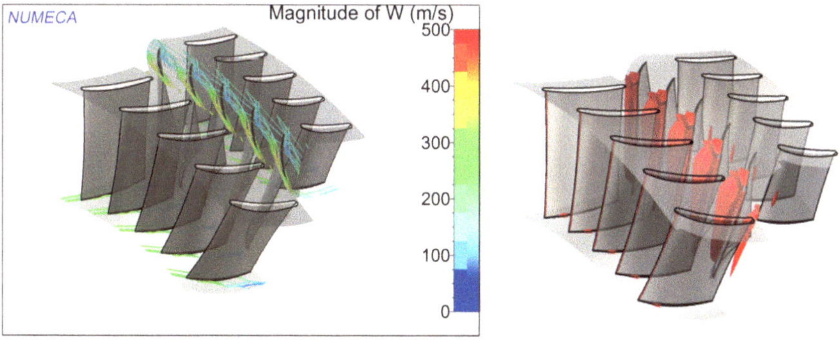

Fig. 5 Flow visualization: (left) velocity vectors across the different gaps (highlights the vortex created through the gap) and (right) iso-surface of Mach = 1 (highlights the shock location)

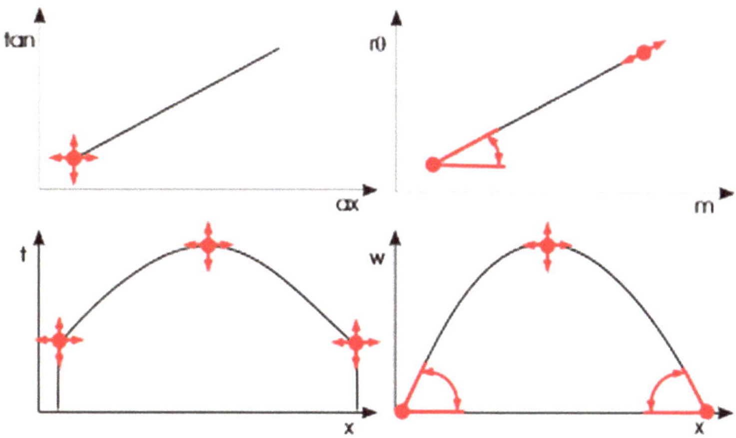

Fig. 6 Parameters describing a section [3]

parameterization of a set of blade sections (cf. Fig. 6). Each of them is described by two parameters for its position (axial and tangential), the stagger angle and the chord length, allowing to define the chord location for each section. The camber line, described by its leading and trailing edge angle and by the value and the position of the maximal camber, is determined based on the leading and trailing edge position (extrema of the chord). Finally, the thickness law, which is described by the value and the position of the maximal thickness and two additional points defining the value and the position of the leading and trailing edge thickness, is added to the camber line in order to define the profile section. In this work, six sections are used to describe the geometry. The fillet radius between the blade and the hub is also considered as a parameter.

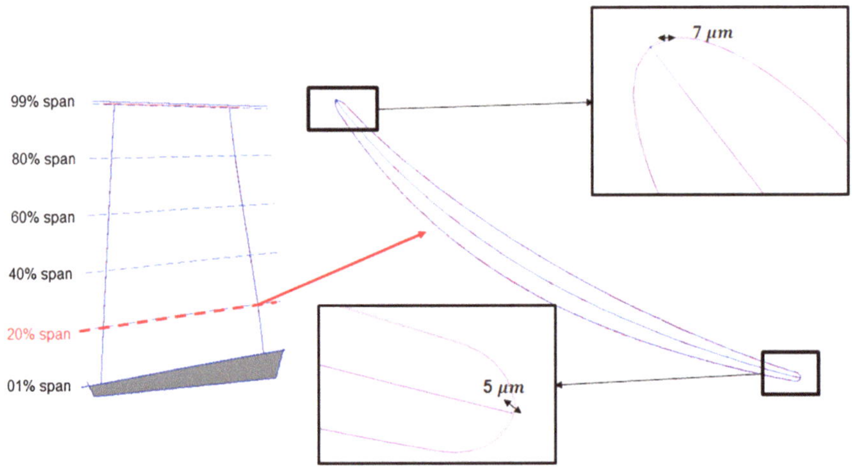

Fig. 7 Differences between the design geometry and the parameterized one

In total, there are 85 parameters: 14 parameters per sections multiplied by 6 sections, which lead to 84 parameters and the fillet radius. Each of these parameters is a potential uncertainty and can theoretically be correlated by another one.

As the geometry is parameterized, the geometry used as nominal design for the UQ computation (i.e. the parameterized one) is not identical to the initial CAD geometry. As the UQ should be performed on the original geometry, the parametric model must be fitted with a high accuracy to the defined geometry. Figure 7 shows that the maximal absolute difference between the initial and parameterized geometry is inferior to 7 μm, which is acceptable for the purpose of this study as shown in [4].

Treatment of the Uncertainties

The parametric model is used to describe the geometry of each manufactured blade: first, for each measured blade, the parameters of the model are determined to match the optical measurements. This set of parameter values is then used to determine the probability density function of each parameter and their correlation.

Lange et al. [1] have shown that the parameters describing the blade section are highly correlated in the spanwise direction. This is due to the manufacturing process used for these blades: forging. The high spanwise correlation allows to limit the number of uncertainties by considering only one uncertainty for each section parameter. This means that the uncertainty describing the maximal thickness is defined as a deformation value, which will be applied on all the blade sections, leading to a thinner or a thicker blade. This observation obtained by means of the Spearman's rank coefficient in [1] allows to reduce the number of uncertainties from 85 to 15.

The correlated parameters are treated as described in chapter "Non-intrusive Probabilistic Collocation Method for Operational, Geometrical and Manufacturing Uncertainties in Engineering Practice" in order to obtain a set of uncorrelated uncertainties which can be used with the NIPColM: the PCA [5] is applied to the correlation matrix to obtain a set of 15 uncorrelated uncertainties describing a Gaussian random field. The truncation of modes usually performed with the PCA is not possible in this case as a huge reduction of the dimensionality has already been applied a priori through the analysis of the correlation matrix and the high spanwise correlation.

A set of collocation points is obtained for the NIPColM coupled with the Smolyak sparse grids. Then, an inverse PCA transform allows to obtain the collocation points in terms of parametric Gaussian uncertainties. However, as the input uncertainties are described with different types of PDF, the Nataf transform [6] is applied to have a deformation corresponding to the true PDF shape. The geometry variability represented with the resulting collocation points corresponds to the random field described with the correlated uncertainties and their defined PDF shape.

An example of the resulting deformations is given for the second eigenmode in Fig. 8. Every parameter, including the fillet radius, is modified simultaneously for each collocation point. A specific treatment based on a butterfly topology of the mesh in the fillet region allows to keep a good mesh quality for different fillet radii (see Fig. 9).

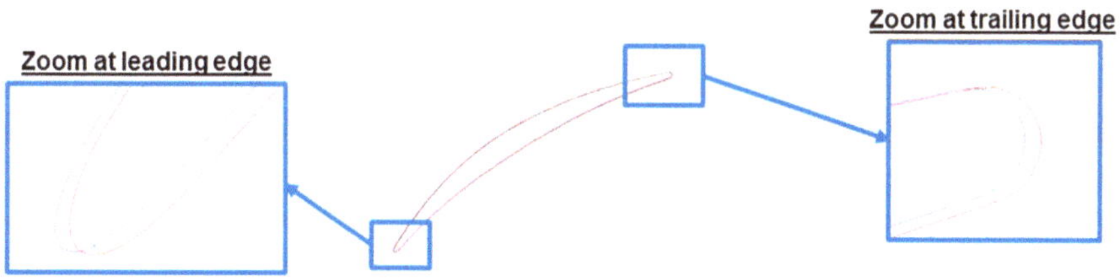

Fig. 8 Example of deformation based on the extremum values of the second eigenmode on the section at 20% span

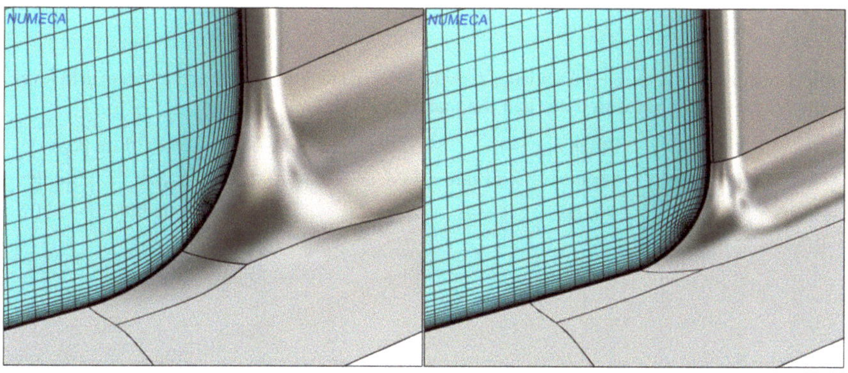

Fig. 9 Mesh near the hub for two different geometries: the mesh quality is ensured even with a varying fillet radius

The application of this technique to the test case described in section "Description of the IC-09: 1.5 Stage High Pressure Compressor" allows to accurately represent the geometrical uncertainties and therefore to obtain meaningful UQ results.

Analysis of the Uncertainty Quantification Result

As the NIPColM accuracy depends on the number of levels used, the first analysis performed determines that in the case of this industrial test case, the sparse grid level 1 is sufficient to determine the mean and the standard deviation as already shown in [7] for an academic test case.

A second analysis is performed to assess the capability of this method to determine the influence of the input correlated parameters on the quantities of interest such as the efficiency based on the scaled sensitivity derivatives as used in [7].

Influence of the Level of Sparse Grid and Corresponding Computation Cost

The convergence test is performed on the two first levels of sparse grids. With the 15 uncertainties defined for this test case, the total number of collocation points (i.e. CFD runs) is 31 for the first level and 511 for the second one. The higher levels are not computationally affordable (5921 CFD runs for a level 3 and 53 921 CFD runs for a level 4).

The comparison of the two first statistical moments (mean and standard deviation) between the two different levels shows that the mean value does not change more than 0.1% for the efficiency, the pressure ratio and the mass flow, while the standard deviation of these three quantities has a maximal variation of 8.3% (cf. Table 1). Therefore, it can be determined that the mean and the variance can be

Table 1 Difference between the statistics found with a sparse grid level 1 and a sparse grid level 2 for the efficiency, the pressure ratio and the mass flow

	Efficiency (–)		Pressure ratio (–)		Mass flow (kg/s)	
	Mean	Std dev	Mean	Std dev	Mean	Std dev
Difference between SGL1 and SGL2 (%)	−0.039	8.344	−0.004	6.922	−0.099	−2.202

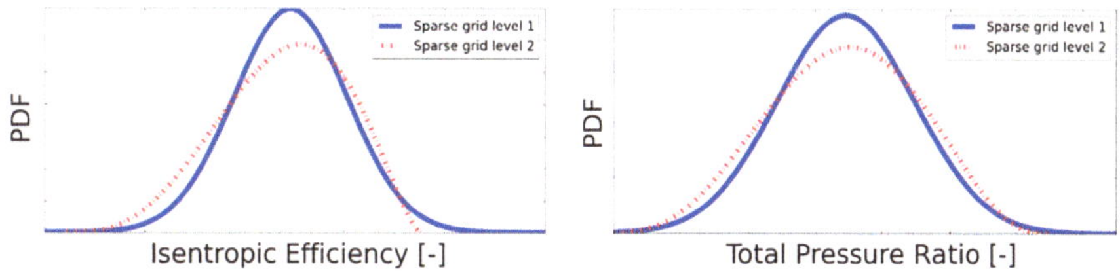

Fig. 10 PDFs for the isentropic efficiency and the total pressure ratio of the 1.5 stage as a function of the sparse grid level

accurately estimated with a sparse grid level 1 (SGL1). This conclusion was also found on another compressor and with a different set of uncertainties [4].

As the sparse grid level 1 only gives an accurate estimation of the mean and the variance, the PDF can be reconstructed using a Gaussian approximation (i.e. imposing that the PDF has a Gaussian distribution with the mean and the variance determined from the UQ computation). This reconstruction can be compared with the PDF reconstructed thanks to the Pearson method [8] based on the first four moments determined with the sparse grid level 2.

The sparse grid level 1 gives valuable information on the approximate shape of the PDF for a reduced cost. Indeed, the PDF obtained with a level 2 has a computational overhead of 16.5 in comparison with the level 1 and gives roughly the same interval of variation for the quantities of interest. The evaluation of a PDF shape with a sparse grid level 1 is accurate when the output PDF is symmetric, which is almost the case for this problem (Fig. 10).

Based on the cost of one single CFD run, which is 4.1 CPU * Hours, the level 1 sparse grid is compliant with an industrial use as it has a computational cost of 128.1 CPU * Hours, meaning that with only 16 cores, the total time is 7.75 h. The sparse grid level 2 has a total cost of 2112 CPU * Hours. As the limit value defined within the UMRIDA project for an efficient UQ method in industry is 10 h on 100 cores with ten simultaneous uncertainties, it can be seen that a level 1 sparse grid fully meets the objective in terms of computational time (17.75 h on 16 cores with 85 possible uncertainties reduced to 15) (Table 2).

Sensitivities of the Performances with Respect to the Input Uncertainties

As it has been proven that the sparse grid level 1 is sufficient to obtain accurate UQ results and that it meets the computational time requirements inherent to the industry, the sensitivities can be computed with a sparse grid level 1 (only 31 CFD runs).

The scaled sensitivities are used to determine the uncertainties with the most influence on the global quantities such as the pressure ratio and the efficiency as

Table 2 Computational cost for the UQ computation for the sparse grid level 1

	#CFD runs	Time for 1 CFD run on 16 cores (h)	Cost for 1 CFD run (CPU * Hours)	Total time on 16 cores (h)	Total cost (CPU * Hours)
Level 1	31	0.25	4.1	7.75	128.1

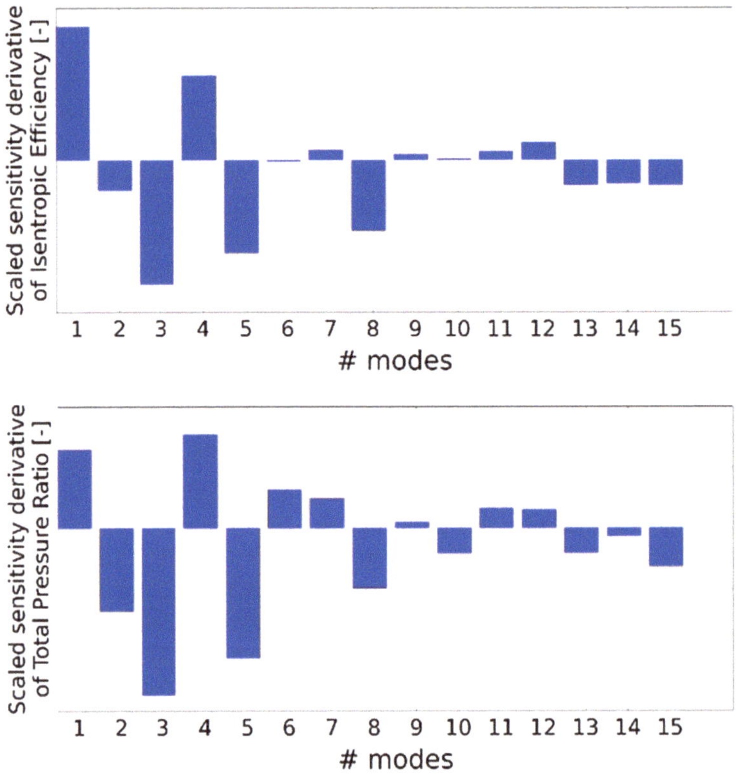

Fig. 11 Scaled sensitivity derivatives of the efficiency (top) and the pressure ratio (bottom) with respect to the eigenmodes

seen in Fig. 11. One important information is that the most influential mode in terms of geometry deformation (i.e. the first mode) is not the mode with the highest impact on the quantities of interest (here, the pressure ratio and the efficiency). However, the first five modes are the most influential. Even if the first few modes have an important influence on the quantities of interest, the last modes still have a non negligible impact. Therefore, a truncation of the number of modes cannot be performed without a loss of accuracy on the UQ result.

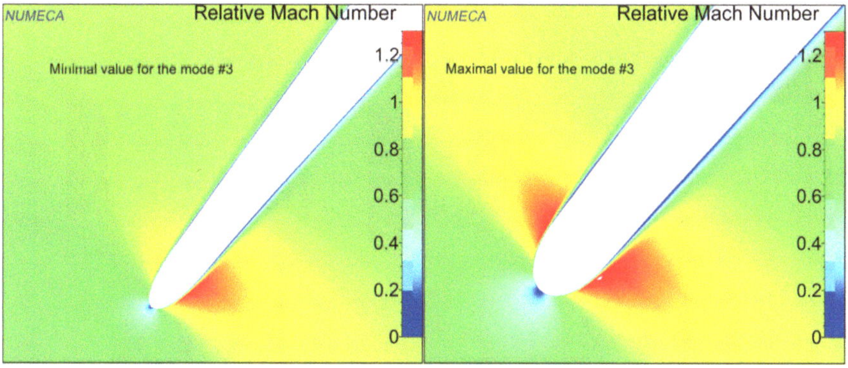

Fig. 12 Comparison of the relative Mach number for two different collocation points: the one with the smallest value of the third mode (left) and the one with the highest value of the third mode (right)

An important information based on the scaled sensitivity derivatives of the quantities of interest with respect to the eigenmodes is to obtain the sensitivities of the output quantities with respect to the input uncertainties (i.e. the parameters of the parametric model).

A qualitative way to obtain such sensitivities is to display the results for two collocation points with extremum value for a given eigenmode and check the geometry deformation and its influence on the flow as given in Fig. 12. The application of this method to the third eigenmode, which has the most influence on the efficiency, shows that the variation of the overall blade thickness is important. This thickness variation leads to an important modification of the flow field. Figure 12 shows that the acceleration at leading edge is higher for high values of the third eigenmode. The boundary layer near the pressure side is also wider for high values of the third eigenmode. This boundary layer leads to more important losses and thus a lower efficiency.

The scaled sensitivity derivatives can be computed with respect to the input correlated uncertain parameters, allowing a quantitative comparison of the influence of the different uncertain parameters. Indeed, from the derivatives with respect to each eigenmodes, which are computed to determine the scaled sensitivity derivatives shown previously, it is possible to compute the derivatives with respect to each input parameter using the directional derivatives. Once these derivatives are computed, the scaled sensitivities are obtained by multiplying them with the standard deviation of the parameter. The result of this treatment is shown in Fig. 13.

The leading edge thickness has the highest influence on the efficiency. As the sensitivity is negative, it means that a reduction of the thickness leads to an increase of the efficiency, which is an expected result. The fillet radius seems to have a high impact on the efficiency, meaning that it cannot be neglected during the UQ analysis.

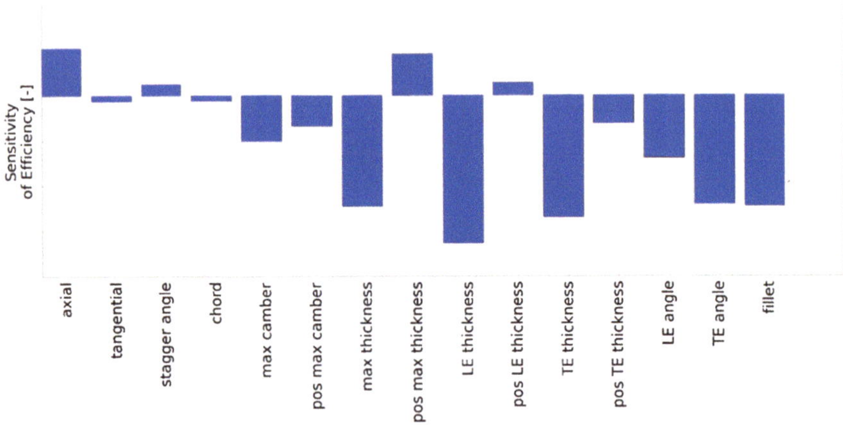

Fig. 13 Scaled sensitivity derivative of the efficiency with respect to the initial set of uncertain parameters

The influence of the leading edge thickness can be validated through an Ant-Hill plot determined from the 511 CFD runs performed for the sparse grid level 2 (cf. Fig. 14) where the efficiency for each collocation point is shown as a function of the leading edge thickness, and a linear correlation is clearly seen. This confirms the results found with the scaled sensitivity derivatives. The Ant-Hill plot of the efficiency as a function of the chord length does not show a clear influence of this parameter, confirming the small sensitivity of the efficiency with respect to the chord length.

The Spearman correlation coefficients determined by Lange et al. [1] show the same trend as the ones found with the scaled sensitivities: the leading edge thickness has the highest impact on the efficiency, and all the thickness-related parameters have also a high impact on this quantity of interest. The chord length parameter has a small impact on the efficiency in comparison with the thickness. As, qualitatively, the same sensitivities are found with the two methods, the use of scaled sensitivity derivatives to assess the influence of the input parameters on the

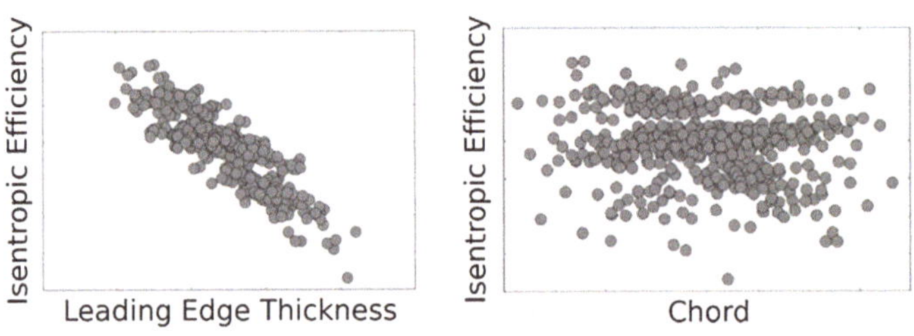

Fig. 14 Ant-Hill plot of the efficiency as a function of the leading edge thickness (left) or as a function of the chord length (right) based on the 511 CFD runs performed for a sparse grid level 2

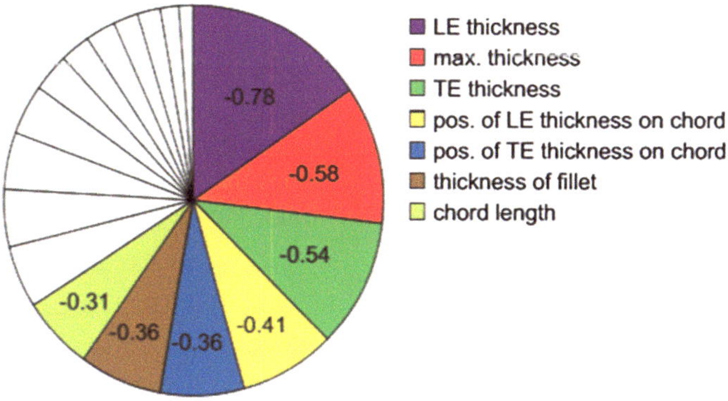

Fig. 15 Spearman rank coefficient of the efficiency for each uncertain parameter [1]

quantities of interest can be validated and is thus a possible alternative to Spearman correlation coefficients and other Ant-Hill plots (Fig. 15).

The computational cost needed to obtain the sensitivities and the mean and standard deviation for quantities of interest is highly reduced in comparison with a Monte Carlo simulation consisting of 500 samples, which was the method used by Lange et al. [1]. Indeed, only 31 samples are needed to obtain the same trend for the sensitivity, which means the cost has been reduced by a factor 16 with the NIP-ColM coupled with PCA in comparison with a Monte Carlo simulation.

Conclusions

The method for the treatment of simultaneous correlated uncertainties developed in chapter "Non-intrusive Probabilistic Collocation Method for Operational, Geometrical and Manufacturing Uncertainties in Engineering Practice" has been successfully applied to an industrially relevant test case which is part of the UMRIDA database. This test case is a 1.5-stage high-pressure compressor in which the parameters are the uncertainties described with a PDF and the correlations between them. A total of 15 correlated uncertainties are used.

It has been shown that a level 1 sparse grid consisting of 31 CFD runs is sufficient to obtain an accurate estimation of the two first statistical moments (mean and standard deviation) of some global quantities such as the pressure ratio and the efficiency. Moreover, the output PDF shape can be estimated with this level 1 and a Gaussian shape approximation based on these two first moments. The PDF shape can be estimated with only two moments since the response is symmetric.

The scaled sensitivity derivatives are computed with respect to the input uncertainties with a sparse grid level 1 as it is sufficient for the convergence of the method. The thickness-related parameters are shown to have an important impact on the efficiency, which is also found with Ant-Hill plots based on a sparse grid level 2 or with Spearman correlation coefficients computed based on a Monte Carlo sampling of 500 samples [1].

Therefore, a sparse grid level 1 gives accurate UQ results with only 31 CFD runs while a Monte Carlo sampling needs 500 samples to give the same qualitative sensitivity, and a sparse grid level 2 needs 511 samples to give a similar interval of variation for the quantities of interest. This means that the computational cost has been reduced by a factor 16 in comparison with a Monte Carlo sampling.

References

1. Lange, A., Voigt, M., Vogeler, K., Gümmer, G., Johan, E., Schrapp, H.: Probabilistic CFD simulation of a high-pressure compressor stage taking manufacturing variability into account. In: ASME paper n° GT2010-22484 (2010)
2. Numeca: User Manual FINE^{TM}/Turbo v11.1 (2016)
3. Lange, A., Vogeler, K., Gümmer, V., Schrapp, H., Clemen, C.: Introduction of a parameter based compressor blade model for considering measured geometry uncertainties in numerical simulation. In: ASME paper n° GT 2009-59937 (2009)
4. Wunsch, D., Nigro, R., Coussement, G., Hirsch, C.: Quantification of combined operational and geometrical uncertainties in turbo-machinery design. In: Proceedings of the ASME GT2015, GT2015-43399 (2015)
5. Loève, M.: Probability Theory I, 4th ed. Graduate Texts in Mathematics, vol. 45. Springer, New York-Heidelberg (1977)
6. Nataf, A.: Détermination des distributions de probabilités dont les marges sont données. Comptesrendus de l'académie des sciences **225**, 42–43 (1962)
7. Nigro, R., Wunsch, D., Coussement, G., Hirsch, C.: Uncertainty quantification in internal flows. NATO CSO report AVT-191, chapter 14 (2015)
8. Pearson, K.: 1895, 1901, 1915. Philos. Trans. R. Soc. Lond. 8 laws variant

Part IV
Robust Design Optimization (RDO) and Applications

Formulations for Robust Design and Inverse Robust Design

Alberto Clarich and Rosario Russo

Introduction

Multi-disciplinary Robust Design Optimization (RDO) is achieving more and more popularity in the aerospace community: Many optimization methodologies have been developed to produce solutions that are interesting for practical industrial cases. In fact, most of the industrial processes are permeated by uncertainties: The numerical design is generally different, from a geometric point of view, from the manufactured product because of the dimensional tolerances, and, more frequently, the working point is not fixed, but is characterized by some fluctuations in the operating variables. Normally, in these conditions, the traditional single-point optimization methodologies tend to over-optimize the final design, giving excellent performances in the design point but limiting the off-design characteristics. Consequently, the application of Robust Design methodologies to improve the performances stability seems to be the most correct approach in the research activities (Fig. 1).

The classical formulation of a RDO problem consists in the definition of two different objectives for each criterion to be optimized: the optimization of the average performance of the system, and the minimization of the standard deviation of the same performance, which guarantee the stability of the same. This problem is normally solved following a multi-objective optimization approach, as it will be discussed in section "Multi-objective RDO Optimization," while in section "Single-Objective RDO Optimization (Min-Max Approach)" a different approach, based on a single-objective reliability-based optimization method, will be proposed as a valid and particularly efficient alternative.

A. Clarich (✉) · R. Russo
ESTECO SpA, Area Science Park, Padriciano 99, 34149 Trieste, Italy
e-mail: engineering@esteco.com; engineering@esteco.it

C. Hirsch et al. (eds.), *Uncertainty Management for Robust Industrial Design in Aeronautics*, Notes on Numerical Fluid Mechanics and Multidisciplinary Design 140, https://doi.org/10.1007/978-3-319-77767-2_28

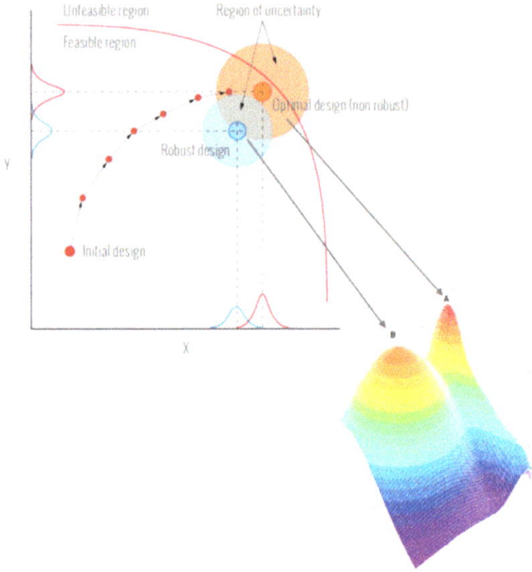

Fig. 1 Robust Design Optimization

In conclusion, the advantages of Inverse RDO, or Tolerance Design, will be introduced at the end of this chapter, to answer to the industrial need of containing manufacturing and maintenance costs while guaranteeing the required performances.

Multi-objective RDO Optimization

The first approach for RDO that we propose in this chapter is based on the definition of a multi-objective optimization problem, consisting generally in the optimization of the mean value of the performances and on the minimization of their standard deviation. As an example of formulation, the objective functions in the RDO of an airfoil for (lift-constrained) drag reduction in case of uncertainties in the cruise Mach number [1] will become:

$$\min_{\mathrm{D}, \alpha(\mathrm{M})} \left(E(c_d), \sigma^2(c_d)\right) \tag{1}$$

The mean and variance C_d are defined as follows:

$$E(c_d) = \int_{M_{min}}^{M_{max}} c_d(D, \alpha, M) p(M) dM \tag{2}$$

$$\sigma^2(c_d) = \int_{M_{min}}^{M_{max}} [c_d(D, \alpha, M) - E(c_d)]^2 p(M) dM \tag{3}$$

where *p(M)* is the probability density function of the Mach Number defined in the interval $Mmin < M < Mmax$.

This approach guarantees the definition of a complete Pareto frontier as trade-off of the optimal solutions, in terms of mean performance and in terms of their stability or robustness. This means that at the end of the optimization the designer has the freedom to select the best solutions having a large variety of possibilities depending on which criteria should be privileged.

The problem of this approach is that a multi-objective optimization algorithm is to be chosen, since the definition of a single objective as weighted sum of the different criteria cannot be proposed for the impossibility of knowing a priori the proper weights for the particular optimization problem. Multi-objective optimization algorithms are in fact generally very robust, but they require a higher number of simulations than a single-objective optimization case, and for a RDO problem the number of simulations may be not feasible from a practical point of view (this number being multiplied by the sampling size for each design to obtain the overall number of simulations required).

A very efficient and robust class of optimization algorithms is the one of the Evolutionary, and in particular the Genetic Algorithms (GA) [2, 3]. This class of algorithms is based on the definition of an initial set of designs, the population of individuals, whose performance evaluations are used by the algorithm to propose a new population or set of designs, by following predefined operations—like Selection, Crossover, Mutation—inspired by Natural Evolution, that guarantee the generation of new designs, which keep the best pieces of information of their parents. The evolution of a defined number of generations produces as result the improvement of the objectives until a convergence to the ideal Pareto frontier is obtained.

The population size and therefore the overall number of simulations required is determined by the number of variables and objectives, in such a way that it is possible to define a "rule of thumb" to link linearly the overall number of simulations to the optimization problem size.

Generally, the population size should be given by $2 \cdot n \cdot m$, where n is the number of variables and m the number of objectives; therefore, for a problem of two objectives, like the one defined for a RDO (maximize mean performance and minimize standard deviation), this number becomes equal to $4 \cdot n$. Considering in addition that at least 20–30 generations should be evaluated to accurately predict the Pareto frontier, we would need about $100 \cdot n$ designs to complete a RDO optimization.

To understand the complexity of the problem, we may consider an RDO problem characterized by 10 input variables, 2 objectives, and 3 uncertainties. Applying a polynomial chaos (PCE) approximation of second order, 10 samples for each design may be enough. Accordingly, to the rule of thumb exposed above, we might need a total of $100 \cdot 10 \cdot 10$, i.e., 10,000 simulations, that for a model of industrial relevance could correspond to a non-affordable CPU time making the application unfeasible.

For this reason, often it has been proposed in the literature a different optimization strategy, using Game Theory-based algorithms [4] which decomposes the input variable space between two players (if there are two objectives, or more in other cases) each one in charge of a single objective. Each objective is then minimized by each player using a fast and efficient single-objective algorithm, such as the simplex [5], and after some iteration, the best solutions found by each player are shared by the same to start a new step, until a convergence is obtained.

The advantage of this algorithm is that it does not search for a complete Pareto frontier, but for few compromise solutions between the contrasting objectives, therefore the computational time is much lighter than a cooperative strategy like the Genetic Algorithm.

As another rule of thumb, the overall number of designs required by the MOGT algorithm [4] is on the order of $10 \cdot n \cdot m$, where n is the number of variables and m of the objectives (considering at least 10 steps where each player perform a reduced simplex execution of about n designs, the double of the variables charged on it).

Single-Objective RDO Optimization (Min-Max Approach)

In order to reduce more significantly the overall number of simulations for an RDO problem, we propose in this section another approach, followed also in one of our previous works [6].

The basic idea is to reduce the number of objectives, so that a single-objective algorithm can be applied.

To achieve this purpose, the so-called *min-max* or *max-min* approach is followed. The idea is to maximize the minimum or worst performance of a distribution function to be maximized (for instance, the aerodynamic efficiency of a wing), or to minimize the maximum or worst limitation to be minimized (for instance, the drag coefficient of a wing).

The effect of this approach is the "shift" of the performance distribution in the desired direction, so in a certain sense both the average performance and the stability at the uncertainties are optimized. Looking for instance at Fig. 2 taken from the NODESIM-CFD Project [6], the drag coefficient distribution of an optimized airfoil configuration under this approach will be shifted below the baseline distribution, since we minimize the maximum value of the distribution or its higher tail.

Besides this objective, other criteria shall be considered (like lift and momentum), but if they can be expressed as constraints, a single-objective algorithm could still be applied.

Generally speaking, the idea is to constraint the worst (maximum or minimum) value of the distribution to be under or above a defined limit (for instance, the

Fig. 2 Example of min-max RDO

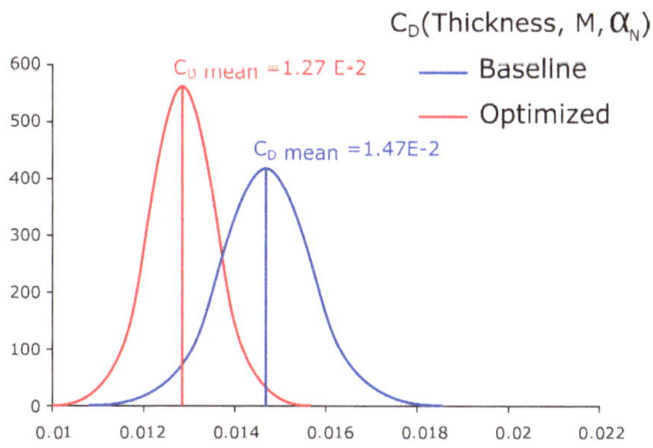

minimum lift should be higher than the desired limit, so that it could be guaranteed even in the worst uncertain conditions).

In formulas, the RDO problem defined previously by Eq. 1 will become:

$$\min_{D,\alpha(M)} \max(c_d(M)) \tag{4}$$

Before analyzing the possible single-objective algorithms that may be chosen, it is opportune to discuss about the definition of maximum and minimum values of a distribution.

In the case of a normal distribution of the performance, since it is unlimited the concept of the extremes may be replaced by a given percentile of the distribution, for instance, 95 or 99%. Usually, the reference value is 99.73% because for a normal distribution it corresponds to the 3-sigma level (see Fig. 3).

This analysis is also called six sigma, since six times the standard deviation corresponds to the 99.73% of the complete distribution, a value that can be assumed enough representative of the whole distribution.

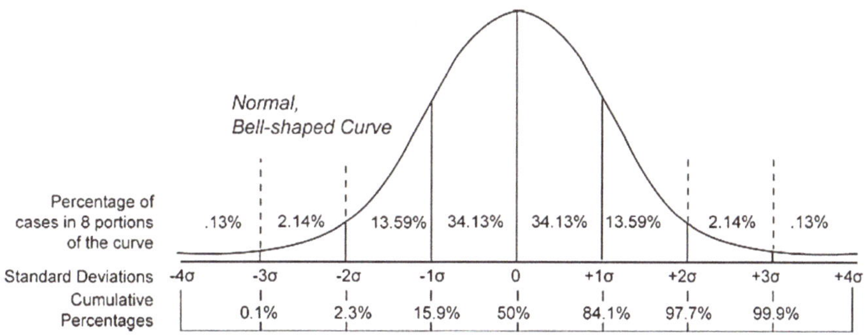

Fig. 3 Six sigma correlation for a normal distribution

Since by Polynomial Chaos analysis, we can compute mean and standard deviation with high accuracy, and the computation of the maximum or minimum value with the expression MEAN $\pm$ 3σ can be evaluated for each design of the RDO optimization; therefore, the objective function can be defined this way.

The limitation of this approach occurs when the performance does not follow a normal distribution: in this case, the six-sigma formulation may not correspond exactly to the same percentile of the distribution, so from design to design the computation of the objective function could be not accurate. This problem is even more evident for a particular class of RDO problems, the reliability-based design optimization, where any constraint may be defined on a given percentile of the distribution.

In order to solve the problem of the accuracy of the min-max approach and make the RDO optimization more efficient, a new methodology, based on the application of Polynomial Chaos on the Reliability RDO, will be introduced in section "Reliability-Based RDO Optimization."

There are several single-objective algorithms which may be applied to solve the max-min RDO problem.

A large variety of algorithms is offered by the gradient-based class, which is actually based on the computation of the gradient of the objective function (by finite difference approximation), and on the iterative movement in the direction of objective improvement. This class includes algorithms such as BFGS, SQP, NLPQLP.

Even though these algorithms are very accurate, they cannot be proposed as a Global Research solution, because they are not robust: Depending on the starting point, they usually converge very soon to the closest local maximum, which is not necessarily the global maximum (Fig. 4, left).

For this reason, we prefer to introduce a different single-objective algorithm, less accurate but more robust and in particular very fast in convergence: the *Nelder* and *Mead simplex* [5]. The algorithm is based on the definition of a geometrical figure of $n+1$ vertices, the *simplex*, in the variable space of n-dimensions (n is the number of variables, and each vertex represents one different design solution). At each

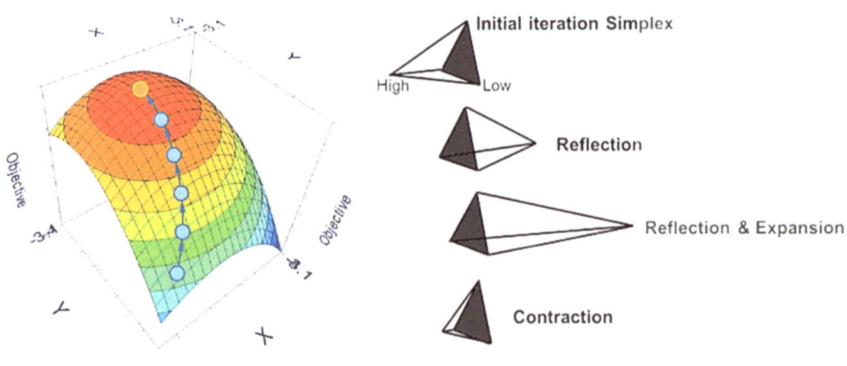

Fig. 4 Convergence to local optima for gradient-based algorithms (left); simplex (right)

iteration, the worst point with respect to the objective function is replaced by a new point, whose variables are obtained by geometrical operations in the n-variables space (Fig. 4 right): The first operation is a *Reflection* through the plane made by the other points of the simplex, then accordingly to the fitness of the new point there could be an *Expansion* in the same direction of the Reflection (if the new solution is particularly good) or a *Contraction* (in the opposite case).

The operations are repeated iteratively until convergence to the global optimum.

To introduce in this case a rule of thumb based on experience, we could affirm that the convergence of simplex in a single-objective optimization problem may be reached by $5 \cdot n$ iterations, which is a number much lower than the two-objective optimization cases proposed in the previous chapter.

The advantage of the approach presented, the single-objective min-max, is particularly evident.

Reliability-Based RDO Optimization

In most of the aeronautical or industrial applications, the need of defining constraints or objectives on the percentiles of some performances distribution, like the minimum value of lift or momentum for an airfoil, emerges clearly, and the way these constraints are generally treated at the state of the art (six sigma) is accurate only in particular cases (normal distribution of the performances).

In most real application cases, the fact that the real distribution of the performances is different produces an error in the definition of the constraint function, which gives as effect a wrong estimation of the failure probability of each proposed design (the probability that this design does not respect the constraint), and most important it can bias the optimization toward a less efficient solution.

For this reason, the accurate estimation of the constraint functions becomes fundamental in a generic problem of Robust Design Optimization in industrial practice.

Generally, an optimization problem in which there is the requirement of the satisfaction of constraints or limits, which should be achieved for a specified percentage of the performance distribution, or for which the percentage of solutions not satisfying the limits (failure probability) must be minimized, is called reliability-based design optimization.

In the literature, there exist already some approaches to solve this kind of problems, in particular First-Order Reliability Method (FORM) or Second-Order Reliability Method (SORM) [7] that evaluate the failure probability of any candidate design on the basis of its uncertainties distribution and of the given limits to be respected.

Approximation methods, such as FORM and SORM, have been developed in the area of structural reliability. These methodologies have been described in detail for instance by Dr. Xiaoping Du in Ref. [8].

Reliability analysis aims at estimating the probability that any structure or mechanical component will fail to meet a predefined criterion. A reliability problem is often formulated in terms of a vector of random input variables $X = (X_1, X_2, \ldots, X_N)$ representing uncertain quantities, such as loads, material properties, structure dimensions, environmental factors, and a performance function $g(X)$, which describes the limiting state of the structure in terms of X. By convention, $g(X) \leq 0$ denotes the failure domain, whereas $g(X) > 0$ denotes the safe set. Since the boundary between the safe and failure set is given by $g(X) = 0$, the performance function is also referred to as the *limit state function* (LSF).

Instead of computing the reliability $R = P\{g(X) > 0\}$, one usually computes its complement, i.e., the *probability of failure*, which reads

$$p_f = P\{g(X) \leq 0\} = \int_{g(X) \leq 0} f_X(X) dX \tag{5}$$

with $f_X(X)$ denoting the joint probability density function (PDF) of the random variable vector X.

In order to overcome problems which might arise in the calculation of the failure probability integral in real-world applications, approximation methods like FORM and SORM have been developed over the years.

Two principal steps are involved in these approximation methods. The first step consists in simplifying the integrand $f_X(X)$ so that its contours become more regular and symmetric; the second consists in simplifying the integration domain. After the execution of these two steps, an analytical solution to the probability integration will be easily found.

The first simplification is achieved by transforming the original random variables X into independent random variables $U = (U_1, U_2, \ldots, U_N)$ that follow a standard normal distribution (i.e., a Gaussian distribution with zero mean and unit standard deviation). In the literature, the original variable space is called X-space, while the standard normal space is called U-space.

After the variable transformation, the failure probability becomes

$$p_f = P\{g(U) \leq 0\} = \int_{g(U) < 0} \Phi_U(u) du \tag{6}$$

where $\Phi_U(u)$ is the joint PDF of U.

The second simplification is achieved by approximating the limit state function (constraint satisfaction). The name First-Order Reliability Method comes from the fact that the LSF is linearized with a first-order Taylor expansion. The linearization is performed at a convenient point, i.e., the point that has the highest probability density on the frontier $g(U) = 0$ between safe and failure set. This point is named *Most Probable Point* (MPP) and it is the LSF shortest distance point from the U-space origin, as shown in Fig. 2. The distance between the MPP and the origin represents the so-called *reliability index* β.

The model for the MPP search can be rewritten as a constrained single-objective optimization problem:

$$\begin{cases} \min_{U} \|u\| \\ \textit{subject to } g(u) = 0 \end{cases} \tag{7}$$

where $\|u\|$ stands for the norm (length or magnitude) of a vector. The solution to the model given in Eq. 8 is the MPP (Fig. 5).

After the two steps described above, the failure probability is easily computed as

$$p_f = \Phi(-\beta) \tag{8}$$

with Φ denoting the standard normal *cumulative distribution function* (CDF).

Second-Order Reliability Methods account for second-order nonlinearities of the performance function. Further details about the MPP search in FORM and SORM approaches are available in the literature [7].

Here, we only stress the following facts. In industrial applications, the calculations of derivatives might be very expensive, assuming that the performance function is differentiable at all. Besides, for particular problems, FORM might not even converge, or locating the MPP might be very difficult.

An important limit of these methodologies can therefore be represented by the high number of evaluations that may be required by the algorithm to compute the failure probability with accuracy, which makes often practically unfeasible its application to optimization problems of industrial relevance, in particular when the computational time required for each evaluation is very expensive.

For these reasons, we propose in next section "Innovative Reliability RDO Methodology Based on PCE" a new methodology to deal efficiently with a reliability-based design optimization problem of industrial relevance, which conjugates accuracy and low number of needed evaluations.

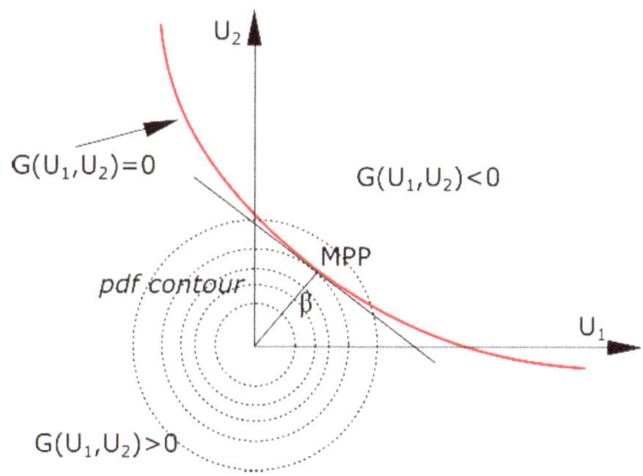

Fig. 5 Reliability problem definition

The methodology is derived from the Robust Design Optimization applying the Polynomial Chaos methodology. The polynomial coefficients are here also used to evaluate the complete cumulative distribution function of the performances of the design, from which it is possible to retrieve accurately the failure probability for the prescribed limits/constraints or objectives of the problem.

Innovative Reliability RDO Methodology Based on PCE

Polynomial Chaos Expansion (PCE) [9] is one of the most efficient methodologies employed for uncertainty quantification, and it is commonly used to solve RDO problems [6, 10].

The main advantage of this methodology relies in the possibility of computing the mean and standard deviation of a given distribution with high accuracy by means of a reduced number of sampling points. In particular, it can be proved that the convergence to exact statistical moments follows an exponential rate with respect to the number of sampling points [11], whereas sampling techniques, such as Monte Carlo (MC) or Latin-Hypercube sampling (LHS), converge much more slowly (as the inverse square root of the sampling size and as the inverse sampling size, respectively).

The basic idea of Polynomial Chaos methodology for UQ is that, under specific conditions [9], a stochastic process can be expressed as a spectral expansion based on suitable orthogonal polynomials, with weights associated to a particular probability density function.

The idea is to project the variables of the problem onto a stochastic space spanned by a set of complete orthogonal polynomials ψ that are functions of random variables $\xi(\theta)$, where θ is a random event [10]. For example, the variable ϕ has the following spectral infinite dimensional representation:

$$\Phi(x,t,\theta) = \sum_{i=0}^{\infty} \Phi_i(x,t)\Psi_i(\xi(\theta)) \tag{9}$$

Now, we want to apply PCE expansion to define the criteria required by a generic reliability-based RDO problem, which is basically the definition of a percentile of the performance distribution or a failure probability.

Equation 10 can in fact be used as a meta-model for the $\Phi(x,t,\theta)$ function, which represents basically the output performance function of the input variables X and t and of the random event θ or uncertain parameters $\xi(\theta)$. The PCE can be used to determine, by means of a MC or LHS, an accurate CDF of the function Φ.

The evaluation of the performance function in industrial use cases can be very demanding, since it often involves expensive CFD or structural simulations. In the strategy proposed, instead, these expensive evaluations are required only to determine the coefficients of the PCE. Once found, the evaluation of the meta-model on any sample is practically free in terms of CPU. Once the CDF is accurately obtained, from the given constraint value we can easily retrieve the corresponding percentage in the distribution, i.e., the failure probability (Fig. 6).

In this way, a Robust Design Optimization problem can be defined, using as criteria for the optimization the minimization of the *failure probability*: In other words, we search for a new design for which the failure probability for the given uncertainties distribution is minimized, either for a new design for which a given percentile (e.g., 99%) of its distribution is minimized.

The big advantage of this approach with respect to using FORM/SORM methodologies is the reduced number of sampling points needed to obtain the Polynomial Chaos-based meta-model (Eq. 10), if compared to the iterations needed to compute the reliability index for each design required by FORM/SORM methodologies.

Tolerance Design: Background

The motivation of Reverse-Multi-Objective Robust Design Optimization (R-MORDO) methodology, called also Tolerance Design, comes basically from economic reasons in industrial production.

To improve the quality of a product, it is necessary to avoid excessive warranty costs, derived mainly by the failures typical when the product is not designed taking into account any uncertainty in the process, i.e., without performing a Robust Design analysis (RDO). At the same time, however, an excessive care of the uncertainties could drive to very restrictive solutions, i.e., solutions that does not allow the input parameters to vary but within very small tolerances, causing to enhance production costs.

Figure 7 expresses in fact the two extreme situations that may occur in a design optimization: The area on the left is the one characterized by excessive warranty costs, where for instance Robust Design is not applied in the Design phase

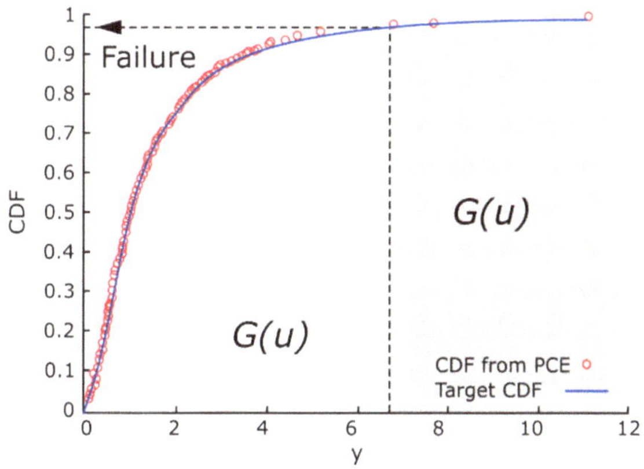

Fig. 6 CDF predicted by Polynomial Chaos used to retrieve failure probability

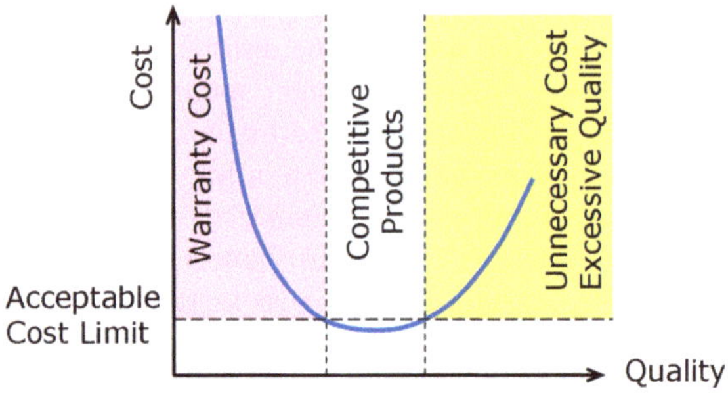

Fig. 7 MORDO and R-MORDO costs

(low Quality), and as a consequence, high Warranty Costs are expected, since the uncertainties not taken into account will produce recurrent failures.

On the other side, the area on the right represents a case in which a RDO methodology is applied with excessive quality; i.e., the minimization of the performance instability is pushed far below any reasonable necessity or constraint, for instance considering very small manufacturing tolerances in the design phase, with a big raise in the production costs.

A best solution is a trade-off design (central area in the picture) that can be obtained following two different objectives: optimize the mean performances under the given constraints (RDO objective), and minimize the standard deviation or tolerance of the input variables at the same time.

In fact, if we are able to optimize the performance distribution of the design configuration and respect the given constraints by a solution whose manufacturing or operational tolerances are higher than the baseline model, it means that we are able to obtain the needed performance by a lower production cost.

As an example of application, we could consider the RDO optimization of an airfoil subjected to manufacturing tolerances (uncertainties on the thickness), in which one objective is the minimization of the mean drag coefficient and a constraint is represented by the satisfaction of a minimum given lift coefficient and a maximum given momentum coefficient, by a high percentage of the distribution (for instance, 99.97%). This typical RDO problem becomes a R-MORDO problem if we consider as an additional objective the maximization of the thickness tolerance (maximization of standard deviation of a normal distribution of the thickness).

In formulas, the R-MORDO problem becomes:

$$\begin{gathered}\min_{D,\alpha(M)} \max(c_d(M)) \\ \max_{D,\alpha(M)} \mathrm{tol}(M)\end{gathered} \tag{10}$$

At the end of the optimization, we might select among a trade-off of different solutions, characterized by respecting the given constraints by minimizing the mean drag coefficient or by maximizing the manufacturing tolerance of the airfoil, i.e., respecting the given constraints with the lowest manufacturing costs as possible.

Figure 8 reports an example of the trade-off solutions that may be obtained by a R-MORDO problem. In the abscissa, a single uncertain parameter, which is also a design variable of the optimization problem, is reported while the curve represents the corresponding performance function.

Three different possible solutions of the R-MORDO problem are represented in the figure, each one characterized by a different tolerance (standard deviation of the input variable normal distribution, centered on each nominal design point). As we can see the performance distribution of the three solutions is practically the same, the Normal distribution of the performance has in fact a standard deviation which is about the same in three cases; this means that the three solutions, from a RDO point of view, are as robust in the same way.

By the way, solution 2 is characterized by a larger input tolerance than solutions 0 and 1: From a R-MORDO or tolerance Design point of view, this solution is then to be preferred to the other two, since it guarantees the same RDO performances with a much lower manufacturing or quality costs.

The R-MORDO methodology is fully implemented in the multi-objective optimization platform modeFRONTIER from ESTECO [12].

modeFRONTIER is a multi-disciplinary optimization platform in which any CAE simulation software can be integrated to define an optimization loop, selecting a proper optimization algorithm, including a suite dedicated to RDO problems.

In the main workflow of the platform, which is used to describe the optimization process, each input variable or uncertainty is represented by a node where to define range of variation and distribution parameters of that variable.

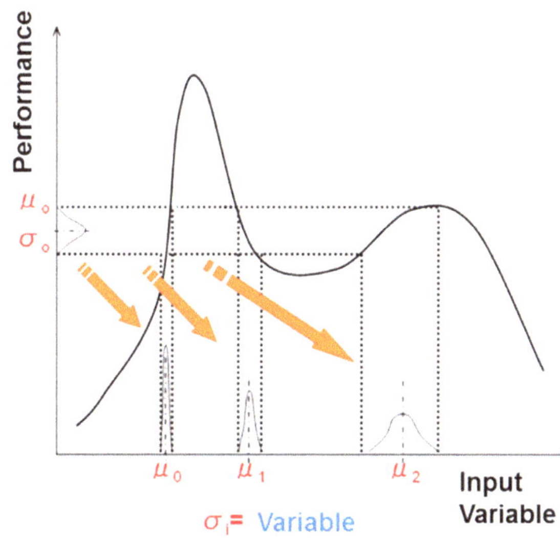

Fig. 8 R-MORDO solutions

Fig. 9 Example of R-MORDO implementation in modeFRONTIER

In the same way, output variables coming as result from any process node of the workflow can be used to define the objectives/constraints of the optimization problem.

As an example, Fig. 9 reports a very simple modeFRONTIER workflow, where on the left a basic optimization (vertical) chain is defined: On the top, the green node represents an input variable characterized by its Normal distribution (represented by the σ symbol), and the variable is then connected to a process node, in this case MATLAB, which is the application used to evaluate the performances of each design solution, and from the application node an output variable, represented by the cyan node, can be extracted.

Also the output variable is represented with a σ symbol since in a RDO problem it is also characterized by a distribution, and therefore the objective function (the arrow node at the bottom) may be defined for instance on the mean value of the distribution or to a given percentage (for instance, six sigma) of it.

The optimization chain on the left represents therefore a simple RDO problem definition in modeFRONTIER workflow.

By the way, the introduction of the other vertical chain on the right makes the problem become a R-MORDO: An additional input variable is introduced, to represent the standard deviation of the main input variable distribution: In other words, this new parameter controls the tolerance of the main input variable. To this variable, a second objective is assigned, which is simply the maximization of this standard deviation or tolerance.

Figure 10 reports a detail of the two variables in modeFRONTIER workflow: If we call the first input variable X, the second variable may be called Sigma_x. The X node main panel, at the right of the figure, reports then in the distribution properties as value for the Standard Deviation of the selected (Normal) distribution the variable Sigma_x instead of a fixed number, as usually occurs in a RDO problem.

Once the workflow is defined this way, the optimization can be performed by modeFRONTIER.

For the sampling phase, an opportune Polynomial Chaos expansion is selected, and an opportune Optimization algorithm can be selected.

The algorithm will then launch the execution of several different design configurations, each one sampled at the variation of the uncertain variable, until the

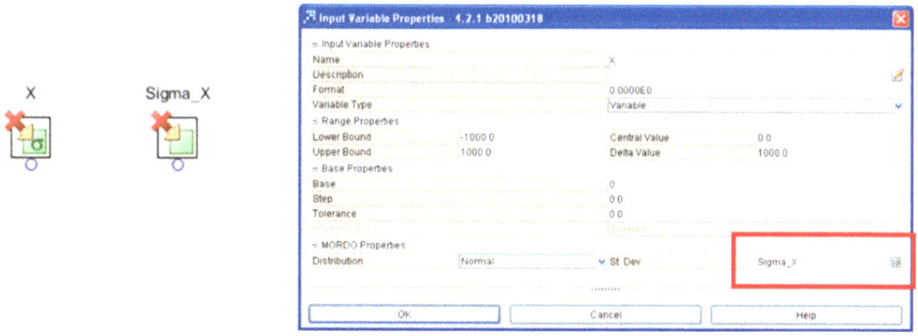

Fig. 10 Main input variable (X) and its standard deviation input variable (Sigma_X) in modeFRONTIER

prescribed objectives will be satisfied by the reaching of the Pareto frontier, or set of not-dominated solutions.

This set represents the trade-off of the two objectives considered, i.e., from one side the maximization of the mean performances, and from the other side the maximization of the input tolerance, so the reduction of manufacturing costs. The user can then finally select the definitive solution accordingly own preference.

References

1. Li, W., Padula, S.: Performance trades study for robust airfoil shape optimisation. In: 21th Applied Aerodynamics Conference, AIAA-2003-3790
2. Deb, K.: Multi objective genetic algorithm: problems and construction of test problem. Evol. Comput. J. **7**(3), 205–230 (1999)
3. Poloni, C., Pediroda, V.: GA coupled with computationally expensive simulation: tool to improve efficiency. In: Genetic Algorithm and Computer Science. Wiley and Sons (1997)
4. Clarich, A., Periaux, J., Poloni, C.: Combining Game Strategies and Evolutionary Algorithms for CAD Parameterisation and Multi-point Optimisation of Complex Aeronautic Systems, Barcelona, acts of Eurogen (2003)
5. Nelder, J.A., Mead, R.: A simplex method for function minimization. Comput. J. **7**, 308–313 (1965)
6. Clarich, A., Pediroda, V.: Robust design applications with modeFRONTIER, applying NODESIM-CFD tools. In: NODESIM-CFD Workshop on Quantification of CFD Uncertainties, Bruxelles, 29–30 October 2009
7. Der Kiureghian, A.: First- and second-order reliability methods. In: Nikolaidis, E., Ghiocel, D.M., Singhal, S. (eds.) Chapter 14 in Engineering Design Reliability Handbook. CRC Press, Boca Raton, FL (2006)
8. Du, X.: First order and second reliability methods. In: Probabilistic Engineering Design, chapter 7. Missouri S&T, ME 360, Fall (2009)
9. Loeven, G., Witteveen, J., Bijl, H.: Probabilistic collocation: an efficient non-intrusive approach for arbitrarily distributed parametric uncertainties. In: 45th AIAA Aerospace Science Meeting and Exhibit, Reno, Nevada (2007)

10. Parussini, L., Pediroda, V., Poloni, C.: Prediction of geometric uncertainty effects on fluid dynamics by polynomial chaos and fictitious domain method. Comput. Fluids **39**(2010), 137–151 (2009)
11. Cao, Y.: On Convergence Rate of Wiener-Ito Expansion for Generalized Random Variables (2000). Mathematical Subject Classifications: 60H15, 60H30, 65C10
12. http://www.esteco.com

Robust Design of Initial Boundary Value Problems

Jan Nordström and Markus Wahlsten

Introduction

Real-world systems based on partial differential equations are affected by a wide range of uncertainties. Common areas include climatology [1], turbulent combustion [2], flow in porous media [3], electromagnetics [4], and seismology [5] to name a few examples.

Non-intrusive methods use information from multiple runs of existing deterministic codes for a particular statistical input [6, 7]. Standard quadrature techniques, often in combination with sparse grid techniques, can be used to obtain the statistics of interest. Intrusive methods are based on polynomial chaos expansions leading to a system of equations for the expansion coefficients [8, 9]. This implies that new non-deterministic codes must be developed. The statistical properties are obtained by a single run for a larger system of equations. Examples of semi-intrusive methods also exist.

In this investigation, we take a step back from the technical developments mentioned above and focus on fundamental questions for the governing initial boundary value problem (IBVP) and in particular on the influence of boundary conditions. Our aim is to minimize the uncertainty or variance of the solution for a given stochastic input. The variance reduction technique in this paper is closely related to well-posedness of the governing initial boundary value problem. In particular, it depends on the sharpness of the energy estimate, which in turn depends on the given boundary conditions.

J. Nordström (✉) · M. Wahlsten
Department of Mathematics, Linköping University,
581 83 Linköping, Sweden
e-mail: jan.nordstrom@liu.se

M. Wahlsten
e-mail: markus.wahlsten@liu.se

C. Hirsch et al. (eds.), *Uncertainty Management for Robust Industrial Design in Aeronautics*, Notes on Numerical Fluid Mechanics and Multidisciplinary Design 140, https://doi.org/10.1007/978-3-319-77767-2_29

The technique used in this analysis is directly applicable to linear problems such as Maxwell's equations, the elastic wave equations, the linearized Euler equations, and the linearized Navier–Stokes equations where the uncertainty is known and limited to the data in the problem. The theoretical derivations are for simplicity and clarity done in one space dimension and for one stochastic variable. The extension to multiple space dimensions and stochastic variables is straightforward and would add more technical details but no principal problems.

The Continuous Problem

Consider the following incompletely parabolic system of equations,

$$\begin{aligned} u_t + Au_x + Bu_y + Cu_z &= F_x + G_y + H_z \quad x \in \Omega, \quad t \geq 0 \\ Hu &= g(x, t, \xi) \qquad x \in \partial\Omega, t \geq 0 \\ u &= f(x, \xi) \qquad\quad x \in \Omega, \quad t = 0, \end{aligned} \tag{1}$$

where

$$\begin{aligned} F &= D_{11}u_x + D_{12}u_y + D_{13}u_z, \\ G &= D_{21}u_x + D_{22}u_y + D_{23}u_z, \\ H &= D_{31}u_x + D_{32}u_y + D_{33}u_z. \end{aligned} \tag{2}$$

The solution is represented by the vector $u = u_k(x, t, \xi)$, $k = 1, \ldots, M$, where $x = (x, y, z)$ and $\xi = (\xi_1, \xi_2, \ldots, \xi_L)$ is the vector of variables describing the stochastic variation of the problem. The $M \times M$ matrices A, B, C, D_{ij} are constant and symmetric. H is the boundary operator defined on the boundary $\partial\Omega$. $f(x, \xi)$ and $g(x, t, \xi)$ are the data to the problem. With a proper form of the matrices involved, the problem (1) is the linearized and symmetrized compressible Navier–Stokes equations [10].

The Energy Method

We multiply the equation in (1) by u^T and integrate over the domain. By rearranging and defining $\|u\|^2 = \int_\Omega u^T u \, dx\, dy\, dz$, we obtain

$$\|u\|_t^2 + 2DI = BT, \tag{3}$$

where

$$\begin{aligned} DI &= \int_\Omega \begin{bmatrix} u_x \\ u_y \\ u_z \end{bmatrix}^T \begin{bmatrix} D_{11} & D_{12} & D_{13} \\ D_{21} & D_{22} & D_{23} \\ D_{31} & D_{32} & D_{33} \end{bmatrix} \begin{bmatrix} u_x \\ u_y \\ u_z \end{bmatrix} dx\, dy\, dz \\ BT &= -\oint_{\delta\Omega} u^T \bar{A} u - 2u^T \bar{D} \nabla u \, ds. \end{aligned} \tag{4}$$

In (4), $ds = \sqrt{dx^2 + dy^2 + dz^2}$, $\nabla u = [u_x, u_y, u_z]^T$ and

$$\begin{aligned} \bar{A} &= n_1 A + n_2 B + n_3 C, \qquad \bar{D} = [D_1, D_2, D_3], \\ D_1 &= n_1 D_{11} + n_2 D_{21} + n_3 D_{31}, D_2 = n_1 D_{12} + n_2 D_{22} + n_3 D_{32}, \\ D_3 &= n_1 D_{13} + n_2 D_{23} + n_3 D_{33} \end{aligned} \tag{5}$$

where $n = (n_1, n_2, n_3)^T$ is the outward pointing unit normal to $\delta\Omega$. The matrices $\bar{A}$, D_1, D_2, and D_3 are of size $M \times M$. By diagonalizing BT, we can write the boundary conditions in the following general way

$$Hu = (H^- - RH^+)u = g \tag{6}$$

where R is a matrix satisfying

$$R^T \Lambda^- R + \Lambda^+ > 0. \tag{7}$$

In (6) and (7), H^+ and H^- are eigenvectors corresponding to the positive (Λ^+) and negative (Λ^-) ones, respectively.

We can prove, (see [11] for more details).

Proposition 1 *The problem (1) with boundary conditions (6) and the condition (7) is strongly well-posed.*

Proof Time integration of (3) subject to (3) with boundary conditions leads to the following estimate

$$\|u(T)\|^2 + 2\int_0^T DI\,dt = \|f\|^2 + \int_0^T BT_c + g^T(|\Lambda^-_{\mathbf{M}}| + |G|)g\,dt, \tag{8}$$

where BT_c is given by

$$BT_c = -\begin{bmatrix} W^+ \\ g \end{bmatrix}^T \begin{bmatrix} R^T \Lambda^-_{\mathbf{M}} R + \Lambda^+_{\mathbf{M}} & (\Lambda^-_{\mathbf{M}} R)^T \\ \Lambda^-_{\mathbf{M}} R & -G \end{bmatrix} \begin{bmatrix} W^+ \\ g \end{bmatrix}. \tag{9}$$

The term BT_c in (9) is negative semi-definite when G is chosen appropriately and R is chosen such that (7) holds. By providing these choices, a bound on (8) is obtained. By imposing the correct number of boundary conditions, also uniqueness and existence are guaranteed.

The Semi-discrete Problem

In this section, we formulate a semi-discrete finite difference scheme based on summation-by-parts operators (SBP) [12] and simultaneous approximation terms (SAT) [13].

The semi-discrete version of (1) on SBP-SAT form can be written as

$$
\begin{aligned}
v_t &+ (D_x \otimes I_y \otimes I_z \otimes A)v + (I_x \otimes D_y \otimes I_z \otimes B)v \\
&+ (I_x \otimes I_y \otimes D_z \otimes C)v \\
&= (D_x \otimes I_y \otimes I_z \otimes I_M)F + (I_x \otimes D_y \otimes I_z \otimes I_M)G \\
&+ (I_x \otimes I_y \otimes D_z \otimes I_M)H \\
&+ (P_x^{-1}E_{N_x} \otimes I_y \otimes I_z \otimes I_M)(\tilde{\Sigma}(\tilde{H}^- - \tilde{R}\tilde{H}^+)v - e_{N_x} \otimes g) \\
v(0) &= f.
\end{aligned} \tag{10}
$$

The discrete solution is arranged as

$$
v = \begin{bmatrix} v_0 \\ v_1 \\ \vdots \\ [v_i] \\ \vdots \\ v_{N_x} \end{bmatrix}, \quad [v_i] = \begin{bmatrix} v_0 \\ v_1 \\ \vdots \\ [v_j] \\ \vdots \\ v_{N_y} \end{bmatrix}_i \quad [v_j]_i = \begin{bmatrix} v_0 \\ v_1 \\ \vdots \\ [v_k] \\ \vdots \\ v_{N_z} \end{bmatrix}_{ij} \quad [v_k]_{ij} = \begin{bmatrix} v_1 \\ v_2 \\ \vdots \\ v_l \\ \vdots \\ v_M \end{bmatrix}_{ijk},
$$

where v_{ijkl} approximates $u_l(x_i, y_j, z_k, t, \xi)$.

For simplicity, we only study the boundary at $x = 1$, since the analysis of the remaining boundaries is completely analogous. In (10), $D_{x,y,z} = P^{-1}_{x,y,z}Q_{x,y,z}$ is the discrete approximation of the first derivative in space in the x, y, and z direction, respectively. I_x, I_y, I_z, and I_M are the identity matrices of size $N_x + 1$, $N_y + 1$, $N_z + 1$, and M. The matrices E_{0_x} and E_{N_x} are zero except for the element $(1, 1)$ and $(N_x + 1, N_x + 1)$, respectively, which is equal to 1. Similarly, the vector elements in e_{N_x} are zero with the exception of the last element which is 1. $P_{x,y,z}$ is a positive definite diagonal matrix, and $Q_{x,y,z}$ is an almost skew-symmetric matrix satisfying $Q_{x,y,z} + Q^T_{x,y,z} = E_{N_{x,y,z}} - E_{0_{x,y,z}} = \mathbb{B} = diag[-1, 0, \dots, 0, 1]$.

We have also introduced the numerical fluxes

$$
\begin{aligned}
F &= (\tilde{I} \otimes D_{11})v_x + (\tilde{I} \otimes D_{12})v_y + (\tilde{I} \otimes D_{13})v_z \\
G &= (\tilde{I} \otimes D_{21})v_x + (\tilde{I} \otimes D_{22})v_y + (\tilde{I} \otimes D_{23})v_z \\
H &= (\tilde{I} \otimes D_{31})v_x + (\tilde{I} \otimes D_{32})v_y + (\tilde{I} \otimes D_{33})v_z,
\end{aligned} \tag{11}
$$

where the notation $\tilde{I} = (I_x \otimes I_y \otimes I_z)$ and

$$
\begin{aligned}
v_x &= (D_x \otimes I_y \otimes I_z \otimes I_M)v, \\
v_y &= (I_x \otimes D_y \otimes I_z \otimes I_M)v, \\
v_z &= (I_x \otimes I_y \otimes D_z \otimes I_M)v,
\end{aligned} \tag{12}
$$

have been used. The boundary data g is the original boundary data vector of the problem (1) at $x = 1$.

The discrete boundary operators $\tilde{H}^+$ and $\tilde{H}^-$ are decomposed as

$$
\begin{aligned}
\tilde{H}^{+} &= (I_x \otimes I_y \otimes I_z \otimes H_0^{+}) + (D_x \otimes I_y \otimes I_z \otimes H_{D_x}^{+}) \\
&\quad + (I_x \otimes D_y \otimes I_z \otimes H_{D_y}^{+}) + (I_x \otimes I_y \otimes D_z \otimes H_{D_z}^{+}) \\
\tilde{H}^{-} &= (I_x \otimes I_y \otimes I_z \otimes H_0^{-}) + (D_x \otimes I_y \otimes I_z \otimes H_{D_x}^{-}) \\
&\quad + (I_x \otimes D_y \otimes I_z \otimes H_{D_y}^{-}) + (I_x \otimes I_y \otimes D_z \otimes H_{D_z}^{-}).
\end{aligned} \tag{13}
$$

In a similar fashion as in the continuous problem, the matrix $\tilde{R}$ is defined as

$$
\tilde{R} = (\tilde{I} \otimes R), \tag{14}
$$

and finally $\tilde{H}$

$$
\tilde{H} = \tilde{H}^{-} - \tilde{R}\tilde{H}^{+}. \tag{15}
$$

The penalty matrix $\tilde{\Sigma}$ will be chosen such that stability is achieved.

Stability

To prove stability of the numerical scheme (10), we use the discrete energy method [14, 15], and the SBP properties described above.

We can now prove, (see [11] for further details),

Proposition 2 *The numerical approximation (10) with boundary operators (13) and penalty coefficients*

$$
\begin{aligned}
\tilde{\Sigma} &= (\tilde{H}^{-})^T(\tilde{I} \otimes \Lambda_{\mathbf{M}}^{-}) = ((\tilde{I} \otimes H_0^{-}) + (D_x \otimes I_y \otimes I_z \otimes H_{D_x}^{-}) \\
&\quad + (I_x \otimes D_y \otimes I_z \otimes H_{D_y}^{-}) + (I_x \otimes I_y \otimes D_z \otimes H_{D_z}^{-}))^T(\tilde{I} \otimes \Lambda_{\mathbf{M}}^{-}),
\end{aligned} \tag{16}
$$

is strongly stable.

Proof Applying the discrete energy method to (10) and integrating in time gives us

$$
\|u\|_{P_{xyz}}^2 + 2\int_0^T \overline{DI}\, dt = \|f\|_{P_{xyz}}^2 + \int_0^T \left[BT + g_{N_x}^T(|\Lambda_{\mathbf{M}}^{-}| + |G|)g_{N_x} \right] dt, \tag{17}
$$

with

$$
BT = -\begin{bmatrix} \tilde{W}^{+} \\ \tilde{W}^{-} \\ g \end{bmatrix}_{N_x}^T \left(P_{yz} \otimes \tilde{T}_2\right) \begin{bmatrix} \tilde{W}^{+} \\ \tilde{W}^{-} \\ g \end{bmatrix}_{N_x} \leq 0,
$$

which concludes the proof.

Remark 1 Note the similarity between the semi-discrete estimate (17) and the continuous counterpart (8).

Stochastic Formulation

In this section, we focus on the stochastic properties of (1). Due to the fact that we have a provably stable scheme which converges to the continuous solution under mesh refinement, we only need to study the stochastic properties of the problem in the continuous setting.

Consider the following form of (1) where we highlight the stochastic nature of the data.

$$\begin{aligned} u_t + Au_x + Bu_y + Cu_z &= F_x(u) + G_y(u) + H_z(u) \\ Hu &= g(x,t,\xi) = \mathbb{E}[g] + \delta g(x,t,\xi) \\ u &= f(x,\xi) \quad = \mathbb{E}[f] + \delta f(x,\xi). \end{aligned} \tag{18}$$

In (18), we denote $F_x = F_x(u)$, $G_x = G_x(u)$, and $H_x = H_x(u)$ in order to stress their corresponding dependence on u. By taking the expected value of (18) and defining $v = \mathbb{E}[u]$, we obtain

$$\begin{aligned} v_t + Av_x + Bv_y + Cv_z &= F_x(v) + G_y(v) + H_z(v) \\ Hv &= \mathbb{E}[g] \\ v &= \mathbb{E}[f], \end{aligned} \tag{19}$$

where we have used the linearity property of F_x, G_y, and H_z.

Remark 2 Equation (19) is equivalent to the equation determining the first term in a polynomial chaos expansion; see [16].

Next, the difference between (18) and (19) together with the definition $e = u - v$ gives

$$\begin{aligned} e_t + Ae_x + Be_y + Ce_z &= F_x(e) + G_y(e) + H_z(e) \\ He &= \delta g(x,t,\xi) \\ e &= \delta f(x,\xi). \end{aligned} \tag{20}$$

The Mean Value

Regarding the relation between boundary conditions and the expected values, we need Theorem 1 below. To this end, let $P(\cdot)$ denote a general linear differential operator, H_1 and H_2, two different boundary operators, and w the exact solution when solving the deterministic problems (21) and (22) (for $\delta g_u = \delta g_v = 0$)

$$\begin{aligned} u_t + P(u) &= F, \\ H_1 u &= H_1 w + \delta g_u, \\ u(x,0) &= w(x,0), \end{aligned} \tag{21}$$

$$\begin{aligned} v_t + P(v) &= F, \\ H_2 v &= H_2 w + \delta g_v, \\ v(x,0) &= w(x,0). \end{aligned} \tag{22}$$

Theorem 1 *The expected value of the solutions to the two problems (21) and (22) is the same (i.e., $\mathbb{E}[u] = \mathbb{E}[v]$) if the added randomness has zero mean (i.e., $\mathbb{E}[\delta g_u] = \mathbb{E}[\delta g_v] = 0$).*

Proof See [11].

Theorem 1 implies that the boundary and initial data are extracted from the exact deterministic solution. The fact that w is the solution to both problems implies that it must satisfy the relation $w_t + P(w) = F$.

Remark 3 This situation is very common in practice since the boundary and initial data are extracted from an exact solution (i.e., from reality), together with a small stochastic error (assumed in this scenario to be unbiased (i.e., have zero mean)) coming from measurements or other sources.

Remark 4 The result regarding equal means is important and makes it possible for us to focus entirely on minimizing the variance. The expected value is left unchanged for any choice of well-posed boundary conditions (provided that the data is extracted from the exact solution, and the added uncertainty has zero mean).

Variance Formulation

We apply the energy method to (20) while using the same structure of boundary conditions as in (6)

$$He = (H^- - RH^+)e = E^- - RE^+ = \delta g = \delta g^- - R\delta g^+. \tag{23}$$

This procedure gives (see the previous analysis leading up to (8)) the following equation

$$\|e\|_t^2 + 2DI_e = -\oint_{\delta\Omega} \begin{bmatrix} E^+ \\ \delta g \end{bmatrix}^T \begin{bmatrix} R^T \Lambda_{\mathbf{M}}^- R + \Lambda_{\mathbf{M}}^+ & (\Lambda_{\mathbf{M}}^- R)^T \\ \Lambda_{\mathbf{M}}^- R & \Lambda_{\mathbf{M}}^- \end{bmatrix} \begin{bmatrix} E^+ \\ \delta g \end{bmatrix} ds, \tag{24}$$

where

$$DI_e = \int_{\Omega} \begin{bmatrix} e_x \\ e_y \\ e_z \end{bmatrix}^T \begin{bmatrix} D_{11} & D_{12} & D_{13} \\ D_{21} & D_{22} & D_{23} \\ D_{31} & D_{32} & D_{33} \end{bmatrix} \begin{bmatrix} e_x \\ e_y \\ e_z \end{bmatrix} dx\,dy\,dz.$$

By taking the expected value of (24) and noting that

$$\mathbb{E}[\|e\|^2] = \mathbb{E}[\|u - \mathbb{E}[u]\|^2] = \int_{\Omega} \mathbb{E}[(u - \mathbb{E}[u])^2]\,dx\,dy\,dz = \|Var[u]\|_1, \tag{25}$$

we obtain the final result

$$\begin{aligned} \|Var[u]\|_{1,t} + 2\mathbb{E}[DI_e] = -\oint_{\delta\Omega} & \mathbb{E}[(E^+)^T \Lambda^+_{\mathbf{M}}(E^+)] + \mathbb{E}[(\delta g^-)^T \Lambda^-_{\mathbf{M}}(\delta g^-)] \\ &+ \mathbb{E}[(R\delta g^+)^T \Lambda^-_{\mathbf{M}}(R\delta g^+)] - 2\mathbb{E}[(R\delta g^+)^T \Lambda^-_{\mathbf{M}}(\delta g^-)] \\ &+ \mathbb{E}[(\delta g^- - R\delta g^+ + E^-)^T \Lambda^-_{\mathbf{M}}(RE^+)]\, ds. \end{aligned} \tag{26}$$

Remark 5 Note that (26) is a complete description of how the variance in the solution develops in time.

Next, we will consider (26) in three different scenarios: zero variance on the boundary, decaying variance on the boundary, and large variance on the boundary.

Zero Variance on the Boundary By assuming no uncertainty in the boundary data ($\delta g = \delta g^+ = \delta g^- = 0$), (26) yields

$$\|Var[u]\|_t + 2\mathbb{E}[DI_e] = -\oint_{\delta\Omega} \mathbb{E}[(E^+)^T(\Lambda^+_{\mathbf{M}} + R^T \Lambda^-_{\mathbf{M}} R)(E^+)]\, ds. \tag{27}$$

From (27), we clearly see that the optimal (in terms of variance reduction) choice of R is zero and that all nonzero elements in R will lead to an increase in the L_1-norm of the variance of the solution.

Decaying Variance on the Boundary We now assume time-decaying variances in the boundary data. By considering only the terms depending on the matrix R in (26), we get

$$\begin{aligned} &-\oint_{\delta\Omega} \mathbb{E}[(R\delta g^+)^T \Lambda^-_{\mathbf{M}}(R\delta g^+)] + 2\mathbb{E}[(R\delta g^+)^T \Lambda^-_{\mathbf{M}}(\delta g^-)] \\ &- \mathbb{E}[(\delta g^- - R\delta g^+ + E^-)^T \Lambda^-_{\mathbf{M}}(RE^+)]\, ds. \end{aligned} \tag{28}$$

As can be seen in (28), the optimal choice of R depends on the various correlations between the data g^+ and g^- and the variables E^+ and E^-. However, in the long run, the problem is well-approximated by the zero variance on the boundary case. The optimal choice also in this case is $R = 0$.

Large Variance on the Boundary When having large non-decaying uncertainty on the boundary, we cannot draw any general conclusions of the best choice of R since the correlation terms in (28) will dominate.

Applications

An Application to Inviscid Flows

We study different subsonic outflow boundary conditions for the Euler equations with random boundary data. The linearized one-dimensional symmetrized form of the Euler equations with frozen coefficients is, see [17],

$$U_t + \bar{A} U_x = 0. \tag{29}$$

In (29),

$$U = \left[\frac{\bar{c}}{\sqrt{\gamma \bar{\rho}}} \rho, u, \frac{1}{\bar{c}\sqrt{\gamma(\gamma-1)}} T \right]^T, \quad \bar{A} = \begin{bmatrix} \bar{u} & \frac{\bar{c}}{\sqrt{\gamma}} & 0 \\ \frac{\bar{c}}{\sqrt{\gamma}} & \bar{u} & \sqrt{\frac{\gamma-1}{\gamma}}\bar{c} \\ 0 & \sqrt{\frac{\gamma-1}{\gamma}}\bar{c} & \bar{u} \end{bmatrix}, \tag{30}$$

and $\bar{A} = X \Lambda X^T$ where

$$X = \begin{bmatrix} -\sqrt{\frac{\gamma-1}{\gamma}} & \frac{1}{\sqrt{2\gamma}} & \frac{1}{\sqrt{2\gamma}} \\ 0 & \frac{1}{\sqrt{2}} & -\frac{1}{\sqrt{2}} \\ \frac{1}{\sqrt{\gamma}} & \sqrt{\frac{\gamma-1}{2\gamma}} & \sqrt{\frac{\gamma-1}{2\gamma}} \end{bmatrix}, \quad \Lambda = \begin{bmatrix} \bar{u} & 0 & 0 \\ 0 & \bar{u}+\bar{c} & 0 \\ 0 & 0 & \bar{u}-\bar{c} \end{bmatrix}.$$

The perturbation variables ρ, u, p, T, c, and γ represent the normalized density, velocity, pressure, temperature, speed of sound, and the ratio of specific heat. The overbar denotes variables at the constant state where $\bar{A}$ is calculated. We have used $\bar{u} = 1$, $\bar{c} = 2$, $\bar{\rho} = 1$, and $\gamma = 1.4$.

The boundary conditions are of the type (6). Together with the characteristic case, we study the following two settings of R_0 and R_1 in the non-characteristic case

Characteristic	$R_0 = [0, 0]^T$,	$R_1 = [0, 0]$,	
Pressure	$R_0 = [0, 0]^T$,	$R_1 = [0, -1]$,	(31)
Velocity	$R_0 = [0, 0]^T$,	$R_1 = [0, +1]$.	(32)

Note that the non-characteristic boundary condition (31) corresponds to specifying the *pressure* p and (32) corresponds to specifying the *velocity* u.

For completeness, we introduce the characteristic variables

$$X^T U = \left[\frac{1}{\sqrt{\gamma-1}\bar{\rho}\bar{c}}(p - \bar{c}^2 \rho), \frac{1}{\sqrt{2}\bar{\rho}\bar{c}}(p + \bar{c}\bar{\rho}u), \frac{1}{\sqrt{2}\bar{\rho}\bar{c}}(p - \bar{c}\bar{\rho}u) \right]^T. \tag{33}$$

The non-decaying randomness in the boundary data is given by $\rho = u = p = 0.1 + 3\cos(2\pi t)\xi^3$. In all calculations below, we use the characteristic boundary conditions on the inflow boundary. On the outflow boundary, three different boundary conditions are considered. We either specify the ingoing characteristic variable or the pressure or the velocity. We restrict ourselves to the study of large variance on the boundary.

We show results for $\xi \sim \mathcal{N}(0, 1)$ and $\xi \sim \mathcal{U}(-\sqrt{3}, \sqrt{3})$ and compare the L_1-norm of the variance for the characteristic, pressure, and velocity boundary condition, respectively. Figure 1 shows that even without a decaying variance on the

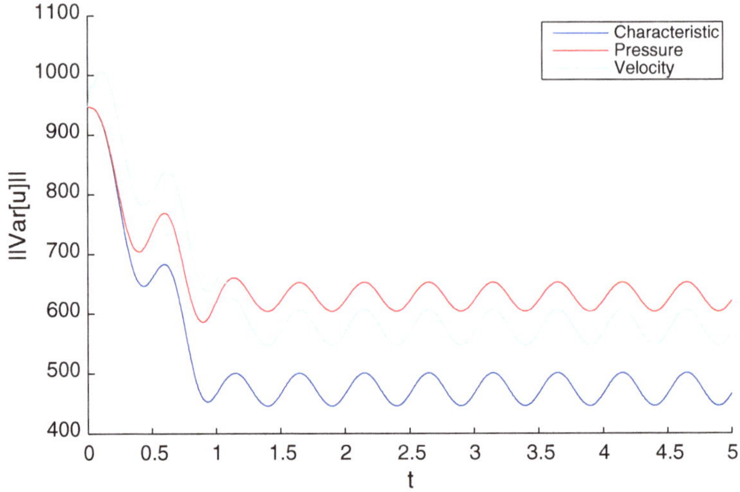

Fig. 1 L_1-norm of the variance as a function of time for $\xi \sim \mathcal{N}(0, 1)$ and characteristic, pressure, and velocity boundary condition

Table 1 Integral of the L_1-norm of the variance for different values of the matrices R_0 and R_1 for $\xi \sim \mathcal{N}(0, 1)$ and $\xi \sim \mathcal{U}(-\sqrt{3}, \sqrt{3})$

	Case	Characteristic	Pressure	Velocity
$\int_0^T \|\|Var[u]\|\| \, dt$	Normal	2579.0	3264.0	3143.0
	Uniform	663.9	840.3	809.1

boundary, the characteristic boundary conditions give us the smallest variance. Figures for the cases with uniformly distributed randomness are omitted due to the resemblance to the normally distributed cases. We also compare the different boundary conditions by integrating the L_1-norm of the variance from zero to T. Table 1 shows the total L_1-norm of the variance of the solution for the different cases. As seen from Table 1, the characteristic boundary condition gives in total the lowest L_1-norm of the variance.

An Application to Viscous Flows

Next, we add on viscosity and consider the linearized and symmetrized Navier–Stokes equations in one dimension; see [18, 19]

$$U_t + AU_x = \varepsilon B U_{xx}, \tag{34}$$

where U and A are given in (30), while

$$B = diag\left(0, \frac{\lambda + 2\mu}{\bar{\rho}}, \frac{\gamma\mu}{\text{Pr}\bar{\rho}}\right). \tag{35}$$

We also use $A = X\Lambda X^T$, $\tilde{B} = X^T BX$, where

$$\tilde{B} = \frac{1}{2}\begin{bmatrix} \tilde{\phi}\alpha^2 & \tilde{\phi}\alpha & \tilde{\phi}\alpha \\ \tilde{\phi}\alpha & \tilde{\phi} + \tilde{\theta} & \tilde{\phi} - \tilde{\theta} \\ \tilde{\phi}\alpha & \tilde{\phi} - \tilde{\theta} & \tilde{\phi} + \tilde{\theta} \end{bmatrix}, \quad \tilde{\phi} = \frac{\lambda + 2\mu}{\bar{\rho}}, \quad \tilde{\theta} = \frac{(\gamma - 1)\mu}{\text{Pr}\bar{\rho}}, \quad \alpha = \sqrt{\frac{2}{\gamma - 1}}.$$

The variables and parameters are μ, λ, Pr, ε and represent the dynamic and second viscosity, the Prandtl number, and the inverse Reynolds number. The overbar denotes the constant state around which the Navier–Stokes equations are linearized.

In our calculations, we use the following numerical values

$$\begin{aligned} &\bar{\rho} = 1, \quad \bar{u} = 1, \quad \bar{c} = 2, \quad \bar{p} = 1, \lambda = -\tfrac{2}{3}, \\ &\gamma = 1.4, \varepsilon = 0.01, \text{Pr} = 0.7, \mu = 1. \end{aligned} \tag{36}$$

The boundary conditions are of the type (6), where R_0 and R_1 are matrices of sizes 3×2 and 2×3. The boundary cases which we compare are the following

$$\begin{aligned} \text{Characteristic: } R_0 &= \begin{bmatrix} 0 & 0 & 0 \\ 0 & 0 & 0 \end{bmatrix}^T, R_1 = \begin{bmatrix} 0 & 0 & 0 \\ 0 & 0 & 0 \end{bmatrix}, \\ \text{Specifying } u\text{: } R_0 &= \begin{bmatrix} 0 & 0 & 0 \\ 0 & 0 & 0 \end{bmatrix}^T, R_1 = \begin{bmatrix} 0 & -1 & r_{13} \\ 0 & 0 & r_{23} \end{bmatrix}, \\ \text{Specifying } p - \varepsilon\left(\tfrac{\theta}{\bar{\rho}}\right)u_x\text{: } R_0 &= \begin{bmatrix} 0 & 0 & 0 \\ 0 & 0 & 0 \end{bmatrix}^T, R_1 = \begin{bmatrix} 0 & 1 & \tilde{r}_{13} \\ 0 & 0 & \tilde{r}_{23} \end{bmatrix}, \\ \text{Inflow 1: } R_0 &= \begin{bmatrix} 0 & s_{21} & 0 \\ 0 & s_{22} & s_{32} \end{bmatrix}^T, R_1 = \begin{bmatrix} 0 & 0 & 0 \\ 0 & 0 & 0 \end{bmatrix}, \\ \text{Inflow 2: } R_0 &= \begin{bmatrix} \tilde{s}_{11} & 0 & 0 \\ \tilde{s}_{12} & \tilde{s}_{22} & \tilde{s}_{32} \end{bmatrix}^T, R_1 = \begin{bmatrix} 0 & 0 & 0 \\ 0 & 0 & 0 \end{bmatrix}. \end{aligned} \tag{37}$$

Note that specifying u and $p - \varepsilon(\theta/\bar{\rho})u_x$ are both well-posed boundary conditions at a subsonic outflow boundary; see [20].

The choices of R_0 and R_1 in (37) are made such that condition (7) is satisfied. The randomness imposed in the boundary data is given by

$$\begin{aligned} \rho = u \ = p \ &= 1 + 3\cos(2\pi t)\xi^3 + 3\sin(2\pi t)\xi_2^3, \\ \rho_x = u_x = p_x &= \quad\ 3\cos(2\pi t)\xi_1^3 + 3\sin(2\pi t)\xi_2^3. \end{aligned}$$

The characteristic variables in this problem are the same as the ones in (33). The boundary conditions used in the comparison are two well-known outflow boundary

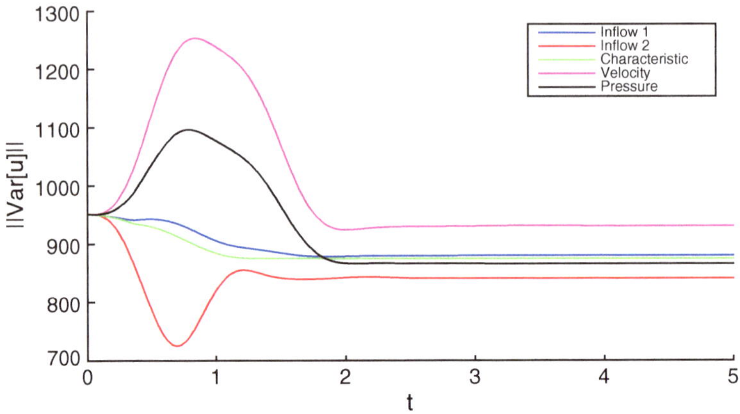

Fig. 2 L_1-norm of the variance as a function of time for normally distributed ξ_1 and ξ_2 for characteristic and non-characteristic boundary conditions when having large non-decaying boundary data for the Navier–Stokes equations

Table 2 Integral of the L_1-norm of the variance for different values of R_0 and R_1 for $T = 5$ and different distributions

	BC	Charact.	Velocity	Pressure	Inflow 1	Inflow 2
$\int_0^T \|\|Var[u]\|\| \, dt$	Uniform	1501.8	1687.9	1558.6	1514.8	1423.8
	Normal	4425.8	4974.3	4593.2	4464.2	4196.0

conditions (*specifying* u and *specifying* $p - \epsilon(\frac{\theta}{\bar{\rho}})$ (with characteristic boundary conditions as inflow) and two artificially constructed inflow boundary conditions (with characteristic boundary conditions as outflow).

Figure 2 shows the L_1-norm of the variance for the different cases stated in (37). Table 2 illustrates the integrated L_1-norm of the variance of the solution for the five different boundary conditions. From Table 2, we note that the generalized characteristic ($R = 0$ for all boundaries) gives in this case the second lowest total variance. In this case, only the boundary condition *inflow 2* gives a smaller total variance. From Fig. 2, we observe that the inflow boundary conditions (*inflow 1* and *inflow 2*) provide a significant smaller variance than the outflow boundary conditions (*specifying* u and *specifying* $p - \epsilon(\frac{\theta}{\bar{\rho}})$).

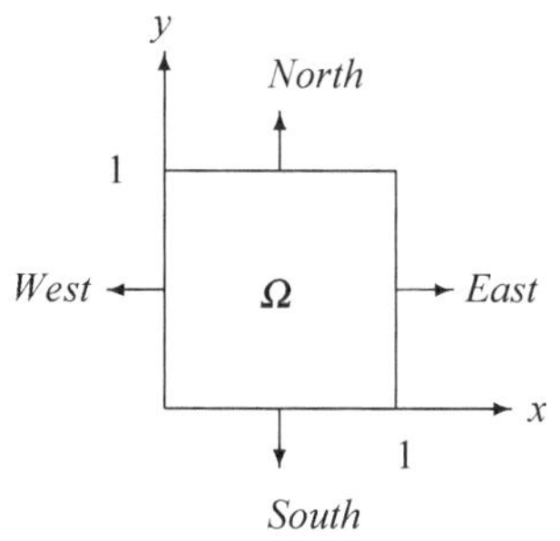

Fig. 3 Schematic of the domain

An Application in Electromagnetics

The relation between the electric and magnetic fields is given by, see [21],

$$\begin{aligned} \mu\frac{\partial H}{\partial t} &= -\nabla \times E, & \varepsilon\frac{\partial E}{\partial t} &= \nabla \times H - J, \\ \nabla \cdot \varepsilon E &= \rho, & \nabla \cdot \mu H &= 0, \end{aligned} \tag{38}$$

where $E, H, J, \rho, \varepsilon$, and μ represent the electric field, magnetic field, electric current density, charge density, permittivity, and permeability. In this example, we let $\rho = 1$, $\varepsilon = 1$, and $\mu = 1$. By also letting $J = 0$, we can write (38) in matrix form as

$$Su_t + Au_x + Bu_y = 0,$$

where $u = \left[H_z, E_x, E_y\right]^T$ and

$$S = \begin{bmatrix} \mu & 0 & 0 \\ 0 & \varepsilon & 0 \\ 0 & 0 & \varepsilon \end{bmatrix}, \qquad A = \begin{bmatrix} 0 & 0 & 1 \\ 0 & 0 & 0 \\ 1 & 0 & 0 \end{bmatrix}, \qquad B = \begin{bmatrix} 0 & -1 & 0 \\ -1 & 0 & 0 \\ 0 & 0 & 0 \end{bmatrix}.$$

Furthermore, we introduce the eigendecomposition

$$\Lambda_A = \Lambda_B = \begin{bmatrix} 1 & 0 & 0 \\ 0 & 0 & 0 \\ 0 & 0 & -1 \end{bmatrix}, \qquad X_A = \begin{bmatrix} -\frac{1}{\sqrt{2}} & 0 & -\frac{1}{\sqrt{2}} \\ \frac{1}{\sqrt{2}} & 0 & -\frac{1}{\sqrt{2}} \\ 0 & 1 & 0 \end{bmatrix}, \qquad X_B = \begin{bmatrix} \frac{1}{\sqrt{2}} & 0 & \frac{1}{\sqrt{2}} \\ 0 & -1 & 0 \\ \frac{1}{\sqrt{2}} & 0 & -\frac{1}{\sqrt{2}} \end{bmatrix}.$$

The boundary conditions are of the type (6), where in this case R is represented at the North, South, East, and West boundary (seen in Fig. 3). Together with the characteristic case (*CHA*), we study the following cases

Table 3 Integral of the L_1-norm of the variance for different values of the matrices R_N, R_S, R_E, and R_W for $\xi \sim \mathcal{N}(0,1)$ and $\xi \sim \mathcal{U}(-\sqrt{3},\sqrt{3})$

	Case	CHA	BC1	BC2
$\int_0^T \|Var[u]\| \, dt$	Normal	74.9	87.1	106.6
	Uniform	921.1	1070.7	1311.8

$$\begin{aligned} \text{BC1 } R_N &= [+\tfrac{1}{2},0]^T, \ R_S = [0,-\tfrac{1}{2}], \\ R_E &= [+\tfrac{1}{2},0]^T, R_W = [0,+\tfrac{1}{2}], \\ \text{BC2 } R_N &= [-\tfrac{1}{2},0]^T, \ R_S = [0,+\tfrac{1}{2}], \\ R_E &= [-\tfrac{1}{2},0]^T, R_W = [0,-\tfrac{1}{2}], \\ \text{CHA } R_N &= [0,0]^T, \quad R_S = [0,0], \\ R_E &= [0,0]^T, \quad R_W = [0,0]. \end{aligned} \tag{39}$$

In (39), both *BC1* and *BC2* specify a linear combination of H_z, E_x, E_y at the North and East boundaries and H_z, E_x at the South and West boundaries. In *CHA*, the characteristic variables

$$X_A^T u = \left[\frac{-H_z + E_x}{\sqrt{2}}, E_y, \frac{-H_z - E_x}{\sqrt{2}}\right]^T, \qquad X_B^T u = \left[\frac{H_z + E_y}{\sqrt{2}}, -E_x, \frac{H_z - E_y}{\sqrt{2}}\right]^T$$

are specified.

When constructing boundary and initial data, we assume randomness in H_z, E_x, and E_y given by

$$H_x = E_x = E_y = 1 + \sin(2\pi x)\sin(2\pi y) + 3\cos(2\pi t)\xi.$$

In Fig. 4, we compare the L_1-norm of the variance for *BC1*, *BC2*, and *CHA* for $\xi \sim \mathcal{N}(0,1)$. We also compare the total variance for the three different cases in Table 3. As can be seen, the trend from the previous sections remains, namely that the characteristic boundary condition (*CHA*) gives the smallest variance.

Summary and Conclusions

We have studied how the boundary conditions affect the uncertainty of the solution when the initial and boundary data are uncertain. A general incompletely parabolic system of equations has been studied, and general well-posed boundary conditions have been derived. The boundary conditions lead to strongly well-posed and strongly stable continuous and discrete problems, respectively.

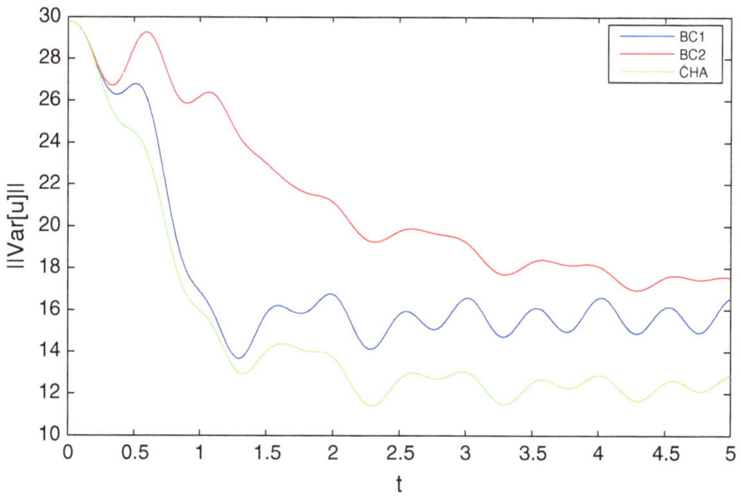

Fig. 4 L_1-norm of the variance as a function of time for $\xi \sim \mathcal{N}(0,1)$ and *BC1*, *BC2*, and *CHA*

Next, randomness was added to the analysis and a formulation relating the variance of the solution with the boundary condition imposed was derived. Hence, a formulation describing the relation between the variance of the solution and the imposed boundary conditions was obtained. It was proven that the expected value of the solution is independent of the choice of boundary condition in the linear case given that the boundary and initial data are taken from the exact solution.

The technique was applied to the one-dimensional linearized Euler, Navier–Stokes and Maxwell's equations, where the case of having large non-decaying variance on the solution was shown for different types of boundary conditions. The results were in line with the model problem showing that the generalized characteristic boundary condition was a good choice, however, the boundary condition *Inflow 2* provided a smaller total variance for the Navier–Stokes case. For the Euler and Maxwell cases, the characteristic boundary condition provided the smallest total variance.

With deterministic boundary data, we conclude that the optimal choice of boundary operator is the generalized characteristic boundary conditions. When having decaying variance in the boundary data, the optimal choice is determined by the different correlations between the inflow and outflow boundary data. However, for long times, the optimal choice is again the generalized characteristic boundary conditions. With large non-decaying variance in the boundary data, no general conclusions can be drawn, although the numerical results suggest that the generalized characteristic boundary conditions are generally a good choice.

References

1. Poroseva, S., Letschert, J., Hussaini, M.Y.: Uncertainty quantification in hurricane path forecasts using evidence theory. APS In: Meeting Abstracts, pp B1+ (2005)
2. Reagan, M.T., Najm, H.N, Ghanem R.G., Knio, O.M.: Uncertainty quantification in reacting-flow simulations through non-intrusive spectral projection. In: Combustion and Flame, pp. 545–555 (2003)
3. Ghanem, R., Dham, S.: Stochastic finite element analysis for multiphase flow in heterogeneous porous media. Porous. Media **32**, 239–262 (1998)
4. Zenga, Z., Jina, J.-M.: Efficient Calculation of Scattering Variation due to Uncertain Geometrical Deviation, vol. 27 (2007)
5. Sayers, C., den Boer, L., Nagy, Z., Hooyman, P.: Well-constrained seismic estimation of pore pressure with uncertainty. Lead. Edge **25**, 1524–1526 (2006)
6. Acharjee, S., Zabaras, N.: A non-intrusive stochastic galerkin approach for modeling uncertainty propagation in deformation processes. Comput. Struct. **85**, 244–254 (2007)
7. Knio, O.M., Le Matre, O.P.: Uncertainty propagation in cfd using polynomial chaos decomposition. Fluid Dyn. Res. **38**, 616–640 (2006)
8. Najma, H.N., Valoranib, M.: Enforcing positivity in intrusive PC-UQ methods for reactive ODE systems. J. Comput. Phys. **270**, 544–569 (2014)
9. Pettersson, P., Iaccarino, G., Nordström, J.: Polynomial chaos methods for hyperbolic partial differential equations. Springer Math Eng. **10**(1007), 978–973 (2015)
10. Nordström, J., Lönn, B.: Energy decay of vortices in viscous fluids: an applied mathematics view. J. Fluid Mech. **709**, 209–232 (2012)
11. Wahlsten, M., Nordström, J.: Robust boundary conditions for stochastic incompletely parabolic systems of equations (In production)
12. Mattsson, K., Almquist, M., Carpenter, M.H.: Optimal diagonal-norm SBP operators. J. Comput. Phys. **264**, 91–111 (2014)
13. Berg, J., Nordström, J.: Superconvergent functional output for time-dependent problems using finite differences on summation-by-parts form. Technical Report 2012-004, Feb 2012
14. Lindström, J., Nordström, J.: A stable and high-order accurate conjugate heat transfer problem. J. Comput. Phys. **229**(14), 5440–5456 (2010)
15. Nikkar, S., Nordström, J.: Fully discrete energy stable high order finite difference methods for hyperbolic problems in deforming domains. J. Comput. Phys. **291**, 82–98 (2015)
16. Pettersson, P., Iaccarino, G., Nordström, J.: A stochastic Galerkin method for the Euler equations with Roe variable transformation. J. Comput. Phys. **257**, 481–500 (2014)
17. Abarbanel, S., Gottlieb, D.: Optimal time splitting for two- and three-dimensional Navier-Stokes equations with mixed derivatives. J. Comput. Phys. **41**, 1–43 (1981)
18. Berg, J., Nordström, J.: Duality based boundary conditions and dual consistent finite difference discretizations of the Navier-Stokes and Euler equations. J. Comput. Phys. **259**, 135–153 (2014)
19. Nordström, J.: The use of characteristic boundary conditions for the Navier-Stokes equations. Comput. Fluids **24**, 609–623 (1995)
20. Nordström, J.: The influence of open boundary conditions on the convergence to steady state for the Navier-Stokes equations. J. Comput. Phys. **85**, 210–244 (1989)
21. Nordström, J., Gustafsson, R.: High order finite difference approximations of electromagnetic wave propagation close to material discontinuities. J. Sci. Comput. **18**(2), 215–234 (2003)

Robust Optimization with Gaussian Process Models

Krzysztof Marchlewski, Łukasz Łaniewski-Wołłk, Sławomir Kubacki and Jacek Szumbarski

Introduction

An optimization process of a certain engineering solution is usually a complex task which should be tackled by appropriate methods and using criteria defined specifically for a given problem. In this chapter, we focus on problems which admit description in terms of sufficiently regular objective functions. Typically, at least continuity with respect to all variables is demanded.

A topic of effective evaluation of such objective functions remains constant in the area of interests of scientists and engineers. The systematic progress in numerical methods and growing availability of high-performance computers are additional factors contributing to further intensification of this research.

In order to solve particular optimization problems, various specialized algorithms have been developed in the last few decades. Still, the main challenge is to apply them routinely to problems of industry-relevant complexity. In this context, it is crucial to design optimization algorithms which need as few as possible objective function evaluations, typically very costly both in terms of computational time and hardware resources.

One of the approaches which enjoy continuously increasing popularity is the response surface method (RSM), also called the method of meta-modelling or

K. Marchlewski (✉) · Ł. Łaniewski-Wołłk · S. Kubacki · J. Szumbarski
The Faculty of Power and Aeronautical Engineering, Warsaw University of Technology, Nowowiejska 24, 00-665 Warsaw, Poland
e-mail: kmarchlewski@meil.pw.edu.pl

Ł. Łaniewski-Wołłk
e-mail: llaniewski@meil.pw.edu.pl

S. Kubacki
e-mail: slawomir.kubacki@meil.pw.edu.pl

J. Szumbarski
e-mail: jasz@meil.pw.edu.pl

C. Hirsch et al. (eds.), *Uncertainty Management for Robust Industrial Design in Aeronautics*, Notes on Numerical Fluid Mechanics and Multidisciplinary Design 140, https://doi.org/10.1007/978-3-319-77767-2_30

surrogate modelling. The essential idea of the surrogate-based optimization is to approximate the parametric variations (response) of real optimized model by their multidimensional algebraic approximation. Such approximation consists basically of two main ingredients: a set of basic functions used to span the response surface and a selection algorithm for samples of a real model response used to *train* a surrogate. Different approaches have been proposed within the RSM framework: application of radial basis functions or thin-plate splines approximation, approximation by Gaussian processes or using artificial neural networks. A comprehensive overview of the response surface method and their application in engineering design and optimization can be found in [1, 2].

Despite their popularity, the response surface methods suffer the same fundamental difficulty—all of them tend to be prohibitively expensive when applied to optimization problems with large number of design/control variables. In such circumstances, the number of samples necessary to fit the data (to train the RSM-based meta-model) is rising exponentially with a number of variables. This effect is commonly referred to as the curse of dimensionality (e.g. [3]).

To create a meta-model, interpolation between the sample points in a design space of an optimized model is needed. However, in practice the response evaluated at these sample points is affected by random errors. Hence, to properly estimate the model response one needs to take uncertainty quantification into account. This is why the Gaussian process (GP) regression (Kriging) has become a popular approach to surrogate-based optimization. Due to its underlying probability framework, it allows us to tell how uncertain different parts of the Kriging model are. Knowing uncertainty of the model, we can design a sampling criterion which targets improvement of the GP model in selected subset of sampling points. These points are indicated as a compromise between search for the global minimum and accuracy in locating it (search for the local minimum). That approach is called the efficient global optimization (EGO) [4].

In recent years, increasing pressure on shortening the design cycle and implementation to production has stimulated growing interest in robust optimization methods. In this context, augmentation of optimization procedures with uncertainty quantification (UQ) has become mandatory. Generally, sources of uncertainties can be divided into two main groups. Epistemic (or systematic) uncertainties appear due to such factors like imperfection or oversimplification of the physical models, insufficiently accurate identification/measurements of the model parameters, lack of perfection in manufacturing procedure or other factors which are currently out of control, but—at least in principle—depend on our knowledge and solicitude. Aleatory (or statistical) uncertainties appear due to factor which is essentially out of control, typical example being operating conditions. Uncertainties which are not controllable are particularly important and often crucial for both a design process and safety operation of a technical system (e.g. icing problem in aircraft [5]).

The need of performing simultaneously the optimization procedure combined with uncertainty quantification is the essence of the idea of the robust design optimization (RDO). The most straightforward implementation of the RDO concept is to perform sequential steps, i.e. optimize and assess uncertainty in two separate sub-steps of the iterative process. Such approach is reliable but usually very time- and resource-consuming. To reduce the time needed for each step, the robust optimization is frequently based on the RSM [6].

In this chapter, we propose a method capable of performing the optimization and UQ procedures simultaneously. The proposed method is based completely on analytical approach. Hence, it allows one to perform the RDO very efficiently. First attempts to develop such approach were presented in [7]. The method is based on the GP regression supplemented with the sampling criterion stemmed from the expected improvement [4] idea. The modification consists in taking into account a correlation between sampled objective function and the mean response of the Kriging model. We called this approach the relative expected improvement (REI) method [8].

This chapter is organized as follows. In the next section, we present a formal definition of the optimization problem. Details of the proposed algorithm are described in the section "Kriging-Based Optimization and Uncertainty Quantification". The section "Example of Application" presents the results of the optimization of the duct channel delivering air to the turboprop engine. In the last section, we briefly summarized the optimization methodology.

Problem Statement

In order to thoroughly describe the developed methodology, two spaces have to be introduced. First one is a space of all possible values of the design variables. The extreme values of these variables should be determined by an experienced engineer. This set of values is called a design space and denoted as $\mathbb{X}$. The points in this space are $\mathbf{x} = (x_1, \dots, x_L)$, where L is a dimension of the problem. A second space contains values of the design variables which arise from considering influence of the uncertainties. While introducing an objective function, a detailed description of these variables will be presented. This set is called a sampling space $\mathbb{S}$ and contains points $\mathbf{s} = (s_1, \dots, s_L)$. It should be noticed that $S \subseteq X$.

Let us assume that an objective function $F(\mathbf{x})$ is known. Its domain is the $\mathbb{X}$ space. We also assume that a variable $\mathbf{x} = (x_1, \dots, x_L)$ is influenced by additive uncertainties. It means that the real value of the objective is a random variable $F(\mathbf{s}) = F(\mathbf{x} + \xi)$, where ξ is the random variable with a probability density function $g(\mathbf{x})$ and the zero mean. We can define a mean response as:

$$F_{err}(\mathbf{x}) := \mathbb{E}_{\xi} F(\mathbf{x} + \xi) \tag{1}$$

Symbol $\mathbb{E}_{\xi}$ denotes an expected value with respect to the uncertainty ξ. A domain of the function F_{err} is the $\mathbb{X}$ space.

In the similar way, we define a variation of the function with respect to the uncertainty:

$$\sigma^2_{err}(\mathbf{x}) := \mathbb{VAR}^2_{\xi} F(\mathbf{x} + \xi) \tag{2}$$

These two statistical moments can be used to quantify the basic information about the objective function influenced by the uncertainties.

Kriging-Based Optimization and Uncertainty Quantification

The evaluation of the objective function F is in most cases very expensive. Therefore, one would like to evaluate quantities $F_{err}(\mathbf{x})$ and $\sigma^2_{err}(\mathbf{x})$ by means of RSM. In our method, we use the Gaussian process prediction (commonly known as Kriging) for that.

More precisely, a technique which will be presented in the following section consists of the following steps:

1. Performing a Design of Experiment (DoE) procedure and evaluation of the objective function for prescribed combination of input parameters.
2. Fitting the Kriging model to the obtained data and validating it through the leave-one-out cross-validation (LOOCV).
3. Applying a sampling criterion to improve the initially fitted model. Particularly, finding the maximum value of the relative expected improvement and sampling the objective function if necessary.
4. Checking a stop condition and returning to the point 2 if convergence requirement is not fulfilled.

Design of Experiment

A data set is required to fit the first model. In the beginning, our knowledge of the deterministic function is very limited; therefore, the initial sample points had to be chosen randomly and evenly at the same time. Their number has to be large enough to capture main features of the objective function. We cannot perform the full factorial experiment because it would imply an extensive evaluation of the objective function. That is why we used a Latin hypercube sampling (LHS) method [9, 10].

Assume that N samples $F(\mathbf{x}^i)$ from the objective function are taken. We use $N = 2 \cdot L + 3$. Such formula is dictated by the number of unknown Kriging parameters which should be estimated to fit the model. In the process, we get the set of N points $X = \left\{\mathbf{x}^1, \ldots, \mathbf{x}^N\right\}$ and the values of the objective function in these points $f(\mathbf{x}^i) = f_i$.

Set X is often augmented with a point corresponding to a base design for the following reasons. First, the objective function value for such design is probably already evaluated. Secondly, this step allows one to assess if the designs found by the algorithm are better than the base design.

Kriging

The real objective function is unknown, but we can estimate its values taking advantage of the Gaussian process. We assume that the objective function is a realization of a random field

$$F(\mathbf{x}) = \varepsilon(\mathbf{x}) + \mu(\mathbf{x})$$

The term $\mu(\mathbf{x})$ is a trend function of the form

$$\mu(\mathbf{x}) = \sum_{j=1}^{M} \beta_j h_j(\mathbf{x})$$

where $h_1, h_2, \ldots, h_M$ are basis functions and β_j are unknown coefficients. It is common to use a linear trend function

$$\mu(\mathbf{x}) = \beta_0 + \sum_{j=1}^{L} \beta_j x_j$$

The term $\varepsilon(\mathbf{s})$ is a centred Gaussian process

$$\varepsilon \sim \mathbb{N}(0, \mathbf{C}(X, Y))$$

where $\mathbf{C}(X, Y)$ is a covariance matrix, and X and Y are random vectors. Properties of functions generated with such GP depend strongly on the covariance function $c(x, y)$. The process ε is assumed stationary, so that $c(x, y) = c(x - y)$. One of the basic functions is the Gaussian kernel which for one-dimension is given by formula

$$g_g(x - y; \theta) = \exp\left(-\frac{(x - y)^2}{2\theta^2}\right)$$

In many dimensions, it is convenient to take

$$c\left(\mathbf{x} - \mathbf{y}, \sigma^2, \tau^2, \theta\right) = \sigma^2 \prod_{k=1}^{L} g_g(x_k - y_k; \theta_k) + \tau^2 \delta_{\mathbf{x},\mathbf{y}}$$

where symbol $\delta_{\mathbf{x},\mathbf{y}}$ denotes the Kronecker delta

$$\delta_{\mathbf{x},\mathbf{y}} = \begin{cases} 1, & \mathbf{x} = \mathbf{y} \\ 0, & \mathbf{x} \neq \mathbf{y} \end{cases}$$

Parameter θ is a length scale for a case, τ^2 is a so-called *nugget effect* introduced to deal with a noise of the solver, and σ^2 parameter improves estimation of variance.

These quantities are called the Kriging parameters. They are estimated with the maximum likelihood estimation (MLE) method. The choice of the covariance kernel will be discussed in detail in section "Utilizing Different Covariance Kernels".

Next, we can define estimators of the expected value and variance of the process

$$\hat{F}(\mathbf{x}) = \mathbb{E}_F F(\mathbf{x}) \tag{3}$$

$$\hat{\sigma}^2(\mathbf{x}) = \mathbb{VAR}_F^2 F(\mathbf{x}) \tag{4}$$

Following the usual Kriging approach, we assume that the expected value estimator has the properties:

- It is a linear combination of known objective function values

$$\hat{F}(\mathbf{x}) = \sum_{i=1}^{N} a^i(\mathbf{x}) F\left(\mathbf{x}^i\right)$$

- It is unbiased

$$\mathbb{E}\left(\hat{F}(\mathbf{x})\right) = \mathbb{E}\left(F(\mathbf{x})\right), \text{and}$$

- The mean squared error (MSE) is minimal

$$MSE(\mathbf{x}) = \mathbb{E}\left(\hat{F}(\mathbf{x}) - F(\mathbf{x})\right)^2 = \min$$

These assumptions allow us to write the Kriging estimators in the forms

$$\hat{F}(\mathbf{x}) = h^T \beta + r^T \mathbf{R}^{-1}\left(F - \mathbf{H}\beta\right) \tag{5}$$

and

$$\hat{\sigma}^2(\mathbf{x}) = \mathrm{c}(\mathbf{x} - \mathbf{x}) - r^T \mathbf{R}^{-1} r + \left(h - \mathbf{H}^T \mathbf{R}^{-1} r\right)^T \left(\mathbf{H}^T \mathbf{R}^{-1} \mathbf{H}\right)^{-1} \left(h - \mathbf{H}^T \mathbf{R}^{-1} r\right) \tag{6}$$

where

$$\beta = \left(\mathbf{H}^T \mathbf{R}^{-1} \mathbf{H}\right)^{-1} \mathbf{H}^T \mathbf{R}^{-1} F$$

- h—a vector of basis function values for $\mathbf{x}$; $h_j = h_j(\mathbf{x})$,
- $\mathbf{H}$—a matrix of transposed vectors h for each $\mathbf{x}^i$; $H_{ij} = h_j\left(\mathbf{x}^i\right)$,
- r—a vector of covariance function values for each $\mathbf{x} - \mathbf{x}^i$; $r_i = c\left(\mathbf{x} - \mathbf{x}^i\right)$,
- F—a vector of known values of the objective function; $F_i = F\left(\mathbf{x}^i\right)$,
- R—a matrix of covariance function values for all $\mathbf{x}^i - \mathbf{x}^j$; $R_{ij} = c\left(\mathbf{x}^i - \mathbf{x}^j\right)$.

The result of applying the described approach can be seen in Fig. 1. The black dots are the samples of the objective function. The grey lines are the samples from the GP in the presence of the measured values. The black line is the mean value, and the dashed lines constrict the 95% confidence intervals.

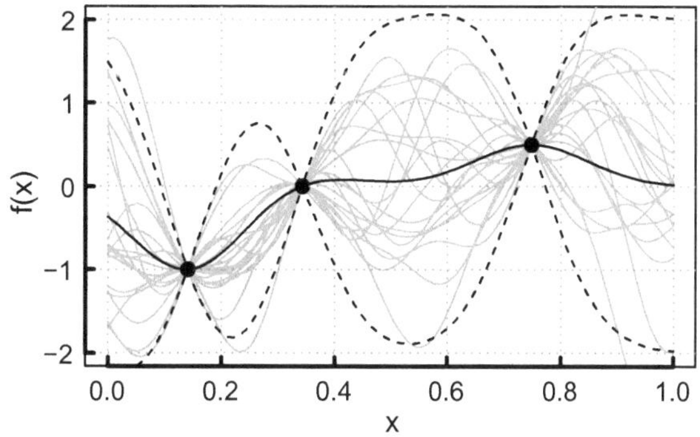

Fig. 1 Twenty samples from a GP. Black dots are measurements points, black line is the mean value, and dashed lines are limits for 95% confidence intervals

Having the expected value and variance estimators, we can define the mean value of the mean response as:

$$\hat{F}_{err}(\mathbf{x}) := \mathbb{E}_{\xi}\hat{F}(\mathbf{x}+\xi) \tag{7}$$

which can be expressed as the following convolution

$$\hat{F}_{err}(\mathbf{x}) = \left(\hat{F} * g\right)(\mathbf{x}) \tag{8}$$

Similarly, the variance of the mean response is defined as

$$\hat{\sigma}^2_{err}(\mathbf{x}) := \mathbb{VAR}^2_{\xi}\hat{F}(\mathbf{x}+\xi) \tag{9}$$

and using convolution properties, we can write

$$\hat{\sigma}^2_{err}(\mathbf{x}) = \left(\hat{F}^2 * g\right) - \left(\hat{F} * g\right)^2 \tag{10}$$

Finally, we can derive explicit formulas for the mean response values

$$\hat{F}_{err}(\mathbf{x}) = \left(\hat{F} * g\right)(\mathbf{x}) = (h * g)^T\beta + (r * g)^T\mathbf{R}^{-1}(F - \mathbf{H}\beta) \tag{11}$$

and variance

$$\begin{aligned}\hat{\sigma}^2_{\text{err}}(\mathbf{x}) = {} & (g * \text{c} * g) - (r * g)^T\mathbf{R}^{-1}(r * g) + \\ & \left((h * g) - \mathbf{H}^T\mathbf{R}^{-1}(r * g)\right)^T \left(\mathbf{H}^T\mathbf{R}^{-1}\mathbf{H}\right)^{-1} \left((h * g) - \mathbf{H}^T\mathbf{R}^{-1}(r * g)\right)\end{aligned} \tag{12}$$

The estimation accuracy of the obtained model should be checked. To this aim, a leave-one-out cross-validation is performed. If results indicate a poor fitting quality, the MLE method is performed or the Kriging parameters are tuned manually.

The crucial feature of the proposed approach is that all convolutions can be calculated analytically. To do so, we have to assume:

1. The error ξ has a normal density function $\left(\xi \sim N\left(0, \sigma_g^2\right)\right)$ of a form

$$g(\xi) = \prod_{k=1}^{L} \frac{1}{\sqrt{2\pi\sigma_{g,k}^2}} \exp\left(-\frac{1}{2}\frac{\xi_k^2}{\sigma_{g,k}^2}\right)$$

2. The covariance function has the form

$$c\left(\mathbf{x}^i - \mathbf{x}^j, \sigma^2, \tau^2, \theta^2\right) = \sigma^2 \prod_{k=1}^{L} \exp\left(-\frac{\left(x_k^i - x_k^j\right)^2}{2\theta_k^2}\right) + \tau^2 \delta_{\mathbf{x}^i,\mathbf{x}^j}, \text{ where}$$

   The variable θ_k is a shape parameter for our function $c\left(\mathbf{x}^i - \mathbf{x}^j, \sigma^2, \tau^2, \theta\right)$. The parameters σ^2 and θ^2 are introduced to improve Kriging model adjustment to the known data. Values of those parameters are found by maximal likelihood estimate (MLE).
3. The trend function is the polynomial

$$\mu(\mathbf{x}) = \beta_0 + \sum_{j=1}^{L} \beta_j x_j$$

Utilizing Different Covariance Kernels

Functions sampled from the described GP approximate well a certain class of objective functions. A quality of this approximation depends on a kernel type used as a covariance function. By *quality*, we mean the ability to estimate an actual objective function using a minimal number of samples. There are many different kernels used in various cases. We prefer using kernels belonging to the Matern family [11]. Particularly, the kernel with smoothness parameter $\nu = 5/2$ is our choice which means that the random process based on such kernel is twice differentiable. The reason for such choice is twofold. First, we would like to optimize only the continuous objective functions. Second, we can measure directly only a value of the objective function. Thus, we do not want to assume too much about the properties of the objective function. According to our experience, there are cases in which additional assumptions may lead to poor estimation results. The Matern kernel is of the form

$$g_{mat}(x^i, x^j, \theta) = \left(1 + \frac{\sqrt{5}|x^i - x^j|}{\theta} + \frac{5(x^i - x^j)^2}{3\theta^2}\right) \exp\left(-\frac{\sqrt{5}|x^i - x^j|}{\theta}\right)$$

In described analytical approach, only the Gaussian functions or their linear combinations can be used. That is why the Matern 5/2 kernel is approximated by Gaussian functions. To this aim, the least squares method is used. It is important to ensure the following conditions:

- The approximation is positive-definite.
- The maximal discrepancy between approximation and Matern 5/2 function (residual) is low.
- The approximation is independent of the Kriging parameters, particularly the covariance radius θ.

In effect of imposed conditions, the approximation takes the form

$$\bar{g}_{mat}(x^i, x^j, \theta) = \sum_{m=1}^{M} a_m \cdot \exp\left(-\frac{((x^i - x^j)/\theta)^2}{2\lambda_m}\right) = \sum_{m=1}^{M} a_m \cdot g_m(x^i, x^j, \theta)$$

Coefficients a_m are determined by minimization of sum of square differences between $\bar{g}_{mat} - g_{mat}$ with the nonnegative least squares method (NNLS). This technique ensures that the approximation is positive-definite. Values of parameters λ_m of the Gaussian functions are determined by a nonlinear optimization of the squared residuals' sum. The multidimensional Matern kernels are defined in the following form

$$\bar{g}_{mat}(\mathbf{x}^i, \mathbf{x}^j, \theta) = \sum_{m=1}^{M} a_m \prod_{l=1}^{L} g_m(x_l^i, x_l^j, \theta_l)$$

In this chapter, the uncertainties of the variables are considered Gaussian. Sometimes, different distributions are better suited. Following the described method of utilizing various covariance kernels, one can take advantage of other probability distributions (e.g. beta distribution). Again, the following conditions should hold:

- The approximation is positive.
- The maximum difference between the cumulative distribution function of the probability function and its approximation is low.

Relative Expected Improvement

In the beginning of the optimization process, our knowledge about the objective function is limited. That is why the process of choosing initial points (Latin hypercube sampling) does not use any information of this function. In contrast to it, the procedure of adding points to the existing set should utilize the previously constructed surrogate model. That step is one of the most important parts in the optimization based on the RSM. The commonly used approach (e.g. EGO algorithm [4]) is called the expected improvement (EI) method. It uses the fact that the objective function value can be modelled as a realization of a Gaussian field with the mean value and the covariance equal to the Kriging estimators. As the best point to measure the objective function, we choose point $\mathbf{x}$ for which the expression

$$EI(\mathbf{x}) = \mathbb{E}\left(\max_{\mathbf{x}\in D}\left(f_{min} - \hat{F}(\mathbf{x}), 0\right)\right)$$

is maximized. This function is called the expected improvement and can be expressed as

$$EI(\mathbf{x}) = \left(f_{\min} - \hat{F}(\mathbf{x})\right)\Phi\left(\frac{f_{\min} - \hat{F}(\mathbf{x})}{\hat{\sigma}(\mathbf{x})}\right) + \hat{\sigma}(\mathbf{x})\phi\left(\frac{f_{\min} - \hat{F}(\mathbf{x})}{\hat{\sigma}(\mathbf{x})}\right) \quad (13)$$

where ϕ is the density and Φ is the cumulative distribution function of the standard normal distribution $\mathbb{N}(0, 1)$.

We use an extension of this method proposed by Laniewski [8]. Two main differences should be noticed. First, we cannot relate improvement of our model to sampled optimum. The function which is optimum we are looking for $\hat{F}_{err}(\mathbf{x})$ (mean performance) is different from the function which value we can calculate $F(\mathbf{x})$ (performance for one, deterministic design). For this reason, we take the minimum of our mean performance as a reference value. Second, a variance of the model should be calculated with respect to the new sample in the points $\mathbf{s}$. We want to know how the model changes if the deterministic function is measured. This important distinction is taken into account and leads to the method called the relative expected improvement (REI) which makes use of the formula (13). The only modifications appear in the expressions for f_{min}, $\hat{F}(\mathbf{x})$ and $\hat{\sigma}(\mathbf{x})$. They can be expressed in the following steps:

1. The values of the design parameters $\mathbf{x}_{min}$ for which the mean response reaches its minimum value should be found. Since we use the estimator rather than deterministic mean response, reaching this goal needs taking into account also the variation of the response. For example, one can search for:

$$\min_{\mathbf{x}_{\min}\in D}\left(\hat{F}_{\text{err}}(\mathbf{x}_{\min}) + 2\hat{\sigma}_{\text{err}}(\mathbf{x}_{\min})\right)$$

2. The formula for the Kriging model variance at the point $\mathbf{x}$ should take into account the relation between the objective function evaluation at the $\mathbf{s}$ point and the model. In consequence, one should obtain the expression for:

$$\hat{\sigma}^2_{\text{err, rel}}(\mathbf{x}, \mathbf{s})$$

   which measures the relative variance of the $\hat{F}_{err}(\mathbf{x})$ if the objective function is sampled in $\mathbf{s}$.
3. The quantity obtained in the step (1) and the formula derived in the step (2) should be used to calculate REI $(\mathbf{x}, \mathbf{s})$. The values of $\mathbf{x}$ and $\mathbf{s}$ should be found in order

to maximize the expression. A domain of **x** is a design region, and a domain of **s** is a sampling space. The obtained parameters **s** are these for which we should evaluate the objective function.

Example of Application

The optimization procedure based on the Gaussian process is not efficient for high-dimensional objective functions. Such property is characteristic for every RSM. Performing the optimization would require using a lot of sampling points (the course of dimensionality). Conveniently, there are engineering problems which can be formulated in terms of relatively small number of design variables. Needless to say, a selection of such variables may not be an easy task which necessitates a great deal of engineering experience.

Let us imagine that an objective function is linear in most dimensions. It means that only in few dimensions the function should be sampled densely. There exist methods of reducing a dimension of the problem (e.g. [12]). They are capable of indicating the dimensions significant in the process of optimization but with the cost of additional sampling. We will show in this example that in case of optimization based on GP such methods are unnecessary.

Formulation of the Test Case

The technical object which we would like to optimize is a duct channel delivering air to the engine of the small aircraft $I-23$ (see Fig. 2). This aircraft is powered by the turboprop engine in tractor configuration: air is drawn in the channel through the inlet located in the nacelle below the spinner and then transported to the engine compressor.

The aim of the optimization was to reduce a total pressure drop between the inlet to the channel and its outlet. To achieve this goal, the shape of the channel had to be modified.

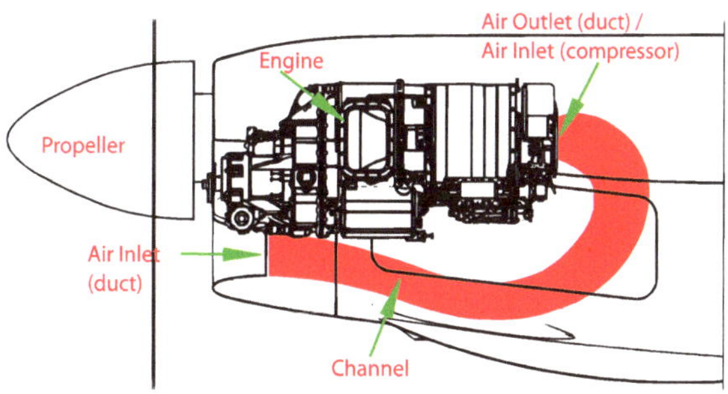

Fig. 2 Engine location and the channel

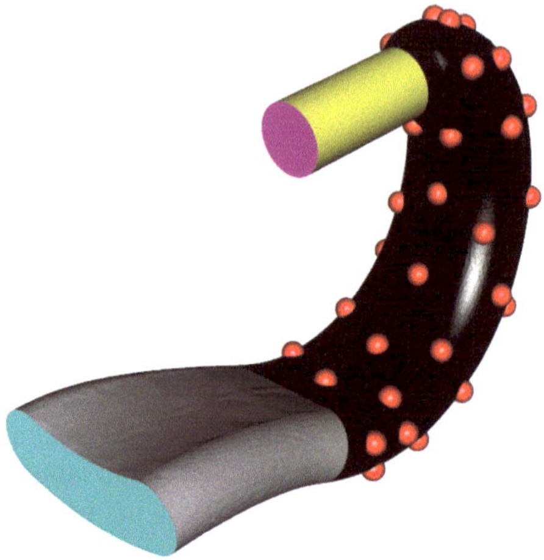

Fig. 3 Channel duct. The red dots are the morphing points. The grey surface cannot be deformed. The yellow cylinder is a dummy block substituting the engine. The black surface is the only area which can be deformed

The maximum deformation was determined basing on a computational mesh. The appearance of negative volumes was the factor limiting the deformation. The deformation was applied in the morphing points and distributed over the whole computational mesh by means of the radial basis function method, with parts of the domain fixed (Fig. 3).

The uncertainties of the design variables were arbitrary. It was assumed that each of them is a random variable with the Gaussian density function, the zero mean and the standard deviation equal to 10% of the maximum deformation.

To perform the optimization, a computational set-up was prepared. The hybrid mesh consisting of the boundary layer mesh and unstructured mesh elements was prepared. The basic conditions of the simulation are summarized in Table 1.

Table 1 Computational set-up summary

Condition	Value
Flow	Turbulent ($Re = 3.8 \cdot 10^5$)
Flight attitude	3000 (m)
Inlet gauge pressure	72,426 (Pa)
Pressure jump (Fan)	8768 (Pa)
Solver	ANSYS fluent (density-based)
Flow	Compressible
Scheme	Second-order upwind
Turbulence model	SST $k - \omega$

Optimization Summary

We started the optimization process with the DoE. There were $N = 3 \cdot L + 3 = 63$ points generated with means of the LHS method. Then, the objective function evaluations (calculation of the pressure drops) were performed. Two evaluations failed due to mesh issues. This fact forced us to do two additional iterations. The values of the objective function in these cases were assumed to be equal to those predicted by the Kriging model, which prevented the REI algorithm from sampling in the neighbourhood of the unconverged samples.

The sampling space was determined to be $\mathbb{S} = [-0.4, 0.4]^L$. The initial model fitting procedure showed that the optimum will be located near the limits of the sampling space. It is a situation leading to poor estimation results near the limits. Because of that, the design space was restricted to $\mathbb{X} = [-0.32, 0.32]^L$. With this step, we ensure that the influence of the Kriging extrapolation outside the boundaries would not affect the optimization procedure.

In the process of optimization, 150 samples have been generated in the parallel mode. More specifically, our algorithm allows us to generate new points where the objective function should be calculated without waiting for all of the previous computations to finish. In order to generate the new point, our method does not change the Kriging model but only modifies underlying probabilities. These modifications consist of assuming zero variance of the Kriging model in points corresponding to ongoing calculations and proceeding with the REI algorithm on such model. It allowed us to run five simulations simultaneously.

Results

The results of the performed optimization process can be seen in Fig. 4. The blue triangles are indicating the values of the objective function obtained with the LHS method. The red squares are the deterministic results of REI sampling. The calculations which failed to converge are indicated with the orange trapezoids. Our reference point is the green star which is the base design to be improved. We can see that some of the randomly generated samples (with the LHS method) are already better than the reference value. It means that relatively big improvement of the objective function should be achieved. The black dots indicate actual value of the optimum estimated by the Kriging model, and the vertical bars are the confidence intervals.

The improvement of the objective function is obtained early during the optimization process (66 iterations), but the confidence interval for the optimum point is too high. After adding new points, the minimum value changes rapidly. It is the moment when the algorithm is focused more on improvement of the objective function value than on the UQ. When the iteration number reaches 90, we can see that changes are

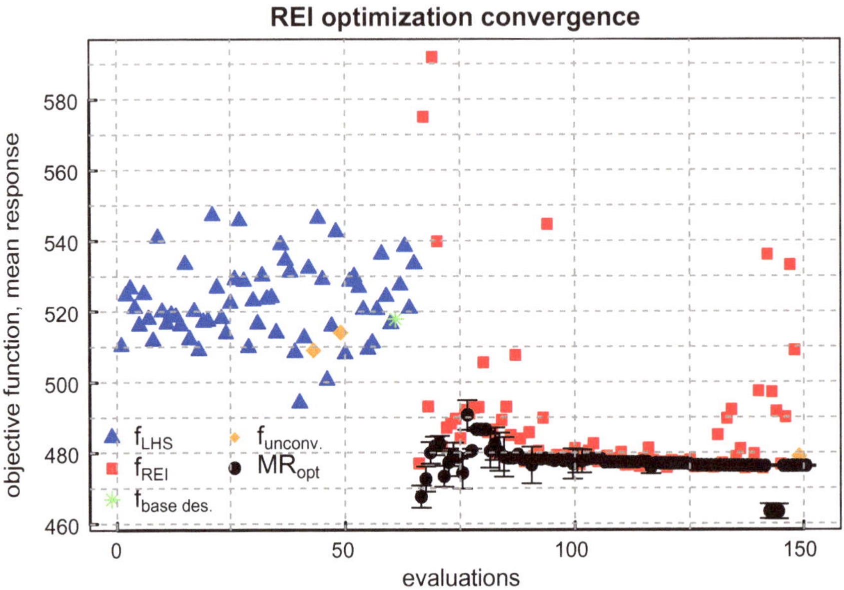

Fig. 4 Results of the optimization process of the channel delivering air to the turboprop engine

slower. Of course, the improvement of the objective function value and the UQ are done simultaneously but we can say that in that moment UQ dominates the search for the optimum. That situation lasts till iteration 155 when the optimum point variance is low and the algorithm starts to check other points.

To sum up, the optimum value is found early during the optimization process (improvement of 8.04%). The optimum is located on the boundaries of 23 design variables. Its location changes only in seven dimensions during the optimization. It means that the objective function should be checked only within these seven dimensions. The optimization algorithm behaves like it would optimize $L = 7$ dimensional objective function not $L = 30$ which substantially lowers amount of iterations necessary to find the objective function optimum.

Summary

The method described in this chapter allows one to find—in a limited number of steps—the optimum of the objective functions first moment with respect to uncertain parameters. It is achieved by using the properties of the Kriging approximation and the new method of sampling. The method is based on the relative expected improvement (REI) criterion. The key advantage of the method is that it allows for the function evaluations only in points which improve our knowledge on the locations of the robust minimum. The properties of the method allow also for the efficient parallelization of the optimization process. The underlying probability makes the proposed approach robust to solver failures and mesh issues. In effect, the described method is particularly suitable for cooperation with external objective function evaluators. The quality of the response surface is constantly checked by means of leave-one-out cross-validation. Such approach allows us to decide if the surrogate model is sufficiently accurate. The present approach facilitates incorporation of uncertainties of design variables by analytical convolutions of Gaussian functions. It does not mean, however, that the proposed method is restricted only to uncertainties with normal distributions. Other distributions can be incorporated providing that they can be approximated by a weighted sum of Gaussian functions. Any other distribution can be used assuming that it can be approximated by a weighted sum of Gaussian functions.

The REI criterion indicates the new points improving the knowledge about the optimum. It means that the new points are added also in order to minimize the variation of the optimum. The algorithm is performing simultaneously the search for an optimal design and its UQ.

References

1. Forrester, A., Sobester, A., Keane, A.: Engineering Design via Surrogate Modelling: A Practical Guide, Wiley (2008)
2. Myers, R.H., Montgomery, D.C., Anderson-Cook, C.M.: Response Surface Methodology. Process and Product Optimization Using Designed Experiments, 4th edn. Wiley (2016)
3. Houle, M.E., Kriegel, H.P., Kroger, P., Schubert, E., Zimek, A.: Can shared-neighbor distances defeat the curse of dimensionality? In: SSDBM 2010: Scientific and Statistical Database Management, pp. 482–500
4. Jones, D.R., Schonlau, M., Welch, W.J.: Efficient global optimization of expensive black-box functions. J. Global Opt. **13**, 455–492 (1998)
5. Ratvasky, T.P., Barnhart, B.P., Lee, S.: Current methods for modeling and simulating icing effects on aircraft performance. Stab. Control, NASA/TM-2008-215453
6. Giunta, A.A., Mcfarland, J.M., Swiler, L.P., Eldred, M.S.: The promise and peril of uncertainty quantification using response surface approximations. Struct. Infrastruct. Eng. **2**(3–4), 175–189 (2006)
7. Janusevskis, J., Le Riche, R.: Simultaneous kriging-based sampling for optimization and uncertainty propagation. J. Global Opt. **55**(2), 313–336 (2013)
8. Łaniewski-Wołłk, Ł.: Relative Expected Improvement in Kriging Based Optimization, Cornell University Library (2009). arXiv:0908.3321
9. McKay, M.D., Beckman, R.J., Conover, W.J.: A comparison of three methods for selecting values of input variables in the analysis of output from a computer code, technometrics (JSTOR Abstract). Am. Stat. Ass. **21**(2), 239–245, May 1979
10. Eglajs, V., Audze, P.: New approach to the design of multifactor experiments. Prob. Dyn. Strengths, Riga **35**, 104–107 (1977)
11. Abramowitz, M., Stegun, I.A.: Handbook of mathematical functions with formulas, graphs, and mathematical tables. National Bureau of Standards, Applied Mathematics Series 55 (1972)
12. Constantine, P.G., Dow, E., Wang, Q.: Active subspace methods in theory and practice: applications to kriging surfaces. arXiv:1304.2070

13. Jones, D.R.: Taxonomy of global optimization methods based on response surfaces. J. Global Opt. **21**, 345–383 (2001)
14. Roustant, O., Ginsbourger, D., Deville, Y.: DiceKriging, DiceOptim: two R packages for the analysis of computer experiments by kriging—based metamodeling and optimization. J. Stat. Softw. **51**(1) (2012)

Robust Design in Turbomachinery Applications

Rémy Nigro, Dirk Wunsch, Grégory Coussement and Charles Hirsch

Introduction

Nowadays, the aerodynamic design procedures are still overwhelmingly deterministic in character, neglecting the influences of uncertainties. There are two major classes of uncertainty-based design problems that can be identified: robust design and reliability-based design problems. A reliability-based design seeks a design with a probability of failure less than an acceptable value, while robust design aims at finding a design insensitive to small changes of uncertainties. A robust design optimization will therefore reduce the level of reduction of performance when deviating from the nominal conditions.

Figure 1 illustrates the principle of a robust design optimization at the example of a function with two minima. In a deterministic optimization formulation, one would seek to minimize the cost function and an adequate optimizer will identify the minimum on the left as the global optimum, since it shows the lowest value. If the design parameter is now uncertain and represented by a PDF as illustrated by the red-colored PDF on the abscissa, the system response varies significantly over the range of variability of this design parameter. This is illustrated by the red-colored PDF on the ordinate. This distribution could, for example, represent the

R. Nigro · D. Wunsch · C. Hirsch (✉)
Numeca International, Brussels, Belgium
e-mail: charles.hirsch@numeca.be

R. Nigro
e-mail: remy.nigro@umons.ac.be

D. Wunsch
e-mail: dirk.wunsch@numeca.be

R. Nigro · G. Coussement
University of Mons, Mons, Belgium
e-mail: gregory.coussement@umons.ac.be

C. Hirsch et al. (eds.), *Uncertainty Management for Robust Industrial Design in Aeronautics*, Notes on Numerical Fluid Mechanics and Multidisciplinary Design 140, https://doi.org/10.1007/978-3-319-77767-2_31

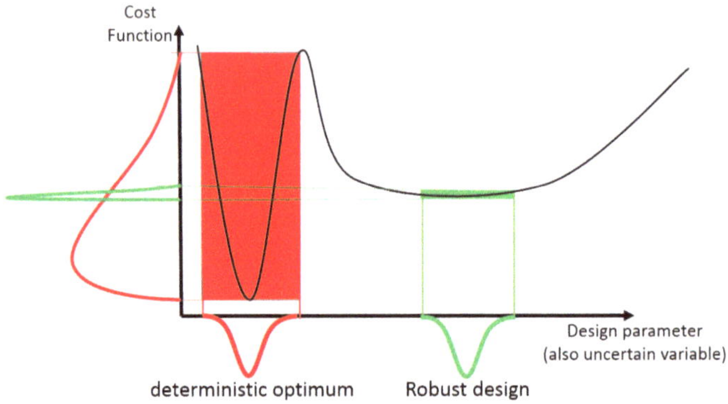

Fig. 1 Principles of robust design optimization: differences between a deterministic (red) and a robust optimum (green)

variation of the efficiency. A robust design optimization formulation, on the other hand, will not only seek to minimize the mean of the cost function, but also to minimize the variability of the cost function. In mathematical terms, this is expressed by minimizing the mean value of the cost function and the standard deviation of the same cost function. The optimization problem becomes thus multi-objective with several optimal solutions. It is clearly seen from Fig. 1 that the minimum on the right of the figure illustrated by PDFs in green color shows a much more narrow distribution of the cost function for the same variability of the input quantity (the design parameter on the abscissa)! This is thus a robust optimum.

In order to calculate the statistical output quantities (mean and variance), a design optimization strategy needs to be coupled with a method for the uncertainty quantification. In the following, a coupling of the non-intrusive probabilistic collocation method as described in chapter "Non-intrusive Probabilistic Collocation Method for Operational, Geometrical and Manufacturing Uncertainties in Engineering Practice" and a multi-objective surrogate-based optimization is applied. This coupling is based on the construction of a surrogate model for both the design parameters and the uncertain variables. The method is then applied to the Rotor 37 test case, and the differences between deterministic and robust optima are discussed.

Surrogate-Assisted Optimization

Deterministic design optimization is often a process based on surrogate-assisted genetic algorithms. This process can be summarized as follows. First, a Design of Experiments (DoE) samples the design space by running full 3D CFD simulations. For each design, the parameters are modified, the geometry is adapted, and the mesh is regenerated. Various DoE such as Latin Hypercube Sampling (LHS) [5],

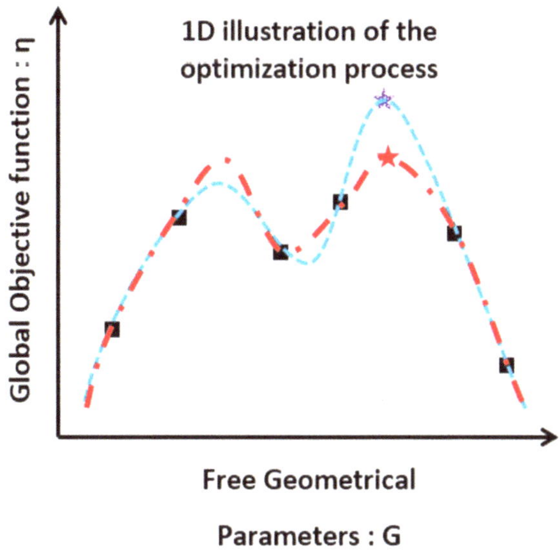

Fig. 2 Principles of a surrogate-based optimization

Plackett-Burmann [4] and many others can be applied; the common goal is to sample the design space with as little samples as possible. Second, based on the DoE, a surrogate model (or response surface model) approximates the solution across the design space. In the following, an artificial neural network [2] is used. Other surrogate modelling techniques such as kriging [6] can also be applied. This allows to access values between the DoE points and runs the optimization on the surrogate model, which is significantly cheaper than running full 3D CFD simulations for every design. The error between the surrogate and the real system response needs to be controlled. Third, a genetic algorithm searches the design space on the surrogate model and finds an optimal design. This optimal design is run as a 3D CFD simulation, and the simulation result is used to update the surrogate. Figure 2 illustrates this process on a 1D example. The black squares represent the DoE, the red dash-dotted line is the initial surrogate approximation, the optimizer finds an optimal solution on the surrogate (red star), then a 3D CFD computation is performed, and the surrogate is updated, which results in the blue dashed line. This is an iterative process that is repeated until convergence on an optimal solution is reached.

Robust Design Optimization Method

For robust design optimization, the UQ method presented in chapter "Non-intrusive Probabilistic Collocation Method for Operational, Geometrical and Manufacturing Uncertainties in Engineering Practice" needs to be coupled with the optimization procedure. As seen above, the optimization objectives and constraints are not single values in robust optimization, but the mean value and standard deviation of the objective functions. The statistical moments, which are the output of the UQ method, need to be calculated for every single design optimization. The most straightforward approach would be to run full 3D CFD simulations for every point in the DoE and calculate a surrogate for the statistical moments. This is, however, very costly, since a database usually contains hundreds of points. At the example of 10 simultaneous uncertainties, which requires 21 CFD simulations per point in the DoE, it is easy to see that for 200 points in the DoE, the computation cost would increase by a factor of 21, which is not feasible in industrial design practice.

The proposed solution is a mixed Design of Experiments (DoE) comprising of both the design variables and the uncertainties. To build this mixed DoE, the uncertainties are added as additional dimensions to the DoE, which requires harmonizing the way the dimensions are sampled independently from their type of distribution. Generally, design parameters in traditional DoE-based optimization are sampled uniformly from a given interval [a, b], such as depicted in Fig. 3. If the sampling of uniformly distributed design parameters must be harmonized with the sampling of arbitrary probability distributions of the uncertain variables, the uniform sampling of the design parameters must be expressed in terms of a PDF. It is clear that in the case of design parameters, the uniform PDF is the equivalent distribution type. The sampling procedure, as described in Fig. 3, is now the following. The PDF (uniform) is represented by its Cumulative Distribution Function (CDF), which on its ordinate ranges from 0 to 1. If the ordinate is divided into equal probable sections, the values of the design parameters corresponding to these sections account for the same probability. As an example, if four sections are chosen, each of the design parameter intervals accounts for 25% of the PDF distribution.

For an arbitrary PDF shape, such as sketched in Fig. 4, the principle remains the same. The PDF is expressed as CDF, and it is divided into equal probable intervals, which are mapped onto the uncertain parameter. The probability interval spacing is

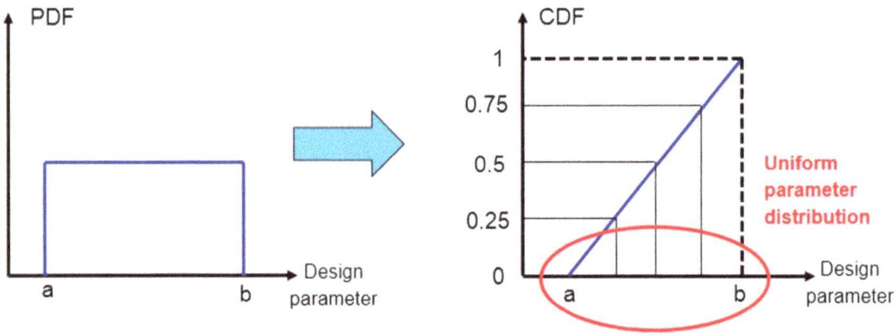

Fig. 3 Sampling of a design parameter from a uniform distribution in interval [a, b]

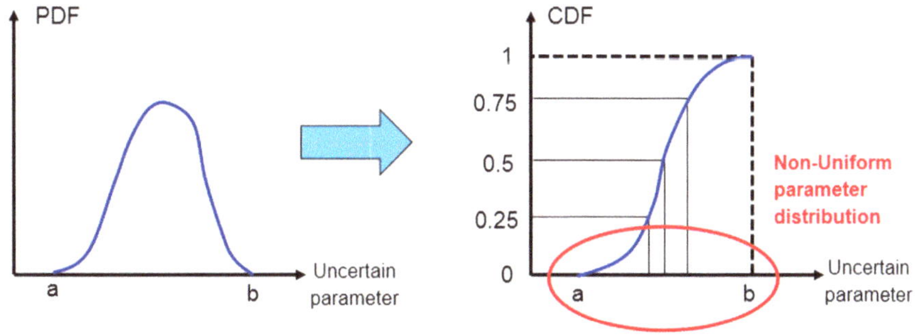

Fig. 4 Sampling of an uncertain variable from an arbitrary probability density function in interval [a, b]

uniform in both cases of a uniform and arbitrary PDF, but the design and uncertain parameter spacing vary in function of the PDF shape.

The DoE sampling is performed in both cases over uniform intervals in terms of probability with the help of the CDF, which allows using the same parametric manager for design and uncertain variables, available in the optimization software. The mixed design variable and uncertainty space can thus be integrated easily with a traditional DoE optimization technique. The main advantage lies in the computational costs compared to an approach where every point in the DoE is run with a UQ simulation. At the hand of the above-cited example of 10 simultaneous uncertainties, which require 21 CFD simulations per point in the DoE and a DoE size of 200 points, the computation cost decreases from 4200 CFD runs (20 design variables * 10 DoE samples * 21 UQ points) to 300 CFD runs for the mixed design and uncertain space ((20 design variables + 10 uncertainties) * 10 DoE samples). This gain in computation time is achieved at the expense of not having the statistical moments directly available in the surrogate. Therefore, at each optimization cycle, a full UQ computation is performed, which allows to have an accurate evaluation of the statistical moments (i.e., the mean and the standard deviation). These results are used to update the surrogate model.

Robust Design Optimization Definition of the BC-01: Rotor 37

The detailed description of the geometry and the experimental setup of the Rotor 37 can be found in [1]. Figure 5 shows the measurement sections and geometry.

Simulations are performed with the flow solver FINE^{TM}/Turbo [3] and the uncertainty quantification methodology developed in [7]. RANS equations and Spalart–Allmaras turbulence model are solved at a constant rotating hub speed of 17188 rpm on a mesh with approximately 2.8 million mesh points.

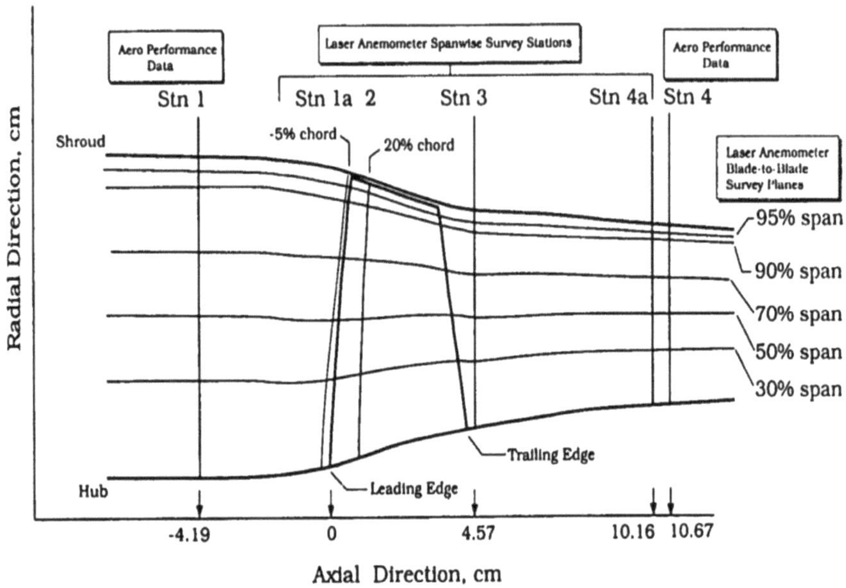

Fig. 5 Rotor 37 measurement sections [1]

The geometry of the compressor blade is parameterized with six spanwise sections. The design space, which consists of 49 dimensions, is formed by the following geometrical parameters:

- The tangential stacking law (relative to the leading edge)
- The leading and trailing edge blade angles in each section
- The axial stacking law (relative to the leading edge)
- The meridional chord (axial position of the trailing edge)
- The maximal thickness in each section (each section keeps its thickness law shape but the maximal thickness can be modified).

The objective of this robust optimization is to maximize the mean value of the isentropic efficiency while minimizing the standard deviations of the isentropic efficiency, of the absolute pressure ratio, and of the mass flow. An additional constraint is included on the mean value of the absolute total pressure ratio: It cannot decrease more than 10% of the original design value (i.e., it cannot be below 1.9).

Uncertainties are considered on the inlet total pressure, the outlet static pressure, leading edge radius, leading edge blade angle, and tip clearance. They are all defined with a symmetric Beta PDF following the test case description in the UMRIDA database, as described in chapter "UMRIDA Test Case Database with Prescribed Uncertainties."

Robust Design Optimization Analysis and Comparison with Deterministic Optimization

The difference between deterministic and robust optimization lies mainly in the definition of the optimization objective functions. While in the deterministic optimization the efficiency is maximized, the robust design optimization maximizes the mean value of the efficiency and minimizes its standard deviation as well as the standard deviation of other quantities. The deterministic optimization is here a single objective optimization, while the robust design optimization is multi-objective by inclusion of the standard deviations in the objectives. Consequently, Pareto fronts form the optimal robust design space. In order to compare the deterministic optimum with the robust optima, a UQ simulation is run for the design parameter values of the deterministic optimum. This allows computing the mean value and standard deviations of the objective functions and includes the deterministic optimal design into the Pareto plots.

Figure 6 shows the two-dimensional Pareto plot, where the standard deviation of efficiency is plotted over the mean value of efficiency. Three optimal robust designs are retained in comparison with the original design, and the deterministic optimal design is rerun with a UQ simulation as explained above. The yellow star indicates the position of the original design. It is seen that all optima increase the mean efficiency and that the robust designs decrease the standard deviation of efficiency in comparison with the original design, whereas the deterministic design increases the standard deviation of efficiency. The largest increase in mean efficiency of 3.63% compared to the original design is achieved by design 3, while it reduces the standard deviation by 34.00%. The robust optimal design 1 leads to the largest reduction in standard deviation of efficiency by 62.23%, at expense of a slightly smaller increase in mean efficiency of 2.48% compared to the original design. It is noteworthy that the mean value of efficiency of the deterministic optimal design

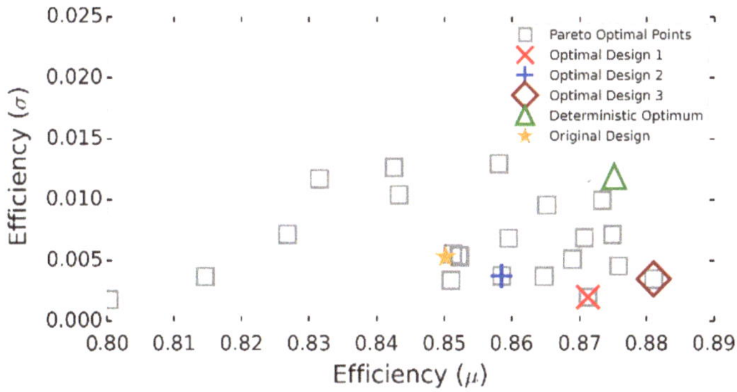

Fig. 6 Pareto plot for standard deviation of efficiency plotted over the mean value of efficiency for the original and optimal designs

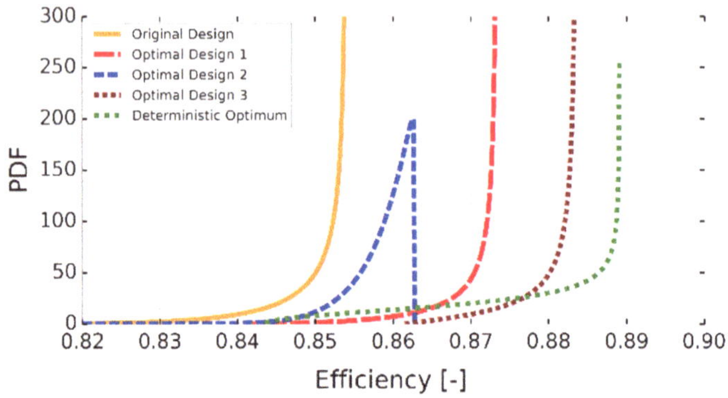

Fig. 7 Probability density function of the isentropic efficiency for the original, deterministic optimum, and robust optimal designs

(green triangle in Fig. 6) is below the mean value of efficiency of the robust design 3 (purple diamond in Fig. 6).

Given the multi-objective character of the robust design optimization and from a purely conceptual point of view, the deterministic optimization should lead to the largest efficiencies. This is also the case in this study, where the absolute value of efficiency predicted by the deterministic optimization is 0.886, but its mean value as a result of the UQ study on the found optimum is 0.875, as seen in Fig. 6. Figure 7 shows the reconstructed PDFs of the original and optimal designs. The reconstructed PDF of the deterministic optimal design shows clearly the highest deterministic value, but also the highest efficiency variability, which varies from approximately 0.84 to 0.89 and thus over a range of 5 points in efficiency, while the robust optimal design 3 varies from approximatively 0.86 to 0.88 and thus over a range of 2 points in efficiency. This is an indication that the variance of the robust optima is smaller than the one for the deterministic optimum, which proves the interest of the robust optimization.

Globally, all PDFs show an improvement in efficiency for all optimal designs in comparison with the original design, while only the robust optimal designs reduce the standard deviation in efficiency and are thus less sensitive to uncertainties compared with the original design.

As a robust design optimization is a multi-objective optimization, also the standard deviation of the pressure ratio was set as objective for minimization. It is recalled that the pressure ratio was also attributed with a constraint on its absolute value. The Pareto plot in Fig. 8 shows that it was not possible to minimize the standard deviation in pressure ratio and that the standard deviations increase between 5.13% for the robust optimal design 2 and 49.07% for the deterministic optimum compared to the original design.

The reconstructed PDFs of the pressure ratio in Fig. 9 show an increase of its mean value for all optimal designs, although the mean value of the pressure ratio is almost unchanged for the deterministic design. It is interesting to note that the

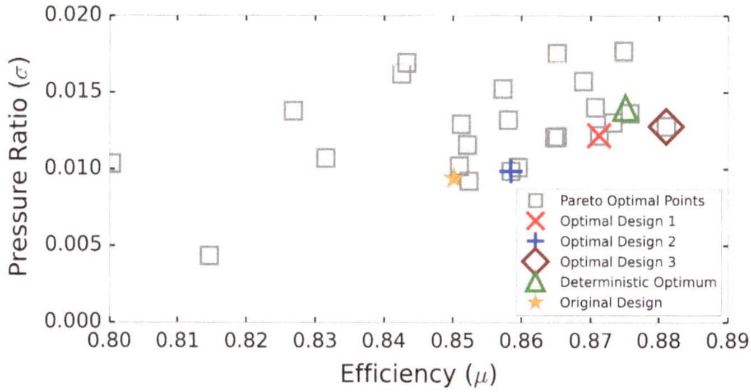

Fig. 8 Pareto plot for standard deviation of pressure ratio plotted over the mean value of efficiency for the original and optimal designs

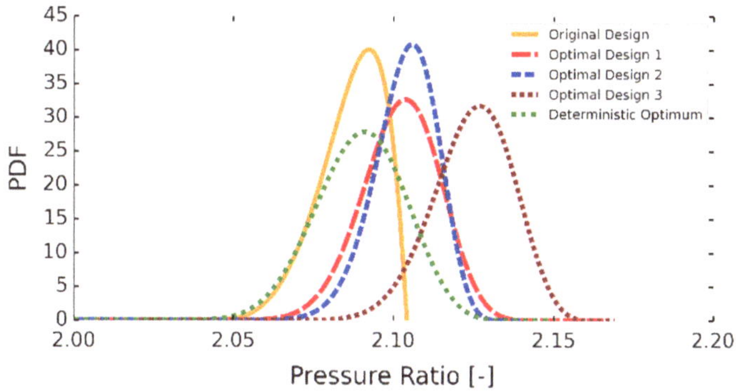

Fig. 9 Probability density function of the pressure ratio for the different designs selected

optimal designs lost the slight skewness of the original design in distribution of the pressure ratio. The smallest standard deviation in pressure ratio is achieved by the robust optimal design 2, which has only a minor increase of 5.13% compared to the original design.

All mean values and standard deviations for the optimal designs are summarized in Table 1, comparing the optimal designs with the original design.

The calculation of scaled sensitivity derivatives allows to assess the influence of the individual uncertainties on the solution across the retained optimal designs. Figures 10 and 11 show the scaled sensitivity derivatives for the isentropic efficiency and total pressure ratio, respectively. All designs show a rather small sensitivity to the leading edge radius, which means that the performances are almost not influenced by this uncertain parameter given its probability distribution. The tip gap has a non-negligible influence on these two quantities.

Table 1 Variations of mean value and standard deviation of efficiency and pressure ratio for the optimal designs in comparison with the original design

	Efficiency (−)		Pressure ratio (−)	
	Mean (%)	Std deviation (%)	Mean (%)	Std deviation (%)
Deterministic	+2.94	+124.46	+0.17	+49.07
Design 1	+2.48	−62.23	+0.79	+29.85
Design 2	+0.98	−29.10	+0.83	+5.13
Design 3	+3.63	−34.00	+1.84	+36.21

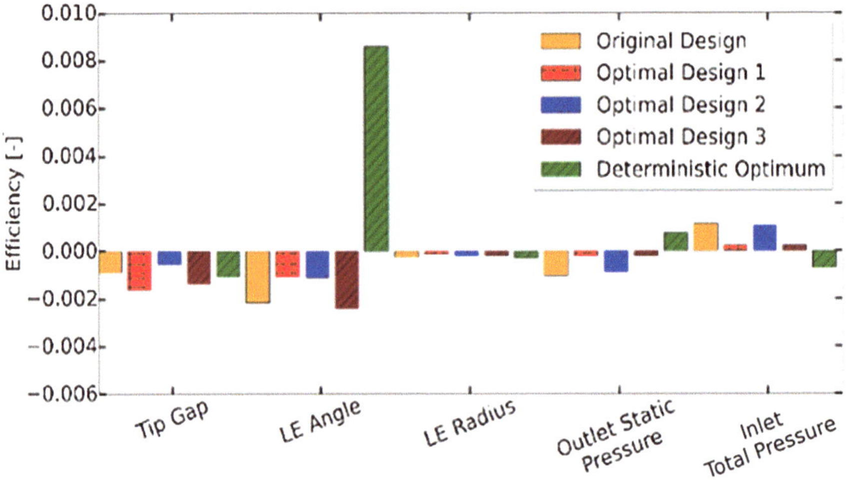

Fig. 10 Scaled sensitivity derivatives of the isentropic efficiency with respect to the input uncertainties for the original and optimal design

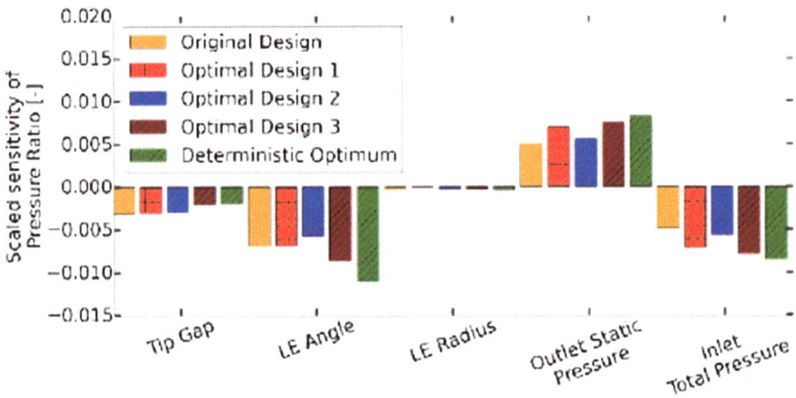

Fig. 11 Scaled sensitivity derivatives of the total pressure ratio with respect to the input uncertainties for the original and optimal design

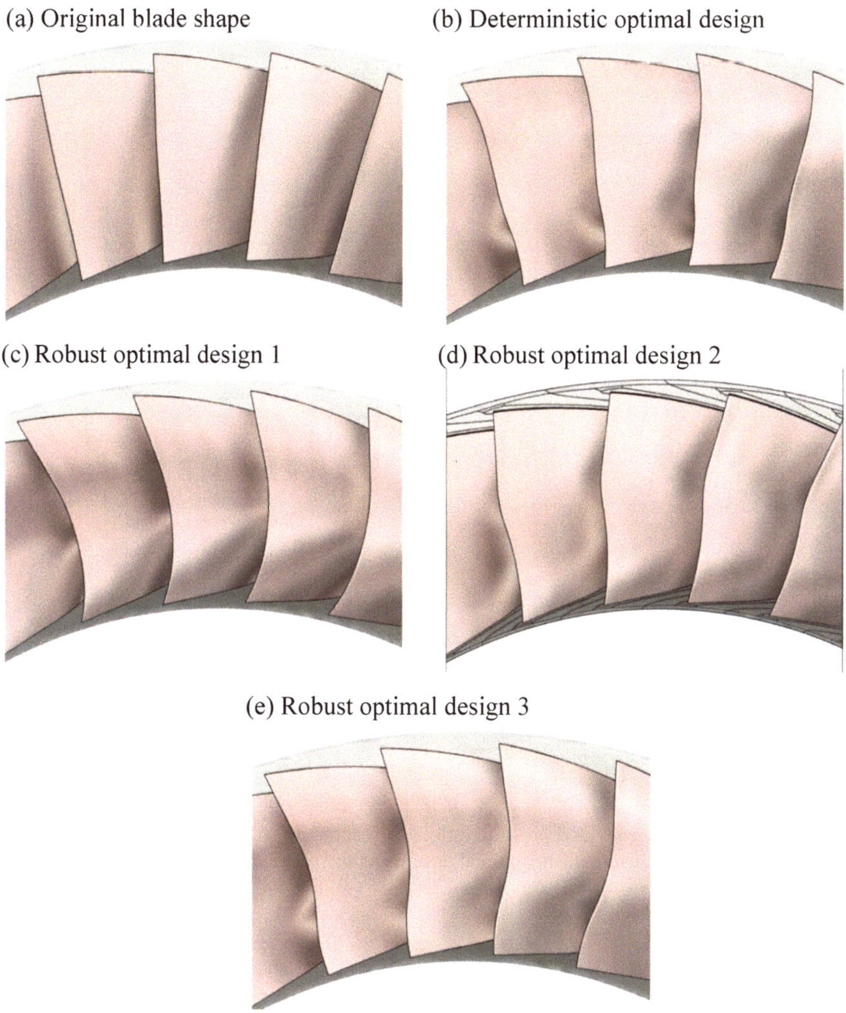

Fig. 12 Blade shape for original and optimal designs; **a** original R37 blade, **b** deterministic optimal design, **c** robust optimal design 1, **d** robust optimal design 2, **e** robust optimal design 3

The leading edge blade angle shows the largest sensitivities and the largest variations between the designs on the efficiency. It is by far the highest for the deterministic design, which shows also the highest standard deviation for the efficiency. *This shows the importance of including objectives on the standard deviation into the optimization formulation, i.e., the importance of a robust optimization formulation.*

Design 2 shows overall a smaller sensitivity on the pressure ratio, even if it is slightly higher than the original design.

Figure 12 shows a comparison of the blade shapes of the original and optimal blade designs. While the original blade has a straight leading edge, all optimal

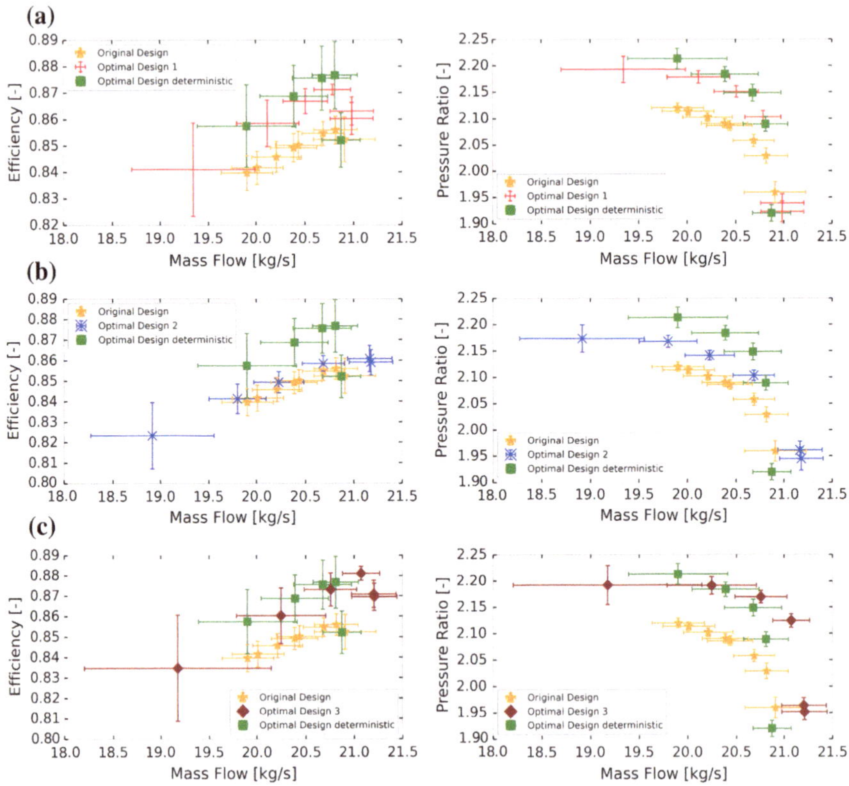

Fig. 13 Isentropic efficiency and pressure ratio as a function of mass flow, comparing the three robust designs with the deterministic optimal design and the original design. **a** Design 1, **b** design 2, **c** design 3

designs show a curved leading edge. The robust optimal design 2 shows a back sweep of the tip region of the blade in contrast to the other designs.

Full UQ performance curves, including all the uncertainties for all the designs, have been obtained, as shown in Fig. 13, which provides information on off-design performances of the selected compressor blade designs. As seen in Fig. 13, the choke mass flow is increased for every robust optimum, while it is slightly decreased for the deterministic optimum in comparison with the original design. The mean efficiency is increased for every optimal design and for all mass flow rates analyzed. One important point to note is that the standard deviation of the efficiency is only improved at the design point for the robust optima, while it is increasing at off-design points, for all optimal designs. This means that multi-point robust design optimizations are needed in order to reduce the performance variability across the entire operating range. In this case, the standard deviation of the off-design points needs to be added to the robust objective formulation.

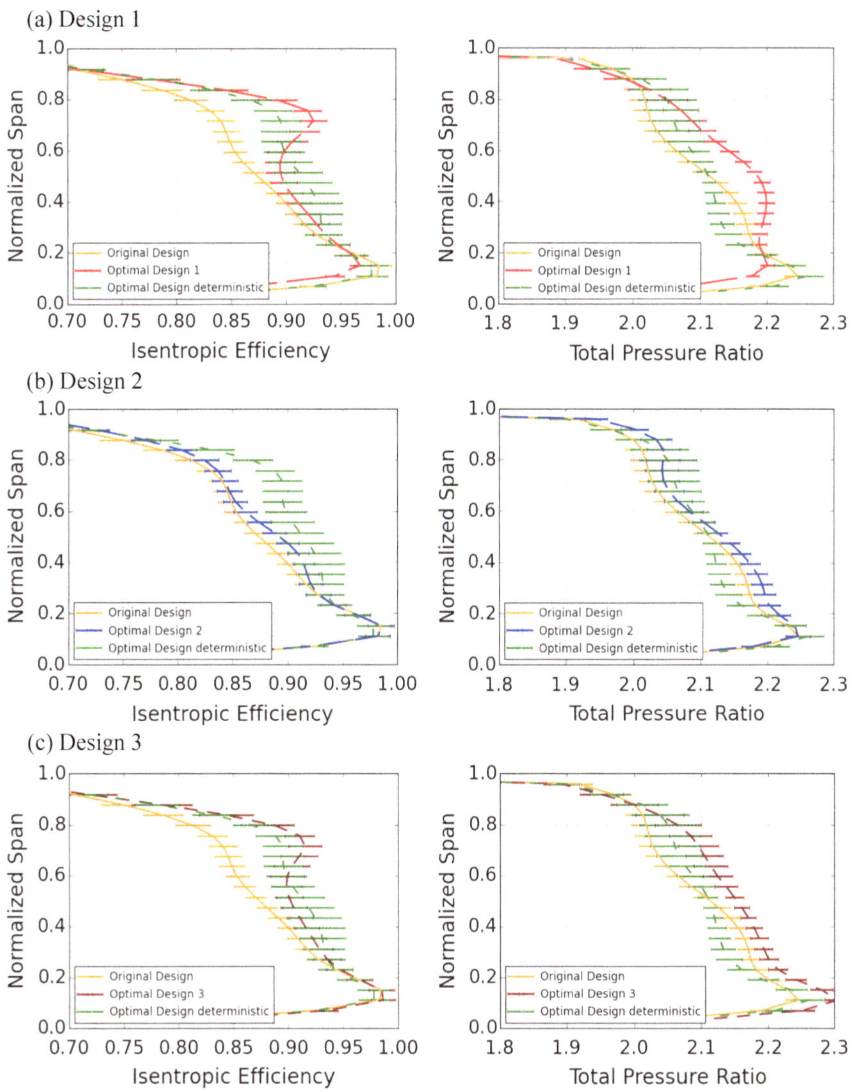

Fig. 14 Radial profiles of isentropic efficiency and pressure ratio at the station 4 comparing the three robust optimal designs with the deterministic optimal design and the original blade. **a** Design 1, **b** design 2, **c** design 3

Figure 14 displays the spanwise profiles of efficiency and pressure ratio at station 4, and at 98% of the choking mass flow, for the three robust designs, compared to the original and deterministic UQ results.

The standard deviation of both the pressure ratio and the efficiency is smaller for the three robust designs than for the deterministic optimal design all along the span-height. Design 2 has a distribution, which is very close to the original blade, with a small and almost constant increase of both the efficiency (0.01 point) and the pressure ratio (0.02) in comparison with the original design. Designs 1 and 3 have similar deviations on the efficiency profiles from the original one, with an important increase of efficiency from 60% span up to the blade tip. However, while design 3 keeps the same level of efficiency as the original design from the hub up to 20% span, design 1 shows a small drop of efficiency (2 points) in this area. The pressure ratio profile for design 1 has a maximum at 40% of span-height, while the other designs have this maximum around 10% of span-height. As the geometry and the radial profiles are considerably modified for the various designs, differences are also expected for the off-design conditions.

Figure 15 compares the flow fields for the smallest and largest leading edge blade angles within the variation range of the LE PDF, for each of the three selected designs at the flow conditions used during the optimization.

The influence of the leading edge blade angle on the shock intensity and its location, for the optimized designs, is shown, comparing with the original design. The large effect of the leading edge blade angle on the flow field can be considered as the main cause of the variation of the efficiency and of its related large standard deviation.

Computational Costs of Robust Design Optimization Applied to Case BC-01: Rotor 37

The computational cost of a robust design optimization scales directly with the number of full 3D CFD simulations that need to be performed. This cost is compared with the cost for the deterministic optimization in Table 2. In both cases, 10 points per dimension of the DoE are used, and the deterministic DoE has 49 dimensions for the design variables, while the robust design optimization DoE has 54 dimensions (49 design variables and 5 uncertainties). This results in 490 and 540 3D CFD runs for the deterministic and robust DoE, respectively. In both cases, a total of 40 global optimization cycles were used and after every global optimization cycle, the surrogate was updated with a CFD run. The computational overhead amounts thus almost entirely from the surrogate update, where 40 CFD runs are needed for the deterministic simulation against 440 for the robust optimization. A total of 530–980 CFD runs mean a computational overhead of 1.85 for the robust design optimization.

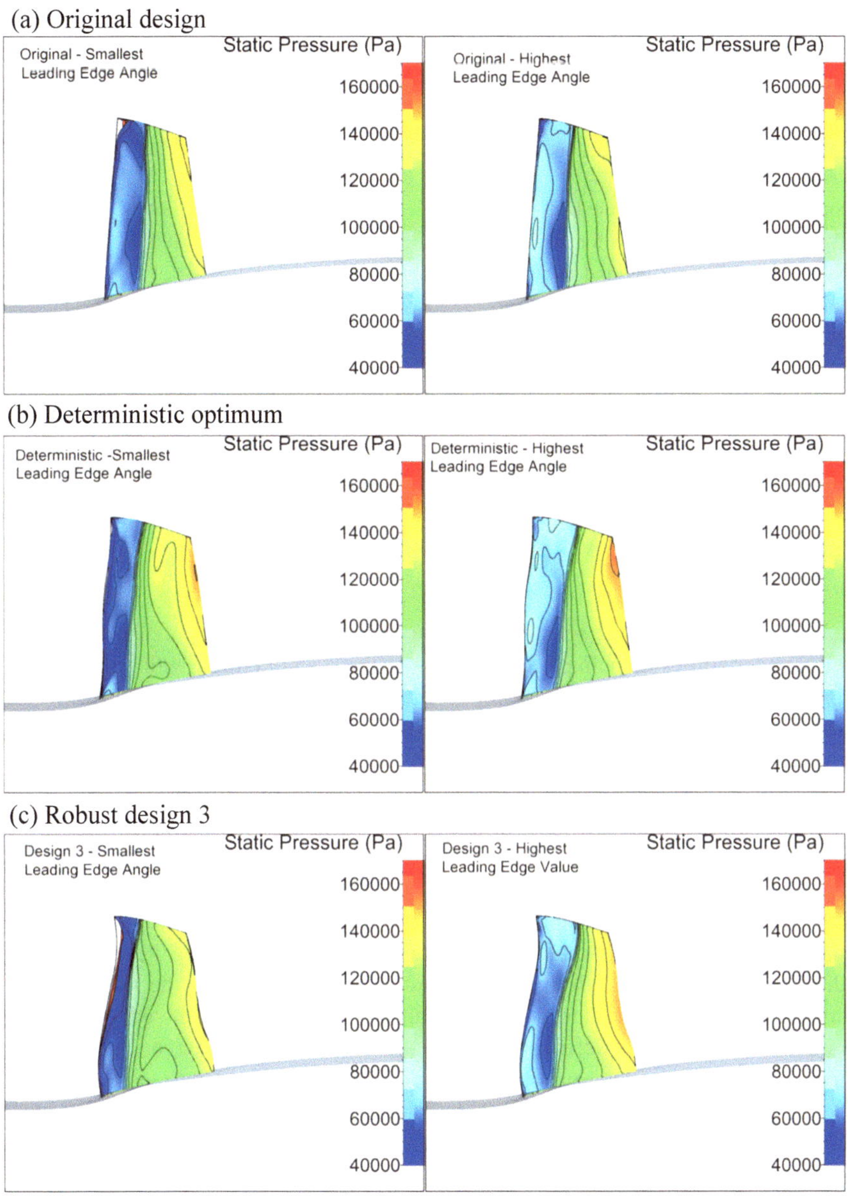

Fig. 15 Static pressure on the suction side at low mass flow rates, showing the smallest and largest leading edge blade angle configuration per design

Table 2 Detail of the computational cost comparing a robust design optimization with a deterministic optimization

	Robust design optimization	Deterministic optimization
Number of points per dimensions in the initial DoE	10	10
Number of dimensions in the DoE	54	49
Total number of CFD runs for the DoE	540	490
Number of CFD runs per updates	11	1
Number of updates	40	40
Total number of CFD runs for the optimization updates	440	40
Total number of CFD runs	980	530
Computational cost[a] (CPU hours)	2617	1415

[a]Considering that 1 CFD run has a cost of 2.67 CPU Hours (10 min on 16 cores)

Conclusions

A strategy for robust design optimization (RDO) is proposed, i.e., optimization under uncertainties reducing the variability of the system output with respect to the input uncertainties. This strategy relies on the non-intrusive probabilistic collocation method for the uncertainty propagation and a surrogate-assisted optimization strategy. In order to allow for RDO within reasonable turnaround times, a mixed Design of Experiments (DoE) is built, which comprises of design variables and uncertainties as individual dimensions. This reduces the cost by one order of magnitude compared to an approach where every point in the DoE is run with a UQ simulation. The robust design optimization problem is formulated as a simultaneous maximization of the mean efficiency and minimization of standard deviations of efficiency and of other global output quantities at the example of the Rotor 37. In addition, a deterministic design optimization is performed and the obtained performance improvements and geometries are compared and discussed. In order to compare the optimized deterministic design with the robust designs, a UQ simulation is done for the deterministic design variable set and added to the Pareto plots showing, for example, the standard deviation of isentropic efficiency over mean value of isentropic efficiency. The deterministic optimal design increases the standard deviation of the isentropic efficiency when the uncertainties are considered. It becomes thus more sensitive to variations in the uncertain input space compared with the original design. This shows the importance of robust design optimization, not only that it allows to reduce the standard deviation of the response, but a deterministic optimization of a problem attached with uncertainties might lead to the selection of sub-optimal design choices.

The influence of scaled sensitivity derivatives is discussed. The computational overhead amounts in this study to a value of 1.85, allowing to reach an increase in mean efficiency for one of the robust optimal designs of 3.63%, while reducing the standard deviation of efficiency by 34.00%. This is nearly 0.7% more in mean value of efficiency compared with the deterministic design, which in addition leads to an increase of 124% in standard deviation of efficiency. The comparison of non-deterministic performance curves for the deterministic and robust optimal design shows that a multi-point formulation is needed, where the standard deviations of the objective function at off-design points are included into the robust objective formulation.

Overall, the significant differences in optimal blade shape between the deterministic and robust designs are a clear indication that robust optimization must be addressed for the future of industrial turbomachinery and global optimization.

References

1. Dunham, J.: CFD validation for propulsion system components, AGARD-AR-355 (1998)
2. Gurney, K.: An Introduction to Neural Networks. Taylor & Francis Inc, Bristol, PA, USA (1997)
3. Numeca, User Manual FINE^{TM}/Turbo v11.1 (2016)
4. Plackett, R.L., Burman, J.P.: The design of optimum multifactorial experiments. Biometrika **33** (4), 305–325 (1946)
5. Shields, M.D., Zhang, J.: The generalization of Latin hypercube sampling. Reliab. Eng. Syst. Safety **148**, 96–108 (2016)
6. Stein, M.L.: Statistical Interpolation of Spatial Data: Some Theory for Kriging. Springer, New York (1999)
7. Wunsch D., Nigro R., Coussement G., Hirsch C.: Quantification of combined operational and geometrical uncertainties in turbo-machinery design. In: Proceedings of the ASME GT2015, Paper GT2015-43399 (2015)

Robust Design Measures for Airfoil Shape Optimization

Daigo Maruyama, Stefan Görtz and Dishi Liu

Introduction

Deterministic aerodynamic optimization results in designs that are optimal under nominal conditions. Robust design optimization, on the other hand, accounts for variability in the design process and results in designs that are less sensitive to small random perturbations. Here, robust design measures are discussed in the context of airfoil shape optimization. Generally, in deterministic aerodynamic design of airfoils, the quantities of interest (QoI) are the drag coefficient (C_d), the lift coefficient (C_l), and the pitching moment coefficient (C_m). The drag coefficient or the lift-to-drag ratio (L/D) is often used to measure the aerodynamic performance and is used to formulate objective functions to be minimized or maximized in shape optimization. The pitching moment coefficient and the lift coefficient are often used to constrain the optimization problem. The pressure coefficient (C_p) is also considered for a variety of design purposes such as shock location/strength control or natural laminar flow design. In inverse design optimization, the objective function is the error (e.g., root mean square error) between the current and a target C_p distributions, which needs to be specified by an experienced designer.

In a nondeterministic setting, statistics of the QoI are introduced to formulate "robustness measures." Here, we considered the mean and standard deviation of aerodynamic coefficients to formulate what is typically called an expectation or

D. Maruyama (✉) · S. Görtz · D. Liu
German Aerospace Center (DLR), Institute of Aerodynamics and Flow Technology, Braunschweig, Germany
e-mail: Daigo.Maruyama@dlr.de

S. Görtz
e-mail: Stefan.Goertz@dlr.de

D. Liu
e-mail: Dishi.Liu@dlr.de

C. Hirsch et al. (eds.), *Uncertainty Management for Robust Industrial Design in Aeronautics*, Notes on Numerical Fluid Mechanics and Multidisciplinary Design 140, https://doi.org/10.1007/978-3-319-77767-2_32

mean-risk approach, and their maximum or minimum value for risk management. These two formulations are generally called robust design optimization (RDO) and reliability-based design optimization (RBDO), respectively. C_d is considered as an important aerodynamic coefficient in this context. The robustness of C_l may be regarded to be less important than C_d as long as the minimum C_l satisfies the lift constraint.

In this chapter, QoI based on the drag coefficient (C_d) are used as robustness measures unless otherwise noted. The lift coefficient C_l is used as a constraint to maintain a constant value. Another constraint used here is to keep the maximum thickness-to-chord ratio constant since it directly influences C_d. These constraints are maintained at the nominal conditions.

Next, the uncertainty factors in airfoil aerodynamics are introduced. The statistical variation of C_d may be caused by random variations of the flight conditions, such as the angle of attack α, the free-stream Mach number M, and Reynolds number Re, due to changing atmospheric conditions. The airfoil shape may also be subject to manufacturing tolerances and to degradation during operations. These factors are irreducible and considered as realistic uncertainties (aleatory uncertainties). On the other hand, numerical and modeling errors are an additional source of uncertainty that has an impact on the robust design results. These are reducible uncertainties (epistemic uncertainties). Figure 1 shows a simple depiction of these uncertainties in numerical flow simulations. Here, we focus on simulating the effect of irreducible uncertainties on shape design by using stochastic methods. The angle

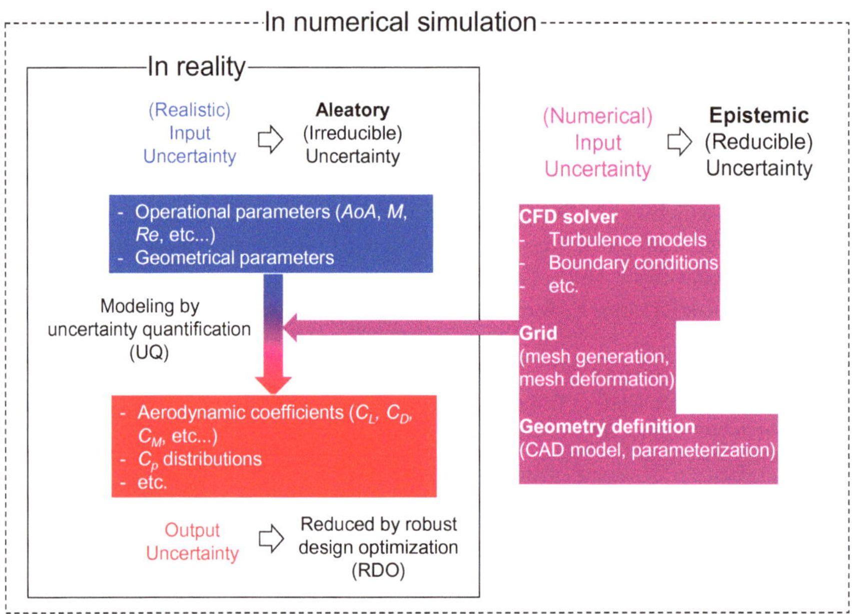

Fig. 1 Overview of aleatory and epistemic uncertainties in numerical simulation

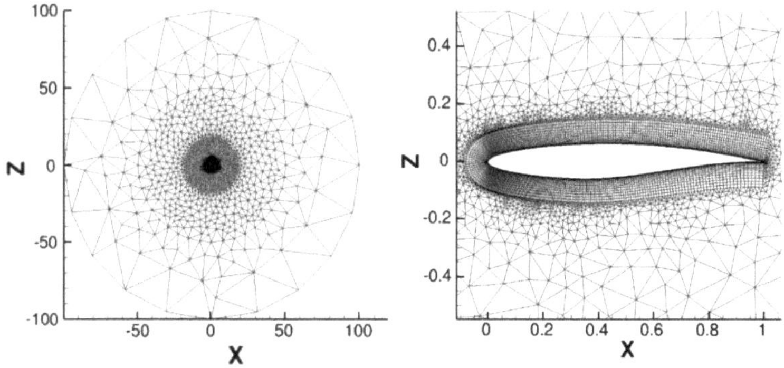

Fig. 2 Hybrid unstructured grid around RAE2822 airfoil. The number of grid nodes is around 29,000. Far-field (left), near-field (right)

of attack and the free-stream Mach number are considered as operational uncertainties with a mean and standard deviation as defined by the UMRIDA BC-02 test case description. The geometrical uncertainties were parameterized by using a truncated Karhunen–Loève expansion (tKLE) [1, 2].

To evaluate the aerodynamic performance of the airfoil, DLR's flow solver TAU [3–5] was used. Fully turbulent computations were performed with the negative Spalart–Allmaras turbulence model [6]. We opted for a central flux discretization, matrix dissipation, the lower/upper symmetric Gauss–Seidel (LU-SGS) implicit method for time integration in a backward Euler solver, and a "3v" multigrid cycle. A quasi two-dimensional hybrid unstructured grid with prisms and tetrahedral elements was used for the RANS simulations. The geometry changes caused by the nominal shape optimization or by the geometrical uncertainties are accomplished by using a radial basis function-based mesh deformation tool developed at DLR [7]. Figure 2 (left) shows the baseline grid around the unperturbed RAE2822 airfoil. A detailed view of the grid close to the surface of the airfoil is shown on the right-hand side.

Deterministic Design Optimization

First of all, the results of deterministic design optimization (DDO) are presented. This section also includes a brief discussion of errors due to epistemic uncertainties as shown in Fig. 1. Figure 3 shows the flowchart of the deterministic optimization framework. This framework was developed based on a DLR in-house aerodynamic shape optimization toolbox named Pyranha [8]. A generalized optimization problem for both deterministic design optimization and robust design optimization is described as follows:

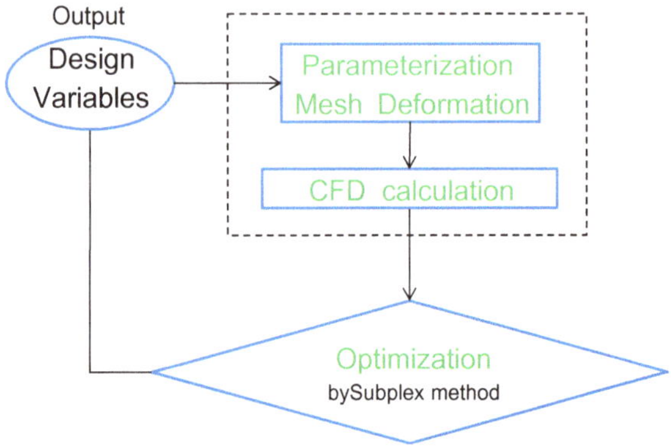

Fig. 3 Overview of deterministic design optimization (DDO) framework

$$\left.\begin{array}{ll}\min\limits_{\chi} & f(\boldsymbol{\chi}) \\ s.t. & t/c = const. \\ & C_l = const.\end{array}\right\} \tag{1}$$

where $\boldsymbol{\chi}$ denotes the 10 deterministic design variables, t and c are maximum thickness and chord length respectively, and C_l is the lift coefficient. Here, Bernstein polynomials were used to define 10 design parameters and applied to the camber line to satisfy the thickness constraints. In DDO, the cost function $f(\boldsymbol{\chi})$ can be considered deterministic under the assumption that epistemic uncertainties can be neglected. In this case, $f(\boldsymbol{\chi})$ in Eq. (1) can be expressed as:

$$f(\boldsymbol{\chi}) = C_d(\boldsymbol{\chi}) \tag{2}$$

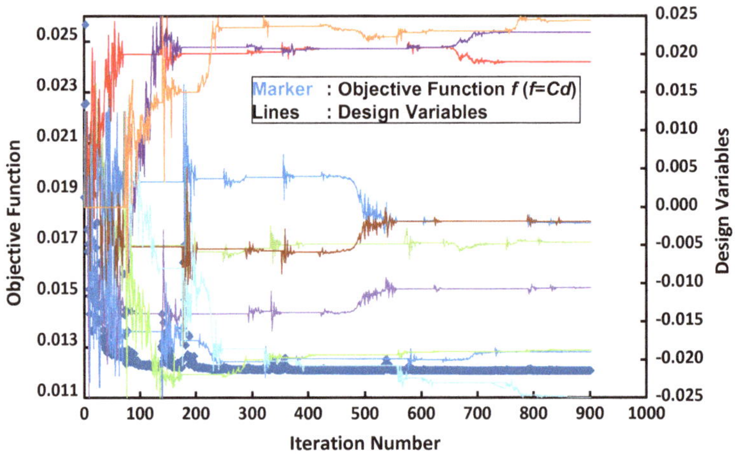

Fig. 4 Histories of the objective function and design variables in deterministic design optimization (DDO)

Figure 4 shows the history of the objective function and the design variables during the course of the optimization. The Subplex algorithm [9] was used to optimize the cost function $f(\chi)$. It is a gradient-free optimization method. It can be observed in Fig. 4 that after 500 iterations of the optimizer, the oscillations in the objective function are much less than 1 drag count (1 drag count = 10^{-4}). We conclude that epistemic errors due to the CFD solver, the mesh deformation, and the parameterization are negligible in our approach.

Robust and Reliability-Based Design Optimization

Robust design is now introduced based on the optimization problem formulation defined in Eq. (1). We distinguish between robust design optimization (RDO) based on the "expectation measure and mean-risk approach" and reliability-based design optimization (RBDO) based on the "worst-case risk measure" as briefly introduced above. The cost function $f(\chi)$ is evaluated by means of uncertainty quantification (UQ) methods and is regarded as stochastic as denoted by $f(\chi, \mathbf{a}, \xi)$. Once the input probability density function (pdf) is defined, the stochastic cost function $f(\chi, \mathbf{a}, \xi)$ can be regarded as deterministic, i.e., as $f(\chi)$, assuming that the statistics of the QoI can be accurately evaluated. This also motivates the use of deterministic optimizers in RDO and RBDO. The following two functions are introduced for robust and reliability-based design optimization, respectively:

$$f(\chi) \equiv \mu_{C_d}(\chi) + \sigma_{C_d}(\chi) \tag{3}$$

$$f(\chi) \equiv \max_{\mathbf{a}, \xi}(C_d(\chi, \mathbf{a}, \xi)) \tag{4}$$

where χ, $\mathbf{a}$, and ξ denote the design variables, the uncertain operational parameters, and the uncertain geometry parameters furnished by tKLE, resulting in 10 independent standard Gaussian variables, respectively. Concerning the geometrical uncertainties, initially we assume a random perturbation $\theta(x)$ of each surface grid node in the direction normal to the airfoil surface on both the upper and lower surfaces with zero-mean normal distributions, i.e., $\theta(x) \sim N(0, \sigma_\theta(x))$. Here, it is assumed that the random perturbations are spatially correlated by a Gaussian correlation function. Then, the KLE can decompose these correlated variables into uncorrelated ones. Only a relatively small number of the first several significant modes is retained. Finally, the correlated random perturbation $\theta(x)$ is transformed into a smaller number of independent standard Gaussian variables as $\xi \sim N(0, 1)$. In our test case, the number of random geometrical variables is reduced from 380 to 10 by using the tKLE. Figure 5 displays σ_θ distributions, and the baseline airfoil and three instances of the randomly deformed airfoil with the deformation 10 times exaggerated for better illustration. This setting means that perturbations are the largest at the mid-chord and zero at the leading and trailing edges. The maximum σ_θ

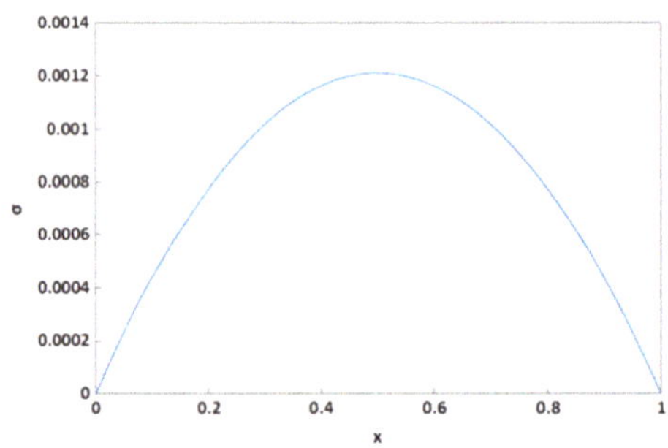

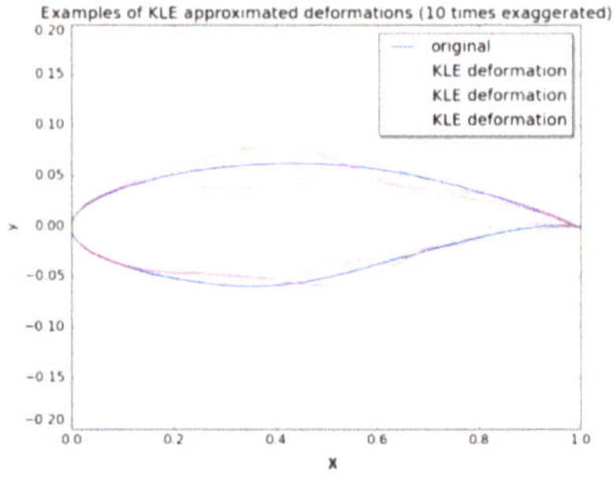

Fig. 5 Distribution of the standard deviation of each surface node (left), and examples of the original and perturbed RAE2822 airfoil (right)

of each surface is 1% of the maximum thickness-to-chord ratio of RAE2822, which is 0.1211.

The UQ method used here to compute the stochastic objective functions (also called the robustness measure) $f(\chi)$ expressed by Eqs. (3) and (4) is a combination of design of experiments (DoE) and surrogate modeling, which are implemented in DLR's surrogate modeling for aerodynamic data toolbox in Python (SMARTy) [10–12]. Details of the methods used to evaluate Eqs. (3) and (4) are described in [13]. In that chapter, gradient-enhanced Kriging (GEK) [10–12] is shown to produce more accurate estimates of the stochastic quantities for a given computational cost than Kriging or direct integration. As shown in [13], in practice, the maximum C_d in Eq. (4) is evaluated by the quantity at the 99th percentile of the cumulative density function (cdf) expressed as $Cd_{99\%\max}$ since we do not use optimization approaches to find the maximum value but use random search methods based on QMC on the constructed surrogate model for further improvement of efficiency of the process. In this approach, the cases with overestimated maximum values caused by extrapolations of the scattered sample points can be more evitable.

Besides, accuracy cannot be guaranteed in an acceptable accuracy level when the maximum value of quite high accuracy of the surrogate model is required, which may cause a drastic increase of the computational process.

Next, the optimization process for RDO and RBDO is shown in Fig. 6. The process is basically the same for RDO and RBDO. The robustness measure f is calculated at each iteration of the optimization by the procedure shown in Fig. 6 (right) and described in [13]. This leads to a unique value of f at each iteration.

The uncertain input parameters are assumed to have a normal distribution with a given mean and standard deviation except for the mean of the angle of attack α. The mean of the angle of attack is determined by an additional CFD computation at each iteration of the optimizer to satisfy the constant C_l constraint according to Eq. (1). Note that once the input probability density function (pdf) is defined, the stochastic

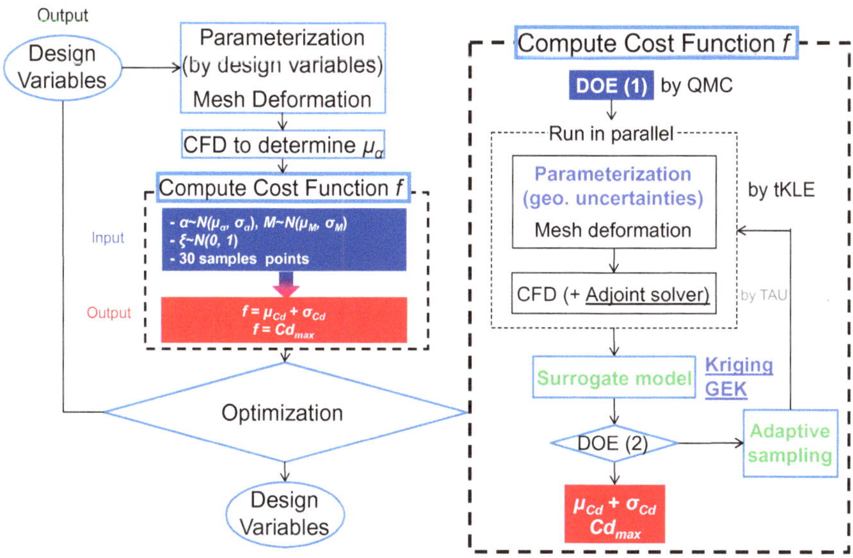

Fig. 6 Overview of framework for robust design optimization (RDO) and reliability-based design optimization (RBDO)

cost function *f* is almost uniquely determined. Then, any errors in *f* are considered as epistemic uncertainties.

The Subplex algorithm [9] was used to optimize the stochastic cost function *f*. In this algorithm, initially the simplex algorithm is assigned into several decomposed subspaces. Thus, it is generally more efficient than the Simplex algorithm for cases with many design variables. Moreover, the Subplex method is applicable to noisy cost functions [9]. It should be noted that less noise of the output design variables would result in a more accurate cost function value, which may also lead to a different optimal solution, details of which can be found in [14]. It is also important to note that there is no guarantee that the Subplex algorithm converges to the global optimum when the cost function *f* has more than one local optimum. However, one of the advantages of this algorithm is that it is a gradient-free optimization algorithm. Therefore, it is easy to implement when the gradients of the stochastic cost functions such as the maximum C_d are difficult and/or expensive to compute. Accuracy issues including the impact of a noisy objective function due to epistemic errors on the optimal solution are presented in [14].

Therefore, in this chapter, the influence of different measures of robustness on the optimal solution is discussed based on the optimization results. As a prerequisite, the robustness measure *f* should be evaluated accurately. We used a fixed number of 30 samples. Under this condition, the sampling techniques used to construct a surrogate model for the UQ part in every iteration of RDO and RBDO are as follows [13, 14]:

- RDO: GEK with 30 QMC samples whose distribution is the same as the input pdf (normal distributions).
- RBDO: GEK with 24 uniformly distributed QMC samples and 6 additional samples selected with the EI-based adaptive sampling technique.

The robustness measures established above were applied to design optimization using the UMRIDA BC-02 test case, which is based on the RAE2822 airfoil. The test case, mesh and flow conditions, setting of uncertainty factors, and all of the other conditions are common with those in the previous section of this chapter.

Figure 7 shows the history of the stochastic objective function f ($f \equiv \mu_{Cd} + \sigma_{Cd}$) and of the design variables in RDO (expectation measure and mean-risk approach) using a normal distribution of 30 samples to build the GEK surrogate model which in turn is used to estimate the statistics in every optimization cycle. It can be seen that there are more oscillations in the objective function f after the design variables almost converged than in the case of the deterministic design optimization (DDO) shown in Fig. 4. These additional oscillations can be attributed to the epistemic uncertainties which are due to a lack of accuracy of the stochastic quantities (more details of these errors can be seen in [14]). The same can be observed in the case of RBDO (the worst-case risk measure) in Fig. 8.

Figure 8a shows the history of the stochastic objective function f ($f \equiv Cd_{99\%max}$) and the evolution of the design variables in the case of RBDO (worst-case risk measure) using a uniform distribution of the initial samples and the adaptive sampling technique. Additionally, the optimization history obtained by a uniform distribution of all 30 samples (without any adaptive sampling) is shown in Fig. 8b since this approach led to a better final result. According to [13, 14], the adaptive sampling

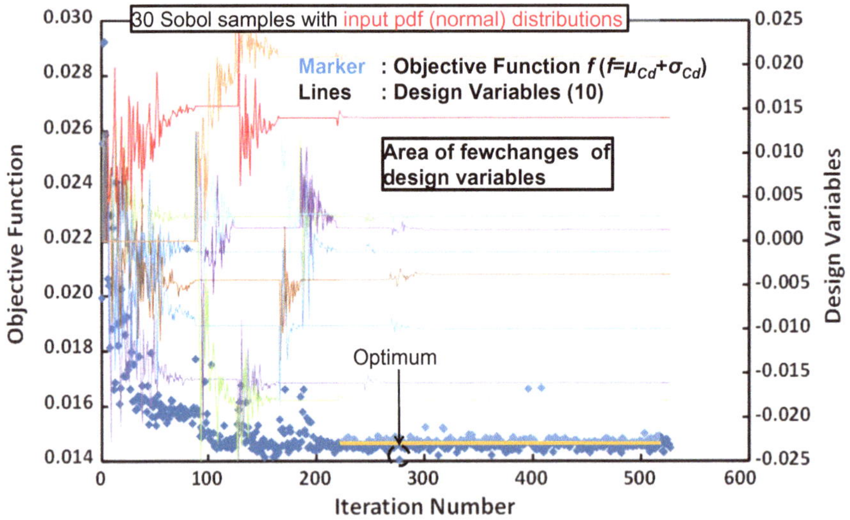

Fig. 7 Histories of the objective function and design variables in robust design optimization (RDO)

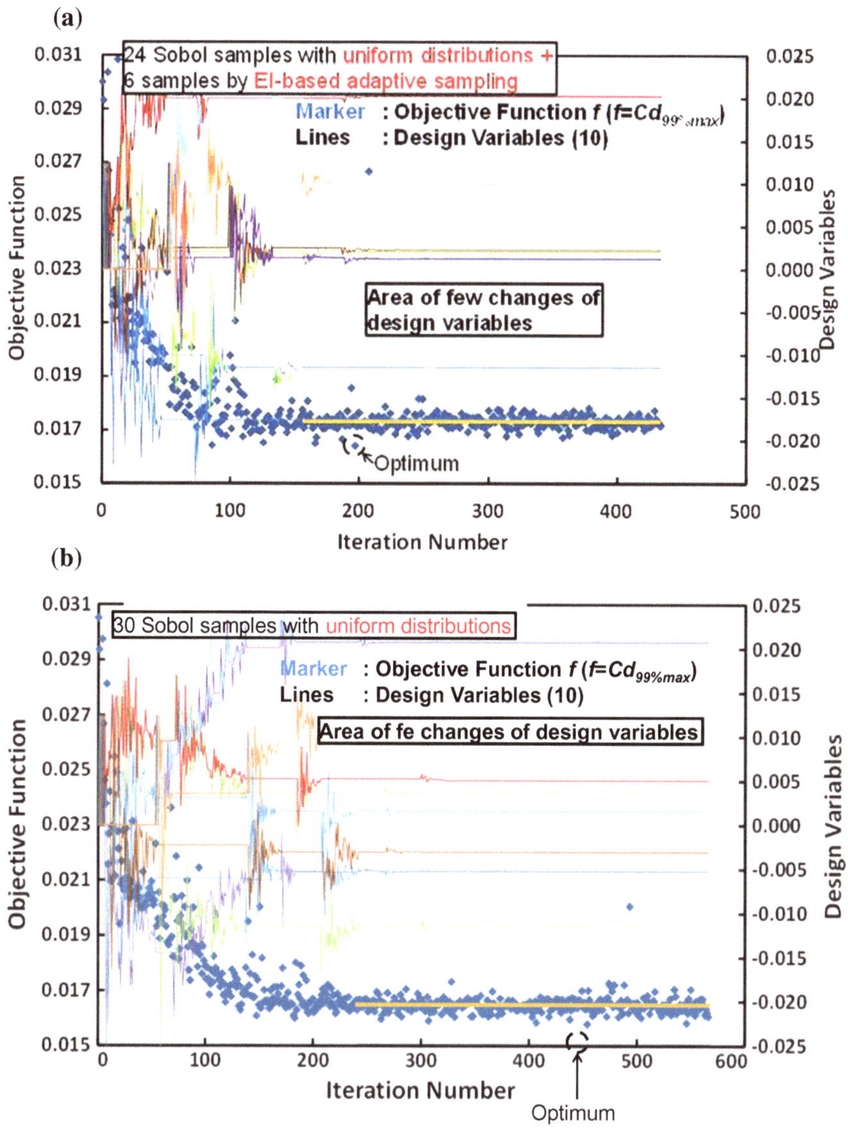

Fig. 8 Histories of the objective function and design variables in reliability-based design optimization (RBDO)—**a** surrogate models constructed by scattered samples with adaptive sampling technique and **b** surrogate models constructed by scattered samples without adaptive sampling technique

strategy typically leads to more accurate estimation of $Cd_{99\%max}$, which was tested by using the initial RAE2822 airfoil. However, Fig. 8 indicates that the optimal solution f_{opt} found without adaptive sampling is better in terms of the cost functions f than the one designed with adaptive sampling. The optimal design variables χ_{opt}, which are

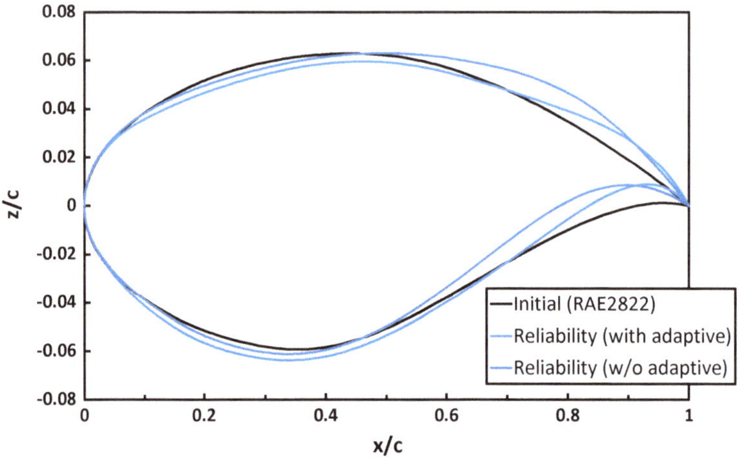

Fig. 9 Comparison of airfoils obtained by reliability-based design optimization (RBDO) based on two different UQ approaches with and without adaptive sampling

deterministic in this procedure, also reveal the differences between these two approaches (with and without adaptive sampling).

Figure 9 shows a comparison of the optimal airfoils. From the viewpoint of the performance improvement, we infer that the cost function f evaluated based on the surrogate model constructed with adaptive sampling converged to a "local optimum." The two optimal airfoils are further investigated by "re-evaluating" the statistics. This is done by constructing 100 surrogate models based on 100 different sets of 30 samples used to construct the surrogate model for each of the airfoil. $Cd_{99\%max}$ is then computed on each of the 100 surrogate models, and the average $Cd_{99\%max}$ is evaluated. The author showed this strategy also in [13–15]. Table 1 shows a comparison of $Cd_{99\%max}$ for the airfoils designed by using the GEK-based UQ method with and without adaptive sampling. The airfoil obtained by the method without adaptive sampling achieves around 10 drag counts better performance in terms of $Cd_{99\%max}$ than the airfoil obtained with adaptive sampling. Then, an interesting result can be presented related to this fact. Figure 10 compares the initial airfoil (RAE2822), the deterministically designed airfoil, the robustly designed airfoil, and the airfoil resulting from reliability-based design "without" adaptive sampling. They are termed "initial airfoil," "deterministic airfoil," "robust airfoil," and "reliability airfoil" as in Fig. 10, respectively. It can be observed that there are much fewer differences between the "robust airfoil" and the "reliability airfoil" than between the airfoils resulting from reliability-based design with and without adaptive sampling as shown in Fig. 9.

Each of the statistics of interest for the four airfoils was fairly and accurately "re-evaluated" as described above. They are tabulated in Table 2. C_d of the nominal airfoils is shown as well. As expected, the three optimized airfoils feature the best objective function value for their respective objective function. This indicates that

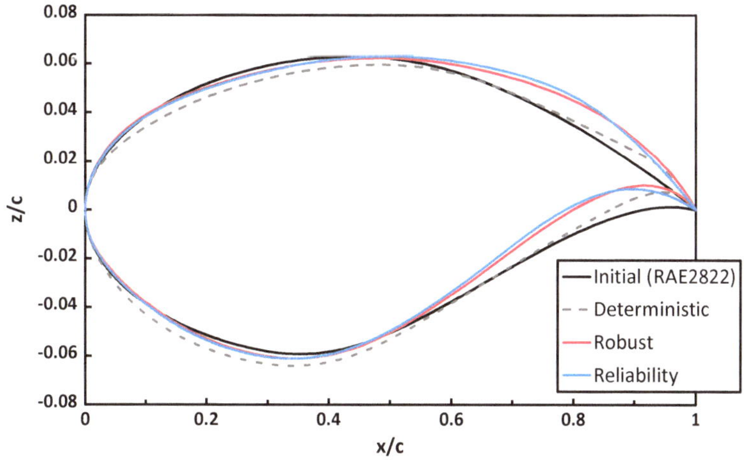

Fig. 10 Comparison of initial airfoil (RAE2822) and the airfoils obtained by deterministic design optimization (DDO), robust design optimization (RDO), and reliability-based design optimization (RBDO)

Table 1 Comparison of maximum C_d (quantity at the 99th percentile of the cumulative density function) of designed airfoils, UQ based on GEK surrogate w/o adaptive sampling and with EI-based adaptive sampling

RBDO	w/o adaptive sampling (cts.)	With adaptive sampling (cts.)
$Cd_{99\%\max}$	167.2	176.0

Table 2 Values of the objective function/robustness measures of the initial and optimized airfoils

Objective	Initial (cts.)	Deterministic (cts.)	Robust (cts.)	Reliability (cts.)
μ_{Cd}	192.2	137.2	**135.9**	138.5
σ_{Cd}	27.4	11.2	9.9	**8.2**
$\mu_{Cd}+\sigma_{Cd}$	219.6	148.4	**145.8**	146.7
$Cd_{99\%\max}$	257.7	179.1	169.1	**167.2**
Cd	190.6	**122.9**	127.7	134.4

during the shape optimization the stochastic objective functions ($f = \mu_{Cd} + \sigma_{Cd}$, $Cd_{99\%max}$) were accurately evaluated by our UQ approach, which combines effective sampling techniques with gradient-enhanced surrogate models. The fact that the "robust airfoil" and the "reliability airfoil" are relatively similar in shape is reasonable since minimizing $Cd_{99\%max}$ could also result in minimizing σ_{Cd} in most practical cases. In fact, the minimum σ_{Cd} (in addition to the minimum $Cd_{99\%max}$) is featured by the "reliability airfoil," not by the "robust airfoil." Of course, the minimums of the original objective function $f = \mu_{Cd} + \sigma_{Cd}$ itself and the mean value μ_{Cd} are achieved by "robust airfoil." Minimizing μ_{Cd} implicitly contains the

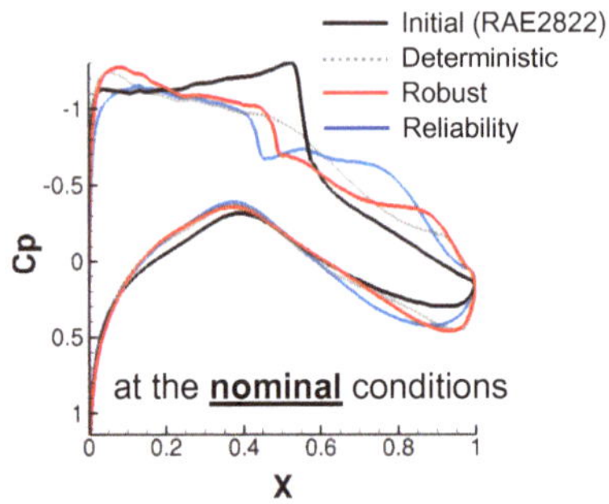

Fig. 11 Surface pressure coefficient (C_p) distributions at the nominal conditions of initial airfoil (RAE2822), deterministic, robust, and reliability-based designed airfoils

effect of reducing C_d at the nominal condition. Therefore, roughly to say, the "robust airfoil" is a kind of airfoil that has a performance in between the "deterministic airfoil" and the "reliability airfoil." C_d of that airfoil at the nominal condition is in fact in between that of the "deterministic airfoil" and the "reliability airfoil," but the differences between them are within a few drag counts.

The aerodynamic characteristics of these four airfoils are summarized in Figs. 11, 12, and 13. Figures 11 and 12 describe aerodynamic characteristics at the nominal conditions by showing the pressure coefficient (C_p) distributions of the four designed airfoils, respectively. It can be observed that the strength of the shock wave was directly reduced in DDO. The shock wave of both of the robust airfoils, on the other hand, is not reduced as much, but its location moved upstream to around $x/c = 0.5$ where the largest random geometry deformations may occur (see also the left hand of Fig. 5).

Figure 13 represents C_p distributions of 95% highest density intervals of the four airfoils to show the statistics of the aerodynamic characteristics. These statistics were simply evaluated by using the CFD results of the 30 samples but taking the mean value of 100 different sets of sample points as done for statistics of C_d. These figures show that the location of the shock wave of the "initial airfoil" is not sensitive to random perturbations. On the other hand, the strength of the shock wave of the "deterministic airfoil" tends to be smaller than that of the "initial airfoil." However, the location of the shock wave is much more sensitive to the random perturbations and may vary between 0.2 and 0.6 of the chord line. We could say that both the "robust airfoil" and the "reliability airfoil" have the advantages of both the "initial airfoil" and the "deterministic airfoil"; i.e., the range of the possible shock locations of the robust and reliable airfoils is smaller than that of the deterministic one and the 95% highest density intervals is also smaller than that of the "initial airfoil."

The fact that the optimized airfoils achieve the best aerodynamic performance in terms of the two robustness measures demonstrates that the UQ methods for evaluating the robustness measurements and the developed robust design framework successfully worked. The framework can also be applied to other test cases.

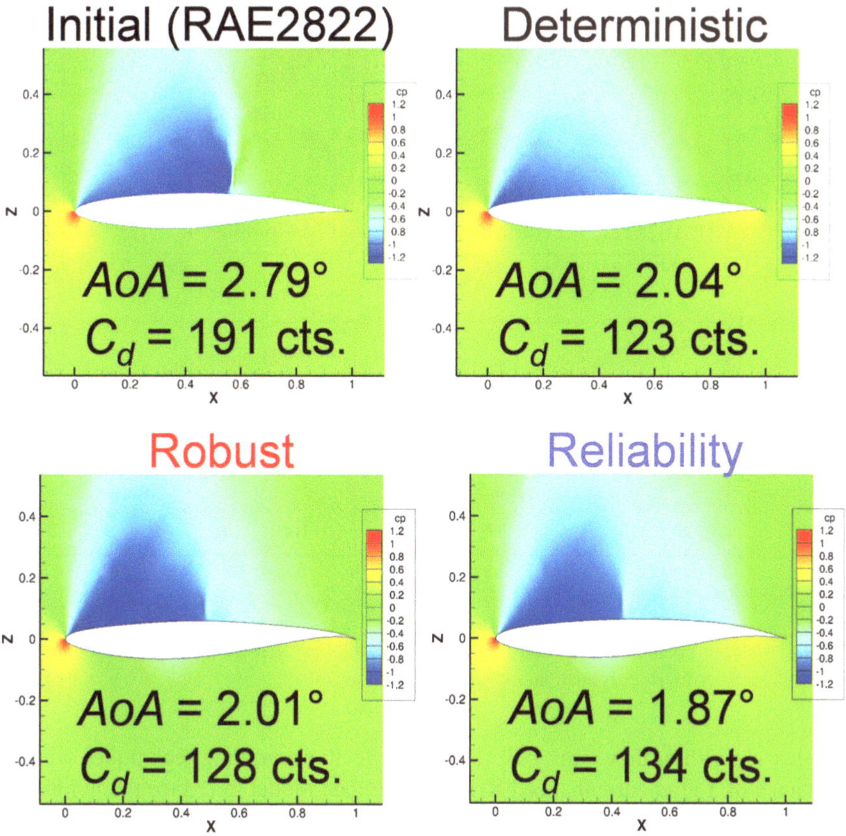

Fig. 12 Pressure coefficient (C_p) of initial airfoil (RAE2822), deterministic, robust, and reliability-based designed airfoils at the nominal conditions

Summary

Two types of robustness measurements were introduced to perform robust aerodynamic shape optimization based on the UMRIDA BC-02 airfoil test case. The airfoil was assumed to be subject to random geometry perturbations, which were represented by 10 independent standard Gaussian parameters furnished by a truncated Karhunen–Loève expansion. In addition, two operational uncertainties were considered. One robustness measure is focusing on the expectation measure and mean-risk approach. The other one is for risk management where we focus on the worst-case risk, which is the so-called reliability analysis. These robustness measures were formulated by the sum of mean and standard deviation of the drag coefficient, and the maximum value was evaluated by using the quantity at the 99th percentile of the cumulative density function of C_d, respectively. At each optimization cycle of the nominal shape optimization, which was conducted by the

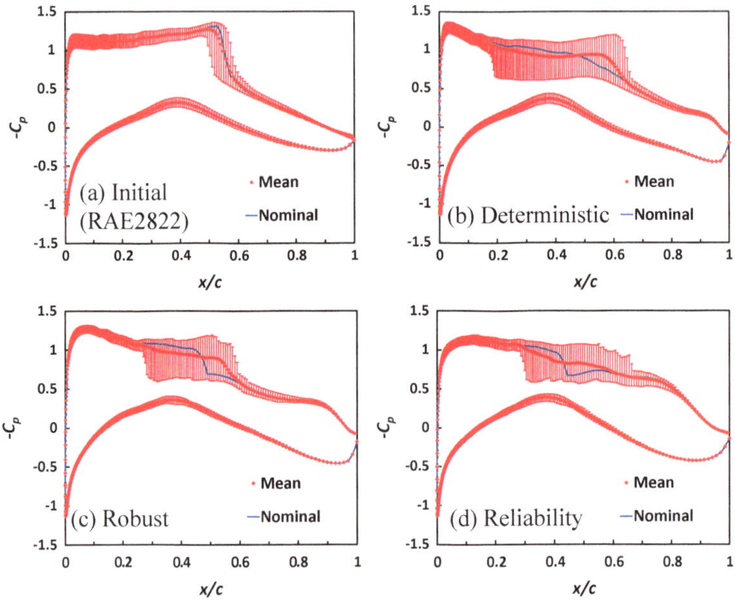

Fig. 13 Pressure coefficient (C_p) distributions of 95% highest density intervals of initial airfoil (RAE2822), deterministic, robust, and reliability-based airfoils

Subplex algorithm, they were calculated by effective sampling techniques with an assistance of a gradient-enhanced Kriging surrogate model to reduce the number of CFD computations, which is the most time-consuming part. The gradients were efficiently computed by an adjoint solver. The robustness measures were efficiently calculated by constructing a surrogate model based on 30 CFD and adjoint computations in the 12-dimensional uncertain input parameters space, and several hundred of optimization iterations were performed.

The robustly designed airfoils were compared with each other and with the initial airfoil and a deterministically designed airfoil. Each optimized airfoil successfully showed the best performance in terms of its objective function. The differences between the airfoils designed based on the two different robustness measurements are smaller than those between them and the deterministically designed one. The key differences are in the shock location and its strength. According to the 95% confidence intervals of the pressure distributions, both of the robustly designed airfoils feature the advantages of both the initial and the deterministically designed airfoils, which are robust to the shock location and its strength to random perturbations.

The developed framework was demonstrated for a generalized robust airfoil shape optimization process. It can also be applied to practical robust design applications such as considering the transition location as the quantity of interest, i.e., robust natural laminar airfoil design, and also to 3D aircraft cases.

References

1. Liu, D., Görtz, S.: Efficient quantification of aerodynamic uncertainty due to random geometry perturbations. In: New Results in Numerical and Experimental Fluid Mechanics IX, volume 124 of the Series Notes on Numerical Fluid Mechanics and Multidisciplinary Design. Springer (2014), pp. 65–73. ISBN 978-3-319-03157-6. ISSN 1612-2909
2. Liu, D., Litvinenko, A., Schillings, C., Schulz, V.: Quantification of airfoil geometry-induced aerodynamic uncertainties—comparison of approaches. SIAM/ASA J. Uncertain. Quant. **5**(1) (2016)
3. Galle, M., Gerhold, T., Evans, J.: Parallel computation of turbulent flows around complex geometries on hybrid grids with the DLR-TAU code. In: Ecer, A., Emerson, D.R. (Eds.) Proceedings of 11th Parallel CFD Conference Williamsburg, VA, North-Holland, 23–26 May 1999
4. Gerhold, T., Hannemann, V., Schwamborn, D.: On the validation of the DLR-TAU code. In: Nitsche, W., Heinemann, H.-J., Hilbig, R. (Eds.) New Results in Numerical and Experimental Fluid Mechanics. Notes on Numerical Fluid Mechanics, vol. 72, Vieweg (1999) ISBN 3–528-03122-0, pp. 426–433
5. Schwamborn, D., Gerhold, T., Heinrich, R.: The DLR TAU-code: recent applications in research and industry, invited lecture. In: Wesseling, P., Oate, E., Priaux, J. (Eds.) Proceedings of the European Conference on Computational Fluid Dynamics (ECCOMAS CFD 2006), The Netherlands (2006)
6. Allmaras, S.R., Johnson, F.T., Spalart, P.R.: Modifications and clarifications for the implementation of the Spalart-Allmaras turbulence model. In: Seventh International Conference on Computational Fluid Dynamics (ICCFD7), ICCFD7-1902, Hawaii, July 2012
7. Heinrich, R., Reimer, K., Michler, A.: Multidisciplinary simulation of maneuvering aircraft interacting with atmospheric effects using the DLR TAU code. In: RTO AVT-189 Specialists' Meeting on Assessment of Stability and Control Prediction Methods for Air and Sea Vehicles, Portsdown West, Oct 12–14, 2011
8. Brezillon, J., Abu-Zurayk, M.: Aerodynamic inverse design framework using discrete adjoint method. In: Notes on Numerical Fluid Mechanics and Multidisciplinary Design, vol. 121. Springer (2013) pp. 489–496
9. Rowan, T.: Functional Stability analysis of numerical algorithms, Ph.D. thesis, Department of Computer Sciences, University of Texas at Austin (1990)
10. Han, Z.H., Görtz, S., Zimmermann, R.: Improving variable-fidelity surrogate modeling via gradient-enhanced Kriging and a generalized hybrid bridge function. J. Aerosp. Sci. Technol. **25**(1) (2013)
11. Liu, D.: Efficient quantification of aerodynamic uncertainies using gradient-employing surrogate methods. In: Management and Minimisation of Uncertainties and Errors in Numerical Aerodynamics, volume 122 of the Series Notes on Numerical Fluid Mechanics and Multidisciplinary Design. Springer (2013) pp. 283–296. ISBN 978-3-642-36184-5. ISSN 1612-2909
12. Han, Z.H., Görtz, S., Zimmermann, R.: Improving variable-fidelity surrogate modeling via gradient-enhanced kriging and a generalized hybrid bridge function. J. Aerosp. Sci. Technol. **25**(1), 177–189 (2013)
13. Maruyama, D., Liu, D., Görtz, S.: Comparing surrogates for estimating aerodynamic uncertainties of airfoils. In: Hirsch, C. et al. (eds.), Uncertainty Management for Robust Industrial Design in Aeronautics Chapter 13 (2017)
14. Maruyama, D., Liu, D., Görtz, S.: Surrogate model based approaches to UQ and their range of applicability. In: Hirsch, C. et al. (eds.), Uncertainty Management for Robust Industrial Design in Aeronautics Chapter 43 (2017)
15. Maruyama, D., Liu, D., Görtz, S.: An efficient aerodynamic shape optimization framework for robust design of airfoils using surrogate models. In: Proceedings of the European Congress on Computational Methods in Applied Sciences and Engineering (ECCOMAS 2016), Crete, Greece (2016)

Robust Design with MLMC

Michele Pisaroni, Fabio Nobile and Penelope Leyland

Introduction

The ever-increasing demand for aircrafts with better performance, higher reliability, and robustness at lower cost requires nowadays optimization techniques seeking optimality under uncertain conditions that may arise during design, manufacture, and operation of the vehicle. The vast majority of problems in aircraft production and operation require decisions made in the presence of large number of uncertainties such as the geometrical and operational parameters that characterize aerodynamic systems. Designs obtained with deterministic optimization techniques seeking only optimality in a specific set of conditions may have very poor off-design performances.

Optimization under uncertainty (OUU) is a broad class of methodologies that address mainly the following two problems:

- Robust Design Optimization (RDO): focuses on the performances of a system under perturbations of the design conditions. Prescribed probabilistic measures of robustness (involving mean, variance, or higher moments) as objective functions are used to robustify the design. In this framework, the optimal design should be most insensitive to perturbations of the design.
- Reliability-Based Design Optimization (RBDO): focuses on safety-under-uncertainty aspects of the system and deals generally with the optimization of a

M. Pisaroni (✉) · F. Nobile · P. Leyland
Scientific Computing and Uncertainty Quantification, Ecole Polytechnique Fédérale de Lausanne, 1015 Lausanne, Lausanne, Switzerland
e-mail: michele.pisaroni@epfl.ch

F. Nobile
e-mail: fabio.nobile@epfl.ch

P. Leyland
e-mail: penelope.leyland@epfl.ch

C. Hirsch et al. (eds.), *Uncertainty Management for Robust Industrial Design in Aeronautics*, Notes on Numerical Fluid Mechanics and Multidisciplinary Design 140, https://doi.org/10.1007/978-3-319-77767-2_33

deterministic objective function subject to probabilistic constraints (failure probability or reliability indices). The optimal design has, in this framework, a higher degree of confidence and guarantees a prescribed minimum level of reliability under uncertain conditions.

We consider transonic airfoils under geometric uncertainties due to manufacturing tolerances and operating uncertainties due to atmospheric turbulence. In the RDO framework, we wish not only to minimize shock drag losses at a given speed, as in deterministic optimization, but also require that the drag and its dispersion are minimized under a set of operating uncertain conditions.

Non-intrusive uncertainty propagation techniques and derivatives-free approaches have gained a lot of interest in the recent years as they simply require multiple solutions of the original model and can use industrial CFD solvers as black box. Polynomial Chaos (PC) [1–3] and Stochastic Collocation (SC) [4] methods have been successfully employed in different engineering fields to built response surfaces in the stochastic variables while Kriging regression [5] has been mainly employed to build meta-models in the design variables. All these approaches have a strong potential for RDO but they are known to suffer the so-called curse of dimensionality (dramatic computational cost increases with the number of uncertain/design variables). In addition, those approaches based on global basis functions that span the entire stochastic domain may see their performances and accuracy jeopardized if the problem under investigation presents strong discontinuities as in the case of transonic/supersonic flows.

On the other hand, Monte Carlo (MC) methods have a dimension independent convergent rate and have been proven to be robust and accurate for non-smooth problems. Nevertheless, their very slow convergence rates make them chimerical for practical applications. As introduced in chapter "General Introduction to Monte Carlo and Multi-level Monte Carlo Methods", the Multi-Level Monte Carlo (MLMC) method can be employed to approximate the expectation (and higher moments) of an output quantity of interest (QoI) with a limited computational budget. The robustness and accuracy of the classical MLMC implementation strongly rely on the problem-dependent convergence rates of the output quantity of interest over the hierarchy of meshes, and the corresponding rate of cost increases to dictate the number of levels and the number of realizations per level. For many engineering problems, such parameters are generally estimated through a computational expensive screening procedure performed prior to the actual uncertainty analysis. In chapter "Continuation Multi-level Monte Carlo" and in [6], we have adapted the Continuation Multi-Level Monte Carlo (C-MLMC) algorithm presented in [7] to propagate operational and geometrical uncertainties in aerodynamic flow problems. As shown in [6], the C-MLMC is robust and efficient and leads to considerable speedup of few orders of magnitude over a traditional MC method for both inviscid and viscous flow problems.

In this chapter, we present a RDO technique based on the combination of evolutionary algorithms (EAs) and the Continuation Multi-Level Monte Carlo (C-MLMC) methodology to treat robust optimization problems in aerodynamic design. In particular, we focus on flows around an airfoil with uncertain operating conditions (angle of attack, Mach number) and aim at obtaining a shape to minimize the drag or maximize the lift-to-drag ratio.

Methodology

We consider a general mathematical formulation of a single objective robust design optimization problem (RDO):

$$\text{RDO}: \begin{cases} \min\limits_{x} \ \mathscr{R}(f(x)) \\ \ s.t \ \mathscr{C}(g(x)) \leq 0 \\ \quad x_L \leq x \leq x_U \end{cases} \tag{1}$$

where x is the vector of design variables.

We denote with $\mathscr{R}(f(x))$ the robust loss function (or fitness in case of maximization) which is typically a function of mean (denoted with μ which is function of the vector of design variables x and the system parameter p) and the standard deviation (denoted with σ which is also function of the vector of design variables x and the system parameter p) of some problem-specific quantity of interest Q_r that has to be optimized:

$$\mathscr{R}(f(x)) = F(\mu_{Q_r}(x, p), \sigma_{Q_r}(x, p)) \tag{2}$$

$\mathscr{C}(g(x))$ denote a set of deterministic or probabilistic constraints which typically depend on some constant reference values C^*, mean and standard deviation of some problem-specific quantity of interest Q_c that have to be constrained:

$$\mathscr{C}(g(x)) = G(C^*, \mu_{Q_c}(x, p), \sigma_{Q_c}(x, p)) \tag{3}$$

x_L and x_U are the lower and upper bounds for the design variables.

We employ evolutionary algorithms (EAs) in conjunction with the Continuation Multi-Level Monte Carlo (C-MLMC) presented in chapter "Continuation Multi-level Monte Carlo." C-MLMC is used to compute efficiently the robust loss function $\mathscr{R}(f(x))$ and the set of probabilistic constraints $\mathscr{C}(g(x))$ that involve mean and standard deviations. EAs are known to be well suited for solving RDO problems because they ideally do not make any assumption about the underlying fitness landscape and are noways employed in different engineering fields.

Combination of EAs and C-MLMC

Evolutionary algorithms (EAs) are a broad class of optimization approaches that aim at simulating the evolution of a population through successive generations of better performing individuals [8]. A generic EA uses mechanisms inspired by biological evolution, such as reproduction, mutation, and selection. The candidate solutions are treated as individuals in a population, and the fitness function determines the quality of the solutions.

We denote with Θ_j^n the population of j individuals at the generation n:

$$\Theta_j^n = \{\theta_1, \theta_2, \dots, \theta_j\} \tag{4}$$

Each individual θ_i is defined by a set of properties (its genotype) which in our application is a set of k geometrical parameters defying the shape of an airfoil.

$$\theta_i = \{g_{(i)1}, g_{(i)2} \dots, g_{(i)k}\} \tag{5}$$

Namely we employ the PARSEC coefficients as genotype for the individuals considered hereafter (Fig. 1).

At each generation, the fitness ϕ of each individual in the population is evaluated. The fitness can be seen as a figure of merit and can be interpreted, in our setting, as the value of the robust objective function in the optimization problem that we want to solve.

$$\phi_i = \mathscr{R}\left(f(\theta_i, p)\right) \tag{6}$$

We indicate with Φ_j^n the set of the fitness values of all j individuals at generation n:

$$\Phi_j^n = \{\phi_1, \phi_2, \dots, \phi_j\} \tag{7}$$

The structure of a generic EA is presented in Algorithm 1.

In the generic EA, we can distinguish between problem-specific and algorithm-specific components.

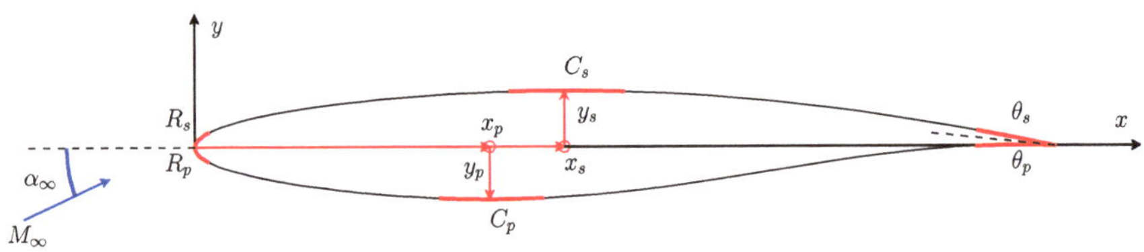

Fig. 1 Geometry of the RAE 2822 transonic airfoil and PARSEC parameters that define the geometry of the airfoil

Algorithm 1: Generic EA

GENERATOR → Θ_j^0
EVALUATOR (Θ_j^0) → Φ_j^0
while TERMINATOR (Φ_j^i, N^i) *not True* **do**
  SELECTOR ($\widetilde{\Theta}_j^{i-1}, \Phi_j^{i-1}$) → $\widetilde{\Theta}_j^i$
  CROSSOVER ($\widetilde{\Theta}_j^i$) → $\widetilde{\Theta}_j^i$
  MUTATOR ($\widetilde{\Theta}_j^i$) → Θ_j^i
  EVALUATOR (Θ_j^i) → Φ_j^i
  REPLACER (Θ_j^i, Φ_j^i) → $\widetilde{\Theta}_j^i$
  update $i = i + 1$
  update $N^i = N^{i-1} + j$

Problem-Specific Components:

- **GENERATOR**: defines how individuals are created. In the present airfoil problem, it is a set of routines that generates PARSEC parameters [9] that satisfy the geometrical constraints.
- **EVALUATOR**: dictates how fitness values are computed from individuals. In case of:
  - Deterministic Optimization: the fitness is computed by interrogating a black box CFD solver that provides the airfoil coefficients (C_L, C_D or L/D).
  - Robust Optimization: the fitness is computed by estimating $\mu_{Q_r}(x, p)$ and $\sigma_{Q_r}(x, p)$ using the C-MLMC methodology presented in the following section.

Algorithm-Specific Operators:

- **SELECTOR**: determines which individuals should become parents. The probability for an individual to be selected is weighed by its fitness. The better the fitness, the higher the probability to be selected.
- **CROSSOVER**: dictates how offspring is created from the parents. The genotypes of the parents are randomly cut and re-assembled to generate the offspring.
- **MUTATOR**: performs a random substitution of some "chromosomes" in the genotype representing an individual. Mutation enhances the probability of exploring untouched areas of the design space avoiding premature convergence in local minimum.
- **REPLACER**: determines which individuals survive into the next generation. The *elitist selection* is used to pick the best individuals from the current generation to carry over to the next, unaltered.
- **TERMINATOR**: determines when the evolution should end.

Starting from an initial population Θ_j^0 of randomly generated individuals, new generations are iteratively developed by applying the above-mentioned operators to the individuals of the previous. In each generation, the fitness of every individual in the population is evaluated. The individuals with highest fitness are selected from the current population, and each individual's genotype is recombined and possibly randomly mutated to form a new generation. The algorithm terminates when either a maximum number of evaluations has been computed, or a satisfactory fitness level has been reached for the whole population.

Computation of $\mathscr{R}(f(x,p))$ and $\mathscr{C}(g(x,p))$ using C-MLMC

Suppose we are interested in the expected value $\mu_Q = \mathbb{E}[Q]$ and variance $\sigma_Q^2 = \mathbb{V}ar[Q]$ of a QoI $Q = f(u)$ of the solution u of a fluid dynamics problem computed using a numerical approximation with a discretization parameter M (number of spatial degrees of freedom).

The MLMC estimator uses a sequence of approximations $Q_{M_0}, Q_{M_1}, \ldots, Q_{M_L} = Q_M$ of increasing accuracy; hence, $M_0 < M_1 < \cdots < M_L = M$ and evaluate the expectation $\mathbb{E}[Q]$ as:

$$\mathbb{E}[Q] \approx \mathrm{E}^{\mathrm{MLMC}}[Q_M] := \sum_{l=0}^{L} \mathrm{E}^{\mathrm{MC}}[Q_{M_l}] - \mathrm{E}^{\mathrm{MC}}[Q_{M_{l-1}}], \tag{8}$$

where each term is a Monte Carlo estimation of $\mathbb{E}[Q_{M_l} - Q_{M_{l-1}}]$ independent of the other levels.

Similarly, a MLMC estimator for $\mathbb{V}ar[Q]$ can be easily constructed as:

$$\mathrm{V}^{\mathrm{MLMC}}[Q_M] := \sum_{l=0}^{L} \mathrm{V}^{\mathrm{MC}}[Q_{M_l}] - \mathrm{V}^{\mathrm{MC}}[Q_{M_{l-1}}], \tag{9}$$

where $\mathrm{V}^{\mathrm{MC}}[Q_{M_l}]$ and $\mathrm{V}^{\mathrm{MC}}[Q_{M_{l-1}}]$ are the MC variance estimators for $\mathbb{V}ar[Q_M]$ computed on level l and $l-1$:

$$\mathrm{V}^{\mathrm{MC}}[Q_M] := \frac{1}{N-1} \sum_{i=1}^{N} \left(Q_M(\omega^{(i)}) - \mathrm{E}^{\mathrm{MC}}[Q_M]\right)^2. \tag{10}$$

using the same samples, whereas estimations on different levels are independent. To overcome some of the limitations of classical MLMC approach and to further reduce the computational burden required to accurately compute the problem-dependent parameters needed to estimate the optimal number of levels and samples per level to achieve a prescribed tolerance, we employ in this work the C-MLMC approach [6, 7]. Additionally, we leverage on the self-tuning nature of C-MLMC approach presented in chapter "Continuation Multi-level Monte Carlo" to effectively accommodate its application in RDO framework.

Thanks to the rigorous mean square error estimators that exist for the MLMC estimators for mean and variance [10] and we are able to approximate, with a controlled accuracy, in just one run of our C-MLMC algorithm the following quantities:

- mean and standard deviation of the QoI that has to be optimized: $\mu_{Q_r}(x,p)$, $\sigma_{Q_r}(x,p)$
- mean and standard deviation of the QoI that has to be constrained: $\mu_{Q_c}(x,p)$, $\sigma_{Q_c}(x,p)$

For the airfoil problem presented next, we compute in one C-MLMC run the mean and standard deviation of lift and drag coefficients (C_L and C_D) as well as the ratio of lift and drag (L/D).

Robust Transonic Airfoil Shape Optimization

We consider hereafter the transonic RAE2822 airfoil [11, 12] affected by operating uncertainties modeled as truncated Gaussian random variables (Table 1).

The initial geometry of the RAE2822 airfoil is defined with a set of PARSEC parameters [9] and is illustrated in Fig. 1.

We employ the PARSEC parametrization of the airfoil shape, which defines the set of design variables and corresponding ranges (geometrical constraints) for the design and geometric constraints variable set. The allowed design range for the geometrical parameters defining the shape of the airfoil is presented in Table 2.

MLMC hierarchy In this work, we employ a 5-levels structured grid hierarchy for the C-MLMC simulations. The features of the grid levels are presented in Table 3 along with the average computational time required to compute one deterministic simulation (on one CPU) using the MSES collection of programs for the analysis of airfoils [13]. MSES solves the steady Euler equations with a finite volume discretization over a streamline grid and is coupled via the displacement thickness,

Table 1 Operating uncertainties for the RAE2822 problem

	Quantity	Reference (r)	Uncertainty $\mathcal{TN}(\mu,\sigma,X_{LO},X_{UP})$
Operating	α_∞	2.31	$\mathcal{TN}(r,2\%,-2\%,+2\%)$
	M_∞	0.729	$\mathcal{TN}(r,5\%,-5\%,+5\%)$
	Re_c	$6.5\cdot 10^6$	–
	p_∞ $[Pa]$	101325	–
	T_∞ $[K]$	288.5	–

Table 2 Geometric design parameters for the RAE2822 problem

	Quantity	Reference (r)	Design range (%)
Geometric design parameters	R_s	0.00839	90–115
	R_p	0.00853	90–115
	x_s	0.431	80–125
	x_p	0.346	80–125
	y_s	0.063	80–125
	y_p	−0.058	80–125
	C_s	−0.432	80–125
	C_p	0.699	80–125
	θ_s	−11.607	–
	θ_p	−2.227	–

Table 3 MLMC 5-levels grid hierarchy for the RAE2822 problem

Level	Airfoil nodes	Cells	*CTime* (s)
*L*0	47	1739	1.9
*L*1	71	2627	3.2
*L*2	107	3959	5.7
*L*3	161	5957	7.5
*L*4	243	8991	14.7
*L*5	365	13505	17.9

with a two-equation integral solver for the viscous regions of the boundary layer and trailing wakes.

RDO problem formulations We now present the following RDO problems:
Transonic Airfoil RDO1: we wish to minimize the weighted sum of the mean and the standard deviation of the drag coefficient C_d of the airfoil (C_d^* denote the drag coefficient of the original RAE2822 airfoil). We constrain the shape of the airfoil to the range of design parameters in Table 2. Additionally, we prescribe the mean of the lift coefficient μ_{C_L} to be above the 90% of the original RAE 2822 airfoil (denoted as C_L^*).

$$\text{RDO1(a)} : \begin{cases} \min\limits_{x} \ \dfrac{\mu_{C_d}(x,p)}{C_d^*} + \dfrac{\sigma_{C_d}(x,p)}{C_d^*} \\ s.t \ \ \mu_{C_L}(x,p) \geq 0.9 * C_L^* \\ \quad x_L \leq x \leq x_U \end{cases} \tag{11}$$

The geometrical constraints are handled by the **GENERATOR** that selects the appropriate genotype of the individuals to accommodate the prescribed bounds. The lift coefficient however is not available before the evaluation of the individual and

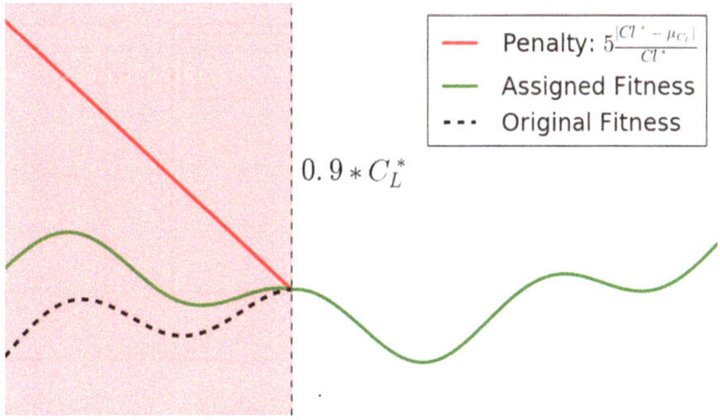

Fig. 2 Application of a linear penalty function to the fitness in case of violation of the lift coefficient constraint

requires a different treatment. We employ a penalty function following the approach of Coello [14]. The penalty function gives a fitness disadvantage to individuals violating a constraint, proportional to the amount of constraint violation.

More specifically the RDO1(a) is rewritten as a penalization problem as:

$$\text{RDO1(a)} : \begin{cases} \min\limits_{x} \dfrac{\mu_{C_d}(x,p)}{C_d^*} + \dfrac{\sigma_{C_d}(x,p)}{C_d^*} + \dfrac{5(C_L^* + \mu_{C_L}(x,p))}{C_L^*} \\ s.t \;\; x_L \leq x \leq x_U \end{cases} \tag{12}$$

Figure 2 shows the fitness (dotted in the region where the constraint is violated) and the penalty function (in red). The continuous green line represents the fitness that is actually assigned to the individuals.

In Fig. 3, we compare the performance of the original RAE2822 airfoil with those of the robust optimized shape (obtained by solving the RDO1(a) presented above) and the deterministic optimal shape. The latter is obtained by solving the following problem:

$$\text{DO1} : \begin{cases} \min\limits_{x} \dfrac{C_d(x,p)}{C_d^*} \\ s.t \;\; C_L(x,p) \geq 0.9 * C_L^* \\ \quad\;\; x_L \leq x \leq x_U \end{cases} \tag{13}$$

As it might be observed by looking at the C_d dispersion plot in Fig. 3 and Table 4, the deterministic optimal DO1 shape achieves the best reduction in drag coefficient at the operating conditions ($M = 0.729$, $\alpha = 2.31$, $Re = 6.5e6$, diamond symbol in the plot). However, the RDO1(a) robust shape presents a lower variability of C_d in the operating range with the same mean drag coefficient of DO1.

Transonic Airfoil RDO2: we wish to maximize the mean of the airfoil lift-to-drag ratio L/D and minimize its standard deviation. As in the previous problem, the thickness of the airfoil is constrained:

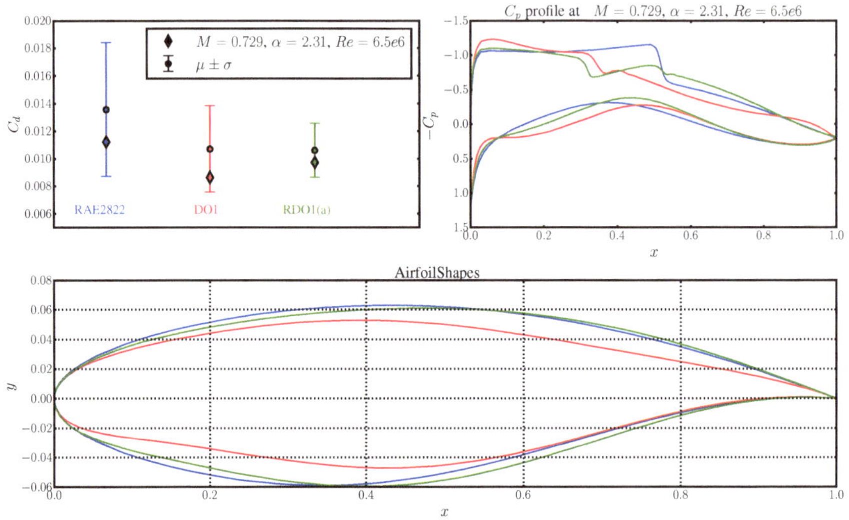

Fig. 3 RDO1(a) robust optimal shape and DO1 deterministic optimal shape compared with the initial RAE2822 airfoil. Top left plot: C_d of the airfoils. The diamond symbols correspond to the C_d of the airfoils at deterministic design conditions ($M = 0.729$, $\alpha = 2.31$, $Re = 6.5e6$). Top right plot: pressure coefficient profiles of the airfoils at deterministic design conditions. Lower plot: airfoil shapes

Table 4 Deterministic (DO1) and Robust (RDO1(a)) optimal airfoil shape performances compared to the RAE2822

	RAE2822	DO1	RDO1(a)
$M = 0.729$ $\alpha = 2.31$ $Re = 6.5e6$	$C_d = 0.0112$	$\mathbf{C_d = 0.0086}$	$C_d = 0.0097$
$M = \mathcal{TN}(0.729, 2\%, -2\%, +2\%)$ $\alpha = \mathcal{TN}(2.31, 5\%, -5\%, +5\%)$ $Re = 6.5e6$	$C_d = 0.0135 \pm 0.0049$ (36%)	$C_d = 0.0107 \pm 0.0031$ (29%)	$\mathbf{C_d = 0.0107 \pm 0.0020}$ **(18%)**

$$\text{RDO2(a)} : \begin{cases} \max\limits_{x} \ \mu_{L/D}(x,p) - \sigma_{L/D}(x,p) \\ x_L \leq x \leq x_U \end{cases} \tag{14}$$

In Fig. 3, we compare the performances of the original RAE2822 airfoil with those of the robust optimized shape (obtained by solving the RDO2(a) presented above) and the deterministic optimal shape. The latter is obtained by solving the following problem:

$$\text{DO2} : \begin{cases} \max\limits_{x} \ L/D(x,p) \\ x_L \leq x \leq x_U \end{cases} \tag{15}$$

Looking at the L/D dispersion plot in Fig. 4 and Table 5, the deterministic optimal DO2 shape achieves the highest L/D ratio at the operating conditions ($M = 0.729$,

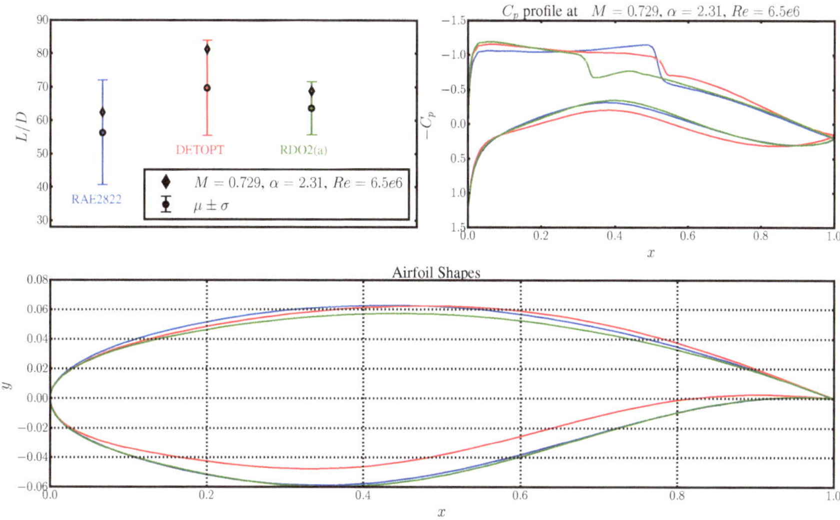

Fig. 4 Performances of the RAE2822 airfoil, RDO2(a) robust optimal shape and DO2 deterministic optimal shape. RDO2(a) robust optimal shape and DO1 deterministic optimal shape compared with the initial RAE2822 airfoil. Top left plot: L/D of the airfoils. The diamond symbols correspond to the L/D of the airfoils at deterministic design conditions ($M = 0.729,\ \alpha = 2.31,\ Re = 6.5e6$). Top right plot: pressure coefficient profiles of the airfoils at deterministic design conditions. Lower plot: airfoil shapes

Table 5 Deterministic (DO2) and Robust (RDO2(a)) optimal airfoil shapes performances compared to the RAE2822

	RAE2822	DO2	RDO2(a)
$M = 0.729$ $\alpha = 2.31$ $Re = 6.5e6$	$L/D = 62.42$	**L/D = 81.25**	$L/D = 68.79$
$M = \mathcal{TN}(0.729, 2\%, -2\%, +2\%)$ $\alpha = \mathcal{TN}(2.31, 5\%, -5\%, +5\%)$ $Re = 6.5e6$	$L/D = 56.32 \pm$ 27.84 (28%)	$\mathbf{C_d = 69.71} \pm$ 14.20 (20%)	$L/D = 63.64$ **±7.90** (**12**%)

$\alpha = 2.31$, $Re = 6.5e6$, diamond symbol in the plot). The RDO2(a) robust shape presents a lower variability of L/D in the operating range at the cost of lower mean L/D ratio compared to DO2.

References

1. Tootkaboni, M., Asadpoure, A., Guest, J.K.: Topology optimization of continuum structures under uncertainty-a polynomial chaos approach. Comput. Methods Appl. Mech. Eng. **201**, 263–275 (2012)
2. Dodson, M., Parks, G.T.: Robust aerodynamic design optimization using polynomial chaos. J. Aircraft **46**(2), 635–646 (2009)

3. Ho, S., Yang, S.: A fast robust optimization methodology based on polynomial chaos and evolutionary algorithm for inverse problems. IEEE Trans. Magnet. **48**(2), 259–262 (2012)
4. Sankaran, S., Audet, C., Marsden, A.L.: A method for stochastic constrained optimization using derivative-free surrogate pattern search and collocation. J. Comput. Phys. **229**(12), 4664–4682 (2010)
5. Simpson, T.W., Mauery, T.M., Korte, J.J., Mistree, F.: Kriging models for global approximation in simulation-based multidisciplinary design optimization. AIAA J. **39**(12), 2233–2241 (2001)
6. Pisaroni, M., Nobile, F., Leyland, P.: A continuation multi level monte carlo (c-mlmc) method for uncertainty quantifcation in compressible aerodynamics. Technical Report (2016)
7. Collier, N., Haji-Ali, A.L., Nobile, F., von Schwerin, E., Tempone, R.: A continuation multilevel monte carlo algorithm. BIT Num. Math. 1–34 (2014)
8. Bäck, T., Fogel, D., Michalewicz, Z.: Handbook of evolutionary computation. Release **97**(1), B1 (1997)
9. Sobieczky, H.: Parametric airfoils and wings. In: Fujii, K., Dulikravich, G.S. (eds.) Notes on Numerical Fluid Mechanics, vol. 68 pp. 71–88 (1998)
10. Bierig, C., Chernov, A.: Convergence analysis of multilevel monte carlo variance estimators and application for random obstacle problems. Numerische Mathematik **130**(4), 579–613 (2015)
11. Haase, W., Brandsma, F., Elsholz, E., Leschziner, M., Schwamborn, D.: EUROVALAn European Initiative on Validation of CFD Codes: Results of the EC/BRITE-EURAM Project EUROVAL, 1990–1992. vol. 42. Springer-Verlag (2013)
12. V.A.: Experimental Data Base for Computer Program Assessment Report of the Fluid Dynamics Panel Working Group. AGARD-AR-138 (1979)
13. Drela, M.: A users guide to MSES 3.05. Massachusetts Institute of Technology (MIT), Cambridge (2007)
14. Coello, C.A.C.: Theoretical and numerical constraint-handling techniques used with evolutionary algorithms: a survey of the state of the art. Comput. Methods Appl. Mech. Eng. **191**(11), 1245–1287 (2002)

Value-at-Risk and Conditional Value-at-Risk in Optimization Under Uncertainty

Domenico Quagliarella

Introduction

Many industrial optimization processes must take account of the stochastic nature of the system and processes to design or redesign and consider the random variability of some of the parameters that describe them. Thus, it is necessary to characterize the system that is being studied from various points of view related to the treatment of uncertainty. In particular, it is necessary to consider the sensitivity of the system to the uncertain parameters and assess its reliability. Having established the ability to characterize the system from this point of view, it is necessary to build an optimization loop that can improve its reliability, or that is capable of providing a robust optimum, or that could withstand acceptably random perturbations of design parameters or operating conditions. The classical approach to this problem is the so-called "robust design optimization" (RDO), which tries to maximize the expected performance and simultaneously to minimize the performance sensitivity with respect to random parameters (variance). Instead, the "reliability-based design optimization" (RBDO) tries to find the optimum design by explicitly assigning a specific level of risk and a given level of reliability. In financial engineering, this corresponds to the concept of minimizing the value-at-risk (VaR). In terms of optimization problems, this means to assign a specific quantile of the probability distribution corresponding to the function to be optimized as the actual objective function and to minimize its value. Therefore, if the goal is that the value of a given objective function f is less than a specific value $\bar{f}$ in 75% of possible cases, the following constraint on the corresponding quantile should be imposed: $\nu^{0.75} \leq \bar{f}$. Alternatively, the problem can be set as the minimization of $\nu^{0.75}$ and a function f is thus obtained which is less than or equal to the value given by the optimization of the quantile in 75% of cases. If,

D. Quagliarella (✉)
Fluid Mechanics Department, C.I.R.A.—Italian Aerospace Research Centre,
Via Maiorise, 81043 Capua, Italy
e-mail: d.quagliarella@cira.it

C. Hirsch et al. (eds.), *Uncertainty Management for Robust Industrial Design in Aeronautics*, Notes on Numerical Fluid Mechanics and Multidisciplinary Design 140, https://doi.org/10.1007/978-3-319-77767-2_34

instead, the objective is $\min v^1$, then the purpose of the optimization procedure is to improve the worst-case scenario, as it happens when the problem is of minimax type.

Although conceptually very attractive, and despite its official adoption as a standard measure of risk in the industry regulations, VaR is often a difficult measure to use and not very tractable numerically, outside of cases in which the loss distributions are normal. An alternative measure of risk is the conditional value-at-risk (CVaR), also called superquantiles. CVaR retains all the desirable features of VaR, but solves many of the problems associated with the use of VaR and makes optimization problems more tractable. One of the advantages of CVaR is that it is a consistent measure of risk [1]. Also, it is possible, in the risk optimization process, to avoid the explicit evaluation of a given level of CVaR by using a special function (F_α), introduced in [2, 3], which reduces the problem to the evaluation of expected values and to the minimization over a scalar. Moreover, a result of fundamental importance for optimization is that function F_α preserves convexity. However, it should be stressed that not always the use of CVaR is better than VaR; in particular, well-modeled probability distribution tails are required to obtain an accurate estimate of CVaR, and this is not always easily achievable. The approach to optimization under uncertainty here described is an evolution of what presented in [4, 5], with the introduction of different risk function, namely VaR and CVaR, that can be used in the optimization loop and the capability of evaluating estimation error end confidence intervals within the optimization process.

Risk Measures

In optimization under uncertainty, inevitably, we have to deal with random events, modeled by random variables. This means that the first thing to do to treat the problem from a mathematical point of view is to agree on a way to measure the risk. Therefore, we have to introduce a functional of a random variable vector that quantifies the level of risk:

$$\mathscr{R}(X) \tag{1}$$

Subsequently, we have to decide a level of risk that we consider acceptable, let us say C, bearing in mind that there will inevitably be adverse events. Hence, an inequality is defined

$$\mathscr{R}(X) \leq C \tag{2}$$

that bounds the risk measure chosen. Then, if the random variables representative of the costs depend on a deterministic decision vector $\mathbf{z}$, we are led naturally to define a constrained minimization problem:

$$\min_{\mathbf{z}\in Z\subseteq\mathbb{R}^n} \mathscr{R}_0(X(\mathbf{z})) \quad \text{s. to:} \quad \mathscr{R}_i(X(\mathbf{z})) \leq c_i,\; i = 1, \ldots, m \tag{3}$$

In the next sections, we will further develop this approach and we will introduce multi-objective problems too.

Here, instead, we will concentrate on the nature and definition of risk measures mainly following the approach described in [6]. Let us start with the simplest possible example. Indeed, the most immediate and familiar way to measure risk is to say that, on average, one should have $X \leq C$. In symbols

$$\mu(X) \leq C \quad \longrightarrow \quad \mathscr{R}(X) = \mu(X) = EX \tag{4}$$

Subsequently, if we want to be more stringent, we can add a condition on standard deviation

$$\mu(X) + \lambda\sigma(X) \leq C \quad \longrightarrow \quad \mathscr{R}(X) = \mu(X) + \lambda\sigma(X) \tag{5}$$

or on variance

$$\mu(X) + \lambda\sigma^2(X) \leq C \quad \longrightarrow \quad \mathscr{R}(X) = \mu(X) + \lambda\sigma^2(X) \tag{6}$$

Although attractive for their simplicity, these risk measures have several problems, both from the practical point of view and from the theoretical one. From the theoretical point of view, these last two measures do not meet the criteria of coherence defined in [1]. From the practical point of view, instead, the addition to the mean of a term constituted by the standard deviation or the variance is equivalent to define a constraint on these statistics quantities following a Lagrangian approach. This implies that, to work properly, the (μ, σ) space should be convex. Furthermore, also in this case, imposing a constraint on a given σ level requires the assignment a specific value to λ that is not known a priori.

Value-at-Risk and Conditional Value-at-Risk

Alternatively, we can explicitly assign that the random variable of interest is less than or equal to the desired limit:

$$v^\alpha(X) \leq C, \;\; \alpha \in (0, 1) \quad \longrightarrow \quad \mathscr{R}(X) = v^\alpha(X) \tag{7}$$

where v^α is the α-quantile or value-at-risk (VaR) in the terminology used in financial engineering. More formally, according to [7–9], let X be a random variable and $F_X(y) = \Pr\{X \leq y\}$ be the cumulative distribution function (CDF) of X. Then, the inverse CDF of X can be defined as $Q_X(\gamma) = F_X^{-1}(\gamma) = \inf\{y : F(y) \geq \gamma\}$. Thus, for any $\alpha \in (0, 1)$, the α-VaR of L is univocally defined as

$$v^\alpha = F_X^{-1}(\alpha), \tag{8}$$

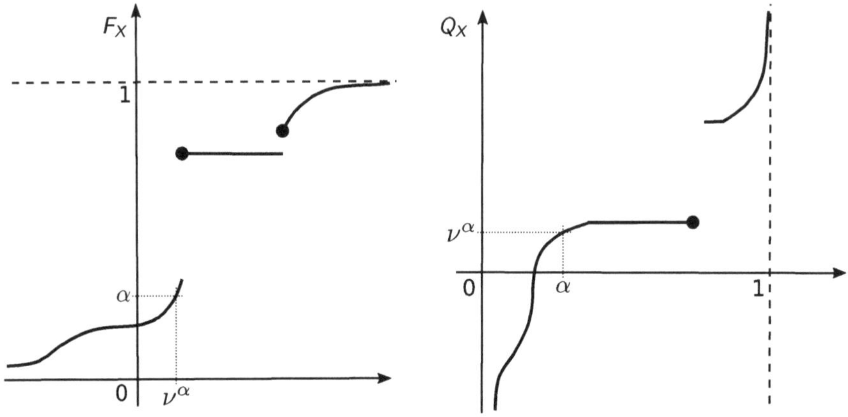

Fig. 1 F_X and its inverse Q_X

and it can be considered the maximum loss that can be exceeded only in a fraction of cases equal to $(1-\alpha)$. The infimum is used in the definition of the inverse CDF because cumulative distribution functions are, in general, weakly monotonic and right-continuous (see [10]). Figure 1 shows a cumulative distribution, F_X, and its inverse, Q_X, in the presence of discontinuities. It is clear that, once known and uniquely defined the inverse, the quantile (VaR) is immediately defined as a function of its α-probability. From a practical standpoint, VaR offers several advantages compared to the measures defined using mean and variance. In particular, the use of quantiles (VaR) allows us to define the risk functional directly in terms of the cumulative distribution function. This, as we shall see in the following sections, allows us to approach with ease and immediacy reliability-based optimization problems. However, from a theoretical point of view, VaR, despite being superior in many ways to traditional risk measures, does not respect all the axioms necessary to qualify as a coherent measure of risk [1, 11].

On the contrary, the conditional value-at-risk (CVAR), while maintaining the positive features of VaR, also respects all theoretical criteria to qualify as a coherent measure of risk. The α-CVaR of X can be thought of as the conditional expectation of losses that exceed the ν^α level and can be expressed as

$$c^\alpha = \frac{1}{1-\alpha}\int_\alpha^1 \nu^\beta \mathrm{d}\beta \tag{9}$$

as it was first outlined by Acerbi in [12]. In Fig. 2, the area representing the integral in formula (9) is highlighted. References [2, 3, 13] show that c^α can also be written as the following stochastic program:

$$c^\alpha = \inf_{t\in\mathbb{R}} \left\{ t + \frac{1}{1-\alpha} E[X-t]^+ \right\} \tag{10}$$

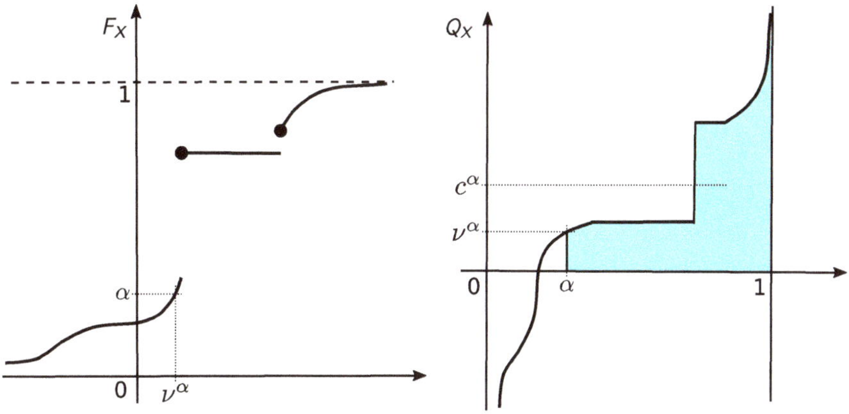

Fig. 2 Quantile and α-tail

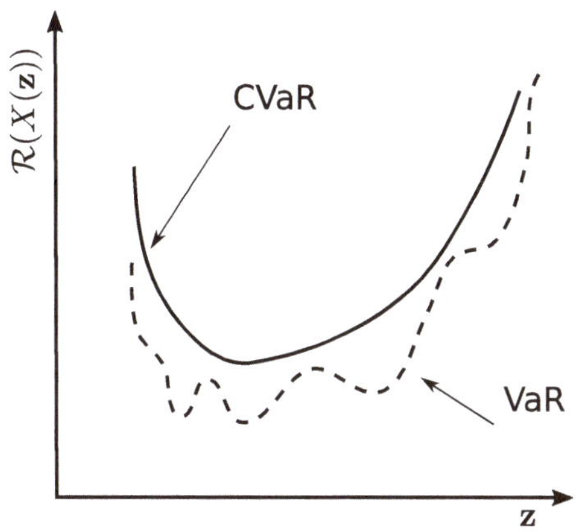

Fig. 3 CVaR preserves convexity

with $[a]^+ = \max\{0, a\}$. The set of optimal solutions to the stochastic program (10) is, according to [9], $T = [\nu^\alpha, u^\alpha]$ with $u^\alpha = \sup t : F(t) \leq \alpha$. In particular, $\nu^\alpha \in T$, so

$$c^\alpha = \nu^\alpha + \frac{1}{1-\alpha} E[X - \nu^\alpha]^+ \tag{11}$$

When X has a positive density in the neighborhood of ν^α, then $\nu^\alpha = u^\alpha$. Therefore, the stochastic program defined in Eq. (10) has a unique solution, and

$$c^\alpha = E[X | X \geq \nu^\alpha] \tag{12}$$

with $E[X | X \geq \nu^\alpha]$ also known as expected shortfall or tail conditional expectation.

CVaR satisfies all the axioms that define a coherent measure and, above all, preserves convexity, as it can be qualitatively appreciated from Fig. 3.

This implies that if the original optimization problem is convex, its transformation obtained using CVaR is still convex and, hence, all the powerful machinery developed for the solution of convex optimization problems can still be exploited.

VaR and CVaR Estimation

In most cases of our interest, the value of the risk functions must be estimated from a finite number of samples. In this work, the *Empirical Cumulative Distribution Function* (ECDF) and the *Weighted Empirical Cumulative Distribution Function* (WECDF) are used for this purpose. The definitions of ECDF [14] and WECDF are reported here for the sake of completeness.

Empirical Cumulative Distribution Function Definitions

We preliminarily recall that, as usual, the probability space is defined using the classical tern $(\Omega, \mathscr{F}, \mu)$ consisting of a sample space Ω of possible outcomes, a σ-algebra $\mathscr{F} \subseteq 2^{\Omega}$, collection of all the considered events (where each event is a set containing zero or more outcomes), and a probability measure function $\mu : \mathscr{F} \rightarrow [0, 1]$ that assigns probabilities to the events.

Definition 1 ECDF, Multivariate Let $X : \Omega \mapsto \mathbb{R}^d$ a random variable, $\mathbf{x}^i = (x_1^i, \dots, x_d^i)$ a random sample of X, μ the defined probability measure, and $\boldsymbol{\alpha} = (\alpha_1, \dots, \alpha_d)$ a generic vector in $\mathbb{R}^d$. The multivariate "empirical distribution function" is defined, for n samples $\{\mathbf{x}^1, \dots, \mathbf{x}^n\}$, as

$$\hat{F}_{\mu}^{n}(\boldsymbol{\alpha}) = \frac{\text{number of elements in the sample} \leq \boldsymbol{\alpha}}{n} = \frac{1}{n} \sum_{i=1}^{n} \mathbb{1}\{\mathbf{x}^i \leq \boldsymbol{\alpha}\} \tag{13}$$

where $\mathbb{1}\{A\}$ is the indicator function of event A:

$$\mathbb{1}_A(x) := \begin{cases} 1 & \text{if } x \in A, \\ 0 & \text{if } x \notin A, \end{cases} \tag{14}$$

and $\mathbf{x}^i \leq \boldsymbol{\alpha}$ meaning $x_j^i \leq \alpha_j, \ \forall j = 1, \dots, d$. Note that the last relation defines a partial order and, if it is true, then $\mathbf{x}^i$ is either dominated by $\boldsymbol{\alpha}$ or equal to it.

Definition 2 WECDF, Multivariate As a first step, the previous definition is extended introducing a variable jump w_i so that, instead of having a step function that jumps up of a fixed quantity related to the data point number, we have:

$$\hat{F}^{n}_{\mu;\mathbf{w}}(\alpha) = \sum_{i=1}^{n} w_i \mathbb{1}\{\mathbf{x}^i \leq \alpha\} \tag{15}$$

with

$$\sum_{i=1}^{n} w_i = 1$$

This weighted ECDF definition will be used to change the probability measure (target measure ν) and to approximate the CDF that would have been obtained according to this new measure $F_{\nu}(\alpha)$.

From now on, therefore, when the optimization algorithm will require the calculation of the $F_Q(s)$, it will use instead its estimator $\hat{F}^{n}_{\mu;\mathbf{w}}(s)$, where n indicates the number of samples used to estimate this ECDF (WECDF).

Note that each indicator function, and hence the ECDF (WECDF), is itself a random variable. This is a very delicate issue to consider. Indeed, if the EDCF (WECDF) is used to approximate the deterministic operator $Q(s)$, a direct residual influence of the random variables that characterize the system under investigation remains on $P_{Q(s)}$. In other words, $Q(s)$ behaves as a random variable, but with the important difference that its variance tends to zero when the ECDF (WECDF) approximates the CDF with increasing precision. It is possible to demonstrate that the estimator $\widehat{F}_{Q_n}(s)$ is consistent, as it converges almost surely to $F_Q(s)$ as $n \to \infty$, for every value of s [15]. Furthermore, for the Glivenko–Cantelli theorem [16], the convergence is also uniform over s. This implies that, if the ECDF is calculated with sufficient accuracy, it can be considered and treated as a deterministic operator. On the other hand, if the number of samples, or the estimation technique of the ECDF, do not allow to consider it as such, one can still correlate the variance of the ECDF with the precision of the obtained estimate. Of course, if the ECDF is estimated in a very precise way, it is possible to use for the optimization also an algorithm conceived for deterministic problems, provided that it has a certain resistance to noise. Conversely, if the ECDF is obtained from a coarse sample, its practical use is only possible with optimization algorithms specifically designed.

Value-at-Risk and Conditional Value-at-Risk Computation

For the sake of simplicity, VaR and CVaR estimation procedures will be given only for the univariate distribution case. If we have a finite number of samples, that is, a ECDF, we proceed as described below. If $x_1, x_2, \ldots, x_n$ are n independent and identically distributed (i.i.d.) observations of the random variable X, then the α-VaR of X can be estimated by

$$\hat{v}^{\alpha;n} = X_{\lceil n\alpha \rceil : n} = \hat{F}_n^{-1}(\alpha) \tag{16}$$

where $X_{i:n}$ is the ith order statistic from the n observations, and

$$F_n(t) = \sum_{i=1}^{n} \mathbb{1}\{x_i \leq t\} \tag{17}$$

is the empirical CDF constructed from the sequence $\tilde{X}$ of $x_1, x_2, \dots, x_n$, $\mathbb{1}\{\cdot\}$ is the indicator function, and t is a scalar value (conversely to the case of Eqs. 13 and 15). When a weighted empirical distribution is used, instead, α-VaR can be estimated, for a given probability level α, using the formula reported below

$$\sum_{k=1}^{k_\alpha} w_k \geq \alpha > \sum_{k=1}^{k_\alpha - 1} w_k \tag{18}$$

to compute the index k_α, from which $\hat{v}^{\alpha;n}$ is directly obtained:

$$\hat{v}^{\alpha;n} = x_{(k_\alpha)} \tag{19}$$

To evaluate the α-CVaR of X, instead, Ref. [9] suggests to use the estimator

$$\hat{c}^{\alpha;n} = \inf_{t \in \mathbb{R}} \left\{ t + \frac{1}{n(1-\alpha)} \sum_{i=1}^{n} [x_i - t]^+ \right\} \tag{20}$$

or, equivalently

$$\hat{c}^{\alpha;n} = \inf_{t \in \mathbb{R}} \left\{ t + \frac{1}{1-\alpha} E[\tilde{X} - t]^+ \right\} \tag{21}$$

where the ECDF of $\tilde{X}$ is F_n. Using Eq. (11), we have a direct estimate of c^α:

$$\hat{c}^{\alpha;n} = \hat{v}^{\alpha;n} + \frac{1}{n(1-\alpha)} \sum_{i=1}^{n} [x_i - \hat{v}^{\alpha;n}]^+ \tag{22}$$

When the WECDF is used, instead, we have:

$$\hat{c}^{\alpha;n} = \frac{1}{1-\alpha} \left[\left(\sum_{k=1}^{k_\alpha} w_k - \alpha \right) x_{(k_\alpha)} + \sum_{k=k_\alpha+1}^{n} w_k x_{(k)} \right] \tag{23}$$

Robust Optimization Algorithm

The robust optimization problem (3) has to be redefined in terms of estimates of the risk functions in order to be numerically solved. Therefore, extending also to multi-objective problems, we have:

$$\min_{\mathbf{z}\in Z} \hat{\rho}_{i;n}(\mathbf{z})\ i = 1, \dots, p \quad \text{s. to:}\ \ \hat{\rho}_{i;n}(\mathbf{z}) \leq c_i\ i = p+1, \dots, p+q \tag{24}$$

where $\hat{\rho}_{i;n}$ is an estimator of the generic risk measure $\mathscr{R}_i$ obtained using a sample of size n. The constraints are also assigned considering a set of inequalities defined in terms of q further risk measure estimates.

The quality of the risk function estimate directly influences the optimization results. Therefore, the following guidelines should be used in building a robust optimization procedure:

- try to use the largest number of samples compatible with the computational budget in the estimation of the risk function;
- use advanced sampling techniques such as multi-level Monte Carlo, or other importance sampling methods;
- use optimization algorithms not quite sensitive to noise since, from the point of view of the optimization algorithm, the higher or lower quality of an estimate can be seen, in the broad sense, as noise level that influences the value of objective functions and constraints;
- use of computational statistics methods to evaluate estimate accuracy and confidence intervals; here in particular, the bootstrap method is used, which has the important characteristic of being a nonparametric method [17].

These issues, with varying degrees of importance, have been present in all the literature related to the optimization under uncertainty. The stochastic nature of risk function estimates has led the research related to the multi-objective method to several extensions of the classical Pareto front concept. In [18], for example, the Pareto front exploration in the presence of uncertainties is faced introducing the concept of *probabilistic dominance*, which is an extension of the classical *Pareto dominance*, while in [19], a probabilistic ranking and selection mechanism is proposed that introduces the *probability of wrong decision* directly in the formula for rank computation. An interesting approach, similar in some aspects to the one here described, is found in [20] where a quantile-based approach is coupled with the probability of *Pareto nondominance* (already seen in [19]). Here, contrary to the cited work, the optimization technique introduced relies on direct estimation of the risk functions obtained through the ECDF or the WECDF and of their confidence intervals computed using the nonparametric bootstrap method.

However, it should be emphasized that the use of confidence intervals within an optimization algorithm can be misleading, given that, as is well known, they are not a measure of the degree of probability that an unknown parameter value lies in a specific interval [21]. The confidence intervals are functions of random variables and therefore are themselves random variables. consequently, their use to evaluate the difference of unknown statistical parameters (in our case VaR and CVaR) related to different distributions must be done with great caution and taking into account that two confidence intervals with a large overlap might not provide any information on the actual difference of the unknown parameter in the two different distributions.

Another important point is that the stochastic nature of the sampling for risk function estimation makes sure that an evolutionary optimization algorithm tends to bring out "pathological samples," that is, samples that give an underestimation of the risk function in minimization problems. The simplest remedy, when it is not possible to refine the sampling, is to make sure that a pathological sampling does not extend its effect beyond a single evolutionary algorithm generation. Consequently, if it is possible, elitism should not be used so that the effect of a single estimate is limited to the current generation. If elitism has to be used, it is necessary to change the activation mechanism taking into account the confidence intervals.

Change of Probability Measure

The robust optimization algorithm here presented relies on the use of risk measures, such as VaR and CVaR, as objectives, and one of the main problems posed by this approach is that obtaining an adequate sampling of the random variables that govern the risk function has very often prohibitive computational costs. Therefore, the need often arises to obtain good estimates of the risk function during the optimization process even in the presence of few samples that have not been generated according to the true distribution of the input random variables. Hence, we have to deal with the problem of computing the weights of a Weighted Empirical Cumulative Distribution Function (WECDF) in order to obtain a suitable change of probability measure. The approach here presented is strictly related to what described in [22, 23].

Empirical Importance Weights Computation

The change of probability measure is obtained through an optimization process aimed to assign suitable values to the weights of the WECDF. We will refer this process as "Empirical Importance Weights Computation." A suitable objective function is chosen:

$$\omega_m^2(\mathbf{w}) = \frac{1}{2}\sum_{k=1}^{m}\left(\hat{F}^n_{\mu;\mathbf{w}}(\mathbf{t}^k) - F_\nu(\mathbf{t}^k)\right)^2 \tag{25}$$

with $\{\mathbf{t}^1, \ldots, \mathbf{t}^m\}$ a properly chosen set of sample points. Using the definition of the weighted ECDF, we obtain:

$$\omega_m^2(\mathbf{w}) = \frac{1}{2}\sum_{k=1}^{m}\left(\sum_{i=1}^{n} w_i \mathbb{1}\{\mathbf{x}^i \le \mathbf{t}^k\} - F_\nu(\mathbf{t}^k)\right)^2 \tag{26}$$

that considering also the constraints associated with the weights leads to the following optimization problem:

$$\begin{aligned} \mathbf{w} = \arg\min_{\mathbf{w}} \; & \omega^2(\mathbf{w}) \\ \text{s. to: } & w_i \geq 0, \qquad i = 1, \ldots, n \\ & \sum_{i=1}^{n} w_i = 1 \end{aligned} \tag{27}$$

As $F^n_{\mu;\mathbf{w}}$ is a linear function with respect to the parameter vector $\mathbf{w}$, it can be solved using a nonnegative least squares algorithm, like the Lawson and Hanson algorithm [24], or transforming it into a convex quadratic programming problem [22]. Choosing $\mathbf{t} = \mathbf{x}^i$, we obtain the following linear equation system:

$$\begin{cases} \sum_{i=1}^{n} w_i & = 1 \\ \sum_{i=1}^{n} w_i \mathbb{1}\{\mathbf{x}^i \leq \mathbf{x}^k\} = F_\nu(\mathbf{x}^k) & \forall k \in 1, \ldots, n \end{cases} \tag{28}$$

This formulation, however, may lead to an ill-posed problem when the dimension of the random vector $\mathbf{x}$ increased. Alternatively, the system can be obtained considering a random hyperrectangle H_k strictly contained in the unit cube $H_k \subset U = [0, 1]^n$ and $\nu(\cdot)$ a function that represents the probability that the random variable vector $\mathbf{x}$ takes on a value contained in H_k. So we have

$$\begin{cases} \sum_{i=1}^{n} w_i & = 1 \\ \sum_{i=1}^{n} w_i \mathbb{1}\{\mathbf{x}^i \in H_k\} = \nu(H_k) & \forall k \in 1, \ldots, K \end{cases} \tag{29}$$

Note that if the cuboid is either void or contains all the points, the resulting equation should be discarded as it could lead to nonconsistent results. Alternatively, it is possible to discard the equation with less than, let us say, q points and more than $n - q$ points. Figure 4 reports some random hypercuboid examples that are used to individuate the points needed to build the K rows of system (29). The figure illustrates, in a case in which the random vector has two components, the process by which the random rectangles are generated. Note that the rectangles can intersect.

Problem (29) is characterized by a linear equality, namely the first equation, that should hold. Several methods are available to force this constraint either exactly or approximately. Here, this is approximately (but satisfactorily) accomplished by heavily weighting the first equation and solving the resulting nonnegative least squares system [24].

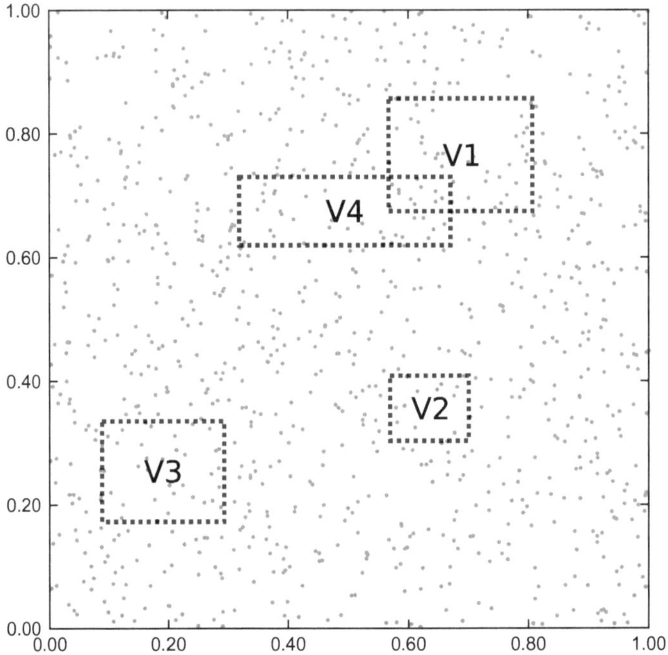

Fig. 4 Example of random cuboids used to build the equations of system (29)

Example

Figure 5 reports the result obtained to restore the proper CDF related to the function:

$$f = \frac{1}{n} \sum_{i=1}^{n} \left(2\pi - u_i\right) \cos\left(u_i - d_i\right) \tag{30}$$

with $\mathbf{u} \in [0, 3]^n$, $\mathbf{d} \in [0, 2\pi]^n$, and $n = 6$. The probability measure was changed to let the random vector $\mathbf{u}$ have components with a uniform distribution function.

100,000 random hyperrectangle constraints were generated (of which 85,548 were usable because the cuboid contained at least one point and not all the available points).

Limits of the WECDF Correction Process

Although very effective, the technique of the empirical importance weights has significant limitations, especially related to the fact that the empirical cumulative distribution is defined by a set of points. Indeed, the demonstration of the well-posedness

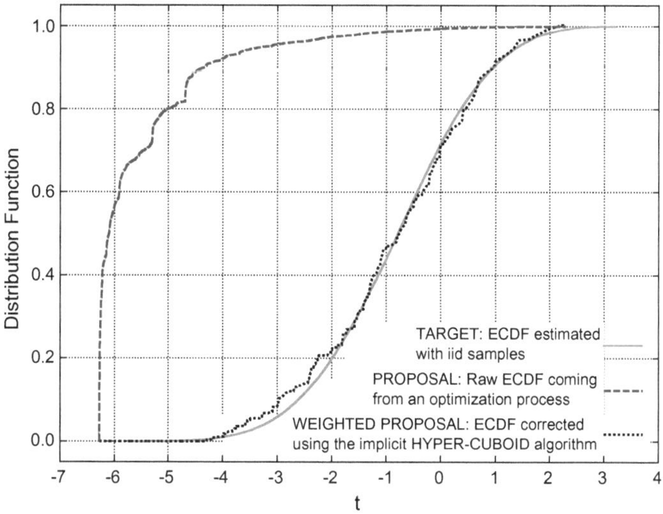

Fig. 5 Change of probability measure example

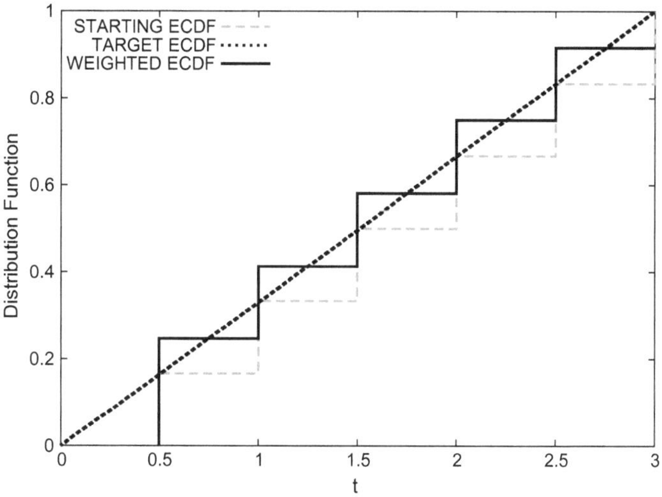

Fig. 6 Univariate uniform distribution approximated using a very coarse WECDF

of the optimization problem is an open question, and if the result seems quite good in the case of empirical distributions defined with many samples, on the other hand a starting distributions very coarse may result in a barely acceptable weighted ECDF. This is the case, for example, reported in Fig. 6. The results obtained with a denser sampler, as those reported in Fig. 7, appear much more acceptable, even though the value of many weights is equal to zero at the end of the process.

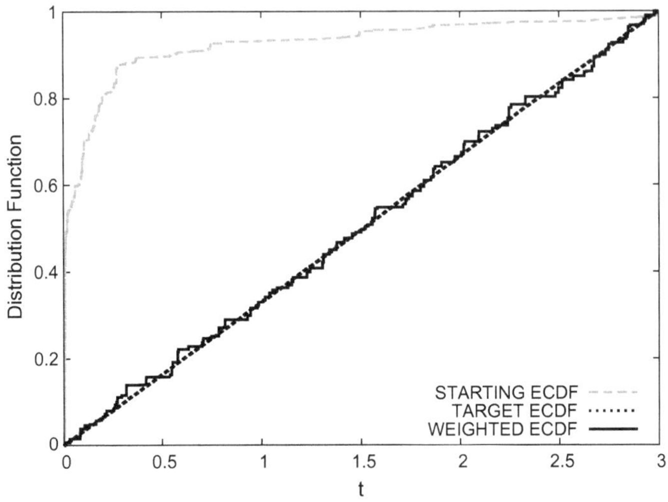

Fig. 7 Uniform distribution approximated using a dense WECDF

Bootstrap Error Analysis

The results of any optimization algorithm for robust or reliable design depend in an essential way on the quality of estimation of the risk measures to be minimized. In the approach here presented, these risk functionals rely on the evaluation of the ECDF or the WECDF estimators. This leads in a natural way to deal with two issues: how to evaluate the quality of the risk functional estimates used in the multi-objective optimization problem, and how to possibly get better estimates with a given computational effort. Related to these two points, however, there are other problems too: Is it possible to conceive an algorithm that can account for the error in the risk estimates? Is it possible to find estimators of ECDF that give the required accuracy in the computation of the risk measures of interest for the optimization problem?

The approach here proposed for error assessment and estimation is based on the bootstrap method introduced by Efron in 1977 [25, 26]. This method represents a major step forward in the statistical practice because it allows to accurately assess the variability of any statistical estimator without making any assumption about the type of distribution function involved. Suppose that a statistic

$$T(X) = T\left(x_1, x_2, \dots, x_n\right) \tag{31}$$

is given, evaluated on a set of data

$$X = \left\{x_1, x_2, \dots, x_n\right\} \tag{32}$$

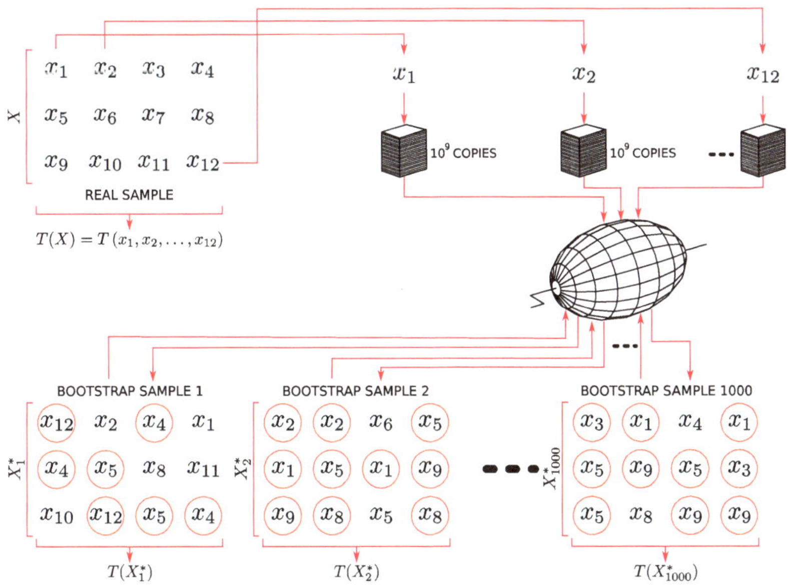

Fig. 8 Bootstrap error analysis

belonging to an assigned space X. The bootstrap essentially consists in repeatedly recalculating the statistic T employing a tuple of new samples

$$X^* = \left\{x_1^*, x_2^*, \ldots, x_n^*\right\} \tag{33}$$

obtained by selecting them from the collection $\left\{x_1, x_2, \ldots, x_n\right\}$ with replacement. The repeated calculation of $T\left(x_1^*, x_2^*, \ldots, x_n^*\right)$ gives a set of values that provides a good indication on the quality of the distribution of T.

Therefore, to calculate the accuracy of a generic risk measure, like v^α or c^α, obtained using the estimators $\hat{v}^{\alpha;n}$ or $\hat{c}^{\alpha;n}$, the bootstrap procedure can be applied to the samples used to compute the estimator.

Figure 8 illustrates graphically the process of bootstrapping. Suppose we have a series of random data made by a set of numbers, such as the realizations of a random variable and another quantity, our quantity of interest θ, related to the first. The original observations related to these pairs appear in the table at the top left of the figure. Let us suppose then to make many copies of each of these random variable realizations data and to put them all in a lottery number shaker from which then we will extract our couples randomly, taking care to return the numbers in shaker at the end of each series of extractions. Each new set must contain the same number of samples of the original, but, by virtue of the sampling mechanism chosen, some realizations will be present more than once, while others will be absent in the new series. The tables at the bottom in the figure represent this process and the realizations selected more than once are highlighted using a circle.

Hence, if T is the estimator of the quantity of interest θ, we say $T(X) = \hat{\theta}$, for a given sample X. For each bootstrap sample X_i^*, it is possible to compute the related estimator value $T(X_1^*), \dots, T(X_n^*)$. In this way, we obtain n different estimates of θ that can be used to calculate the "bootstrap mean," the "bootstrap variance," the "percentiles bootstrap," etc., which are approximations of the corresponding unknown values of the quantity of interest, and carry information about the distribution of $T(x)$. Starting then from these estimated quantities, it is possible to calculate standard error, confidence intervals, to test hypotheses, etc.

Different methods and algorithms are available to compute the confidence intervals from the bootstrap distribution. We report here, for the sake of completeness, the two simpler and more commonly used approaches, namely the "basic bootstrap" and the "percentile bootstrap." The basic bootstrap uses in a straightforward way the empirical quantiles obtained from the bootstrap distribution related to the estimator of the quantity of interest:

$$\left(2\hat{\theta} - \hat{\theta}^*_{(1-\alpha/2)}; 2\hat{\theta} - \hat{\theta}^*_{(\alpha/2)}\right) \tag{34}$$

where $\hat{\theta}^*_{(1-\alpha/2)}$ is the $1 - \alpha/2$ percentile of the bootstrapped coefficients $\hat{\theta}^*$. The percentile bootstrap uses again the quantiles, but with a different formulation:

$$\left(\hat{\theta}^*_{(\alpha/2)}; \hat{\theta}^*_{(1-\alpha/2)}\right) \tag{35}$$

For further details and more advanced techniques, see [17, 27, 28].

VaR and CVaR Error Estimation Example

In the following section, an example of application of the bootstrap technique is illustrated. This example is related to the characterization of the standard error of VaR and CVaR calculated on an empirical distribution quite densely defined (21105 samples), but with a few jumps in the upper tail. This distribution is shown in Fig. 9.

The values of VaR and CVaR are calculated according to formulas (18) and (23) and are reported in Table 1.

Figures from 10, 11, 12 and 13 report the bootstrap probability density distributions for VaR and CVaR at $\alpha = 0.90$ and $\alpha = 0.99$ alongside the quantile–quantile

Table 1 VaR and CVaR values at some α values in the upper CDF tail

α	VaR	CVaR
0.90	0.101067	0.1268328
0.99	0.203824	0.2441841

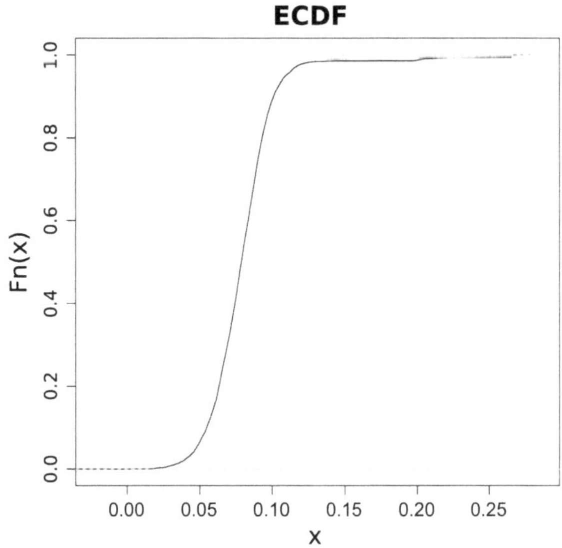

Fig. 9 Reference empirical distribution for the VaR and CVaR error estimation example

plot that compares their cumulative distribution with the normal cumulative distribution. 10,000 bootstrap iterations were used to generate these data.

The analysis of these figures shows that, at least for this particular distribution, CVaR has a more regular, smooth and next to the normal distribution than VaR. However, this apparent advantage, especially evident in the tails, not always corresponds to more favorable values for bias and standard error.

Figure 14 compares the VaR and CVaR plot related to the example distribution. Finally, Fig. 15 is probably the most significant for our purposes. It was obtained by extracting 1,000 samples randomly from the original data set and performing the bootstrap procedure on these samples for $\alpha = 0.99$. The values of VaR and CVaR obtained from these subdistributions are then shown on the charts alongside with the related bootstrap standard error reported as errorbar. The comparison of these intervals with the estimate obtained from the full sample is indicative of the error related to a very coarse sampling. The comparison between the results obtained for VaR and CVaR shows, on the whole, less variability and greater consistency of results obtained with CVaR when the number of available samples decreases.

Optimization Example

The function reported in Table 2, taken from [29], is used as a benchmark to test the VaR-/CVaR-based approach to robust optimization. With respect to the function reported in the reference, the following changes have been introduced: The ranges of design and uncertain parameters have been changed as reported in table, and a

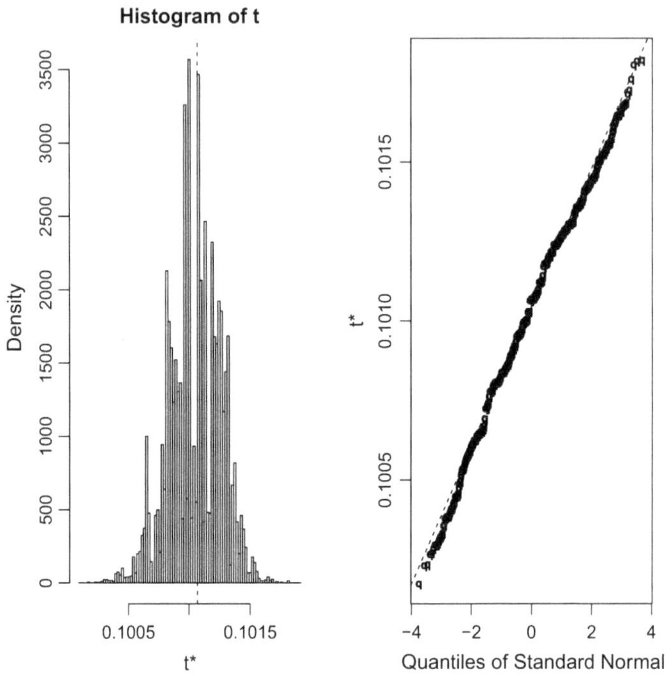

alpha	VaR	bias	std. error
0.90	0.101067	-2.515364e-05	0.00022092

Fig. 10 Bootstrap distribution for VaR risk function at $\alpha = 0.90$ and related quantile–quantile plot

multiplicative factor equal to $1/n$ has been introduced to make easier the result comparison when the dimension of the parameter space changes. The random variables $\mathbf{u}$ have a uniform distribution function. Table 3 reports the solutions to the optimization problems

$$\min \nu^0 = \min_{\mathbf{d}\in D, \mathbf{u}\in U} f(\mathbf{d}, \mathbf{u})$$

$$\min \nu^1 = \min_{\mathbf{d}\in D} \max_{\mathbf{u}\in U} f(\mathbf{d}, \mathbf{u})$$

over the Cartesian product of D and U. The first problem represents the best possible solution obtainable if the $\mathbf{u}$ are considered as design parameters varying in U. The second one, instead, minimizes the maximum possible loss or, alternatively, maximizes the minimum gain, according to the framework of decision theory [30]. These solutions have been obtained analytically and verified by exhaustive search for $n = 1$. It is worth to note that these particular optimal solutions are the same whatever is the dimension of the search space.

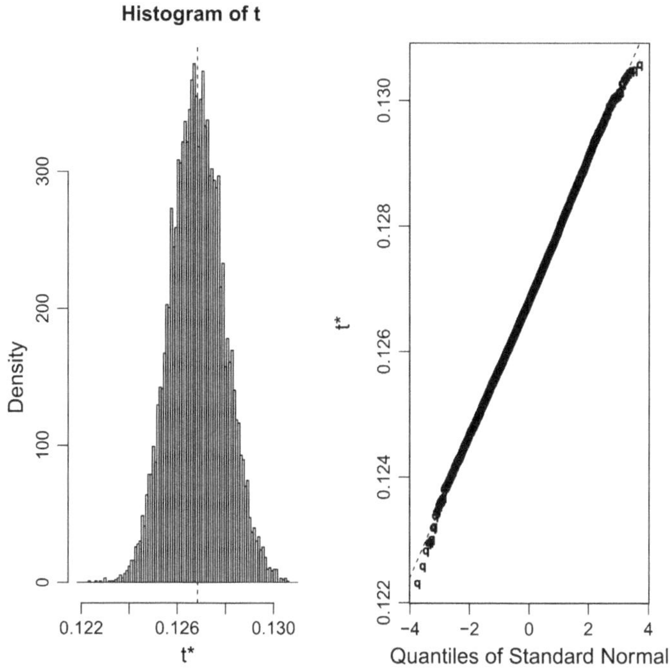

alpha	CVaR	bias	std. error
0.90	0.1268328	1.153863e-05	0.00109707

Fig. 11 Bootstrap distribution for CVaR risk function at $\alpha = 0.90$ and related quantile–quantile plot

The optimization examples reported here are instead related to the minimization of $\nu^{0.9}$ and $c^{0.9}$, whose numerical solutions are quite near to the solution of $\min \nu^1$, although not identical. The optimization method used here is a simple genetic algorithm based on local random walk selection [31, 32]. Crossover operator is the classical *one-point* crossover which operates at bit level, and also mutation operator works at bit level. Therefore, the real variable vector is transformed into a bit string in which each real variable is encoded using 63 bits. Gray coding is used to smooth the effect of bit mutation and crossover. Finally, no elitism is used in the optimization runs. Two different optimization runs have been performed, the first one using $\nu^{0.9}$ as objective function and the second one using instead $c^{0.9}$. The dimension of design variable vector and uncertain parameter vector was set to $n = 6$. Each run was divided into two stages: In the first one, the ECDF used to estimate the risk function was computed using 100 Monte Carlo samples, while in the second stage 1000 samples were used. The population size was set to 64 elements for each stage, and the last population of the first stage was used to initialize the population of the second one. The number of generations was set to 10 for the first stage, while it was set to 20 for the second one.

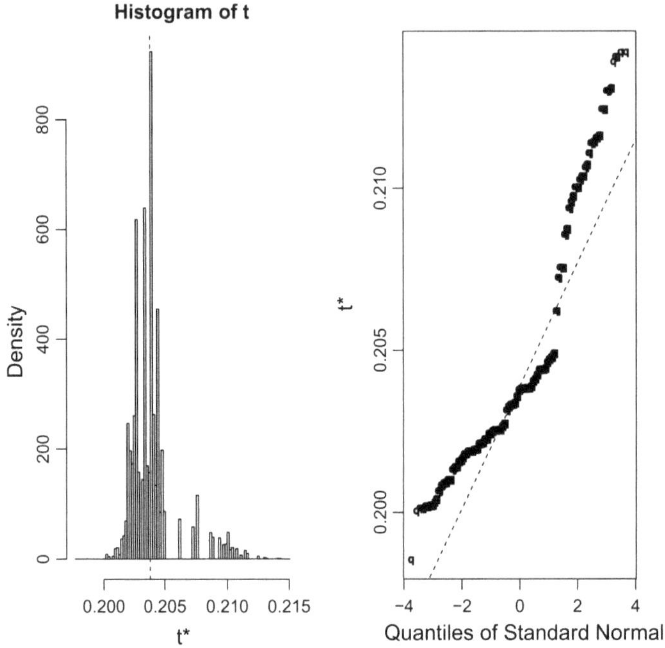

alpha	VaR	bias	std. error
0.99	0.203824	6.6676e-06	0.0008973185

Fig. 12 Bootstrap distribution for VaR risk function at $\alpha = 0.99$ and related quantile–quantile plot

In every stage of all the optimization runs, the crossover probability was set to 80% and the mutation rate to 1%. Figure 16 reports the evolution history related to the $v^{0.9}$ (VaR)-based optimization run, while Fig. 17 is related to the $c^{0.9}$ (CVaR) run. As can be easily observed, the trend of the two runs is very similar, and similar results are obtained in terms of optimal values, as reported in Table 4.

Of course, two single runs cannot be used to draw statistically significant conclusions about the quality of one approach over the other, and an in-depth analysis of the advantages of a risk function with respect to another must necessarily consider a systematic benchmark activity on many different objective functions and problem sizes. This aspect will be further investigated in future work.

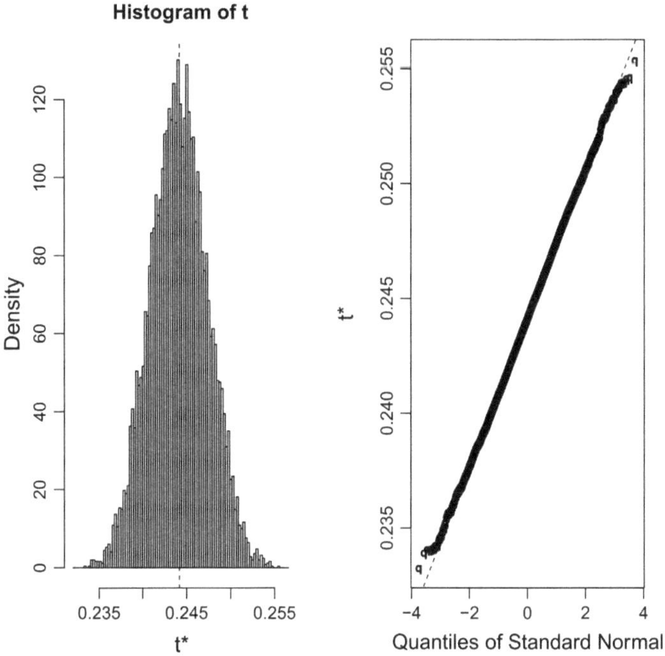

alpha	CVaR	bias	std. error
0.99	0.2441841	0.001955025	0.004689159

Fig. 13 Bootstrap distribution for CVaR risk function at $\alpha = 0.99$ and related quantile–quantile plot

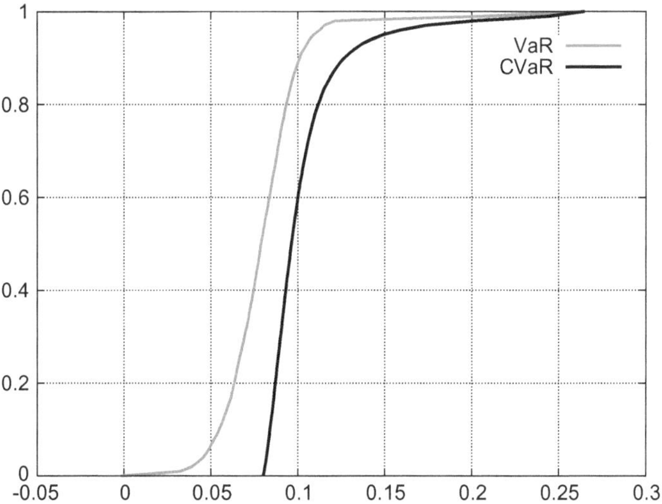

Fig. 14 VaR versus CVaR plot comparison

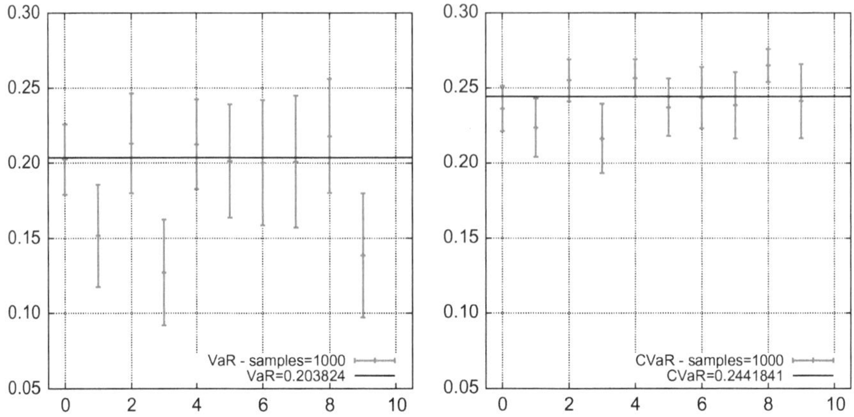

Fig. 15 VaR versus CVaR bootstrap standard error comparison

Table 2 Benchmark function table

ID	Function	Ranges
MV4	$f = \frac{1}{n}\sum_{i=1}^{n}(2\pi - u_i)\cos(u_i - d_i)$	$\mathbf{u} \in [0,3]^n, \mathbf{d} \in [0, 2\pi]^n$

Table 3 Benchmark functions table v^0 and v^1 optimal values

ID	$v^0 = \min_{\mathbf{d}\in D, \mathbf{u}\in U} f(\mathbf{d},\mathbf{u})$			$v^1 = \min_{\mathbf{d}\in D} \max_{\mathbf{u}\in U} f(\mathbf{d},\mathbf{u})$		
	d	**u**	f	**d**	**u**	f
MV4	$[3.1416]^n$	$[0]^n$	−6.283185 ...	$[4.6638]^n$	$[0]^n$	−0.305173 ...

Conclusions

Robust-based optimization and reliability-based optimization are very challenging from the computational point of view. Consequently, the introduction of these techniques in the industrial practice cannot be separated from finding new methods that significantly enhance computational efficiency and the ability to exploit with maximum effectiveness the information produced during the optimization process. The philosophy of the analysis carried out in this work can be effectively summarized in what written above. The approach proposed here has its roots in the introduction of risk measures appropriate to the problem in question and, above all, with characteristics, in terms of information conveyed, more favorable than most classical and well-established techniques.

Alongside to this "rational" and, in a sense, axiomatic definition of the concept of risk, very present in the literature to date, there are also in this work a whole series of techniques designed to improve the computational efficiency of the opti-

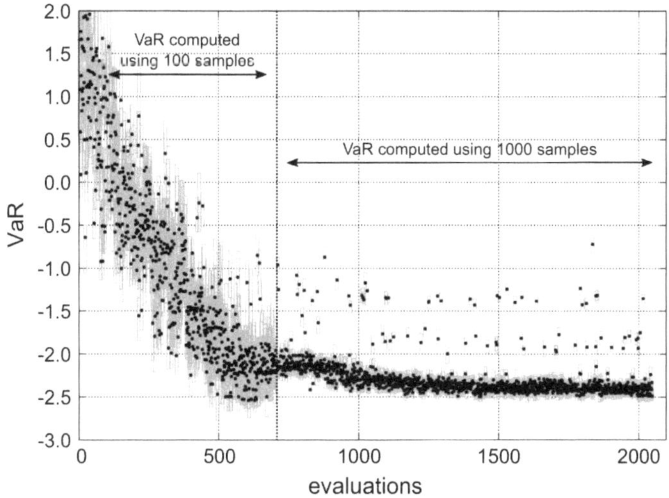

Fig. 16 Evolution history for the VaR-based two stage optimization run

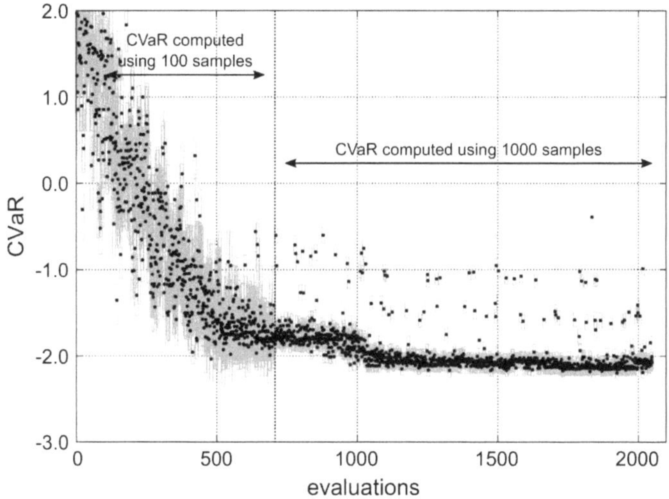

Fig. 17 Evolution history for the CVaR-based two stage optimization run

Table 4 VaR and CVaR runs optimal results

ID	**d**						$\nu^{0.9}$	$c^{0.9}$
VaR run	4.460	4.617	4.546	4.473	4.657	4.565	−2.495	−2.156
CVaR run	4.618	4.528	4.497	4.473	4.645	4.557	−2.461	−2.197

mization process. We started from the extensive adoption of the cumulative probability distribution function paradigm and the introduction of two empirical estimators, namely ECDF and WECDF, which are the basis for the calculation of risk functions in the optimization process. Then, we introduced two risk measures, namely VaR and CVaR, whose benefits are reported and illustrated, with reference to current literature. Subsequently, we dealt with the technique of change of probability measure which allows, with relative ease, to estimate correctly the risk functions, even when it is not possible to use the correct distributions in the input random parameters, because, for example, the space of random parameters was explored using importance sampling or optimization techniques. Finally, we introduced the "bootstrapping" computational statistics technique to interactively estimate the standard errors and confidence intervals during the optimization process, even when the input distributions are not known or not specified, or when they are too coarse. All of the techniques described above were blended and integrated into an optimization algorithm whose preliminary results, here reported as example, seem to confirm the advantages of this approach compared to more traditional methods. Further developments are underway to better assess and demonstrate the benefits and possible limitations of this methodology.

References

1. Artzner, P., Delbaen, F., Eber, J.M., Heath, D.: Coherent measures of risk. Math. Financ. **9**(3), 203–228 (1999)
2. Rockafellar, R.T., Uryasev, S.: Optimization of conditional value-at-risk. J. Risk **2**, 21–41 (2000)
3. Rockafellar, R.T., Uryasev, S.: Conditional value-at-risk for general loss distributions. J. Bank. Financ. **26**, 1443–1471 (2002)
4. Quagliarella, D., Petrone, G., Iaccarino, G.: Optimization under uncertainty using the generalized inverse distribution function. In: Fitzgibbon, W., Kuznetsov, Y.A., Neittaanmäki, P., Pironneau, O. (eds.): Modeling, Simulation and Optimization for Science and Technology of Computational Methods in Applied Sciences, June 2014, vol. 34, pp. 171–190. Springer, Netherlands, Print ISBN 978-94-017-9053-6, Online ISBN 978-94-017-9054-3 (2014)
5. Quagliarella, D., Petrone, G., Iaccarino, G.: Reliability-based design optimization with the generalized inverse distribution function. In Greiner, D., Galván, B., Périaux, J., Gauger, N., Giannakoglou, K., Winter, G. (eds).: Advances in Evolutionary and Deterministic Methods for Design, Optimization and Control in Engineering and Sciences of Computational Methods in Applied Sciences. vol. 36, pp. 77–92. Springer International Publishing, ISBN 978-3-319-11540-5 (2015)
6. Rockafellar, R.T., Uryasev, S.: The fundamental risk quadrangle in risk management, optimization and statistical estimation. Surv. Oper. Res. Manag. Sci. **18**(1–2), 33–53 (2013)
7. Hong, L.J., Liu, G.: Monte carlo estimation of value-at-risk, conditional value-at-risk and their sensitivities. In: Proceedings of the 2011 Winter Simulation Conference (WSC), Dec 2011, pp. 95–107 (2011)
8. Hong, L.J., Hu, Z., Liu, G.: Monte carlo methods for value-at-risk and conditional value-at-risk: a review. ACM Trans. Model. Comput. Simul. **24**(4), 22:1–22:37
9. Trindade, A.A., Uryasev, S., Shapiro, A., Zrazhevsky, G.: Financial prediction with constrained tail risk. J. Bank. Financ. **31**(11), 3524–3538 (2007)

10. Wikipedia: Cumulative distribution function
11. Ying, L.: A study on risk measures and their applications in the real-world. In: 14th NUROP Congress (2009)
12. Acerbi, C., Tasche, D.: Expected shortfall: A natural coherent alternative to value at risk. Economic Notes **31**(2), 379–388 (2002)
13. Pflug, G.C.: Some remarks on the value-at-risk and the conditional value-at-risk. In: Uryasev, S.P. (ed.) Probabilistic Constrained Optimization: Methodology and Applications, pp. 272–281. Springer, US, Boston, MA (2000)
14. Woo, C.: Empirical distribution function (version 4)
15. van der Vaart, A.W.: Asymptotic Statistics. Cambridge University Press (1998)
16. Serfling, R.J.: Approximation Theorems of Mathematical Statistics. Wiley (2008)
17. Efron, B., Tibshirani, R.J.: An Introduction to the Bootstrap, 1st edn. (Chapman & Hall/CRC Monographs on Statistics & Applied Probability). Chapman and Hall/CRC May 1994
18. Teich, J.: Pareto-front exploration with uncertain objectives. In: Evolutionary multi-criterion optimization, pp. 314–328. Springer (2001)
19. Hughes, E.: Evolutionary multi-objective ranking with uncertainty and noise. In: Evolutionary multi-criterion optimization, pp. 329–343. Springer (2001)
20. Filomeno Coelho, R., Bouillard, P.: Multi-objective reliability-based optimization with stochastic metamodels. Evol. Comput. **19**(4), 525–560 (2011)
21. Mayo, D.G.: In defense of the neyman-pearson theory of confidence intervals. Philos. Sci. **48**(2), 269–280 (1981)
22. Amaral, S.: A Decomposition-Based Approach to Uncertainty Quantification of Multicomponent Systems. Ph.D. thesis, MIT, Cambridge, MA, June 2015
23. Amaral, S., Allaire, D., Willcox, K.: Optimal L_2-norm empirical importance weights for the change of probability measure. Stat. Comput. 1–19 (2016)
24. Lawson, C.L., Hanson, R.J.: Linear least squares with linear inequality constraints. classics in applied mathematics. In: Solving Least Squares Problems. SIAM—Society for Industrial and Applied Mathematics, pp. 158–173. Philadelphia (1995) ISBN: 978-0-89871-356-5
25. Efron, B.: Bootstrap methods: another look at the jackknife. Ann. Stat. **7**(1), 1–26 (1979)
26. Diaconis, P., Efron, B.: Computer intensive methods in statistics. Technical Report No. 83, Division Of Biostatistics, Stanford University, January 1983
27. Davison, A.C., Hinkley, D.V.: Bootstrap Methods and their Application, 1st edn. Number 1 in Cambridge Series in Statistical and Probabilistic Mathematics, October 1997. Cambridge University Press. ISBN-13: 9780521574716 (1996)
28. Davison, A.C., Hinkley, D.V., Young, G.A.: Recent developments in bootstrap methodology. Statist. Sci. **18**(2), 141–157 (2003)
29. Vasile, M., Minisci, E.: An evolutionary approach to evidence-based multi-disciplinary robust design optimisation. In Poloni, C., Quagliarella, D., Periaux, J., Gauger, N., Giannakoglou, K. (eds.): EUROGEN 2011 Proceedings—Evolutionary and Deterministic Methods for Design, Optimization and Control with Applications to Industrial and Societal Problems. ECCOMAS Thematic Conference, Capua, Italy, ECCOMAS, CIRA, September 14–16 2011
30. von Neumann, J., Morgenstern, O.: Theory Of Games And Economic Behavior. Princeton University Press, Princeton (1953)
31. Quagliarella, D., Vicini, A.: GAs for aerodynamic shape design II: Multiobjective optimization and multi-criteria design. In: Lecture Series 2000–07, Genetic Algorithms for Optimisation in Aeronautics and Turbomachinery, May 2000. Belgium, Von Karman Institute (2000)
32. Vicini, A., Quagliarella, D.: Inverse and direct airfoil design using a multiobjective genetic algorithm. AIAA J. **35**(9), 1499–1505 (1997)

Combination of Polynomial Chaos with Adjoint Formulations for Optimization Under Uncertainties

Dinesh Kumar, Mehrdad Raisee and Chris Lacor

Introduction

Over the last decade, design optimization is receiving more and more interest in engineering applications. Conventional optimization deals with the problem of finding numerically the minimum (or maximum) of an objective function. In aeronautics, an example of optimization can be considered as finding the optimal shape of an aircraft wing that minimizes the drag while maintaining the prescribed lift and geometry constraints.

For optimization, sampling-based random search algorithms (gradient free, such as Genetic algorithm) [1, 2] and gradient-based methods [3, 4] are common methodologies. Evolutionary algorithms are less dependent to the starting point, as the whole design space is explored. Therefore, there is less risk of ending up in a local minimum. Gradient-based methods show fast convergence to the optimum but strongly dependent on the starting point and hence higher risk of ending up in a local minimum. The derivatives needed in gradient-based optimization can be computed efficiently by using the adjoint approach. Adjoint methods are nowadays widely used for optimization with a large number of design variables.

Robust optimization is an extension of conventional optimization where uncertainties are also included in the design procedure. In computational modeling, there

D. Kumar (✉) · C. Lacor
Department of Mechanical Engineering, Vrije Universiteit Brussel,
Pleinlaan 2, 1050 Brussels, Belgium
e-mail: dkumar@vub.ac.be

C. Lacor
e-mail: clacor@vub.ac.be

M. Raisee
School of Mechanical Engineering, College of Engineering, University of Tehran,
P. O. Box: 11155-4563, Tehran, Iran
e-mail: mraisee@ut.ac.ir

C. Hirsch et al. (eds.), *Uncertainty Management for Robust Industrial Design in Aeronautics*, Notes on Numerical Fluid Mechanics and Multidisciplinary Design 140, https://doi.org/10.1007/978-3-319-77767-2_35

are always numerous input parameters such as material properties, geometry, boundary conditions, initial conditions, whose values are often inaccurate or uncertain.

The polynomial chaos method (PCM)-based uncertainty quantification (UQ) approach offers a large potential for non-deterministic simulations. In PCM, stochastic variables with different distributions can be handled both with intrusive and non-intrusive approaches. The properties of the input random variables and the output stochastic solution can be described in terms of its statistical moments and probability density functions. The polynomial chaos methodology was originally formulated by Wiener [5] in an intrusive framework and was further expanded by Xiu and Karniadakis [6]. Several non-intrusive polynomial chaos methods (NIPCM) were also developed during recent years [7–9]. NIPCM are sampling-based methods and can be implemented easily into any in-house and commercial CFD code. For these reasons, here, non-intrusive polynomial chaos is used.

The presence of uncertainties brings several difficulties to the optimization process. Introducing uncertainties in the design process, the objective becomes non-deterministic and can be characterized by its mean, variance, and higher-order moments. It means, the optimization problem becomes multi-objective in presence of uncertainties. In case of a multi-objective optimization, one can either combine all objective functions into one single objective function by a weighted sum, or, alternatively, keep the different objectives which usually leads to a so-called Pareto front. These are different designs, where, if one compares design x on the front with design y on the front, x is not better than y in all objectives. With the increase in computational power and resources, optimization under uncertainties has become a very active area of research.

Several methods for robust optimization have been proposed independently and applied in different scientific fields. Wang et al. [2] applied non-intrusive polynomial chaos with multi-objective genetic algorithm for the robust optimization of rotor blades. Palar et al. [10] recently applied robust optimization in the aerodynamic shape optimization of a transonic airfoil using non-intrusive polynomial chaos with gradient-free evolutionary algorithm. Sriram and Jameson [3] proposed a robust optimization framework to optimize the mean value of a given objective function using polynomial chaos and adjoint-based gradient methods. Schillings and Schulz [11] proposed a polynomial chaos and gradient-based robust optimization framework to optimize the objective function defined as a combination of the mean and the variance. They applied it to aerodynamic shape optimization under uncertainties in 2D Euler flows. Recently, Maruyama et al. [12] proposed a robust optimization framework using surrogate models and applied it for robust design of a 2D airfoil under uncertainties.

In this chapter, a new approach for robust optimization is presented. Our goal is to combine the non-intrusive polynomial chaos uncertainty quantification method with adjoint methods for robust optimization and apply it in CFD applications.

Polynomial Chaos for UQ

The polynomial chaos expansion is the spectral representation of a random variable, which decomposes the random variable into separable deterministic and stochastic parts. Let J be the objective function of a stochastic problem with n input random variables. In polynomial chaos expansion, the uncertain objective $J(x;\boldsymbol{\xi})$ is decomposed in a basis of complete orthogonal polynomials as:

$$J(x;\boldsymbol{\xi}) = \sum_{i=0}^{P} J_i(x)\psi_i(\boldsymbol{\xi}) \tag{1}$$

where J_i are deterministic PC coefficients. ψ_i are the multi-dimensional orthogonal polynomials which are function of the independent random variables $\boldsymbol{\xi} = \{\xi_1, \xi_2 \ldots, \xi_n\}$ and are used to span the n-dimensional stochastic space. Here ξ_j is a random variable with a specific probability density function.

If p is the highest order of polynomial considered to approximate the stochastic response and n is the number of random variables, the total number of terms $P + 1$ in Eq. 1 depends on the order of polynomial (p) and on the number of random dimensions (n) as:

$$P + 1 = \frac{(n+p)!}{n!p!} \tag{2}$$

The polynomials ψ_i are orthogonal with respect to the probability distribution function $\boldsymbol{W}$ of the random variables $\boldsymbol{\xi}$. Orthogonality means that:

$$<\psi_i\psi_j> = \int_{\boldsymbol{\xi}} \psi_i(\boldsymbol{\xi})\psi_j(\boldsymbol{\xi})\boldsymbol{W}(\boldsymbol{\xi})d(\boldsymbol{\xi}) = <\psi_i^2>\delta_{ij} \tag{3}$$

where δ_{ij} is the Kronecker delta and $<\psi_i\psi_j>$ is the inner product.

The expectation or the mean of u can be written as:

$$\begin{aligned} E[J] &= \int_{\boldsymbol{\xi}} J(x;\boldsymbol{\xi})\boldsymbol{W}(\boldsymbol{\xi})d(\boldsymbol{\xi}) \\ &= \int_{\boldsymbol{\xi}} \{\sum_{i=0}^{P} J_i(x)\psi_i(\boldsymbol{\xi})\}\boldsymbol{W}(\boldsymbol{\xi})d(\boldsymbol{\xi}) \\ &= J_0 \end{aligned} \tag{4}$$

Further, the variance of the uncertain parameter u can be written as:

$$\begin{aligned}\sigma_J^2 &= E[(J - E[J])^2] \\ &= \int_{\xi} \{\sum_{i=1}^{P} J_i(x)\psi_i(\xi)\}^2 W(\xi)d(\xi) \\ &= \sum_{i=1}^{P} J_i^2(x) < \psi_i^2 > \end{aligned} \tag{5}$$

The PC expansion coefficients (J_i) can be computed using sampling-based non-intrusive methods such as regression, projection, or collocation methods. Currently, the quadrature-based projection method is considered to compute the PC expansion coefficients of the objective and the gradient of the objective.

The Adjoint Method

The adjoint methods have long been considered as a preferable choice for gradient-based optimization. In a design optimization problem, an objective function $J(U, \alpha)$ is a function of the state variables U and the design variables α. Based on the derivation of the adjoint equations, adjoint solvers can be classified as discrete and continuous. In a continuous adjoint solver, the adjoint equations are derived from the governing equation and then they are discretized. However, in a discrete adjoint solver, the governing equations are first discretized and then the adjoint equations are obtained. Both approaches have their positive and negative aspects. However, the discrete adjoint formulations are explained below. The adjoint method aims at computing the gradient of the objective function with respect to the design variables α. The total derivative of the objective function, $G = \frac{dJ}{d\alpha}$ can be written as:

$$G = \frac{dJ}{d\alpha} = \frac{\partial J}{\partial \alpha} + \frac{\partial J}{\partial U}\frac{dU}{d\alpha} \tag{6}$$

The partial derivatives in the above equation can be evaluated directly by varying the design variables and re-evaluating the objective function in the numerator. However, evaluation of the term $\frac{dU}{d\alpha}$ requires the solution of the governing equations. If $R(U, \alpha) = 0$ represents the residuals of the flow equations of an aerodynamic problem, the total derivative of the flow equations with respect to the design variables α can be expressed as:

$$\frac{dR}{d\alpha} = \frac{\partial R}{\partial \alpha} + \frac{\partial R}{\partial U}\frac{dU}{d\alpha} = 0 \tag{7}$$

or

$$\frac{dU}{d\alpha} = -\left(\frac{\partial R}{\partial U}\right)^{-1} \frac{\partial R}{\partial \alpha} \tag{8}$$

which gives the change in the state variables U with respect to the design variables α. By combining Eqs. 6 and 8, we obtain the following expression for the total gradient:

$$G = \frac{dJ}{d\alpha} = \frac{\partial J}{\partial \alpha} - \frac{\partial J}{\partial U}\left(\frac{\partial R}{\partial U}\right)^{-1} \frac{\partial R}{\partial \alpha} \tag{9}$$

Now let assume λ satisfies the following linear equation:

$$\left(\frac{\partial R}{\partial U}\right)^{T} \lambda = \left(\frac{\partial J}{\partial U}\right)^{T} \tag{10}$$

which is also known as the adjoint equation and λ is the vector of the adjoint variables or the adjoint solution. Combining Eqs. 9 and 10, one finally obtains the following expression for the gradient G:

$$G = \frac{dJ}{d\alpha} = \frac{\partial J}{\partial \alpha} - \lambda^{T} \frac{\partial R}{\partial \alpha} \tag{11}$$

Polynomial Chaos with Adjoint Formulations for Robust Optimization

When introducing uncertainties in a design process, the objective function is no longer deterministic and can be characterized by its mean and variance, i.e., in a robust design the optimization becomes multi-objective. Gradient-based optimization of the mean objective and of the variance of the objective therefore requires the gradient of both quantities. The gradient of the mean objective is combined with the gradient of its variance using weights.

Stochastic Objective and Its Gradient

In stochastic applications, the objective function (J) will also be stochastic and can be written in terms of polynomial chaos expansion as:

$$J = \sum_{i=0}^{P} J_i \psi_i(\boldsymbol{\xi}) \tag{12}$$

where the mean value of J is $\bar{J} = J_0$ and the variance of J is $\sigma_J^2 = \sum_{i=1}^{P} J_i^2 <\psi_i^2>$.

The gradient of the objective function J with respect to the design variable can be expressed as:

$$G = \nabla J = \sum_{i=0}^{P} \nabla J_i \psi_i(\boldsymbol{\xi}) = \sum_{i=0}^{P} \frac{dJ_i}{d\alpha} \psi_i(\boldsymbol{\xi}) = \sum_{i=0}^{P} G_i \psi_i(\boldsymbol{\xi}) \tag{13}$$

In stochastic applications, the stochastic objective function is usually written as the weighted sum of its statistical moments. Considering a multi-objective optimization, a new objective function can be defined as a linear combination of the mean and the standard deviation of the original objective function as:

$$I = K_1 \bar{J} + K_2 \sigma_J \tag{14}$$

In order to optimize (or minimize) the objective function I using gradient-based methods, one needs to compute the gradient of the objective function I with respect to the design variable as:

$$\nabla I = K_1 \nabla \bar{J} + K_2 \nabla \sigma_J \tag{15}$$

From Eq. 13

$$\nabla \bar{J} = \nabla J_0 = G_0 \tag{16}$$

where $\nabla \bar{J}$ is the gradient of the mean and G_0 is the mean of the objective gradient.

The gradient of the variance can be expressed as:

$$\nabla \sigma_J^2 = \sum_{i=1}^{P} \nabla J_i^2 <\psi_i^2> \tag{17}$$

or

$$2\sigma_J \nabla \sigma_J = 2 \sum_{i=1}^{P} J_i G_i <\psi_i^2> \tag{18}$$

The above equation can further simplified as:

$$\nabla \sigma_J = \frac{1}{\sigma_J} \sum_{i=1}^{P} J_i G_i \langle \psi_i^2 \rangle \tag{19}$$

where J_i and G_i are the PC expansion coefficients of the objective and its gradient, respectively.

Non-intrusive Calculation of PC Expansion Coefficients

For a deterministic sample *j*:

$$G^j = \left(\frac{dJ}{d\alpha}\right)^j = \left(\frac{\partial J}{\partial \alpha} - \lambda^T \frac{\partial R}{\partial \alpha}\right)^j = \frac{\partial J^j}{\partial \alpha} - \lambda^T \frac{\partial R^j}{\partial \alpha} \tag{20}$$

Define the gradient G of the stochastic objective function J. The polynomial chaos expansion of the gradient G can be written as:

$$G = \frac{dJ}{d\alpha} = \nabla J = \sum \left(\frac{dJ}{d\alpha}\right)_i \psi_i(\boldsymbol{\xi}) = \sum G_i \psi_i(\boldsymbol{\xi}) \tag{21}$$

The Galerkin projection of the above expansion with ψ_i:

$$G_i = \langle G\psi_i \rangle / \langle \psi_i^2 \rangle \tag{22}$$

Similarly, the Galerkin projection of the stochastic objective function can be written as:

$$J_i = \langle J\psi_i \rangle / \langle \psi_i^2 \rangle \tag{23}$$

Using the numerical quadrature method, PC coefficients of G_i and J_i can be written in terms of deterministic solutions G^j and J^j as:

$$G_i = \left(\frac{dJ}{d\alpha}\right)_i = \frac{1}{\langle \psi_i^2 \rangle} \sum_{j=1}^{N} G^j \psi_i(\boldsymbol{\xi}^j) w^j \tag{24}$$

and

$$J_i = \frac{1}{<\psi_i^2>} \sum_{j=1}^{N} J^j \psi_i(\xi^j) w^j \tag{25}$$

where G^j and J^j are from deterministic sample j and N is the total number of deterministic simulations. Once the PC coefficients G_i and J_i are computed using above equations, the gradient of the stochastic objective I (Eq. 14) can be formulated in order to optimize the stochastic function.

Aerodynamic Shape Optimization of the RAE2822 Airfoil Under Uncertainties

To validate the developed approach, here the robust optimization approach is applied to a 2D transonic airfoil, the RAE2822 under operational uncertainties. The adjoint solver and the flow solver of SU2 (an open-source CFD solver) are coupled with the polynomial chaos method for the optimal shape design under uncertainties. The optimization procedure is performed using a Python-based optimizer SciPy. The SciPy is a gradient-based optimizer for solving nonlinear optimization problems with constraints using the sequential least square programming (SLSQP) algorithm [13].

Test Case Description

In this section, the robust design methodology is applied to one of the basic test case of the UMRIDA project, the RAE2822 airfoil in transonic viscous flow. The airfoil geometry is described in Cook et al. [14]. The nominal flow conditions correspond to Mach number $M_\infty = 0.729$, angle of attack $\alpha = 2.31°$, and Reynolds number $Re_\infty = 6.5 \times 10^6$. The operating flow conditions for the test case correspond to case 6 (AGARD report, [14]) with corrected wall interference. The Mach number and the angle of attack are considered as uncertain parameters with standard deviations of $\sigma_M = 0.01$ and $\sigma_\alpha = 0.4°$ respectively. Both uncertain inputs are considered as uniformly distributed parameters. The stochastic information of the input data is tabulated in Table 1.

Table 1 UQ data for RAE2822 airfoil

Variables	Mean	Std	Distribution
Mach number	0.729	0.01	Uniform
Angle of attack	2.31°	0.4°	Uniform

For optimization, the objectives considered are the mean drag coefficient and its standard deviation with a constraint in the mean lift. For uncertainty quantification, the quadrature-based second-order polynomial method is used. The quadrature points for PC order 2 are the roots of the third-order Legendre polynomial (for uniformly distributed uncertainties). Two uncertainties and second-order polynomial requires $(p+1)^n = 9$ individual runs of the flow and the adjoint solver in each iteration of the optimization process. The optimization problem becomes as:

$$\begin{aligned} min \;\; & K_1 \bar{C}_d + K_2 \mu_{C_d} \\ & subjected\ to: \\ & \bar{C}_l \geq C_o \end{aligned} \tag{26}$$

where $\bar{C}_d$ is the mean drag coefficient and μ_{C_d} is the standard deviation of the drag coefficient. $\bar{C}_l$, the mean lift coefficient is an inequality which is maintained greater than a constant C_o during the optimization process.

Deterministic Solution

For the RAE2822 airfoil, the unstructured hybrid mesh with 13,937 cells, used for the CFD simulation, is shown in Fig. 1. The computational domain and grid cells near the airfoil surface are depicted in the zoomed area (see Fig. 1). A verified and validated solver SU2 is used for deterministic CFD solutions. The Spalart–Allmaras (SA) model is used for modeling of turbulence. In addition to solving the RANS equations, the adjoint equations can be also solved in SU2 to calculate the surface

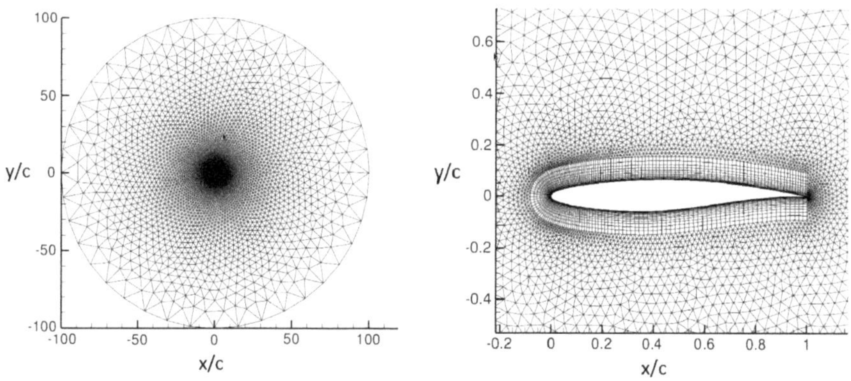

Fig. 1 RAE2822: computational domain (left) and zoomed mesh near to airfoil (right)

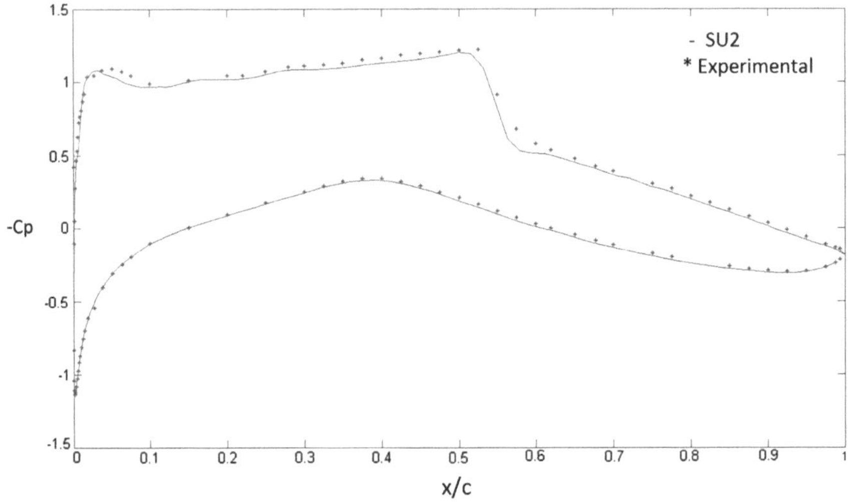

Fig. 2 RAE2822: pressure coefficient comparison (SU2 and experimental results)

sensitivities for the optimization purposes. In Fig. 2, the pressure coefficient for the test case is depicted. The pressure coefficient distribution obtained with the SU2 solver is also compared with experimental results (see Fig. 2). It can be observed that the numerical predictions of the C_p using the SA turbulence model are in close agreement with the measurements.

In a gradient-based aerodynamic shape optimization model, the chosen objective function is optimized by computing its gradient with respect to the shape or design parameters. The gradient of the objective function with respect to the deformation in the geometry is computed using the adjoint approach. The adjoint solver of SU2 computes the surface sensitivities at each node of the airfoil surface.

Shape Parameterization

For aerodynamic shape optimization problems, design variables are related to the surface boundaries. The design variables are important aspect of an optimization process. A simple way to construct the design variables is to consider surface mesh points. During the optimization process, the independent movement of these surface points may lead to a non-smooth configuration. The shape parameterization ensures smooth geometry deformation. Therefore, in an optimal shape design process, it is essential to parameterize the geometry using an appropriate parameterization technique. In this way, the design parameters can be defined and geometry changes with respect to the design parameters can be provided to the optimizer. Shape

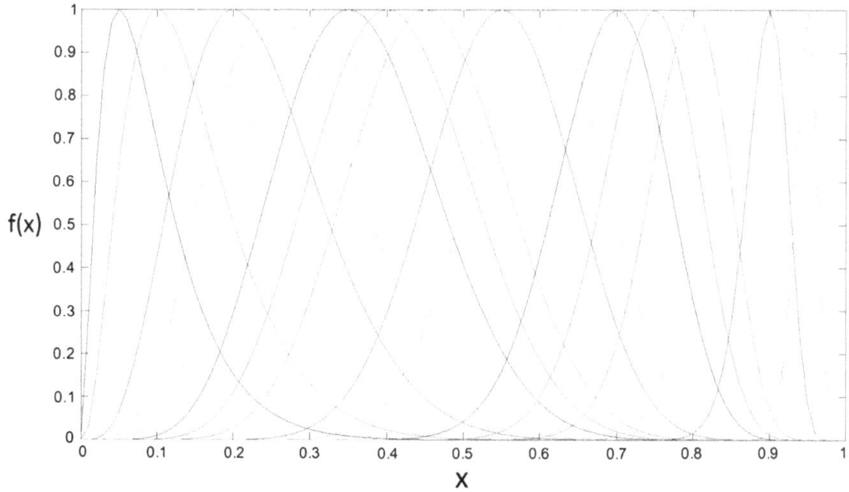

Fig. 3 Hicks–Henne bump functions for $t_2 = 10$ and $t_1 = 0.05, 0.10, 0.15, \ldots, 0.95$

parameterization is needed for evolutionary and genetic algorithms as well as for gradient-based optimization.

For robust optimization, the shape of the airfoil is parameterized and modified using a set of Hicks–Henne bump functions. Hicks–Henne bump functions recently received popularity in modeling small perturbations in many shape optimization problems [15].

The Hicks–Henne functions are defined as:

$$f(x) = [sin(\pi x^{\frac{log0.5}{logt_1}})]^{t_2}; 0 \leq x \leq 1 \tag{27}$$

where t_1 is the x location of the maximum and t_2 is the width of the bump.

The modified airfoil shape can be expressed as a weighted sum of sin functions (Hicks–Henne):

$$y = y_{base} + \sum_{1}^{N} a_i f_i(x) \tag{28}$$

where $(a_1, a_2, \ldots a_N)$ are N design variables which are used to control the magnitude of the shape functions. In Fig. 3, the Hicks–Henne bump function for $t_2 = 10$ and $t_1 = 0.05, 0.10, 0.15, \ldots, 0.95$ is shown. The bump functions reach maximum at the given values of $t_1 = 0.05, 0.10, 0.15, \ldots, 0.95$.

The bump functions are applied to the baseline airfoil to modify the shape of the airfoil. A total of 38 Hicks–Henne bump functions are applied to the upper and the lower surface of the airfoil at 5, 10, ..., 95% of the chord length. In this way, 38 coefficients of the bump functions are considered as the design variables. The leading and trailing edges are kept fixed in the design process by not adding any shape function to those points.

Robust Optimization

In this section, the initial geometry of the RAE2822 airfoil is taken for the shape optimization in order to minimize the drag as the objective function. A built-in gradient-based optimizer using SLSQP algorithm in SU2 framework combines the flow solver, the adjoint solver, and the mesh deformation tools to minimize the objective function. By using a simple chain rule, the surface sensitivities (computed with adjoint solver) are projected into the design space. For uncertainty quantification at each iteration of the optimization process, the input values of the Mach number and angle of attack for all CFD runs are shown in Table 2.

In Fig. 4, the convergence history and CFD solutions are plotted for all CFD runs. As it can be observed, the CFD solutions and residuals for all the samples are converged.

In the robust optimization process, the objective function (a weighted sum of the mean and its standard deviation) and the gradient of the objective function are provided to the optimizer. By changing the weights, the results for three different test cases are obtained. The first case is where a higher weight is given to the mean (i.e., $K_1 = 0.9$ and $K_2 = 0.1$), for the second test case, more weight is given to the standard deviation (i.e., $K_1 = 0.1$ and $K_2 = 0.9$), and the last case is where equal weights are given to the mean and the standard deviation.

In Fig. 5, the optimization history of the mean and the standard deviation for the three test cases is shown. From the optimization history, one can see that for all the test cases, the mean and the standard deviation of the drag coefficients are optimized after a few iterations only.

In Table 3, the mean and the standard deviation of the drag coefficient are shown for the baseline RAE2822, deterministic optimized airfoil, and for the robust design of the three test cases. From Table 3, it can be seen that for the case 1, where more weight is given to the mean, the optimized mean drag is lower than in all other designs. Similarly for case 2, where more weight is given to the standard deviation, the standard deviation is lower than in all other designs. Case 3, where equal

Table 2 Mach number and AoA samples for RAE2822 airfoil

	S_1	S_2	S_3	S_4	S_5	S_6	S_7	S_8	S_9
M	0.715	0.729	0.742	0.715	0.729	0.742	0.715	0.729	0.742
AoA	1.773°	1.773°	1.773°	2.31°	2.31°	2.31°	2.846°	2.846°	2.846°

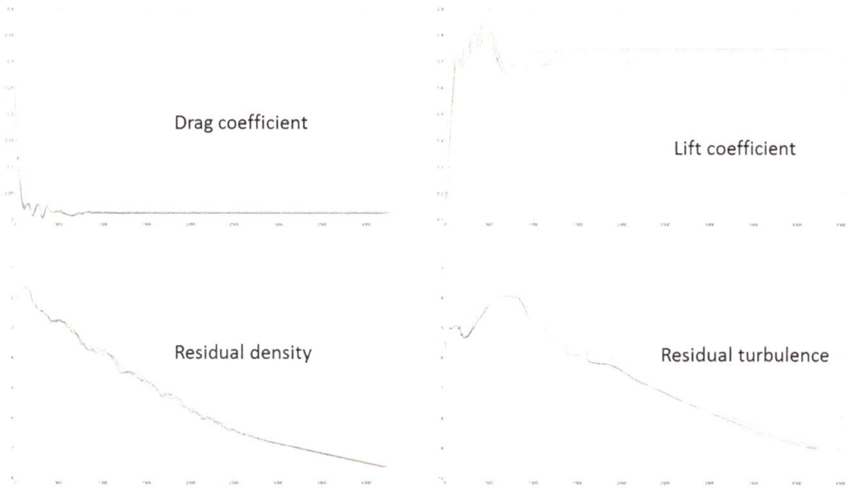

Fig. 4 CFD solution and residuals for all CFD runs of RAE2822 airfoil

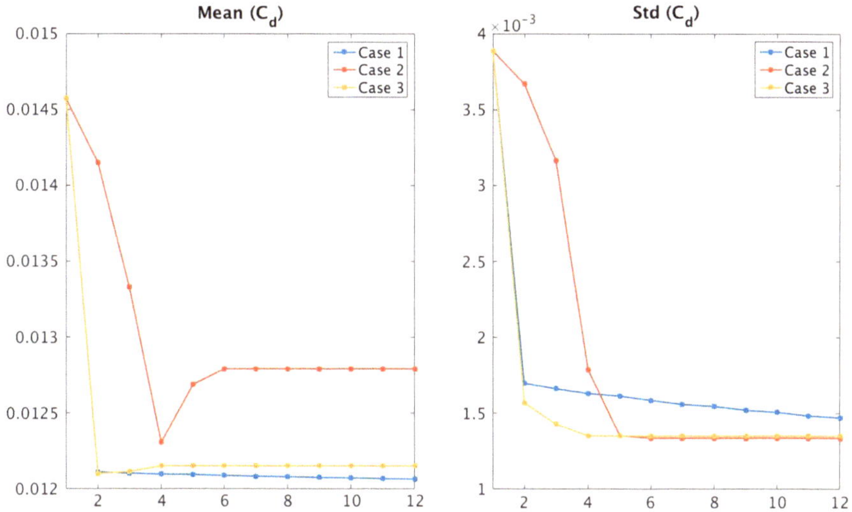

Fig. 5 Optimization history for the mean and the standard deviation of the three cases

Table 3 Mean and std of the drag coefficient for baseline RAE, optimized and robust airfoils

Case	Mean C_d	Std C_d	CoV
RAE2822 baseline	1.45729679e−02	3.88569535e−03	26.66% mean
Det. optimized	1.24683603e−02	2.51665720e−03	20.18% mean
Case 1, K = (0.9, 0.1)	1.20657193e−02	1.47263413e−03	12.20% mean
Case 2, K = (0.1, 0.9)	1.27927945e−02	1.33563551e−03	10.44% mean
Case 3, K = (1, 1)	1.21520489e−02	1.34977932e−03	11.11% mean

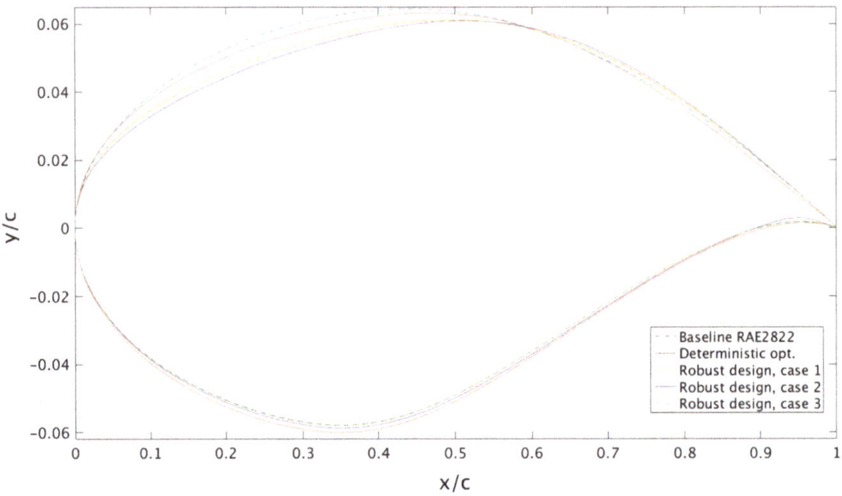

Fig. 6 Airfoil shape: RAE2822 baseline and optimized

weights are given to the mean and the standard deviation, optimizes both values, i.e., the mean and the standard deviation simultaneously. For case 3, the optimized mean value is very close to the mean value of case 1, and the optimized standard deviation is very close to the standard deviation of case 2.

When the uncertainties are applied to the airfoil obtained with a deterministic optimization, the mean drag coefficient is reduced by approximately 15% and the coefficients of variance by approximately 25% compared to the original design (see Table 3). When the uncertainties are included in the design process (case 1 and case 2), the coefficients of variance are reduced by approximately 50%, making the final designs robust with respect to the uncertainties. In Fig. 6, the airfoil designs for all cases, i.e., the baseline RAE2822, optimal shape (without uncertainties) and robust shape for both the test cases (under uncertainties) are shown. It can be seen that during the optimization process, mostly the area near the shock region is deformed. Three different robust airfoil designs (for the cases under uncertainties) are obtained. The shapes of these airfoils are quite different from the one without uncertainties. In Fig. 7, the standard deviation of C_d is plotted against the mean C_d for these three cases of robust designs making a Pareto front.

It is clear that this is just a part of the full Pareto front. This is related to the use of a gradient-based method where only part of the design space is explored in the neighborhood of the original airfoil. Putting it in another way: A gradient-based method will allow shape optimizations of, e.g., a VW car (for drag reduction) but it will not change it into a Ferrari (with even lower drag). For this an evolutionary method would be needed exploring the design space to a large extent.

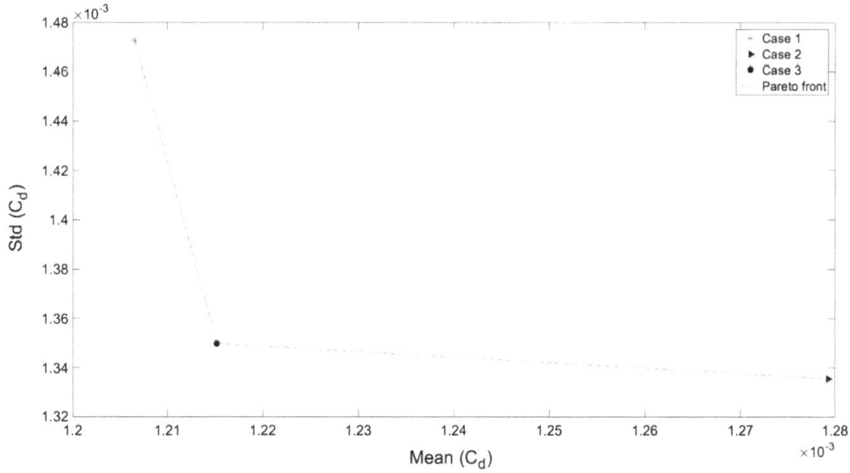

Fig. 7 Three points on the Pareto front

Conclusion

In this chapter, the non-intrusive polynomial chaos methods are combined with the adjoint-based gradient methods. The developed robust optimization method is applied to the optimal shape design of the RAE2822 airfoil using the flow solver and the adjoint solver of SU2. For the RAE2822 airfoil, three optimal designs are obtained by considering three different points on the Pareto front. Currently, the quadrature-based polynomial chaos is used for uncertainty quantification as only two operation uncertainties are considered in the robust optimization process. However, for a large number of uncertainties (e.g., geometrical uncertainties), an efficient non-intrusive UQ method such as sparse quadrature and reduced basis can be employed easily to the robust optimization approach.

References

1. Poloni, C., Mosetti, G.: Aerodynamic shape optimisation by means of hybrid genetic algorithm. Z. Angew. Math. Mech. **76**, 247–250 (1996)
2. Wang, X., Hirsch, C., Liu, Z., Kang, S., Lacor, C.: Uncertainty-based robust aerodynamic optimization of rotor blades. Int. J. Numer. Meth. Eng. **94**, 111–127. https://doi.org/10.1002/nme.4438 (2013)
3. Sriram, Jameson, A.: Robust optimal control using polynomial chaos and adjoints for systems with uncertain inputs. In: 20th AIAA Computational Fluid Dynamics Conference, Honolulu, Hawaii, 27–30 June 2011
4. Congedo, P.M., Corre, C., Martinez, J.-M.: Shape optimization of an airfoil in a BZT flow with multiple-source uncertainties. Comput. Methods Appl. Mech. Eng. **200**, 216–232 (2011)
5. Wiener, N.: The homogeneous chaos. Am. J. Math. **60**(4), 897–936 (1938)

6. Xiu, D., Karniadakis, G.E.: The wiener-askey polynomial chaos for stochastic differential equations. SIAM J. Sci. Comput. **24**(2), 619–644 (2002)
7. Eldred, M.S., Webster, C.G., Constantine, P.: Evaluation of non-intrusive approaches for Wiener-Askey generalized polynomial chaos. In: Proceedings of the 10th AIAA Nondeterministic Approaches Conference, No. AIAA-2008-1892, Schaumburg, IL, 7–10 April 2008
8. Babuska, I., Nobile, F., Tempone, R.: A stochastic collocation method for elliptic partial differential equations with random input data. SIAM J. Numer. Anal. **45**(3), 1005–1034 (2007)
9. Raisee, M., Kumar, D., Lacor, C.: A non-intrusive model reduction approach for polynomial chaos expansion using proper orthogonal decomposition. Int. J. Numer. Meth. Eng. **103**, 293312 (2015)
10. Palar, P.S., Tsuchiya, T, Parks, G.: Decomposition-based evolutionary aerodynamic robust optimization with multi-fidelity point collocation non-intrusive polynomial chaos. In: 17th AIAA Non-Deterministic Approaches Conference, AIAA SciTech, AIAA paper 2015-1377 (2015)
11. Schillings, C., Schulz, V.: On the influence of robustness measures on shape optimization with stochastic uncertainties. Optim. Eng. **16**, 347 (2015)
12. Maruyama, D., Liu, D., Gortz, S.: An Efficient Aerodynamics Shape Optimization Framework for Robust Design of Airfoils Using Surrogate Models. ECCOMAS congress, Crete Island (2016)
13. Boggs, P.T., Tolleb, J.W.: Sequential quadratic programming for large-scale nonlinear optimization. J. Comp. Appl. Math. (2000)
14. Cook, P.H., McDonald, M.A., Firmin M.C.P.: Aerofoil RAE 2822—pressure distributions, and boundary layer and wake measurements. In: Experimental Data Base for Computer Program Assessment, AGARD Report AR 138 (1979)
15. Masters, D.A., Taylor, N.J., Rendall, T., Allen, C.B., Poole, D.J.: Review of aerofoil parameterisation methods for aerodynamic shape optimisation. In: 53rd AIAA Aerospace Sciences Meeting, AIAA SciTech (2015)
16. Schittkowski, K.: The nonlinear programming method of Wilson. Han, and Powell with an augmented Lagrangian type line search function, Numerische Mathematik **38**(1), 83–114 (1982)

Robust Multiphysics Optimization of Fan Blade

K. Vinogradov, G. Kretinin and I. Leshenko

Introduction and Task Statement

One of the most relevant tasks is the fan blade robust multiphysics optimization under geometrical uncertainties. A modern civil aircraft fan blade is considered as an investigation object. The blade must provide high level of the aerodynamic efficiency (adiabatic coefficient of efficiency) and necessary structural properties. The fan blade flutter phenomenon sensitivity is also considered. Computational model of the fan blade for test case IC-04 includes blade solid domain and air path. CAD model of the blade profile is presented in Fig. 1. From enormous number of calculations necessitates to solve multiphysics robust optimization problem task statement with one blade was chosen. In aerodynamic calculations only blade profile is modeled. In the strength calculations, full blade (with foot) is considered. Blade foot has no geometrical uncertainties and variable parameters and is not involved in optimization procedure (nominal geometry). Computational aerodynamic model of the fan blade is presented in Fig. 1. NUMECA AutoBlade 8.9.1 was used as parameterization software to construct computational mesh and to automatize meshing and aerodynamic calculations procedures. Simplified parameterized blade model in NUMECA AutoBlade is also present in Fig. 1. The next cross sections were considered: 0, 30, 50, 70, 85, 100% of the normalized blade height.

Computational mesh provided for this test case IC-04 consists of the two parts. For CFD aerodynamic calculations, hexahedral structured mesh is used. To generate this type of mesh, NUMECA AutoGrid5 is used. The number of nodes for aerodynamic calculations is 1 300 000 nodes. Aforementioned number of nodes was chosen based on preliminary grid dependency test at the same aerodynamic

K. Vinogradov (✉) · G. Kretinin · I. Leshenko
NPO Saturn, Lenin av., 163, 152903 Rybinsk, Yaroslavl region, Russia
e-mail: keryisdex@mail.ru

C. Hirsch et al. (eds.), *Uncertainty Management for Robust Industrial Design in Aeronautics*, Notes on Numerical Fluid Mechanics and Multidisciplinary Design 140, https://doi.org/10.1007/978-3-319-77767-2_36

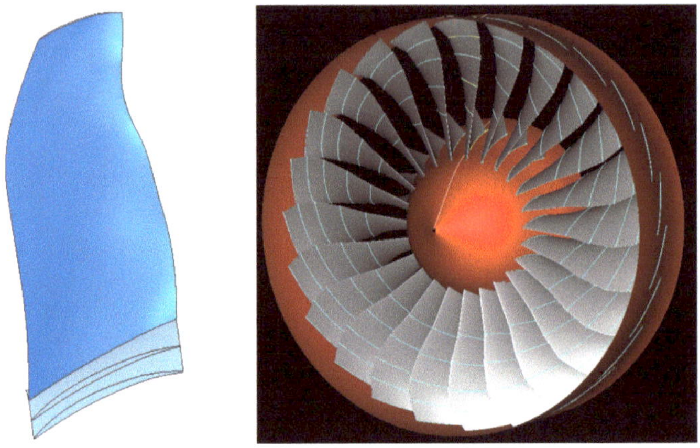

Fig. 1 Computational model for aerodynamic calculations

calculations for other fan and compressor blades. Chosen mesh has growth ratio 1.5 and the first cell size 5–6 m. “H–O–H” topology of the computational mesh was used because this mesh type provides high mesh quality level and necessary computational time.

To carry out strength analysis and to determine mode of deformation hexahedral combined (structured/unstructured) mesh also was used. To generate this type of mesh, ANSYS Mechanical Meshing was used. The number of nodes for strength analysis is approximately 100 000 nodes. Aerodynamic mesh of the fan blade is presented in Fig. 2. Mesh for structural analysis is also presented in Fig. 2.

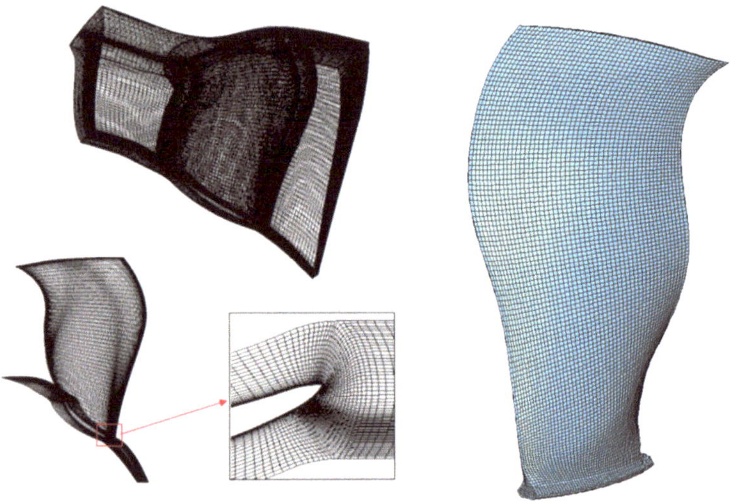

Fig. 2 Computational mesh for aerodynamic and structural calculations

Total pressure and total temperature in stationary frame were used as inlet boundary conditions. The flow direction is set by dimensionless angle components. Boundary conditions for aerodynamic calculations are set in accordance with standard atmosphere conditions. To couple parameterized blade profile and constant blade foot specify technology in ANSYS Mechanical APDL was developed. By means of additional curves and surfaces sketching, we can construct intermediate blade part between profile and foot in hub blade cross section for different stagger angles. Visualization of this procedure is present in Fig. 3.

Analysis of fluid flow and film cooling has been performed using NUMECA FINE/Turbo [1] which employs a structured grid system. The solutions have been obtained using the finite volume method to discretize the compressible RANS equations. The Spalart–Allmaras turbulence model is used as a turbulence viscosity equation solving and closure system of equations. The boundary conditions at the outlet of the computational model (radial distributions of pressure were accepted according to the preliminary aerodynamic calculations at the operating point, near the stall margin and in other areas of interest. One of the key features of aerodynamic calculations and optimization task statement is the one optimization iteration corresponds to five aerodynamic calculations (operating point-95% of Speed, 80% of speed to flutter sensitivity, and three points to determine stall margin in automatic mode). Deterministic optimization task statement of the FSI multiphysics problem is presented in the section "General Robust Design Optimization Task Statement".

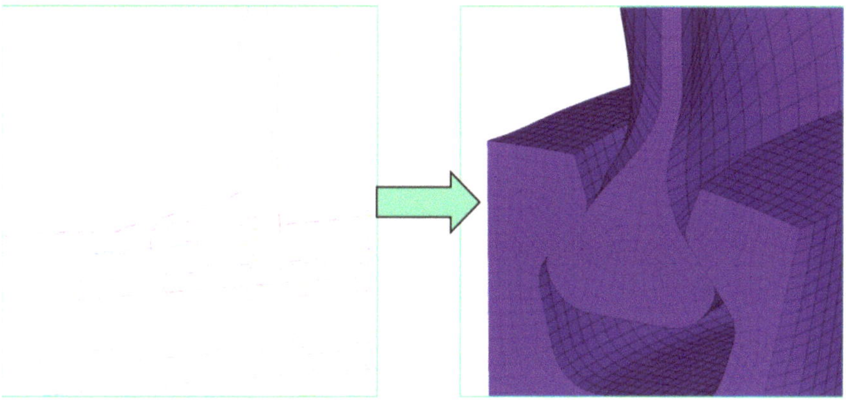

Fig. 3 Intermediate blade part between profile and foot

Geometrical Uncertainties Considered

In the proposed test case, huge number of geometrical uncertainties was considered. There are geometrical uncertainties from fan blade manufacture tolerances and deviations. Scheme of the considered blade sections location and scheme of the main geometrical uncertainties considered (red circles) in blade section are presented in Fig. 4.

Red circles show cross sections chose for providing uncertainties and further UQ investigations. The main uncertainties considered as it is shown in Fig. 5 are blade thicknesses in different profile locations (at 3 mm from leading and trailing edges (e1, e2), and 20 mm from leading and trailing edges (e3, e4) and maximal blade thickness in section (E). Profile angle of incidence is also taking into consideration.

General information example for the considered geometrical uncertainties is presented in Table 1. In this table, main statistic parameters (mean value, variance, and distribution law) for probability density function description are presented. Geometrical uncertainties are presented by means of deviations from nominal fan blade dimensions. Nominal fan blade dimensions were obtained from CAD model of the fan.

Example of the theoretical graphs of the probability density functions in comparison with experimental bar graphs for presented uncertainties is shown in the Fig. 5.

Analysis of probability density function graphs for considered geometrical uncertainties showed most of experimental stochastic parameters can be described by means of Gaussian pdf distribution law with acceptable accuracy level. Some parameters will have to be described by means of lognormal and beta distribution.

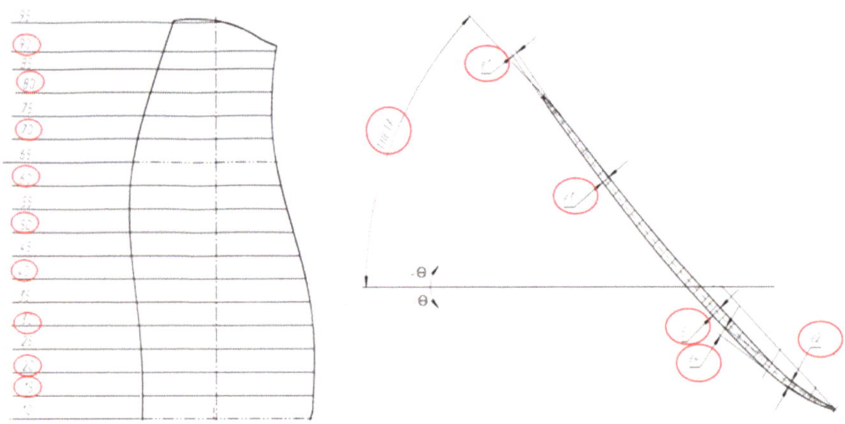

Fig. 4 Scheme of the considered blade geometrical uncertainties

Table 1 Example of general information about geometrical uncertainties

Name of the uncertainty	Mean value	Variance	Law of distribution
ΔE15	−0.0896	0.079	Gaussian
ΔTHETA15	−0.4239	4.9772	Gaussian
Δe2_15	μ = −1.199	σ = 0.302	Lognormal

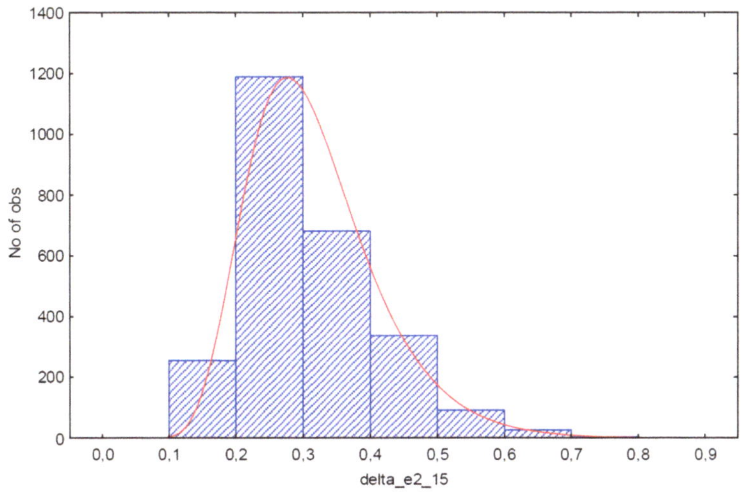

Fig. 5 Theoretical pdf example in comparison with experimental bars

General Robust Design Optimization Task Statement

Let us consider mathematical formalization of the RDO problems. While creating a technical system, the designers are to form the vector of values of system efficiency $y = (y_1, y_2, \ldots, y_m)$, which are to be maximized, minimized, and constrained, to form the vector of variable parameters $x = (x_1, x_2, \ldots, x_n)$, varying of which leads to the variation of the efficiency, and to form the vector of external conditions $e = (e_1, e_2, \ldots, e_k)$. The correlation between these vectors as $y = f(x, e)$ forms the mathematical model of the system under investigation. The existence of a mathematical model makes it possible to formulate a design problem as an optimization task, which lies in the search of one or several vectors $x^* \in D$ that ensure the best (in some way) efficiency. Here

$$D = \{x \in R^n | x_{j-} \leq x_j \leq x_{j+1}, j = 1, \ldots, n;\ g_i(x, e) \leq 0, i = 1, \ldots, w\} \tag{1}$$

is the search region, $g_i(x, e)$ is the constrained efficiency values. Such an "ideal" design problem statement was regarded, until recently, as a necessary and sufficient condition to obtain an optimal design. In practice, however, such an approach of solving real-life tasks deals with serious problems connected with impossibility to implement optimal project solutions. The main reason for this lies in the existence of a large number of uncertainties, which are not taken into account while modeling the system, optimization problem statement, and solving procedure.

The attempt to include uncertainties in design problem formalization results in the necessity to consider relations: $x = x(\bar{x}, \xi_x)$; $e = e(\bar{e}, \xi_e)$, where $\bar{x}$, $\bar{e}$ are the ideal vectors of variable parameters and environmental conditions; $\xi = (\xi_x, \xi_e)$ corresponds to the vector of random values including the uncertainties in implementation of variable parameters and environmental conditions. Generally, to solve a RDO problem one must be able to determine system efficiency values $y = f(x, e)$ for given values of $\bar{x}$, $\bar{e}$, and hence to know the laws of distribution of vector components ξ [2]. In our situation, we consider aforementioned geometrical uncertainties (blade thickness) in five cross sections of the blade. These parameters are the stochastic values, which distribution laws were obtained as a result from the experimental data and are shown (as example) in Fig. 5. Blade leading edge and trailing edge angles as to stagger angle were set as variable parameters in parameterization model. Total number of variable parameters and uncertainties are 42. As a probabilistic criteria were used efficiency values with probability no less than one given (P = 90%).

The main problem occurring while solving robust design optimization problem is determining probabilistic criteria values. The simplest and the most universal method of evaluation of probabilistic criteria is the Monte Carlo method. The main advantage of this method, as applied to RDO problems, is no necessity of setting of any a priori assumptions about the goal function peculiarities (smoothness, monotony, continuity, differentiability, and so on). However, the efficiency of the Monte Carlo method when solving real-life problems to a great extent depends on the required accuracy of definition of probabilistic criteria. Particularly, applying the gradient methods of optimization, requiring high accuracy of definition of probabilistic criteria, resulting in high computational expense (required number of tests at each iteration of extremum search makes up $\approx 10^6$–10^9). The second approach includes a number of methods which are based on different approximation techniques (Taylor's series, response surfaces, and so on). In this approach, results of response surface modeling algorithms (surrogate models) are used for probabilistic criteria evaluation. When solving the task under consideration, we used the Monte Carlo method along with the method of multicriteria optimization IOSO [3] as well as the procedure of multilevel optimization involving surrogate models [1–7].

Deterministic Optimization Results

To solve deterministic optimization task, aerodynamic and stress/flutter sensitivity computations in NUMECA FINE/Turbo and ANSYS Mechanical APDL were carried out in one software loop. Deterministic optimization results were used as initial DoE to further robust optimization process (to construct surrogate model for

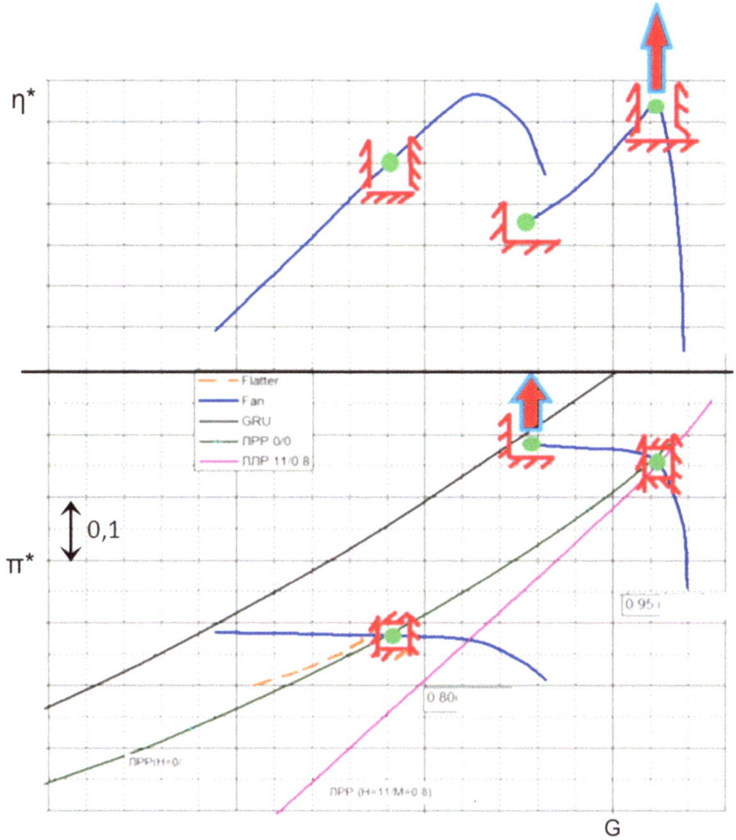

Fig. 6 Task statement

robust optimization in iterative procedure). Necessary information about deterministic optimization task statement is presented in Fig. 6.

Some additional details about deterministic optimization are presented below.

Objective functions:

- increasing of the aerodynamic efficiency at rotating speed n = 0.95;
- increasing of the stall margin at rotating speed n = 0.95 (no less than 0.1);
- decreasing maximal static stresses in blade profile;
- decreasing flutter sensitivity across TBC criterion (Torsion-bending coupling).

Constraints: compressor pressure ratio, mass flow at rotating speed n = 0.95, n = 0.8

Variables: camber line form (blade LE and TE angles), stagger angle in six sections.

TBC criterion is the experimental-based dependence between torsion and bending displacement by first flexion mode for leading edge and trailing edge control points.

Scheme of the TBC criterion calculations and necessary mathematical equations is presented in Fig. 7.

Based on aforementioned task, statement and details more than 2500 iterations of the deterministic optimization were carried out. Essential improvement for all criteria considered was obtained as a result of the optimization. Aerodynamic efficiency was improved on 1%, stall margin improvement was 0.75, and equivalent von Mises stress (sstt) was decreased on 42%. We selected four points with maximal levels of criteria considered. Comparison between aerodynamic and multiphysics optimization is presented in Fig. 8. Figure 8 shows that the results of aerodynamic optimization have a much higher level of aerodynamic efficiency and stall margin (D_Ky) but poor structural properties (static stresses in the blade profile and flutter sensitivity are too high (TBC > 0.3)). Such structural properties make the blade non-viable. The results of deterministic multiphysics optimization are represented with yellow triangles and show worse aerodynamic characteristics (by 0.9% of aerodynamic efficiency) than the results of aerodynamic optimization. At the same time, they show much higher structural characteristics. In particular, level of von Mises stress on 35–40% less, flutter sensitivity is TBC < 0.22. The results of

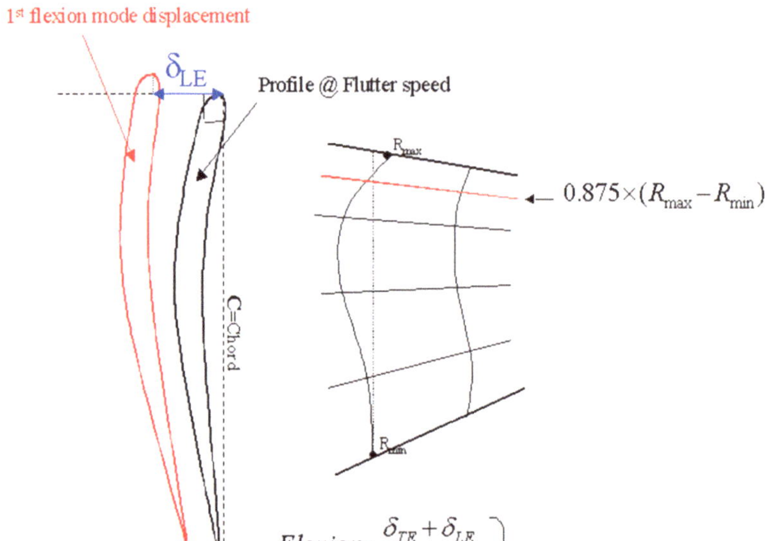

$$Flexion = \frac{\delta_{TE} + \delta_{LE}}{2}$$

$$Torsion = \frac{\delta_{TE} - \delta_{LE}}{C}$$

$$TBC = \frac{C}{2} \times \frac{Torsion}{Flexion} = \frac{\delta_{TE} - \delta_{LE}}{\delta_{TE} + \delta_{LE}}$$

Fig. 7 TBC criterion formulation

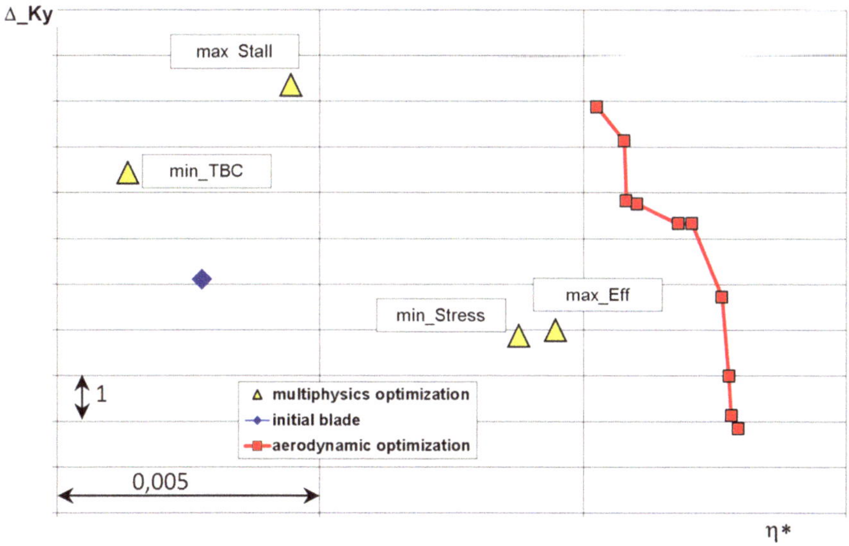

Fig. 8 Comparison between aerodynamic and multiphysics optimization

the multiphysics optimization have the essential difference in all criteria considered (aerodynamic efficiency difference is 0.8%, stall margin difference is 4–5%, and different stress level (max difference 15–19%). The results of this comparison prove the necessity of multiphysics optimization. Some of the aerodynamic characteristics of the results of deterministic multiphysics optimization are presented in Fig. 9. Structural characteristics of the fan blades (distribution of the von Mises stress) for the results of deterministic multiphysics optimization are presented in Fig. 10 that show that substantial differences in the level of maximal stress are present between all of represented blades.

Thus, the problem of multiphysics optimization of the fan blade (aerodynamics + strength + flutter sensitivity) in the deterministic approach has been solved.

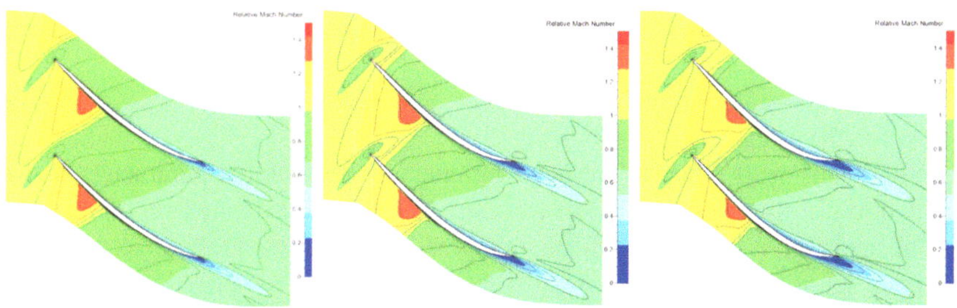

(a) Initial blade (b) Max efficiency (c) Min stresses

Fig. 9 Flow patterns in span 50% in operating point for some results

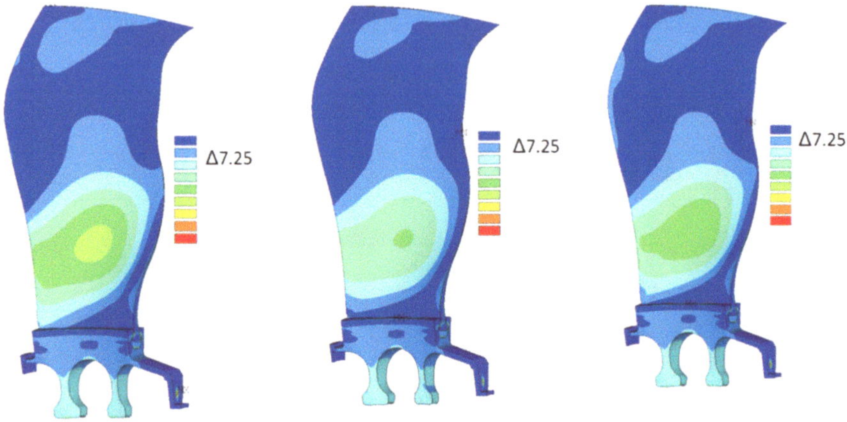

(a) Max efficiency (b) Min static stress (c) Min flutter sens.

Fig. 10 3D stress code results (static stress in the fan) for some multiphysics optimization results

Pareto sets for all four criteria under consideration were obtained. Results of deterministic optimization were used as initial DoE for the robust optimization problem.

Multilevel Robust Design Optimization Using Surrogate Models and IOSO Technique

In the present work, Approx software [8] is used for the surrogate model construction. This software allows us to obtain different types of the surrogate models: from full-square regression models [9] to different types of artificial neural networks.

The central point during construction of the surrogate model is the choice of the structure of the approximating function. Models based on linear regression with the function parameters defined as:

$$Par = K_0 + \sum_{i=1}^{N} K_i \cdot m_i \tag{1}$$

where K0, Ki—regression coefficients.

Within this work, due to the large number of the problem parameters, it was decided to use modified method of least squares with an extended set of variables.

The approach is based on full-square regression with the regressors formed on an extended set of variables. The extended set of variables is comprised not only of the variables themselves but also of their functional dependencies. Number of regressors might end up very large, so the algorithm uses adaptive selection of only those regressors that represent the response surface the most fully. Tuning of the

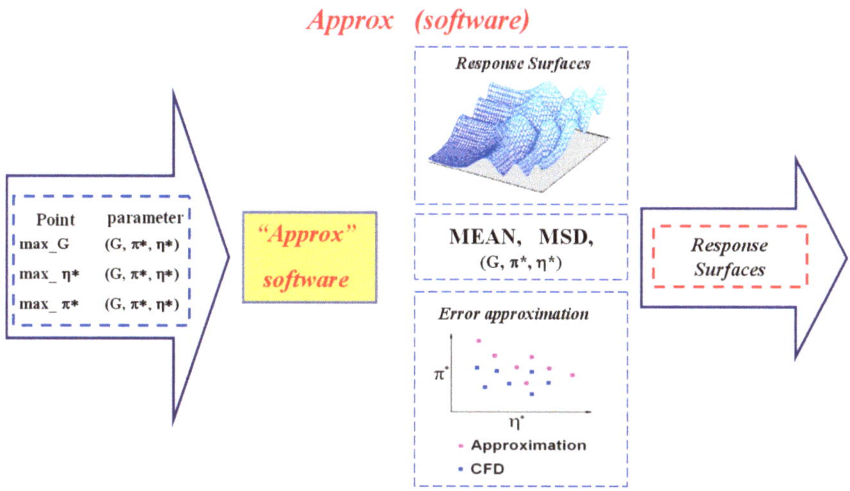

Fig. 11 Surrogate model creating scheme in Approx software

parameters of the response surface of this kind takes into account as many of the regression coefficients as possible, as well as the relative accuracy achieved by the adaptive selection procedure. The higher is the number of the regression coefficients the more accurately the starting points can be described. Scheme of the surrogate model construction in Approx software is presented in Fig. 11. The typical situation, while solving a problem of optimization of complex engineering systems, is that the user has several tools of various degree of fidelity to perform the analysis. These tools differ according to their levels of complexity of modeling the actual physical phenomena and their different levels of numerical accuracy. The high-fidelity tools could be represented by detailed nonlinear mathematical models of the researched systems or even by the experimental samples of such systems. However, the use of such tools in optimization is associated with significant time expenditures.

The low-fidelity (surrogate) models also allow carrying out optimization search, but the reliability of the obtained results can be rather low. Therefore, within the framework of the development of RDO methodology for complex systems, the methods based on the combination of various fidelity analysis tools are widely practiced.

The objective here is to offer the procedure of multiobjective optimization of complex systems based on the adaptive use of analysis tools of various levels of complexity. The intention is to minimize the use of high-fidelity time-consuming tools for the analysis. This approach ensures the possibilities to search Pareto-optimal set of solutions and also ensures improving the surrogate mathematical model.

The simplified scheme of work for the multilevel optimization procedure can be represented as follows (Fig. 12).

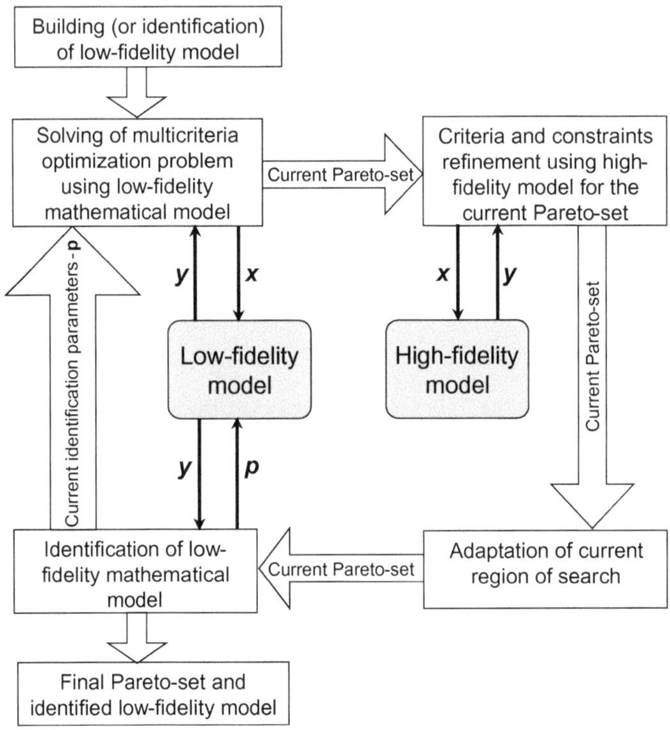

Fig. 12 Scheme of multilevel optimization procedure via IOSO NM software

- Generation of surrogate model on the basis of the data set previously obtained.
- Solution of the multiobjective optimization problem based on a surrogate model. Updating of the objectives and constrained parameters obtained for the Pareto set using the high-fidelity analysis tool.
- Refinement of the surrogate model.
- Replacement of the surrogate model and return to step II.

The particular features of the problem define the number of iterations for such a multilevel procedure. The number of applications of high-fidelity analysis tools is limited to the product of the number of iterations and the number of Pareto-optimal solutions. Interaction between Approx software and IOSO optimization algorithm is presented in Fig. 13.

The information stored during the search is used to improve the surrogate models. However, this model is correct not for the entire initial search area but only for a certain neighborhood of the obtained Pareto set. This ensures purposeful improvement of approximating properties only in the area of optimal solutions that noticeably reduce the computing effort to construct surrogate models. In this approach, results of response surface modeling algorithms (surrogate models) are

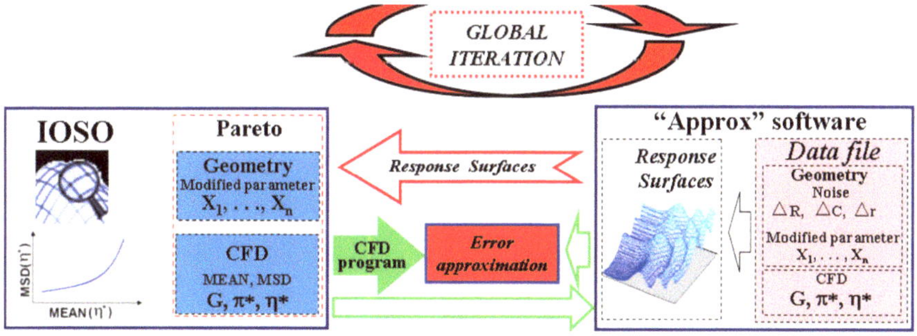

Fig. 13 IOSO NM interaction algorithm with surrogate model

used for probabilistic criteria evaluation. In our case, uncertainty quantification and robust optimization tasks are solving together.

RDO Results for Fan Blade

Surrogate model construction for RDO procedure.

At the first step, initial DoE based on deterministic optimization results was constructed. To generate initial combinations of the variable parameters deterministic optimization task was solved. Within the applied procedure of multilevel optimization, 15 global iterations were carried out. Initial design of experiments has 1400 calculations based on deterministic optimization results. Additional 25–50 calculations (high-fidelity CFD simulation) at every multilevel iteration of the optimization were carried out. Total number of calculations in database to construct surrogate model was ≈500. Variations of accuracy of low-fidelity models via iterations are shown in Fig. 14. The tendencies for all criteria considered are noticeable toward increase of low-fidelity model accuracy under the growth of the number of iterations. It should be noted that the worse accuracy of approximation has the maximal value less than 2% (for TBC criterion).

Fig. 14 Surrogate model accuracy modifications (aerodynamic efficiency)

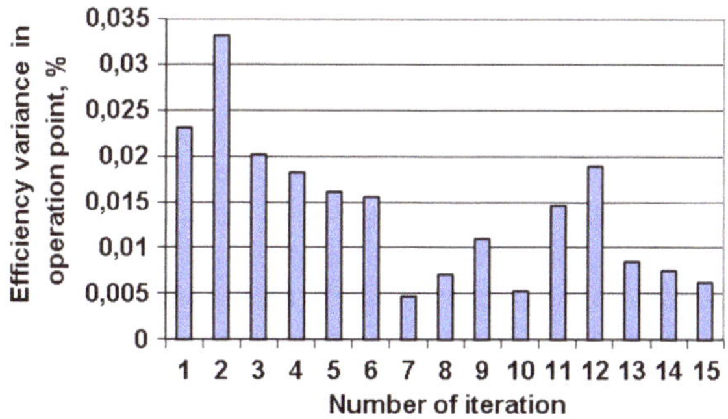

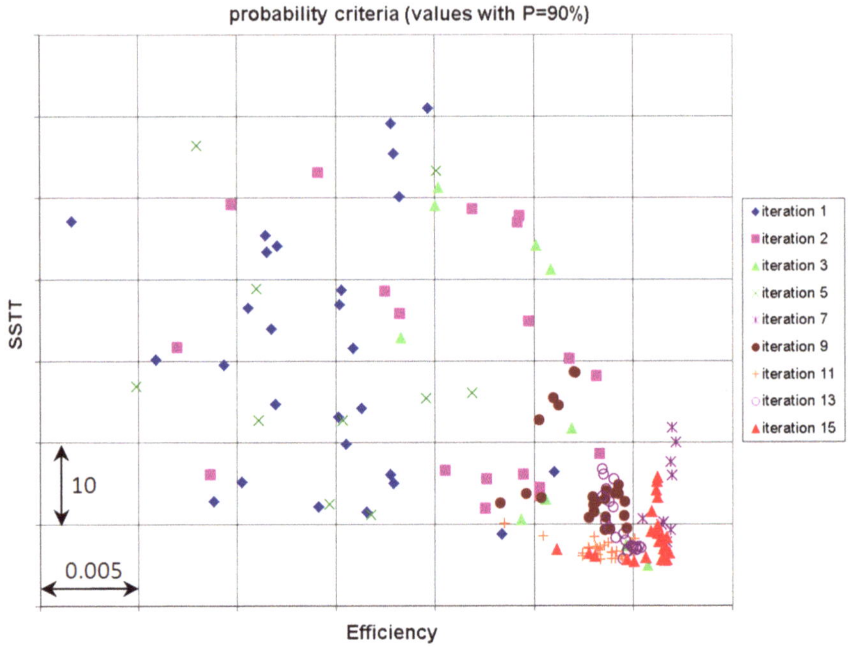

Fig. 15 Surrogate model database modifications for aerodynamic efficiency and static stress

The next accuracy levels were obtained:

- 0.3% for mass flow rate$
- 0.006% for aerodynamic efficiency
- 0.015% for stall margin
- 1% for static stress
- 1.8% for TBC criterion.

During iteration refinement of the surrogate model, substantial nonlinearity in the change of the model accuracy was observed which was caused by the high dimensionality of the problem (42 parameters). In spite of aforementioned problems, obtained surrogate model can provide necessary level of prediction accuracy to carry out robust optimization based on these results. Overview of the optimizing search during robust optimization on probabilistic criteria (efficiency and strength with 90% confidence) for some of 15 multilevel iterations is shown in Fig. 15. The figure shows clear tendency for the aerodynamic efficiency to increase and for the maximal stress to decrease during the robust optimization.

As a result, evident Pareto set between aerodynamic efficiency and structural properties has formed on the 15 iterations. It has a pronounced corner point. In addition, the search dynamic demonstrates the influence of complexity and multi-physicality of the problem on the obtained result.

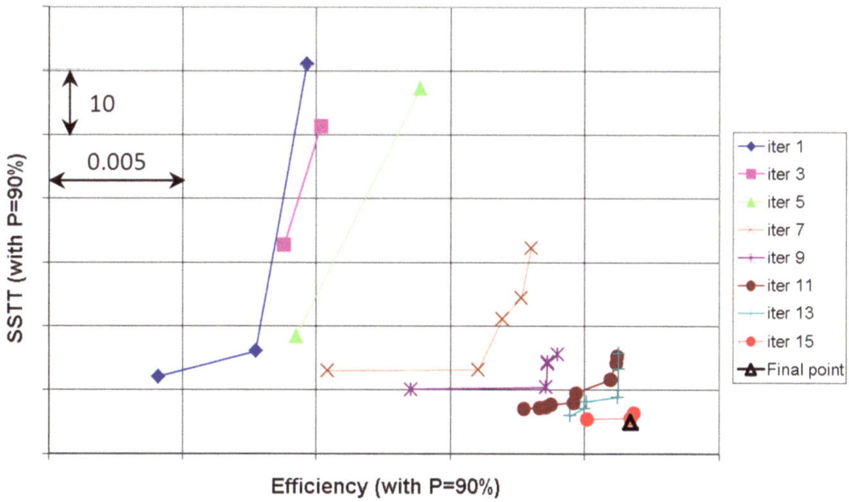

Fig. 16 Pareto set (between aerodynamic efficiency and static stress) modifications via multilevel iterations of robust optimization

Dynamic of the change of the Pareto set between aerodynamic efficiency and structural properties during multilevel iteration of robust optimization is presented in Fig. 16. Figure 16 shows tendency for the values of probabilistic criteria to improve. For example, for efficiency with P = 90% the gain was about 1.4% and for maximal stress the reduction was 19%. This result demonstrates high efficiency of the developed and employed technology for the robust optimization problems of high dimensionality. As a final result of the optimization, according to the traditional rules of the Pareto set analysis, the "angle point" corresponding to the best aerodynamic efficiency and structural properties was used.

The probability density distribution of the aerodynamic efficiency for deterministic and robust optimization results is presented in Fig. 17. The chart clearly demonstrates the achieved improvement. During the initial stages of the robust optimization, the deterministic and the mean values of a criterion had worsened but eventually (on the fifteenth iteration) there was a significant gain in the mean value of efficiency, decrease in its variance (from $\sigma = 0.08$ to $\sigma = 0.059\%$), as well as an increase in the probabilistic criterion up to $\Delta\eta_{90\%} = 0.15\%$.

The probability density distribution of the structural properties (static von Mises stress) for deterministic and robust optimization results is presented in Fig. 18. The tendencies for the distributions are similar to the ones for the aerodynamic characteristics. The results of the deterministic optimization have the highest mean value and probabilistic criterion ($SSTT_{90\%}$) for the static stress. During the initial stages of the robust optimization, the mean value of the stress as well as its variance had decreased. On the fifteenth iteration, further significant decrease of the mean value of the stress, its variance (from $\sigma = 1.362$ MPa to $\sigma = 0.982$ MPa), and a decrease in the probabilistic criterion were obtained. The probability density

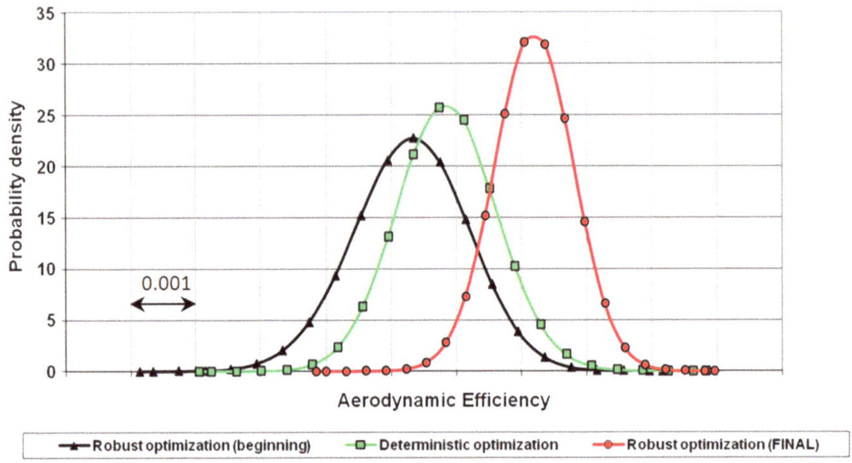

Fig. 17 Pdf for deterministic and robust optimization results (aerodynamic efficiency)

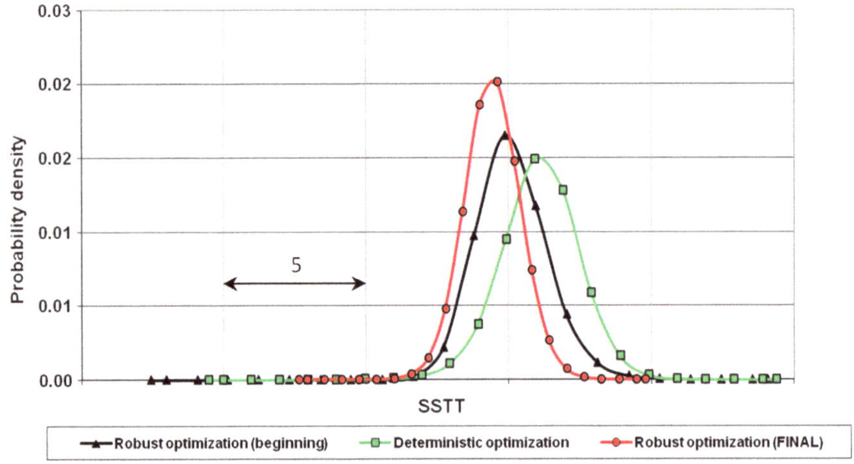

Fig. 18 Pdf for deterministic and robust optimization results (structural properties)

distribution of the stall margin for deterministic and robust optimization results is presented in Fig. 19. The results of the deterministic optimization have the highest mean value and probabilistic criterion ($D\_Ky_{90\%}$) for the stall margin, which was caused differences in the task statement of the deterministic and robust optimization. The deterministic optimization was supposed to maximize the stall margin but for the robust optimization it was used as a constraint. During the initial stages of the robust optimization, the mean value had decreased while the variance had increased significantly. By the fifteenth iteration, the mean value had not changed

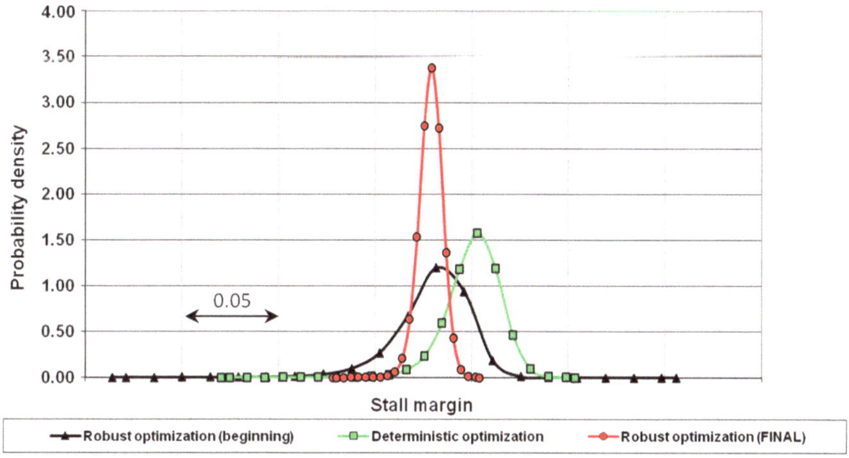

Fig. 19 Pdf for deterministic and robust optimization results (stall margin)

much while the variance had decreased significantly (from $\sigma = 1.3$ to $\sigma = 0.6\%$). Thus, the results of robust optimization can guarantee a probabilistic constraint on the stall margin.

Conclusions

1. One of the most promising techniques to solve RDO problems in coupling is usage of approximate assessments of probabilistic criteria under Monte Carlo combined with direct optimization techniques IOSO.
2. Application of the multilevel optimization procedure offers a significant reduction of the computing time expenditures for the solution of complex real-life problems while maximizing the probability of manufacturing the object under study.
3. Deterministic multiphysics optimization of a fan blade has been carried out. The fan aerodynamic efficiency has been increased, the maximal stress in the blade profile, and flutter sensitivity has been decreased. Pareto sets were constructed from the results of the deterministic optimization.
4. The problem of the robust optimization of a fan blade under the influence of geometrical uncertainties (deviations of manufacturing) has been stated and solved by means aforementioned technique. The total number of parameters was 42.
5. From the results of the robust optimization a point from a Pareto set has been obtained which can provide maximal efficiency, minimal stress in the blade profile and the necessary level of the stall margin and the flutter sensitivity with the 90% probability. Furthermore, during the robust optimization the variance of

the aerodynamic and the structural properties caused by the geometrical uncertainties has decreased. The variance has decreased on average from 1.3 to 0.6%.
6. The results show high efficiency of using the developed approach for the robust optimization problems with a high number of geometrical uncertainties.
7. The objectives of the further research will be the increase in the number of the optimization criteria and the uncertainty quantification in the blade foot and also the direct simulation of vibro-stress in the blade (two-way FSI) during the robust optimization.

References

1. Egorov, I.N., Kretinin, G.V., Leshchenko, I.A.: Stochastic optimization of parameters and control laws of the aircraft gas-turbine engines—a step to a robust design. Inverse Prob. Eng. Mech. III, Elsevier Science Ltd, 345–353 (2002)
2. Egorov, I.N., Kretinin, G.V., Leshchenko, I.A., Babiy, Y.I.: Optimization of Complex Engineering Systems Using Variable-Fidelity Models. In: Proceedings of the 1st ASMO UK/ ISSMO Conference on Engineering Design Optimization, MCB University Press, ISBN 0-86176-650-4, pp. 143–149 (1999)
3. Egorov, I.N., Kretinin, G.V., Kostiuk, S.S., Leshchenko, I.A., Babi, U.I.: The methodology of stochastic optimization of parameters and control laws for the aircraft gas-turbine engines flow passage components. ASME J. Eng. Gas Turbines Power **123**, 495–501 (2001)
4. Egorov, I.N., Kretinin, G.V., Leshchenko, I.A.: How to execute robust design optimization. In: Proceedings of 8th AIAA/USAF/NASA/ISSMO Symposium on Multidisciplinary Analysis and Optimization, AIAA Paper 2002–4328, Atlanta, GA, USA (2002)
5. Egorov, I.N.: Indirect Optimization Method on the Basis of Self-Organization. Curtin University of Technology, Optimization Techniques and Applications (ICOTA'98), vol. 2, pp. 683–691, Perth, Australia, (1998)
6. Egorov, I.N.: Optimization of a multistage axial compressor. stochastic approach. ASME Paper 92-GT-163 (1992)
7. Egorov, I.N., Kretinin, G.V.: Optimization of Gas Turbine Engine Elements by Probability Criteria. ASME Paper 93-GT-191 (1993)
8. EU FP6 Program NODESIM-CFD. http://www.nodesim.eu/
9. Polynkin, Andrey, Toropov, Vassili V.: Mid-range metamodel assembly building based on linear regression for large scale optimization problems. Struct. Multidisc. Optim. **45**(4), 515–527 (2012)

UQ Sensitivity Analysis and Robust Design Optimization of a Supersonic Natural Laminar Flow Wing-Body

Domenico Quagliarella and Emiliano Iuliano

Introduction

Within the UMRIDA project an aerodynamic configuration of a supersonic business jet wing-body was defined and proposed as baseline for a robust aerodynamic shape design problem. This configuration, while being of industrial interest, is not covered by copyright or confidentiality clauses and can be used for benchmarks and comparisons even outside of the UMRIDA consortium.

The present chapter reports the robust optimization task that was carried out to enhance the natural laminar flow (NLF) on the wing of the aforementioned supersonic business jet under geometrical uncertainties. The reference wing-body configuration has been already optimized by CIRA within the SUPERTRAC EU project by using deterministic methods: Hence, robust optimization has been used to refine the aerodynamic design in order to produce an optimal wing-body shape which offers a better behavior with respect to uncertainties in wing geometry.

In order to take into account the propagation of the uncertain input variables distribution onto the objective function and establish criteria for robust optimization, various risk measures can be defined in the context of robust and reliability-based optimization, and different risk functional definitions can lead to different approaches to the problem of optimization under uncertainty. The optimization framework here adopted is based on the value-at-risk (VaR) and conditional value-at-risk (CVaR) risk measures, also called quantile and super-quantile, respectively, which were originally conceived in the area of financial engineering. Very coarse VaR and CVaR estimations are used in the optimization process and, as this could lead to inaccurate

D. Quagliarella (✉) · E. Iuliano
Fluid Mechanics Department, C.I.R.A.—Italian Aerospace Research Centre, Via Maiorise, 81043 Capua, Italy
e-mail: d.quagliarella@cira.it

E. Iuliano
e-mail: e.iuliano@cira.it

C. Hirsch et al. (eds.), *Uncertainty Management for Robust Industrial Design in Aeronautics*, Notes on Numerical Fluid Mechanics and Multidisciplinary Design 140, https://doi.org/10.1007/978-3-319-77767-2_37

estimation of the objective function cumulative distribution, the “bootstrap” computational statistics technique is used to get an estimate of standard error and confidence intervals of the risk function.

After a brief recall of the theory behind the risk measures and their inclusion within a robust optimization process, the baseline description and the design problem definition, available in the database of the UMRIDA project, are reported. Then, the adopted shape parameterization, the definition of the uncertain variables and a rough sensitivity analysis are presented. Finally, the results obtained by using the VaR and CVaR risk measures are reported and related issues are discussed.

Theory Recall

The approach to robust and reliability-based design here adopted relies on the use of various risk measures, and it is described in detail in chapter “Value-at-Risk and Conditional Value-at-Risk in Optimization Under Uncertainty” of this collection and in the UMRIDA deliverable reports [1, 2]. In a very concise outline, the starting point is the formal definition of a risk measure, together with the consideration that different risk functional definitions can lead to different approaches to the problem of optimization under uncertainty and, hence, to different results. Here, in particular, the value-at-risk (VaR) and conditional value-at-risk (CVaR) measures are adopted to evaluate the wing-body aerodynamic performance in the risk measure-based optimization algorithm. Coherently, with what exposed in chapter “Value-at-Risk and Conditional Value-at-Risk in Optimization Under Uncertainty” and in [1] the “bootstrap” computational statistics technique is used to get an estimate of the standard error on VaR and CVaR.

Value-at-Risk and Conditional Value-at-Risk

The solution of the NLF design optimization problem here illustrated relies on the application of two risk measures, namely the value-at-risk (VaR) and the conditional value-at-risk (CVaR), also called quantiles and superquantiles. These risk measures originated in the area of financial engineering, but they are very well and naturally suited to reliability-based design optimization problems and they represent a viable alternative to more traditional robust design approaches. The definitions of VaR and CVaR, taken from [3–5] and adopted in [2], are here briefly reported for the sake of completeness. Let X be a random variable and $F(y) = \Pr\{X \leq y\}$ be the cumulative distribution function (CDF) of X. Then the inverse CDF of X can be defined as $F^{-1}(\gamma) = \inf\{y : F(y) \geq \gamma\}$. Following the definitions of [5], and coherently with what reported in [6], for any $\alpha \in (0, 1)$, the α-VaR of X is defined as

$$v^{\alpha} = F^{-1}(\alpha), \tag{1}$$

and it can be considered the maximum loss that can be exceeded only in $(1-\alpha)100\%$ of cases. Correspondingly, the α-CVaR of X can be thought of as the conditional expectation of losses that exceed the v^α level, and can be expressed as

$$c^\alpha = \frac{1}{1-\alpha}\int_\alpha^1 v^\beta \mathrm{d}\beta \tag{2}$$

as it was first outlined by Acerbi in [7]. References [8–10] show that c^α can also be written as the following stochastic program:

$$c^\alpha = \inf_{t\in\mathbb{R}}\left\{t + \frac{1}{1-\alpha}E[X-t]^+\right\} \tag{3}$$

with $[a]^+ = \max\{0, a\}$. The set of optimal solutions to the stochastic program (3) is, according to [5], $T = [v^\alpha, u^\alpha]$ with $u^\alpha = \sup t : F(t) \le \alpha$. In particular, $v^\alpha \in T$, so

$$c^\alpha = v^\alpha + \frac{1}{1-\alpha}E[X-v^\alpha]^+ \tag{4}$$

In the case where we have a finite number of samples, that is a ECDF, we proceed as described below. If $X_1, X_2, \dots, X_n$ are n independent and identically distributed (i.i.d.) observations of the random variable X, then the α-VaR of X can be estimated by

$$\hat{v}^{\alpha;n} = X_{\lceil n\alpha\rceil:n} = F_n^{-1}(\alpha) \tag{5}$$

where $X_{i:n}$ is the i-th order statistic from the n observations, and

$$F_n(y) = \sum_{i=1}^n \mathbb{1}\{X_i \le y\} \tag{6}$$

is the empirical CDF constructed from the sequence $\tilde{X}$ of $X_1, X_2, \dots, X_n$, and $\mathbb{1}\{\cdot\}$ is the indicator function. The estimation of α-CVaR of X, according to reference [5], can be directly obtained using Eq. (4):

$$\hat{c}^{\alpha;n} = \hat{v}^{\alpha;n} + \frac{1}{n(1-\alpha)}\sum_{i=1}^n [X_i - \hat{v}^{\alpha;n}]^+ \tag{7}$$

Robust Optimization Algorithm

A robust optimization problem can be defined in terms of an arbitrary number of risk measures of different types within the framework of multi-objective optimization:

$$\min_{\mathbf{x}} \hat{\rho}_{i;n}(\mathbf{x}) \quad i = 1, \dots, p \tag{8}$$

being p the number of objectives chosen, $\mathbf{x} \in X \subseteq \mathbb{R}^m$ the vector of deterministic design parameters, and $\hat{\rho}_{i;n}$ an estimator of the risk measure ρ_i obtained using a sample of size n. The problem is easily extended to the constrained case considering the following set of inequalities defined in terms of q further risk measures:

$$\hat{\rho}_{i;n}(\mathbf{x}) \leq K_i \quad i = p+1, \ldots, p+q \tag{9}$$

Regardless of the algorithm used, it is evident that the result of the optimization will be affected by the quality of the risk function estimate and, therefore, by the number of samples and by the algorithm used to obtain it. We note in this regard that advanced importance sampling algorithms can significantly increase the quality of the risk function estimation. From the point of view of the optimization algorithm, the higher or lower quality of an estimate can be seen, in the broad sense, as noise (more or less random) that influences the value of objective functions and constraints. Therefore, the first thing to do, in the absence of other possibilities, is to use an optimization algorithm scarcely sensitive to noise. In this work, we chose to use a gradient-free algorithm, in particular, the ADG multi-objective genetic algorithm developed in-house [11–13]. In any case, even an algorithm not quite sensitive to noise, but not explicitly designed for robust optimization problems will respond in a non-ideal way when applied to this class of problems. In particular, the most critical phase of the optimization process occurs when the magnitude of the improvement obtainable in a single step of the algorithm is of the same order of magnitude than the uncertainty in the estimation of the risk function. Therefore, it is of fundamental importance to adapt the optimization algorithms so that they can operate effectively in these situations. They must therefore be equipped with the ability to evaluate the error related to a sample of given size and to the algorithm used to obtain the estimate.

It is important to note that the error estimate must be applied to each set of samples that constitute an ECDF, since this, similarly to the value of the risk functions (which are the building blocks of our robust design and optimization system) also depends on the particular value assumed by the deterministic design parameters. Once the error value is available, it can be used at various levels in the algorithm. In the present work, in particular, it is simply used to decide when to stop the and possibly switch to a more refined sampling level in the estimation of risk functions. Here we adopt the bootstrapping technique to measure the accuracy of our estimates [14] in terms of variance, confidence intervals, and prediction errors. The details of the non-parametric bootstrap method used are, again, reported in [1] and in chapter "Value-at-Risk and Conditional Value-at-Risk in Optimization Under Uncertainty".

Design Problem Description

The reference configuration is the optimized wing-body shape produced by CIRA within the SUPERTRAC EU project [15]. This shape was optimized for natural laminar flow and was obtained from a baseline configuration produced by Dassault Aviation within the framework of the Supersonic Business Jet project and made available within SUPERTRAC project. The inboard wing has a 65° leading-edge sweep angle, while the outboard-wing sweep is 56°. The wing semi-span is 9.35 m and the aspect ratio is 3.5. The cruise flight Mach number is 1.6.

Within SUPERTRAC project, wing section airfoils and twist angle were optimized in order to maximize the laminar flow region while monitoring and controlling the pressure (vortex and wave) drag. Compared to SUPERTRAC baseline, the optimized wing showed an improved extent of laminar flow. The airfoil wing sections optimized by CIRA are not covered by copyright, so the test case is usable without non-disclosure agreements.

Geometry and Design Problem Definition

Figure 1 reports the UMRIDA baseline in isometric and orthographic projection, while wing-body geometric features are summarized in Table 1.

Fig. 1 UMRIDA baseline view

Table 1 Wing-body geometric features

Parameters	Values
Inboard sweep	65
Outboard sweep	56
Semi-span length	9.35 m
Aspect ratio	3.5
Wing area	50 m^2

Table 2 Design flow conditions

Parameters	Values
Mach	1.6
Reynolds	51.8×10^6
L_{ref}	6.27 m
AOA	3.65
C_L	0.182

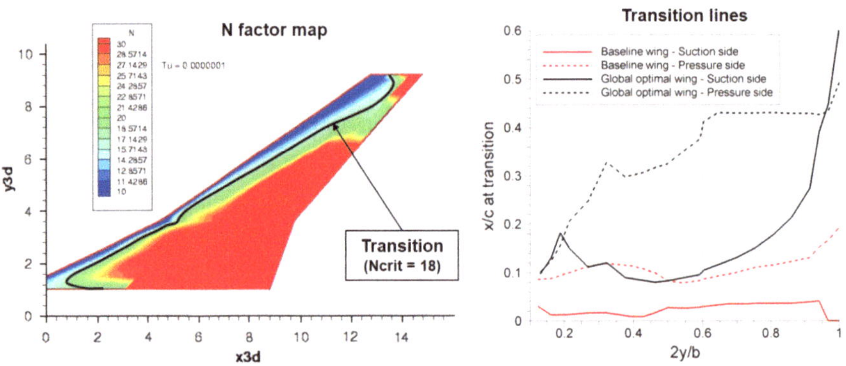

Fig. 2 UMRIDA optimized baseline features

The flow conditions for the optimization problem are those related to the main cruise design point and are reported in Table 2.

Figure 2 shows the N factor map on the wing suction side (left) and the transition locations (suction and pressure side) along the wingspan (right) as compared to the baseline wing. The deterministic optimization managed to damp the N factor levels so that not only the laminar flow is naturally enhanced (as observable from the right-hand side picture) but also the application of active flow control (e.g., suction) would require much less energy to further decrease the friction drag by increasing the laminar flow portion.

Optimization Problem

The detailed description of the original optimization problem is reported in [15]. A synoptic view of design condition, constraints, and objective as set up by CIRA is summarized in Table 3. The objective function here proposed has been changed to account for the new characterization of Mach and lift coefficient as uncertain parameters, and it is defined as:

Table 3 Problem definition

Design variables	
Wing twist	−3; +3
Wing section shape	User choice
Design point	
Mach number	1.6
Reynolds number	51 millions
Reference chord	6.27 (m)
Altitude	44,000 (ft)
Lift coefficient	0.182
Design constraints	
Lift coefficient	$C_L \geq 0.180$
Pitching moment	$C_M \geq -0.05$
Trailing edge angle	$tea \geq \tilde{tea} = 0.050$ (rad)
Leading-edge radius	$lea \geq \tilde{lea} = 0.0020$ (m)
Objective	
$G(C_L, C_D, C_M, ler, tea, S_{lam,u}, S_{lam,l})$	To be minimized

$$G = K_a \frac{C_D + C_{D,M} + C_{D,L}}{C_L} \frac{\tilde{C}_L}{\tilde{C}_D} - K_u S_{lam,u} - K_l S_{lam,l} + \\ + K_r P\left(1 - \frac{ler}{\tilde{ler}}\right) + K_t P\left(1 - \frac{tea}{\tilde{tea}}\right)$$

with the terms below aimed at accounting the contribution to drag due to trim:

$$C_{D,M} = \max[0, 0.05(\tilde{C}_M - C_M)]$$

$$C_{D,L} = \max[0, 1.0(\tilde{C}_L - C_L)]$$

and where $\tilde{C}_L$, $\tilde{C}_D$, and $\tilde{C}_M$ are, respectively, the lift, drag, and pitching moment coefficients of the baseline, and $K_{a,u,l,r,t}$ are constant values that define the relative importance of the corresponding aerodynamic/geometric performance component. Used values are: $K_a = 0.2$, $K_u = 0.3$, $K_l = 0.5$, $K_r = 100$, and $K_t = 100$. The quadratic penalty function is activated only if its argument is positive. Hence, P has the following expression:

$$P(x) = \begin{cases} x^2 & \text{if } x > 0 \\ 0 & \text{if } x \leq 0 \end{cases}$$

The function S_{lam} is introduced to estimate the transition position along the whole wingspan, and it is defined as

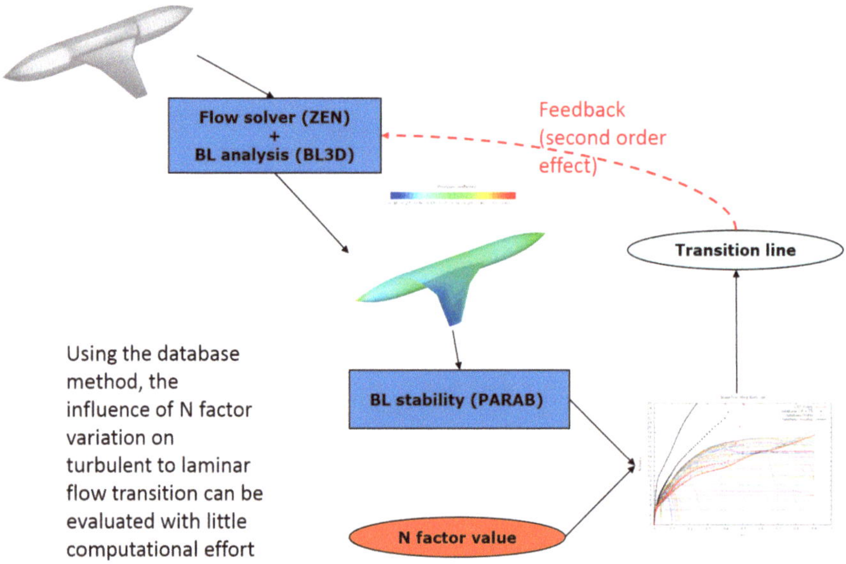

Fig. 3 Computational setup for the objective function computation

$$S_{lam} = \frac{1}{S} \int_0^{b/2} X_{tr}(y) dy$$

where S is the wing area, $b/2$ is the half-span length, and $X_{tr}(y)$ is the chordwise laminar to turbulent transition location at spanwise section y. The right-hand side picture in Fig. 2 provides a graphical representation of $X_{tr}(y)$. The function S_{lam} may assume any value between zero and one, with $S_{lam} = 0$ meaning that the flow is fully turbulent and $S_{lam} = 1$ meaning that the flow is fully laminar. In other words, it gives an estimate of the fraction of the wing area along which the flow runs laminar.

Computational Model

The physical and computational model adopted for objective function computation is thoroughly described in [15]. A scheme of the computational setup is reported in Fig. 3 for the sake of completeness.

Definition of Uncertainties

The definition of uncertain parameters must take into account the design problem at hand. The aim of the designer is to obtain a configuration that has a satisfactory level of laminar flow even in the presence of parameters and working conditions not completely deterministic and controllable. A further difficulty is due to the epistemic uncertainty inherently included in the computational model for the transition from laminar to turbulent flow. This can significantly affect the performance calculation of the new configurations. Consequently, a good design should be quite robust even with respect to this latter type of uncertainty sources.

Geometrical Uncertainties

Natural laminar flow is mostly sensitive to the shape of the leading edge region. This is due to its effect on pressure coefficient gradient which, in turn, is one of the factors that have most influence on the transition. Here, the deterministic parameterization (i.e., parameters to obtain the best performance in an optimization process) of the wing shape is univocally defined and the uncertain parameters (i.e., parameters to control the shape change due to geometrical uncertainties) are seen as additive terms whose values are extracted from known statistical distribution. In other words, given $\{x_1, x_2, \dots, x_n\}$ the set of deterministic design variables whose realization $\{x_1^*, x_2^*, \dots, x_n^*\}$ univocally defines the three-dimensional wing surface $S^*(x_1^*, x_2^*, \dots, x_n^*)$, the uncertainty in the geometry definition is introduced by adding a set of random design variables $\{\xi_1, \xi_2, \dots, \xi_n\}$ to the realization of the deterministic set, so that the wing surface is represented by the statistical function $S^*(x_1^* + \xi_1, x_2^* + \xi_2, \dots, x_n^* + \xi_n)$. In the present case, uniform probability distributions are considered for all geometric uncertain parameters. Section "Shape Parameterization and Sensitivity Analysis" will give details about the chosen parameterization.

Operational Uncertainties

Operational uncertainties are the classical ones related to Mach number and lift coefficient (C_L). Mach and C_L are modeled as four parameter beta distributions. For the sake of completeness, we recall that the probability density function f for $a \leq x \leq b$ and shape factors α, β is given by

$$f(x; \alpha, \beta, a, b) = \frac{(y - a)^{\alpha-1}(b - y)^{\beta-1})}{(b - a)^{\alpha+\beta-1}B(\alpha, \beta)}$$

and by $f = 0$ otherwise, with $B(\alpha, \beta) = \int_0^1 u^{\alpha-1}(1 - u)^{\beta-1}du$. The Table 4 summarizes the parameters that define the Mach and CL random variables.

In the present context, operational uncertainties are taken into account only in the sensitivity analysis phase in order to assess their first-order effect on the performance.

Table 4 Uncertain operational parameters

Parameter	α	β	a	b
Mach	4	4	1.55	1.65
C_L	2.5	2.5	0.180	0.184

Model Uncertainties (Epistemic)

One of the challenges that have to be faced when approaching the numerical design of natural laminar flow wings is the reliable estimation of the point of transition from laminar to turbulent flow. A significant uncertainty in the determination of transition location is inherent to the methods for numerical transition prediction and in particular to the e^N method. A robust design approach should take into account this epistemic uncertainty source. If the e^N method is chosen, then the uncertain parameter to be considered is the N_{critical} factor. Its nominal value has been fixed at 18 for deterministic runs, while it will be modeled by a uniform distribution in the interval $[16, 20]$ when considered as uncertain parameter. As for the operational uncertainties, the N factor uncertainty are considered only in the sensitivity analysis phase.

Shape Parameterization and Sensitivity Analysis

Parameterization Approach

The criticality of a parameterization that effectively models the uncertainty in the form of the leading edge of a wing lies mainly in the fact that these parameters must simultaneously act on a small portion of the wing and, at the same time, should not produce discontinuities such to prevent the use of the fluid dynamic solver or interfere with the parameterization of the deterministic variables that control the shape of the whole wing. The method here chosen is based on NURBS. It uses a grid of 21×7 NURBS control points (CPs) defined on the whole wing surface (see Fig. 4) and a subset of it (21 CPs) is used to control the wing leading-edge shape. As depicted in Fig. 5, seven control points are employed to modify the wing leading edge shape in the streamwise direction using a third-order basis functions, and three control points, with second-order basis functions, operate spanwise. The design variables are identified with the vertical displacements of the control points.

As aforementioned, the uncertainty in the wing shape geometry is represented by a uniform random perturbation $\xi_i, i = 1, \ldots, n$ that is added to the vertical displacements of the deterministic set of NURBS control points. Table 5 summarizes the deterministic and uncertain design variables together with their ranges and distributions.

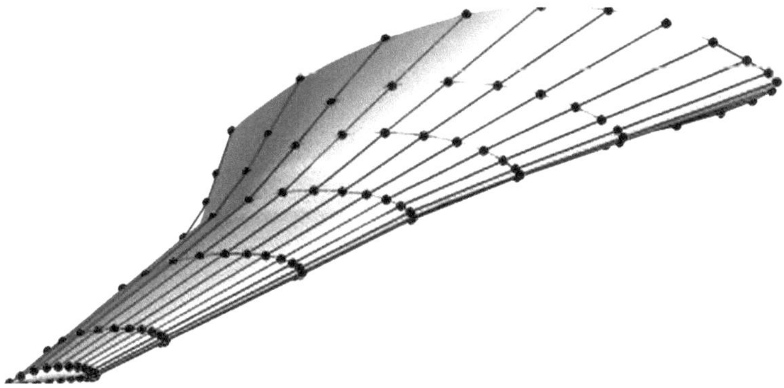

Fig. 4 NURBS control points to control the whole wing shape (21 × 7)

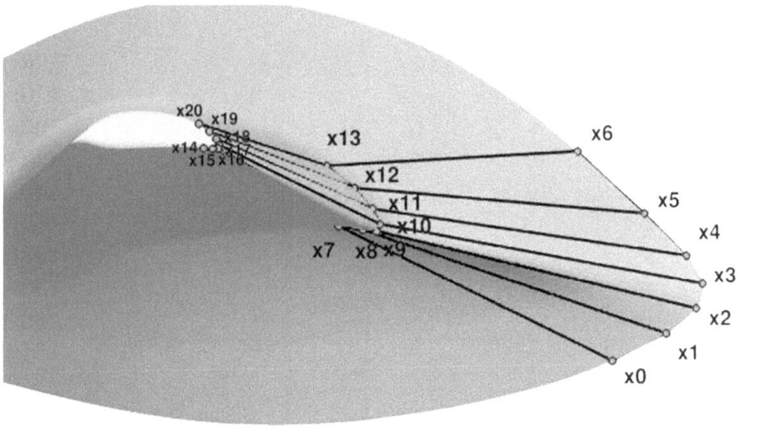

Fig. 5 Active NURBS control points to control local leading-edge shape modifications (7 × 3)

Sensitivity Analysis

A sensitivity analysis has been performed to preliminarily identify, using the analysis of variance, the main dependencies and interactions of the uncertain variables.

The combined effects of geometry and operating condition uncertainties are here considered using sensitivity analysis. This task is aimed to discover the global sensitivities of the objective function related to the baseline configuration with respect to the above-reported uncertainty sources.

The sensitivities are computed in the hypothesis that uncertain variables are characterized by an uniform distribution on their whole range. The expected goal of this screening activity is the identification of the most important uncertain variables, relative to their effect on the complete objective function.

Table 5 Uncertain design variables details

Design variable	Lower bound	Upper bound	Random variable	Lower bound	Upper bound	Distribution
x_0	−35	−15	ξ_0	−1	1	UNIFORM
x_1	−30	−10	ξ_1	−1	1	UNIFORM
x_2	−25	−5	ξ_2	−1	1	UNIFORM
x_3	−2	3.5	ξ_3	−0.275	0.275	UNIFORM
x_4	20	40	ξ_4	−0.5	0.5	UNIFORM
x_5	50	90	ξ_5	−2	2	UNIFORM
x_6	90	160	ξ_6	−3.5	3.5	UNIFORM
x_7	−20	0	ξ_7	−1	1	UNIFORM
x_8	−20	0	ξ_8	−1	1	UNIFORM
x_9	−20	0	ξ_9	−1	1	UNIFORM
x_{10}	−2	3.5	ξ_{10}	−0.275	0.275	UNIFORM
x_{11}	5	30	ξ_{11}	−1.25	1.25	UNIFORM
x_{12}	20	50	ξ_{12}	−1.5	1.5	UNIFORM
x_{13}	40	80	ξ_{13}	−2	2	UNIFORM
x_{14}	−20	10	ξ_{14}	−1.5	1.5	UNIFORM
x_{15}	−20	10	ξ_{15}	−1.5	1.5	UNIFORM
x_{16}	−20	10	ξ_{16}	−1.5	1.5	UNIFORM
x_{17}	−2	3.5	ξ_{17}	−0.275	0.275	UNIFORM
x_{18}	5	30	ξ_{18}	−1.25	1.25	UNIFORM
x_{19}	10	30	ξ_{19}	−1	1	UNIFORM
x_{20}	15	50	ξ_{20}	−1.75	1.75	UNIFORM

Variance-Based Decomposition via Sobol Indices

The parametric study was performed computing the Sobol indices via variance-based decomposition which is a form of global sensitivity analysis that decomposes the variance of the model output into fractions which can be attributed to single input variables and into other parts related to sets of interacting input variables. This kind of analysis is able to deal with nonlinear responses and to evaluate the effect of interactions in non-additive systems. A basic assumption is that inputs are independently and uniformly distributed within the unit hypercube.

The total variance of the sampled response function is decomposed as:

$$\mathrm{Var}(Y) = \sum_{i=1}^{d} V_i + \sum_{i<j}^{d} V_{ij} + \cdots + V_{12\ldots d} \tag{10}$$

where V_i is the variance due to the pure variation of factor X_i:

$$V_i = \mathrm{Var}_{X_i}\left(E_{X_{\sim i}}\left(Y|X_i\right)\right) \tag{11}$$

and V_{ij} is the variance due to the pure variation of both factors X_i, X_j:

$$V_{ij} = \mathrm{Var}_{X_{ij}}\left(E_{X_{\sim ij}}\left(Y|X_{ij}\right)\right) \tag{12}$$

with $\sim i$ notation indicating the effect of all variables except X_i.

The Sobol index S_i is defined as

$$S_i = \frac{V_i}{\mathrm{Var}(Y)}, \tag{13}$$

and it measures the main (first order) effect, e.g., the effect of varying X_i alone, but averaged over variations in the other input parameters. The following relations hold:

$$\sum_{i=1}^{d} S_i \leq 1, \qquad \sum_{i<j}^{d} S_i + \sum_{i<j}^{d} S_{ij} + \cdots + S_{12\ldots d} = 1 \tag{14}$$

The index S_{Ti} measures instead the total effect, e.g., the contribution of all terms in the variance decomposition which do include X_i. Thus, the first-order effect plus the interactions are accounted by:

$$S_{Ti} = \frac{E_{X_{\sim i}}\left(\mathrm{Var}_{X_i}\left(Y|X_{\sim i}\right)\right)}{\mathrm{Var}(Y)} = 1 - \frac{\mathrm{Var}_{X_{\sim i}}\left(E_{X_i}\left(Y|X_{\sim i}\right)\right)}{\mathrm{Var}(Y)} \tag{15}$$

The interaction effect between, e.g., X_i and X_j is counted in both S_{Ti} and S_{Tj}, hence we have

$$\sum_{i=1}^{d} S_{Ti} \geq 1 \tag{16}$$

Design Space Sampling

The goal of this sensitivity analysis is to understand the contribution of the uncertainty sources on the whole design problem and, hence, the full objective function value is considered (see section "Optimization Problem"), which is a weighted mixture of aerodynamic efficiency and of a measure of laminar portions on the wing surfaces.

An Orthogonal Array Latin Hypercube sampling strategy [16] was chosen as it has both orthogonality and stratification features and is therefore a good choice for computing Sobol indices. The Sobol indices were computed using a corrected version of Saltelli's formula [17, 18] which is implemented within DAKOTA package. The sampling size chosen to fill the hypercube required 21,025 evaluations.

Failure Handling

The geometry modification process has been set up by empirically fixing the variation ranges of the uncertain geometry variables in order to obtain a global uncertainty in the description of the leading edge shape that was within the order of 5% of the wing section shapes. This has made unavoidable having mesh generator/CFD solver failures for particular combinations of some design variables The observed probability of failure was 0.006%: Although quite limited, failed samples may have the capacity to impair the computation of variance contributions and, hence, they have to be handled appropriately to not reduce the reliability of the computed sensitivity indices.

In this work, an exploratory approach was adopted to the problem of handling failed samples, by imposing a fictitious value to the objective function of the candidate sample in case of failures within the computational chain. Three different failure modes were experimented here:

1. assign a large response penalization to failed samples;
2. assign a response value close to the mean of the objective function values of the samples database;
3. assign a response value slightly larger than the maximum objective function value of the samples database;

In the subsequent sections, the results obtained with each of these approaches are reported and discussed. Thanks to DAKOTA restart capabilities, assigning new response values to particular samples is very easy and does not require the recomputation of the objective function database.

Sensitivity Analysis Results

The S_i (main effect) and S_{Ti} (total effect) indices are reported in Figs. 6 and 7, respectively. Variables from 0 to 20 represent the shape variables, while variables 21 and 22 identify Mach and C_L variables. It can be observed from Fig. 6 that the main effect is predominantly concentrated on variables 8, 12, 13, and 15 that control the pressure side shape on the mid-outboard wing. These variables have the highest influence on the objective function, but the main effect of Mach and C_L (variable 21 and 22) is also significant. These results are in good agreement with preliminary, coarse grid CFD analyses not reported here.

It appears also evident that adding a large penalty to failed computations might lead to mask the true influence of the design parameters on the objective function. To avoid this problem, different strategies to assign a score to failed computations were tried. In particular, we found that assigning an objective value equal to the average or to the maximum value computed from the successful computations does not seem to affect the results quality in terms of relative sensitivities. Figure 7 is related to the variable total effect, e.g., including the interactions with the other variables, and it tells a different story. Here the mean value and the maximum value curves

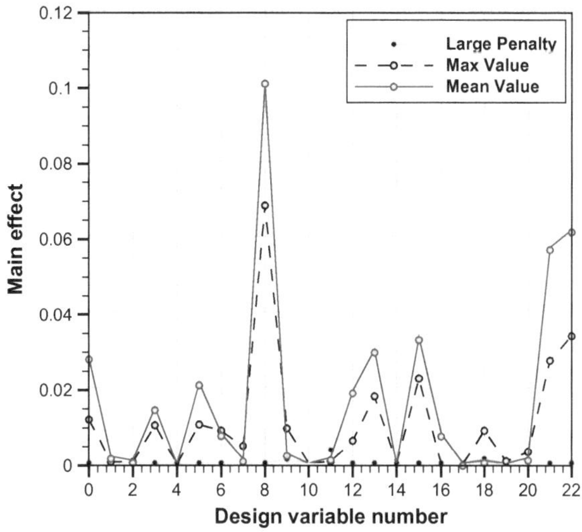

Fig. 6 Main effect of uncertain variables

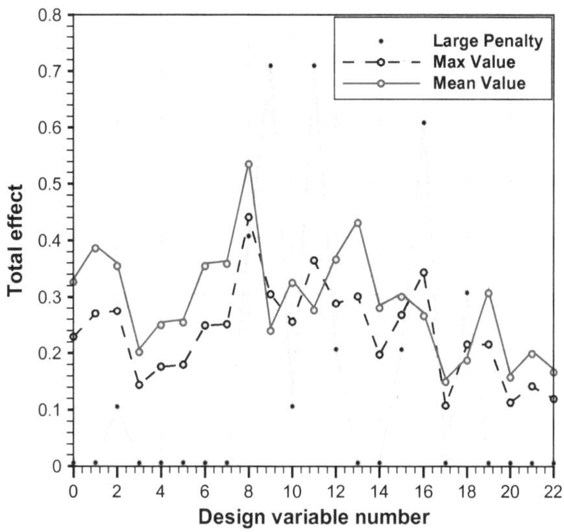

Fig. 7 Total effect of uncertain variables

evidence that all the parameter shape values show significant interactions. Inboard wing variables (from 0 to 6) and Mach/C_L variables have statistically less importance with respect to mid-outboard ones. The large penalty curve shows how the interaction between variables 9, 11, 16, and 18 (controlling the leading edge radius and shape on the mid-outboard wing) plays a major role in generating failed samples. The sum of main and total index values are reported below to allow a comparison of the importance order of these effect when the objective function values related to failed configurations are changed:

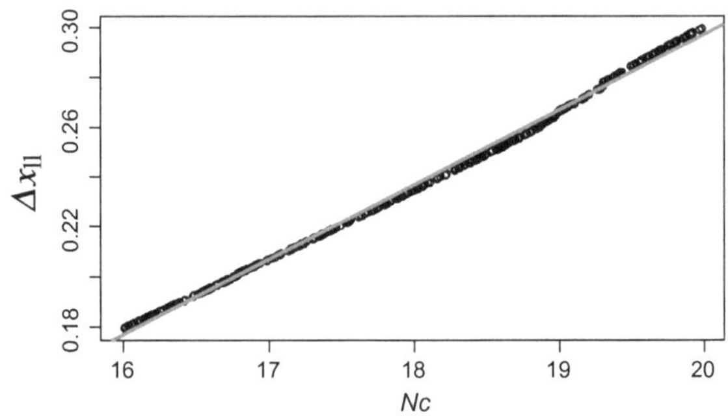

Fig. 8 Influence of N_{critical} (Nc) on lower surface laminar flow extension

$$\sum_{i=1}^{d} S_i = \begin{cases} 0.006 \text{ (large penalty)} \\ 0.39 \quad \text{(mean value)} \\ 0.24 \quad \text{(max value)} \end{cases} \tag{17}$$

$$\sum_{i=1}^{d} S_{Ti} = \begin{cases} 3.32 \text{ (large penalty)} \\ 6.68 \quad \text{(mean value)} \\ 5.34 \quad \text{(max value)} \end{cases} \tag{18}$$

The sensitivity analysis suggests that, at least with the chosen parameterization, it is not straightforward to obtain clear indications toward the reduction of the number of variables or simplification of the model. However, it clearly shows that the correct management of computational chain failures has a crucial role in the setup of the optimization problem. In other words, while some variables have a main effect considerably higher than other ones, the interaction effects are significant for each variable, and this implies that a reduction of the number of uncertain variables is not advisable here.

Sensitivity Analysis with Respect to Model Uncertainties

The most important parameter in determining the behavior of the computational chain modeling the laminar to turbulent flow transition is, undoubtedly, the critical value N_{critical} of the amplification factor N, when the e^N transition method is adopted [19]. Consequently, it was decided to focus the sensitivity analysis to just this parameter, the value of which, among other things, must be obtained from experimental correlations. In our case, in order to simplify the interpretation of the numerical test data, it was decided to study its effect exclusively on the extension of laminar zone on the lower surface of the wing, namely $S_{lam,l}$. The parameter N_{critical} was, hence, sampled according to an uniform distribution in [16, 20]. The population sample is composed of 400 members and includes the baseline.

Table 6 Analysis of variance for the N_{critical} linear model fit

Response: Δx_{ll}

	Df	Sum Sq	Mean Sq	*F* value	*Pr*(> *F*)
N_{critical}	1	0.48266	0.48266	126,694	$<2.2 \times 10^{-16}$
Residuals	398	0.00152	0.00000		

The results, summarized in Table 6 and reported in Fig. 8, show that, at least for the limited region of parameter space in proximity to the baseline, the overall effect of N_{critical} is almost perfectly linear. Therefore, if the same behavior would be confirmed also for the optimized configuration, its effect may be easily taken into account using a linear correlation which can be applied even after the optimization phase.

Wing-Body Robust Optimization

Problem Setup

The objective of this section is to perform a robust optimization by using both VaR and CVaR risk measures. A genetic algorithm is launched with a population size of 32 with 1% activation probability of the classical bit mutation (with binary strings represented using Gray code) and 80% activation probability of one point crossover. A total number of 32 generations have been computed; however, significant results have been obtained after just five generations as it will be made clear afterward. In order to partially mitigate the drawback of using only 16 samples within the uncertainty quantification internal loop, the bootstrap technique is used to estimate the confidence intervals. In the following, the VaR results will be shown first together with a discussion on the lack of accuracy in the tail prediction due to coarse sampling; then, the adoption of the CVaR measure will be introduced as a potential remedy of such issues and CVaR-based optimization results will be finally shown.

Uncertainty Quantification of the Baseline Configuration

Prior to getting in the optimization phase, the baseline configuration has been analyzed and the propagation of the input uncertainties over the objective function has been quantified. A random plan of 160 samples has been used to compute the VaR and CVaR levels from the empirical cumulative distribution function (ECDF). Moreover, a non-parametric bootstrap analysis with 10,000 repetitions (extracted from the 160 samples with sample re-injection) has been performed to estimate the standard error of both VaR and CVaR measures. Results are shown in Fig. 9. The plot indi-

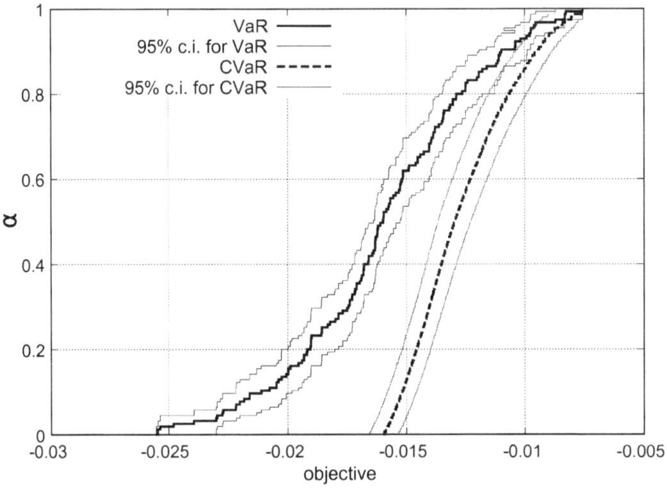

Fig. 9 Var and CVaR curves and their confidence intervals computed using non-parametric bootstrap

cates that the spread of the objective function due to the uncertain variables is quite large (from about −0.025 to −0.0075) and that a big role is played by the tails of the distribution, especially the upper one where an increase in the CVaR gradient is clearly observed. As a matter of fact, the following optimization phase is expected to achieve two main objectives: The first is to move the curves toward the left-hand side of the plot in order to minimize the target response in average; the second is to minimize the variation along the x-axis, thus making the response distribution more compact.

VaR-Based Optimization: Results and Discussion

Value-at-risk response function at $\alpha = 0.9$ is estimated with 16 Monte Carlo samples during the optimization phase, while in the post-run analysis phase results are validated using 160 samples. Figure 10 shows the convergence history of the optimization run: As the response value scale is quite large, a zoom around the optimal values is proposed in Fig. 11. The simulation has been stopped after 32 generations; however, a kind of convergence is observable by looking at the plateau after 800 evaluations.

Figure 12 depicts a comparison of ECDF curves as obtained with the VaR criterion at selected generations 5, 15, 21, and 32. Two basic features can be observed:

1. the ECDF curve is globally and progressively shifted toward the left region as the problem is cast in a minimization fashion;

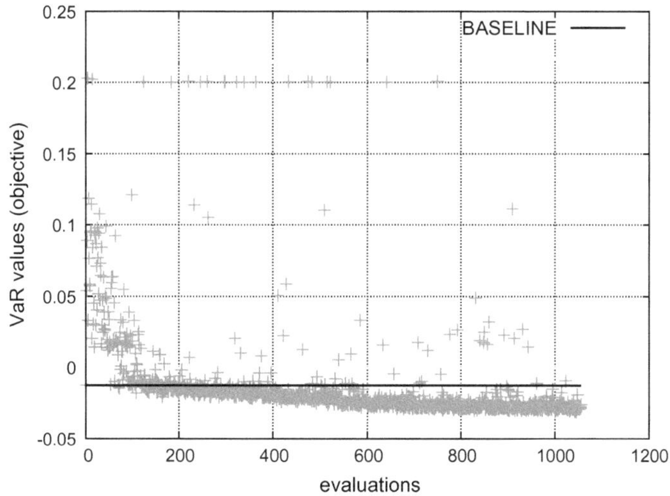

Fig. 10 Evolution history after 32 generations for the VaR optimization run

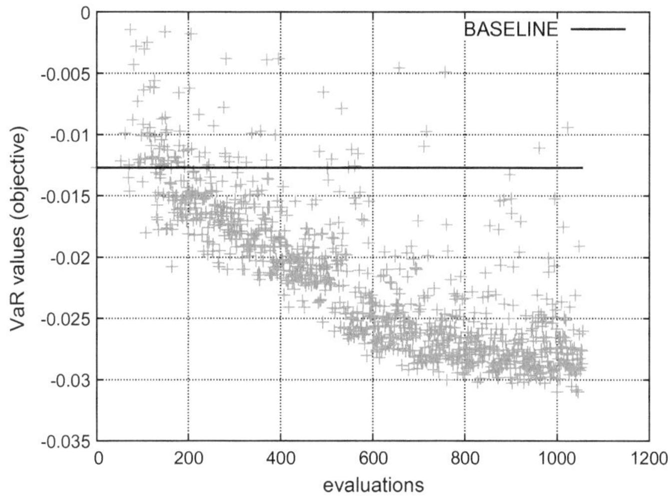

Fig. 11 Evolution history after 32 generations (zoom) for the VaR optimization run

2. the local slope of the ECDF is becoming steeper and steeper with the progress of the optimization simulation.

This behavior allows to reduce both the mean and the variance of the uncertain response. However, in order to evaluate the effect of the coarse sampling within the uncertainty quantification step and to estimate the noiseness of $\nu^{0.9}$ during the optimization process, a bootstrap analysis is performed on the ECDF of the baseline shape and of the computed optima at generations 5, 15, 21, and 32: Basically,

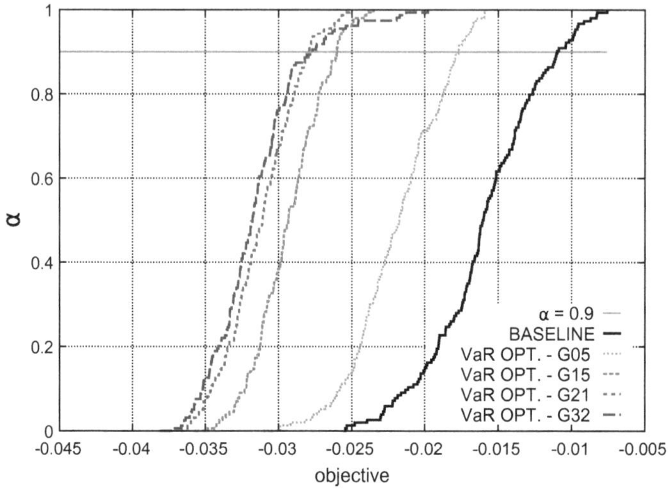

Fig. 12 Comparison of the ECDF obtained with VaR at different generations

given a set of 16 (as during the optimization run) random samples involving only the uncertain variables, a high number of sub-sampling sets of dimension 16 are drawn from the initial set with replacement and confidence intervals are estimated. This has been done by using 5 different initial sampling sets and results are shown in Fig. 13. It can be observed that the general trend is in line with the progress of the simulation, as the mean values decrease with increasing number of generations. However, the bootstrap analysis highlights that the noise level in the estimation might not allow a correct ranking of the population elements: As a matter of fact, it is not clearly distinguishable which is the best design candidate between generations 21 and 32 optima by looking at the confidence intervals. Indeed, dealing with an optimization under uncertainty problem, we cannot refer only to mean values, as those values may be strongly dependent on the sample size that has been used to compute them. This is what Fig. 13 suggests: The combined effect of value-at-risk measure and coarse sample size may lead to confusion in candidate ranking in advanced stages of the optimization process.

Another evidence of this behavior is reported in Fig. 14 where a similar bootstrap analysis is repeated over one (instead of 5) sampling set of size 160 (ten times the previous size): Moreover, both value-at-risk and conditional value-at-risk measures at $\alpha = 0.9$ have been computed and shown with the related confidence interval. The grey curve (i.e., VaR values) basically confirms the onset of possible problems in comparing solutions with good ranking, i.e., the optimum candidate at generation 32 presents larger confidence intervals with respect to generation 21: This means that it has been selected only by chance due to the fact that the VaR-based criterion is not able to detect such issues. On the other hand, the black dotted curve (i.e., CVaR values) provides better information as, according to such criterion, the generation 21 optimum is definitely preferable to generation 32. This provides the final information

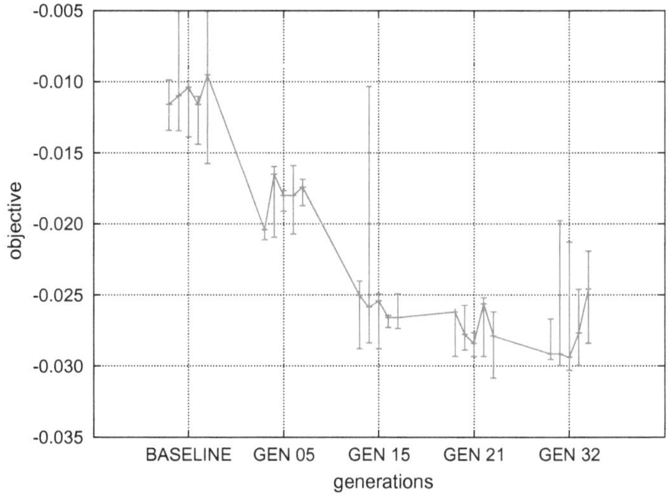

Fig. 13 Bootstrap over repeated coarse samples of the ECDF (16 samples)

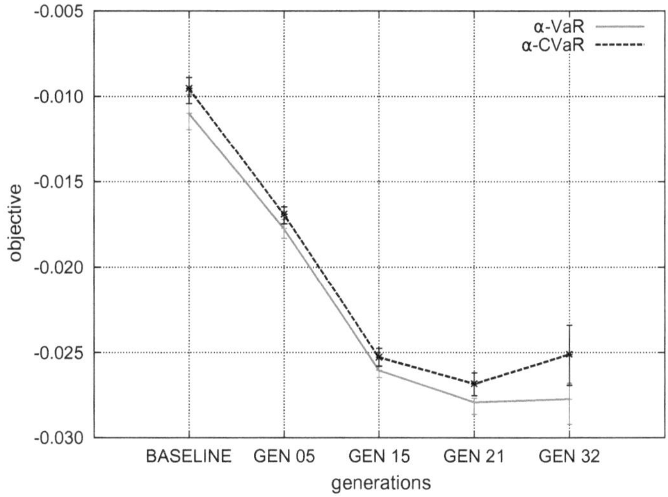

Fig. 14 VaR, CVaR at $\alpha = 0.9$ and related confidence intervals for a 160 sample ECDF

about the capability of both approaches; however, it should be recalled that the results in Fig. 14 have been obtained over a sample size of 160, which is computationally not practical when dealing with an optimization under uncertainty. Hence, the next question is naturally: Is CVaR able to properly rank candidates even by decreasing the sample size? The answer is condensed in Figs. 15, 16 and 17: Here, the objective function values, both VaR and CVaR, are computed by repeating 31 times the random drawing of samples of respectively 16, 32, and 64 size from the 160-size sampling

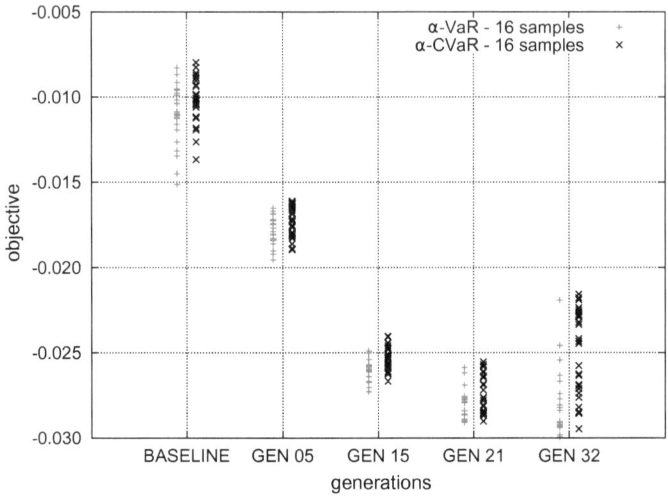

Fig. 15 VaR and CVaR at $\alpha = 0.9$ over 16-size repeated samples drawn from a 160 sample ECDF

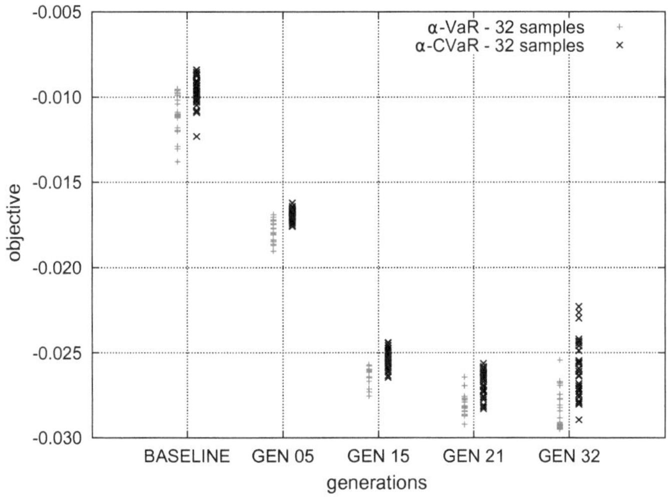

Fig. 16 VaR and CVaR at $\alpha = 0.9$ over 32-size repeated samples drawn from a 160 sample ECDF

set that has been mentioned above. With this approach, by taking the sampling size to its limit of 160, the curves in Fig. 14 would be obtained.

Of course, by increasing the sample size, the spread tends to narrow as more and more samples are common to the repetitions, but the interesting point lies in observing that, starting from 32 samples, the CVaR measure is able to detect the mentioned issue on generation 32, as the response distribution is located higher (i.e., it is worse on average) and is wider (i.e., it has larger variability) than generation 21.

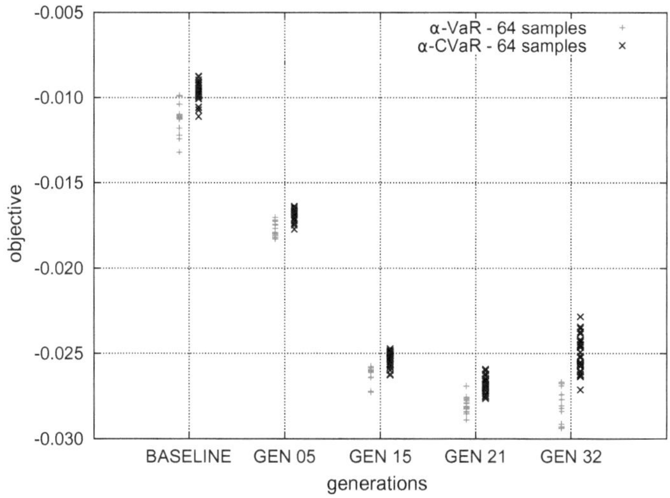

Fig. 17 VaR and CVaR at $\alpha = 0.9$ over 64-size repeated samples drawn from a 160 sample ECDF

As a result, a new optimization using the CVaR risk measure as objective function is envisaged and detailed in the following section.

CVaR-Based Optimization: Results and Discussion

The CVaR-based optimization is performed by using nearly the same setup as the previous one, except for the following differences:

- the genetic algorithm is stopped after 40 generations (32 in the VaR-based optimization);
- the inner sampling to quantify the response uncertainty for each design candidate is enlarged to 32 (instead of 16). This reflects the need to have a reliable estimation of the risk measure (as observed in the final part of the previous section).
- the starting point is the deterministic optimum as found by running the genetic algorithm without any uncertain variable. In the previous optimization, the baseline shape has been used instead.
- the first random population is initialized by considering a hypercube whose size is 10% the original design space size: This should avoid a large spread of the objective function data in the initial stages of the process. In the previous case, no compression has been used.

Figure 18 reports the ECDF of five design, namely: the baseline shape, the deterministic optimum, the optima at generations 21 and 32 of the VaR-based optimization and the new CVaR-based optimum obtained at the final generation 40. The plot well

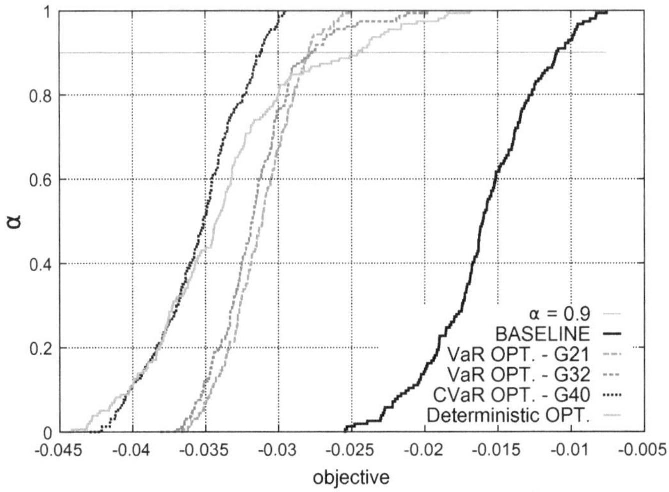

Fig. 18 Comparison of the ECDF obtained with CVaR-based optimization versus VaR based, deterministic, and baseline

explains how the CVaR optimization improves both the VaR value and the tail of the ECDF, even with respect to the VaR-based optima. As a matter of fact, the performance degradation in the tail is much more acceptable with respect to other solutions. Moreover, if compared to the ECDF built on the deterministic optimum, a kind of "counter-clockwise rotation" can be noticed which improves both the response distribution mean and the variance. As a consequence, dealing with a robust optimization problem, the drawback consists in having the lowest objective function value located on the deterministic optimum curve. Such a result of the optimization process is further shown in Figs. 19 and 20, where the VaR and CVaR curves are respectively depicted with the same approach as in Fig. 9. The CVaR-based and the deterministic optima are considered together with the baseline configuration. The comparison clearly confirms what has been already observed about the effect of the robust optimization and allows to further notice that the confidence intervals on the CVaR-based optimum are significantly reduced throughout the whole ECDF curve. This is not a trivial improvement as it underlines that not only the CVaR solution is more robust, but it is also much less sensitive to the sampling adopted to compute the ECDF.

A comparison of laminar to turbulent transition prediction is proposed in Figs. 21 and 22, respectively, for the suction (upper) and pressure (lower) sides of the wing surface. On the left-hand side, the solution on the baseline shape is shown, while the CVaR optimum is depicted on the right-hand side. The plots are obtained by projecting the transition locus along the wingspan for each individual along the ECDF presented above. Hence, a kind of "transition band" is obtained: The more it is located far from the leading edge, the better is the response value and the larger is the band, the higher is the response spread with uncertainty. It can be noticed that the solution does not change too much on the upper side, where a larger sensitivity to the uncer-

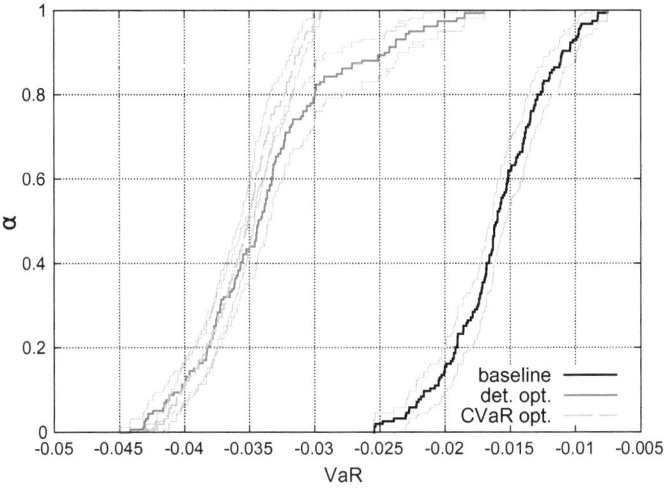

Fig. 19 Comparison of VaR curves with bootstrap confidence intervals for baseline, deterministic optimum, and CVaR-based optimum

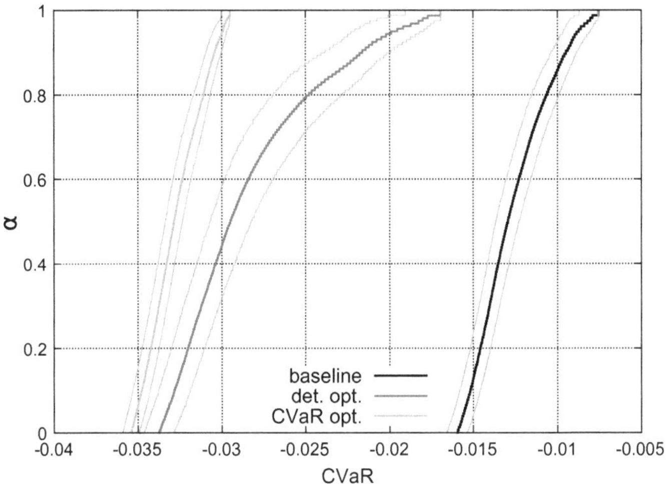

Fig. 20 Comparison of CVaR curves with bootstrap confidence intervals for baseline, deterministic optimum, and CVaR-based optimum

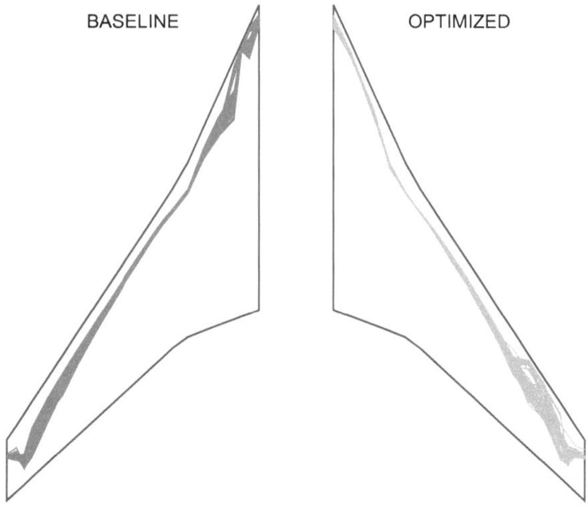

Fig. 21 Transition line envelopes comparison for the wing upper surface with $\alpha \leq 1$

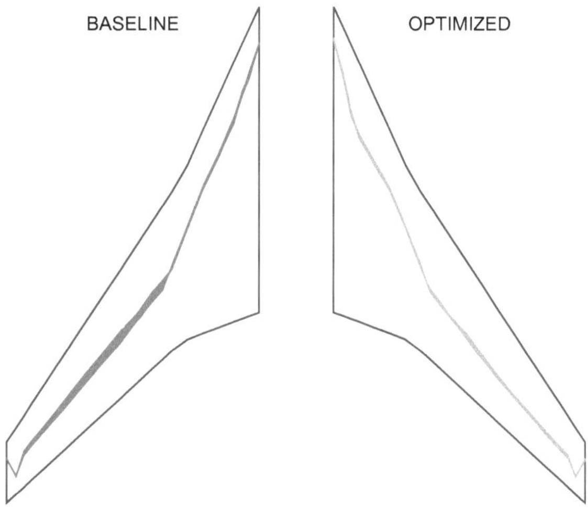

Fig. 22 Transition line envelopes comparison for the wing lower surface with $\alpha \leq 1$

tain variables is observed on the outboard wing. However, the bulk of the response function is linked to the transition prediction on the lower side, where a sensible increase of the overall performance along the inboard wing and a significant reduction of the uncertainty along the outboard wing are evident.

Conclusions

The use of an optimization approach based on VaR and CVaR as risk measures has been successfully applied to an aerodynamic robust design problem of industrial interest. Within this context, VaR and CVaR risk measures appeared to be, in terms of information conveyed, more favorable than more classical and well-established techniques based on expected value and variance. The problem at hand was characterized by a rather high number of design variables and uncertain parameters. Furthermore, the evaluation of the aerodynamic characteristics of a given configuration required a rather high computational effort. This imposed the use of a very rough estimate of VaR and CVaR risk functions in the optimization process, and, consequently, the adoption of an optimization algorithm relatively insensitive to noise and the use of a statistical technique capable to estimate the error and the confidence intervals related to a given sample. To this purpose, the bootstrapping method was chosen. The bootstrap analysis allowed to verify up to what stage of the optimization process the VaR estimates were accurate, stable and consistent enough to allow the optimization algorithm to progress toward the optimal solution. When this limit was reached, the introduction of CVaR, computed using a larger sample, allowed further progress in the optimization. The adoption of this strategy made possible that a simple genetic algorithm achieved significant results in a few generations of evolution.

The baseline and optimal solutions obtained were validated using a more refined sampling which, together with bootstrap analysis for the calculation of confidence intervals, confirmed the reliability of the solutions obtained through the optimization process.

Further work is needed to enhance the ranking capabilities of the algorithm near the optimum as the noise introduced by the coarseness of the samples could in some circumstances lead to misleading conclusions even when used in conjunction with confidence interval estimation methods. Therefore, it is likely that we will have to resort to a hierarchy of computational models of increasing complexity and fidelity, as well as to advanced statistical sampling techniques such as importance sampling and multi-level Monte Carlo methods.

A final consideration is devoted to the need to better define the well posedness of a robust design optimization problem. Indeed, it is very important to accurately define the margins of uncertainty that characterize the problem parameters and the design and operating conditions, because a definition not consistent with the reality of these margins is likely to make the optimal solution to the robust design problem either inefficient if the uncertainty margins are unnecessarily restrictive or not really robust if too optimistic and reduced margins are given.

Acknowledgements This work is related to a research activity that has been carried out within the framework of the UMRIDA (Uncertainty Management for Robust Industrial Design in Aeronautics) research project that has received funding from the European Union's Seventh Framework Programme for research, technological development, and demonstration under grant agreement n° ACP3-GA-2013-605036.

References

1. Quagliarella, D.: Definition of novel robustness criteria and multiobjective formulations in optimization under uncertainty. UMRIDA project deliverable D4.1-24(12), CIRA (2016)

2. Quagliarella, D., Iuliano, E.: Multiobjective/multifidelity algorithms in optimization under uncertainty with applications to aerodynamic shape design. UMRIDA project deliverable D4.1-30(12), CIRA (2016)
3. Hong, L.J., Liu, G.: Monte Carlo estimation of value-at-risk, conditional value-at-risk and their sensitivities. In: Proceedings of the 2011 Winter Simulation Conference (WSC), pp. 95–107. (Dec 2011)
4. Hong, L.J., Hu, Z., Liu, G.: Monte Carlo methods for value-at-risk and conditional value-at-risk: a review. ACM Trans. Model. Comput. Simul. **24**(4), 22:1–22:37 (2014)
5. Trindade, A.A., Uryasev, S., Shapiro, A., Zrazhevsky, G.: Financial prediction with constrained tail risk. J. Bank. Finance **31**(11), 3524–3538 (2007)
6. Quagliarella, D., Petrone, G., Iaccarino, G.: Optimization under uncertainty using the generalized inverse distribution function. In: Fitzgibbon, W., Kuznetsov, Y.A., Neittaanmäki, P., Pironneau, O. (eds.) Modeling, Simulation and Optimization for Science and Technology. Computational Methods in Applied Sciences, vol. 34. Springer Netherlands, pp. 171–190 (2014). Print ISBN 978-94-017-9053-6, Online ISBN 978-94-017-9054-3
7. Acerbi, C., Tasche, D.: Expected shortfall: a natural coherent alternative to value at risk. Econ. Notes **31**(2), 379–388 (2002)
8. Pflug, G.C.: Some remarks on the value-at-risk and the conditional value-at-risk. In: Uryasev, S.P. (ed.) Probabilistic Constrained Optimization: Methodology and Applications, pp. 272–281. Springer, US, Boston, MA (2000)
9. Rockafellar, R.T., Uryasev, S.: Conditional value-at-risk for general loss distributions. J. Bank. Finance **26**, 1443–1471 (2002)
10. Rockafellar, R.T., Uryasev, S.: Optimization of conditional value-at-risk. J. Risk **2**, 21–41 (2000)
11. Vitagliano, P.L., Quagliarella, D.: A hybrid genetic algorithm for constrained design of wing and wing-body configurations. In: Bugeda, G., Désidéri, J.A., Périaux, J., Schoenauer, M., Winter, G. (eds.) Evolutionary Methods for Design. Optimization and Control Applications to Industrial and Societal Problems, International Center for Numerical Methods in Engineering (CIMNE), Barcelona, Spain (2003)
12. Quagliarella, D., Vicini, A.: A genetic algorithm with adaptable parameters. In: 1999 IEEE International Conference On Systems, Man, And Cybernetics, Institute of Electrical and Electronic Engineers (IEEE), Tokyo, Japan, Oct 1999
13. Quagliarella, D., Vicini, A.: A parallel multiobjective genetic algorithm for applications to wing design. Technical Note, CIRA-TN-98-111, CIRA—Centro Italiano Ricerche Aerospaziali, Via maiorise, 81043, Capua, Italy, June 1998
14. Efron, B., Tibshirani, R.J.: An Introduction to the Bootstrap. Chapman & Hall/CRC Monographs on Statistics & Applied Probability, 1st edn. Chapman and Hall/CRC (1994)
15. Iuliano, E., Quagliarella, D., Donelli, R.S., El Din, I.S., Arnal, D.: Design of a supersonic natural laminar flow wing-body. J. Aircr. **48**(4), 1147–1162 (2011). July
16. Tang, B.: Orthogonal array-based latin hypercubes. J. Am. Stat. Assoc. **88**(424), 1392–1397 (1993)
17. Saltelli, A., Ratto, M., Andres, T., Campolongo, F., Cariboni, J., Gatelli, D., Saisana, M., Tarantola, S.: Global Sensitivity Analysis: The Primer. Wiley (2008)
18. Saltelli, A., Annoni, P., Azzini, I., Campolongo, F., Ratto, M., Tarantola, S.: Variance based sensitivity analysis of model output. design and estimator for the total sensitivity index. Comput. Phys. Commun. **181**(2), 259–270 (2010)
19. van Ingen, J.L.: The e^N method for transition prediction. historical review of work at TU Delft. In: 38th Fluid Dynamics Conference and Exhibit. Fluid Dynamics and Co-located Conferences. American Institute of Aeronautics and Astronautics, June 2008

Robust Compressor Optimization by Evolutionary Algorithms

Dirk Büche

Introduction

Robust design optimization (RDO) seeks for designs that are insensitive with respect to uncertainties. Uncertainties are always present in the real-world applications resulting especially from manufacturing processes, operating conditions, thermo-mechanical deformation and machine degradation over the lifetime. A prerequisite for robust design optimization is that the uncertainties are known and quantified, as the magnitude and type of the uncertainties influence the optimization result. Otherwise, the result of the robust design optimization is of no practical use [1].

The magnitude and thus the effect of uncertainties in a real machine can typically be reduced at the expense of additional cost and time, but is never zero. This applies, for example, to manufacturing processes as more expensive machining equipment and increased machining time typically increase precision. Furthermore, operational degradation can be reduced by shortening service intervals.

Roughly 15 years ago, automated design optimization of aerodynamic shapes became a viable alternative to the manual design thanks to the increase in computational power and more efficient CFD solvers. The rapid increase in computational power allowed for the evaluation of a larger set of design alternatives which is necessary for automated optimization, especially if robust methods such as evolutionary algorithms are employed. The roots of evolutionary algorithms (EAs) are found back in the 1960s with early publications, e.g. by Rechenberg [2]. At those times, parallel computing was not considered feasible. EAs became popular in the 1990s when conferences such as GECCO, CEC and PPSN were

D. Büche (✉)
MAN Diesel & Turbo Schweiz AG, Hardstr. 319, 8005 Zurich, Switzerland
e-mail: Dirk.Bueche@man.eu

C. Hirsch et al. (eds.), *Uncertainty Management for Robust Industrial Design in Aeronautics*, Notes on Numerical Fluid Mechanics and Multidisciplinary Design 140, https://doi.org/10.1007/978-3-319-77767-2_38

established and are even more interesting in today's parallel computing environments due to their population-based parallel and robust search.

EAs imitate the process of natural evolution. They evolve a population of individuals, where each individual is one possible solution to the design problem. For example, an individual might represent a parameter set for a compressor blade or an aircraft design. Each individual is evaluated using numerical or experimental methods such as computational fluid dynamics (CFD) and/or computational structural mechanics (CSM). The result is then an evaluated solution to the design problem. On average, better solutions are selected as parent population for the next iteration of the EA. From the selected parents, an offspring population is generated by selecting properties of different parents (recombination) and adding some random variation (mutation). Advanced algorithms such as the CMA-ES [3] extract statistical information over several iterations to learn about optimal mutation strength. This highly accelerates the convergence speed of the algorithm to the optimum.

The early design optimizations found in the literature typically considered only a single design condition (operating point) due to computational cost constraints. This may lead to poor results as outlined by Huyse and Lewis [4]. Their first optimization considered an airfoil for a single free flow Mach number: The design attained improvements only over a very narrow Mach number range close to the design condition, while off-design behaviour was poor. Similarly, an optimization with multiple but fixed Mach numbers was insufficient as the parameterization of the airfoil allowed for generating several localized optima around the specified points. Their solution was to define the Mach number as random variable. This resulted in a true improvement over the considered Mach number range.

Others such as Büche [5] extended the optimization to a multidisciplinary optimization problem including also structural aspects such as mechanical stresses and eigenfrequencies as this is the only way to obtain a solution that fulfils all requirements concurrently and can be integrated in an actual machine. The analyses considered multiple operating points and the design parameterization prohibited localized adaptation.

Today's computing power allows for either larger numerical models such as 3D unsteady simulations or more cost-intensive extensions such as robust design. Including uncertainty quantification in the automated optimization can increase computational cost by a factor of up to 100 [6]. This paper discusses how robust design optimization could be handled in a very cost-effective manner by extracting further information from the population for uncertainty quantification. The proposed method is analysed using simple test functions and then applied to the optimization of radial compressor designs.

Evolution Strategy with Covariance Matrix Adaptation (CMA-ES)

The CMA-ES is only briefly described herein as several very detailed descriptions and tutorials exist provided by the main developer, Nikolaus Hansen. We refer to the implementation of Hansen et al. [6] and define a simple test problem. The problem is a quadratic function $f(\mathbf{x})$ with two design variables x_1 and x_2:

$$f_{ell} = x_1^2 + 10 \cdot x_2^2 \quad (1)$$

The contour plot is given in Fig. 1. Due to the elliptic contour lines, the function is also referred to as the ellipsoid function.

The optimization starts by generating an initial solution. Let this solution be set to $\mathbf{x}_\mathrm{p} = [0.5, 0.5]$. The solution is the parent for the following iteration by means of the CMA-ES algorithm. A new population of λ solutions is generated as copies of the parent and by adding random mutation as illustrated in Fig. 1. In mathematical terms, mutation adds to each solution $\mathbf{x}_{j, j=1,\cdots,\lambda}$ a random vector $\mathbf{z}$, taken from a scaled, multivariate normal distribution:

$$\mathbf{x}_j = \mathbf{x}_\mathrm{p} + \mathbf{z}, \quad \mathbf{z} \sim \sigma \cdot N(0, \mathbf{C}) \quad (2)$$

The normal distribution $N(0, \mathbf{C})$ has zero mean and a symmetric, positive definite covariance matrix $\mathbf{C}$ as well as a multiplierσ denoted as step size. The random vector $\mathbf{z}$ can be generated by a vector of normally distributed random numbers $\mathbf{n}$ with unit variation $n_i \sim N(0, 1)$ and the following multiplication:

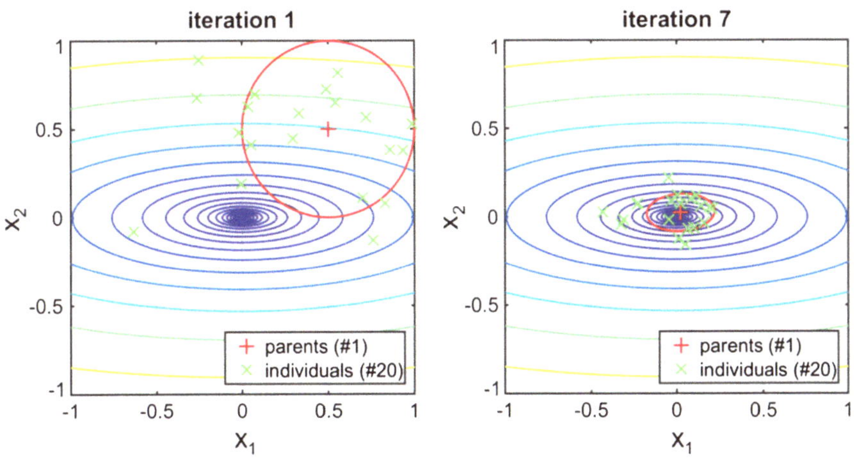

Fig. 1 Left: Start of an evolutionary optimization on a simple test function: Initial parent (red plus sign), mutation distribution (red circle) and generated population of solutions (green x). Right: Iteration no. 7 where the mutation distribution is adapted to the local curvature of the function and reduced in magnitude

$$\mathbf{z} = \sigma \cdot \mathbf{B} \cdot \mathbf{D} \cdot \mathbf{n} \tag{3}$$

where **B** is the matrix of ortho-normal eigenvectors of **C** and **D** is a diagonal matrix comprising the square root of the eigenvalues of **C** as diagonal elements. Both matrices result from an eigenvalue decomposition. The matrix **C** can be regenerated by

$$\mathbf{C} = \mathbf{B} \cdot \mathbf{D} \cdot \mathbf{D} \cdot \mathbf{B}^{\mathrm{T}} \tag{4}$$

As initialization for this optimization, we set the covariance matrix to the identity matrix $\mathbf{C} = \mathbf{I}$ and the step size to $\sigma = 0.5$. In Fig. 1, the red circle represents the distance of one standard deviation from the parent $\mathbf{x}_\mathrm{p}$. Therefore, the probability that a new solution is within this circle is 68%.

The population of solutions is evaluated on the problem $f_{ell}(\mathbf{x})$ and then ranked according to the resulting objective values f_{ell}. The μ best solutions are selected and recombined by a weighted sum, giving solutions of higher rank a higher weight w as defined in [6]:

$$\mathbf{x}_\mathrm{p} = \sum_{i=1}^{\mu} w_i \cdot \mathbf{x}_i \tag{5}$$

The recombined solution $\mathbf{x}_\mathrm{p}$ is the new parent and used for the next iteration. An included procedure updates the step size σ and covariance matrix **C** according to the distance and direction of the best solutions compared to the previous parent solution as outlined in [6]. This algorithm allows adjusting the mutation distribution very efficiently such that the covariance matrix approximates the local curvature matrix

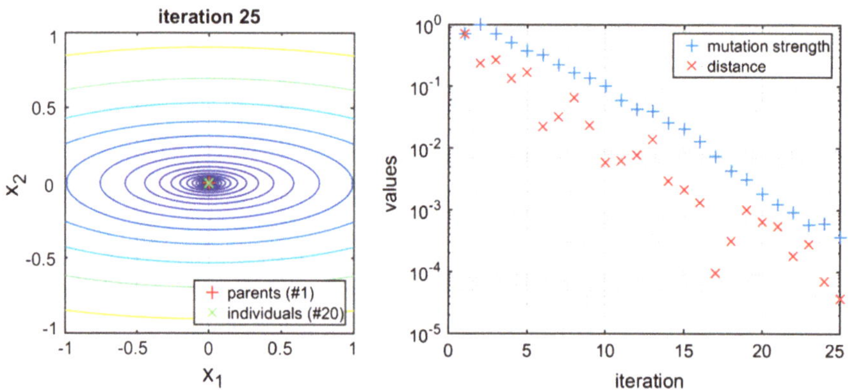

Fig. 2 Left: Iteration no. 25 where the algorithm has converged to the optimum. Right: The mutation strength is successfully adjusted such that exponential convergence is obtained

of the function to be optimized. This is illustrated in the right part of Fig. 1, where the mutation ellipsoid agrees well with the local contour line.

Thanks to adaptation of the mutation distribution, the CMA-ES converges exponentially towards to the optimum. Figure 2 shows the population of iteration no. 25 which collapsed in the optimum and also displays the evolution of the mutation strength and distance of the parent to the optimum over the iterations.

Robust Design Optimization Problem

Robust design optimization (RDO) is not straightforward as there are already different understandings about the term itself. In classical optimization, *robust optimization* refers to the convergence of an algorithm. In that case, a robust algorithm finds the true optimum with a high probability and is unlikely get stuck by premature convergence or in local optima.

In the context of uncertainties, *robust optimization* refers to the goal of finding an optimum solution to a problem that is insensitive with respect to the uncertainties. In the following, we refer to the latter definition. According to [7], uncertainties can be divided into four classes: (1) noise in the evaluation, (2) uncertainties in the design variables and parameters, (3) approximation errors by model building, and (4) time varying functions as illustrated in Fig. 3.

The main focus of this work is on uncertainties $\boldsymbol{\xi}$ in the design variables or other input parameters $\mathbf{x} = \{x_{i,}, i = 1 \ldots D\}$ resulting, e.g., from manufacturing tolerances or operational uncertainties. This can be modelled by:

$$F(\mathbf{x}) = f(\mathbf{x} + \boldsymbol{\xi}), \quad \boldsymbol{\xi} \sim \mathrm{N}(0, \boldsymbol{\Theta}) \tag{6}$$

where N is a normal distribution with zero mean and covariance matrix $\boldsymbol{\Theta}$. An illustration is given in the middle of Fig. 3.

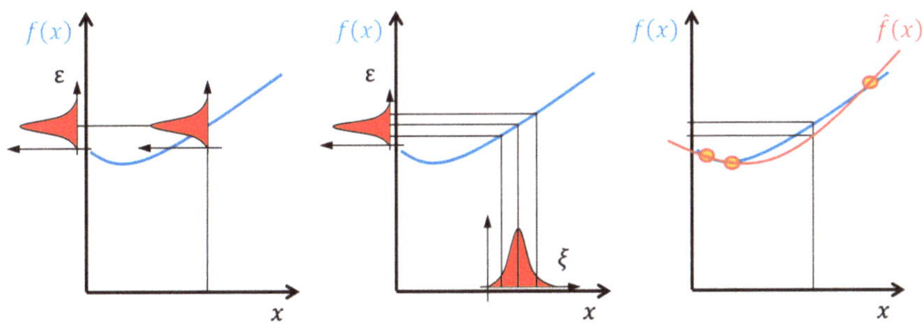

Fig. 3 Possible uncertainties in optimization which are noise in the evaluation (Left), uncertainties in the design variables (Middle) and approximation error by models (Right)

A solution that is insensitive to such uncertainties is referred to as robust solution. Evaluating the robustness of a solution can significantly increase computational cost as more simulations are required. Finding an efficient tool is a key focus of this paper.

Given $f(x+\boldsymbol{\xi})$ as the stochastic simulation result, the interest is in different robust quantities such as the mean E and variance var of the result:

$$\mathrm{E}(f(x)) = \int_{\Omega} f(x+\boldsymbol{\xi}) \cdot w(\boldsymbol{\xi}) d\boldsymbol{\xi} \tag{7}$$

$$\mathrm{var}(f(x)) = \int_{\Omega} (f(x+\boldsymbol{\xi}) - \mathrm{E}(f(x)))^2 \cdot w(\boldsymbol{\xi}) d\boldsymbol{\xi} \tag{8}$$

where w is the weight function belonging to the probability distribution of the uncertainties $\boldsymbol{\xi}$. In general, these integrals cannot be solved analytically and only be approximated by a finite set of samples. To do so, different uncertainty quantification methods were described in the previous chapters of this book. For the simplest method, the Monte Carlo method, the result converges with $1/\sqrt{N}$, where N is the number of simulations. The robust optimization problem can then be formulated as

$$\text{find } \mathbf{x}^* = \operatorname{argmin}(E(f(\mathbf{x}+\boldsymbol{\xi}))), \text{ given } \xi_i \sim N(0, \Theta_i) \tag{9}$$

Evolutionary algorithms evolve a population of solutions over several iterations to converge towards the optimum. The question is how to integrate UQ into an evolutionary optimization in the most cost-efficient manner. In the literature [7], different approaches exist.

The first approach is denoted as **explicit averaging**. For each solution in the population, an UQ is conducted. Each UQ requires a set of solver evaluations. The mean objective value may be considered as a robust result. Thus, compared to a deterministic optimization, the total computational cost of a robust optimization multiplies with the number of evaluations N_{UQ} that are required for one UQ. This approach is very expensive, increasing the total cost by a factor of 3 to 100 [6].

Explicit averaging could also make use of response surfaces. Response surfaces are cost-efficient surrogates of the real numerical solver but introduce further uncertainties. They are trained on already evaluated solutions of the actual or previous iterations, typically located in the neighbourhood of the solution to be analysed. However, these surrogates are less effective in higher dimension [6] due to the curse of dimensionality. Furthermore, the simulation results must be sufficiently smooth and well behaved such that an approximation with lower order functions is sufficiently representative. As we are also interested in off-design compressor operation, the behaviour of the objective function can become highly nonlinear or even discontinuous in design space.

Implicit averaging suggests adding random perturbation to all solutions in a population. Statistical information for UQ can then be extracted from the population, given that the population is sufficiently spread over the design space. This approach is more cost-efficient as the solutions are not resampled as in explicit averaging.

CMA-ES with Uncertainty Quantification

The CMA-ES has several properties that make it an ideal candidate for robust optimization. The algorithm is well-established and analysed in the literature. It is population based and uses a direct search (no gradient information is used). In the selection operator, individuals in the population are simply ranked; thus, no detailed information about the objective function is needed. The non-elitist selection operator in CMA-ES reduces the risk of getting stuck in local minima. The algorithm converges fast to the optimum by learning the optimal mutation parameters.

Other evolutionary algorithms such as genetic algorithms use recombination as their main variation operator. This makes it difficult to analyse the actual variation strength. In contrast, CMA-ES uses a multivariate normal distribution which allows direct comparison of the variation strength with the given uncertainty level. Thus, implicit averaging can be used for UQ.

We follow two approaches in order to generate a robust optimization. The approaches are based on the implicit averaging (Implicit-CMA-ES) and a modification of CMA-ES (MinMut-CMA-ES).

Implicit-CMA-ES

Implicit averaging is used as described above. In each iteration, CMA-ES generates a new population of solutions with design variables **x**. Before evaluating the designs with the given solver, uncertainty is added by a random perturbation vector $\boldsymbol{\xi}$:

$$F = f(x + \boldsymbol{\xi}), \xi_i \sim N(0, \theta_i^2) \tag{10}$$

The main assumption is that the selection operator in CMA-ES picks on average robust solutions as these solutions, by definition of robustness, are less prone to deterioration due to the perturbation vector. This implicit averaging requires a sufficiently large population size. The advantage of this approach is its simplicity. The CMA-ES is unchanged as perturbation is added for the evaluation of the solution and the perturbed vectors are not passed on to the CMA-ES.

MinMut-CMA-ES (Minimum Mutation)

The second approach requires modifying the CMA-ES source code. The mutation in CMA-ES is a multivariate, correlated normal distribution $N(0, \sigma^2 \cdot \mathbf{C})$. During each iteration, a deterministic algorithm adjusts the mutation distribution. The underlying idea for the novel robust optimization is to keep the mutation strength above a lower limit with the effect that the mutation strength is always larger or equal to the uncertainty. The assumption is that more solutions are picked from robust areas of the design space than from sensitive areas with steep function value gradient.

Hansen et al. [6] propose to adjust the step size σ in order to guarantee a minimum mutation strength. However, the covariance matrix $\mathbf{C}$ might degenerate leading to highly correlated mutation. Then, mutation occurs mainly in one (correlated) space direction (i.e. the main eigenvector of $\mathbf{C}$), such that the actual uncertainty is poorly represented. In order to prevent degradation, we propose to adjust the covariance matrix $\mathbf{C}$ instead of the step size σ. The mutation strength $\mathbf{n}$ in each design space direction x_i can be obtained by solving the equation:

$$\sigma \cdot \mathbf{B} \cdot \mathbf{D} \cdot \mathbf{n} = e_i \Rightarrow \mathbf{n} = 1/\sigma \cdot \mathbf{D}^{-1} \cdot \mathbf{B}^T \cdot e_i \tag{11}$$

where e_i is a unit vector in direction i. The actual mutation strength (standard deviation) is then

$$p_i = \frac{1}{\|\mathbf{n}_2\|} \tag{12}$$

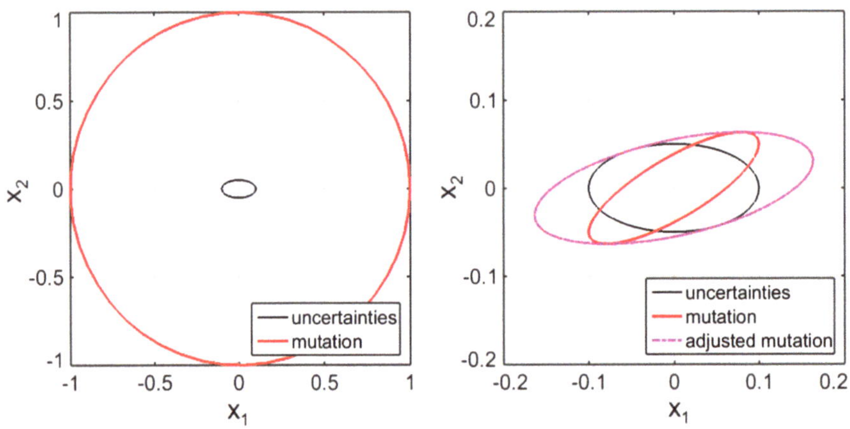

Fig. 4 Comparison of the standard deviation of the mutation distributions (red ellipse) with the uncertainty strength (black ellipse). In the beginning of the mutation (left), the mutation strength is larger than the uncertainty, whereas at a later generation (right), it can be below the uncertainty effect and needs to be adjusted in some space dimensions (magenta ellipse)

In each iteration, the mutation strength is compared with the uncertainty and adjusted if required. Figure 4 illustrates the procedure by plotting a contour line in a distance of one time the standard deviation for the mutation and for the uncertainty. The left part of Fig. 4 shows the beginning of the optimization, where the mutation strength is $p_1 = p_2 = 0.5$, while the uncertainties for x_1 and x_2 are $\Theta_1 = 0.1, \Theta_2 = 0.05$ and thus much smaller than the mutation strength. The mutation distribution is not adjusted, and the uncertainty has no effect on the convergence.

The right part of Fig. 4 illustrates a later iteration, where the mutation strength is smaller than the uncertainties. The mutation distribution is correlated such that the ellipse has a certain aspect ratio. When looking exactly into each axis direction, the mutation strength is smaller than the uncertainty as the mutation ellipsoid is within the uncertainty ellipsoid such that the diagonal elements of the covariance matrix are therefore adjusted by

$$C_{ii} \leftarrow C_{ii} \cdot \frac{\Theta_i^2}{p_i^2}, \text{ if } p_i < \Theta_i \tag{13}$$

The result of the adjustment is added to Fig. 4. This adjustment does not guarantee that the mutation strength in each axis direction is exactly met. However, as the mutation is adjusted gradually over the generations, the adjustments are on average small.

It has to be noted that this is only one possible approach to adjust the covariance matrix. Alternatively, any of the $\frac{D \cdot (D+1)}{2}$ elements of the matrix $\mathbf{C}$, specifically the non-diagonal elements, could be adjusted as well.

Experimental Analysis on Simple Test Functions

A simple test function is the so-called sphere function, which is a non-scaled quadratic function with

$$f_{\text{sph}} = \sum_{i=1}^{D} x_i^2 \tag{14}$$

The optimum is at $x_i = 0 \forall i = 1, \ldots, D$. The optimum is robust as the vicinity is smooth and continuous. By adding a step function at the optimum with the form

$$f_{\text{sph, rob}} = \sum_{i=1}^{D} x_i^2 + \alpha_i, \quad \alpha_i = \begin{cases} 0.1 & \text{if} x_i > 0 \\ 0 & \text{else} \end{cases} \tag{15}$$

the location of the optimum is unchanged but becomes very sensitive towards small uncertainties. A contour plot is given in Fig. 6. For the first test, the number of

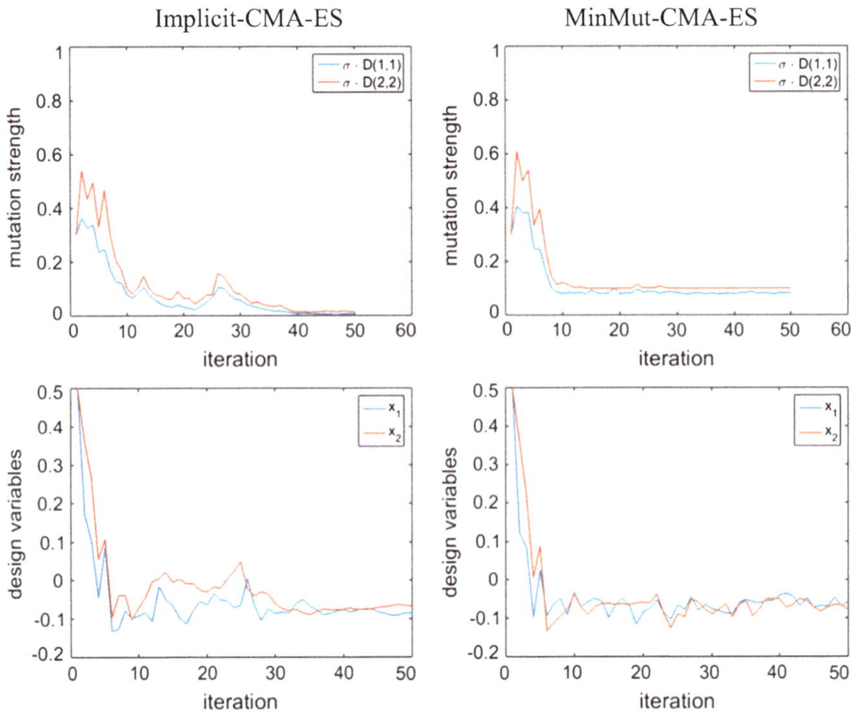

Fig. 5 Convergence of the Implicit-CMA-ES and the MinMut-CMA-ES on the 2D test function plotted over the number of iterations. The mutation strength represented by lengths of the eigenvectors of the mutation distribution (upper figure) and the design variables of the parent (lower figure) are shown

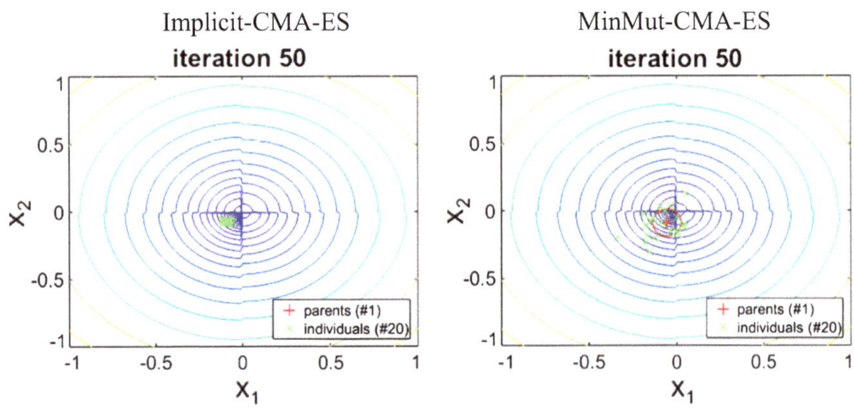

Fig. 6 Contour plots illustrating the final population of the two robust variants of the CMA-ES

parameters is set to $D = 2$, $x_{i,i=1,2} \in [-1, 1]$ and uncertainties are normally distributed with zero mean and standard deviation $\theta_{i,i-1,2} = 0.1$.

An optimization run is started with the two robust approaches Implicit-CMA-ES and MinMut-CMA-ES. As starting point, the point $x_{i,i=1,2} = 0.5$ is chosen for both approaches. Figures 5 and 6 illustrate the convergence of these two approaches.

Figure 5 illustrates that roughly at iteration 10, the mutation strength is about 0.1 and thus similar to the level of uncertainty. From here on, the two approaches behave differently. While Implicit-CMA-ES further converges and reduces the mutation strength, the implemented modification in MinMut-CMA-ES prevents a further decrease of the mutation strength as shown in the lower contour plots in Fig. 6 and lower graphs in Fig. 5. Both algorithms converge to a point of roughly $x_{i,i=1,2} = -0.08$, thus a distance of roughly the standard deviation of the uncertainty ξ_i from the step function in each dimension.

In order to test the behaviour of the two approaches for a typical number of design variables, which often lies in the range of 6–30, we increase the number of design variables to $D = 20$ with design space $x_{i,i=1,\dots,D} \in [-1, 1]$ and set the population size of the CMA-ES to $\lambda = 50$. The uncertainty is set to a realistically

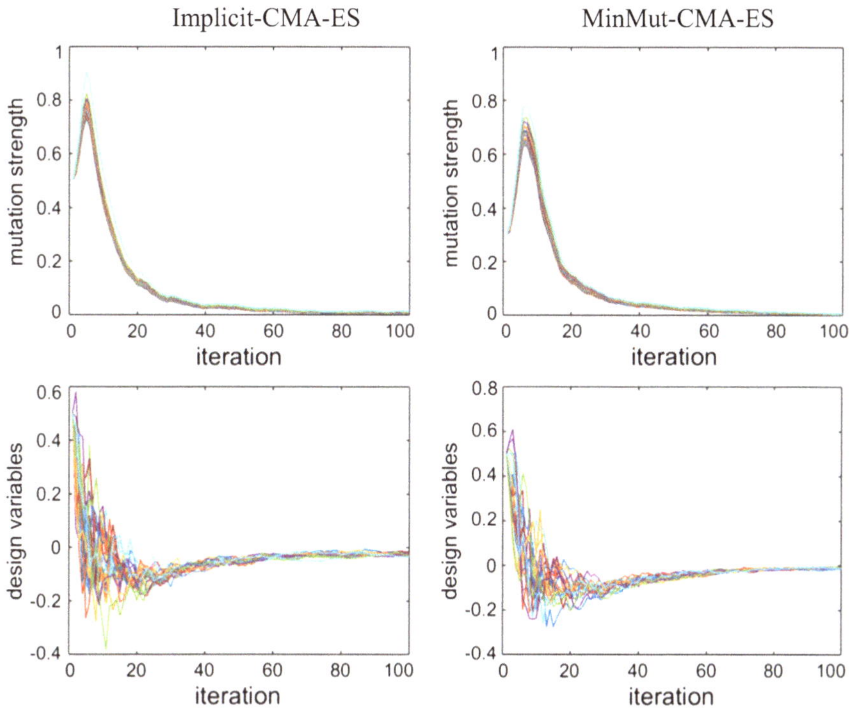

Fig. 7 Convergence of the Implicit-CMA-ES and the MinMut-CMA-ES on the 20-dimensional test function. The mutation strength represented by lengths of the eigenvectors of the mutation distribution (upper figure) and the design variables of the parent (lower figure) are shown

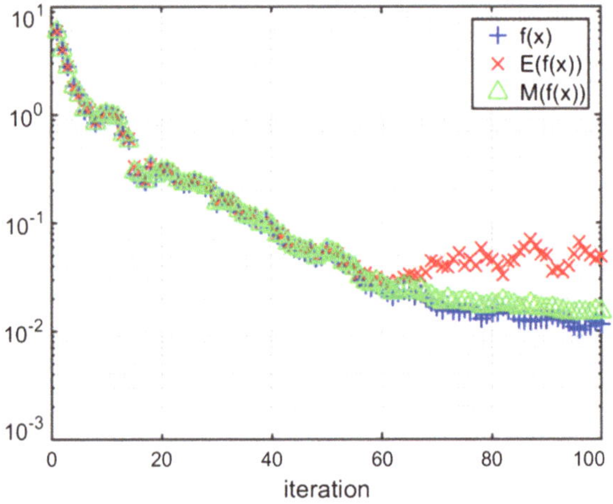

Fig. 8 Convergence of the Implicit-CMA-ES on the 20-dimensional test function over the iterations. For the parent solution, the deterministic function value $f(\mathbf{x})$ as well as the mean $E(\mathbf{x})$ and median value $M(\mathbf{x})$ for the given uncertainties is plotted

small value of $_{i,i=1,\ldots,N}=0.01$, while the step in the test function $f_{\mathrm{sph,rob}}$ is kept at 0.1.

Figure 7 illustrates the results of the two investigated methods. The adaptation of mutation distribution is more important in this case compared to the previous one, because the uncertainty is significantly smaller and allows for a higher convergence towards a point in the close vicinity of the optimum.

The adaptation of the mutation works successfully which can be seen from the decrease in mutation strength with increasing iteration number. Furthermore, the robustness in the final solution is demonstrated as both approaches converge to negative design variable values with a certain distance to the step in the function.

Implicit-CMA-ES seems to converge roughly to $x_i=-0.024$, i.e. about 2.4 times the standard deviation of the uncertainties of $\Theta_{\mathrm{i}}=0.01$ from the step. MinMut-CMA-ES converges closer towards the optimum with about $x_i=-0.009$ which corresponds roughly to 1 time standard deviation of the uncertainties and agrees with the result of the 2-dimensional test function, but with a different level of uncertainty.

The final result of the two approaches differs. The Implicit-CMA-ES delivers a more robust solution than the MinMut-CMA-ES as seen by the about 2.5 times larger distance of the final solutions to the step in the function.

Further work is needed to analyse these effects more precisely. For example, three different statistical quantities for the parent solution of the Implicit-CMA-ES are plotted in Fig. 8 over the number of iterations. The quantities are the deterministic function value $f(\mathbf{x})$, the mean value for the given uncertainties $E(\mathbf{x})$ and the robust mean (median) $M(\mathbf{x})$. The optimization reduces all quantities to a certain level. It has to be discussed in future work how these quantities correlate with the two different approaches for the robust optimization.

Robust Compressor Optimization

The proposed robust optimization approach MinMut-CMA-ES is applied to the design optimization of a high-pressure radial compressor stage. Radial compressor stages typically comprise of a radial impeller, a diffuser, a bend and a return channel. An example is given by Büche et al. [8]. Here, we focus on the impeller and especially on the impeller blades.

First, the geometry is parametrized and the uncertainties are set. Then, two different optimization approaches are executed independently which are the original CMA-ES (deterministic optimizer) that does not consider uncertainties and the proposed MinMut-CMA-ES (robust optimizer) including uncertainties.

Geometrical Uncertainties

The manufactured shape of the blades always differs from the nominal, generating geometrical uncertainties. Two examples are given in Fig. 9. The actual leading edge can deviate in thickness (shape tolerance) or can be shifted or rotated (position tolerance). Actual values for the geometrical uncertainties can be extracted, e.g., from actual manufacturing measurements of the quality department. While drawings usually only define deviation limits, actual measurements provide statistical information as probability distributions, given a sufficiently high number of measured samples.

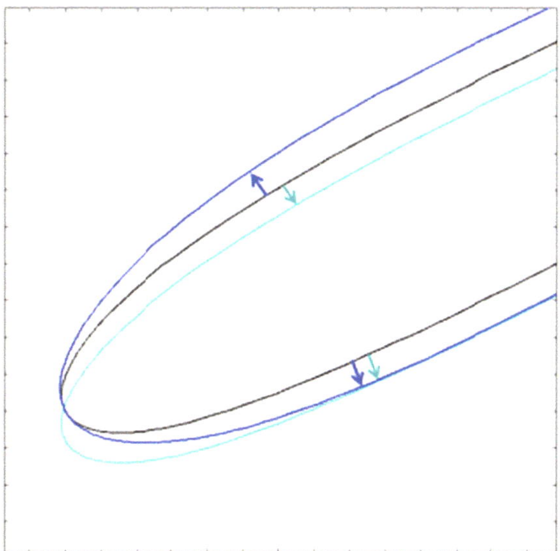

Fig. 9 Leading edge of a blade profile (black line) with two different shape deviations: increased thickness (blue) and meridional shift (cyan)

Impeller blade profiles can be parameterized by subdividing the profile into the four regions: leading edge (LE), trailing edge (TE), suction side (SS) and pressure side (PS). Each of the four regions is defined using limited number parameters.

Given statistical information for overall geometric uncertainty σ_{tot} of the leading edge, the question is how to transfer this into uncertainties of the individual parameters. As an example, the leading edge is assumed to be described by three parameters, which are nose radius r_{LE} , metal angle α_{LE} and radial position R_{LE}. Assuming that the individual effects of these parameters are equal, normally distributed and pointing into the same space axis, the total effect of the uncertainties is

$$\left(0, \sigma_{\mathrm{tot}}^2\right) = \mathrm{N}\left(0, \sigma_{\mathrm{r_{LE}}}^2\right) + \mathrm{N}\left(0, \sigma_{\mathrm{\alpha_{LE}}}^2\right) + \mathrm{N}\left(0, \sigma_{\mathrm{R_{LE}}}^2\right) \tag{16}$$

The total variance σ_{tot}^2 is then given by

$$\sigma_{\mathrm{tot}}^2 = \sigma_{\mathrm{r_{LE}}}^2 + \sigma_{\mathrm{\alpha_{LE}}}^2 + \sigma_{\mathrm{R_{LE}}}^2 \tag{17}$$

and the standard deviations are

$$\sigma_{\mathrm{r_{LE}}} = \sigma_{\mathrm{\alpha_{LE}}} = \sigma_{\mathrm{R_{LE}}} = \frac{1}{\sqrt{3}} \cdot \sigma_{\mathrm{tot}} \tag{18}$$

For the other three regions, PS, SS and TE, the same approach is applied as for the LE. For PS and SS, the LE and TE parameters have to be considered too as they also influence their shape. In total, 14 uncertainties are considered.

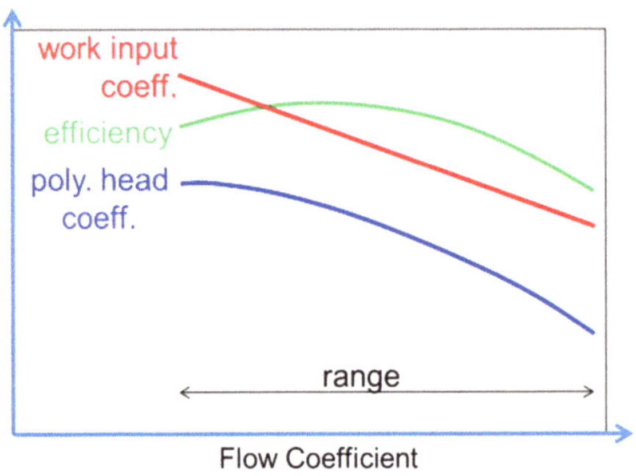

Fig. 10 Typical compressor characteristics

Comparison Between Deterministic and Robust Design Optimization

The optimization goal is to maximize the efficiency of multiple operating points with constraints on the minimum required range in flow from surge to choke and a minimum required pressure rise expressed by means of polytropic head (see Fig. 10).

The deterministic optimization does not consider uncertainties and thus uses the standard implementation of the CMA-ES without adding any uncertainties to the solution. The robust design optimization uses the MinMut-CMA-ES with the uncertainties given in Chapter "Numerical Uncertainties Estimation and Mitigation by Mesh Adaption".

The convergence of both algorithms is illustrated in Fig. 11. In the beginning of the optimization, the mutation strength is much larger than the uncertainties. Thus, deterministic optimization and robust optimization do not differ as shown by roughly identical evolution of the objective function and design variables up to evaluation 2500. From there on, the deterministic algorithm reduces further the mutation strength. In contrast, the robust optimization keeps a minimum mutation

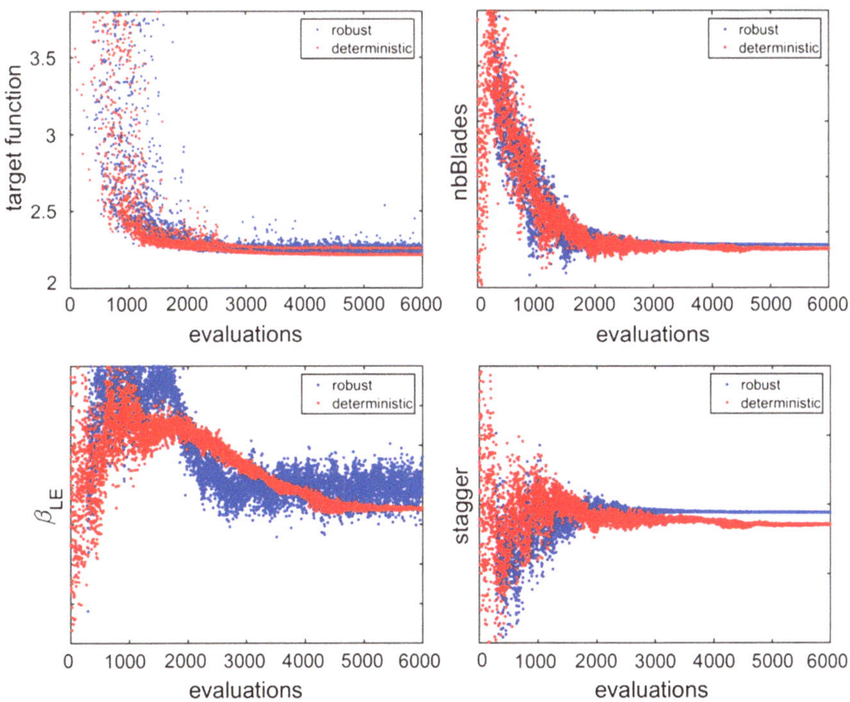

Fig. 11 Comparison of the deterministic and the robust optimization for the target function (upper left) and three design variables as function of the number of evaluations

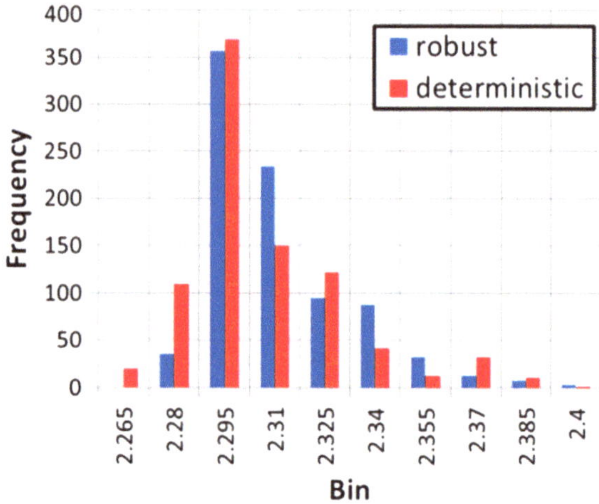

Fig. 12 Histogram to illustrate the result of the uncertainty quantification for the final result of the deterministic and robust optimization

strength which is especially visible for the leading edge angle β_{LE}. After 6000 evaluations, both algorithms have converged as no improvement in the objective function is visible over the last 1'000 evaluations.

As best solution for the deterministic optimization, the solution with minimum objective value is selected. For the robust optimization, the best solution is computed as the mean of the design variable values of the last 500 evaluations.

For both best solutions, a deterministic evaluation and an UQ are performed. The best solution of the deterministic optimization has a slightly lower deterministic target value as the best solution of the robust optimization. Furthermore, when running an UQ for both best solutions, the mean and standard deviation of the target function are very similar, as can be seen by the histogram in Fig. 12. What is surprising is that the influence of the uncertainties is low in this example. When computing the influence of a single flow coefficient instead of multiple operating points, there is a clear effect. However, computing the influence over the entire operating range, the effects cancel partially. As an example, increasing metal angles at LE or TE typically improves the efficiency towards choke while deteriorating the efficiency towards stability limit.

Overall, it can be concluded that for the considered case, the uncertainties are sufficiently low such that the influence is minor. Thus, the manufacturing deviations are adequate to guarantee a product fit for the intended purpose.

Summary and Outlook

Robust design optimizations require additional data and methods compared to deterministic ones. First, quantitative information about the uncertainties must be available and modelled by a probability distribution. Then, the optimization algorithm has to be modified to integrate uncertainty quantifications in a cost-efficient manner.

A robust evolutionary optimization algorithm was proposed and tested on a simple test function and for a radial compressor design case. The robust optimization needed roughly at the same computational effort. The best solution of the deterministic and robust optimization differs as the uncertainties influence the optimization, even though to a marginal extent in the compressor example. This allows for concluding that the actual manufacturing tolerances are sufficiently low for given requirements.

As an outlook, more robust design optimizations are required to confirm these first findings and to get more insight into the effect on the final design. Furthermore, other uncertainties, especially from the compressor assembly, need to be included.

References

1. Kipouros, T., Pachidis, V., Ghisu, T., Templalexis, I.: Characterisation of the uncertainties of the operating conditions in turbomachinery design. In: Proceedings of ASME Turbo Expo 2010, Glasgow, UK. GT201023253 (2010)
2. Rechenberg, I.: Evolutions strategie: Optimierungtechnischer Systemenach Prinzipien der biologischen Evolution, Technical University of Berlin, Department of Process Engineering (1971)
3. Hansen, N., Ostermeier, A.: Completely derandomized self-adaptation in evolution strategies. Evolution. Comput. **9**(2), 159–195 (2001)
4. Huyse, L., Lewis, R.M.: Aerodynamic shape optimization of two dimensional airfoils under uncertain operating conditions. ICASE No. 2001-1. Technical report, NASA Langley Research Center Hampton, VA (2001)
5. Büche, D.: Multi-objective evolutionary optimization of gas turbine components. Dissertation imBereich Technische Wissenschaften, Eidgenössische Technische Hochschule ETH Zürich, 15240, Shaker, ISBN 3-8322-2800-4 (2004)
6. Hansen, N., Niederberger, A., Guzzella, L., Koumoutsakos, P.: A method for handling uncertainty in evolutionary optimization with an application to feedback control of combustion. IEEE Trans. Evolution. Comput. **13**(1) (2009)
7. Jin, J., Branke, J.: Evolutionary optimization in uncertain environments. In: IEEE Trans. Evolution. Comput. **9**(3) (2006)
8. Büche, D., Beetz, M., Ribi, B.: Uncertainty analysis for large-scale industrial radial compressors. In: Proceedings of ASME Turbo Expo 2010: Power for Land, Sea and Air, June 14–18, 2010, Glasgow, UK, GT2010-22918 (2010)

Robust Optimization of Acoustic Liners

N. Magnino

Introduction

Reduction of aeronautical noise pollution is a technical challenge for designer with social, commercial implications because of repercussions on the acceptance of new airports and potential reduction of fees and taxes imposed by agencies and local governments on airliners. In modern turbofan engines, the fan contributes greatly to the total noise because of the predominant portion of the fan thrust, hence the great attention given to the design of the acoustic liners installed inside the engine nacelle just in front of the fan. In general, these devices have a sandwich construction consisting at least of a porous external facing sheet, to allow the passage of sound waves into one or more layers of honeycombs that act as Helmholtz resonators closed by a rigid backing skin to allow reflection. As acoustic liners have to be effective in all the certification conditions prescribed by authorities, the design process is by nature a multi-objective optimization between eventually conflicting requirements. Especially in the preliminary phase, where a lot of scenarios are to be considered, that can be done manually via traditional trial-and-error legacy procedures, or with various algorithms [1], but at the risk of obtaining local or just ideal optima whose real performance falls short in the field. In this chapter, we will document an innovative methodology developed within UMRIDA by Leonardo Finmeccanica in cooperation with ESTECO to do automatic multi-objective robust design optimization based on efficient single-objective reliability-based formulation [2], taking into account the uncertainties due to manufacturing tolerances experimentally measured and analyzed with the efficient adaptive sparse polynomial chaos expansion used for UQ (Fig. 1).

N. Magnino (✉)
Leonardo S.p.A., Via P. Foresio 1, 21040 Venegono Superiore, VA, Italy
e-mail: nicola.magnino@leonardocompany.com

C. Hirsch et al. (eds.), *Uncertainty Management for Robust Industrial Design in Aeronautics*, Notes on Numerical Fluid Mechanics and Multidisciplinary Design 140, https://doi.org/10.1007/978-3-319-77767-2_39

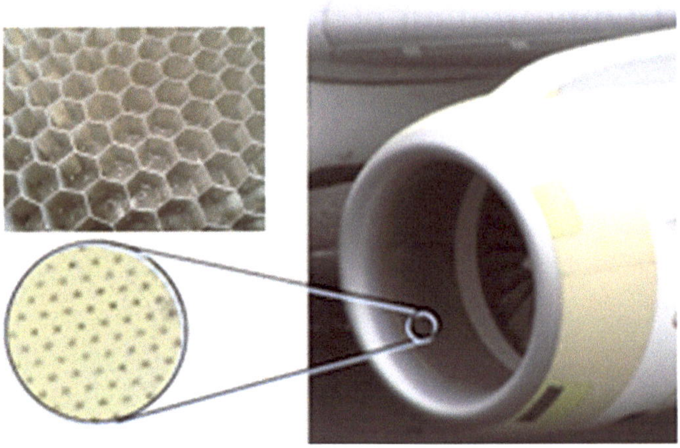

Fig. 1 Full engine nacelle (right) and detailed views of the acoustic panel interior (upper left) and perforated facing skin (lower left)

Definition of MORDO Test Case

The industrial validation case targeted in this study is a typical regional jet engine nacelle integrating an acoustic liner consisting of N Helmholtz resonating cavities. Each one is essentially composed of a honeycomb cell of a specific height targeted to attenuate the design frequency of the impinging noise, being closed at the bottom by a rigid backing skin and on top by a porous face sheet to dissipate energy as the air passes through. A consolidated (all metal single degree o freedom, i.e., with each honeycomb cell being undivided) acoustic liner technology has been selected (Fig. 3) from vast Leonardo Finmeccanica production, with the objective function being represented by the noise (attenuated by the liner) computed at the prescribed certification flight points. Please note that the name of the program and the customer as well as engine data and liner impedance together with some technological process details will not be disclosed, as they are strictly company confidential. Acoustic requirements are prescribed by customers and aviation agencies for three typical flight conditions: approach, sideline (or takeoff), and flyover (or cut-back), as illustrated in Fig. 2.

Since these requirements can be in contrast to each other and an acoustic liner providing the best attenuation in one flight condition not necessarily performs well in the other conditions, liners shall be designed through a multi-objective optimization procedure. The geometrical uncertain parameters considered with their main influencing factors are:

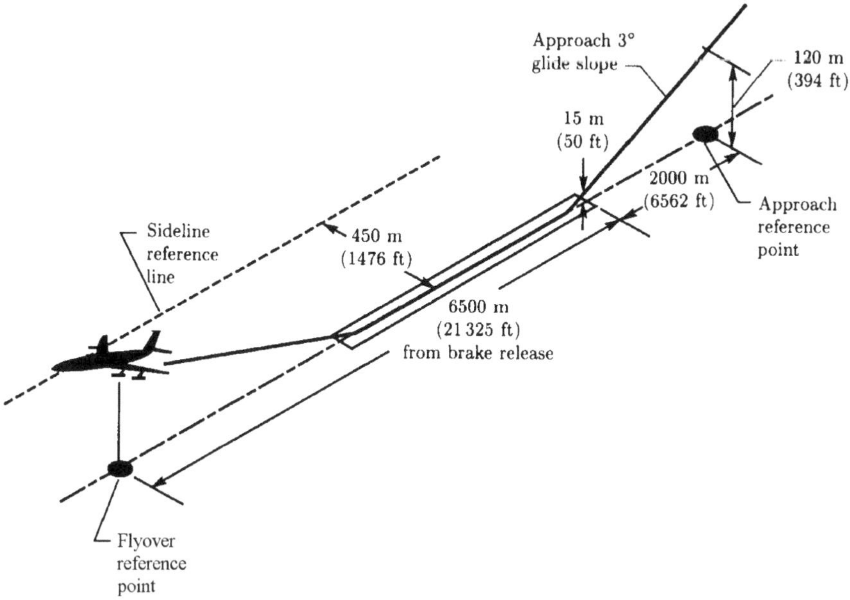

Fig. 2 FAR Part 36 noise certification conditions

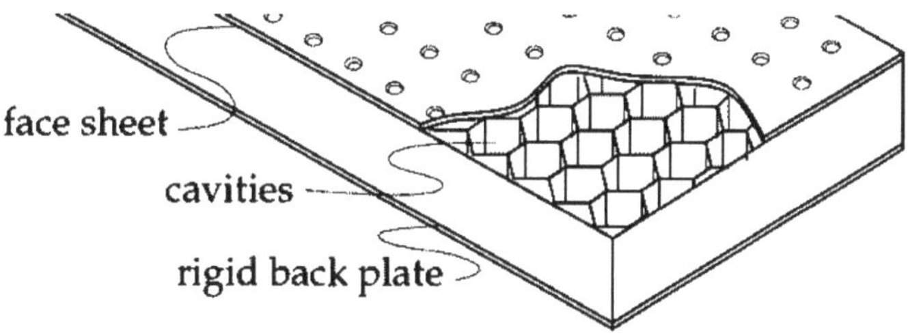

Fig. 3 Nacelle SDOF liner acoustic panel

• Honeycomb height (h)	f(Material)
• Face sheet thickness (t_{fs})	f(Material, Manufacturing process)
• Face sheet open area (PoA_{eff})	f(Machinery Tool)
• Holes diameter (d_{eff})	f(Nacelle Geom., Machinery Tool)

These geometrical parameters are input into a proprietary semi-empiric impedance model [3] together with values depending on flight condition to produce the admittance to be applied to the boundary condition representing the acoustic liner in the finite element model used by computational acoustic code utilized. Indeed numerical simulations have been performed with MSC Actran software developed by Free Field Technology, on an 2D axisymmetric model consisting of about 30000 quadratic elements; the level of refinement of the mesh being enough to resolve the highest frequencies present considering the maximum speed of the flow in critical areas like the lip. As a cost function, it should be used ideally the quite realistic effective perceived noise level (EPNL), but in order to calculate it, we would need to have the whole aircraft's noise sources modeled, (not only the engine as it is the case normally). Moreover, all the 1/3 octave band center frequencies should be considered in the calculation, thus leading to unacceptable computational cost. In this work, we opted for another simpler metric, more easily computed, which is normally used in preliminary design and is capable to prove the MORDO methodology, i.e., the overall sound pressure level (OASPL) defined as:

$$OASPL = 10 \log \sum_j 10^{\frac{SPL_j}{10}} \tag{1}$$

$$SPL(f) = 10 \log \left(\frac{\sum_{40^\circ}^{90^\circ} \sum_{m,n} p^2_{rms\,(\vartheta, f, m, n)}}{p^2_{ref}} \right) \tag{2}$$

The above-mentioned objective function uniformly sums over the directivity angles and over frequencies the sound pressure level (SPL) which is a function of the frequency, the root mean square of the acoustic pressure of the duct radial and azimuthal modes computed, and directivity angle at a distance of 150 ft according to the scheme in Fig. 4.

All the Actran simulations were performed at two frequencies the blade passage frequency (BPF) and its first harmonic 2BPF and for a range of angles that varies depending on flight condition and in accordance with customer preferences. In fact, noise radiating from an inlet is mainly tonal, and therefore, at the BPF and its harmonics, higher sound levels are propagated. For this reason, a liner should be optimized to attenuate the noise primarily at these frequencies. The simulation time to complete one configuration design was about 2 min running in parallel on 8 cores of one of the computational nodes present in FNM's AAHPC cluster.

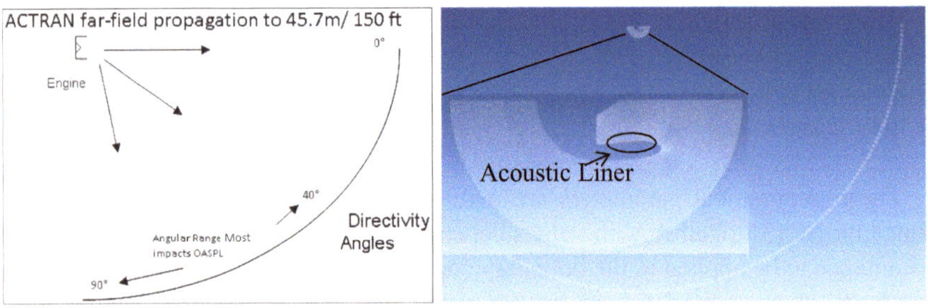

Fig. 4 Schematic of far-field propagation and ACTRAN mesh with acoustic panel circled

Setup of MORDO Case in modeFRONTIER

The workflow used to set up the automatic optimization in modeFRONTIER is similar to the one already prepared and reported in chapter "Manufacturing Uncertainties for Acoustic Liners" used to perform the tests needed to find the optimal number of sampling points and methodology for an accurate UQ of the system response OASPL (Fig. 5).

The main components, including the impedance model and the Actran model interface, remain unaltered, as well as the nodes that define the input parameters of the system, i.e., the geometrical variables *h*, *t*, *d*, and *PoA*. This time, however, besides the distribution definition (as obtained from the experimental data), we have to select the range of variation of these parameters, based on experience and wider enough to let the optimization algorithm explore the variables space. All the data are summarized in Table 1 here.

From the Actran interface, all the OASPL signal of interest is extracted, corresponding to the different flight conditions that are evaluated. Finally, the objectives are applied to the responses of interest: Accordingly to the approach considered and described in the following chapters, they can be applied to mean value, standard

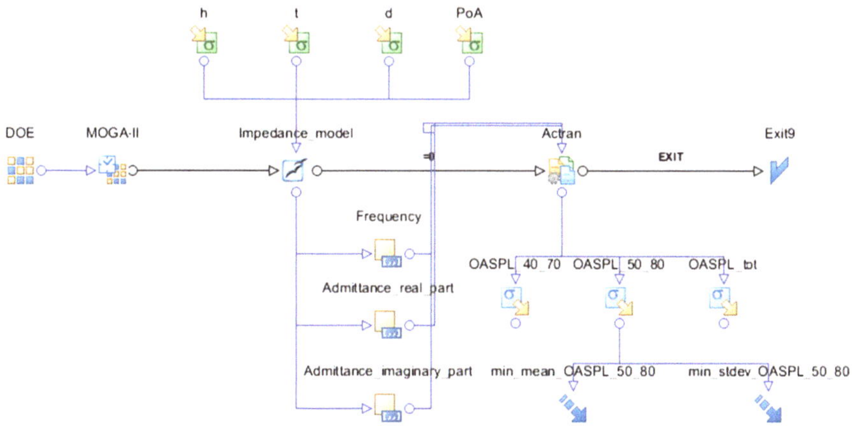

Fig. 5 modeFRONTIER workflow for aeroacoustic robust design optimization

Table 1 Summary of uncertain parameters and their range of variation

Uncertainty	Distribution	σ	Range
Cell height	Normal	1.5069E−2	[10–18 mm]
Face sheet thickness	Normal	2.2625E−2	[0.7–1.35 mm]
Hole diameter	Normal	1.2803E−2	[1–2.5 mm]
Percentage of open area	Normal	1.0815E−3	[3–15%]

deviation, or percentile, as computed by modeFRONTIER following the procedures based on polynomial chaos and described in chapter "Manufacturing Uncertainties for Acoustic Liners."

Preliminary RDO on Single Condition

In order to verify also on the IC-06 the conclusions derived within UMRIDA on the airfoil application of BC-02, a simplified version of the aeroacoustic model has been tested, considering a single flight condition, i.e., takeoff because of its higher frequencies that imply more propagating acoustic modes and require finer grid, hence a higher computational cost. We had to confirm that a single-objective approach, based on the minimization of an extreme percentile of a quantity to be reduced (min-max approach) and combined with the application of polynomial chaos approximation for the accurate definition of the percentile, has revealed to be much more efficient than the classical approach, based on the minimization of the mean value of the performance and the minimization of its standard deviation, in terms of quality of results and number of simulations spent. In order to obtain a comparison based on the same number of simulations, it has been decided to set to 50 the maximum number of designs for the optimization approach, and to 15 the number of samples for each design (minimum for an accurate UQ using a second-order polynomial chaos with four uncertainties). Figure 6 shows the result obtained following the first approach, i.e., a single-objective optimization using Simplex algorithm [4] and defining as objective the minimization of the 99.97% percentile of the OASPL ($P_{99.97}$(OASPL), i.e., the value corresponding to the 99.97% cumulative percentage of the cumulative distribution function of the OASPL signal). In order to compare this result with the others, in Fig. 6 are plotted the mean and standard deviation of OASPL, each point corresponding to a different design evaluated. It can be observed that most of the 50 designs evaluated are

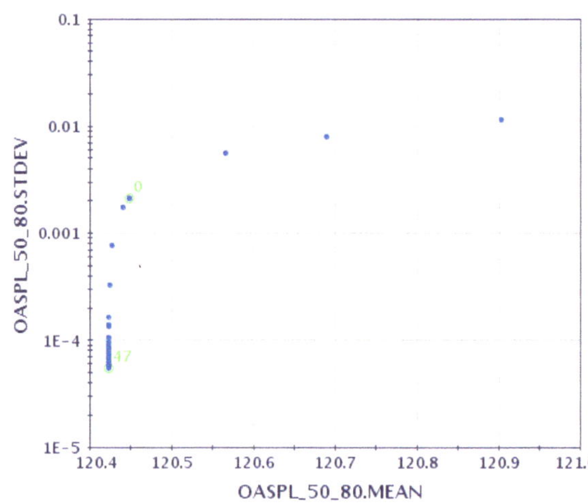

Fig. 6 1-obj optimization results (Simplex 50 designs): OASPL mean versus st.dev

Table 2 Comparison of best designs performances obtained in different approaches

RDO	RID	PoA [%]	h [mm]	d [mm]	t [mm]	$P_{99.97}$(OASPL) [dB]	σ(OASPL) [dB]	Time [CPUh]	Time* [CPUh]
Baseline	0	6.01	12.54	1.32	0.81	120.45	2.14E-03	4	-
S.O. SIMPEX	47	6.59	15.55	1.02	1.03	120.42	5.55E-05	240	960
M.O. MOGA	26	3	18	1.01	1.23	120.42	**6.79E-05**	240	960
M.O. MOGA	196	3.08	12.39	1.02	1.19	120.42	4.95E-05	960	3840

placed around the optimal region of the space, corresponding to low value of mean and standard deviation. In particular, design #47 is selected as the best one, and its detailed values are summarized for comparison in Table 2. The next approach considered is based on two-objective methodology and the results plotted in Fig. 7. In this case, a genetic algorithm (MOGAII of modeFRONTIER [5]) is being applied. Due to the low total number of designs allowed, only five generations of ten designs each have been evaluated, which is normally a too small quantity to expect good results. In fact, using the same chart, we can see that the designs found are more distributed in the variable space, and just a few are really close to the optimal region, confirming that this classical methodology is not efficient when a low number of simulations are a key factor to choose the optimization. More in detail, even if the mean value of OASPL of the best design RID #27 is practically the same, 120.4 dB (Table 2), the best standard deviation is higher (6.79E−5 vs. 5.55E−5) than the one obtained by the single-objective approach. As a further confirmation, we also plot in Fig. 8 the results obtained letting the GA algorithm run up to 20 generations, for a total of 200 designs. At the end, the final values are similar to the one obtained by the Simplex with more designs gathering around the optimum. The absolute best design with a standard deviation value of 4.95E−5 represents an improvement too small w.r.t. 5.55E−5 for the Simplex to justify an increase that large in computational cost.

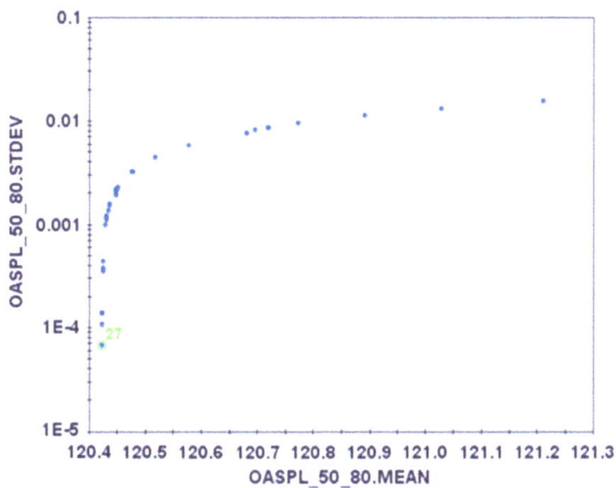

Fig. 7 2-obj optimization results (MOGAII 50 designs): OASPL mean versus st.dev

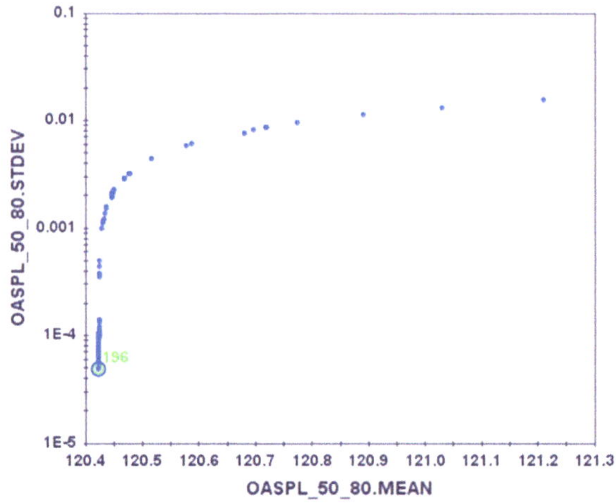

Fig. 8 2-obj optimization results (MOGAII 200 designs): OASPL mean versus st.dev

In Table 2 here above are presented the values for the best results obtained following the three different approaches executed, in terms of mean/standard deviation/percentile of the OASPL, and in terms of CPU hours spent to obtain the results. For 50 designs of 15 samples simulations, each taking 2 min on 8 cores, the total computational effort for the single-objective approach is equal to 240 CPU hours. If the two-objective optimization is stopped after the same number of evaluations, the computational cost is the same, but the performance is worse. In order to improve slightly the performance, the GA approach is to be extended to at least 200 designs, which requires a total of 960 CPUh, 4 times more! Please note that the global UMRIDA target is being able to handle a large number of uncertainties (>10) including geometrical uncertainties, in a turnaround time in the order of 10 h on a cluster of 100 cores size: 10 uncertainties × 10 CPUh × 100 cores. That means 1000 CPUh is a limit that cannot be passed, even with a number of uncertainties larger (10) that the one considered in this case. We can add that passing from four to ten uncertainties, it would almost multiply by 4 the possible number of samples, considering that the minimum number N of samples for a polynomial chaos expansion of order 2 is given by:

$$N = \frac{(n+2)!}{n!*2!} \tag{3}$$

It follows then that if by 4 uncertainties the minimum number of samples (eventually reduced by adaptive sparse method) is equal to 15, by 10 uncertainties that number would be possibly as big as 66. It follows the number of CPUh in the last column of Table 2 as an estimation of the global computational effort, to the reference case of UMRIDA project. It appears that the GA two-objective classical approach cannot be competitive w.r.t. UMRIDA goal; conversely, the single-objective approach remains competitive also if the number of uncertainties is increased to 10. In Table 2 are detailed also the geometrical values of the best designs obtained by the different methodologies, for a further analysis of the results. It is interesting to note that the designs found by MOGAII are quite different especially in terms of *PoA* and honeycomb cell depth *h*; in particular, the SIMPLEX finds quickly an optimum design quite similar to the baseline with a slightly higher *PoA* and slightly higher *h,* while all the designs the MOGA-II finds are with the lowest admissible *PoA,* and apart from the design #26 that has an *h* at the upper bound, all the others have lower values. The hole diameter *d* for all the designs is lower than the baseline and close to the limit. Moving from these observations, a few general considerations can be done to comment briefly on preliminary results. Very low values (below 5%) of *PoA* in general make the impedance of nonlinear acoustic liners too sensible to incident SPL which is dependent on engine RPM that may be slightly changed by pilot input especially during takeoff, approach, and landing maneuvers from nominal values, thus making the solution inherently non-robust. Also, from a manufacturing point of view, without getting too much into company confidential details, it can be asserted in general that too low values of *PoA* are critical because it is difficult to guarantee required uniformity over the liner (ideally, there should be at least two holes per honeycomb cell); small holes are difficult to realize with high accuracy and are subject more easily to clogging.

Multi-criteria Optimization Problem

After the execution of the tests described in previous section, the optimization methodology has been applied to a multiple flight conditions (approach, cut-back, and takeoff) problem, using the same numerical model developed but computing now a different OASPL for each flight condition as the angle range for integration of SPL changes according to experience and customer preferences as follows:

- 1: OASPL_app, approach condition, 40–70° range for SPL integration
- 2: OASPL_cbk, cut-back condition, 50–80° range for SPL integration
- 3: OASPL_tko, takeoff condition, 50–80° range for SPL integration

Figure 9 above illustrates the main workflow that has been implemented in modeFRONTIER to define this optimization problem. The four geometrical parameters are directly connected to a sub-system node, here called "flight condition," and a fourth input called "flight_cond" is added. For any design configuration and sample point proposed by the robust design optimization algorithm, the sub-system node will be executed in a loop of three times, one for each value of the flight_cond parameter. For each execution of the sub-system (Fig. 10), an Actran analysis is performed to evaluate the OASPL performance at the corresponding flight condition, and a vector node called OAPSL in Fig. 9 is extracted, contacting for each component the corresponding OASPL parameter. The last calculator node of Fig. 9 workflow is then used to extract from the vector the single responses, and then, assign these values to the optimization objectives/constraints, later described.

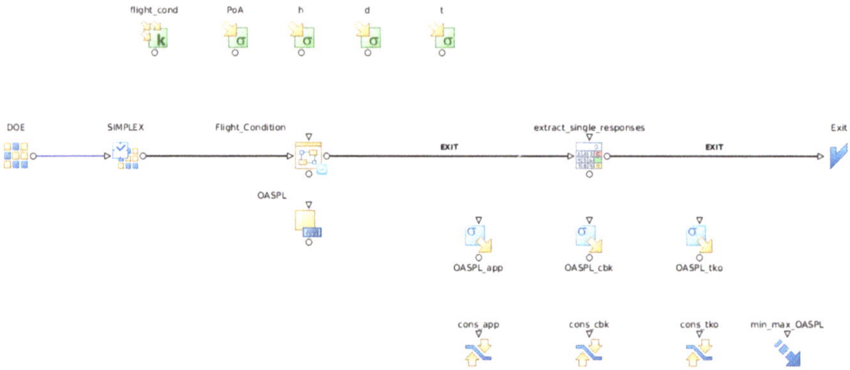

Fig. 9 Main MORDO workflow in modeFRONTIER for multiple flight conditions problem

It is worth noting that this workflow could be easily generalized for any number of flight conditions. The process flow used for the execution of Actran simulations is illustrated in Fig. 10. A first OpenOffice node is used to extract, in function of the different flight condition, the corresponding values to be used in the Actran model. The parameters, together with the admittance table function of the design parameters and flight condition, are then transferred by the corresponding nodes to the Actran node, which executes Actran analysis and Python script to let PLTViewer utility calculate the OASPL from raw solution files.

The main optimization workflow is completed by the definition of constraints/ objectives. As explained in order to avoid a drastic increment of design simulations, we decided to define a single-objective function: the average of the three single OASPL performance parameters (i.e., the 99.97% percentile of OASPL signal at each flight condition). To prevent penalization of any of the flight conditions, a constraint on each OASPL is added: to be lower than the baseline value. In conclusion, the multi-criteria (but single-objective constrained) optimization problem is defined by:

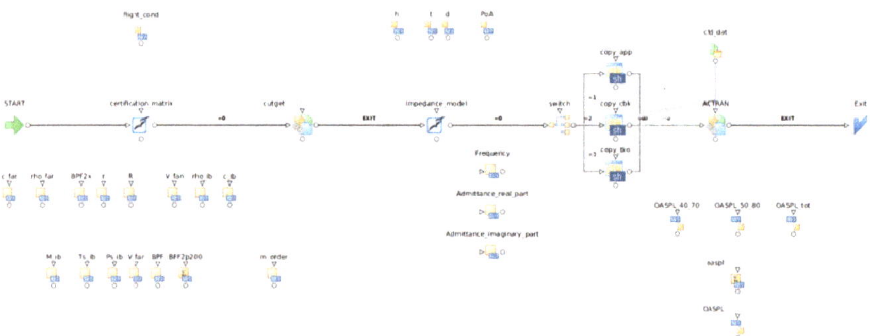

Fig. 10 Subprocess workflow in modeFRONTIER to execute Actran simulations

$$\wedge P_{99.97}(OASPL) \quad P_{99.97}(OASPL) = \frac{1}{3}\sum_{c} P_{99.97}(OASPL_c) \tag{4}$$

$$P_{99.97}(OASPL_c) < P_{99.97}\left(OASPL_c^{base}\right) \quad c = app, cbk, tko \tag{5}$$

Also in this case, it is worth to note that the procedure can be extended to any number of flight conditions, without impacting on the optimization, apart from the obvious additional constraints and for the additional CPU time of CAA simulations. Conversely, a multi-objective approach would raise heavily the number of simulation required by the number of objectives considered. We have therefore retained also the same 15 samples and second-order sparse polynomial chaos required for the UQ of the single-objective approach proposed in previous chapter. However, we are solving the single-objective problem using independently two different algorithms, the SIMPLEX and a genetic algorithm (MOGAII in modeFRONTIER), to determine which approach is the most efficient. In Fig. 11, it is offered a visual comparison between SIMPLEX, and MOGAII in terms of convergence speed. As expected, the SIMPLEX is faster to approach a convergence trend after only 20 designs, while the MOGAII requires at least 40 designs that correspond to four

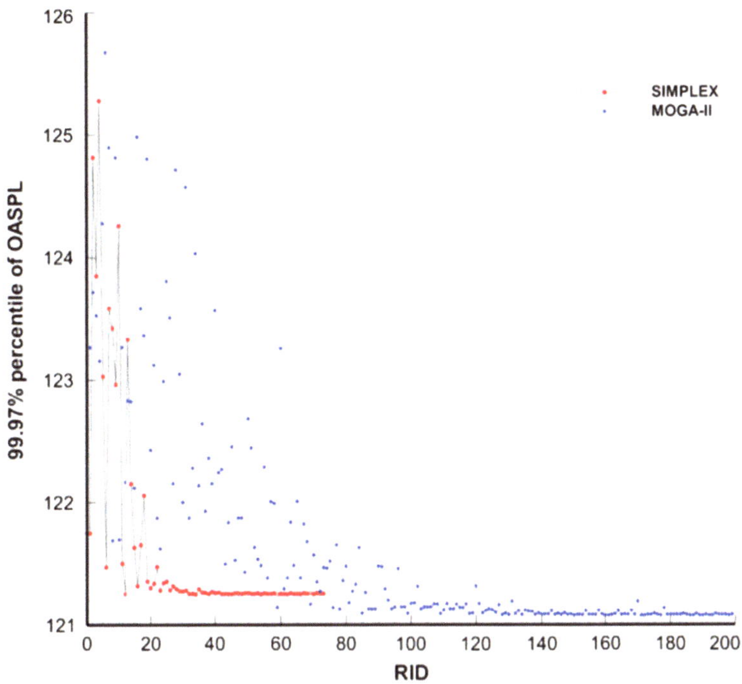

Fig. 11 $P_{99.97}$(OASPL) convergence history of SIMPLEX versus MOGA-II

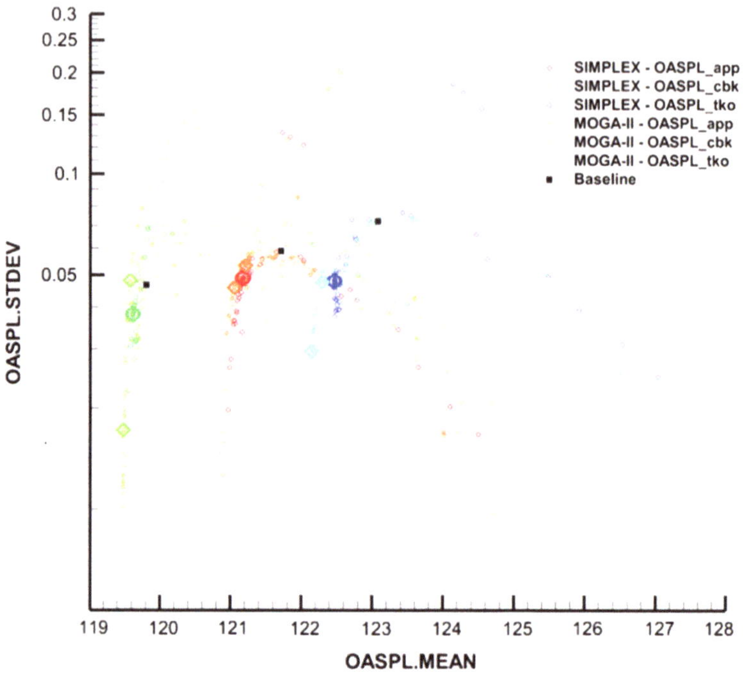

Fig. 12 μ,σ(OASPL) plots for app, cbk, and tko conditions of SIMPLEX versus MOGA-II

generations. Nevertheless, the gain in terms of objective function is limited, i.e., 121.06 for MOGAII @ RID #150 versus 121.25 for SIMPLEX @ RID #54

Figure 12 offers a visual comparison between designs explored by SIMPLEX (hollow diamonds) and MOGAII (hollow circles) algorithms in terms of standard deviation of OASPL plotted versus the mean of the OASPL for all the three certification conditions. It can be observed that SIMPLEX as well as MOGAII improve w.r.t. baseline (black solid square) not only in terms of the percentile, as already shown in Fig. 11, but also both the mean and the standard deviation of AOASPL; with *takeoff* being the most improved and *approach* the less and the *takeoff* being the only one for which even MOGAII RID #59 is better than SIMPLEX RID #54. Eventually, it would be possible to favor one condition (typically according to customer requirements) or another selecting a different solution design following any user criteria; for this application, we have selected the design RID #150 which represents the best tradeoff being the minimum average of all the three conditions.

In Fig. 13, a 3D Pareto front visualization of the results is offered to ease comprehension of the single contribution of the three components to the global $P_{99.97}$(OASPL). On the three axes are the components of the $P_{99.97}$(OASPL) for each condition, i.e., approach, cut-back, takeoff. The SIMPLEX series is represented by spheres, while tetrahedrons identify MOGAII. The contour plot is colored

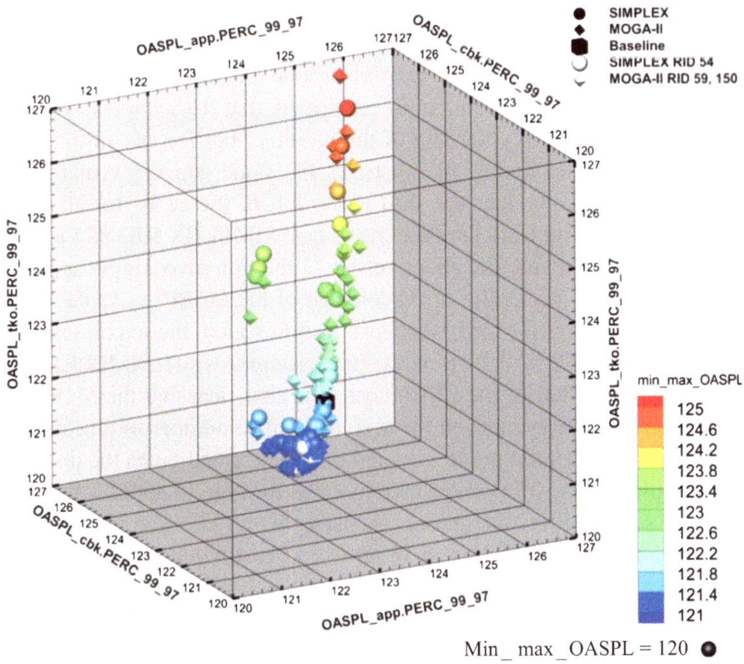

Fig. 13 3D Pareto optimization results of SIMPLEX versus MOGA-II

to the absolute value of the global $P_{99.97}$(OASPL) in dB as well as the size with the reference sphere for 120 dB at the bottom. The best robust designs found by SIMPLEX, RID #54 and MOGAII RID #59 and #150 are distinguishable in white, while the value for the baseline design is marked with a black cube. It can be noted that MOGAII has two Pareto fronts the first one being quite similar to SIMPLEX's one, while the second extends down further improving all the performance criteria. The values reported in Table 3 allow for quantitative comparison of the optimal design versus the baseline.

Apart from the exception represented by the cut-back, it is proved by raw numbers that minimizing the average of the 99.97% percentile of the OASPL with the constrain for each $P_{99.97}$ to be lower than the respective value at the baseline

Table 3 Comparison of best design mean and st. dev. for all flight conditions

Optimization	RID	$P_{99.97}$(OASPL)	$\mu(OASPL_{app})$	$\sigma(OASPL_{app})$	$\mu(OASPL_{cbk})$	$\sigma(OASPL_{cbk})$	$\mu(OASPL_{tko})$	$\sigma(OASPL_{tko})$
Baseline	0	121.75	121.71	5.89E-02	119.81	4.67E-02	123.08	7.23E-02
SIMPEX	54	121.25	121.17	4.90E-02	119.62	3.82E-02	122.47	4.79E-02
MOGA-II	59	121.14	121.21	5.34E-02	119.49	1.72E-02	122.30	4.76E-02
MOGA-II	150	121.08	121.06	4.58E-02	119.58	4.81E-02	122.15	2.96E-02

indeed improves also the mean and the standard deviation of each OASPL. Please note that the values related to optimal result are reported after the validation obtained running a series of 150 samples around the optimal design. Since the results are practically coincident, it is confirmed that 15 samples were enough accurate to estimate not only the UQ of the baseline, but for any design proposed during the optimization. Figure 14 illustrates the same concept visually showing how in these plots it is evident the shift to the left of the PDFs for all the certification points progressing from the baseline to best SIMPLEX RID #54 to MOGAII best RID #150, thus proving the effectiveness of the min–max single approach that minimizes the 99.97% percentile of the average of the OASPLs. Table 4 shows in detail the key parameters for the best robust designs found; the maximum total gain by MOGAII RID #150 is 0.7 dB w.r.t. the baseline, while SIMPLEX provides 0.5 dB with less than half of the CPUh cost. It is also clear that the MOGA-II tops at the bottom of the range imposed for *h*, *d*, and *t*. The number of repeated designs increases as the optimization progresses and that is expected given the discrete base.

3D scatter plot of Fig. 15 (same conventions as for Fig. 13 apply here, except for size that is scaled after thickness) is used to try to graphically represent $P_{99.97}$(OASPL) as function of *PoA*, *h*, *d*, and *t*, to visually identify how each one of

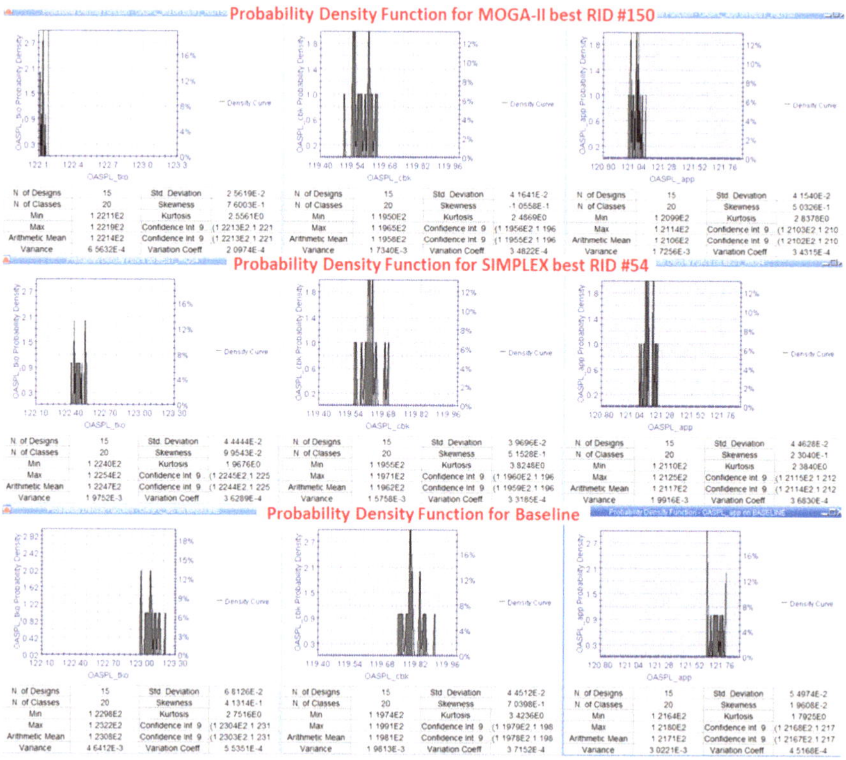

Fig. 14 MORDO probability density function plot of SIMPLEX versus MOGA-II

Table 4 MORDO results summary table

Optimization	RID	Repeated	CPUh	CPUh 4u/100 cores	$P_{99.97}$(OASPL) [dB]	3<PoA<15 (0.1) [%]	10<h<18 (0.01) [mm]	1<d<2.5 (0.01) [mm]	0.7<t<1.4 (0.01) [mm]
Baseline	0	-	14	1	121.75	6.01	12.54	1.32	0.81
SIMPEX	54	0*	738	59	121.25	5.2	13.17	1.18	0.70
MOGA-II	59	3	779	62	121.14	5.2	10.82	1.00	0.70
MOGA-II	150	32**	1626	130	121.08	5.1	10.00	1.00	0.70

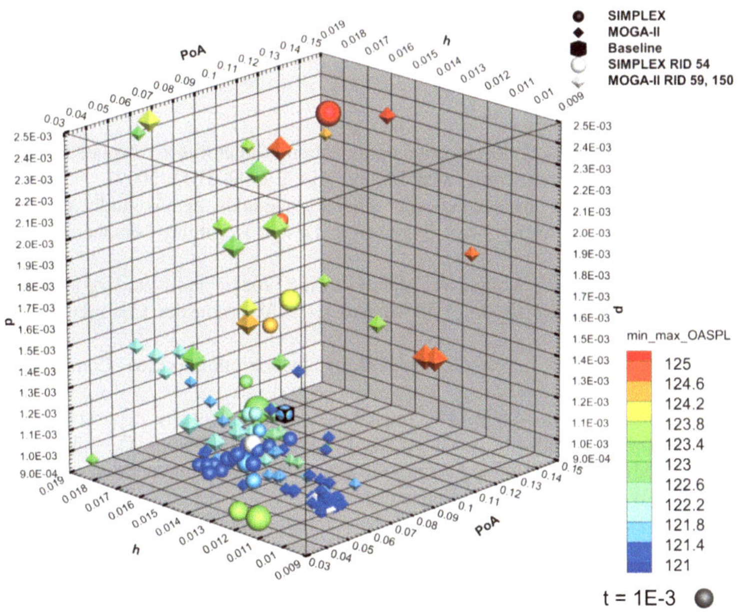

Fig. 15 3D scatter plot of variables for SIMPLEX versus MOGA-II

the uncertainties contributes to the objective. It can be noted that best RIDs of MOGAII group at the lower boundaries of the range for all uncertainties except PoA, while the SIMPLEX's are more centered also for *h* and *d*.

Conclusion

In this chapter is described the implementation and presented the results of an automatic multi-objective robust design optimization methodology applied to a problem of great relevance for the aerospace industry: the acoustic performance of the liner of a typical regional jet engine nacelle produced by Leonardo Finmeccanica. The preliminary tests, performed on a single flight condition, have confirmed the efficiency of the proposed min–max strategy, with the definition of a single-objective function based on a reliability optimization approach (maximization of percentile performances), combined with the efficient adaptive sparse polynomial chaos method for UQ. The same methodology has successfully been extended to the multiple flight conditions problem, reaching a convergence within a small number of iterations (circa 20–40 robust designs), the usage of a genetic algorithm bringing most robust results at the expense of a higher computational effort w.r.t. Simplex.

References

1. Copiello, D., Ferrante, P.: Multi-objective optimization of inlet true zero-splice liners. In: 15th AIAA/CEAS Aeroacoustics Conference, Miami, AIAA 2009–3107 (2009)
2. Der Kiureghian, A.: First- and second-order reliability methods. In: Nikolaidis, E., Ghiocel, D. M., Singhal, S. (eds.) Chapter 14 in Engineering Design Reliability Handbook, CRC Press, Boca Raton, FL p. 200
3. Murray, P.B., Ferrante, P., Scofano A.: Manufacturing process and boundary layer influences on perforate liner impedances. In: 11th AIAA/CEAS Aeroacoustics Conference, Monterey, California (2005)
4. Nelder, John A., Mead, R.: A simplex method for function minimization. Comput. J. **7**, 308–313 (1965)
5. Poloni, C., Pediroda, V.: (1997). GA coupled with computationally expensive simulations: tools to improve efficiency. In: Quagliarella, D., Périaux, J., Poloni, C., Winter, G. (eds.) Genetic Algorithms and Evolution Sategies in Engineering and Computer Science, Chapter 13, pp. 267–288, Wiley, West Sussex, England
6. Copiello, D., Ferrante, P., Iemma, U., Maci, A.: Preliminary design of aero-engine intake acoustic liners by means of the multi-objective approach. AIAA 2010–3828

Application of Robust Design Methodologies to Falcon

G. Rogé and X. Loyatho

Geometry and Design Point

The industrial test case generic Falcon jet (IC-03) deals with the same choice as baseline for UQ study: an unstructured tetrahedral mesh (7.17M nodes), a free stream Mach number = 0.85, an altitude = 40,000 ft, a Spalart–Allmaras turbulence model, and a CL target = 0.35 (Figs. 1 and 2).

The CFD framework includes Dassault Aviation in-house Navier–Stokes solver AETHER (RANS model, finite element approach), gradient by finite difference, direct or adjoint method. The optimization framework is built on local optimization/gradient-based method (Fig. 3).

Geometrical deformation:

Two kinds of geometrical variables are considered. The first set deals with optimization variables for design conception. In our study, we have selected wing trailing edge **camber** angle (Fig. 4).

The second set deals with uncertainty variables. In our study, we have selected wing tip **twist** angle (Fig. 5).

We remain than, using adjoint methodology, the complexity is **independent** of the number of variables.

G. Rogé (✉) · X. Loyatho
Dassault Aviation, 78 quai Marcel Dassault, 92552 Saint-Cloud, France
e-mail: gilbert.roge@dassault-aviation.com

X. Loyatho
e-mail: ximun.loyatho@dassault-aviation.com

C. Hirsch et al. (eds.), *Uncertainty Management for Robust Industrial Design in Aeronautics*, Notes on Numerical Fluid Mechanics and Multidisciplinary Design 140, https://doi.org/10.1007/978-3-319-77767-2_40

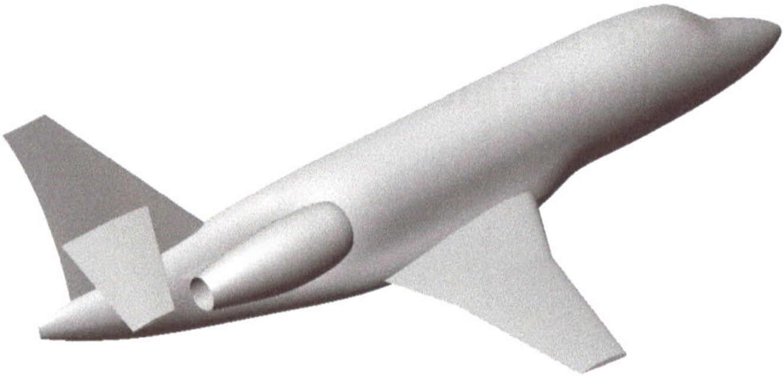

Fig. 1 Generic Falcon jet (IC-03)

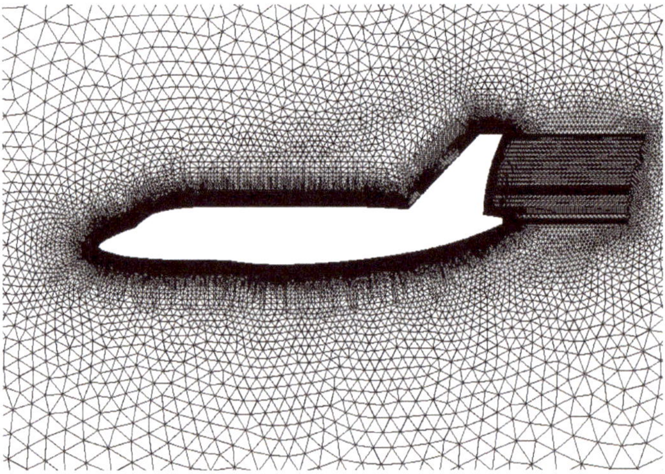

Fig. 2 Mesh. symmetry plane y = 0. Zoom

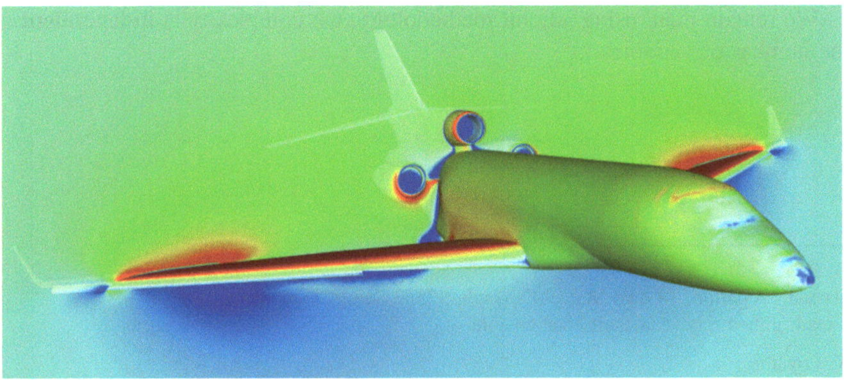

Fig. 3 RANS results

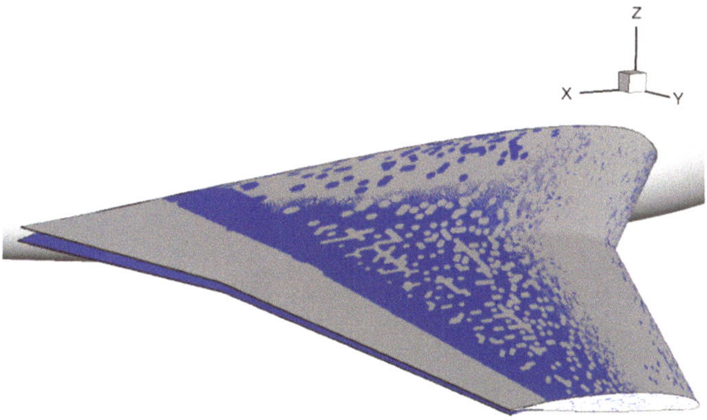

Fig. 4 Camber = −15°

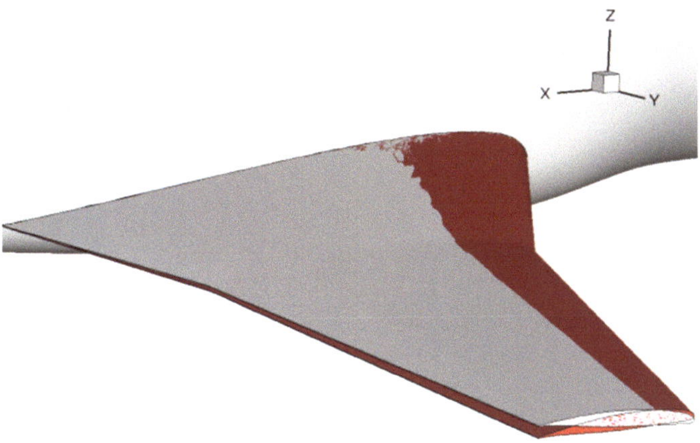

Fig. 5 Twist = +5°

Deterministic Optimization

For the initial shape, we have camber = 0° and twist = 0°.

For Mach = 0.85 and CL = 35 cts, we have AoA = 1.76° and CD = 390.5 cts (Figs. 6 and 7).

CPU time (on 2048 procs BullX cluster; all the CPUs are given for this cluster) is 20’ for 1 RANS and 20’ for 1 adjoint.

Deterministic optimization problem:

We have to solve the following control problem (Figs. 8 and 9):

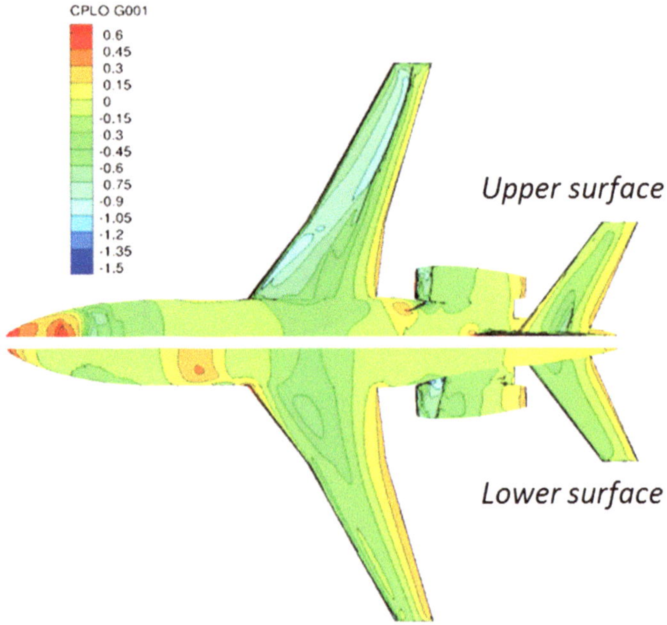

Fig. 6 Baseline Cp

Fig. 7 Baseline polar curve

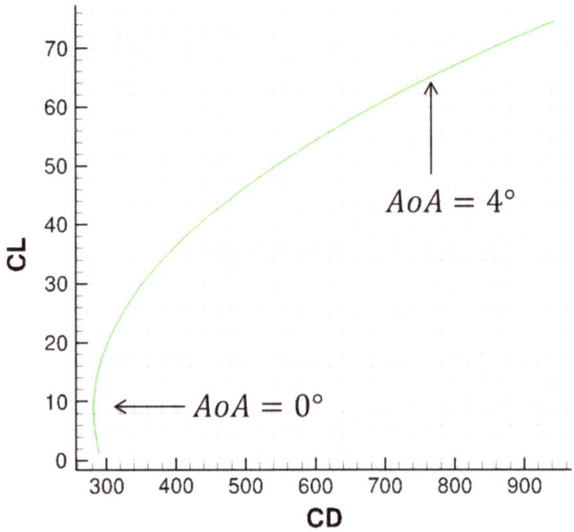

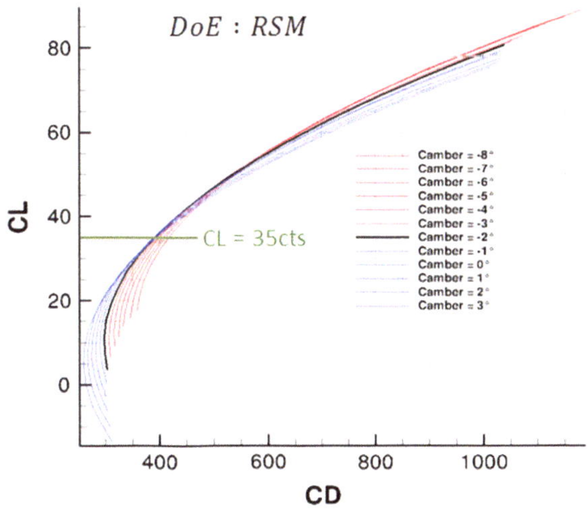

Fig. 8 Polar curves for different TE camber angle

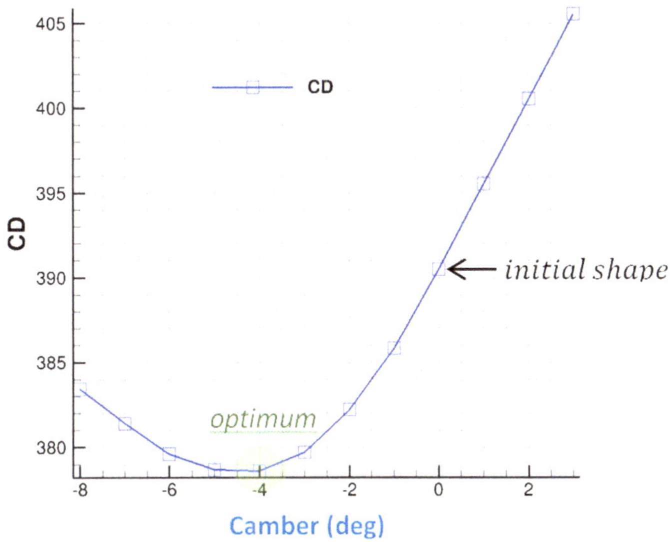

Fig. 9 Drag coefficient versus TE camber angle

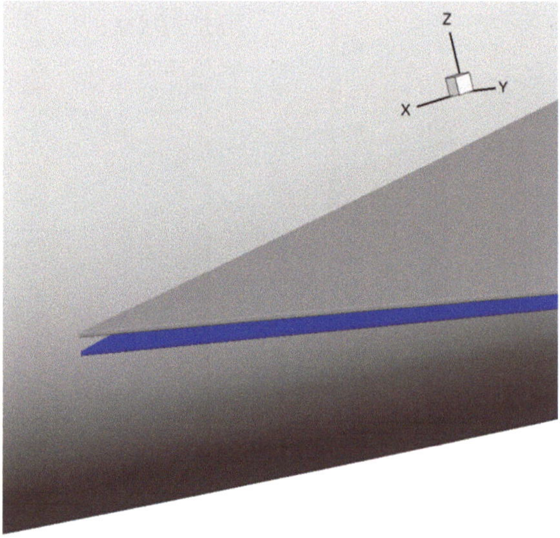

Fig. 10 Baseline versus deterministic optimum (zoom close to TE)

$$\boldsymbol{min}\,CD$$
Variables: $AoA, TE\,\boldsymbol{Camber\,Angle}\,\left(\frac{x}{c} = \mathbf{0.75}\right)$
Constraint: $\mathrm{CL} = 35\,\mathrm{cts}$

Deterministic optimum: The whole optimization problem is solved in ~10 h. We obtain **camber** = −4.1° and CD = 378.6 cts (Gain = 11.9 cts) (Fig. 10).

Robust Optimization

UQ: We first set some notations

$$X_{output} = F\left(X_{input}\right)$$

The Method of Moment—**first-order approach** gives the same probability law for X_{output} than for X_{input}.

We obtain easily the mean value:

$$\mu_{output} = F\left(\mu_{input}\right)$$

and the standard deviation: $\sigma_{output} = |\nabla F| \times \sigma_{input}$

In our example:

$$X_{input} = Twist$$
$$X_{output} = CD$$

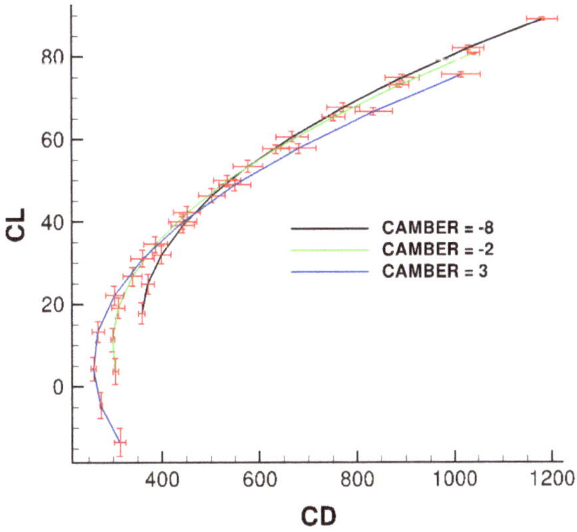

Fig. 11 UQ analysis on polar curves (± standard deviation)

$\nabla \text{CD} = \frac{\partial CD}{\partial Twist}$ is computed by finite differences, direct or adjoint method.

The twist angle uncertainty, i.e., the level of confidence on structural modeling (lack of knowledge) is (Figs. 11 and 12)

$$\boldsymbol{\sigma}_{input} = \boldsymbol{\sigma}_{twist} = \mathbf{1}^{\circ}$$

σ constraint Robust optimization problem:

We have to solve the following control problem (Figs. 13 and 14):

$$\begin{array}{ll} & \boldsymbol{min\, \mu}_{CD} \\ \text{Variables:} & AoA, \boldsymbol{Camber} \\ \text{Constraint:} & CL = 35\,\text{cts}; \quad \sigma_{CD} \leq 15\,\text{cts} \end{array}$$

σ constraint optimum: is obtained in ∼15 h.

We obtain **camber** = −5.2° and CD = 379.0 cts (Gain = 11.5 cts) (Fig. 15).

MinMax (cf chapter "Formulations for Robust Design and Inverse Robust Design" [2]*) Robust optimization problem*:

We have to solve the following control problem (Figs. 16 and 17):

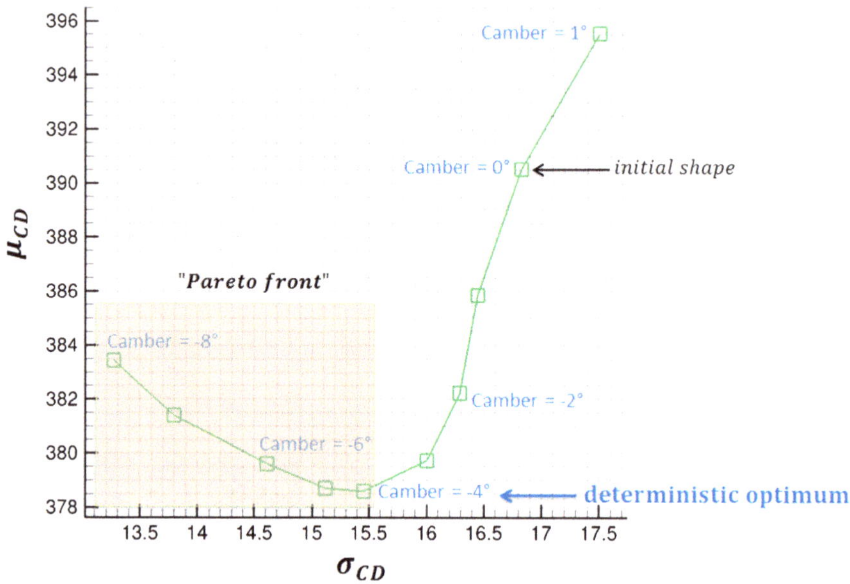

Fig. 12 Mean value versus standard deviation (CD)

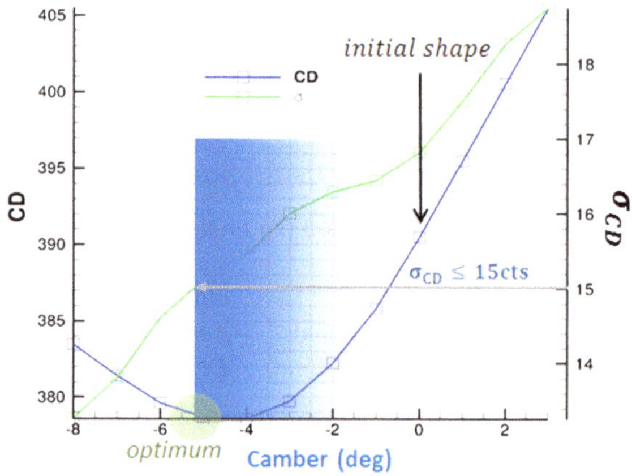

Fig. 13 Robust optimization with constraint on standard deviation. CD versus camber

$$\boldsymbol{min}\ (\boldsymbol{\mu_{CD}} + \boldsymbol{3\sigma_{CD}})$$

$$\text{Variables: } AoA, \boldsymbol{Camber}$$

$$\text{Constraint: } CL = 35\,\text{cts}$$

MinMax optimum is obtained in ∼15 h.

We obtain **camber** = −7.0° and CD = 383.4 cts (Gain: 7.1 cts) (Figs. 18, 19, and 20).

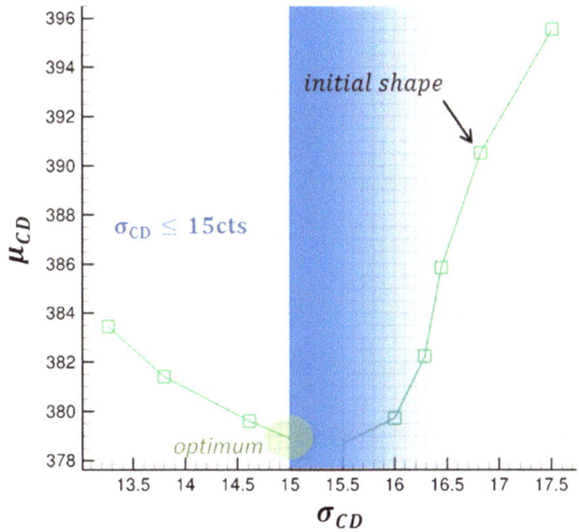

Fig. 14 Robust optimization with constraint on standard deviation. Mean value versus standard deviation

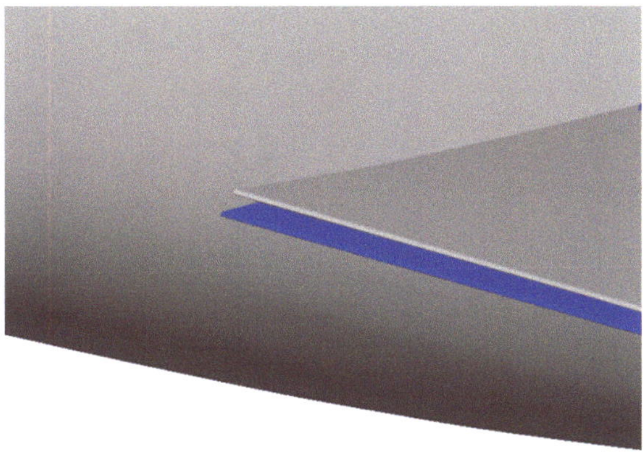

Fig. 15 Baseline versus constraint optimum (zoom close to TE)

Remark: According to MoM order 1 methodology and Pearson approach, for the same price as σ constraint optimization, we are in position to solve Reliability Optimization.

In addition to

X_input probability *law* → X_output: *same law*

Mean value: μ_output = F(μ_input)

Standard deviation: $\sigma_{output} = |\nabla F| \times \sigma_{input}$

We recall that: Third and fourth statistical moment coefficients are unchanged.

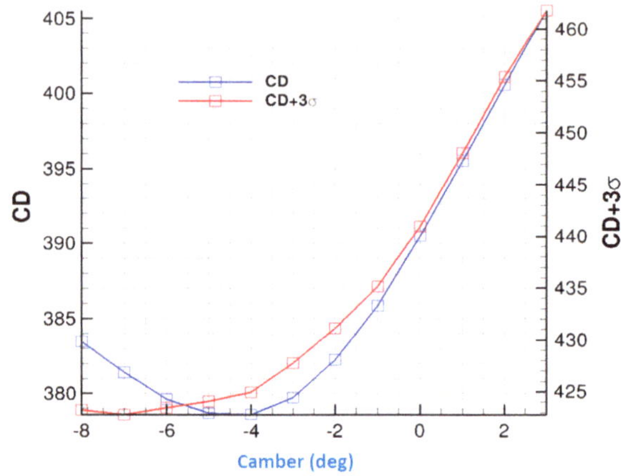

Fig. 16 CD and CD+3 standard deviation versus camber

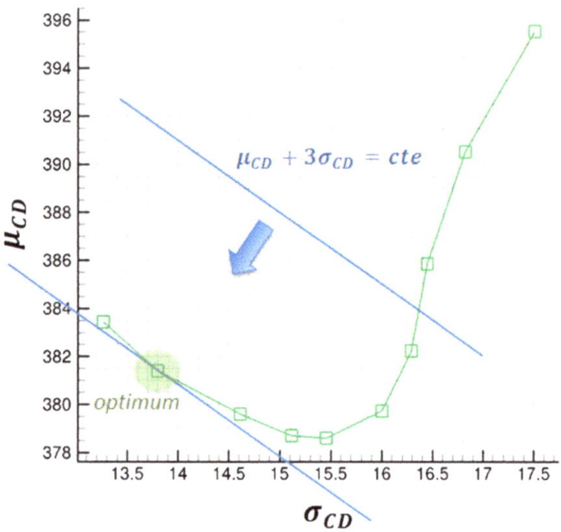

Fig. 17 MinMax optimum

Different objectives, different results!

Discussion about the way to manage both a large number of uncertain variables and a large number of design variables:

$$\text{For Method of moment order 1, we have : } \mu_{CD} = CD, \sigma_{CD} = \left|\frac{\partial CD}{\partial \boldsymbol{Twist}}\right| \times \sigma_{Twist}$$

Local optimization methods need gradients of cost and constraints respect to optimization variables:

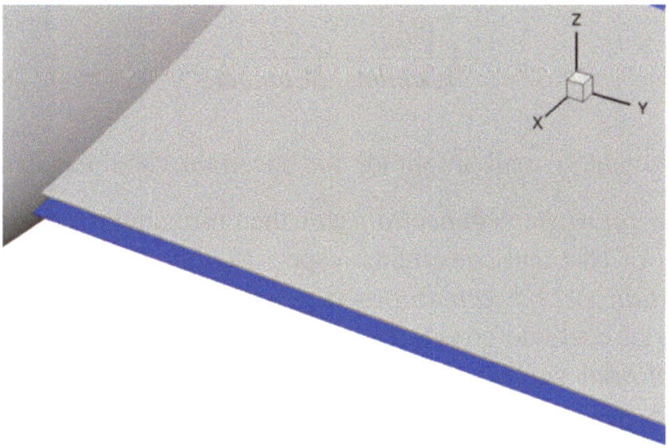

Fig. 18 Baseline versus MinMax optimum (zoom close to TE)

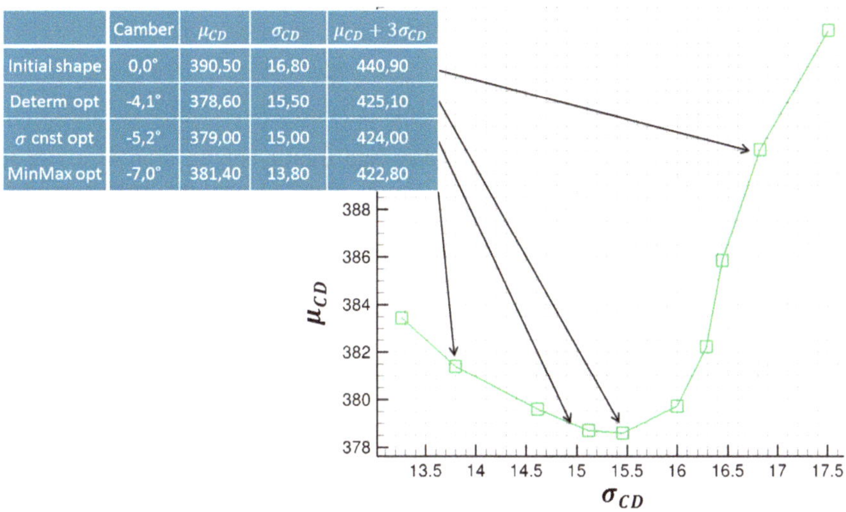

	Camber	μ_{CD}	σ_{CD}	$\mu_{CD} + 3\sigma_{CD}$
Initial shape	0,0°	390,50	16,80	440,90
Determ opt	-4,1°	378,60	15,50	425,10
σ cnst opt	-5,2°	379,00	15,00	424,00
MinMax opt	-7,0°	381,40	13,80	422,80

Fig. 19 Baseline versus robust design solutions

σ constraint optimization :

$\min \mu_{CD}$

Variables : AoA, $Camber$

Constraints :

$CL = 35\ cts$

$\sigma_{CD} \leq 15\ cts$

MinMax optimization :

$\min \mu_{CD} + 3\sigma_{CD}$

Variables : AoA, $Camber$

Constraint :

$CL = 35\ cts$

Robust design optimization

Probability optimization :

$\min \mu_{CD}$

Variables : AoA, $Camber$

Constraints :

$CL = 35\ cts$

$P(CD \geq 380) \leq 1\%$

Reliability design optimization

Fig. 20 Summary: robust design methods

$$\frac{\partial \mu_{CD}}{\partial \boldsymbol{Camber}}, \frac{\partial \sigma_{CD}}{\partial \boldsymbol{Camber}}$$

1. If MoM is suitable (small amplitude for uncertain variables, derivability)
   - *if #optimization var* $\gg$ *#uncertain var*, then use adjoint approach to compute gradients of cost and constraints respect to *optimization var*.
   - *if #uncertain var* $\gg$ *#optimization var*, then use adjoint approach to compute gradients of cost and constraints respect to *uncertain var*.
   - *if #optimization var* $\gg 1$ *and #uncertain var* $\gg 1$, then Hessian of cost and constraints is needed: e.g., $\frac{\partial^2 CD}{\partial \boldsymbol{Camber} \partial \boldsymbol{Twist}}$
2. Else use NIPC/Sparse Grid for UQ and adjoint approach to compute gradients of cost and constraints respect to *optimization var*.

Conclusion/Lesson's Learned

A well-posed formulation both for standard and robust optimization of a Falcon shape (IC-03 test case) has been given. Results obtained with two robust approaches have been compared to standard result. Discussion about impact of robustness concludes this work.

Recommendations:

1. Robustness

- In order to avoid large degradation respect to initial for the cost functional, we need to control his derivatives close to the optimum.
- Probability of failure has to be an eligible constraint for aerodynamics engineer. PDF of uncertain variables is needed.
- An intermediate formulation between a weighted cost functional based on mean value and standard deviation and a formulation based on probability or CDF is to add terms for third and fourth moments (skewness and kurtosis).

2. Multipoints

- Weights have to be chosen through mathematics (Gram–Schmidt …) and physics (flight path statistics …) considerations.
- Independent CFD calculations have to be solved in parallel mode.

References

1. UMRIDA Project: Test Case Description Innovative Database for UQ and RDM: IC-03
2. UMRIDA M4.1-26 (10, 2) Evaluation of methodologies for robust design and identification of the most promising methods

Part V
UMRIDA Best Practices: Methods for Uncertainty Quantification, RDO and Their Applicability Range

Uncertainties Identification and Quantification

Dirk Büche, Sönke Klostermann and Martin Schmelzer

Introduction

The performance of a system such as a turbomachine or an aircraft can only be predicted with limited precision as there are uncertainties. For the performance prediction, all system parameters are typically set to their nominal values; however, in reality they deviate due to various effects as for example manufacturing tolerances and variation in the operational conditions. Thereby, actual values typically deviate in both mean value and distribution from the nominal values. Furthermore, the prediction tool introduces additional uncertainties and limited prediction precision. The tools include simplifications due to model assumptions such as, e.g. given by empirically set model parameters in turbulence models of CFD or assumptions such as steady-state flow.

There are tremendous efforts to identify and measure the most relevant uncertainties. The measurement results are typically converted into a probability distribution. The most important distributions are uniform, Gaussian, and beta distributions.

The probability distributions then serve as input for uncertainty quantification (UQ) and robust design optimization (RDO). UQ follows the goal of predicting the influence of the uncertainties on the system behaviour. RDO searches for a system design that has both high performance and is insensitive to the given uncertainties.

D. Büche (✉)
MAN Diesel & Turbo Schweiz AG, Hardstr. 319, 8005 Zurich, Switzerland
e-mail: Dirk.Bueche@man.eu

S. Klostermann
Airbus Group Innovations, Hein-Saß-Weg 22, 21129 Hamburg, Germany
e-mail: soenke.klostermann@airbus.com

M. Schmelzer
TU Delft, High Speed Laboratory (HSL), Kluyverweg 2, 2629 HT Delft, Netherlands
e-mail: m.schmelzer@tudelft.nl

C. Hirsch et al. (eds.), *Uncertainty Management for Robust Industrial Design in Aeronautics*, Notes on Numerical Fluid Mechanics and Multidisciplinary Design 140, https://doi.org/10.1007/978-3-319-77767-2_41

It is important to note that both UQ and RDO assume that the uncertainties are given. This paper however addresses the process of identifying and of modelling uncertainties. Furthermore, it shows their consequence on a robust design optimization.

Uncertainties in Product Development

In product development, the optimization of entire systems such as aircrafts or system components such as winglets or flaps is a key engineering activity. The goal is to obtain the best system performance given a set of requirements on technology, cost, and manufacturing techniques. Especially in early design phases, many of the key parameters of an aircraft such as engine weight or wingspan are uncertain. These uncertainties may result from the fact that only rough estimates for the component performance are available. However, the total system performance has to be estimated and optimized. When optimizing the system for specific values of the components only, there is a risk of a highly sensitive optimum. Instead, the design should be optimized such that the resulting performance is insensitive to the uncertainties. This minimizes the risk of redesigning in case of larger deviations in some estimated values.

With each design freeze in the development process, the level of uncertainty decreases as more and more parameters are determined. This then allows for more precise and more complex modelling techniques such as 3D CFD on system level and the start of detailed development on component level. The level of uncertainty can often be specified by the component developer and then directly integrated into UQ and RDO tools.

Nevertheless, not all uncertainties can be eliminated: Especially, uncertainties from manufacturing scattering, assembly, and operation remain. These uncertainties are difficult and expensive to quantify. An overview is given in the chapter "Uncertainties in Compressor and Aircraft Design". Uncertainties in the prediction tools are addressed in the chapter "Estimation of Input Uncertainties for Turbulence Modelling".

Designing with Uncertainties

Designers are typically aware of some of the uncertainties in both the design and the simulation models required for running the performance prediction of the design. However, there is often a lack of knowledge about the actual variation magnitude and stochastic distribution as well as the effect of these uncertainties on the system performance. In order to obtain more clarity, two different actions may be performed.

In the first action, the designer may be interested in an analysis that quantifies the effect of uncertainties on the system performance. From the probability distribution of the uncertainty, several samples are drawn by tools such as Monte Carlo or Latin Hypercube and evaluated by the prediction tool. Then, statistical properties such as mean and variance of the result are computed. This is referred to as *uncertainty quantification* (UQ).

The second action is more complex and addresses design optimizations that include uncertainties. The goal is to find a robust design which is both highly performant and is also insensitive to the uncertainties. This action is referred to as *robust design optimization* (RDO). But how can uncertainties be quantified and modelled? Precise knowledge about the uncertainties is a key prerequisite for robust design optimization as otherwise the optimization result is of no practical use [1].

Modelling Uncertainties

Depending on the available data, uncertainties are typically modelled with either *probabilistic models* or *domain (or interval)-based models*.

In *probabilistic models*, each uncertainty is described by a probability density function (PDF) such as uniform, Gaussian, or beta distribution. Inter-correlation between the uncertainties is often represented by covariance matrices. Typically, these PDFs have lower probabilities in the tail regions of the distributions. Setting up PDFs requires a sufficiently large set of measurement data, which can be expensive to generate. Examples for such measurements can be found in [2–4].

Typical examples for applying *domain-based models* are uncertainties that are extracted from manufacturing drawings. A drawing always defines theoretical exact dimensions as well as tolerance intervals. The interval defines a lower and upper limit for the measure, but no probability distribution is provided. Due to the lack of measurements, the actual scattering of the geometry is unknown. For this reason, worst-case scenarios using always the tolerance limits (even that they might be very unlikely to occur) may drive the design into very conservative layouts thus reducing overall performance. However, *domain-based models* can also be converted into probabilistic models by either assuming a uniform probability density between the tolerance limits or by choosing a Gaussian distribution, assuming a certain percentage (e.g. 95%) of all measures within the tolerance interval.

While for statistical analysis such as design of experiments, domain-based models are sufficient, and UQ or RDO such as described in the chapters "General Introduction to Monte-Carlo and Multi-level Monte-Carlo Methods" and "Robust Compressor Optimization by Evolutionary Algorithms" typically require probabilistic models for all uncertainties.

An Example for Input Uncertainties for Turbulence Modelling with Reynolds-Averaged Navier–Stokes

A common problem in CFD based on Reynolds-Averaged Navier–Stokes (RANS) for industrial flow cases is how to deal with the epistemic uncertainties due to the usage of imperfect turbulence models. The non-universality of existing modelling approaches is illustrated by the vast number of existing models such as Spalart–Allmaras, k–ε, k–ω, or SST, varying in complexity and calibrated on canonical flow cases or optimized for specific applications. So far, mainly "expert" experience is used to decide the modelling strategy for the problem at hand and to estimate non-quantitatively the reliability of the output, based on problem knowledge and experience. However, for industrial circumstances in which the goal is to design new and innovative products, for which no or only sparse validation data is available, this expert experience is over-strained. This questions the capability of CFD as a predictive tool in industry. The need for a quantitative measure of the uncertainty on the flow is therefore high and subject to ongoing research.

The field of UQ for industrial problems often deals with efficient propagation of measured or estimated uncertainties on input variables. Using an expensive code to compute the quantity of interest (QoI), the uncertainty due to modelling is of a different ilk. It cannot always be traced back to physical quantities such as geometry, inflow parameters, boundary conditions. This is due to the fact that the mathematical form of the model is the result of modelling assumptions, and its inputs are not necessarily even physical quantities. For the case of RANS-based linear eddy viscosity one-point closures, the models aim to capture the mean effect of the dynamics of the Reynolds stress tensor using a linear stress–strain relation. Transport equations are derived from a combination of physical intuition and fundamental principles. Therefore, an UQ methodology needs to focus on the input parameters to the model (model-parameter uncertainty) and on the underlying modelling assumptions (model-form uncertainty).

The transport equations of any RANS-based model are closed by coefficients; e.g. for the Launder–Sharma k–ε model, we have four coefficients: κ, C_{μ}, $C_{\varepsilon 2}$, σ_k. Commonly, these coefficients are identified by deterministic calibration on canonical and simple flow cases, such as decaying homogeneous isotropic turbulence or turbulent flow in a channel. In industry, the coefficients are subject to application-dependent optimization leading to an extensive library of coefficients per model.

In the chapter "Estimation of Model Error using Bayesian Model-Scenario Averaging with Maximum A Posteriori Estimates" of this book, the methodology of Bayesian Model-Scenario Averaging (BMSA) introduced by Edeling et al. [1, 5] was presented which uses stochastic estimates of the closure coefficients as estimates of the parametric input uncertainties. The uncertainty on the closure coefficients is identified by Bayesian calibration of the models using boundary layer data sets from the AFOSR-IFP-Stanford Conference Proceedings [6]. Using the Markov chain Monte Carlo (McMC) algorithm for Bayesian inverse modelling requires several thousand forward calls of the models in order to achieve converged chains for the coefficients. In order to not propagate the full chains through the code Maximum A Posteriori (MAP) estimates of the underlying probability distributions are used, which lead to a single value per coefficient for each model and flow scenario. This is detailed out in the above-mentioned chapter. The MAP estimates applied within the BMSA methodology serve as our global input uncertainties to the turbulence modelling framework.

An Example for Robust Design Optimization

In this chapter, an example for a RDO is described. The focus is on how to set up the problem as well as on the influence of the uncertainties on the outcome of the optimization. This is illustrated with the help of a simple, bimodal function with two design variables.

Problem Specification

The first step is the *problem specification* that gathers all information required for running a RDO. The optimization in general is defined as the search for a solution $\mathbf{x}$ in the design space X that minimizes an objective function(s) $J(\mathbf{x})$ such as aerodynamic efficiency

$$J(\mathbf{x}), \quad \mathbf{x} = (x_1, x_2, \ldots, x_D) \in X \subseteq \mathbb{R}^D \tag{1}$$

while concurrently fulfilling inequality constraints $\mathcal{G}(\mathbf{x})$ such as on mechanical integrity requirements

$$\mathcal{G}(\mathbf{x}) \leq 0. \tag{2}$$

The design variables $\mathbf{x}$ as well as model parameters may be uncertain and modelled by their nominal values $x_{i,\, i=1,\ldots,D}$ and uncertainties ξ_i, which is given by a probability distribution such as a Gaussian distribution with

$$\xi_i = N\left(0, \sigma_i^2\right) \tag{3}$$

As an example, a test function with the two design variables x_1 and x_2 is selected:

$$\mathcal{J}(x_1, x_2) = 15 \cdot \left(x_1^2 + x_2^2\right) - 100 \cdot e^{-5\left((x_1+1.6)^2 + (x_2+1.6)^2\right)}. \tag{4}$$

The function has two minima (bimodal). The inequality constraints are given by

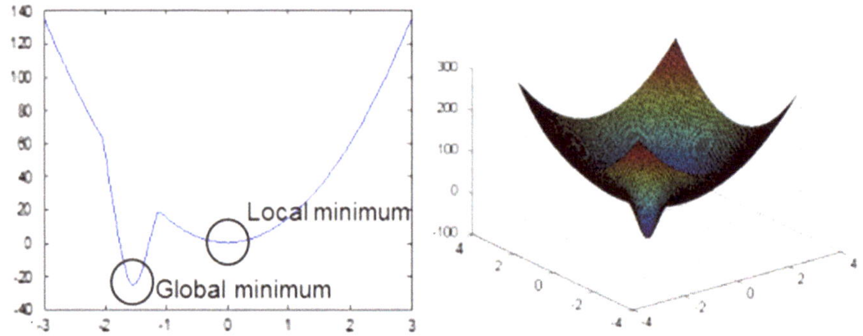

Fig. 1. Non-robust test function with a non-robust global and a robust local optimum

$$\mathcal{G}(x_1,x_2) = \begin{pmatrix} (x_1-0.5)^2 + x_2^2 - 4 \\ (x_1+0.5)^2 + x_2^2 - 4 \end{pmatrix} \leq 0 \tag{5}$$

and could be interpreted as circular bounds of the design space. The test function is shown in Fig.® 1.

The test function is extended by defining both design variables as noisy, leading to the following objective and constraint functions:

$$\mathcal{J}(x_1,x_2,\xi_1,\xi_2) = \mathcal{J}(x_1+\xi_1, x_2+\xi_2). \tag{6}$$

$$\mathcal{G}(x_1,x_2,\xi_1,\xi_2) = \mathcal{G}(x_1+\xi_1, x_2+\xi_2) \tag{7}$$

Robustness and Risk Formulation

This step is denoted as *robustness and risk formulation*. An optimization will lead to a robust solution, if instead of the objective function J a robust measure ρ_J such as the mean $E(J)$ is optimized

$$\rho_J = E(J) = \int_\xi w(\boldsymbol{\xi}) \cdot J(x+\boldsymbol{\xi}) d\boldsymbol{\xi},$$

where $w(\boldsymbol{\xi})$ is the probability density function of the uncertainties $\boldsymbol{\xi}$.

A reliability measure $\rho_{\mathcal{G}}$ representing the constraints is defined in the same manner as

$$\rho_{\mathcal{G}} = E(\mathcal{G})$$

Typically, the two measures cannot be computed by analytical integration but are approximated by a set of sample points that represent the probability space adequately. This is denoted as uncertainty quantification (UQ). The simplest approach is sampling by Monte Carlo methods. Different UQ methods are given in the chapters "General Introduction to Monte-Carlo and Multi-level Monte-Carlo Methods" and "Latin Hypercube Sampling based Monte Carlo Simulation: Method, Extension of the Sample Size and Correlation Control".

Optimization

In the *optimization* step, the robust optimization problem for the robustness measure ρ_J and reliability measure ρ_G is solved. Algorithms for RDOs are given in the chapter "Robust Compressor Optimization by Evolutionary Algorithms and 28 Formulations for Robust Design and Inverse Robust Design".

Analysis

The last step is the *analysis* where the robustness of the solution is verified and it is checked if the solution fulfils the performance target. Special care is taken in case of disjoint constraints.

Figure 2 shows the results of a RDO example for three different noise levels. For the lowest noise level of $\sigma = 0.1$, the optimum is within the vicinity of the global minimum of the function. This optimum is very sensitive and thus not considered as sufficiently robust. For an increase in noise to $\sigma = 0.2$, the optimum is still found in the vicinity of the global minimum but a higher probability exists that the constraint for the chosen α-level is violated.

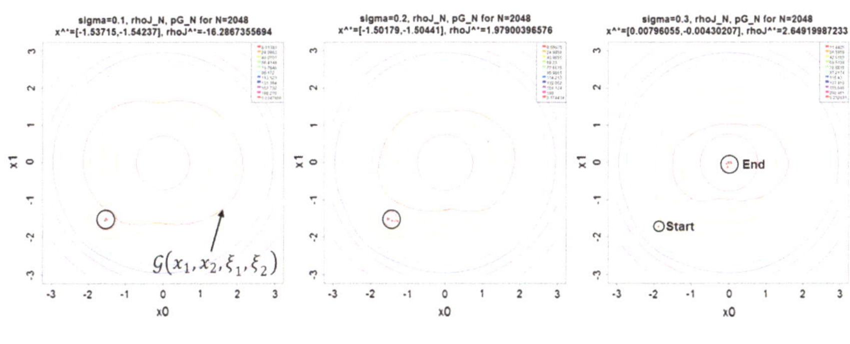

Fig. 2. Contour plot of the objective function with overlaid inequality constraint (red border) for $N_{10} = 2048$ and three different noise levels $\sigma = 0.1, 0.2, \text{and } 0.3$.

With a further increase of the noise level to $\sigma = 0.3$, the optimization result differs. The optimum is now close to the robust region around the local optimum. This emphasizes the importance of sufficient knowledge about the level of uncertainty. Depending on the level, the optimization result changes. Furthermore, the effect of the uncertainties has to be resolved sufficiently in order to guide the optimization to the robust optimum. This may require additional computational effort by sufficiently large sample size to accurately describe the effects of uncertainty.

Conclusions

Uncertainties are present in all engineering optimization tasks. Precise knowledge about the uncertainties with respect to their probability distribution is required in order to be able to specify the task correctly. Otherwise, the UQ and RDO results may be misinterpreted and the wrong optimum solution is chosen.

The dilemma is that data about uncertainties is typically limited and insufficient to generate a statistical representation by probability distribution. Thus, best guesses are required. For manufacturing problems, geometrical tolerances from manufacturing drawings may be considered.

References

1. Edeling, W.N., Cinnella, P., Dwight, R.P.: Predictive RANS simulations via Bayesian model-scenario averaging. J. Comput. Phys. **275**, 65–91 (2014)
2. Garzon, V.E.: Probabilistic Aerothermal Design of Compressor Airfoils (2003)
3. Büche, D., Beetz, M., Ribi, B.: Uncertainty analysis for large-scale industrial radial compressors. In: ASME Turbo Expo 2010, GT2010-22918 (2010)
4. Schmidt, R., Voigt, M., Vogeler, K., Meyer, M.: Comparison of two methods for sensitivity analysis of compressor blades. In: Proceedings of ASME Turbo Expo 2016, GT2016-57378 (2016)
5. Edeling, W.N., Cinnella, P., Dwight, R.P., Bijl, H.: Bayesian estimates of parameter variability in the k-epsilon turbulence model. J. Comput. Phys. **258**, 73–94 (2014)
6. Coles, D., Hirst, E.: Computation of turbulent boundary layers. In: Proceedings of AFOSR-IFP Stanford Conference, vol. 2 (1968)

Polynomial Chaos and Collocation Methods and Their Range of Applicability

Chris Lacor and Éric Savin

Problem Setup

Governing Equations

Let $\mathscr{D} \subset \mathbb{R}^d$ (typically $d = 3$ in physical space) be a fixed domain with a boundary $\partial\mathscr{D}$ and $\mathbf{x} \in \mathscr{D}$ the physical coordinates. Let $(\Omega, \mathscr{A}, \mathscr{P})$ be a probability space where Ω is the abstract set of elementary events, $\mathscr{A}$ is a σ-algebra of subsets of Ω, and $\mathscr{P}$ is a probability measure on $\mathscr{A}$. Our aim is to approximate the random field $u(\mathbf{x};\boldsymbol{\xi}) : \mathscr{D} \times \Gamma \to \mathbb{R}^m$ satisfying the parameterized partial differential equations:

$$\begin{aligned} \mathscr{L}(\mathbf{x},\boldsymbol{\xi};u) &= 0 \quad \text{in } \mathscr{D}, \\ \mathscr{B}(\mathbf{x},\boldsymbol{\xi};u) &= 0 \quad \text{on } \partial\mathscr{D}, \end{aligned} \tag{1}$$

where $\mathscr{L}$ is a linear or nonlinear differential operator, and $\mathscr{B}$ is a boundary operator. Here $\boldsymbol{\xi}(\omega) = (\xi_1(\omega), \xi_2(\omega), \dots, \xi_N(\omega)) : \Omega \to \Gamma \subseteq \mathbb{R}^N$ is a vector of random parameters defined on $(\Omega, \mathscr{A}, \mathscr{P})$ with probability distribution $\mathscr{P}_\Xi$, of which components $\xi_1, \xi_2, \dots, \xi_N$ are mutually independent random variables with values in subsets of $\mathbb{R}$, $\Gamma_1, \Gamma_2, \dots, \Gamma_N$, respectively. We consider without lost of generality that the random field u has scalar values, i.e., $m = 1$. In practice, one may also be interested in quantities:

C. Lacor (✉)
Mechanical Engineering Department, Vrije Universiteit Brussel, Pleinlaan 2, 1050 Brussels, Belgium
e-mail: clacor@vub.ac.be

É. Savin
Onera–The French Aerospace Lab, Computational Fluid Dynamics Department, 29 avenue de la Division Leclerc, F-92322 Châtillon, France
e-mail: eric.savin@onera.fr

C. Hirsch et al. (eds.), *Uncertainty Management for Robust Industrial Design in Aeronautics*, Notes on Numerical Fluid Mechanics and Multidisciplinary Design 140, https://doi.org/10.1007/978-3-319-77767-2_42

$$y = F(u(\cdot;\boldsymbol{\xi})), \tag{2}$$

that are functions of the solution u of the boundary value problem (1), in addition to the solution itself. In computational fluid dynamics for instance, u may be the pressure field satisfying the compressible Navier–Stokes equations about a fixed profile, and y may be the aerodynamic forces (e.g., drag, lift) exerted by the flow on that profile. In this latter case, the differential operator $\mathscr{L}$ may also depend on time t, and the boundary value problem (1) needs be supplemented with initial conditions. We do not consider that more general situation in the following discussion, for its main features basically extend to time-dependent problems.

Probabilistic Framework

The vector of random parameters $\boldsymbol{\xi}$ is representative of variable geometrical characteristics, boundary conditions, loads, physical or mechanical properties, or combinations of them. It can be discrete, continuous, or a combination of both. In the continuous case, it is understood that its probability distribution $\mathscr{P}_\Xi$ admits a probability density function (PDF) $\boldsymbol{\xi} \mapsto W_\Xi(\boldsymbol{\xi})$ with values in $\mathbb{R}_+ = [0, +\infty]$ such that $\mathscr{P}_\Xi(B) = \int_B W_\Xi(\boldsymbol{\xi}) d\boldsymbol{\xi}$ for any subset B of $\mathbb{R}^N$. In addition, $\mathscr{P}_\Xi(d\boldsymbol{\xi}) = \mathscr{P}_1(d\xi_1) \times \mathscr{P}_2(d\xi_2) \times \cdots \times \mathscr{P}_N(d\xi_N)$ owing to the assumption of mutual independence. In the present setting, it is further assumed that the random parameters are exponentially integrable, that is there exists $\beta > 0$ such that:

$$\int_{\mathbb{R}^N} \mathrm{e}^{\beta\|\boldsymbol{\xi}\|} \mathscr{P}_\Xi(d\boldsymbol{\xi}) < +\infty, \tag{3}$$

where $\|\boldsymbol{\xi}\| = (\sum_{n=1}^N \xi_n^2)^{\frac{1}{2}}$ is the usual Euclidean norm in $\mathbb{R}^N$. Any distribution with compact support satisfies the above condition, for example uniform or beta distributions, as well as the gamma or normal distributions. Together with mutual independence, it ensures that each random variable ξ_n possesses finite moments of all orders, that is $\mathbb{E}\{|\xi_n|^k\} = \int_{\mathbb{R}} |\xi|^k \mathscr{P}_n(d\xi) < +\infty$ for all $k \in \mathbb{N}$. This (uniquely) defines a sequence of univariate orthonormal polynomials $\{\psi_j^{(n)}\}_{j\in\mathbb{N}}$ associated with the probability measure $\mathscr{P}_n$ for all $1 \le n \le N$, and a sequence of multivariate orthonormal polynomials $\{\psi_{\mathbf{j}}\}_{\mathbf{j}\in\mathbb{N}^N}$ associated with the probability measure $\mathscr{P}_\Xi$ given by:

$$\psi_{\mathbf{j}}(\boldsymbol{\xi}) = \prod_{n=1}^{N} \psi_{j_n}^{(n)}(\xi_n), \quad \mathbf{j} = (j_1, j_2, \ldots, j_N) \in \mathbb{N}^N, \tag{4}$$

such that $\{\psi_{\mathbf{j}}(\boldsymbol{\xi})\}_{\mathbf{j}\in\mathbb{N}^N}$ constitutes an orthonormal sequence of random variables in the space $L^2(\Omega, \sigma(\boldsymbol{\xi}), \mathscr{P})$ of second-order random variables defined on the probability space endowed with the σ-algebra $\sigma(\boldsymbol{\xi})$ generated by the random parameters $\boldsymbol{\xi}$; see [1, Theorem 3.6]. Alternatively, the polynomial set $\{\psi_{\mathbf{j}}\}_{\mathbf{j}\in\mathbb{N}^N}$ constitutes an orthonormal basis of the functional space $L^2(\mathbb{R}^N, \sigma_B(\mathbb{R}^N), \mathscr{P}_\Xi(d\boldsymbol{\xi}))$ of square-integrable functions with respect to $\mathscr{P}_\Xi$, where $\sigma_B(\mathbb{R}^N)$ is the Borel σ-algebra on $\mathbb{R}^N$.

Consequently, any random variable u in $L^2(\Omega, \sigma(\boldsymbol{\xi}), \mathscr{P})$ can be expanded in a series of multivariate orthonormal polynomials in the random parameters $\boldsymbol{\xi}$ as:

$$u = \sum_{\mathbf{j}\in\mathbb{N}^N} u_\mathbf{j}\psi_\mathbf{j}(\boldsymbol{\xi}), \quad u_\mathbf{j} = \mathbb{E}\{u\psi_\mathbf{j}(\boldsymbol{\xi})\} = \int_{\mathbb{R}^N} u\psi_\mathbf{j}(\boldsymbol{\xi})\mathscr{P}_\Xi(d\boldsymbol{\xi}). \tag{5}$$

This is the so-called generalized polynomial chaos expansion. Likewise, the random field $\mathbf{x} \mapsto u(\mathbf{x};\boldsymbol{\xi})$ satisfying Eq. (1) has finite energy from physical considerations, so it belongs to $L^2(\mathbb{R}^N, \sigma_B(\mathbb{R}^N), \mathscr{P}_\Xi(d\boldsymbol{\xi}))$ and can be expanded as:

$$u(\mathbf{x};\boldsymbol{\xi}) = \sum_{\mathbf{j}\in\mathbb{N}^N} u_\mathbf{j}(\mathbf{x})\psi_\mathbf{j}(\boldsymbol{\xi}), \quad u_\mathbf{j}(\mathbf{x}) = \mathbb{E}\{u(\mathbf{x};\boldsymbol{\xi})\psi_\mathbf{j}(\boldsymbol{\xi})\} = \int_{\mathbb{R}^N} u(\mathbf{x};\boldsymbol{\xi})\psi_\mathbf{j}(\boldsymbol{\xi})\mathscr{P}_\Xi(d\boldsymbol{\xi}). \tag{6}$$

In practical numerical applications, the foregoing expansions are truncated up to a total order p such that $|\mathbf{j}| = j_1 + j_2 + \cdots + j_N \leq p$. Denoting by $\mathbb{P}^p[\cdot]$ the orthogonal projection onto the space of N-variate polynomials of total degree p in $\xi_1, \xi_2, \ldots, \xi_N$, say V_N^p, we seek for an approximate solution $\mathbb{P}^p[u]$ of Eq. (1) in V_N^p as:

$$u(\mathbf{x};\boldsymbol{\xi}) \simeq \mathbb{P}^p[u](\mathbf{x};\boldsymbol{\xi}) = \sum_{|\mathbf{j}|\leq p} u_\mathbf{j}(\mathbf{x})\psi_\mathbf{j}(\boldsymbol{\xi}),$$

which, by reordering the multi-indices $\mathbf{j}$ such that $|\mathbf{j}| \leq p$, also reads:

$$u(\mathbf{x};\boldsymbol{\xi}) \simeq \mathbb{P}^p[u](\mathbf{x};\boldsymbol{\xi}) = \sum_{j=0}^{P} u_j(\mathbf{x})\psi_j(\boldsymbol{\xi}), \quad P+1 = \binom{N+p}{p}. \tag{7}$$

From [1, Theorem 2.2], the sequence $\mathbb{P}^p[u]$ converges to u in the mean-square sense in $L^2(\Omega, \sigma(\boldsymbol{\xi}), \mathscr{P})$ provided that the condition (3) is fulfilled, i.e.,

$$\lim_{p\to+\infty} \mathbb{E}\{|u - \mathbb{P}^p[u]|^2\} = 0.$$

Mean-square convergence classically implies convergence in probability, which in turn implies convergence in distribution, the converse being of course untrue in the general case.

Now the deterministic functional coefficients $\mathbf{x} \mapsto u_j(\mathbf{x})$ in the truncated series remain unknown since the random field u is unknown. Weighted versions of Eq. (1) together with the above approximation are considered in order to determine them.

Stochastic Galerkin Method

Similarly to the weak formulation of deterministic problems, one can form the weak form of Eq. (1) and seek an approximate solution $u^p \in V_N^p$ such that:

$$\begin{aligned} &\mathbb{E}\{\mathscr{L}(\mathbf{x},\boldsymbol{\xi};u^p)v(\boldsymbol{\xi})\} = 0 \quad \forall v(\boldsymbol{\xi}) \in V_N^p, \mathbf{x} \in \mathscr{D}, \\ &\mathbb{E}\{\mathscr{B}(\mathbf{x},\boldsymbol{\xi};u^p)v(\boldsymbol{\xi})\} = 0 \quad \forall v(\boldsymbol{\xi}) \in V_N^p, \mathbf{x} \in \partial\mathscr{D}, \end{aligned} \tag{8}$$

where $\mathbb{E}\{\cdot\}$ stands for the mathematical expectation. The resulting system becomes a deterministic one in the physical domain $\mathscr{D}$ for the coefficients $u_j(\mathbf{x})$ and may be solved by standard discretization techniques, e.g., finite elements, finite volumes, finite differences, boundary elements, etc; see [2] and references therein for an extensive presentation of this method.

Stochastic Collocation Method

Alternatively, one may seek an approximate solution formed by interpolation between solutions of Eq. (1) for Q particular choices of the random parameters $\boldsymbol{\xi}$, namely the set $\{\boldsymbol{\xi}^l\}_{1\leq l\leq Q}$ of so-called nodes, such that:

$$\begin{aligned} &\mathscr{L}(\mathbf{x},\boldsymbol{\xi}^l;u(\mathbf{x};\boldsymbol{\xi}^l)) = 0 \quad \forall l = 1,2,\dots,Q, \mathbf{x} \in \mathscr{D}, \\ &\mathscr{B}(\mathbf{x},\boldsymbol{\xi}^l;u(\mathbf{x};\boldsymbol{\xi}^l)) = 0 \quad \forall l = 1,2,\dots,Q, \mathbf{x} \in \partial\mathscr{D}. \end{aligned} \tag{9}$$

Then the approximate solution $\mathbb{I}^Q[u]$ to Eq. (1) reads as the Lagrange interpolation [3–6]:

$$u(\mathbf{x};\boldsymbol{\xi}) \simeq \mathbb{I}^Q[u](\mathbf{x};\boldsymbol{\xi}) = \sum_{l=1}^{Q} u(\mathbf{x};\boldsymbol{\xi}^l)L_l(\boldsymbol{\xi}), \tag{10}$$

where $\{L_l\}_{1\leq l\leq Q}$ is the set of N-variate Lagrange polynomials based on the nodes $\{\boldsymbol{\xi}^l\}_{1\leq l\leq Q}$ chosen so that uniqueness of the interpolation is ensured. Sufficient conditions for such a proper interpolation are outlined in [7]. Choosing the nodes within a quadrature rule $\boldsymbol{\Theta}(N,Q) = \{\boldsymbol{\xi}^l, w^l\}_{1\leq l\leq Q}$ tailored such that $\sum_{l=1}^{Q} w^l f(\boldsymbol{\xi}^l)$ is a good approximation of the N-dimensional integral $\int_{\mathbb{R}^N} f(\boldsymbol{\xi})\mathscr{P}_{\Xi}(d\boldsymbol{\xi}) = \mathbb{E}\{f(\boldsymbol{\xi})\}$ for sufficiently smooth functions f, and the collocation approach may be used to compute an approximate solution $\mathbb{P}_Q^p[u]$ defined by:

$$
\begin{aligned}
\mathbb{P}^p_Q[u](\mathbf{x};\boldsymbol{\xi}) &= \sum_{j=0}^{P}\left(\sum_{l=1}^{Q} w^l u(\mathbf{x};\boldsymbol{\xi}^l)\psi_j(\boldsymbol{\xi}^l)\right)\psi_j(\boldsymbol{\xi}) \\
&= \sum_{l=1}^{Q} u(\mathbf{x};\boldsymbol{\xi}^l)\left(w^l\sum_{j=0}^{P}\psi_j(\boldsymbol{\xi}^l)\psi_j(\boldsymbol{\xi})\right) \\
&= \sum_{l=1}^{Q} u(\mathbf{x};\boldsymbol{\xi}^l)\tilde{L}_l(\boldsymbol{\xi})
\end{aligned}
\tag{11}
$$

in view of Eq. (7); that is, the quadrature set $\boldsymbol{\Theta}(N,Q)$ is used to evaluate the coefficients $u_j(\mathbf{x})$ in Eq. (7). The latter approach is called pseudo-spectral collocation in [8]. Provided that the quadrature rule $\boldsymbol{\Theta}(N,Q)$ integrates exactly all N-variate polynomials of total order $2p$ and $L_l \in V^p_N$, one has $\tilde{L}_l \equiv L_l$ owing to the orthonormalization of the polynomials $\{\psi_j\}_{0\leq j\leq P}$ which are such that $\mathbb{E}\{\psi_j(\boldsymbol{\xi})\psi_k(\boldsymbol{\xi})\} = \delta_{jk}$, the Kronecker symbol.

Regression

In regression approaches, the $P+1$ expansion coefficients in Eq. (7) are determined on the basis of a set of observations $\{u(\cdot;\boldsymbol{\xi}^l\}_{1\leq l\leq Q}$, obtained by computations or measurements, of the random variable or field u for some particular choices of the random parameters $\boldsymbol{\xi}$, again the set $\{\boldsymbol{\xi}^l\}_{1\leq l\leq Q}$. They consist in solving a weighted least-squares minimization problem:

$$
\mathbf{U} = \arg\min_{\hat{\mathbf{U}}\in\mathbb{R}^{P+1}} \frac{1}{2}\left(\mathbf{y}-\boldsymbol{\Phi}\hat{\mathbf{U}}\right)^{\mathsf{T}}\mathbf{W}\left(\mathbf{y}-\boldsymbol{\Phi}\hat{\mathbf{U}}\right), \tag{12}
$$

where $\mathbf{y} = (u(\cdot;\boldsymbol{\xi}^1), u(\cdot;\boldsymbol{\xi}^2), \ldots, u(\cdot;\boldsymbol{\xi}^Q))^{\mathsf{T}}$ is the vector of observations, $[\boldsymbol{\Phi}]_{lj} = \psi_j(\boldsymbol{\xi}^l)$ is the so-called $Q\times(P+1)$ measurement matrix, $\mathbf{W}$ is a $Q\times Q$ weighting matrix, and $\mathbf{U} = (u_0, u_1, \ldots, u_P)^{\mathsf{T}}$ is the sought vector of coefficients. This is the approach retained in, e.g., [9, 10], for which numerous methods are available to solve this optimization problem whenever $Q > P$. In [11–14], for example, the choice of the observation set $\{\boldsymbol{\xi}^l\}_{1\leq l\leq Q}$ is also discussed. Alternatively, one may consider the situation whereby $Q \leq P$ and more particularly $Q \ll P$, that is, underdetermined systems. This can be achieved thanks to some recent mathematical results pertaining to the resolution of under-sampled linear systems promoting sparsity of the sought solution, known as compressed sensing or compressive sampling [15, 16]. A recent review of the application of this approach to generalized polynomial chaos expansions is proposed in [17]; see also [18] for an application in the context of the UMRIDA project. The compressed sensing approach consists in reformulating the least-squares minimization problem (12) as a convex minimization problem with some sparsity constraint, namely

$$\mathbf{U} = \arg \min_{\hat{\mathbf{U}} \in \mathbb{R}^{P+1}} \{ \|\hat{\mathbf{U}}\|_1 ; \ \|\mathbf{W}(\mathbf{y} - \mathbf{\Phi}\hat{\mathbf{U}})\|_2 \leq \varepsilon \}, \tag{13}$$

for some tolerance $0 \leq \varepsilon \ll 1$ on the polynomial chaos truncation (7). Here the ℓ_m-norm is $\|\mathbf{a}\|_m = (\sum_{j=0}^{P} \left|a_j\right|^m)^{\frac{1}{m}}$ for $m > 0$, and $\|\mathbf{a}\|_0 = \#\{j;\ a_j \neq 0\}$ otherwise. Sparsity means that only a small fraction of the sought coefficients $\mathbf{U}$ are non-negligible. The latter problem is known as basis pursuit denoising (BPDN) [19]. It is uniquely solved thanks to some ad hoc mixing properties of the measurement matrix $\mathbf{\Phi}$ which are briefly reviewed in the chapter "Generalized Polynomial Chaos for Non-intrusive Uncertainty Quantification in Computational Fluid Dynamics" of this Book.

Intrusive Versus Non-intrusive Methods: General Comments

The stochastic Galerkin method is **intrusive** in that it may require significant alterations of the existing deterministic codes used for solving numerically the boundary value problem (1). Generally, it also yields coupled equations for the expansion coefficients of its solutions. Hence, new codes need be developed to handle the larger and coupled systems of equations arising from the Galerkin formulation. A recent application of an intrusive approach to the 3D compressible Navier–Stokes equations can be found in Dinescu et al. [20]. The stochastic collocation method and regression methods are **non-intrusive** in that they require only repetitive executions of the existing deterministic codes for carefully selected parameters values. They are the preferred methodologies in CFD, and for their applicability is not affected by the complexity and high nonlinearity of the existing flow solvers so long as they achieve a reasonable accuracy. However, it is to be expected that, for a fixed total order of the polynomial expansion, say, the non-intrusive methods may require the solutions of a much larger number of equations than that of the intrusive Galerkin method, especially for higher dimensions N of the stochastic space [21]. The aliasing error induced by the numerical quadrature rules used in collocation methods may increase with the stochastic dimension alike, indicating that the intrusive Galerkin method may actually offer the most accurate and least demanding solutions in higher dimensions.

A final remark worth to mention is that most of the techniques developed so far, being either intrusive or non-intrusive, have been tested for elliptic and parabolic equations, of which solutions are known to be smooth with respect to space and time. Hyperbolic systems have been much less studied, one important issue being the consideration of the characteristics of the original equations in their Galerkin weak formulation. This issue is addressed in, e.g., [22–24], though.

Overview of the Methodologies Used in UMRIDA

Within UMRIDA, four different non-intrusive methodologies have been implemented. They are described shortly hereafter, with their main advantages and disadvantages. For more details, the reader is referred to the UMRIDA Book.

Sparse Quadrature Sampling Schemes

NUMECA and ONERA both developed a stochastic collocation (SC) method using possibly sparse numerical quadrature rules $\Theta(N, Q)$. The SC expansion of a time-dependent random field $\mathbf{x}, t \mapsto u(\mathbf{x}, t; \boldsymbol{\xi})$ reads:

$$u(\mathbf{x}, t; \boldsymbol{\xi}) \simeq \mathbb{I}^{Q}[u](\mathbf{x}, t; \boldsymbol{\xi}) = \sum_{l=1}^{Q} u(\mathbf{x}, t; \boldsymbol{\xi}^{l}) L_{l}(\boldsymbol{\xi}), \tag{14}$$

where the expansion coefficients are the random field evaluated at the nodes $\{\boldsymbol{\xi}^{l}\}_{1 \leq l \leq Q}$. In stochastic dimension one ($N = 1$), the nodes ξ^{l} are chosen as the numerical quadrature points associated with the PDF $\xi \mapsto W_{1}(\xi)$ of the single random parameter ξ. The moments are then calculated using numerical quadrature, e.g., for the mean:

$$\begin{aligned}\mathbb{E}\{u\}(\mathbf{x}, t) &= \int_{\Gamma_{1}} u(\mathbf{x}, t; \xi) W_{1}(\xi) d\xi \\ &\simeq \sum_{l=1}^{Q} w^{l} u(\mathbf{x}, t; \xi^{l}),\end{aligned} \tag{15}$$

and similar expressions for the higher-order moments. In higher stochastic dimensions, the numerical quadrature formulas involve tensor products of the one-dimensional quadrature nodes. Such a tensorization quickly becomes very expensive in terms of the number of sampling nodes—the so-called curse of dimensionality. NUMECA therefore used a sparse quadrature sampling scheme, based on Smolyak's algorithm [25]. It significantly reduces the number of samples at the cost of ignoring some higher-order coupling terms in the expansion. Since the expansion coefficients of these terms can supposedly be small, this has only a small impact on the accuracy of the calculated moments. The basic methodology is for scalar random data but by repeating it, it can also be used for random fields.

Adaptive Sparse Polynomial Chaos

This methodology is used by AGI and ESTECO. It is based on the regression approach developed by Blatman and Sudret [26–28]. The basic idea of this approach is that in the classical polynomial chaos (PC) expansion of a time-dependent random field:

$$u(\mathbf{x}, t; \boldsymbol{\xi}) \simeq \mathbb{P}^P[u](\mathbf{x}, t; \boldsymbol{\xi}) = \sum_{j=0}^{P} u_j(\mathbf{x}, t)\psi_j(\boldsymbol{\xi}), \tag{16}$$

the expansion coefficients $u_j(\mathbf{x}, t)$ of many of the higher-order terms are negligible. An iterative procedure is developed to find out which are the important terms to be retained. The procedure is based on least angle regression (LAR, as opposed to numerical quadrature) and is quite efficient. Compared to the classical regression approach by least-squares minimization, where oversampling is used with typically $Q = 2P$ samples, under-sampling is used here. The size of the experimental design (ED) is difficult to select a priori but a sequential one can be chosen. A leave-one-out procedure is used to test the quality of the proposed expansion. Apart from the size of the ED, the choice of the location of the sampling nodes is also important. Depending on the problem at hand, the solution might be more or less sensitive to this choice. The basic methodology is for scalar random data but a procedure has been proposed for random fields as well based on proper orthogonal decomposition (POD) in [29].

Compressive Sampling

This methodology is used by ONERA. As in the adaptive sparse PC expansion of the foregoing section, it exploits the fact that many terms in the PC expansion of Eq. (16) are negligible. The leading expansion coefficients are determined by regression from an underdetermined system with the following additional condition:

$$\|\mathbf{U}\|_1 = \sum_{j=0}^{P} \left| u_j(\mathbf{x}, t) \right| \quad \text{is minimal.} \tag{17}$$

This latter condition expresses the sparsity of the expansion; see Eq. (13). Different software packages are available to solve Eq. (13), such as `SPGL1` [30] or `CVXOPT` [31]. The obtained result is comparable to that of the adaptive sparse PC expansion but there is of course some dependency on the used software. However, the ℓ_1-minimization procedure has the great advantage of being non adaptive, as it identifies both the amplitude and the rank of the leading coefficients in a single run. This is a much desirable feature for practical industrial applications with complex configurations. The size of the ED and the location of the sampling points can also have an effect on the quality of the solution. The basic methodology is for scalar stochastic data but by repeating it, it can also be used for stochastic fields.

Reduced Basis Method

This methodology is developed by VUB. Instead of the classical PC expansion of Eq. (16) with the polynomials $\{\psi_j(\boldsymbol{\xi})\}_{0\leq j\leq P}$ as the basis, an expansion in a reduced basis is used:

$$u(\mathbf{x}, t; \boldsymbol{\xi}) \simeq \sum_{j=0}^{m} \hat{u}_j(\mathbf{x}, t) z_j(\boldsymbol{\xi}), \tag{18}$$

where the new basis is formed by $\{z_j(\boldsymbol{\xi})\}_{1\leq j\leq m}$ and $m \ll P$ so that less expansion coefficients have to be determined. The new basis can be determined via a Karhunen–Loève expansion, also known as POD. This requires however the knowledge of the covariance of the random field $u(\mathbf{x}, t; \boldsymbol{\xi})$. Assuming that this covariance (in stochastic space) is largely independent from the resolution in real space, it can be determined by a PC analysis on a coarse grid. Once the reduced basis is found, the expansion

Table 1 Comparison of different PC methods used in the UMRIDA project

Method	Idea	Methodology	Technique	Application
Sparse quadrature PC	Cheaper numerical quadrature	Sparse instead of tensorial quadrature rule	Numerical quadrature	Scalar data extendable to fields by repeating procedure
Sparse PC	Only most important terms in PC decomposition are kept	Different expansions tried in a systematic and iterative way	Regression LAR or stepwise	Scalar data extendable to fields using POD
Compressive sampling	Only most important terms in PC decomposition are kept	Found in one shot	Regression with constraint	Scalar data extendable to fields by repeating procedure
Reduced basis	Find optimal PC decomposition based on covariance and POD	PC on coarse grid. All above techniques can be used to reduce CPU on coarse mesh	Regression on fine grid	Fields >1 field: use covariance of all fields scalar data: use reduced basis of the field data (scalar data depends on field data)

coefficients are computed with (overdetermined) regression, typically using $Q = 2m$ samples. The size of the reduced basis depends on the number of POD modes taken into account. Normally, only few modes suffice, but if the correlation length is small, more modes are needed for an accurate representation. Nevertheless, $m \ll P$ so that significantly less samples are needed on the fine grid compared to a "full" PC expansion. It is also to be noted that for the coarse grid PC analysis, any of the other methods discussed here—e.g., compressive sampling, sparse adaptive PC expansion—can be used to further increase the efficiency. As the methodology is based on the covariance, it is directly applicable to random fields solely. In case of different random fields, the covariance of all the fields is to be used in principle. However, tests with the RAE2822 basic case considered in the UMRIDA project showed that, using only one random field, e.g., the pressure field, the resulting reduced basis is very similar to that obtained using all random fields. For scalar data, which are usually derived from (a) random field(s), the reduced basis derived from these field(s) can be used. The different methodologies are summarized in Table 1.

Comparison of the Different Methodologies in Terms of Efficiency

We remind here that the total order of the polynomial expansion (6) is denoted by p. If the stochastic dimension is $N = 1$, then the Lagrange polynomials in Eq. (14) are of order p. Hence, $p + 1$ quadrature points $\{\xi^l\}_{1\leq l\leq p+1}$ are needed. Since $u(\mathbf{x}, t; \xi^l)$ in Eq. (14) is the value of $u(\mathbf{x}, t)$ at the quadrature point ξ^l, this implies that $p + 1$ simulations are needed. If the stochastic dimension is $N > 1$, this becomes $(p + 1)^N$ in the case of a fully tensorized scheme based on $p + 1$ nodes in each dimension. Using a sparse quadrature rule instead, different levels k can be considered, where k varies from $k = 1$ up to $k = p + 1$. The number of quadrature points, and hence the number of simulations, for a sparse quadrature of level k and $N \gg 1$ can be approximated as:

$$Q \simeq \frac{(2N)^k}{k!}. \tag{19}$$

If we set to unity the cost of a single simulation, this is also the total cost. It should be noted that, with the full tensorial scheme, polynomials up to a total order $N(2p + 1)$ (i.e., a product of N polynomials each of order $2p + 1$ in each dimension) are integrated exactly, whereas, with the sparse scheme of level k based on an underlying one-dimensional Gauss rule with $p + 1$ nodes, only polynomials up to total order $2k - 1$ are integrated exactly, i.e., maximal order $2p + 1$ for the highest level $k = p + 1$.

In the reduced basis method implemented by VUB, a full PC expansion constructed by regression is first considered on a **coarse** grid. The total number of samples required for the regression is $Q = 2(P + 1)$, where $P + 1$ is again the number of terms in the expansion of Eq. (16). In stochastic dimension N, $P + 1 = \binom{N+p}{p}$.

The total cost is therefore $2c \times \binom{N+p}{p}$ where c takes into account the coarseness of the grid; that is, c is the ratio of coarse to fine grid size. Next, regression is used on the fine grid. Assuming the reduced basis has dimension m, this requires $2m$ fine grid samples, with a cost $2m$. Based on the RAE2822 and ROTOR37 test cases defined in the UMRIDA project where respectively $N = 10$ and $N = 21$ uncertain parameters were considered, a reduced basis of size $m \simeq 20$ can be expected, whereas for the value of the coarse to fine grid size $c = \frac{1}{8}$ is found acceptable. The total cost can therefore be estimated as $\frac{1}{4} \times \binom{N+p}{p} + 40$.

In the adaptive sparse PC expansion and compressive sampling approaches, the number of samples Q is not directly related to p and N. However, as an underdetermined system of equations is considered for the regression analysis, Q is (much) smaller than $P + 1$. ESTECO, for instance, used $Q = 200$ samples for the RAE2822 test case with $N = 13$ uncertainties, going up to the PC order $p > 10$. A similar number can be derived from the literature. Blatman and Sudret [28] mention speedups of 3.3–10.2 compared to full PC expansions of order 2, when the stochastic dimension varies from $N = 30$–70. This leads to an equivalent of $Q = 300$–500 samples. In compressive sampling, ONERA used $Q = 80$ samples for the RAE2822 test case with $N = 3$ uncertainties. As a rule of thumb, the number of samples to be used in compressive sampling is typically four times the sparsity $S \equiv \|\mathbf{U}\|_0$ of the sought solution, that is the number of non-negligible coefficients: $Q \geq 4S$. In Hampton and Doostan [32], errors of less than 1% in the mean (compared to full PC expansion) are obtained with about 200–300 samples for an example with $N = 20$. A choice of 100 samples for $N = 5$, 180 samples for $N = 10$ and 250 samples for $N = 20$ therefore seems appropriate for both methods.

Based on the reasoning above, which admittedly gives only rough cost estimates, the costs of the different methods are compared in Tables 2 through 4 below for $N = 5$, $N = 10$ and $N = 20$, respectively. The first two rows show the number of samples of full PC with respectively regression (assuming $2P$ samples) and a ten-

Table 2 Cost of different PC methods used in UMRIDA for $N = 5$ and varying total order p

Method $N = 5$	$p = 1$	$p = 2$	$p = 3$	$p = 4$
Full PC regression	12	42	112	252
Tensorial quadrature PC	32	243	1024	3125
Sparse quadrature PC	50 (11)	166 (71)	417 (341)	833 (1341)
Adaptive sparse PC	100	100	100	100
Compressive sampling	100	100	100	100
Reduced basis	42	45	54	72

Table 3 Cost of different PC methods used in UMRIDA for $N = 10$ and varying total order p

Method $N = 10$	$p = 1$	$p = 2$	$p = 3$	$p = 4$
Full PC regression	22	132	572	2002
Tensorial quadrature PC	1024	59,049	1,048,576	9,765,625
Sparse quadrature PC	200 (21)	1333 (241)	6666 (1981)	26,666 (12981)
Adaptive sparse PC	180	180	180	180
Compressive sampling	180	180	180	180
Reduced basis	43	47	112	290

Table 4 Cost of different PC methods used in UMRIDA for $N = 20$ and varying total order p

Method $N = 20$	$p = 1$	$p = 2$	$p = 3$	$p = 4$
Full PC regression	42	462	3542	21252
Tensorial quadrature PC	1,048,576	3.486E9	1.099E12	95.367E12
Sparse quadrature PC	800	10,666	106,666	853,333
Adaptive sparse PC	250	250	250	250
Compressive sampling	250	250	250	250
Reduced basis	45	98	483	2697

sorized numerical quadrature with $(p + 1)^N$ samples. For the sparse quadrature sampling and reduced basis approaches, PC total orders up to $p = 4$ are considered. For sparse quadrature sampling, the given cost is for the scheme with the highest level k. According to the results obtained by NUMECA, this highest level is not always required. For the ROTOR37 case, for instance, NUMECA reports differences of only 0.02% and 3%, respectively, on the mean and variance if the level $k = 2$ is used instead of the level $k = 3$. The cost of sparse quadrature sampling using lower levels is found in the columns more to the left. That is, for a PC order $p = 4$ the number of samples required for $k = 4$ is found in the column $p = 4$, those of $k = 3$ in column $p = 3$, those of $k = 2$ in column $p = 2$ etc. (Tables 2 through 4).

It is important to stress once more that the tables above are based on very rough cost estimates. Nevertheless, some conclusions can be drawn. Assuming that, in practice, at least a PC total order of 3 is needed for accurate results, we focus on the last two columns in the tables.

- The adaptive sparse PC and compressive sampling seem the most efficient methods, especially when going to high-stochastic dimensions. Whereas for $N = 5$ the reduced basis is most efficient, this is not the case anymore for $N = 10$ and $N = 20$;
- The sparse quadrature sampling, though much more efficient than the fully tensorized quadrature sampling, seems less efficient than the other methods. When comparing it to the full PC with regression, the sparse quadrature sampling is not competitive for $N = 5$. At the higher stochastic dimensions considered, $N = 10$ and $N = 20$, it can compete with the other methods only if the lowest level is used ($p = 1$). It is stressed that the number of samples mentioned is based on a general formula which gives a rough estimate only. Depending on the numerical quadrature formula used (e.g., with nested or non-nested stencils), the number of samples might be lower. This is illustrated in Tables 2 and 3 for $N = 5$ and $N = 10$ where the number of samples for the sparse numerical formulas used by NUMECA is mentioned in between brackets. In those cases, the sparse quadrature sampling method is competitive with the other methods for the higher stochastic dimensions $N = 5$ and $N = 10$, if one restricts it to the level $k = 2$;
- Looking at the effect of increasing the stochastic dimension N, the efficiency of adaptive sparse PC and compressive sampling increases with it. This is also clearly found in Blatman and Sudret [28], who mention speedups of 3.3–10.2 compared to full PC of order 2, when the stochastic dimension varies from $N = 30$–70. Similarly for the reduced basis, the speedups compared to full PC with regression of the same order p are respectively 3.5, 6.9, and 7.8 (for $p = 4$ and $N = 5, 10$, and 20, respectively). If sparse quadrature sampling is compared to full PC with regression and $p = 4$, there is only a speedup if the formulas of level $k = 1$ or $k = 2$ are used. For level 2, the speedups for $N = 5, 10$, and 20 are respectively 1.5, 1.5, and 2.0—based on the number of samples from the approximate formula (19). With the number of samples used by NUMECA, this becomes 3.5 and 8.3, respectively, for $N = 5$ and 10. We can therefore conclude that, compared to standard PC methods, all considered methods become more efficient with increasing stochastic dimension. Therefore, they allow to partially mitigate the effect of the "curse of dimensionality."

Comparison of Computing Times on UMRIDA Test Case ROTOR37

The NASA ROTOR37 is an axial flow transonic compressor designed and experimentally tested by Reid and Moore [33]. In the UMRIDA project, the working point corresponding to a rotational speed of Ω =17,188 rpm and an outlet static pressure of p_{out} =11,000 Pa is chosen. This corresponds to a mass flow which is 98% of the mass flow at choking conditions.

Only two partners contributed to this test case, using PC methodology: NUMECA with sparse quadrature sampling, and VUB with the reduced basis method.

Table 5 CPU cost in h on 100 cores for the ROTOR37 test case on the actual mesh and corrected for mesh size of 1M cells

Partner	Method	N	Mesh (Mcells)	CPU (h/100 cores)	CPU 1M mesh (h/100 cores)
NUMECA	Sparse quad. $k = 1$	10	2.8	0.56	0.20
	Sparse quad. $k = 2$	10	2.8	6.43	2.29
VUB	Reduced basis	12	0.77	0.23	0.29
	Reduced basis	21	0.77	0.66	0.86

Uncertainties are imposed by both partners on the following operational parameters: the inlet total pressure profile (described with one uncertainty, i.e., assumed fully correlated) and the outlet static pressure. In addition, geometrical uncertainties are introduced on the tip clearance and the blade parametrization: leading and trailing edge angles at different spanwise locations as well as blade half thickness parameters (VUB) and leading edge radius (NUMECA). In total, VUB considered two cases with respectively $N = 12$ and 21 uncertainties, whereas NUMECA used $N = 10$ uncertainties. Both partners used the same simulation code, i.e., the FINE/Turbo solver of NUMECA with RANS and Spalart–Allmaras turbulence model and central scheme with artificial dissipation and multigrid algorithms. NUMECA used a grid with 2.8 million points, VUB with 0.77 million. The PC total order considered was $p = 2$.

Table 5 compares the CPU times. The numbers given are the CPU hours on 100 cores. In the last column, corrected numbers are given, taking into account the difference in grid size: they are rescaled for a grid of 1 million points, hereby assuming the computational cost is linear with grid size (as should be the case when using multigrids). One of the milestones of the UMRIDA project was the ability to handle a large number of uncertainties ($N > 10$), including geometrical uncertainties, in a turnaround time of the order of 10 h on a cluster of 100 cores. This milestone is therefore clearly achieved with the present methodologies. Comparing the corrected times for NUMECA ($N = 10$) with those of VUB ($N = 12$) the estimated results of Table 3 are confirmed, if one takes into account that the actual size of the reduced basis was only $m = 7$ (instead of 20 anticipated in Table 3). According to Table 3, the reduced basis method should be 2.2 times more expensive than the level 1 sparse quadrature method of NUMECA, and 5.1 times more efficient than NUMECA's level 2 method. The actual numbers (from Table 5) are respectively 1.5 times more expensive and 7.9 more efficient, which is thus explained by the smaller reduced basis.

The level 1 results of NUMECA provide only an approximation of the mean and the variance and do not allow for an accurate estimation of the PDF, in contrast to the level 2 results and the reduced basis results. Taking this into account, the reduced basis method seems more efficient, at least for this application. Looking at the VUB results, the effect of the stochastic dimension can also be evaluated. In the case $N = 21$, the number of POD modes considered (i.e., the size of the reduced basis) was such that the sum of the corresponding eigenvalues (in absolute value) was 99.99% of the total sum. In the case $N = 10$, this was only 99.9%. This resulted in a reduced basis of size $m = 21$ for $N = 21$ and $m = 7$ for $N = 10$. If we correct for this effect and assume the same basis size of $m = 21$ for $N = 10$, the CPU time would be 0.38 (instead of 0.23) compared to 0.66 for $N = 21$. This is more or less a linear increase with the number of uncertainties and not an exponential one as observed in classical methods, due to the "curse of dimensionality."

Conclusions

The different methodologies, used in UMRIDA to deal with the "curse of dimensionality," are compared for their efficiency. Based on rough estimates of the number of required simulations, it is found that all methods allow to mitigate the "curse of dimensionality." For the sparse quadrature method and the reduced basis method, this is also confirmed by simulations of the 3D compressible Navier–Stokes test case ROTOR37 with a stochastic dimension of at least $N = 10$. It is also confirmed that the UMRIDA milestone, of a turnaround time of the order of 10 h on a cluster of 100 cores for a case $N = 10$, is clearly achieved. In the ROTOR37 test case, the required simulation time is at least an order of magnitude less.

References

1. Ernst, O.G., Mugler, A., Starkloff, H.J., Ullmann, E.: On the convergence of generalized polynomial chaos expansions. ESAIM. Math. Model. Numer. Anal. **46**(2), 317–339 (2012)
2. Le Maître, O., Knio, O.: Spectral Methods for Uncertainty Quantification: With Applications to Computational Fluid Dynamics. Springer, Dordrecht (2010)
3. Mathelin, L., Hussaini, M., Zang, T.: Stochastic approaches to uncertainty quantification in CFD simulations. Numer. Algorithms **38**(1), 209–236 (2005)
4. Xiu, D., Hesthaven, J.S.: High-order collocation methods for differential equations with random inputs. SIAM J. Sci. Comput. **27**(3), 1118–1139 (2005)
5. Babuška, I., Nobile, F., Tempone, R.: A stochastic collocation method for elliptic partial differential equations with random input data. SIAM J. Numer. Anal. **45**(3), 1005–1034 (2007)
6. Ganapathysubramanian, B., Zabaras, N.: Sparse grid collocation schemes for stochastic natural convection problems. J. Comput. Phys. **225**(1), 652–685 (2007)
7. Chung, K., Yao, T.: On lattices admitting unique Lagrange interpolations. SIAM J. Numer. Anal. **14**(4), 735–743 (1977)
8. Xiu, D.: Efficient collocational approach for parametric uncertainty analysis. Commun. Comput. Phys. **2**(2), 293–309 (2007)
9. Berveiller, M., Sudret, B., Lemaire, M.: Stochastic finite element: a non-intrusive approach by regression. Rev. Européenne de Mécanique Numérique **15**(1–3), 81–92 (2006)
10. Hosder, S., Walters, R.W., Balch, M.: Point-collocation nonintrusive polynomial chaos method for stochastic computational fluid dynamics. AIAA J. **48**(12), 2721–2730 (2010)
11. Gao, Z., Zhou, T.: On the choice of design points for least-square polynomial approximations with application to uncertainty quantification. Commun. Comput. Phys. **16**(2), 365–381 (2014)

12. Arnst, M., Ponthot, J.P.: An overview of nonintrusive characterization, propagation, and sensitivity analysis of uncertainties in computational mechanics. Int. J. Uncertainty Quantif. **4**(5), 387–421 (2014)
13. Hosder, S., Walters, R.W., Balch, M.: Efficient sampling for non-intrusive polynomial chaos applications with multiple uncertain input variables. In: 48th AIAA/ASME/ASCE/AHS/ASC Structures, Structural Dynamics, and Materials Conference, AIAA paper 2007-1939, Honolulu HI, 23–26 April 2007
14. Raisee, M., Kumar, D., Lacor, C.: A non-intrusive model reduction approach for polynomial chaos expansion using proper orthogonal decomposition. Int. J. Numer. Methods Eng. **103**(4), 293–312 (2015)
15. Candès, E.J., Romberg, J.K., Tao, T.: Stable signal recovery from incomplete and inaccurate measurements. Commun. Pure Appl. Math. **59**(8), 1207–1223 (2006)
16. Donoho, D.L.: Compressed sensing. IEEE Trans. Inf. Theory **52**(4), 1289–1306 (2006)
17. Hampton, J., Doostan, A.: Compressive sampling methods for sparse polynomial chaos expansions. In Ghanem, R.G., Higdon, D., Owhadi, H., (eds.) Handbook of Uncertainty Quantification. Springer, Cham, 29 p (2016)
18. Savin, É., Resmini, A., Peter, J.: Sparse polynomial surrogates for aerodynamic computations with random inputs. In: 18th AIAA Non-Deterministic Approaches Conference, AIAA paper 2016-0433, San Diego CA, 4–8 Jan 2016
19. Chen, S.C., Donoho, D.L., Saunders, M.A.: Atomic decomposition by basis pursuit. SIAM J. Sci. Comput. **20**(1), 33–61 (1998)
20. Dinescu, C., Smirnov, S., Hirsch, C., Lacor, C.: Assessment of intrusive and non-intrusive non-deterministic CFD methodologies based on polynomial chaos expansions. Int. J. Eng. Syst. Model. Simul. **2**(1–2), 87–98 (2010)
21. Xiu, D.: Fast numerical methods for stochastic computations: A review. Commun. Comput. Phys. **5**(2–4), 242–272 (2009)
22. Gottlieb, D., Xiu, D.: Galerkin method for wave equations with uncertain coefficients. Commun. Comput. Phys. **3**(2), 505–518 (2008)
23. Poëtte, G., Després, B., Lucor, D.: Uncertainty quantification for systems of conservation laws. J. Comput. Phys. **228**(7), 2443–2467 (2009)
24. Tryoen, J., Le Maître, O., Ndjinga, M., Ern, A.: Intrusive Galerkin methods with unwinding for uncertain nonlinear hyperbolic systems. J. Comput. Phys. **229**(18), 6485–6511 (2010)
25. Smolyak, S.: Quadrature and interpolation formulas for tensor products of certain classes of functions. Sov. Math. Dokl. **4**, 240–243 (1963)
26. Blatman, G., Sudret, B.: Sparse polynomial chaos expansions and adaptive stochastic finite elements using a regression approach. C. R. Méc. **336**(6), 518–523 (2008)
27. Blatman, G., Sudret, B.: An adaptive algorithm to build sparse polynomial chaos expansions for stochastic finite element analysis. Probab. Eng. Mech. **25**(2), 183–197 (2010)
28. Blatman, G., Sudret, B.: Adaptive sparse polynomial chaos expansion based on least angle regression. J. Comput. Phys. **230**(6), 2345–2367 (2011)
29. Blatman, G., Sudret, B.: Sparse polynomial chaos expansions of vector-valued response quantities. In: Proceedings 11th International Conference on Structural Safety and Reliability (ICOSSAR13) (2013)
30. van den Berg, E., Friedlander, M.P.: SPGL1: A Solver For Large-scale Sparse Reconstruction (June 2007)
31. Andersen, M., Dahl, J., Vandenberghe, L.: CVXOPT: Python Software For Convex Optimization (2016)
32. Hampton, J.: Doostan: compressive sampling of polynomial chaos expansions: convergence analysis and sampling strategies. J. Comput. Phys. **280**, 363–386 (2015)
33. Reid, L., Moore, R.: Performance of single-stage axial-flow transonic compressor with rotor and stator aspect ratios of 1.19 and 1.26, respectively, and with design pressure ratio of 1.82. NASA Technical Paper 1338 (1978)

Surrogate Model-Based Approaches to UQ and Their Range of Applicability

Daigo Maruyama, Dishi Liu and Stefan Görtz

Introduction

In the context of aerodynamic design under uncertainty, surrogate modeling is considered as one of the suitable approaches to efficiently calculate statistics of the quantity of interest (QoI) under scattered data. The surrogate model-based approaches to UQ here in this chapter are the method that the statistics ideally computed by a large number of data information are obtained by complementary data by an assistance of surrogate models in the uncertainty parameters space. The scattered data as sample points is produced by using Design of Experiments (DOE) and adaptive sampling if necessary in this chapter. The dependency of the statistics of the QoI on the number and distribution of sample points used to build a surrogate model and on the kind of surrogate model is discussed in [1]. In the case of robust design, the statistical value of interest is the sum of the mean and standard deviation of the QoI, or its maximum value. Each of them is considered as the objective function in optimization processes. Note that QoI here is lift coefficient (C_l) or drag coefficient (C_d) evaluated by a CFD computation.

In this chapter as Best Practice Guide it is discussed which kind of methods are the most efficient for computing the statistics of QoI as the objective function in certain tolerances of accuracy compared to the reference, e.g., one drag count $(=10^{-4})$. The errors less than this order can be sometimes regarded as epistemic

D. Maruyama (✉) · D. Liu · S. Görtz
German Aerospace Center (DLR), Institute of Aerodynamics and Flow Technology, Braunschweig, Germany
e-mail: Daigo.Maruyama@dlr.de

D. Liu
e-mail: Dishi.Liu@dlr.de

S. Görtz
e-mail: Stefan.Goertz@dlr.de

C. Hirsch et al. (eds.), *Uncertainty Management for Robust Industrial Design in Aeronautics*, Notes on Numerical Fluid Mechanics and Multidisciplinary Design 140, https://doi.org/10.1007/978-3-319-77767-2_43

uncertainties due to imperfectness of CFD solvers. To accurately and efficiently compute the statistical values of interest, we focus on the following three aspects:

- Type of surrogate model;
- Number of sample points (used to build the surrogate model); and
- Distribution of sample points (used to build the surrogate model).

The type of surrogate models and a sufficient number of sample points are firstly shown in section "Selection of Surrogate Models and the Number of Samples". Then, efficient sampling techniques considering both the number and distribution of sample points are introduced in section "Sampling Techniques for Different Measures of Robustness" for computing the above-mentioned two kinds of objective functions in the robust design optimization. The CFD solver used to evaluate the aerodynamic coefficients on the sample points is the DLR-TAU-code [2–4]. Fully turbulent computations were performed with the negative Spalart–Allmaras turbulence model [5]. A quasi-two-dimensional hybrid unstructured grid with prisms and tetrahedral elements was used for the RANS simulations.

Selection of Surrogate Models and the Number of Samples

The points to discuss here are which surrogate model is used and how many sample points are selected. The direct integration of quasi-Monte Carlo (QMC) sampling, Kriging, and GEK are compared. The comparison is performed in terms of the accuracy of the statistics for a given number of samples used to build the surrogate model. The influence of different numbers of samples is also studied. The distribution of the sample points is based on the Sobol sequence [6–8], maintaining a high degree of "uniformity" (low-discrepancy) of samples even in high-dimensional cases (≥ 10). Figure 1 shows the distributions of mean and standard deviation of estimated lift coefficient (C_l) [9]. Details of the test case are introduced in [9]. GEK requires the gradients with respect to the input uncertainty parameters, which can be efficiently computed by an adjoint solver. Therefore, $N_c = 2$ N in case of GEK where N is the number of sample points. Note that the input uncertainty space is 26 dimensions in this test case.

As can be observed in Fig. 1, GEK has comprehensively less errors than the others and converges faster than Kriging along with increase of the number of samples. One can observe that the errors of them when the number of sample points is more than around 15 ($N_c \approx 30$ in Fig. 1) nearly converge to the reference. This is one reason why GEK is recommended.

Another reason to use GEK is further efficiency in high-dimensional cases. When GEK is used, the scattered data information to build a GEK surrogate model is efficiently replenished since the computational cost of an adjoint solver is independent of the dimensionality. This could compensate one of the bottlenecks that the number of sample points should be increased with of the dimensionality

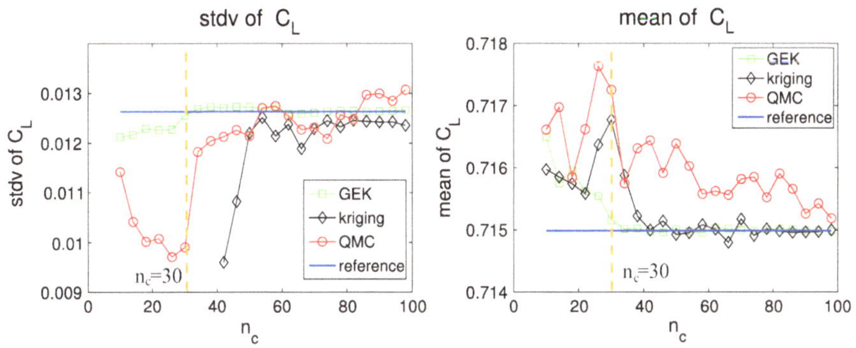

Fig. 1 Convergence of estimate C_l statistics (mean and standard deviation) to the reference statistics by various UQ methods (note that $N_c = 2$ N in case of GEK because only the gradient of C_l was considered, while N_c = N for Kriging and QMC) [10]

(details can be referred in [10]). We judge that the required number of sample points to satisfy the good accuracy does not change so much even if the dimensionality increases.

Because of the above-mentioned reasons, our conclusion of selecting surrogate models is GEK when the gradients of the QoI are able to be calculated efficiently by an adjoint solver. The number of sample points can be more than around 15. More details on the number of sample points and sampling techniques are introduced in the next section.

Sampling Techniques for Different Measures of Robustness

Criteria to Assess the Accuracy of the Statistics of QoI

Here, we focus also on the accuracy of specific statistical values of the QoI by GEK (also Kriging as comparison) with different distributions of the sample points. Two fixed numbers of sample points (12 and 30) are used to be compared with each other. The QoI considered here is the drag coefficient (C_d). The measures of robustness (objective functions) f considered here are:

$$f \equiv \mu_{C_d} + \sigma_{C_d} \tag{1}$$

$$f \equiv \max_{\mathbf{u}}(C_d(\mathbf{u})) \tag{2}$$

where $\mathbf{u}$ denotes input uncertainty parameters whose dimensionality is 12 in the applications to the UMRIDA BC-02 test case. The optimizations of these stochastic quantities are called “expectation measure with mean-risk approach” and "worst-case risk measure," respectively. These statistics expressed by Eqs. (1) and (2) are ideally uniquely determined under fixed probability density functions (pdfs) of the input uncertainty parameters **u**. Note that the pdfs of the input uncertainty parameters are assumed to be normally distributed. Details of these equations on how to calculate f by using surrogate models can be seen in [1].

The accuracy is assessed in terms of the following three criteria:

(1) The expected value (mean) μ_f of f obtained for different distributions of the sample points;
(2) The dispersion (standard deviation) σ_f of f obtained for different distributions of the sample points; and
(3) The influence of the above two values μ_f and σ_f on the result of robust design, f_{opt}, χ_{opt}.

The first and second criteria are to investigate the accuracy of f itself. The third criterion is then for examining the accuracy of the optimal solutions in terms of f_{opt}, χ_{opt} in applications to robust design. The closer the mean μ_f is to the reference f_{ref} and the closer the standard deviation σ_f is to 0, the better the accuracy of the estimated f.

The different sets of the sample points are achieved by consecutive rows in the Sobol sequence where "uniformity" (low-discrepancy) is maintained. Three different sets of 30 sample points in two dimensions are shown in Fig. 2i as an example. Each set of sample points was constructed by extracting 30 consecutive rows in the Sobol sequence. These different sets of sample points are transformed by using the cumulative density function (cdf) of uncertain input parameters (e.g., see Fig. 2ii). The uniformity is conserved for each set of sample points. The reason why different sets of sample points are used is as follows.

In robust design, the robustness measures expressed by Eqs. (1) and (2) have to be evaluated at every iteration of the optimization, corresponding to different design variables. In other words, a new surrogate model needs to be built at each iteration and the statistics are evaluated on the surrogate model. Assuming a fixed number of samples and a low-discrepancy distribution of the samples, ideally the statistics should be insensitive to the sample set used to build the surrogate model. The surrogate models we use here allow for different sample sets to be used. This advantage is used below to study the effect of the sample set on the accuracy of the statistics; see Fig. 3.

Note that in this section, Kriging and GEK with a Gaussian kernel (correlation function) were adopted and the hyperparameters were optimized by a global optimizer (a differential evolution algorithm was used) by maximum likelihood estimation (MLE).

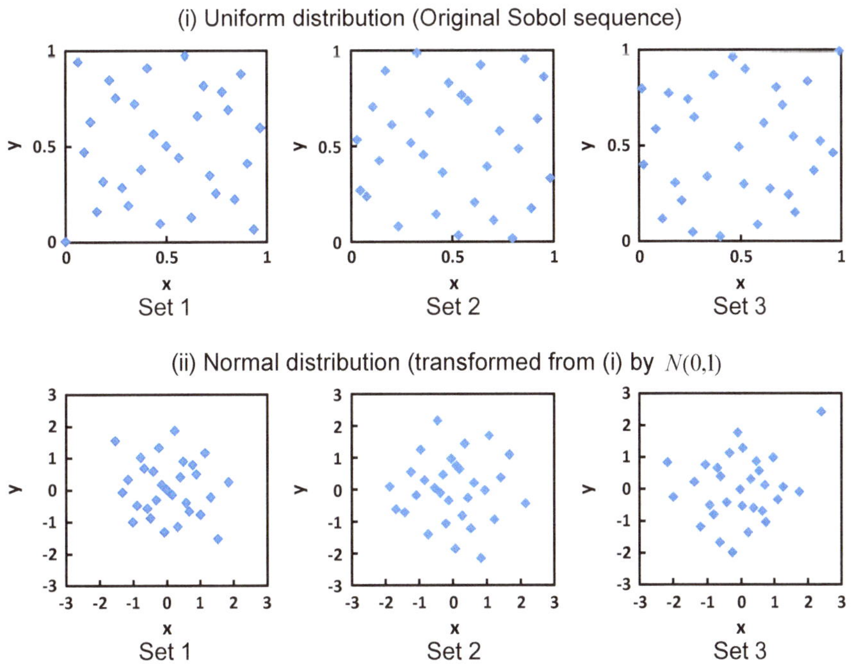

Fig. 2 Three examples of different sets/distributions of sample points (extracted from the Sobol sequence), (i) the number of which is 30 in two-dimensional, and (ii) their transformation into the normal distributions

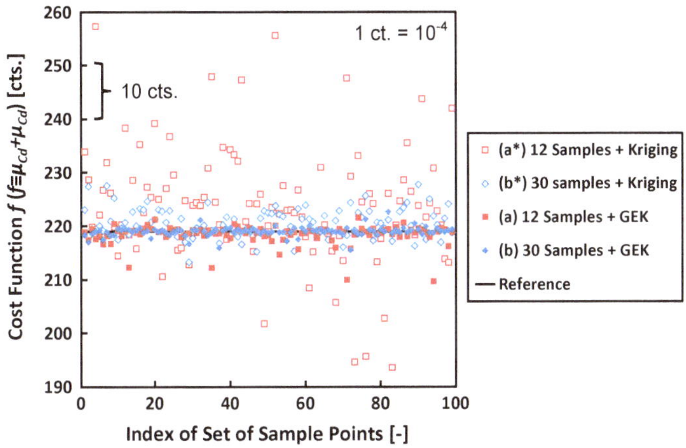

Fig. 3 Cost function $f(f \equiv \mu_{Cd} + \sigma_{Cd})$ distributions evaluated by 100 different sets of sample points for (a) 12 samples with GEK, (b) 30 samples with GEK, (a*) 12 samples with Kriging, (b*) 30 samples with Kriging

Results

Here, two investigations in terms of the criteria (1)–(3) are demonstrated, leading to the best approaches to quantify/optimize the robustness represented by Eqs. (1) and (2).

Investigation of the number of sample points

The first investigation is about influences due to the number of sample points when different sets of sample points of the Sobol sequence are used. The criteria (1)–(3) introduced in the previous sub-section are firstly investigated by using the statistical value f of Eq. (1) as $f \equiv \mu_{Cd} + \sigma_{Cd}$. The following two numbers of sample points are discussed:

(a) 12 samples; and
(b) 30 samples.

Figure 3 shows the cost function f distributions evaluated by 100 different sets of sample points for (a) and (b) extracted from arbitrary consecutive rows of the Sobol sequences. Considering the cases that an adjoint solver is not available, the results obtained by using Kriging are also described as (a*) and (b*) for comparison. The reference value in the figure was evaluated by direct integration of 10^5 Sobol sequence samples. It can be observed that the cost function f evaluated by using GEK has tendency of less dispersions and better agreement with the reference than Kriging. The criteria (1) and (2) in the previous sub-section are discussed here with Fig. 3. Now the mean μ_f and standard deviation σ_f of these distributions can be calculated from the 100 cost functions $f_1 \sim f_{100}$ in Fig. 3.

Figure 4 summarizes μ_f and σ_f, which represent the accuracy of (a) and (b) from the above-mentioned criteria (1) and (2). μ_f is transformed into the absolute error as

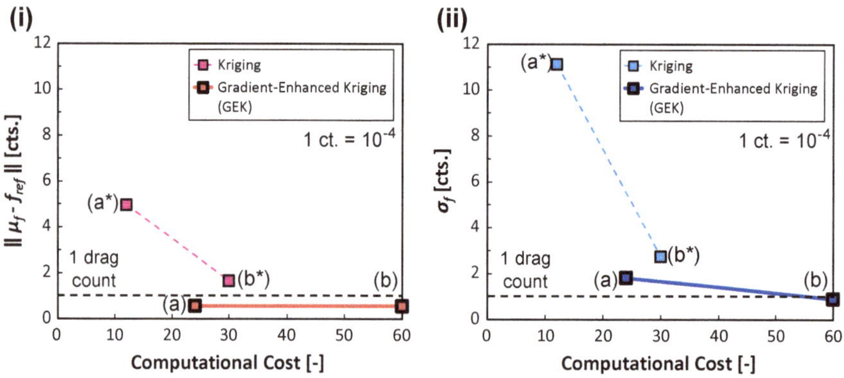

Fig. 4 Comparison between accuracy and computational cost to evaluate a robustness measure, (a) 12 samples with GEK, (b) 30 samples with GEK, (a*) 12 samples with Kriging, (b*) 30 samples with Kriging. The unit of computational cost is one-time state calculation or adjoint calculation in CFD, both of which are assumed to have the same computational time

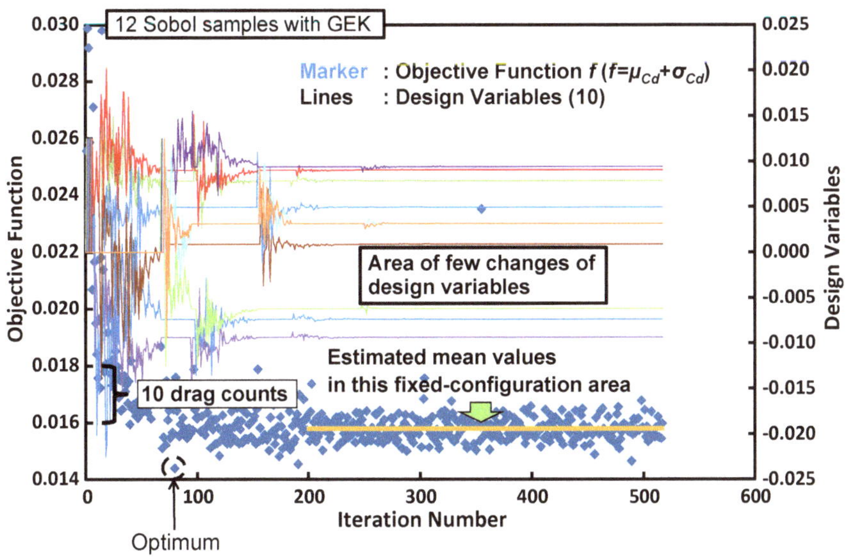

Fig. 5 Histories of objective function evaluated by (a) 12 samples with GEK and design variables in robust design optimization (RDO)

$\|\mu_f - f_{ref}\|$, where f_{ref} is the reference value. The computational cost as the horizontal axis is simply counted by the number of CFD computations with an assumption that the additional computational cost of the adjoint calculation for GEK is identical to that of the state calculation of flow in CFD. Concerning the error tolerances, the errors are discussed with the order of the drag count (where one drag count is 10^{-4} and is denoted as 1 ct.) since the epistemic uncertainties caused by CFD is considered to be not completely negligible in the order of less than 1 ct.

μ_f in (a) and (b) (12 and 30 samples with GEK) have few differences (see Fig. 4i). This fact is in a good agreement with the conclusion in the previous section (see also Fig. 1). On the contrary, σ_f decreases with increase of the number of samples (see Fig. 4ii). The errors of both μ_f and σ_f in (b) are less than 1 ct. How σ_f influences to the optimum solution is introduced next. Note that it can be observed from Fig. 4 that "(a) 12 samples with GEK" is even better than "(b*) 30 samples with Kriging" in terms of both accuracy for μ_f and σ_f, and efficiency.

Now the criterion (3) is discussed with results of applications to robust design optimization. Details of the optimization procedure can be seen in [11]. Figures 5 and 6 show the optimization histories of the cost function $f(f \equiv \mu_{Cd} + \sigma_{Cd})$ and design variables χ. The cost function f of Figs. 5 and 6 at each iteration was computed by (a) and (b), respectively. We can observe that the optimum results f is quite different from each other. This can be confirmed in the areas where the design variables are almost constant; i.e., the configuration is almost fixed. This fact is caused by the difference of σ_f in Fig. 4ii while the mean value μ_f in Fig. 4i is

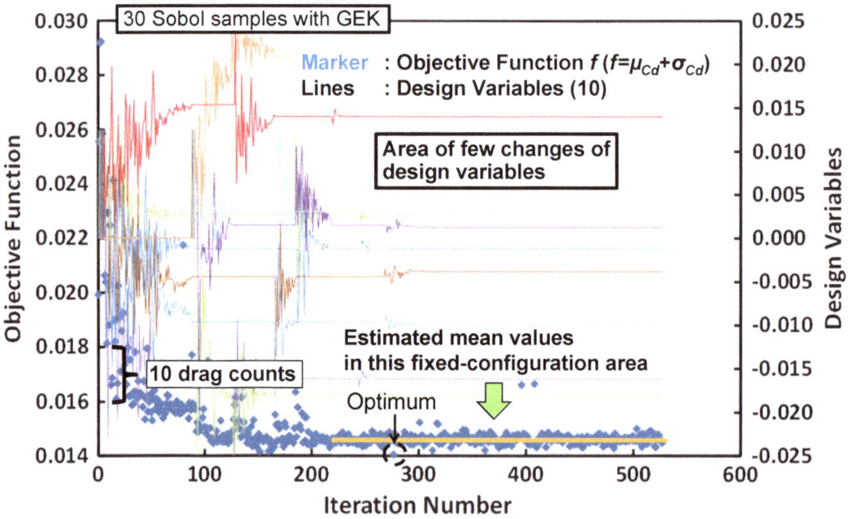

Fig. 6 Histories of objective function evaluated by (b) 30 samples with GEK and design variables in robust design optimization (RDO)

almost constant. Table 1 summarizes the cost function f of the two designed airfoils, which were re-evaluated by the common strategy (b) 30 samples. This table clarifies that more accurate evaluation could lead to an optimal solution with better performance.

Finally as comparison, a fixed set of sample points (see Fig. 2i), i.e. a fixed consecutive row of the Sobol sequence is added as another type of sampling technique for comparison:

(c) 30 samples (by using a fixed set of sample points).

The set of sample points which has the closest f to $\widehat{f}$ was picked up from $f_1 \sim f_{100}$ in Fig. 3 and that set of sample points was fixed and used for each iterative process. The histories of f, χ and the re-evaluated f of the optimum configuration are shown in Fig. 7 and Table 1, respectively. There are few differences between the sampling strategies (b) 30 samples and (c) 30 samples (by using a fixed set of sample points). The conclusion here is that σ_f due to different sets of sample points is important here as "another indicator" to determine the number of sample points.

Table 1 Values of objective function $f(f \equiv \mu_{Cd} + \sigma_{Cd})$ of optimized airfoils obtained by different strategies (a) 12 samples, (b) 30 samples, (c) 30 samples (by using a fixed set of sample points)

	(a) (cts.)	(b) (cts.)	(c) (cts.)
μ_{C_d}	144.4	135.9	136.2
σ_{C_d}	14.1	9.9	10.8
$f(\equiv \mu_{C_d} + \sigma_{C_d})$	158.5	145.8	147.0

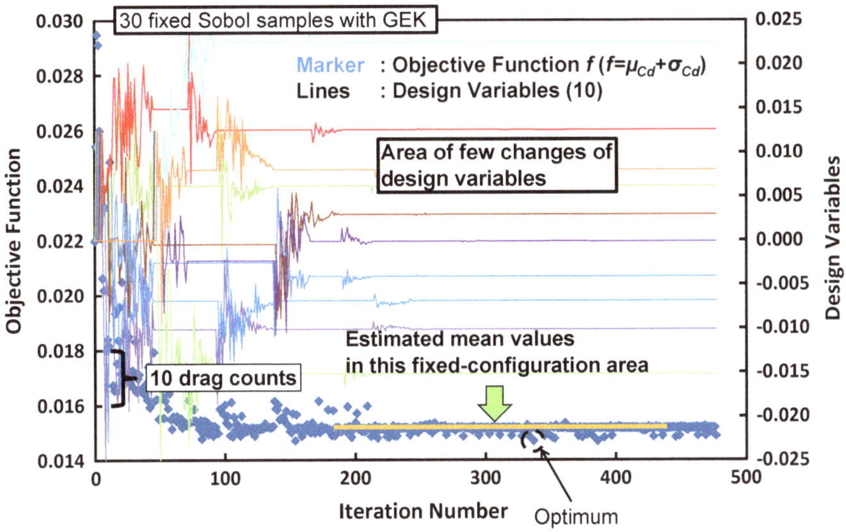

Fig. 7 Histories of objective function evaluated by (c) 30 samples (by using a fixed set of sample points) with GEK and design variables in robust design optimization (RDO)

Investigation of distribution of sample points

In the second investigation, more details of influences of distribution of sample points are demonstrated. That is, the distributions of the sample points are not only by the original Sobol sequence but the one transformed into input pdf (normal distributions here, see Fig. 2ii) and/or the one with dynamic infilled sample points. The number of samples and the surrogate model are fixed at 30 and GEK, respectively. The sampling techniques used are:

(a) input pdf (normal distributions);

(b) uniform distributions;

(c) uniform distributions and an adaptive sampling.

Suitable sample techniques for different measures of robustness (statistical values of QoI) are introduced in [1]. The results obtained here are straightforward as follows:

- For evaluating mean and standard deviation of QoI (expectation measure with mean-risk approach) represented by Eq. (1), the same distributions as the pdf of the input uncertainty parameters (normal distributions are often used) can be applicable.
- For evaluating maximum or minimum value of QoI (worst-case risk measure) represented by Eq. (2), an adaptive sampling technique in the uniform distributions leads to good accuracy.

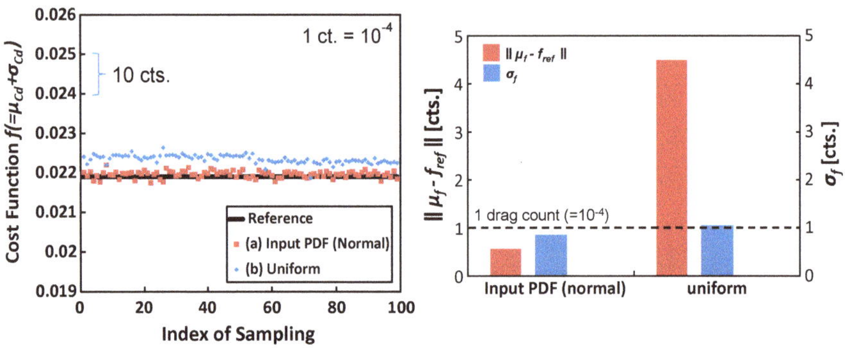

Fig. 8 Cost function $f(f \equiv \mu_{Cd} + \sigma_{Cd})$ distributions evaluated by 100 different distributions of Sobol sample points with (a) input pdf (normal distributions), (b) uniform distributions

Qualitative substantiation of them is demonstrated here. Figures 8 and 9 show cost function $f(f \equiv \mu_{Cd} + \sigma_{Cd}$ and $f \equiv \max_u(C_d(\mathbf{u}))$, respectively) distributions evaluated by 100 different sets of Sobol sample points by (a) input pdf (normal distributions), (b) uniform distributions, (c) uniform distributions with adaptive sampling (only for Fig. 9), and μ_f and σ_f, as with Figs. 3 and 4 for the first approach. The adaptive sampling technique here is an Expected-Improvement (EI)-based approach to search for the maximum or minimum value of QoI on the surrogate model. The initial sample points are the Sobol sequence with 24 points. The surrogate model is updated in stages by an imposed sample point until the total number of sample points reaches to 30. Details of the adaptive sampling technique can be found in [1].

In Fig. 8, for the cost function $f \equiv \mu_{Cd} + \sigma_{Cd}$ by Eq. (1), μ_f and σ_f by the input pdf (normal distributions) are lower than the uniform distributions and are lower than 1 ct, respectively. Note that μ_f and σ_f by the normal distributions correspond to (b) 30 samples in Figs. 2 and 3, and also its optimization result can be seen in Fig. 5.

On the other hand, for the cost function $f \equiv \max_u(C_d(\mathbf{u}))$ by Eq. (2), the expected value μ_f by the input pdf (normal distributions) is lower than 1 ct., whereas σ_f is quite large as can be also confirmed in Fig. 9(a) that f varies widely with different sets of the Sobol sample points. On the contrary, σ_f by the uniform distributions is low but μ_f is overestimated with around 3 cts. as also can be seen in Fig. 9(a). The uniform distributions with an adaptive sampling technique bring the same accuracy as the input pdf for $f \equiv \mu_{Cd} + \sigma_{Cd}$ in terms of both μ_f and σ_f, which are less than 1 ct.

In this chapter, the types of surrogate models, the number and distribution of sample points were discussed. In the worst-case risk measure, dynamic adaptive sampling techniques are necessary to keep the same order of accuracy as the expectation measure with mean-risk approach. Further improvement of the accuracy of evaluating the cost functions could be led by a variety of adaptive sampling techniques to enhance the quality of the Kriging-based surrogate models [12, 13].

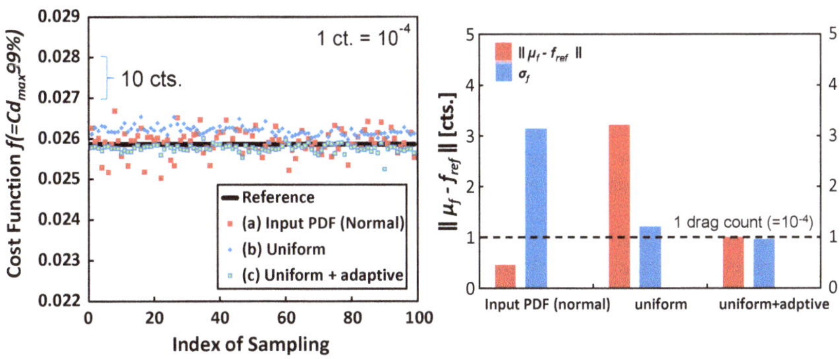

Fig. 9 Cost function $f(f \equiv C_{d99\%max})$ distributions evaluated by 100 different distributions of Sobol sample points with (a) input pdf (normal distributions), (b) uniform distributions, (c) uniform distributions with adaptive sampling

Summary

The accuracy and efficiency of surrogate model-based approaches to UQ and their application to robust design were demonstrated for the UMRIDA BC-02 test case. Twelve uncertain parameters, yielding a 12-dimensional input parameter space for surrogate model construction, were considered. Both Kriging and gradient-enhanced Kriging (GEK) were investigated. GEK was shown to lead to a good agreement of the statistical values such as the mean and standard deviation of the aerodynamic coefficients with reference values when the number of samples is more than around 12. GEK is the best choice when an adjoint solver is available. The accuracy of the statistics was also investigated from the point of view of how the sampling influences the surrogate model used in robust design. It was confirmed that the error dispersions of the statistical cost function is a function of the number of samples, the distribution of the samples. Sampling techniques in accordance with statistics to be evaluated are required to reduce the error dispersion and to achieve good robust design solutions. Different robustness measures can be evaluated accurately to within one drag count by using 30 sample points.

References

1. Maruyama, D., Liu, D., Görtz, S.: Comparing surrogates for estimating aerodynamic uncertainties of airfoils. In: Hirsch, C. et al. (eds.) Uncertainty Management for Robust Industrial Design in Aeronautics, Chap. 13, pp. xx–xx. (2017)
2. Galle, M., Gerhold, T., Evans, J.: Parallel computation of turbulent flows around complex geometries on hybrid grids with the DLR-TAU code. In: Ecer, A., Emerson, D.R. (eds.) In: Proceedings 11th Parallel CFD Conference, Williamsburg, VA, North-Holland, 23–26 May 1999

3. Gerhold, T., Hannemann, V., Schwamborn, D.: On the validation of the DLR-TAU code. In: Nitsche, W., Heinemann, H.J., Hilbig, R. (eds.) New Results in Numerical and Experimental Fluid Mechanics, Notes on Numerical Fluid Mechanics, vol. 72, pp. 426–433. Vieweg (1999). ISBN 3-528-03122-0
4. Schwamborn, D., Gerhold, T., Heinrich, R.: The DLR TAU-code: recent applications in research and industry, invited lecture. In: Wesseling, P., Oate, E., Priaux, J. (eds.) Proceedings of the European Conference on Computational Fluid Dynamics (ECCOMAS CFD 2006), The Netherlands (2006)
5. Allmaras, S.R., Johnson, F.T., Spalart, P.R.: Modifications and clarifications for the implementation of the Spalart-Allmaras turbulence model. In: Seventh International Conference on Computational Fluid Dynamics (ICCFD7), ICCFD7-1902, Hawaii, July 2012
6. Sobol, I.M.: Distribution of points in a cube and approximate evaluation of integrals. Zh. Vychisl. Mat. Mat. Fiz. **7**(4), 784–802 (1967)
7. Joe, S., Kuo, F.Y.: Remark on algorithm 659: Implementing Sobol's quasirandom sequence generator. ACM Trans. Math. Softw. **29**, 49–57 (2003)
8. Joe, S., Kuo, F.Y.: Constructing Sobol sequences with better two-dimensional projections. SIAM J. Sci. Comput. **30**, 2635–2654 (2008)
9. Liu, D., Litvinenko, A., Schillings, C., Schulz, V.: Quantification of airfoil geometry-induced aerodynamic uncertainties—comparison of approaches. SIAM/ASA J. Uncertainty Quant. (2016)
10. Liu, D., Maruyama, D., Görtz, S.: Geometrical uncertainties—accuracy of parametrization and its influence on UQ and RDO. In: Hirsch, C. et al. (eds.) Uncertainty Management for Robust Industrial Design in Aeronautics, Chap. 51, pp. xx–xx. (2017)
11. Maruyama, D., Görtz, S., Liu, D.: Robust design measures for airfoil shape optimization. In: Hirsch, C. et al. (eds.) Uncertainty Management for Robust Industrial Design in Aeronautics, Chap. 32, pp. xx–xx. (2017)
12. Dwight, R., Han, Z.H.: Efficient uncertainty quantification using gradient-enhanced Kriging. AIAA Paper 2009–2276 (2009)
13. Shimoyama, K., Kawai, S., Alonso, J.J.: Dynamic adaptive sampling based on Kriging surrogate models for efficient uncertainty quantification. In: 54th AIAA/ASME/ASCE/AHS/ASC Structures, Structural Dynamics, and Materials Conference. AIAA paper 2013-1470 (2013)

Monte Carlo-Based and Sampling-Based Methods and Their Range of Applicability

Robin Schmidt, Matthias Voigt, Michele Pisaroni, Fabio Nobile, Penelope Leyland, Jordi Pons-Prats and Gabriel Bugeda

Motivation

Uncertainty quantification (UQ) has become nowadays an essential ingredient in aerodynamic robust design and optimization. Despite the advances in computational fluid dynamics (CFD) and the wide availability of modern parallel computer architectures, the efficient propagation of uncertainties from the sources to the quantities of interest (QoI) of the problem under investigation is still a significant challenge, especially when many sources of uncertainties are present and when each deterministic realization requires the solution of high-fidelity models with many degrees of freedom.

R. Schmidt (✉) · M. Voigt
Institute of Fluid Mechanics, Technische Universität Dresden, 01062 Dresden, Germany
e-mail: robin.schmidt@tu-dresden.de

M. Voigt
e-mail: matthias.voigt@tu-dresden.de

M. Pisaroni · F. Nobile · P. Leyland
Scientific Computing and Uncertainty Quantification,
Ecole Polytechnique Fédérale de Lausanne, 1015 Lausanne, Switzerland
e-mail: michele.pisaroni@epfl.ch

F. Nobile
e-mail: fabio.nobile@epfl.ch

P. Leyland
e-mail: penelope.leyland@epfl.ch

J. Pons-Prats · G. Bugeda
CIMNE, Aeronautical Group, 08034 Barcelona, Spain
e-mail: jpons@cimne.upc.edu

G. Bugeda
Universitat Politecnica de Catalunya - BarcelonaTech, 08034 Barcelona, Spain
e-mail: bugeda@cimne.upc.edu

C. Hirsch et al. (eds.), *Uncertainty Management for Robust Industrial Design in Aeronautics*, Notes on Numerical Fluid Mechanics and Multidisciplinary Design 140, https://doi.org/10.1007/978-3-319-77767-2_44

The geometrical and operational parameters that characterize aerodynamic systems in mathematical and computational models are naturally affected by aleatory uncertainties due to the intrinsic variability of manufacturing processes and the surrounding environment. These uncertainties have to be taken into account to achieve and guarantee the highest safety standards and to design aerodynamic systems whose performance is unchanged when exposed to variabilities.

Non-intrusive uncertainty propagation techniques have gained a lot of interest in the recent years as they simply require multiple solutions of the original model and can use the computational flow solver as black box. Polynomial chaos (PC) [1, 2] and stochastic collocation (SC) [3] have been successfully applied in different engineering fields and have strong potential for efficient and accurate UQ but they suffer the so-called curse of dimensionality (computational cost dramatically increase with the number of uncertain variables). Real aerodynamic systems can be affected by an innumerable number of uncertainties, and any accurate UQ analysis will be high dimensional, in particular if geometrical uncertainties are considered.

In addition, if the latter approaches are based on global basis functions that span the entire stochastic domain, their performances and accuracy can be jeopardized if the problem under investigation, governed by nonlinear mathematical models, present strong discontinuities as in the case of transonic, supersonic, and hypersonic flows.

On the other hand, sampling-based approaches like Monte Carlo (MC) methods have a dimension independent convergence rate and have been proven to be robust and accurate for non-smooth problems; nevertheless, their very slow convergence rate makes them chimerical for practical aerodynamics applications that require the solution of expensive computational fluid dynamics (CFD) models.

The Multi-Level Monte Carlo (MLMC) method has been introduced by Heinrich [4] in the context of parametric integration and thereupon extended by Giles [5] to approximate stochastic differential equations (SDEs) in financial mathematics as a way to improve the efficiency of MC simulations. The key idea of MLMC algorithms is that one can draw MC samples simultaneously and independently on several approximations of the problem under investigation on a hierarchy of computational grids (levels). The expectation of an output quantity of interest (QoI) is computed as a sample average of the coarsest solutions corrected by averages of the differences of solutions computed on two consecutive levels in the hierarchy. By this way, most of the computational effort can transport from the finest level (as in a standard MC approach) to the coarsest one.

In this chapter, the applicability issues of Monte Carlo (MC) and sampling-based methods with special focus on the Multi-Level-Monte Carlo (MLMC) method is taken into account. The aim is to provide a set of comments and conclusions from the perspective of the applications developed and the experience gained.

Remarks on the Selection of the Stochastic Variables

The selection of the stochastic variables and their stochastic definition is of prime importance not only on the definition of the MC/MLMC analysis, but to any other UQ methodology. Correlation between variables, sensitivity of the problem to their variability, number of stochastic variables, and typology of the variables must be taken into consideration before starting any single computation. If in addition the UQ method should be used along a robust design optimization analysis, all this applies as well to the design variables.

Regarding UQ analysis, the number of uncertain parameters will determine the efficiency of the method. As stated in section "Motivation," the convergence of Monte Carlo methods is independent to the number of uncertain parameters. Therefore, Monte Carlo methods are the one suffering less from the curse of dimensionality. However, the larger the number of uncertain parameters, the more difficult the definition of the problem will be. Due to the fact that the computational cost, and computational time will be considerable, it is better to spend a sufficient amount of time taking care of a good definition of the initial conditions. That means, for example, to have a sufficiently large number of experimental measurements in order to be able to deduce the distributions of and the correlations between uncertain parameters. Correlated variables should be properly defined from the perspective of this correlation. Otherwise, the results can mask variability effects.

Range of Application of MC Methods

MC methods are suitable for any number of stochastic input variables. As already mentioned in section "Motivation," they are particularly recommendable for a high number of input variables. MC simulation does not make any assumptions about the investigated system (e.g., uncorrelated, continuous, smooth, no branching system, etc.) and thus consistently realizes the work with black box models.

By using variance reduction methods such as Stratification or Latin Hypercube Sampling (LHS), a reduction in the sample size by many times is possible. When applying Quasi-Monte Carlo methods, especially for high number of stochastic variables $d > 10$, special care is to be taken to achieve good space filling. Here, optimized Latin hypercube samples can show better discrepancy and space filling.

Finally, MC and also MLMC have demonstrated their ability to deal with any kind of uncertain parameter. The aeronautical applications have been developed in the UMRIDA project described both geometrical and operational uncertainties, which MC and MLMC method have been able to deal without any problems.

Random Number Generation

We recommend the use of already existing random number generators of different programming languages. For example, the mentioned Mersenne Twister with very good performance (see chapter "General Introduction to Monte Carlo and Multi-level Monte Carlo Methods") is already included in the C++ Standard 2011.

Guide Values

Note, the values given here serve only as a rough guide. The quality of an MC simulation must always be ensured by means of suitable criteria such as, for example, confidence intervals of the various statistical result quantities. For various objectives of a Monte Carlo analysis, a value for Crude Monte Carlo (CMC) with random sampling and a value for optimized LHS (OLHS) is given. The sample size N depends on the nature of the result quantity, their variability and confidence level.

- Estimate the mean and coefficient of variation
  $N_{CMC} \approx 500, \ldots, 1000$; $N_{oLHS} \approx 50, \ldots, 100$
- Estimate robustness (without rare events)
  $N_{CMC} \approx 500, \ldots, 1000$; $N_{oLHS} \approx 50, \ldots, 100$
- Estimate sensitivities (depends on the used sensitivity measure)
  $N_{CMC} \approx 500, \ldots, 2000$; $N_{oLHS} \approx 50, \ldots, 200$
- Estimate probability of occurrence P_o
  (assumption: magnitude of probability is small,
  coefficient of variation of the standard error ≈ 0.1)
  $N_{CMC} \approx 100/P_o$; $N_{oLHS} \approx 10/P_o$ (see [6], p. 54)
- Estimate robustness with rare events
  $N_{CMC} \approx 100/P_o$; $N_{oLHS} \approx 10/P_o$.

Assessment of MLMC Applicability

Sampling and Levels of Discretization

MLMC method is based on the definition of a hierarchical evaluation of the problem. Course mesh defines the first level to increase the refinement of the mesh and the accuracy of the levels. From the first level, the successive levels are the difference of the two discretizations, which means the number of evaluations is double. The number of evaluations in each level and the number of levels greatly depend on the defined final accuracy. The final evaluation cost of each individual is the total cost of the MLMC evaluation, so the two mentioned factors, namely the number of samples per level, and the number of levels is really relevant. Although they cannot be reduced to nothing, a good definition of the levels of accuracy and the tolerances on the mean and variance values the method uses will create a difference. A trade-off analysis should be necessary in order to keep the cost under control, but a simple rule is to consider the following factors: (i) Accuracy of the solver; (ii) Physical meaning of the QoI and the capability to measure this magnitude. Do not define a MLMC accuracy larger than these two parameters. It has no sense and can lead to a never-ending evaluation. If the analyzed problem is about aeronautics, for example, and the QoI is the lift and the drag coefficients (C_l and C_d), the baseline values should be assessed, so tolerance around 10^{-3}, for the C_l, and 10^{-4}, for the C_d, could be more than enough. Of course, if the defined levels define a fine mesh which is not able to get this accuracy, then the values should be increased to avoid looking for something which is not feasible with the setting you are working with.

Solver Coupling

The black box strategy is a nice option in order to facilitate the integration of any kind of optimizer and solver with the MLMC code. The black box strategy enables to deal with each solver almost independently. Only some connection should be created to transfer information from one point to another. But those parameters only used in one of the black boxes should not be transferred along the chain, so a simplified procedure can be defined. It is certain that the strategy should be adjusted to the specific problem the user is dealing with, which means it could not be so transparent when transferring the parameters. Anyway, from a general perspective and saving particular cases, it is a good solution.

Computational Cost

Whenever the computational resources or the time is limited, the computational cost can be an issue. Since limiting the computations to a certain time can lead to a mismatching of the accuracy criteria, the optimal use of the computational resources is mandatory. It is clear that the cost of each individual MLMC analysis should be analyzed carefully. Limiting the number of samples and the number of levels for each analysis is feasible, but requires a lot of effort. It is not easy to define an universal setting for the tolerance which works for all the cases. It can happen that some of the analysis can have a lack of convergence. It can be caused because the tolerances on mean and variance, or the tolerance of the method itself have been defined too small, or on the contrary too relaxed. In this second case, which is the one the user should avoid, the number of samples does not stop to increase in each iteration. In some cases, it could be advisable to define a time-out instruction. Linux systems enable to define the time-out instruction when calling the solver. If the run of the solver is longer than the predefined time, it will be stopped and an error message will be returned. Windows systems require a customized programming. Anyway, whenever the time-out instruction should be used, the user can be sure that the final results will have no sense, so it should be always avoided to halt the analysis. Due to the fact that the computational cost and time are always to keep under control, and the parallelization of the code is fully advisable. Another option could use a grid computing environment. There are several open source libraries which help to define and implement the grid computation. The requirement is to have some spare computers, with administration privileges. Usually, a daemon should be installed in the slave computers, while the solver and calculation files can both be directly transferred by the master machine only when required, or can be installed beforehand on the slave computer, and only the parameters should be transferred by the master. The experience with HTCondor, a high-throughput computing open source library, is really positive. It enables to save time using the whole computational power in the room or department. The grid of computation can be easily enlarged or decreased according to the needs and availability. The only modification the code requires is to automate the call of the evaluations, creating them according to the grid requirements.

References

1. Knio, O., Le Maitre, O.: Uncertainty propagation in CFD using polynomial chaos decomposition. Fluid Dyn. Res. **38**(9), 616–640 (2006)
2. Hosder, S., Walters, R.W., Perez, R.: A non-intrusive polynomial chaos method for uncertainty propagation in CFD simulations. AIAA paper **891**, 2006 (2006)
3. Witteveen, J.A., Doostan, A., Chantrasmi, T., Pecnik, R., Iaccarino, G.: Comparison of stochastic collocation methods for uncertainty quantification of the transonic RAE 2822 airfoil. In: Proceedings of Workshop on Quantification of CFD Uncertainties (2009)
4. Heinrich, S., Sindambiwe, E.: Monte Carlo complexity of parametric integration. J. Complex. **15**(3), 317–341 (1999)
5. Giles, M.B.: Multilevel Monte Carlo path simulation. Oper. Res. **56**(3), 607–617 (2008)
6. Bucher, C.: Computational Analysis of Randomness in Structural Mechanics. Structures and Infrastructures Series, vol. 3. CRC Press (2009)

Introduction to Intrusive Perturbation Methods and Their Range of Applicability

Alain Dervieux

Introduction

Perturbation methods generally refer to methods in which the knowledge of some quantities is made easier by introducing exact or approximate derivatives of them. Starting from a given black box simulation tool, perturbation methods involve to get some derivative of it, which cannot be performed without an intrusion into the black box, that is without some effort in the transformation and further development of the application-dependent software.

Intrusive perturbation methods can be of two sorts, in relation with two different problematics.

(a) In uncertainty propagation by perturbation or moments, the probabilistic behavior of the physical process output is evaluated by a two-step scheme: (1) derive a surrogate black box by derivation of the initial black box, (2) use the surrogate black box in a sampling-based integrator such as Monte Carlo.

(b) In robust optimization, the uncertainty propagation in state system can be applied in a non-intrusive manner, but the research of an optimum may then be addressed with an extended gradient method, which relies on an intrusive adjoint-based functional derivation. The robust/probabilistic optimization then can use for example the following three steps: (1) build probabilistic state and functional with for example a non-intrusive polynomial chaos method, (2) compute the gradients of the mean and of the variance of the functional, and (3) update the design variable with a combination of the two gradients.

In this section, we concentrate on problematics (a) relying on second derivatives of the functionals. More precisely, we recall in short the essential information on

A. Dervieux (✉)
Ecuador INRIA, 2002 route des lucioles, 06902 Sophia Antipolis, France
e-mail: alain.dervieux@inria.fr

C. Hirsch et al. (eds.), *Uncertainty Management for Robust Industrial Design in Aeronautics*, Notes on Numerical Fluid Mechanics and Multidisciplinary Design 140, https://doi.org/10.1007/978-3-319-77767-2_45

the algorithms for differentiating in order to examine in which range the different options can be applied more efficiently. By range, we clearly refer to the number of parameters.

Engineering studies often deal with *scalar outputs*. By the way, many methods like response surface method are too computer costly when applied to mechanical fields themselves. Let us rely on the *Optimal Control* context for the definition of main problematics and methods. Developing an Optimal Control-based analysis starts from an existing simulation application and reuses key parts of it. Let us identify these parts. Given a set of parameters, a simulation software gives a prediction of a physical process. For instance, a "numerical wind tunnel" predicts the flow of air around a plane shape, using only computation. The predicted flow is the unique solution of a (set of) mathematical equation which we call the *state equation*, according to Optimal Control terminology. Numerical resolution of the discretized state equation involves in fact two important parts.

- One is the assembling part: for given arbitrary values of the state variables W and using values of external parameters γ (e.g., the geometry), it computes a *residual* array, which reflects how the state variables satisfy the state equation. It is therefore written:

$$\textbf{state residual}: (\gamma, W) \mapsto \Psi(\gamma, W)$$

  where Ψ is in fact the left-hand side of the N-dimensional discretized state equation (N can be millions):

$$\Psi(\gamma, W) = 0.$$

- The other one is the solution algorithm: for a given fixed external parameters γ, it uses (possibly hundreds times) the residual returned by the state residual to (finally) produce the state solution $W(\gamma)$ that nullifies the residual (or at least makes it sufficiently small).

$$\textbf{state-resolution}: \gamma \mapsto W(\gamma) \quad \text{such that } \Psi(\gamma, W(\gamma)) = 0.$$

  This is most often done iteratively, by incremental modifications of an arbitrary initial state, each modification driven by the residual for the current state.

To turn a simulation application into an Optimal Control application requires an additional ingredient, the *objective functional* that will evaluate a scalar cost for any possible parameters and state. Given a set of parameters γ, which we now view as *control parameters*, and given the corresponding solution state $W(\gamma)$, it computes one (or several) optimization criterion, i.e., the value of the objective functional for these control parameters and state. This objective functional takes into account all industrial targets and constraints for a given process or product.

$$\textbf{objective-assembler}: (\gamma, W) \mapsto J(\gamma, W)$$

The goal of Optimal Control is to find control parameters γ which will make the objective functional smaller.

When the size of the parameter array is small, many methods can find the minimum easily enough. However for large numbers of control parameters, approaches that use analytic gradients become necessary. We advocate the following strategy to obtain these gradients:

- from the state residual, develop a **gradient-assembler** that computes the residual corresponding to derivatives of the state equation. This part of the strategy relies on *Automatic Differentiation* (AD).
- develop an adequate resolution algorithm for the gradient and second derivatives.

Therefore use the gradient-assembler to iteratively find the requested gradient.

In order to compute the gradient of a function using AD, we can choose between two modes: Tangent and Reverse mode differentiation. Now we want to compute the gradient j' of the constrained functional using the differentiation of explicit parts implementing Ψ and J with the two differentiation modes.

Tangent Mode First-order Differentiation

It consists in computing the Gâteaux-derivatives of j with respect to each component direction e_i, $i = 1, \ldots, n$ ($e_i = (0, \ldots 0, 1, 0, \ldots, 0)^T$, where 1 is at the ith component):

$$\frac{dj}{d\gamma_i} = \frac{dj}{d\gamma}e_i = \frac{\partial J}{\partial \gamma}e_i + \frac{\partial J}{\partial W}\frac{dW}{d\gamma}e_i \tag{1}$$

where $\frac{dW}{d\gamma}e_i$ is the solution of the linear system:

$$\frac{\partial \Psi}{\partial W}\frac{dW}{d\gamma}e_i = -\frac{\partial \Psi}{\partial \gamma}e_i \, . \tag{2}$$

In order to get the gradient, (2) must be solved and (1) has to be evaluated at the point (γ_h, W_h) for each vector e_i of the canonical basis, i.e., n times and the main cost is due to the solution of n linearized N-dimensional systems.

If we choose to solve the single system (2) with an iterative method (that could be performed using AD without store the matrix $\frac{\partial \Psi}{\partial W}$ into memory, see [5]), and the solution is obtained after $n_{\text{iter},T}$ step, the total runtime cost will be of the order of $\alpha_T n_{\text{iter},T}$, i.e., $n_{\text{iter},T}$ evaluation of the matrix-by-vector operation $\frac{\partial \Psi}{\partial W}x$, where we assume that each evaluation costs α_T times the evaluation of the state residual $\Psi(\gamma, W)$ and the cost of the state residual is taken as reference equal to 1. Therefore, the runtime cost of the full gradient will be $n\alpha_T n_{\text{iter},T}$.

Reverse Mode First-order Differentiation

The complete gradient is given by the equation

$$\left(\frac{dj}{d\gamma}\right)^T = \left(\frac{\partial J}{\partial \gamma}\right)^T - \left(\frac{\partial \Psi}{\partial \gamma}\right)^T \Pi \tag{3}$$

where $\Pi : (\gamma, W) \mapsto \Pi(\gamma, W) \in \mathbb{R}^N$ (the adjoint state) is the solution of the linear system

$$\left(\frac{\partial \Psi}{\partial W}\right)^T \Pi = \left(\frac{\partial J}{\partial W}\right)^T . \tag{4}$$

It is important to note that the above formulation permits us to obtain all the derivatives needed by (3)–(4) using only Reverse mode differentiation of the programs implementing $J(\gamma, W)$ and $\Psi(\gamma, W)$.

To compute the gradient j' with this approach, we thus need to solve only one linearized N-dimensional system, the adjoint system (4). If we choose to solve the adjoint system with an iterative method, we can apply the same estimate as in the case of the Tangent mode differentiation but, this time, the overhead associated with the evaluation of the matrix-by-vector operation $\left(\frac{\partial \Psi}{\partial W}\right)^T x$ with respect to the state residual evaluation will be α_R (and usually $\alpha_R > \alpha_T$) and the number of iterations $n_{\text{iter},R}$ for the convergence of the solution could be different from $n_{\text{iter},T}$ of the previous case (in our experience, the number of iterations needed by, e.g., GMRES; see [6], to solve the system $Ax = b$ and $A^T x = c$ are of the same order; see [5]): therefore, the total runtime cost for the gradient will be $\alpha_R n_{\text{iter},R}$.

From the previous arguments, it clearly appears that if we need to compute the gradient j' only, the Reverse mode is cheaper in terms of CPU time than the Tangent mode if $n \gg 1$. Nevertheless, the Tangent mode algorithm in section "Tangent Mode First-order Differentiation" will be used in the following because it is the basis for the Hessian computation with the Tangent-on-Tangent approach.

Tangent-on-Tangent Second-order Differentiation

In the same manner as the computation of the gradient to compute second derivatives, we have different possibilities, which are theoretically equivalent, but they differ in the computational cost (and the choice of the best strategy depends, in the end, on the number n of the control variables for the given functional). Moreover, the best strategy depends on which use of the second-order derivative we need, i.e., the best strategy to obtain the full Hessian matrix could be different from the best strategy to obtain its diagonal part or the multiplication of the Hessian matrix by a vector.

The first method to obtain the second-order differentiation of a constrained functional performs two successive Tangent mode differentiations for both the functional and the state residual and use the adjoint state to compute every single element in the Hessian matrix [4]: we call this approach Tangent-on-Tangent (ToT). The second approach (Tangent-on-Reverse, ToR) performs a Tangent mode differentiation of the gradient (3) obtained with Reverse differentiation (section "Reverse Mode First-order Differentiation").

This method was initially investigated by [7] along with various other algorithms, but the publication does not go into the implementation details for a generic fluid dynamic code. Here we present the mathematical background behind the idea and the efficient AD implementation of [4] but with a different analysis of the computational cost.

Starting from the ith element of the gradient (1), we perform another differentiation with respect to the variable γ_k obtaining the i-k element of the Hessian matrix

$$\left(\frac{d^2 j}{d\gamma^2}\right)_{i,k} = \frac{d^2 j}{d\gamma_i d\gamma_k} = D^2_{i,k} J + \frac{\partial J}{\partial W}\frac{d^2 W}{d\gamma_i d\gamma_k} \tag{5}$$

where

$$D^2_{i,k} J = \frac{\partial}{\partial \gamma}\left(\frac{\partial J}{\partial \gamma} e_i\right) e_k + \frac{\partial}{\partial W}\left(\frac{\partial J}{\partial \gamma} e_i\right)\frac{dW}{d\gamma_k} + \frac{\partial}{\partial W}\left(\frac{\partial J}{\partial \gamma} e_k\right)\frac{dW}{d\gamma_i} + \frac{\partial}{\partial W}\left(\frac{\partial J}{\partial W}\frac{dW}{d\gamma_i}\right)\frac{dW}{d\gamma_k}. \tag{6}$$

Differentiating the Eq. (2), we get

$$D^2_{i,k} \Psi + \frac{\partial \Psi}{\partial W}\frac{d^2 W}{d\gamma_i d\gamma_k} = 0 \tag{7}$$

where

$$D^2_{i,k} \Psi = \frac{\partial}{\partial \gamma}\left(\frac{\partial \Psi}{\partial \gamma} e_i\right) e_k + \frac{\partial}{\partial W}\left(\frac{\partial \Psi}{\partial \gamma} e_i\right)\frac{dW}{d\gamma_k} + \frac{\partial}{\partial W}\left(\frac{\partial \Psi}{\partial \gamma} e_k\right)\frac{dW}{d\gamma_i} + \frac{\partial}{\partial W}\left(\frac{\partial \Psi}{\partial W}\frac{dW}{d\gamma_i}\right)\frac{dW}{d\gamma_k} \tag{8}$$

and $e_i (e_k)$ is the usual vector of the canonical basis with 1 at the ith (kth) component and zero otherwise. Substituting the second derivatives of the state with respect to the control variables $\frac{d^2 W}{d\gamma_i d\gamma_k}$ in Eq. (5) from Eq. (7), we get

$$\begin{aligned} \frac{d^2 j}{d\gamma_i d\gamma_k} &= D^2_{i,k} J - \frac{\partial J}{\partial W}\left(\frac{\partial \Psi}{\partial W}\right)^{-1} D^2_{i,k} \Psi \\ &= D^2_{i,k} J - \Pi^T D^2_{i,k} \Psi \end{aligned} \tag{9}$$

where Π is the solution of the adjoint system (4). The i-k element of the Hessian matrix $\left(\frac{d^2 j}{d\gamma^2}\Big|_{\gamma_h}\right)$ is then obtained evaluating the (9) at the point (γ_h, W_h) solution of the state equation $\Psi = 0$, namely

$$\left(\frac{d^2j}{d\gamma^2}\bigg|_{\gamma_h} \right)_{i,k} = (D^2_{i,k}J)|_{(\gamma_h,W_h)} - \Pi_h^T (D^2_{i,k}\Psi)|_{(\gamma_h,W_h)} \tag{10}$$

where $\Pi_h \in \mathbb{R}^N$ is the solution of the adjoint linear system

$$\left(\frac{\partial \Psi}{\partial W}\bigg|_{(\gamma_h,W_h)} \right)^T \Pi_h = \left(\frac{\partial J}{\partial W}\bigg|_{(\gamma_h,W_h)} \right)^T . \tag{11}$$

The n derivatives $\dfrac{dW}{d\gamma_i}$ in the formulas for $D^2_{i,k}J$ and $D^2_{i,k}\Psi$ should be computed (and stored) solving the linear systems

$$\left(\frac{\partial \Psi}{\partial W}\bigg|_{(\gamma_h,W_h)} \right) \frac{dW}{d\gamma_i} = -\left(\frac{\partial \Psi}{\partial \gamma}\bigg|_{(\gamma_h,W_h)} \right) e_i \qquad i = 1, \ldots, n \tag{12}$$

and this task can be performed using an iterative matrix-free method [5]. We assume the number of iterations needed for the iterative linear solver to the convergence of the solution to be $n_{\text{iter},T}$, and each iteration calls a tangent-differentiated routine implementing the matrix-by-vector multiplication $\frac{\partial \Psi}{\partial W}x$ whose cost is α_T times the cost of the original routine implementing the evaluation of the state residual $\Psi(\gamma, W)$.

Description of the Algorithm for the Hessian Matrix with the ToT Approach.
The algorithm to compute the Hessian matrix with the ToT approach can be summarized as follows:

1. compute the state W_h such that $\Psi(\gamma_h, W_h) = 0$;
2. compute $\bar{W}_J = \left(\frac{\partial J}{\partial W}\bigg|_{(\gamma_h,W_h)} \right)^T$;
3. compute the adjoint state Π_h solving the linear system $\left(\frac{\partial \Psi}{\partial W}\bigg|_{(\gamma_h,W_h)} \right)^T \Pi_h = \bar{W}_J$;
4. for each element of the canonical basis e_i, $i = 1, \ldots, n$
   (a) compute $\dot{\Psi}^{(i)}_\gamma = \left(\frac{\partial \Psi}{\partial \gamma}\bigg|_{(\gamma_h,W_h)} \right) e_i$;
   (b) compute (and store) the vector $\theta^{(i)}_h$ solution of the linear system

$$\left(\frac{\partial \Psi}{\partial W}\bigg|_{(\gamma_h,W_h)} \right) \theta^{(i)}_h = -\dot{\Psi}^{(i)}_\gamma; \tag{13}$$

5. for $i = 1, \ldots, n$ and $k = i, \ldots, n$
   (a) compute $(D^2_{i,k}\Psi)|_{(\gamma_h,W_h)}$;
   (b) compute $(D^2_{i,k}J)|_{(\gamma_h,W_h)}$;
   (c) compute $\left(\frac{d^2j}{d\gamma^2}\bigg|_{\gamma_h} \right)_{i,k} = (D^2_{i,k}J)|_{(\gamma_h,W_h)} - \Pi_h^T (D^2_{i,k}\Psi)|_{(\gamma_h,W_h)}$.

From Eq. (10), we see that the ToT approach gives us a single element *i-k* of the Hessian matrix at time, then using the symmetry property of the Hessian we can compute the full $n \times n$ matrix applying $\frac{n(n+1)}{2}$ time the steps 5(a)–5(c) in the algorithm above.

For each element *i-k*, we need to know the vectors $\theta_h^{(i)} = \frac{dW}{d\gamma_i}$ and $\theta_h^{(k)} = \frac{dW}{d\gamma_k}$ obtained solving iteratively (e.g., using GMRES) the linear system (13) whose cost is $\alpha_T n_{\text{iter,T}}$ (for simplicity we assume that the number of iterations needed to solve the linear system is independent from the right-hand side).

Moreover, the quantity $\left(D^2_{i,k}\Psi\right)\big|_{(\gamma_h,W_h)}$ (step 5a) can be obtained with a single invocation of the differentiated-twice subroutine of `state_residual (psi, gamma, w)`, and its cost is α_T^2 times the cost of the evaluation of the residual $\Psi(\gamma, W)$ (that it is assumed to be unitary). The analogous quantity relative to the functional $\left(D^2_{i,k}J\right)\big|_{(\gamma_h,W_h)}$ (step 5b) can be obtained with a single invocation of the differentiated-twice subroutine of `objective-assembler (j, gamma, w)` and its cost is negligible with respect to running the subroutine `state_residual(psi, gamma, w)`, being negligible the cost to evaluate the subroutine `objective-assembler(j, gamma, w)` with respect to `state_residual (psi, gamma, w)`.

Therefore, assuming the adjoint state Π_h to be available, the evaluation of the full Hessian with the ToT approach costs

$$n\alpha_T\left[n_{\text{iter},T} + \frac{(n+1)}{2}\alpha_T\right] \tag{14}$$

and we note that the cost is quadratic respect to the dimension of the control variables but, if we have $n_{\text{iter},T} \gg n$, the main contribution could be from the cost to solve the n linear systems (13). Therefore, if the dimension of the control variables n is small, the cost is dominated by the solution of the linear systems; otherwise, (and assuming the number of iterations n_{iter} to be independent from n), the main cost is due to the differentiated-twice subroutines.

Another important thing to note is the fact that with ToT we can compute the diagonal of the Hessian without computing the extra-diagonal values, due to the fact that the Hessian is built element-by-element: this fact results in a cost for the entire diagonal of

$$n\alpha_T\left[n_{\text{iter},T} + \alpha_T\right] \tag{15}$$

(i.e., one linear system (13) and one evaluation of the differentiated-twice routines in the steps 5(a)–5(c) for each element of the diagonal).

Tangent-on-Reverse Option

This consists in the direct derivation in any direction e_i, $i = 1, \ldots, n$ of the (non-scalar) function:

$$\left(\frac{dj}{d\gamma}\right)^T = \left(\frac{\partial J}{\partial \gamma}\right)^T - \left(\frac{\partial \Psi}{\partial \gamma}\right)^T \Pi \tag{16}$$

where $W: \gamma \mapsto W(\gamma)$ such that $\Psi(\gamma, W) = 0$ and $\Pi : (\gamma, W) \mapsto \Pi(\gamma, W)$ is the adjoint state defined as

$$\Pi = \left(\frac{\partial \Psi}{\partial W}\right)^{-T} \left(\frac{\partial J}{\partial W}\right)^T. \tag{17}$$

To build the algorithm to compute the Hessian in the present context, we need the following

Lemma *Let $\gamma_h \in \mathbb{R}^n$ and $W_h \in \mathbb{R}^N$ such that $\Psi(\gamma_h, W_h) = 0$ and let*

$$\begin{aligned} j: \mathbb{R}^n &\longrightarrow \mathbb{R} \\ \gamma &\longmapsto j(\gamma) := J(\gamma, W) \end{aligned} \tag{18}$$

then the projection of the Hessian $\left(\left.\frac{d^2 j}{d\gamma^2}\right|_{\gamma_h}\right) \in \mathbb{R}^{n\times n}$ along a direction $\delta \in \mathbb{R}^n$ is given by

$$\boxed{\begin{aligned}\left(\left.\frac{d^2 j}{d\gamma^2}\right|_{\gamma_h}\right)\delta = &\left.\left[\frac{\partial}{\partial \gamma}\left(\frac{\partial J}{\partial \gamma}\right)^T\right]\right|_{(\gamma_h, W_h)} \delta + \left.\left[\frac{\partial}{\partial W}\left(\frac{\partial J}{\partial \gamma}\right)^T\right]\right|_{(\gamma_h, W_h)} \theta_h + \\ &- \left.\left[\frac{\partial}{\partial \gamma}\left(\left(\frac{\partial \Psi}{\partial \gamma}\right)^T \Pi_h\right)\right]\right|_{(\gamma_h, W_h)} \delta - \left.\left[\frac{\partial}{\partial W}\left(\left(\frac{\partial \Psi}{\partial \gamma}\right)^T \Pi_h\right)\right]\right|_{(\gamma_h, W_h)} \theta_h + \\ &- \left(\left.\frac{\partial \Psi}{\partial \gamma}\right|_{(\gamma_h, W_h)}\right)^T \lambda_h\end{aligned}} \tag{19}$$

where $\Pi_h, \theta_h, \lambda_h \in \mathbb{R}^N$ satisfy

$$\left\{\begin{aligned} &\left(\left.\frac{\partial \Psi}{\partial W}\right|_{(\gamma_h, W_h)}\right)^T \Pi_h = \left(\left.\frac{\partial J}{\partial W}\right|_{(\gamma_h, W_h)}\right)^T \\ &\left(\left.\frac{\partial \Psi}{\partial W}\right|_{(\gamma_h, W_h)}\right)\theta_h = -\left(\left.\frac{\partial \Psi}{\partial \gamma}\right|_{(\gamma_h, W_h)}\right)\delta \\ &\left(\left.\frac{\partial \Psi}{\partial W}\right|_{(\gamma_h, W_h)}\right)^T \lambda_h = \frac{\partial}{\partial \gamma}\left(\frac{\partial J}{\partial W}\right)^T \delta + \frac{\partial}{\partial W}\left(\frac{\partial J}{\partial W}\right)^T \theta_h + \\ &\qquad - \frac{\partial}{\partial \gamma}\left[\left(\frac{\partial \Psi}{\partial W}\right)^T \Pi_h\right]\delta - \frac{\partial}{\partial W}\left[\left(\frac{\partial \Psi}{\partial W}\right)^T \Pi_h\right]\theta_h. \square \end{aligned}\right. \tag{20}$$

If we apply this lemma using $\delta = e_i$ (where $e_i = (0, \ldots, 1, \ldots, 0)^T$ is the vector having the only non-zero value at the i-th position, i.e., the i-th element of the canon-

ical basis of $\mathbb{R}^n$*), it means that we are computing the ith column (and, by symmetry, the ith row) of the Hessian, obtaining*

$$\boxed{\begin{aligned}\left(\left.\frac{d^2j}{d\gamma^2}\right|_{\gamma_h}\right)e_i = &\left.\left[\frac{\partial}{\partial\gamma}\left(\frac{\partial J}{\partial\gamma}\right)^T\right]\right|_{(\gamma_h,W_h)}e_i + \left.\left[\frac{\partial}{\partial W}\left(\frac{\partial J}{\partial\gamma}\right)^T\right]\right|_{(\gamma_h,W_h)}\theta_h^{(i)} + \\ &- \left.\left[\frac{\partial}{\partial\gamma}\left(\left(\frac{\partial\Psi}{\partial\gamma}\right)^T\Pi_h\right)\right]\right|_{(\gamma_h,W_h)}e_i - \left.\left[\frac{\partial}{\partial W}\left(\left(\frac{\partial\Psi}{\partial\gamma}\right)^T\Pi_h\right)\right]\right|_{(\gamma_h,W_h)}\theta_h^{(i)} + \\ &- \left(\left.\frac{\partial\Psi}{\partial\gamma}\right|_{(\gamma_h,W_h)}\right)^T\lambda_h^{(i)}.\end{aligned}} \tag{21}$$

Then, to compute the full Hessian, we need to apply the Hessian-by-vector multiplication (21) to each component of the canonical basis of $\mathbb{R}^n$.

For each $i = 1, \dots, n$, Eq. (21) needs the adjoint state Π_h, solution of the adjoint system

$$\left(\left.\frac{\partial\Psi}{\partial W}\right|_{(\gamma_h,W_h)}\right)^T\Pi_h = \left(\left.\frac{\partial J}{\partial W}\right|_{(\gamma_h,W_h)}\right)^T \tag{22}$$

and the arrays $\theta_h^{(i)}$, $\lambda_h^{(i)}$ solutions of the linear systems:

$$\begin{cases}\left(\left.\frac{\partial\Psi}{\partial W}\right|_{(\gamma_h,W_h)}\right)\theta_h^{(i)} = -\left(\left.\frac{\partial\Psi}{\partial\gamma}\right|_{(\gamma_h,W_h)}\right)e_i \\ \left(\left.\frac{\partial\Psi}{\partial W}\right|_{(\gamma_h,W_h)}\right)^T\lambda_h^{(i)} = \left.\left[\frac{\partial}{\partial\gamma}\left(\frac{\partial J}{\partial W}\right)^T\right]\right|_{(\gamma_h,W_h)}e_i + \left.\left[\frac{\partial}{\partial W}\left(\frac{\partial J}{\partial W}\right)^T\right]\right|_{(\gamma_h,W_h)}\theta_h^{(i)} + \\ \qquad - \left.\left[\frac{\partial}{\partial\gamma}\left(\left(\frac{\partial\Psi}{\partial W}\right)^T\Pi_h\right)\right]\right|_{(\gamma_h,W_h)}e_i - \left.\left[\frac{\partial}{\partial W}\left(\left(\frac{\partial\Psi}{\partial W}\right)^T\Pi_h\right)\right]\right|_{(\gamma_h,W_h)}\theta_h^{(i)}\end{cases} \tag{23}$$

where all the derivatives in the Eqs. (21)–(23) are evalued at the final state W_h. Moreover, the second linear system in (23) is of the same type of the adjoint system (22) but with a different right-hand side, so we can use the same matrix-free algorithm and the same preconditioner (but with different right-hand side) for both equations.

Description of the Algorithm for the Hessian Matrix with the ToR Approach.
The algorithm to compute the Hessian matrix with linearized system can the ToR approach can be summarized as follows:

1. compute the state W_h such that $\Psi(\gamma_h, W_h) = 0$;
2. compute $\bar{W}_J = \left(\left.\frac{\partial J}{\partial W}\right|_{(\gamma_h,W_h)}\right)^T$;
3. compute the adjoint state Π_h solving the linear system $\left(\left.\frac{\partial\Psi}{\partial W}\right|_{(\gamma_h,W_h)}\right)^T\Pi_h = \bar{W}_J$;
4. for each element of the canonical basis e_i, $i = 1, \dots, n$:

(a) compute $\dot{\Psi}_\gamma = \left(\frac{\partial \Psi}{\partial \gamma}\Big|_{(\gamma_h, W_h)} \right) e_i$;

(b) compute the vector $\theta_h^{(i)}$ solving the linear system $\left(\frac{\partial \Psi}{\partial W}\Big|_{(\gamma_h, W_h)} \right) \theta_h^{(i)} = -\dot{\Psi}_\gamma$;

(c) compute $\dot{\bar{W}}_J = \left[\frac{\partial}{\partial \gamma} \left(\frac{\partial J}{\partial W} \right)^T \right]\Big|_{(\gamma_h, W_h)} e_i + \left[\frac{\partial}{\partial W} \left(\frac{\partial J}{\partial W} \right)^T \right]\Big|_{(\gamma_h, W_h)} \theta_h^{(i)}$;

(d) compute $\dot{\bar{\gamma}}_J = \left[\frac{\partial}{\partial \gamma} \left(\frac{\partial J}{\partial \gamma} \right)^T \right]\Big|_{(\gamma_h, W_h)} e_i + \left[\frac{\partial}{\partial W} \left(\frac{\partial J}{\partial \gamma} \right)^T \right]\Big|_{(\gamma_h, W_h)} \theta_h^{(i)}$;

(e) compute $\dot{\bar{W}}_\Psi = \left[\frac{\partial}{\partial \gamma} \left(\left(\frac{\partial \Psi}{\partial W} \right)^T \Pi_h \right) \right]\Big|_{(\gamma_h, W_h)} e_i + \left[\frac{\partial}{\partial W} \left(\left(\frac{\partial \Psi}{\partial W} \right)^T \Pi_h \right) \right]\Big|_{(\gamma_h, W_h)} \theta_h^{(i)}$;

(f) compute $\dot{\bar{\gamma}}_\Psi = \left[\frac{\partial}{\partial \gamma} \left(\left(\frac{\partial \Psi}{\partial \gamma} \right)^T \Pi_h \right) \right]\Big|_{(\gamma_h, W_h)} e_i + \left[\frac{\partial}{\partial W} \left(\left(\frac{\partial \Psi}{\partial \gamma} \right)^T \Pi_h \right) \right]\Big|_{(\gamma_h, W_h)} \theta_h^{(i)}$;

(g) compute the vector $\lambda_h^{(i)}$ solving the linear system

$$\left(\frac{\partial \Psi}{\partial W}\Big|_{(\gamma_h, W_h)} \right)^T \lambda_h^{(i)} = \dot{\bar{W}}_J - \dot{\bar{W}}_\Psi;$$

(h) compute $\bar{\gamma}_\Psi = \left(\frac{\partial \Psi}{\partial \gamma}\Big|_{(\gamma_h, W_h)} \right) \lambda_h^{(i)}$;

(i) compute the i-th column (or row) of the Hessian matrix:

$$\left(\frac{dj}{d\gamma}\Big|_{\gamma_h} \right) e_i = \dot{\bar{\gamma}}_J - \dot{\bar{\gamma}}_\Psi - \bar{\gamma}_\Psi\ .$$

From the previous algorithm, we see that for each column of the Hessian matrix we need to solve two linear systems: one is to compute the vector $\theta_h^{(i)}$ (step 4b) and the other is to compute the vector $\lambda_h^{(i)}$ (step 4g). Moreover, the quantities $\dot{\bar{W}}_J$, $\dot{\bar{\gamma}}_J$ (steps 4c and 4d) can be obtained with a single invocation of the twice-differentiated subroutine of `objective-assembler(j, gamma, w)`; while the quantities $\dot{\bar{W}}_\Psi$, $\dot{\bar{\gamma}}_\Psi$ (steps 4e and 4f) can be obtained with a single invocation of the differentiated-twice subroutine of `state_residual`.

The computational cost for a single Hessian-by-vector multiplication, evaluated with the Tangent-on-Reverse approach, can be estimated with the same assumptions made in section "Tangent-on-Tangent Second-order Differentiation" and is due to the following contributes:

- $\alpha_T n_{\text{iter},T}$ for computing the derivatives of the state variables respect to the control $\theta_h^{(i)} = \frac{dW}{d\gamma_i}$ (step 4b), where $n_{\text{iter},T}$ is the number of iterations needed by the matrix-free algorithm to solve the linear system;
- $\alpha_R \alpha_T$ for evaluating the quantities $\dot{\bar{W}}_\Psi$, $\dot{\bar{\gamma}}_\Psi$ (steps 4e and 4f) with the single invocation of the subroutine `state_residual`;

- $\alpha_R n_{\text{iter},R}$ for computing the vector $\lambda_h^{(i)}$ (step 4g), where $n_{\text{iter},R}$ is the number of iterations needed by the matrix-free algorithm to solve adjoint systems.

Therefore, the full Hessian evaluation with ToR costs

$$n\alpha_T\left(n_{\text{iter},T} + \alpha_R + \frac{\alpha_R}{\alpha_T} n_{\text{iter},R}\right) \tag{24}$$

and we note that the major contribution is due to the solution of the linear systems, usually being the number of iterations to the convergence $\gg \alpha_R$.

Another important thing to note is the fact that with ToR, we cannot compute the diagonal of the Hessian without computing the extra-diagonal values, due to the fact that the Hessian is built column-by-column (or, by symmetry, row-by-row) using the above lemma on the elements of the canonical basis.

As minor remark, ToR approach for the full Hessian does not need to store the derivatives of the state variables respect to the control $\theta_h^{(i)} = \frac{dW}{d\gamma_i}$ for all $i = 1, \ldots, n$, but it can use the same memory locations for the various $\theta_h^{(i)}$, resulting in a memory saving and in a serialization of the algorithm, while using a different location for each vector results in an easily parallelizable algorithm (each Hessian-by-vector multiplication is independent from the others, so each evaluation can be run in parallel).

Comparison Between ToT and ToR

At this point, the natural question arising from the previous analysis is about the choice of the method that is less expensive for a given problem. The cost to evaluate the full Hessian, its diagonal part and the Hessian-by-vector multiplication for the two different strategies, is given in Table 1, where we do not take into account the cost to solve the state equation $\Psi = 0$ and to solve the adjoint system (4).

From the algorithms in sections "Tangent-on-Tangent Second-order Differentiation" and "Tangent-on-Reverse option," we note that the two approaches to evaluate the full Hessian share a common part, namely the computation of the derivatives of the state variables respect to the control $\frac{dW}{d\gamma_i}$ $(i = 1, \ldots, n)$ as solution of the linear system

$$\left(\left.\frac{\partial \Psi}{\partial W}\right|_{(\gamma_h, W_h)}\right)\frac{dW}{d\gamma_i} = -\left(\left.\frac{\partial \Psi}{\partial \gamma}\right|_{(\gamma_h, W_h)}\right) e_i\,. \tag{25}$$

This cost appears in Table 1 as the $n\alpha_T n_{\text{iter},T}$ term, therefore the characteristic cost grows as $\frac{n(n+1)}{2}\alpha_T^2$ for the ToT approach and $n\alpha_R\left(\alpha_T + n_{\text{iter},R}\right)$ for ToR. Thus, we can say that, *using a single strategy* to compute the full Hessian, ToT has a lower computational cost with respect to ToR if

Table 1 ToT and ToR comparison. Computational cost for the evaluation of the full $n \times n$ Hessian matrix, only its diagonal part and the Hessian-by-vector multiplication. α_T, α_R are the overheads associated with the Tangent- and Reverse-mode differentiation for the subroutine implementing the evaluation of the state residual. $n_{\text{iter},T}$, $n_{\text{iter},R}$ are the number of iterations needed for the matrix-free algorithm to solve the tangent and adjoint linear system, respectively. The values in the table do not take into account the runtime cost to compute the adjoint state Π_h, that is assumed to be available. The cost to compute the adjoint state Π_h as solution of the adjoint linear system (4) can be estimated as $\alpha_R n_{\text{iter},R}$

	Hessian (full)	Hessian (diagonal)	Hessian-by-vector
ToT	$n\alpha_T\left[n_{\text{iter},T} + \frac{(n+1)}{2}\alpha_T\right]$	$n\alpha_T\left[n_{\text{iter},T} + \alpha_T\right]$	$n\alpha_T\left[n_{\text{iter},T} + \alpha_T\right]$
ToR	$n\alpha_T\left(n_{\text{iter},T} + \alpha_R + \frac{\alpha_R}{\alpha_T}n_{\text{iter},R}\right)$	–	$\alpha_T\left(n_{\text{iter},T} + \alpha_R + \frac{\alpha_R}{\alpha_T}n_{\text{iter},R}\right)$

$$n < 2\frac{\alpha_R}{\alpha_T}\left(1 + \frac{n_{\text{iter},R}}{\alpha_T}\right) - 1. \tag{26}$$

Therefore, ToT is cheaper than ToR if the dimension n of the control γ is small. This last result can be used to build better strategy (i.e., less time-consuming) for the full Hessian *using ToT and ToR for evaluate different parts of the Hessian*. The key idea is to use ToT to build the upper triangular part of the Hessian until the $\bar{n}$th column and then evaluate the remaining $n - \bar{n}$ columns with ToR (using the Hessian-by-vector multiplication). This mixed strategy costs

$$\begin{cases} n\alpha_T n_{\text{iter},T} + \dfrac{n(n+1)}{2}\alpha_T^2 & \text{for } n \leq \bar{n} \\ n\alpha_T n_{\text{iter},T} + \dfrac{\bar{n}(\bar{n}+1)}{2}\alpha_T^2 + (n-\bar{n})\alpha_R\left(\alpha_T + n_{\text{iter},R}\right) & \text{for } n > \bar{n} \end{cases} \tag{27}$$

where optimal values for $\bar{n}$ are found to be $\bar{n} = \frac{\alpha_R}{\alpha_T}\left(1 + \frac{n_{\text{iter},R}}{\alpha_T}\right)$. For a given problem, we assume that the values α_T, α_R can be obtained with not too much effort in a preprocessing phase, using program profiling. Much more difficult could be the estimate of $n_{\text{iter},R}$, the number of iterations needed to solve the adjoint linear system, in fact this number depends on many factors: the dimension of the problem itself, the dimension of the Krylov space, the kind of preconditioner used, etc. A comparison for the cost of the full Hessian using different strategies is given in Fig. 1 where we assumed $n_{\text{iter},R} = n_{\text{iter},T} = 300$ for the number of iterations needed to solve the linear systems (direct and adjoint) and $\alpha_T = 2$, $\alpha_R = 4$ for the overhead associated with the Tangent differentiation and Reverse differentiation (obtaining the corresponding threshold value for the mixed strategy $\bar{n} = 302$).

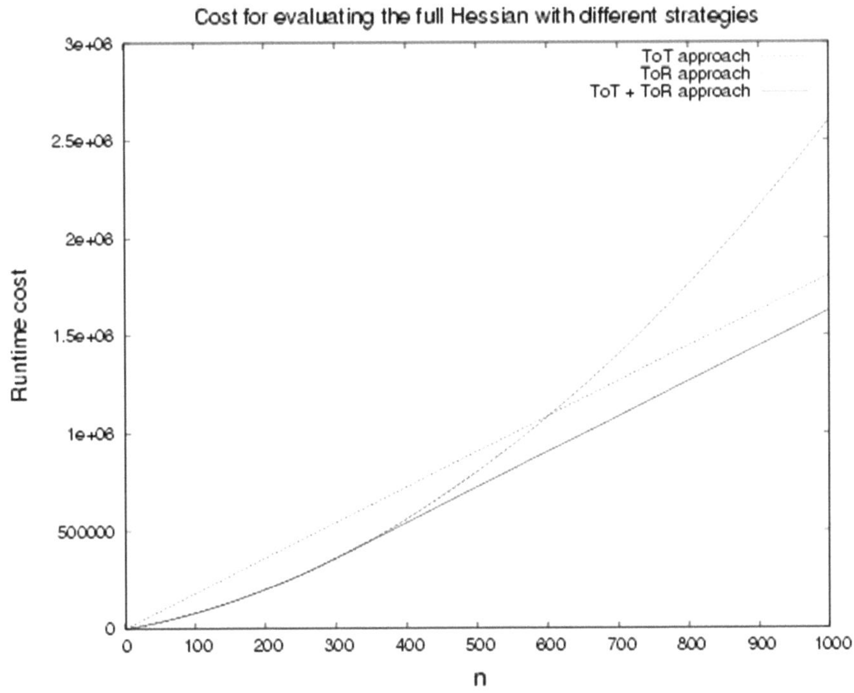

Fig. 1 Comparison for the cost of the full Hessian using different strategies as a function of n, the dimension of the control variable γ. We assumed $n_{\text{iter},R} = n_{\text{iter},T} = 300$ for the number of iterations needed to solve the linear systems (tangent and adjoint) and $\alpha_T = 2$, $\alpha_R = 4$ for the overhead associated with the Tangent differentiation and Reverse differentiation (obtaining the corresponding threshold value for the mixed strategy $\bar{n} = 302$)

Large Number of Uncertainties

An important output of the previous sections is that the ToT approach generally should be chosen for a number N of uncertain variables less than 1000 while the ToR may be preferred larger numbers N. The total cost will be about 1000 linearized systems to solve, for obtaining 1,000,000 second derivatives. For $N \geq 1000$, the total cost will be about 2000 linearized systems to solve, for obtaining 1,000,000 second derivatives. The best of these options is therefore linearly increasing with N, this is an important advantage of the perturbation approach.

We have first to identify why we need so many parameters:

- They can (in most part) correspond to the discretization of a continuous field. In this case, the number of degrees of freedom could be decreased or the derivatives on some intermediate nodes can be obtained by interpolation.
- They can correspond to the discretization of a continuous field which is not available, for example for software reasons (discrete shapes inside Katia).

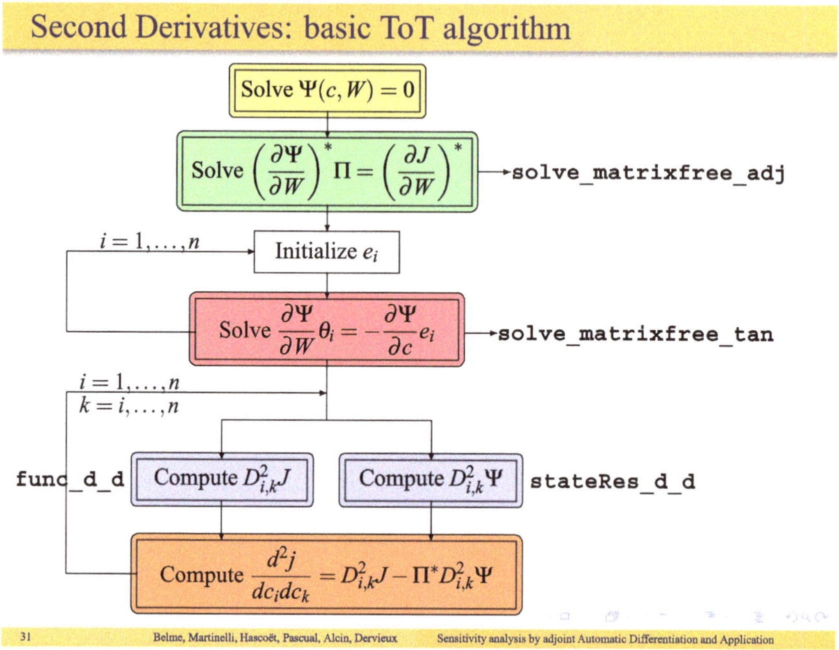

Fig. 2 ToT approach

- The worst case is when these parameters cannot be related to a continuous analog.

Once we are convinced that N cannot be reduced, it is necessary to compute all these derivatives in the shortest wall-clock time. This can be obtained thanks to parallel computing. Of course, each linearized system can be solved in a shorter time with a parallel computer using, for example mesh partitioning. But, also, we observe that the solution of the different systems and the assembly of different right-hand sides can be performed independently, with memory saving.

We restrict to the ToT option. Figure 2 depicts the main part of the algorithm. After solving the state and adjoint, the derivatives θ_i of flow field W with respect to each uncertain variable C_i need be computed. Then N^2 derivatives need be assembled (without system resolution).

It is clear that the derivatives θ_i constitute the main computational effort, and that they can be computed in separate runs.

Further improvements can imagine from the observation that these systems have a common matrix and different right-hand sides. Many published works propose smart solvers for this context; see for example [3]. They use Gram–Schmidt/Krylov basis or/and deflation methods. In the latter case, coarse grid inverse matrices can be built once for all. These coarse grids can be derived either from eigenvalues of the matrix of from coarser meshes.

Static POD methods can also be used both for multiple RHS and for multidisciplinary optimization. We refer for example to the work of Farhat (e.g., [2]). We also point out that a simplified shape modification method can be introduced in the POD model for example as in [1].

Conclusion

The cost of a perturbation approach (PA) for uncertainty propagation combines a sensitivity analysis with Monte Carlo-like post-processing. The implementation of PA is quite intrusive and complex since it involves a sensitivity (first and second order) calculation. Among other methods, Automatic Differentiation can be applied. When the number N of uncertain variables increases, the computational cost of this approach is linear with respect to N. But it can be enormous for large N. Several sources of parallelism can help to reduce the time to compute, therefore using many processors. The reduction of computational complexity can be obtained mainly by two ways:

- the use of preconditioned solvers adapted to multiple RHS,
- the derivation of sensitivity on reduced order models.

References

1. Bourguet, R., Braza, M., Dervieux, A.: Reduced order modeling of transonic flows around an airfoil submitted to small deformations. J. Comput. Phys. **230**(1), 159–184 (2010)
2. Carlberg, K., Farhat, C.: An adaptive POD-krylov reduced-order model for structural optimization. In: Proceedings 8th World Congress on Structural and Multidisciplinary Optimization. Lisbon, Portugal, June 1–5 (2009)
3. Chan, T.F., Wan, W.L.: Analysis of projection methods for solving linear systems with multiple right-hand sides. SIAM J. Sci. Comput. **18**(6), 1698–1721 (1997)
4. Ghate,D., Giles, M.B.: Efficient Hessian calculation using automatic differentiation. In: 25th Applied Aerodynamics Conference. Miami (Florida), AIAA 2007-4059 (2007)
5. Martinelli, M.: Sensitivity evaluation in aerodynamic optimal design. Ph.D. Thesis, Scuola Normale di Pisa, Italy (2007)
6. Saad, Y.: Iterative Methods for Sparse Linear Systems, PWS Publishing Company (1996)
7. Sherman, L.L., Taylor III, A.C., Green, L.L., Newman, P.A.: First and second-order aerodynamic sensitivity derivatives via automatic differentiation with incremental iterative methods. J. Comput. Phys. **129**, 307–331 (1996)

Use of Open Source UQ Libraries

Sönke Klostermann and R. Lebrun

For this section, we aim to demonstrate the use of open source software for aircraft preliminary design at the example of a wing configuration robust design optimization. On airfoil level, we demonstrate the meta-modelling of the relation between the lift and drag coefficients and the parameters used to describe the airfoil shape. The meta-model is built with the open source library OpenTURNS that is dedicated to the treatment of uncertainty [1]. For the mesh construction on airfoil level, the three-dimensional finite element mesh generator GmSH is used [2]. The computations on airfoil level are conducted with the open source CFD code SU2 [3, 4].

Meta-Modelling on Airfoils' Level

Our aim is to construct a meta-model that will replace the CFD computation for a 2D airfoil. Hence, the surrogate model will have as inputs, the different design variables defining the geometry of the airfoil, and the physical parameters linked to the operating conditions and will return the different aerodynamic coefficients.

For the design variables defining the airfoil geometry the so-called PARSEC parameters have been proposed by Sobieczky [5] that define the airfoil shape (Fig. 1) by means of 11 parameters given in Table 1.

One of the drawbacks of the PARSEC parameterization is that, for certain combination of parameters values, a non-geometrical shape (e.g. bumps or crossings between upper and lower surface) can be obtained. For this reason, acceptable

S. Klostermann (✉) · R. Lebrun
Airbus Group Innovations, Hein-Saß-Weg 22, 21129 Hamburg, Germany
e-mail: soenke.klostermann@airbus.com

R. Lebrun
e-mail: regis.lebrun@airbus.com

C. Hirsch et al. (eds.), *Uncertainty Management for Robust Industrial Design in Aeronautics*, Notes on Numerical Fluid Mechanics and Multidisciplinary Design 140, https://doi.org/10.1007/978-3-319-77767-2_46

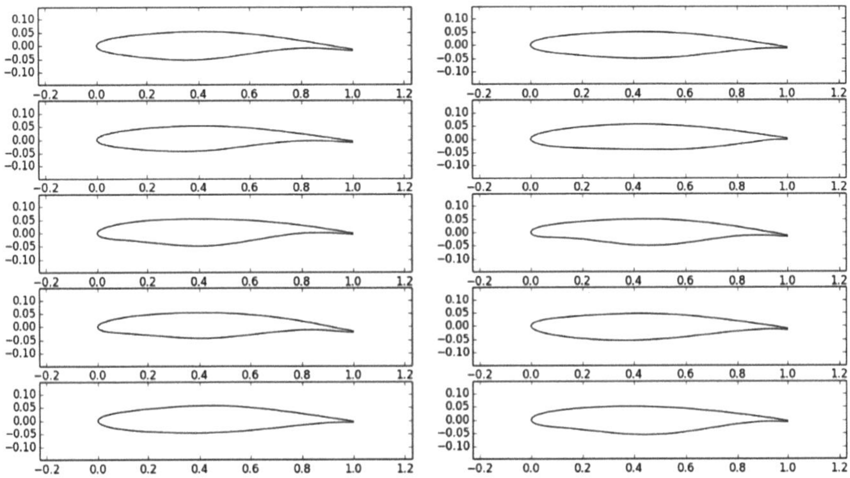

Fig. 1 Illustrations of sample realizations of the airfoils defined by the PARSEC parameters

Table 1 PARSEC-11 parameterization parameters

Parameter	Description	Minimum	Maximum
r_{LE}	Leading-edge radius of curvature	0.005	0.00938
x_u	Upper surface max. thickn. location	0.36	0.45
z_u	Upper surface maximum thickness	0.045	0.057
z_{xxu}	Upper surface curvature at x_u	−0.555	−0.26
x_l	Lower surface max. thickn. location	0.3	0.56
z_l	Lower surface maximum thickness	−0.058	−0.04
z_{xxl}	Lower surface curvature at x_l	0.28	1.1
z_{TE}	Trailing-edge position	−0.02	0.009
Δz_{TE}	Trailing-edge thickness	0.005	0.0082
α_{TE}	Trailing-edge direction angle	−0.13	−0.08
β_{TE}	Trailing-edge wedge angle	0.1	0.29

intervals for the maximum and minimum values according to [6] are given in Table 1.

For a further reduction of the complexity, some of the above parameters have been removed. Similar to the NACA 4-digit airfoils family, the leading-edge radius of curvature is expressed by a relation of the upper and lower surface maximum thickness. Furthermore, the wedge angle and the trailing-edge thickness have been set to constant values as they have very little effect on the actual aerodynamic performance. Including the two operational parameters Angle of Attack (AoA) and Mach number, we have in total 10 input parameters for the meta-model.

For the creation of the meta-model, we use a computation chain as illustrated in Fig. 2 that generates a mesh in GmSH based on the geometrical information

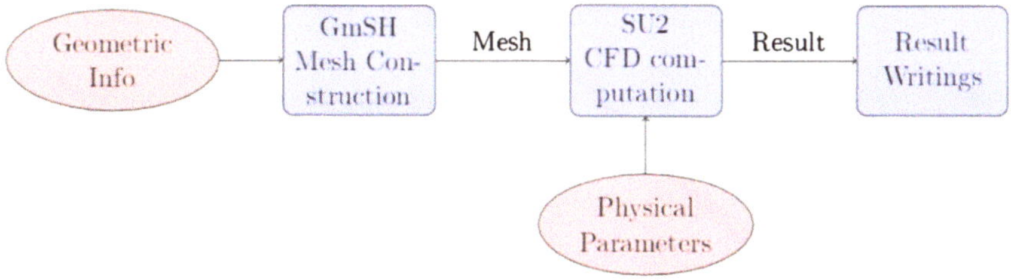

Fig. 2 Computation chain for the creation of the airfoil meta-model

derived from the PARSEC parameters. The computational mesh together with the operational conditions is used to calculate the CFD result inside SU2.

In the current tests, we try to recover a highly nonlinear function from R^{10} to R^2 based on a database of size 49,775, which is split into a learning database of size 44,797 (90% of the database) and a test database of size 4978 (the remaining 10% of the database) using the following strategy:

- Explore the polynomials in 10 variables up to degree 100 using a hyperbolic enumeration function that allows to favour low-dimension interactions upon the full interactions.
- The parametric search includes a search for the best total degree and the best hyperbolicity coefficient.
- For each coefficient (lift and drag), the best approximation is built using an internal cross-validation based on the corrected leave-one-out cross-validation, and the quality is assessed on the test database. The selection is done using the corrected Q2 coefficient (variance of the residual over the initial variance of the quantity to approximate).

The quality of the meta-model resulting from this approached can be assessed, where a good agreement of the meta-model with the model on the test database is shown (Fig. 3).

The meta-model for the lift coefficient is on the left, with a Q2 coefficient of 99.87% and a hyperbolicity exponent of $q = 0.547$, meaning that the coefficient is a function of the geometric and environmental parameters that includes significant interactions between the parameters. The total degree is 30, meaning that the function is significantly nonlinear.

The meta-model for the drag coefficient is on the left, with a Q2 coefficient of 99.96% and a hyperbolicity exponent of $q = 0.5$, meaning that the coefficient is a function of the geometric and environmental parameters that includes less interactions between the parameters than for the lift coefficient. The total degree is 35, meaning that the nonlinearity is more pronounced than for the lift coefficient.

In order to improve the performance of the SPCE algorithm, some improvements have been implemented in the OpenTURNS software (cf. Table 2) such as parallelization of the evaluation of the polynomial basis, OpenBLAS for a parallel

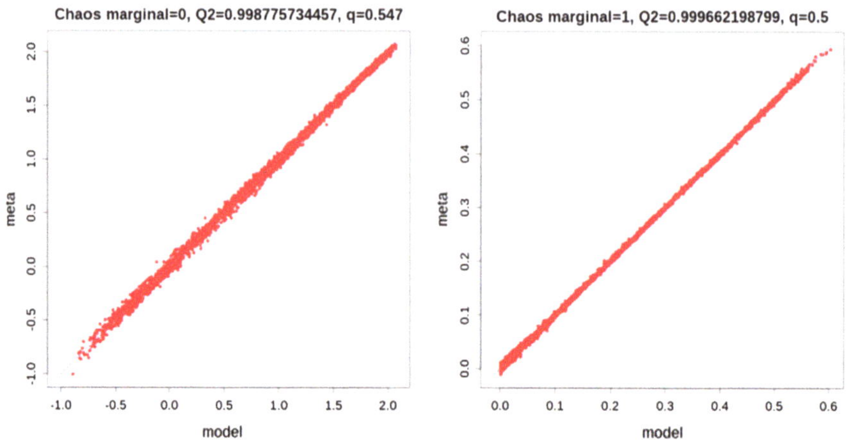

Fig. 3 Assessment of the meta-model quality for the lift coefficient on the left and the drag coefficient on the right

Table 2 Running time and speed-up with respect to the 1.2 version of OpenTURNS for the sparse chaos meta-modelling step

	OpenTURNS 1.2	OpenTURNS 1.3	OpenTURNS 1.4
Time (s)	43,040	19,089	2332
Speed-up	1	2.25	18.45

approach for the linear algebra part of the computations and a switch to the normal equation instead of the QR decomposition to solve the least-squares problems.

Optimization of Wing Configuration

We aim to optimize the wing configuration of a generic passenger transport aircraft sized similar to current generation short- to medium-range narrow-body aircraft, taking into account uncertainties on design parameters and operational conditions to account for the scatter of physical boundary conditions and system characteristics (test case IC-02). Since the aim is to do the optimization in early phases of the development process—where detailed information about the design do not yet exist—a coarse approximation of the system performance based on a simplified wing geometry is deemed sufficient, thus enabling for lower computational costs (Fig. 4).

Several options exist in order to simulate the flow over a three-dimensional wing to compute lift and drag for a given configuration. Three important options can be noted as compressible Euler equations, Navier–Stokes equations for Newtonian fluids, and Prandtl lifting line theory. The first two options necessitate a full three-dimensional simulation process making these options computationally costly.

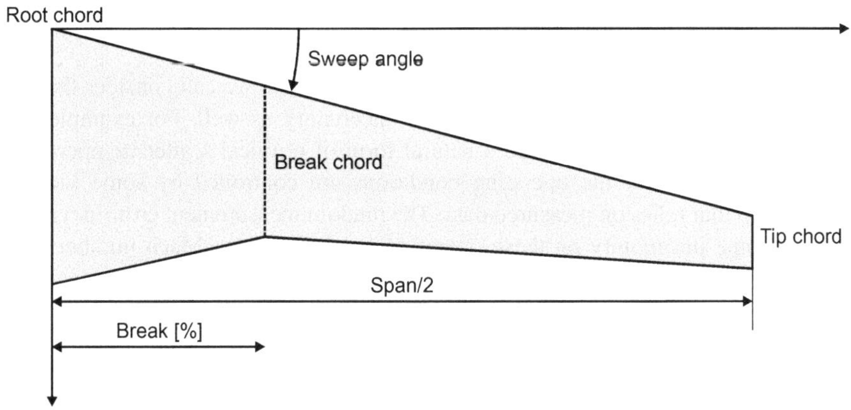

Fig. 4 Illustration of wing parameters for definition of simplified wing geometry

In addition, one needs a detailed three-dimensional design description. As we intend to do the optimization in early phases of the design process, this normally does not yet exist.

The Prandtl lifting line theory is a one-dimensional simulation process that relies on information from two-dimensional information on airfoil level (e.g. from 2D CFD computations or a surrogate model). It is a linearized fluid model that enables for an estimation of the lift and especially the lift-induced drag [7]. The so-called Differentiated Lifting Line Method (DLLM) as a nonlinear lifting line method has been developed that enables for the estimation of the wing's lift, induced drag, wave drag, friction drag, and viscous drag. In the nonlinear lifting line method, the state variables are the induced angles of attack over discrete sections of the wing.

The text user interface of OpenTURNS is accessible via the Python programming language. The implementation of the DLLM is based on pure Python, and therefore, interaction between both codes can be implemented straightforward. The DLLM code is called by the *sequential Monte Carlo algorithm* of OpenTURNS (cf. section BPG I. Best Practices for Input Uncertainty Identification and Quantification) as a "PythonFunction" within a "NumericalMathFunction" that overrides the latter and delivers the results from the specified Python code.

The DLLM code acts as surrogate model to deliver an approximate solution for the aerodynamic performance of the full three-dimensional wing configuration for a given set of input parameters. A more detailed description of interfacing with OpenTURNS can be found in the reference guide [1]. The DLLM code relies on *NumPy* [8] and *SciPy* [9] for some general numerical methods and some visualization tasks, so the whole computation chain is based on open source software.

We choose a probabilistic approach to model the uncertainties as this does not lead to an overestimation of worst-case scenarios which is typical for domain-based uncertainty formulations. We consider that the design variables may exhibit some kind of uncertainty which represents manufacturing tolerances or assembly effects, for example.

In addition to the uncertainties of the design variables, we can consider that the operating conditions exhibit some form of uncertainty as well. For example, the environmental conditions underlie a natural form of physical scatter (temperature, pressure, humidity) or the operating conditions are controlled by some kind of control loop that relies on measured data. The random measurement error inevitably leads to some uncertainty on these operating conditions (e.g. Mach number, altitude, AoA).

Within the limited capability of the geometrical description of our simplified model, the variation of the different sections' twist angle enables for a simplified modelling of manufacturing tolerances regarding the local aerodynamic properties on airfoil level. A more detailed consideration of random geometry by means of random fields (cf. book section "Uncertainties in Compressor and Aircraft Design") would require a fully detailed three-dimensional geometry representation which is out of the scope here.

Our optimization objective is to minimize the drag for a given configuration under cruise conditions as the cruise drag has a major influence on the overall fuel consumption of an aircraft. Table 3 shows the operating conditions for the optimization problem.

For the design variables of the optimization problem that fully describe the simplified geometry according to Fig. 5, the initial values are chosen so they approximately resemble the values of an Airbus A320 as a typical current generation short- to medium-range narrow-body aircraft. For each parameter, an optimization range with lower and upper bound has been assigned as given in Table 4.

For the sequential Monte Carlo algorithm, the parameter setting is given in Table 5. The solving of our optimization problem is done by the COBYLA algorithm [10].

The inequality constraints of the chance-constrained optimization problem are defined by two result quantities of the DLLM: a minimum value for the wings lift and a minimum value for the wings reference surface.

The minimum value for the lift is a typical optimization constraint for aerodynamic optimization problems as the optimized configuration must be able to produce enough lift for level flight during cruise conditions given a mass for the

Table 3 Operating conditions in cruise case

Factor	Quantity
Temperature (sea level)	15 °C
Pressure (sea level)	101,325 Pa
Humidity	0
Altitude	10,000 m
Mach number	0.8
Angle of attack	3.5°

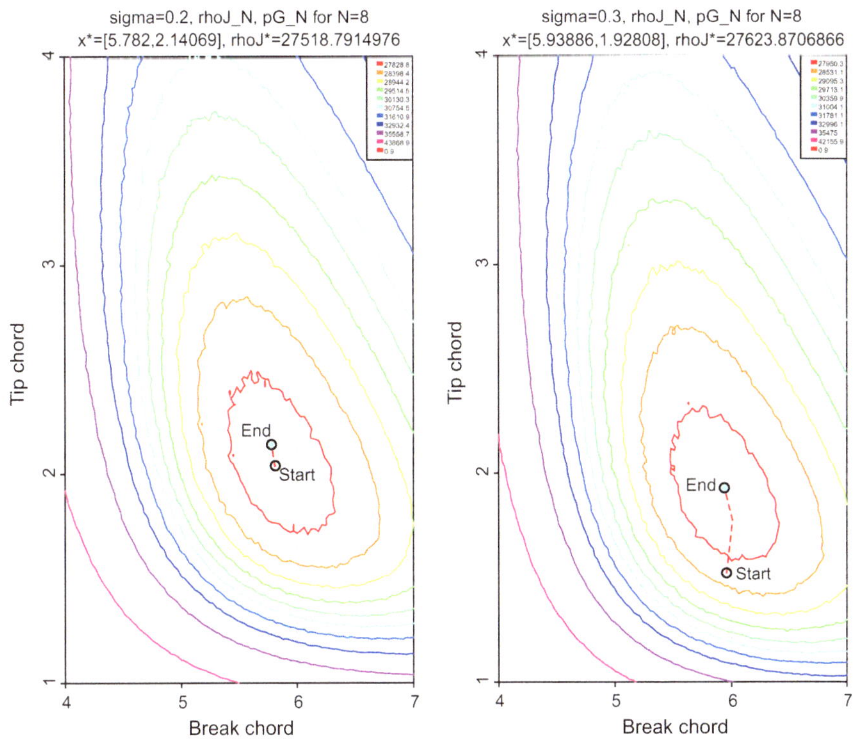

Fig. 5 Optimization result for chord length at break and tip for design variable uncertainty of $\sigma = 0.2$ (left) and $\sigma = 0.3$ (right)

Table 4 Design variables initial values with lower and upper bound for optimization

Design variable	Initial value	Lower bound	Upper bound
Span (m)	34.1	10	50
Sweep angle (°)	34	0	40
Break (%)	33	20	40
Root chord (m)	6.1	5	7
Break chord (m)	4.6	5	7
Tip chord (m)	1.5	1	4
Root height (m)	1.28	1	1.5
Break height (m)	0.97	0.8	1.2
Tip height (m)	0.33	0.2	0.5

Table 5 Parameter setting for sequential Monte Carlo algorithm

Parameter	Setting
N_{LHS}	1000
N_0	2
ε_0	5×10^{-3}
n_{local}	100
n_{robust}	5
α	0.9

aircraft. In our case, the minimum lift is set to $L_{\text{min}} = 850{,}000$ N resembling a typical take of mass for this type of aircraft.

Typically, the aircraft wing is used to store a significant amount of fuel. To account for this volume, we set a minimum value for the reference surface as second inequality constraint with $S_{\text{ref}} = 120\ \text{m}^2$. From a pure aerodynamic view, the resulting optimum will always be located at the maximum allowable span and sweep angle as this configuration will have the lowest drag and the highest lift over drag. From a structural point of view, this type of wing is prohibitive of course as we would not be able to design such a slender wing with a high sweep angle at affordable costs.

Within this study, we do not have access to a corresponding structural computation that we might use to get more sensible optimization results with these design variables. For the optimization—taking into account uncertainties of design variables and operating conditions—we restrict the design space to the three chord lengths and the break value along the wing's longitudinal axis. The other design variables are fixed to their initial values given in Table 4.

The uncertainties are modelled as noise vector $\boldsymbol{\xi} = (\xi_1, \xi_2, \xi_3, \xi_4)$ given by a joint probability density of normally distributed $\xi_1, \xi_2, \xi_3, \xi_4$ for a set of standard deviations $\sigma = [0.1, 0.2, 0.3, 0.4, 0.5]$ for the design variables to illustrate the effect of uncertainty on the location of the optimal design point.

As described above, we can assume that in addition to the uncertainties of the design variables the operating conditions exhibit some form of uncertainty as well. For this reason, we impose some normally distributed noise on the operating conditions. We assume a normally distributed noise vector that is added to the nominal Mach number with a standard deviation of $\sigma_{Mach} = 5 \times 10^{-4}$. For the temperature, we define $\sigma_{Temp} = 1\,°\text{C}$ and for the pressure $\sigma_{Pr} = 100\,\text{Pa}$. In order to take into account local manufacturing tolerances on airfoil level, we define a normally distributed noise vector for the twist angle of each airfoil with a standard deviation of $\sigma_{Twist} = 0.1°$. The effect of increased uncertainty for the design variables with constant uncertainty in the operating conditions is illustrated in Fig. 5.

For an increase in the design variables' uncertainty, the sample size used to model the uncertain parameters plays an important role (as can be seen by the longer optimization path for increasing sample size on the right) and should be taken into account when estimating the computational costs of a robust design optimization study.

Conclusion

In our opinion, the open architecture of open source software like for example OpenTURNS provides a feasible environment to incorporate RDO at early stages of the development into the design loop including different physics of interest without additional licensing issues enabling for easier collaboration between departments and risk share partners. This would enable for an incorporation of additional constraints to achieve more robust design under real-world scenarios as the proposed non-intrusive approach does not exhibit any methodical restriction regarding the design variables or optimization constraints.

References

1. www.openturns.org
2. Geuzaine, C., Remacle, J.-F.: Gmsh: a three-dimensional finite element mesh generator with built-in pre- and post-processing facilities. Int. J. Numer. Methods Eng. **79**(11), 1309–1331 (2009)
3. Economon, T.D., Palacios, F., Copeland, S.R., Lukaczyk, T.W., Alonso, J.J.: SU2: an open-source suite for multiphysics simulation and design. AIAA J. **54**(3), 828–846 (2016)
4. Palacios, F., et al.: Stanford University Unstructured (SU2): an open-source integrated computational environment for multi-physics simulation and design. In: AIAA Paper 2013-0287
5. Sobieczky, H.: Parametric airfoils and wings. In: Fujii, K., Dulikravich, G.S. (eds.) Notes on Numerical Fluid Mechanics, vol. 68, pp. 71–88 (1998)
6. Maginot, J.: Sensitivity Analysis for Multidisciplinary Design Optimisation. Ph.D. thesis, Cranfield University, School of Engineering (2008)
7. Anderson, J.D.: Fundamentals of Aerodynamics. McGraw-Hill, Boston (2001)
8. www.numpy.org
9. www.scipy.org
10. Powell, M.J.D.: A View of Algorithms for Optimization Without Derivatives. Technical Report DAMTP, Cambridge University (2007)

Uncertainty Quantification in an Engineering Design Software System

Dirk Wunsch, Rémy Nigro, Grégory Coussement and Charles Hirsch

Introduction

The UMRIDA project reached significant progress in the field of uncertainty quantification (UQ) and robust design optimization (RDO). This is demonstrated by the successful application of UQ and RDO to industrial challenges from the UMRIDA database and clearly shows that the technology readiness level (TRL) was significantly increased to a level of industrial applicability during the project. However, UQ is still a new field to many design engineers and the complexity and usability of UQ and RDO methods remain a challenge.

In order to reach a widespread application of these methods in industry, it seems mandatory to reduce the user-experienced complexity of the tools as much as possible and automatize the UQ toolchains. The UQ toolchain in FINE^{TM} [1] is designed towards this goal and described in the following, emphasizing aspects related to the user experience and usability.

Selection of Sparse Grid Level

As shown in chapter "Manufacturing Uncertainties in High-Pressure Compressors", multiple simultaneous uncertainties lead to an exponential increase in the number of simulations that need to be run. This is the so-called curse of dimensionality. To overcome this curse of dimensionality to a certain extent and make non-intrusive

D. Wunsch (✉) · R. Nigro · C. Hirsch
Numeca International, Chaussée de la Hulpe 189, Brussels, Belgium
e-mail: dirk.wunsch@numeca.be

R. Nigro · G. Coussement
University of Mons, Mons, Belgium

C. Hirsch et al. (eds.), *Uncertainty Management for Robust Industrial Design in Aeronautics*, Notes on Numerical Fluid Mechanics and Multidisciplinary Design 140, https://doi.org/10.1007/978-3-319-77767-2_47

Table 1 Comparison of collocation points (CFD runs) required in function of needed output

Number of uncertainties	Sufficient for mean and variance		Sufficient for PDF shape	
	Basic	Sparse grid	Basic	Sparse grid
1	3	3	5	5
2	9	5	25	17
3	27	7	125	31
4	81	9	625	49
5	243	11	3125	71
…	…	…	…	…
10	59,049	21	$9.7 * 10^6$	241

Table 2 Ratio of sparse over basic method for estimation of mean value and variance for several output quantities applied to the rotor 37

Sparse/basic	Mean	Variance
Mass flow rate (kg/s)	1.0001	0.9949
Total pressure ratio (–)	1.0000	1.0102
Isentropic efficiency (–)	1.0001	1.0047

collocation methods accessible for multiple uncertainties, a sparse grid technique was introduced. This method chooses the order of the underlying Lagrange interpolating polynomials in function of a "level". The experience gained on the application to several test cases showed that a level 1 sparse grid is sufficient, when only the mean value and its standard deviation (or variance) are required. This is notably the case for manufacturing uncertainties. A level 2 sparse grid is needed if the full probability distributions of output quantities are required. Table 1 shows the number of CFD runs required comparing the standard method with the introduced sparse grid approach. Table 2 proofs at the example of five simultaneous uncertainties applied to the rotor 37 (In this volume, chapter "UMRIDA Test Case Database with Prescribed Uncertainties"), that the sparse grid is sufficiently accurate with respect to the standard method. The selection of the sparse grid level is one of the two required inputs from user side for a UQ simulation in FINE™.

UQ Simulation Set-up for Both Simultaneous Operational and Geometrical Uncertainties

The identification and characterization of input uncertainties rely on expert knowledge. The identification of the range of variability of the operating conditions (boundary conditions) or the determination of the geometric variability must be provided by the software user. The result of experimental measurements is often provided in form of a mean value attached with error bars. This information can be

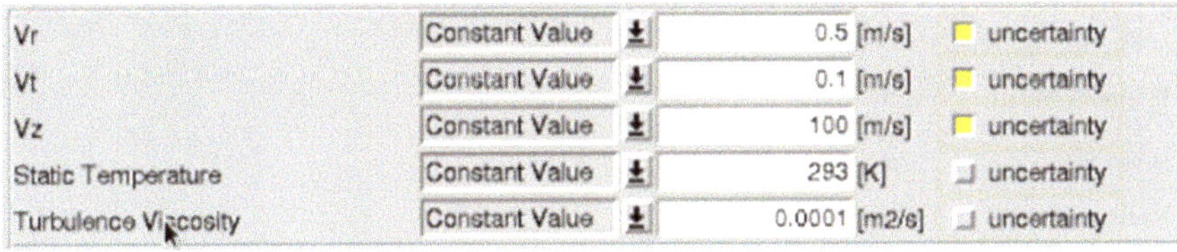

Fig. 1 Selection of operational uncertainties in FINE™

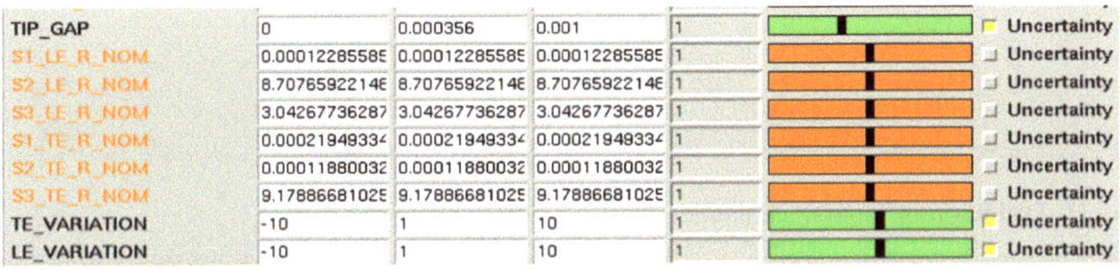

Fig. 2 Selection of geometrical uncertainties in FINE™

used to define the uncertain problem. As explained in chapter "Vision, Objectives and Research Activities" of this volume, the shape of the input distribution matters and has an influence on the solution.

The UQ software should ease the input of limited measurement data. If for example only the minimum, maximum and mean values are known, a beta-PDF can be defined on these grounds, and this process is automatized in FINE™. In addition, FINE™ provides the possibility to enter different input PDF shapes for every boundary condition or geometrical parameter selected. Figure 1 shows, on the example of operational conditions, the selection process of boundary conditions for UQ simulations. A simple check is needed. Figure 2 shows the equivalent selection procedure for geometrical uncertainties based on a parametric model.

Independent of the choice of operational or geometrical uncertainties, the only additional user input required is the definition of the distribution itself, as seen in Fig. 3. A selection of predefined shapes such Gaussian or beta distributions is available. In addition, user-defined distribution in form of profiles can be given as input.

Once all input distributions are defined, the collocation points are generated and the individual computations accounting for simultaneous operational and geometrical uncertainties are generated automatically. This means that the boundary conditions and geometrical parameter values are modified jointly, the geometries are changed and re-meshed, the simulations are run and the UQ post-processing is performed. All of this without any further user action in a fully automated chain.

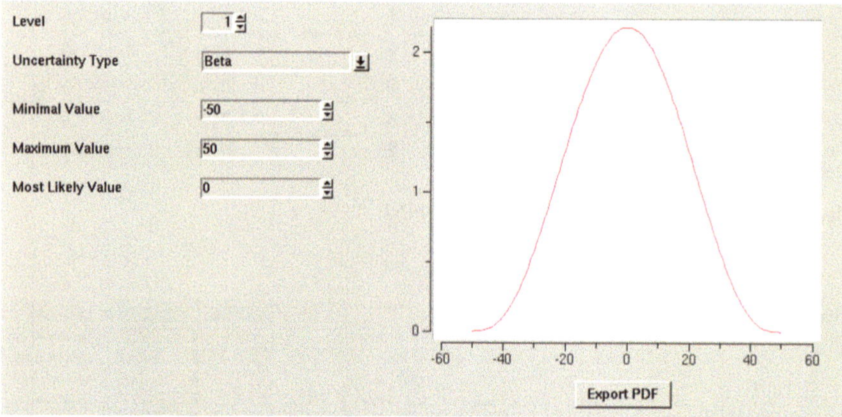

Fig. 3 Definition of input uncertainties

UQ Simulation Set-up for Manufacturing Uncertainties

In the case of manufacturing uncertainties, which are characterized by correlations between the geometrical parameters or surface points, a covariance matrix needs to be provided by the user. Figure 4 shows the selection of the surface that is to be considered as a random field (uncertain surface) and Fig. 5 the definition the covariance matrix. A principal component analysis is performed and the number of modes that are retained are automatically calculated based on the reconstruction accuracy defined by the user as shown in Fig. 5.

It is recommended to retain at least 99% of the surface reconstruction accuracy, since one per cent error in the surface reconstruction translates into larger errors on the prediction accuracy of the simulation output quantities, as described in chapter "Non-intrusive Probabilistic Collocation Method for Operational, Geometrical, and Manufacturing Uncertainties in Engineering Practice" of this volume. It was shown that due to the small variability resulting from the manufacturing process (which

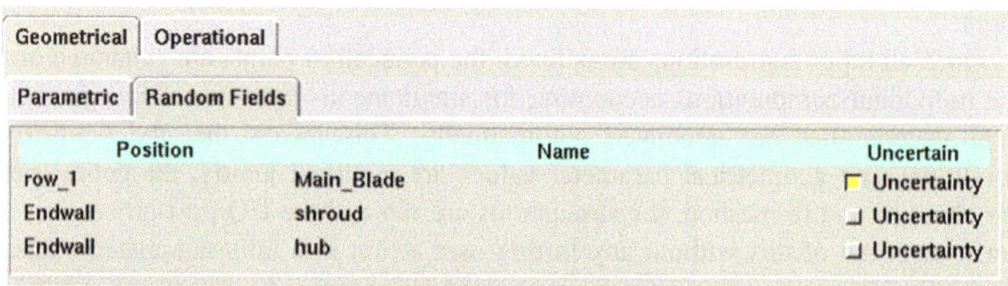

Fig. 4 Selection of surfaces to be considered as random fields

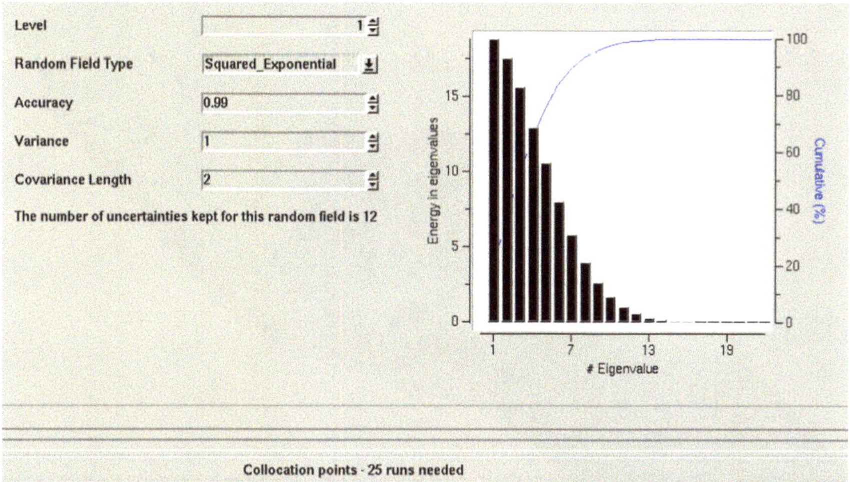

Fig. 5 Set-up page for manufacturing uncertainties

necessarily lies within the defined tolerances) a level 1 sparse grid is usually sufficient for the simulation of manufacturing uncertainties.

Once the correlations and reconstruction accuracy are defined, the process of simulation set-up, geometry modification, simulation execution and post-processing is automated in the same way as for combined operational and geometrical uncertainties.

Application of the Automatized UQ Simulation Chain

Every UQ simulation consists of a given number of independent deterministic simulations, which are set up as described above. For each of these simulations, the correct geometry is built and meshed by keeping the number of cells and the global mesh topology constant. Figures 6 and 7 illustrate, on the example of a varying leading edge blade angle, the automatic mesh generation process. Figure 6 shows a global view of the geometry while Fig. 7(left) shows the mesh generated with the smallest leading edge blade angle and Fig. 7(right) the mesh generated with the largest leading edge blade angle.

The UQ post-processing is automated, such that mean value and variance as well as higher moments of output quantities are calculated without user interference for all output quantities defined in the simulation set-up. This allows, if based on the mean and variance, to plot results with $\pm\sigma$ (standard deviation) around the mean value. The $\pm\sigma$ interval contains 68.3% of the values for a Gaussian distribution, and

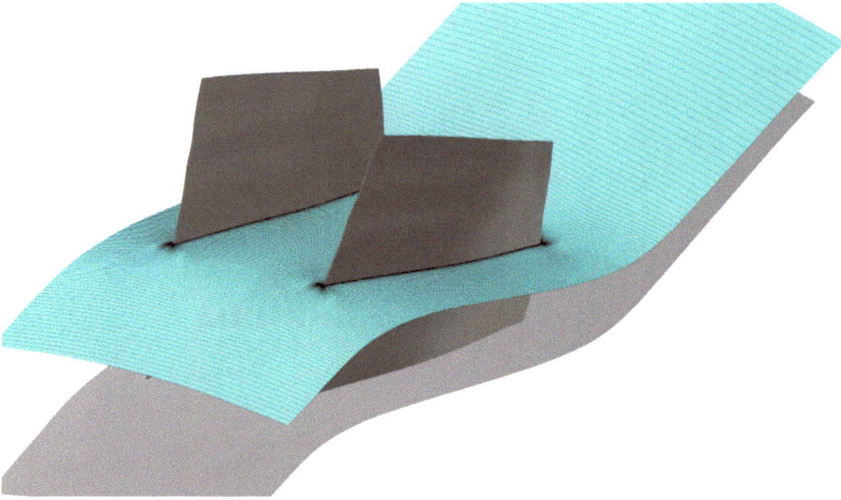

Fig. 6 Global geometry view indicating constant radius cut with mesh visualization. Zoom in Fig. 7

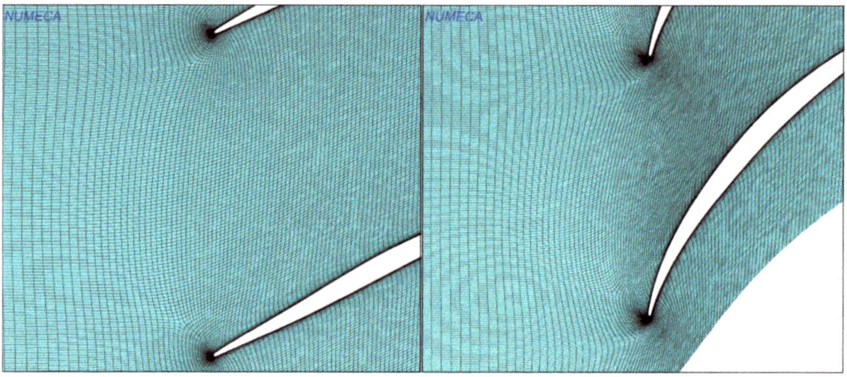

Fig. 7 (Left) Geometry and automatically generated mesh with smallest leading edge blade angle. (Right) Geometry and automatically generated mesh with largest leading edge blade angle

it therefore provides a range of confidence for the CFD results as shown in Fig. 8. Figure 9 shows the automatically reconstructed PDF of the absolute total pressure ratio based on the first for statistical output moments computed. Figure 10 shows scaled sensitivity derivatives, which indicate how a quantity varies with a given uncertainty.

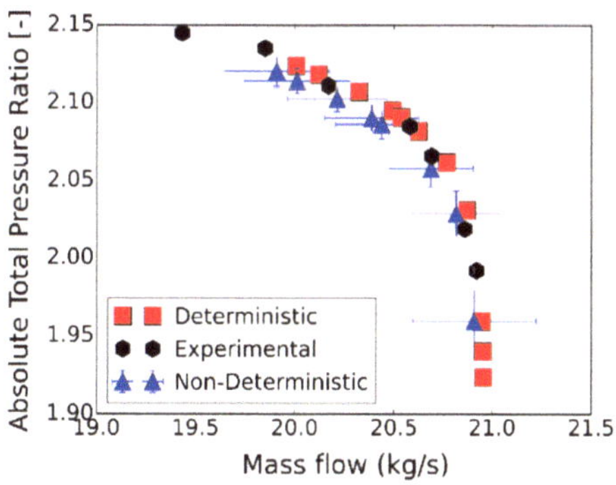

Fig. 8 Absolute total pressure ratio over mass flow comparing deterministic results with results from UQ simulation

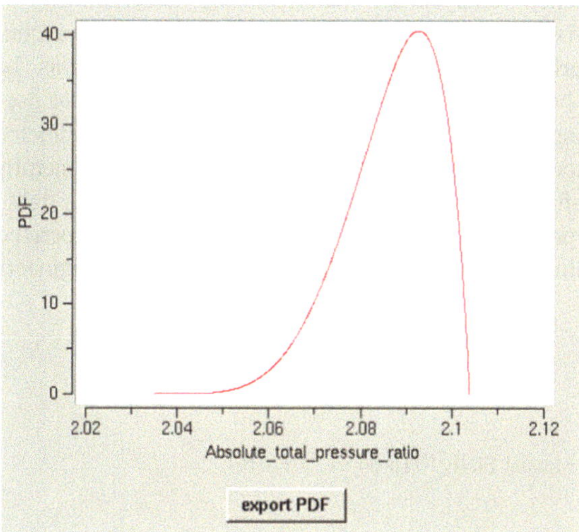

Fig. 9 Reconstructed PDF of absolute total pressure ratio

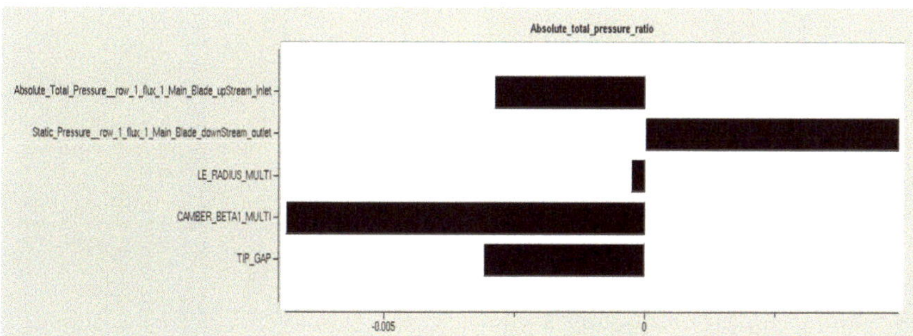

Fig. 10 Scaled sensitivity derivatives of output total pressure ratio for different uncertainties

Conclusions

A fully automated UQ environment for engineering practice is developed in FINE™. It is shown how to account in a very user-friendly way for simultaneous operational, geometrical and manufacturing uncertainties in an engineering design software. The user-friendliness is achieved by fully automating the process for modifying geometry, meshing, CFD and UQ post-processing. The curse of dimensionality is overcome to a certain extent by the use of sparse grids. A level 1 sparse grid is appropriate for estimation of mean value and standard deviation, while a level 2 sparse grid is needed for an appropriate estimation of the PDF shape of output quantities. The use of scaled sensitivities, which are part of the automated post-processing, helps to understand the influence of individual uncertainties on the solution. The reliability of a design can be assessed by building a domain of confidence through PDF reconstruction. The same methodology is applicable to the principal component analysis of manufacturing uncertainties, where only the covariance matrix for the uncertain surfaces and the required reconstruction accuracy of the surface are required. Alternatively, correlations between design parameters can be considered. This results in eigenmodes that need to be propagated. The propagation of these modes and corresponding geometry modifications are equally automated. For manufacturing uncertainties, generally, a level one sparse grid is sufficient since the variability usually lies within tight manufacturing tolerances. The proposed software system reduces the user-experienced complexity of UQ to a minimum and makes it accessible in the daily engineering practice.

Reference

1. Numeca: User Manual FINE™/Turbo v11.1 (2016)

Use of Automatic Differentiation Tools at the Example of TAPENADE

Alain Dervieux

This section recalls what is automatic differentiation (AD) and explains how the AD tool TAPENADE [6] is applied on the routines computing a functional (subject to a state equation) in order to get new routines computing the second derivative of the functional with respect to a specified parameter.

Automatic Differentiation

Given a program `P` computing a function $\boldsymbol{\Phi}$

$$\begin{aligned} \boldsymbol{\Phi} : \mathbb{R}^m &\to \mathbb{R}^n \\ u &\mapsto v \end{aligned} \tag{1}$$

we want to build a program that computes the derivatives of $\boldsymbol{\Phi}$. Specifically, this program should compute the derivatives of the *dependent variables* (i.e., some variables in v) with respect to the *independent variables* (i.e., some variables in u). The program `P` is a *sequence* of instructions $\mathtt{I}_1, \mathtt{I}_2, \ldots, \mathtt{I}_p$ that can be identified as a composition of functions, where each simple instructions $\mathtt{I}_k$ is a function $\phi_k : \mathbb{R}^{q_{k-1}} \to \mathbb{R}^{q_k}$. Thus we see $\mathtt{P} : \{\mathtt{I}_1, \mathtt{I}_2, \ldots, \mathtt{I}_p\}$ as

$$u \mapsto v = \boldsymbol{\Phi}(u) = \phi_p \circ \phi_{p-1} \circ \cdots \circ \phi_1(u)$$

and the derivative is obtained using the chain rule

A. Dervieux (✉)
Ecuador, INRIA, 2002 route des lucioles, 06902 Sophia-Antipolis, France
e-mail: alain.dervieux@inria.fr

C. Hirsch et al. (eds.), *Uncertainty Management for Robust Industrial Design in Aeronautics*, Notes on Numerical Fluid Mechanics and Multidisciplinary Design 140, https://doi.org/10.1007/978-3-319-77767-2_48

$$\frac{\partial \Phi}{\partial u}(u) = (\phi'_p \circ \phi_{p-1} \circ \cdots \circ \phi_1(u))\cdot \\ \cdot (\phi'_{p-1} \circ \cdots \circ \phi_1(u))\cdot \\ \cdot \ \cdots \ \cdot \\ \cdot \phi'_1(u)$$

or in a more compact way

$$\frac{\partial \Phi}{\partial u}(u) = \phi'_p(w_{p-1})\phi'_{p-1}(w_{p-2}) \cdots \phi'_1(w_0)$$

where $w_0 = u$ and $w_k = \phi_k(w_{k-1})$. We note here that if we are not interested in the full Jacobian (that is expensive because its computation would involve matrix-by-matrix multiplications), we can compute the derivatives along a given direction (Gâteaux differentiation) using matrix-by-vector multiplications.

The **Tangent mode** applied to a routine computing Φ produces another routine computing, from u and from an arbitrary direction $\dot{u}$ (of same dimension as u), the derivative in direction $\dot{u}$:

$$u, \dot{u} \mapsto \frac{\partial \Phi}{\partial u}(u)\dot{u} = \phi'_p(w_{p-1})\phi'_{p-1}(w_{p-2}) \cdots \phi'_1(w_0)\dot{u} \tag{2}$$

In order to use matrix-by-vector multiplications, the obtained routine computes the derivatives in the same order as the initial routine computes the original values (right to left in Eq. (2)). We note that tangent mode delivers only one real number if Φ is a real valued functional.

The **Reverse mode** when applied to the previous initial routine computing Φ produces a routine which computes, from u and from an arbitrary direction $\bar{v}$ (of same dimension as v), the following product of same dimension as u:

$$u, \bar{v} \mapsto \left(\frac{\partial \Phi}{\partial u}(u)\right)^* \bar{v} = \phi'^*_1(w_0)\phi'^*_{p-1}(w_{p-2}) \cdots \phi'^*_p(w_{p-1})\bar{v}. \tag{3}$$

To use matrix-by-vector multiplications, we need to store (or to recompute) the intermediate results w_k and using them in reverse order. For a functional of n variables, the routine produced by the reverse mode delivers n numbers and to compute the *gradient* of the functional, it can be (at least theoretically) n times more efficient than the tangent mode.

To better understand how AD works, we introduce a special notation (similar in some way to the notation in [5]) that will help us for a correct implementation. First of all, we assume that the function

$$f\colon \mathbb{R}^{n_x} \times \mathbb{R}^{n_y} \to \mathbb{R}^m \\ (x, y) \mapsto f(x, y)$$

is implemented by the routine `func(f,x,y)`, where `x` and `y` are the variables that should contain the $x \in \mathbb{R}^{n_x}$ and $y \in \mathbb{R}^{n_y}$ respectively, and `f` is the variables where will be stored the result $f(x, y) \in \mathbb{R}^m$ (i.e., `func(f,x,y)` is a "subroutine" in FORTRAN language or a "void function" in C/C++ terminology).

For differentiation, we need to know which are the independent variables (that are input parameters of the function) and which are the dependent variables (output parameters). For the simplicity of our explanation, we assume that no variable is at the same time an input and an output. In our notation, we use an arrow over any parameter containing the independent variable and an arrow under any parameter containing the dependent variable, thus, if we want to specify that the routine `func(f,x,y)` (implementing the function $f(x, y)$) has `f` as output and `x`, `y` as input we write

$$\texttt{func}(\underset{\downarrow}{\texttt{f}},\overset{\downarrow}{\texttt{x}},\overset{\downarrow}{\texttt{y}}) \tag{4}$$

Furthermore, if we would specify that the function is evaluated with some specific values $x = x_0$ and $y = y_0$ we shall write

$$\texttt{func}(\underset{f(x_0,y_0)}{\texttt{f}},\overset{x_0}{\texttt{x}},\overset{y_0}{\texttt{y}})$$

where the value written over a parameter means "value taken by the input variable" and the value under a parameter means "value stored in the output variable."

The last step is to specify which mode, tangent or reverse, we use for differentiation and study the output generated by the AD tool (TAPENADE in our case) that performs the differentiation required. First of all, we must keep in account that for each independent variable with respect to which we differentiate, we shall have a correspondent dual variable that will be of the same kind (input) if we use tangent mode, and of the opposite kind (output) if we use the reverse mode. The same thing happens for the dependent (output) variables: the dual variables will be output variables in the case of tangent mode and input variables for reverse mode.

Using the notation previously introduced, tangent mode differentiation for the case (4) respect all the independent variables `x` and `y` gives us

```
func_d(f,fd,x,xd,y,yd)
```

where the new parameters `fd` (output), `xd` (input) and `yd` (input) are the dual variables of `f`, `x`, and `y` (the `d` character after the variables and function name means "*dot*"). If we give at the new input parameters `xd` and `yd` the values $\dot{x} \in \mathbb{R}^{n_x}$ and $\dot{y} \in \mathbb{R}^{n_y}$, respectively, we obtain

$$\texttt{func\_d}(\underset{f(x_0,y_0)}{\texttt{f}},\underset{\frac{\partial f}{\partial x}\dot{x}+\frac{\partial f}{\partial y}\dot{y}}{\texttt{fd}},\overset{x_0}{\texttt{x}},\overset{\dot{x}}{\texttt{xd}},\overset{y_0}{\texttt{y}},\overset{\dot{y}}{\texttt{yd}})$$

where the output parameter `fd` contains the value $\dot{f} = \frac{\partial f}{\partial x}\dot{x} + \frac{\partial f}{\partial y}\dot{y}$ with $\dot{f} \in \mathbb{R}^m$ (and the derivatives $\frac{\partial f}{\partial x}$ and $\frac{\partial f}{\partial y}$ are both evaluated at (x_0, y_0)).

Reverse mode differentiation of (4) gives us

```
func_b(f,fb,x,xb,y,yb)
```

where now we have a new input parameter `fb` and two new output parameters `xb` and `yb` (the `b` character after the variables and function name means "*bar*"). Storing the value $\bar{f} \in \mathbb{R}^m$ in the parameter `fb`, we obtain

$$\texttt{func\_b}(\underbrace{\texttt{f}}_{f(x_0,y_0)}, \overset{\bar{f}}{\texttt{fb}}, \overset{x_0}{\texttt{x}}, \underbrace{\texttt{xb}}_{\left(\frac{\partial f}{\partial x}\right)^* \bar{f}}, \overset{y_0}{\texttt{y}}, \underbrace{\texttt{yb}}_{\left(\frac{\partial f}{\partial y}\right)^* \bar{f}}),$$

i.e., the output parameters `xb` and `yb` will contain the values $\bar{x} = \left(\frac{\partial f}{\partial x}\right)^* \bar{f}$ and $\bar{y} = \left(\frac{\partial f}{\partial y}\right)^* \bar{f}$, respectively, (with $\bar{x} \in \mathbb{R}^{n_x}, \bar{x} \in \mathbb{R}^{n_y}$ and the derivatives $\frac{\partial f}{\partial x}, \frac{\partial f}{\partial y}$ both evaluated at (x_0, y_0)).

For the case where we are differentiating with respect to only some independent variables (e.g., `x`), we obtain for tangent mode

$$\texttt{func\_dx\_d}(\underbrace{\texttt{f}}_{f(x_0,y_0)}, \underbrace{\texttt{fd}}_{\frac{\partial f}{\partial x}\dot{x}}, \overset{x_0}{\texttt{x}}, \overset{\dot{x}}{\texttt{xd}}, \overset{y_0}{\texttt{y}}) \tag{5}$$

and for reverse mode,

$$\texttt{func\_dx\_b}(\underbrace{\texttt{f}}_{f(x_0,y_0)}, \overset{\bar{f}}{\texttt{fb}}, \overset{x_0}{\texttt{x}}, \underbrace{\texttt{xb}}_{\left(\frac{\partial f}{\partial x}\right)^* v}, \overset{y_0}{\texttt{y}}) \tag{6}$$

where the suffix `_dx` means that the differentiation is performed with respect to only the independent variable `x`.

It interesting to note here that if we need to solve a linear system $A\xi = b$ where $A = \frac{\partial f}{\partial x}$ or $A = \left(\frac{\partial f}{\partial x}\right)^*$ and b is a known vector, we can do it *without* storing the matrix A. This can be done by using the so-called matrix-free iterative algorithms (like GMRES, see e.g. [11]) that do not need to know the matrix A, but only its effect on a given vector ξ_i, i.e., the matrix-by-vector multiplication $A\xi_i$. With the analysis done to study the result of AD on a routine, we could perform this multiplication using the derivatives (5) or (6) obtained by AD. To be more clear, given a general GMRES routine, we need to replace any matrix-by-vector occurrence of the kind $A\xi_i$ with the corresponding routine (5) or (6), depending on the definition of A: we shall use the tangent mode derivative (5) if $A = \frac{\partial f}{\partial x}$ and the reverse mode derivative (6) if $A = \left(\frac{\partial f}{\partial x}\right)^*$.

Second Differentiation with Tangent-on-tangent Option

This method was initially investigated by Ghate and Giles [12] along with various other algorithms, but the publication does not go into the implementation details for a generic fluid dynamic code. Here, we present the mathematical background behind the idea and the efficient AD implementation of [4] but with a different analysis of the computational cost.

Theory

We are interested by obtaining the second derivatives of a functional j depending of $\gamma \in \mathbb{R}^n$ and expressed in terms of a state $W \in \mathbb{R}^N$ as follows:

$$\begin{cases} \psi(\gamma) = \Psi(\gamma, W(\gamma)) = 0 \\ j(\gamma) = J(\gamma, W(\gamma)) \, . \end{cases} \tag{7}$$

We apply twice the differentiation with respect to the variable γ_k obtaining

$$\frac{d^2 j}{d\gamma_i d\gamma_k} = D^2_{i,k} J + \frac{\partial J}{\partial W} \frac{d^2 W}{d\gamma_i d\gamma_k} \tag{8}$$

where

$$D^2_{i,k} J = \frac{\partial}{\partial \gamma}\left(\frac{\partial J}{\partial \gamma} e_i\right) e_k + \frac{\partial}{\partial W}\left(\frac{\partial J}{\partial \gamma} e_i\right)\frac{dW}{d\gamma_k} + \frac{\partial}{\partial W}\left(\frac{\partial J}{\partial \gamma} e_k\right)\frac{dW}{d\gamma_i} + \frac{\partial}{\partial W}\left(\frac{\partial J}{\partial W}\frac{dW}{d\gamma_i}\right)\frac{dW}{d\gamma_k}.$$

Similarly, we get

$$D^2_{i,k} \Psi + \frac{\partial \Psi}{\partial W} \frac{d^2 W}{d\gamma_i d\gamma_k} = 0 \tag{9}$$

where

$$D^2_{i,k} \Psi = \frac{\partial}{\partial \gamma}\left(\frac{\partial \Psi}{\partial \gamma} e_i\right) e_k + \frac{\partial}{\partial W}\left(\frac{\partial \Psi}{\partial \gamma} e_i\right)\frac{dW}{d\gamma_k} + \frac{\partial}{\partial W}\left(\frac{\partial \Psi}{\partial \gamma} e_k\right)\frac{dW}{d\gamma_i} + \frac{\partial}{\partial W}\left(\frac{\partial \Psi}{\partial W}\frac{dW}{d\gamma_i}\right)\frac{dW}{d\gamma_k}.$$

Substituting the second derivatives of the state with respect to the control variables $\frac{d^2 W}{d\gamma_i d\gamma_k}$ in Eq. (8) from Eq. (9) we get

$$\begin{aligned} \frac{d^2 j}{d\gamma_i d\gamma_k} &= D^2_{i,k} J - \frac{\partial J}{\partial W}\left(\frac{\partial \Psi}{\partial W}\right)^{-1} D^2_{i,k} \Psi \\ &= D^2_{i,k} J - \Pi_0^* D^2_{i,k} \Psi \end{aligned} \tag{10}$$

where Π_0 is the solution of the adjoint system evaluated at the point $(\gamma, W(\gamma))$ solution of the state equation $\Psi(\gamma, W) = 0$.

The n derivatives $\dfrac{dW}{d\gamma_i}$ should be computed (and stored) using tangent mode differentiation of the nonlinear solver algorithm, and each derivatives costs $n_{\text{iter},T}\alpha_T$. If we need the full Hessian matrix, we have to evaluate the quantity (10) $n(n+1)/2$ times, i.e., we have to evaluate the terms $D^2_{i,k}\Psi$ and $D^2_{i,k}J$ for $i = 1, \ldots, n$ and $j = i, \ldots, n$ due to the symmetry of the Hessian, and each evaluation of $D^2_{i,k}\Psi$ costs α_T^2 (the evaluation of $D^2_{i,k}J$ is negligible with respect to $D^2_{i,k}\Psi$). Therefore, the full Hessian costs $n\alpha_T[n_{\text{iter},T} + (n+1)\alpha_T/2]$. With similar arguments, if we want to compute only the diagonal part of the Hessian, the cost is $n\alpha_T[n_{\text{iter},T} + \alpha_T]$.

Implementation Details

Let us suppose that the subroutine computing the state residual equation $\Psi(\gamma, W)$ is `state_residuals(psi,gamma,w)`, where the input variables are `gamma` and `w`, and the output variable is `psi`.

```
                    ↓     ↓
state_residuals(psi,gamma,w)
                 ↓
```

If we perform a differentiation in tangent mode with respect to the input variables `gamma` and `w`, we have

```
                             ↓      ↓     ↓  ↓
state_residuals_d(psi,psid,gamma,gammad,w,wd)
                   ↓   ↓
```

where `gammad` $= \dot{\gamma}$, `wd` $= \dot{W}$ are input variables and `psid` $= \left(\frac{\partial \Psi}{\partial \gamma}\right)\dot{\gamma} + \left(\frac{\partial \Psi}{\partial W}\right)\dot{W}$ is an output variable.

Now we differentiate in tangent mode the output variable `psid` with respect to `gamma` and `w`, obtaining

```
                                      ↓       ↓       ↓    ↓  ↓   ↓
state_residuals_d_d(psi,psid,psidd,gamma,gammad0,gammad,w,wd0,wd)      (11)
                     ↓   ↓    ↓
```

where `gammad0` $= \dot{\gamma}_0$, `wd0` $= \dot{W}_0$ are input variables and

$$\texttt{psidd} = \ddot{\Psi} = \frac{\partial}{\partial \gamma}\left(\frac{\partial \Psi}{\partial \gamma}\dot{\gamma}\right)\dot{\gamma}_0 + \frac{\partial}{\partial W}\left(\frac{\partial \Psi}{\partial \gamma}\dot{\gamma}\right)\dot{W}_0 + \frac{\partial}{\partial W}\left(\frac{\partial \Psi}{\partial \gamma}\dot{\gamma}_0\right)\dot{W} + \frac{\partial}{\partial W}\left(\frac{\partial \Psi}{\partial W}\dot{W}\right)\dot{W}_0$$

is an output variable.

In order to evaluate the term $D^2_{i,k}\Psi$, we must call the routine (11) with the right arguments, that is:

$$\texttt{state\_residuals\_d\_d}(\underset{\Psi}{\texttt{psi}}, \underset{\dot{\Psi}}{\texttt{psid}}, \underset{\ddot{\Psi}}{\texttt{psidd}}, \overset{\gamma}{\texttt{gamma}}, \overset{e_k}{\texttt{gammad0}}, \overset{e_i}{\texttt{gammad}}, \overset{W}{\texttt{w}}, \overset{\frac{dW}{d\gamma_k}}{\texttt{wd0}}, \overset{\frac{dW}{d\gamma_i}}{\texttt{wd}})$$

where

$$\texttt{psidd} = \ddot{\Psi} = D^2_{i,k}\Psi$$

and where e_i (e_k) is the usual vectors of the canonical basis with 1 at the i-th (k-th) component and zero otherwise. The derivative of the state variables respect to the control $\frac{dW}{d\gamma_i}$ is obtained as solution of the linear system $\left(\frac{\partial\Psi}{\partial\gamma_i}\right) + \left(\frac{\partial\Psi}{\partial W}\right)\frac{dW}{d\gamma_i} = 0$.

The same previous argument applies to the evaluation of the term $D^2_{i,k}J$. In this case, we perform a tangent-on-tangent derivative of the routine

$$\texttt{functional}(\underset{\downarrow}{\texttt{J}}, \overset{\downarrow}{\texttt{gamma}}, \overset{\downarrow}{\texttt{w}})$$

and we get

$$\texttt{functional\_d\_d}(\underset{J}{\texttt{j}}, \underset{\dot{J}}{\texttt{jd}}, \underset{D^2_{i,k}J}{\texttt{jdd}}, \overset{\gamma}{\texttt{gamma}}, \overset{e_k}{\texttt{gammad0}}, \overset{e_i}{\texttt{gammad}}, \overset{W}{\texttt{w}}, \overset{\frac{dW}{d\gamma_k}}{\texttt{wd0}}, \overset{\frac{dW}{d\gamma_i}}{\texttt{wd}})$$

where the evaluation of $D^2_{i,k}J$ will be in the variable `jdd`.

It is useful to note that the n derivatives of the state with respect to the control $\frac{dW}{d\gamma_i}$ must be evaluated and stored *before* any evaluation of $D^2_{i,k}J$ or $D^2_{i,k}\Psi$. If the number of the state variables N and/or the number of the control variable n are high, the previous strategy could be not applicable. One possible solution for this problem could be to store the vectors $\frac{dW}{d\gamma_i}$ into the hard disk instead of keeping them into the RAM, but this strategy will have some negative impact on the performance of the computation due to the I/O overhead.

Numerical Experiments

The interest of this approach is briefly illustrated by the building of the drag response surface for a particular wing shape of business aircraft (courtesy of Piaggio Aero Ind.), for a transonic regime. See the shape and the mesh in Fig. 1. The nominal operational conditions are defined by the free-stream Mach number $M_\infty = 0.83$ and the incidence $\alpha = 2°$. We suppose that only these two quantities are subject to

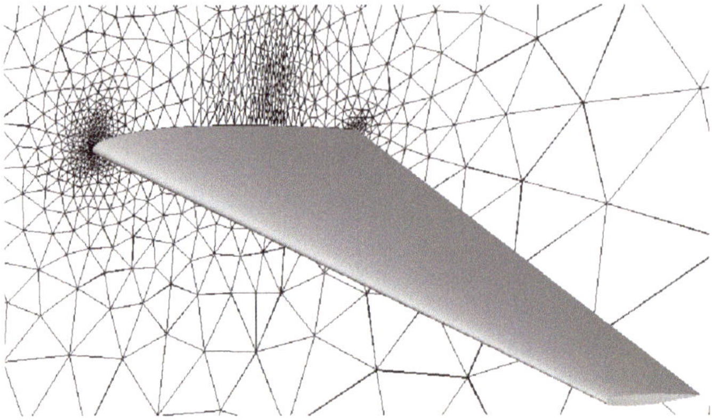

Fig. 1 Wing shape and mesh in the symmetry plane

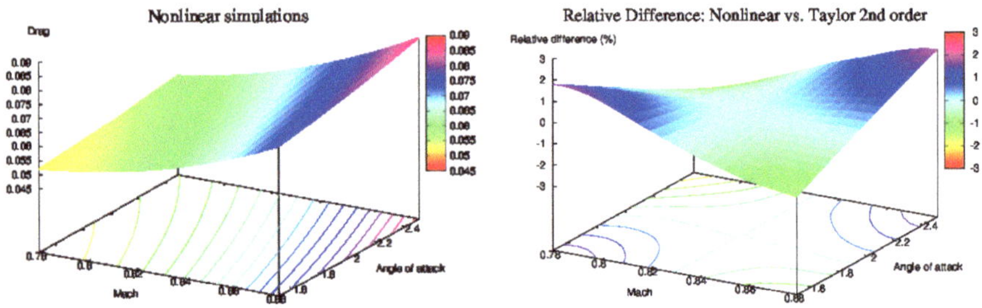

Fig. 2 Drag coefficient versus Mach number and angle of attack (first-order spatial accuracy) for the transonic wing: nonlinear simulations (first image), percentage relative difference between the nonlinear simulations and the second-order Taylor approximation (second image). For the top plot, we have solved 21×21 nonlinear systems $\Psi = 0$

random fluctuations. For simplicity, we assume that their PDF is Gaussian with given mean and variance. The mean values correspond to the nominal values. The section of the initial wing shape corresponds to the NACA 0012 airfoil.

For the present work, due to the fact that we consider only two uncertain variables, we used a ToT approach for the Hessian evaluation. The accuracy of the second-order response surface obtained with the differentiated software is not different from the one obtained in other works, such as those of Ghate and Giles [3] who, by the way, also used TAPENADE, but on a different CFD software and for a different CFD case. We illustrate the accuracy obtained with our example in Fig. 2. The direct evaluation required 21×21 nonlinear simulations. The second-order approximation required only one nonlinear state equation $\Psi = 0$ plus 4 linear systems using ToT. Relative error is less than 2%. Using only first derivatives produces an error of 16%. Let us mention that this method compares also well with Kriging methods as was demonstrated in the comparison paper [8] (Fig. 3).

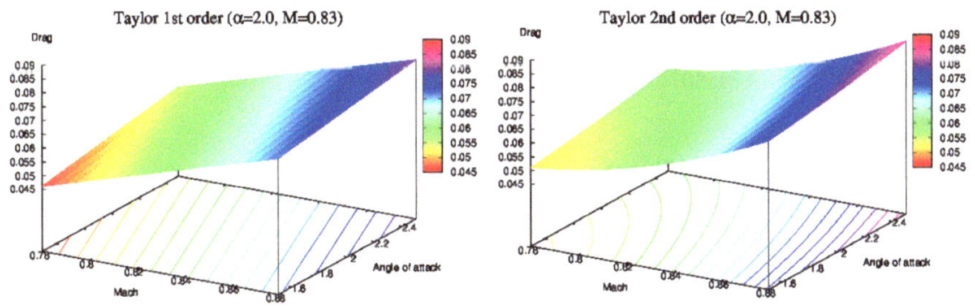

Fig. 3 First-order (first image) and second-order (second image) Taylor approximation around $\alpha = 2°$ and $M = 0.83$; For the first- and second-order Taylor approximation, we have solved only one nonlinear system

Concluding Remarks

This section has presented the main lines of the implementation of the second differentiation of a functional subject to a state equation. Many other informations concerning the method and concerning application to robust optimization can be found in [1, 2, 7–10].

References

1. Belme, A., Martinelli, M., Hascoët, L., Pascual, V., Dervieux, A.: AD-based perturbation methods for uncertainties and errors. Int. J. Eng. Syst. Model. Simul. **2**(1–2) (2010)
2. Duvigneau, R., Martinelli, M., Chandrashekarappa, P.: Uncertainty quantification for robust design. In: Breitkopf, P., Coelho, R.F. (eds.) Multidisciplinary Design Optimization in Computational Mechanics. ISTE, Wiley (2010)
3. Ghate, D., Giles, M.B.: Inexpensive Monte Carlo uncertainty analysis. In: Recent Trends in Aerospace Design and Optimization, pp. 203–210. Tata McGraw-Hill, New Delhi (2006)
4. Ghate, D., Giles, M.B.: Efficient Hessian calculation using automatic differentiation. In: 25th Applied Aerodynamics Conference, Miami (Florida), AIAA 2007-4059 (2007)
5. Griewank, A., Gauger, N., Riehme, J.: Extension of fixed point PDE solvers for optimal design by single-step one-shot method. Eur. J. Comput. Mech. (REMN) **17**, 87–102 (2006)
6. Hascoët, L., Pascual, V.: TAPENADE 2.1 user's guide. Technical Report 0300, INRIA, September 2004
7. Martinelli, M.: Sensitivity Evaluation in Aerodynamic Optimal Design. Ph.D. thesis, Scuola Normale di Pisa, Italy (2007)
8. Martinelli, M., Duvigneau, R.: Comparison of second-order derivatives and metamodel-based Monte-Carlo approaches to estimate statistics for robust design of a transonic wing. In: 10th AIAA Non-deterministic Approaches Conference, Schaumburg, USA, AIAA 2008-2071, April 2008
9. Martinelli, M., Duvigneau, R.: On the use of second-order derivative and metamodel-based Monte-Carlo for uncertainty estimation in aerodynamics. Comput. Fluids **37**(6) (2010)
10. Martinelli, M., Hascoët, L.: Tangent-on-tangent vs. tangent-on-reverse for second differentiation of constrained functionals. Lect. Notes Comput. Sci. Eng. **64**, 151–161 (2008)
11. Saad, Y.: Iterative Methods for Sparse Linear Systems. PWS Publishing Company (1996)

12. Sherman, L.L., Taylor III, A.C., Green, L.L., Newman, P.A.: First and second-order aerodynamic sensitivity derivatives via automatic differentiation with incremental iterative methods. J. Comput. Phys. **129**, 307–331 (1996)

Formulations for Robust Design and Inverse Robust Design

Alberto Clarich and Rosario Russo

Best Practice for Robust Design

The classical approach to solve a Robust Design Optimization problem is based on the definition of a multi-objective optimization problem, consisting generally in the optimization of the mean value of the performances and on the minimization of their standard deviation [1]. The drawback of this methodology is the high number of simulations generally required to solve a multi-objective optimization problem, whatever optimization algorithm is selected.

For this reason, in the course of UMRIDA project we have developed and applied a different methodology, which is based on the Reliability Design Optimization approach [2]. This methodology is based on the typical requirement of aeronautical or industrial practice of defining constraints or objectives on the percentiles of some performances distribution, like the minimum value of lift or momentum for an airfoil, or the maximum value of drag. It is in other words required to limit the worst stochastic performance below the desired limit which may be prescribed by any normative, or elsewhere to improve the worst stochastic performance as an optimization objective.

It emerges clearly that if we are able to convert a two-objective optimization problem (optimize mean and minimize standard deviation of any performance) into a single objective one (optimize worst stochastic performance), we can reduce drastically the number of simulations required to obtain the optimal results.

The way the "worst" stochastic performance can be defined may depend on a prescribed normative (for instance, this is a typical requirement for automotive industry that requires a given percentile of injury criteria during crash-tests to be below a safety threshold), or can be arbitrary defined as an objective (following a

A. Clarich (✉) · R. Russo
ESTECO, Trieste, Italy
e-mail: clarich@esteco.com

C. Hirsch et al. (eds.), *Uncertainty Management for Robust Industrial Design in Aeronautics*, Notes on Numerical Fluid Mechanics and Multidisciplinary Design 140, https://doi.org/10.1007/978-3-319-77767-2_49

Six Sigma rule, we may identify the 99.97‰ of a performance parameter to be minimized, such as a drag coefficient, or a 0.03‰ of a performance parameter to be maximized, such as a lift coefficient).

Normally, in literature the reliability analysis often implements methodologies like FORM or SORM that evaluate the failure probability of any candidate design on the basis of its uncertainties distribution and of the given limits to be respected. One limit of this methodology can be represented by the high number of evaluations that may be required by the algorithm to compute the failure probability with accuracy, which makes often practically unfeasible its application to optimization problems of industrial relevance, in particular, when the computational time required for each evaluation is very expensive.

The methodology that we have developed instead just requires the computation of the needed percentile from a large Monte Carlo population evaluated directly using the Polynomial Chaos expansion of the requested performance. This methodology takes advantage of the small sampling that is required to compute the Polynomial Chaos coefficients, so that the number of numerical simulations is limited to the samples required for its UQ (that as discussed in chapter "Reduced Basis Methods Using Regression Based Polynomial Chaos" can be drastically reduced by the application of Adaptive Sparse Collocation methods).

Finally, the single-objective optimization problem can be efficiently solved applying any optimization algorithm, such as the Nelder–Mead Simplex, or a genetic algorithm which could be more robust but requiring an higher number of design simulations.

For this reason, the proposed reliability optimization approach based on Polynomial Chaos exploitation for the computation of percentiles, guarantees high efficiency both in terms of quality of results, and in terms of computational cost, whatever is the industrial field of application.

In the following paragraph, we illustrate the efficiency of the methodology by analyzing the test case results developed during UMRIDA, giving an indication of the applicability range based on the outputs which may be reachable and the costs (in terms of engineering and CPUh); in addition, an indication of the industrial reproducibility of the methodology outside aeronautics, will also be depicted.

Efficiency and Applicability of Robust Design Formulation

In order to assess the efficiency of the Reliability-based Robust Design Formulation, we report the results of the test case IC-06 elaborated during UMRIDA project [3], and fully described in chapter "Surrogate Model Based Approaches to UQ and their Range of Applicability" of this Appendix.

The object of the optimization problem is an acoustic panel of a typical regional jet engine inlet produced by Alenia Aermacchi (AAEM) integrating an acoustic liner to reduce noise emissions. The objective is the noise attenuation at certified flight conditions (approach, take-off, or cut-back) in the far field surrounding the nacelle.

Geometrical uncertain parameters are indicated in Table 1. The table includes also the distribution type and properties of the uncertainties, as they have been quantified during UMRIDA project using the distribution fitting tool of modeFRONTIER, and the range of variation that has been considered for the optimization problem.

These parameters are entered into a proprietary impedance model together with values depending on flight condition to produce the admittance to be applied to the acoustic liner in the computational acoustic code.

Simulations have been performed with ACTRAN software from FFT, on an axisymmetric model of about 30,000 quadratic elements; the level of refinement of the mesh being enough to resolve the frequencies considered and could be eventually refined to gain accuracy despite increase in computational costs. The simulation time to complete one configuration design was about 2 min for each flight condition running in parallel on eight cores of a computational node of AAEM's AAHPC.

As a cost function, we have used a simple metric function: the overall sound pressure level (OASPL). In fact, noise radiating from an inlet is tonal and therefore at the blade passage frequency (BPF) and its harmonics higher sound levels are propagated. In this view, a liner should be optimized to attenuate the noise mainly at these frequencies. This objective function uniformly sums over the directivity angles and over frequencies the sound pressure level (SPL) which is a function of the frequency, the root mean square of the acoustic pressure of the duct radial and azimuthal modes computed and directivity angle at a distance of 150 ft. All the ACTRAN simulations were performed at two frequencies: the BPF and its first harmonic 2BPF; finally, the range of directivity used in this work is 50°–80°; considered the most critical for the take-off condition.

To set up the complete process flow, from the automatic simulation analysis to the algorithm implementation, a workflow has been implemented in modeFRONTIER software. As a first proof of concept, we first define an optimization problem where only one criterion is considered, i.e., a single OASPL signal, corresponding to the 50°–80° integral of noise pressure levels, related to a single flight condition (take-off).

Taking advantage of the Adaptive Sparse Collocation method (chapter "Reduced Basis Methods Using Regression Based Polynomial Chaos"), we can perform an

Table 1 Summary of uncertain parameters and their range of variation

Uncertainty	Distribution type	Standard deviation	Range of variation
Cell depth	Normal	1.5069E−2	10–18 mm
Sheet thickness	Normal	2.2625E−2	0.7–1.35 mm
Hole diameter	Normal	1.2803E−2	1–2.5 mm
Percentage of open area	Normal	1.0815E−3	3–15%

efficient UQ of the signal using just 15 sample points for each design proposed by the optimization.

In order to compare the proposed methodologies, we have defined a total (target) number of designs to be evaluated equal to 50, which corresponds to a total number of cpu hours of about 240 (50 designs × 15 samples × 2.30/60 h × 8 cpus). We can estimate that in case of 10 uncertainties, due to the additional effort required by the Polynomial Chaos UQ, the total number of cpu hours could be 4 times higher (since the minimum number of samples required by a Polynomial Chaos expansion of order 2 moves from 15 to 66 passing from 4 to 10 uncertainties); therefore, we would reach a total of 1,000 cpuh to solve a 10 uncertainties problem, which is an acceptable requirement for the industrial practice.

The reliability-based approach requires the definition of a single objective, the minimization of the 99.97‰ of the OASPL_50–80 signal response; therefore, we have applied the simplex single-objective algorithm of modeFRONTIER defining 50 total iterations. The performance of the best design is reported in Table 2.

These results are compared with the one obtained by the classical approach, i.e., the multi-objective optimization of mean and standard deviation of OASPL_50–80 (both to be minimized applying MOGAII Genetic Algorithm of modeFRONTIER [2]). In this case, only five generations of ten designs each one has been evaluated, which is generally a small number to expect important results from. In fact, even if the mean value of OASPL is practically the same, 120.4 dB, the best standard deviation is higher (6.79E−5 instead of 5.55E−5) than the one obtained by the single-objective approach, confirming the higher efficiency of the reliability-based approach.

On the other side, if we continue the multi-objective optimization considering a total of 20 generation (i.e., a total of 200 design evaluations), as reported in Table 2 we may expect a relative improvement of the standard deviation objective (about 1.8%), which is however paid by a total cost of cpu hours 4 times greater.

If this additional cost could be acceptable in the test case considered, characterized by 4 uncertainties only, it would not be acceptable in case of an higher number of uncertainties, such as 10, for which the compromise solution obtained by the reliability approach may be more convenient.

To generalize the results, in UMRIDA project also a more demanding optimization problem has been considered, including as objectives the OASPL signals

Table 2 Comparison of best designs performances obtained in different approaches

RDO approach	Mean of OASPL	Standard deviation of OASPL	99.97‰ of OASPL	Cpu hours 4 uncertainties–projection for 10* (cpuh)
1 obj 50 des	120.4	$5.55^{E}-5$	120.42323	240–960
2 obj 50 des	120.4	$6.79^{E}-5$	120.42324	240–960
2 obj 200 des	120.4	$4.95^{E}-5$	120.42309	960–3840

corresponding to three different flight conditions (take-off, approach, and cut-back). In order to keep the advantages of the single-objective reliability-based approach, in such cases when the different criteria to be optimized are homogenous, it is convenient to sum all the criteria into a single objective (in this case, the average of the OASPL signals measured for each flight condition). Since the sum of objectives may penalize one of them, it is advised to include one additional constraint for each objective, for instance, to guarantee a performance not lower than one of the baseline configuration, for each of the criteria considered.

In formulas, we get:

- 99.97‰ of OASPL_app to be lower than 123.35278 (baseline)
- 99.97‰ of OASPL_cbk to be lower than 120.00368 (baseline)
- 99.97‰ of OASPL_tko to be lower than 121.90074 (baseline)
- Minimize (99.97‰ of OASPL_app + 99.97‰ of OASPL_cbk + 99.97‰ of OASPL_tko)/3.

For this application, a simplex algorithm can be used to reach a convergence threshold quite satisfactory even after 20–50 designs (as noted above), either a genetic algorithm could be applied to improve slightly the results (about 0.1%) after 50–100 designs. Reporting the results of the second approach, in Fig. 1 the baseline configuration (ID 0) is highlighted, as well as one of the optimal solutions chose, ID 89, that improves of over 1% all the three criteria with respect to the baseline configuration.

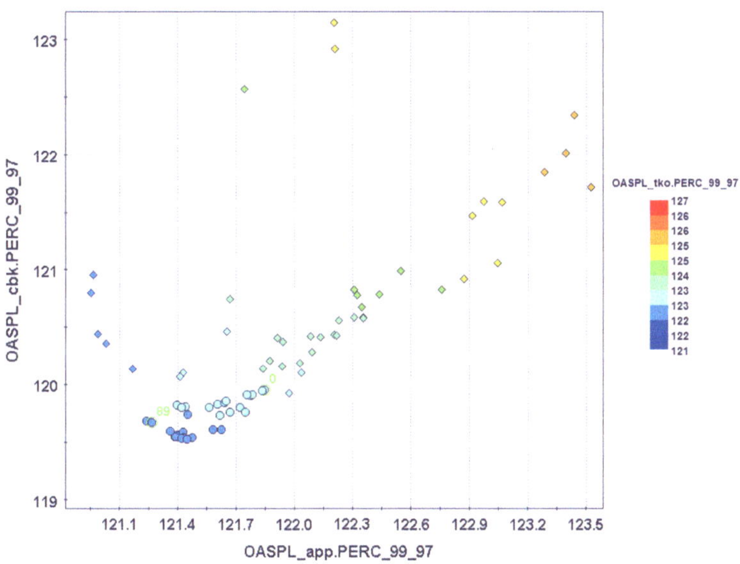

Fig. 1 Optimization results (OASPL percentiles at three flight conditions)

The improvement may be considered small, but it has to be considered that the sensitivity of the parameters to the response function is very small, and that the selection of the best design is always a trade-off between each criterion: The marginal improvement of each criterion separately may in this case reach 6–7%. In this way, if we decide that one criterion is most important than another (for instance, the approach flight condition), we can select the optimal design in order to get a greater improvement of the most important criteria (in this case, of 5%), with a contained loss of performance for the other criteria (in this case, less than 1%).

Also in case of heterogeneous criteria (such as for instance drag coefficient and lift coefficient of an airfoil), it is recommended to reduce as far as possible the number of objective functions, possible to one only (in this case, drag minimization may be more convenient), keeping the other criteria as constraints (lift and momentum coefficients to be respectively greater than and lower than the baseline ones), the same approach that has been followed in [2].

About the applicability of this methodology outside the range of UMRIDA applications (aeronautics), it is worth to remember that Reliability-based Robust Design Optimization is a practice already common in several industrial fields, in particular automotive, as it can be proved by several ESTECO's customer of modeFRONTIER software [4].

In the automotive industry, a typical problem could be, for instance, represented by crash safety analysis: The safety restraint system should guarantee in case of impact an acceleration value on different parts of the human body less than a critical limit (NCAP normative), and of course the lower is the percentage of the crashes for which the accelerations violate this limit (considering a stochastic distribution due to the uncertainties of the system), the better is the reliability of the system [5].

Another typical industrial requirement could be related to manufacturing tolerances: Performance of the product should respect any prescribed limit or threshold (and here the advantage of reliability approach), under the distribution of the manufacturing tolerances [6].

To conclude the analysis in Fig. 2, we report a simplified function, expressing the number of simulations (samples) required at the variation of number of uncertainties and number of variables. The function is based on the assumption that the number of samples for the UQ of the response is given by the minimum number of samples for a Polynomial Chaos expansion of order 2 (and the application of Adaptive sparse methods may reduce further this number), and that to solve a single-objective optimization problem, an empirical rule of $5 \times n$ (with n number of variables) number of design evaluations could be enough to obtain a good convergence.

As for the main target of UMRIDA, to solve a Robust Design Optimization Problem of ten uncertainties by a computational effort of 1,000 cpuh (100 h on 10 CPUs cluster), limiting the number of design variables to 10 (that could coincide with the same uncertain parameters or not necessarily), we may require a number of simulations around 3,000: It is therefore expected to have numerical models enough accurate that can require about 1/3 cpuh. If this requirement is not allowable, the solution is to reduce the number of uncertainties (eventually using one of the

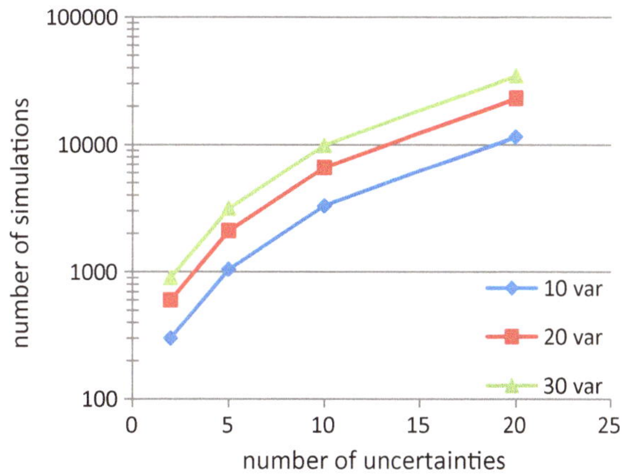

Fig. 2 Number of simulations in function of uncertainties and optimization variables

screening methods described in chapter "Reduced Basis Methods Using Regression Based Polynomial Chaos"), or to reduce the number of optimization parameters, to the ones that are still most important in the process.

Best Practice for Reverse Robust Design

The motivation of R-MORDO (Reverse-Multi-Objective Robust Design Optimization) methodology, called also Tolerance Design, comes basically from economic reasons in industrial production.

To improve the quality of a product it is necessary to avoid excessive warranty costs, derived mainly by the failures typical when the product is not designed taking into account any uncertainty in the process, i.e., without performing a Robust Design Optimization (RDO). At the same time, however, an excessive care of the uncertainties could drive to very restrictive solutions, i.e., solutions that do not allow the input parameters to vary but within very small tolerances, causing the production costs go higher.

A optimal solution can be a trade-off design that can be obtained following two different objectives: optimize the mean performances under the given constraints (RDO objective), and maximize the standard deviation or tolerance of the input variables at the same time.

In fact, if we are able to optimize the performance distribution of the design configuration and respect the given constraints by a solution whose manufacturing or operational tolerances are higher than the baseline model, it means that we are able to obtain the needed performance by a lower production cost.

The way a problem of this kind can be treated is not dissimilar from a Generic Robust Design Problem (for which the best practice is the application of Adaptive Sparse Polynomial Chaos for the uncertainties quantification, and definition of a single-objective function with additional constraints on other criteria [7]), with the difference that one of the criteria to be treated is the value of the tolerances (main parameter of the uncertainties distribution, which can be for instance the standard deviation of a normal distribution, or the range of a uniform or a beta), which has to be optimized or to be constrained (if we want to privilege another criteria to be optimized and keep the tolerances over a desired value).

As an example of application, as it will be discussed in next section, we could define as objective the maximization of the manufacturing tolerances of an airfoil, keeping as constraint the extreme percentile values of the aerodynamic performances (99.97‰ of drag and momentum, and 0.03‰ of lift). The results of the optimization would be read as the reproduction of the desired performances (baseline), by an higher tolerance on the uncertainties, therefore a reduced manufacturing (and maintenance, when, for instance, also operational uncertainties are considered) costs.

Conversely, one might prefer to optimize the percentile performances (therefore, maximizing 99.97‰ of drag with constraints on lift and momentum), keeping a constraint on the tolerances, less tight than the baseline solution.

In every case, a very important factor to consider is how to couple together tolerances coming from several uncertainties, generally coming from heterogeneous quantities, and even different type of distributions. Since the definition of separate objectives (one for each tolerance) may increase drastically the simulation costs, as discussed in previous chapter, it is recommended to define a single-objective function, defining a tolerance factor which could then be used for each separate uncertainty to get the tolerance value from a scale range (from minimum to maximum tolerance). In the next section, we illustrate by an example of application how this procedure could be implemented, and which results can be obtained, also in terms of computational costs and applicability coverage.

Efficiency and Applicability of Reverse Robust Design Formulation

As application case for Reverse Robust Design, we selected the RAE2822 airfoil test case [4].

The baseline airfoil model is a RAE2822, with nominal Mach number equal to 0.734 and angle of attach equal to 2.79°, Reynolds number equal to 6.5E6. The uncertainties considered are three, and precisely:

- Thickness ($\sigma = 0.005$), Normal distribution
- Mach ($\sigma = 0.005$), Normal distribution
- angle ($\sigma = 0.1$), Normal distribution.

Simulations have been performed through FINE/Open software from NUMECA, with a full turbulent (Spalart-Allmaras model) model of about 6 millions of cells; the level of refinement of the mesh is very accurate, because it is important to reduce the effect of numerical uncertainties, which are not considered in the problem. The simulation time to complete one configuration design on a 2-cpu core is about 1 h.

The simulation model has been implemented in modeFRONTIER software from ESTECO, to guarantee an automatic execution of the optimization processes. The airfoil shape is automatically modified using a Bezier curve, a parametric curve that remains continuous and regular at the variation of the control point position. In our case, four control points are used for both the two Bezier curves, to be added respectively on the upper profile and to the lower profile of the RAE2822 airfoil to modify its shape [5].

For each of the aerodynamic performances, a specific constraint is defined, in order to guarantee the performance of the baseline configuration, and more specifically:

- Constraint 1: 00.03‰ of Cl > 0.9
- Constraint 2: 99.97‰ of Cd < 0.0757
- Constraint 3: 99.97‰ of Cm < 0.1305.

Besides the satisfaction of these constraints, the definition of the optimization problem is completed by the maximization objective of the tolerance factor (*tol*). This parameter can arbitrary vary between 0 and 1 (0% means no tolerance, while 100% means maximum tolerance), and it is a common multiplier of the single uncertainties standard deviation parameters.

Basically, a 0 tolerance will leave the baseline uncertainties unaltered, while a 100% value will amplify the standard deviation to the maximum value that we arbitrary considered as feasible for this problem. The standard deviation parameters inside the input variables nodes of the workflow are then defined as follows:

- Standard deviation of *Mach* = 0.005 * (1 − *tol*) + 0.0075 * *tol*
- Standard deviation of *Angle* = 0.1 * (1 − *tol*) + 0.15 * *tol*
- Standard deviation of *Thickness* = 0.005 * (1 − *tol*) + 0.01 * *tol*.

For each design proposed by the optimization, 10 sampling points have been evaluated using the Polynomial Chaos Expansion (deliverable D4.1 for more details); the number of sampling points has been defined after proper tests of convergence (the error of estimation for the output performance main moments are less than 1% for the baseline configuration).

After a small number of design evaluations (25 using Simplex algorithm), it has been possible to reach a satisfactory convergence, to the values indicated in Table 3.

In particular, the optimal tolerance objective has reached a value of 24.7%, which means that we can guarantee at least the same stochastic performance of the

Table 3 Baseline and optimized airfoil configuration performances

Airfoil	Mach (σ)	Angle (σ)	Thickness (σ)	Max 99.97% Cd	Max 99.97% Cm	Min 0.03% Cl
Baseline RAE2822	0.005	0.1	0.005	7.568E−2	1.309E−1	9.019E−1
Optimized	0.0057	0.114	0.00637	6.968E−2	1.305E−1	9.045E−1

baseline, under a significant percentage of tolerance increment for all the uncertainties.

About the computational costs, the results are perfectly in line with the summary chart of chapter (Fig. 2), i.e., 250 simulations for a problem of 8 design variables and 3 uncertainties.

Also in this case, the range of applicability of the methodology goes far over aeronautical applications, being automotive and industrial requirements in general very demanding in terms of manufacturing costs reduction.

In UMRIDA project, we have applied as test the same methodology here described to a heat exchanger manufacturing [6] problem.

A cellular element of a cross-flow heat exchanger (HE), which is repeated along three directions for the air (cold) side and for the liquid (hot) side in the whole heat exchanger, is represented by an analytical function able to define its performances (Nu Nusselt number and Cf friction coefficient), in function of geometrical (height, length, pitch, and thickness of the fins) and operational (Re Reynolds and Pr Prandtl number) parameters. The objective of the optimization is to respect the constraints on the performances (mean value of friction coefficient Cf and 99.97‰ of heat transfer coefficient to be respectively below 0.040 and 4000 W/m^2K, i.e., the baseline performances), with the additional objective to maximize the tolerance on the geometrical and operational parameters. Also in this case, a common tolerance factor *tol* is used to evaluate the tolerance of each parameter and is used to define the objective function to be maximized.

Since the number of uncertainties is equal to six, in this case a number of 50 design samples have been considered (a little higher than the minimum required, since in this case the computational cost is very low).

As a result, we have been able to respect the given constraints on the mean performances, with a tolerance value of 30% of the considered range: for instance, the geometrical dimensions will have a manufacturing tolerance of 50 μm instead of 1 μm, and the Re and Pr number tolerance is ten times higher than the other case.

References

1. Li, W., Padula, S.: Performance trades study for robust airfoil shape optimisation. In: 21th Applied Aerodynamics Conference, AIAA-2003-3790
2. Clarich, A., Pediroda, V.: Robust design applications with modeFRONTIER, applying NODESIM-CFD tools. In: NODESIM-CFD Workshop on Quantification of CFD Uncertainties. Bruxelles, 29–30 Oct 2009
3. Clarich, A., Russo, R., Magnino, N.: Reliability-based robust design optimization using polynomial chaos expansion for aeronautical applications. In: UMRIDA Workshop on RDO. Bruxelles, 20–21 Sept 2016
4. Clarich, A., Marchi, M., Russo, R.: Reliability-based optimization applying polynomial chaos expansion. In: 6th European Conference on Computational Fluid Dynamics (ECFD VI). Barcelona (Spain), July 20–25, 2014
5. Fu, Y.: Robust design framework for IIHS rear impact, modeFRONTIER users'meeting. Trieste, 28 Sept 2006
6. Clarich, A., Carriglio, M.: modeFRONTIER for Virtual Design and Optimization of Compact Heat Exchangers, SAE Conference, 2014-01-2406 (2014)
7. Clarich, A., Russo, R.: Innovative methodologies for robust design optimization with large number of uncertainties using modefrontier. In: Proceedings of the International conference on Evolutionary and Deterministic Methods for Design Optimization and Control with Applications to Industrial and Societal Problems, EUROGEN, pp. 233–240. (2015)

Use of RD in Multiphysics Applications

S. Richard and N. Magnino

Guidelines for UQ Applied to Combustion Instabilities in Aero-engines

Chain of Tools for UQ: Main Steps to Follow

It has been shown in numerous studies from the literature that unsteady CFD simulations can be used to capture combustion instabilities in complex geometries. It is however hardly conceivable having recourse to 3D LES to deal with uncertainties in a reasonable computational timeframe. Nevertheless, CFD is needed for the knowledge of the flame response to acoustic perturbations, accounting for the real geometry and confinement effects. Such flame response can be extracted in the form of a flame transfer function to be imposed on acoustic solvers like 3D Helmholtz tools. These solvers are also not adapted to achieve intensive robust design studies but furnish reference results on unstable modes and associated growth rates accounting for the full annular combustor geometry. The obtained reference results can then be fitted by lower order models like quasi-analytical tools or surrogate models in order to evaluate the risk of occurrence of combustion instabilities (CI) following UQ approaches. This step-by-step procedure constitutes the first element of the best practices to follow regarding RD of stable combustors (Fig. 1). Guidelines associated with each step are provided in the following sections.

S. Richard
Safran Helicopter Engines, Avenue Joseph Szydlowski, 64511 Bordes, France

N. Magnino (✉)
Leonardo S.p.A., Via P. Foresio 1, 21040 Venegono Superiore, VA, Italy
e-mail: nicola.magnino@leonardocompany.com

C. Hirsch et al. (eds.), *Uncertainty Management for Robust Industrial Design in Aeronautics*, Notes on Numerical Fluid Mechanics and Multidisciplinary Design 140, https://doi.org/10.1007/978-3-319-77767-2_50

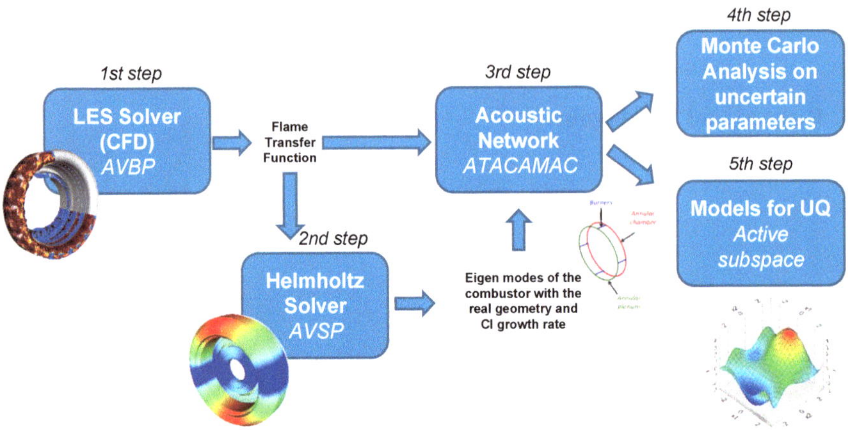

Fig. 1 Full annular Safran Helicopter Engines helicopter combustor (validation case)

CFD to Establish a Reference Flame Transfer Function

In this chapter, it is proposed to perform a UQ study on combustion instabilities. A first step is therefore to define a reference flame transfer function for unsteady CFD simulations (a compressible LES solver is mandatory). The objective of these simulations is not to find and understand possible unstable modes of the combustor, including all scenarios like longitudinal, transverse and azimuthal modes. Thus, it is proposed to minimize the CFD effort by reducing the problem on the base of its symmetries. For that purpose, only one sector of the combustor should be simulated and acoustic forcing at the combustor inlet should be used with amplitudes of the order of a few percentage of the operating airflow rate. Such a forcing allows obtaining heat release fluctuations characteristic of CI and finally to retrieve by post-processing the parameters n and τ of the combustor flame transfer function following:

$$\boldsymbol{q}' \sim \boldsymbol{n}\boldsymbol{u}'(t-\tau)$$

where n is the gain and τ the delay of the FTF, q' is the heat release fluctuation, t the time and u' the velocity fluctuation.

The obtained FTF parameters will be the main variables to be used for the RD study, because they are the most uncertain ones. Indeed, slight manufacturing variations are known to play a role on UQ occurrence. In addition, modelling errors from CFD like turbulence/flame interactions or two-phase flow description may affect the flame dynamics and response to acoustic forcing. For this reason, it highly recommended when possible to perform a small DOE varying the flame LES filter size or the droplet diameter to get an order of magnitude of FTF parameters uncertainty. If this is not possible, values proposed by CERFACS on the base of several experimental data could be used: $\sigma_n/n = 10\%$ and $\sigma_\tau/\tau = 5\%$.

Recommendations for the Use of Acoustic Tools

To avoid expensive 3D Helmholtz simulations, it is proposed to exploit a quasi-analytical tool called ATACAMAC for the UQ study. ATACAMAC is a network-based description of the whole annular combustor and gives access to azimuthal modes which are often the main concern of combustor design engineers. However, the simplified geometrical description of the combustor limits the predictive character of such tools. It is thus recommended to fit 3D results obtained with a 3D Helmholtz computation (here the AVSP code) accounting for the whole complexity of the combustion chamber, by adjusting the geometrical parameters of ATACAMAC presented in Fig. 2.

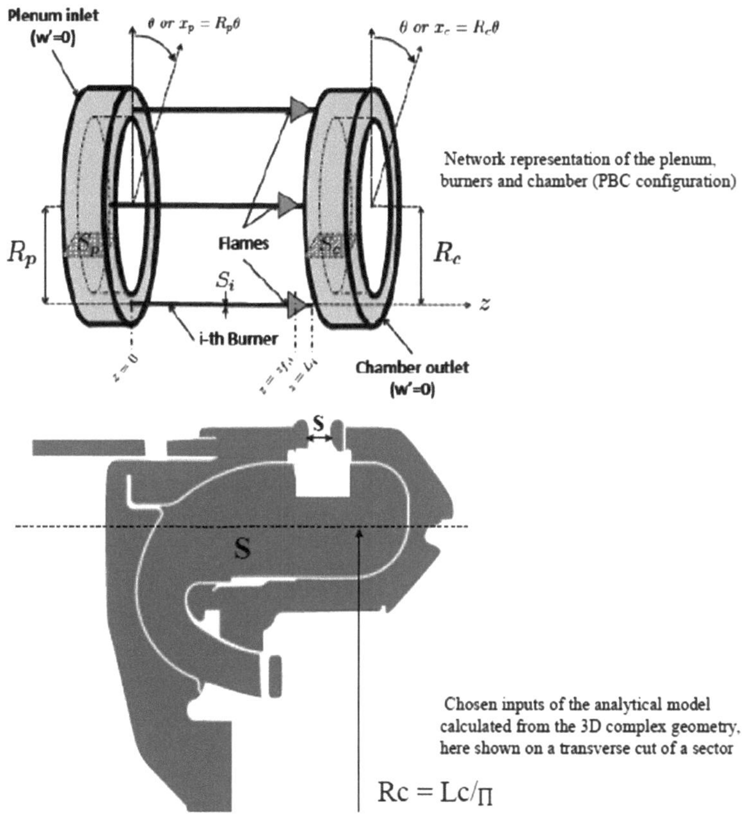

Fig. 2 Network model geometrical parameters

The baseline parameters like the combustor of injector cross sections denoted S are not uncertain and can be directly extracted from the CAD. On the contrary, the injector length L or the mean combustor dome radius can be difficult to define due to the complexity of the combustor shape. These parameters are therefore fitted to ensure a proper description of a targeted unstable mode growth rate when modifying the FTF delay in both ATACMAC and AVSP (see Fig. 3). An important requirement of the fitting procedure is to correctly represent the change of sign of the growth rate. Once ATACAMAC is fitted, a Monte Carlo analysis on the base of 5000–10 000 samples can be performed for a first estimation of the CI probability (Fig. 4).

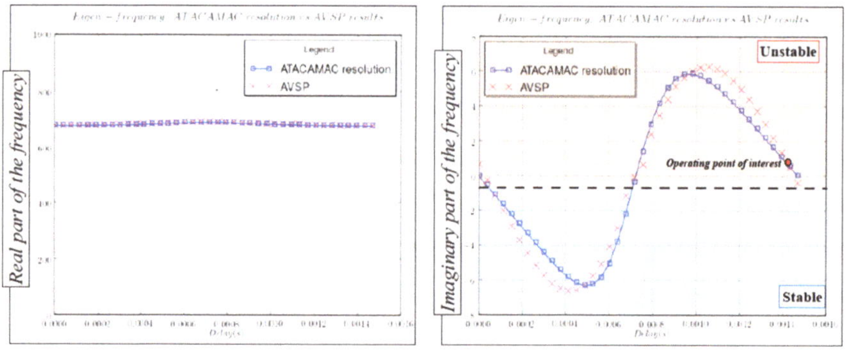

Fig. 3 Real (left) and imaginary (right) parts of a targeted combustor eigen-mode for several FTF delays—comparisons between AVSP and ATACAMAC after adjustment of the burner length parameter

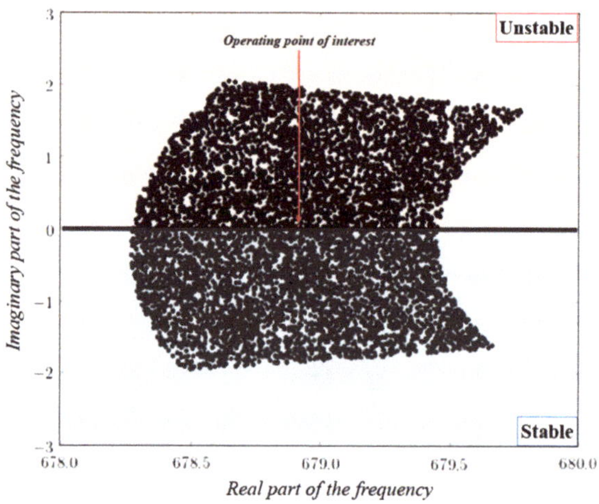

Fig. 4 Monte Carlo results using 8,000 samples: stability map of the first azimuthal mode is displayed by varying n and τ parameters

Guidelines for MORDO Applied to Aero-engine Acoustic Liners

Workflow for MORDO of Acoustic Liners

Acoustic propagation of engine noise is highly influenced by the liner impedance model, fan model and flow field parameters, as shown in Fig. 5, but within UMRIDA project the focus was only on the uncertainties of the geometrical parameters of the liner that have a direct impact on the acoustic impedance. Indeed the motivation behind FNM participation was to investigate the viability of an automatic multiobjective optimization methodology capable to efficiently produce a design of an acoustic liner robust in terms of attenuation of the noise with respect to the real uncertainties of the panel due to the manufacturing tolerances and in a reasonable turnaround time.

So the base ground to build an efficient RDO methodology for acoustic liners is the experimental database of measurements to develop UQ upon. The four uncertainties that influence the most the acoustic impedance of the liner according to the proprietary semi-empiric model are: Percentage of open Area (PoA_{eff}), honeycomb cell Height (h), holes diameter (d_{eff}), face sheet Thickness (t_{fs}) as illustrated in Fig. 6.

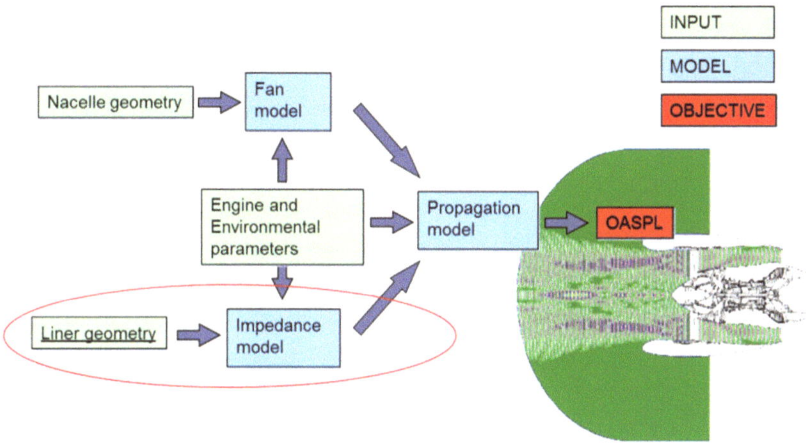

Fig. 5 Nacelle liner acoustic design scheme

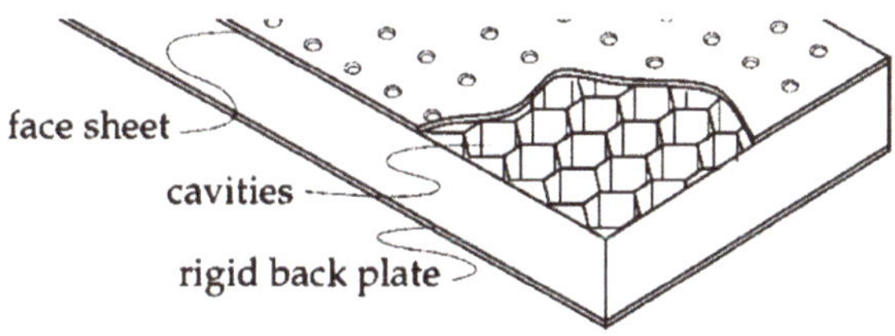

Fig. 6 Nacelle SDOF liner acoustic panel

In Table 1 are reported for each variable the characteristics of the distribution as determined by using the ESTECO Fitting Tool, together with the range necessary for the optimization to explore the space and the step which has to be a trade-off between actual manufacturing values and a reasonable dimension of the set.

Exploiting modeFRONTIER's recent capability to streamline complexity with addition of subprocess nodes, two workflows were built, one nested into the other: Fig. 7 shows at a glance the whole internal workflow to execute Actran's aero-acoustic simulation for each of the certification conditions, while Fig. 8 shows the outer automation loop to cycle through and compute final objective function $P_{99.97}$(OASPL).

Table 1 Liner uncertainties PDF, range and discretization

Uncertainty	Distribution	σ	Range	Step	Base
PoA	Normal	1.08E−03	[3–15%]	1.00E−03	121
h	Normal	1.51E−05	[10–18 mm]	1.00E−05	801
d	Normal	1.28E−05	[1–2.5 mm]	1.00E−05	151
t	Normal	2.26E−05	[0.7–1.35 mm]	1.00E−05	71

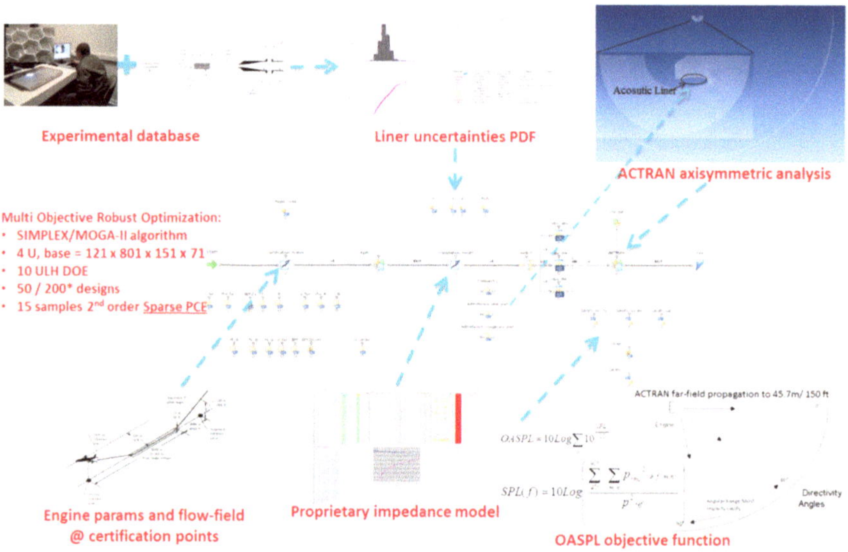

Fig. 7 Engine nacelle acoustic MORDO workflow

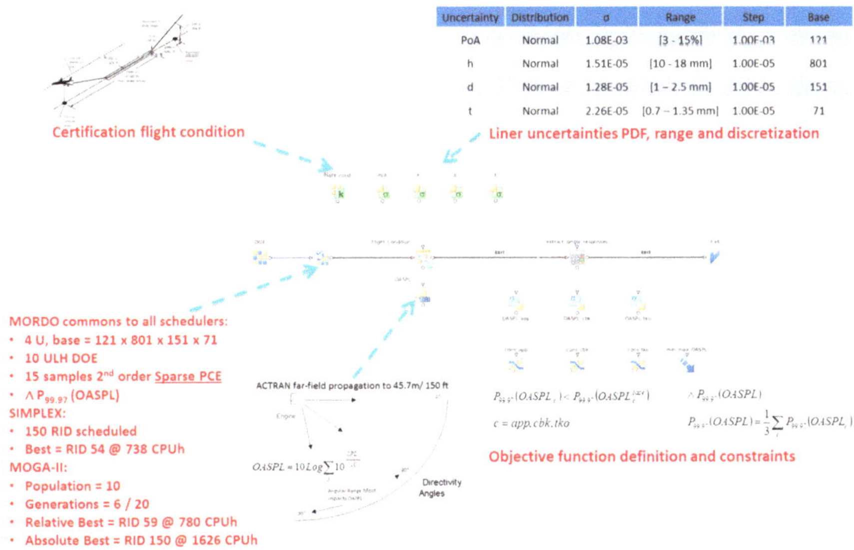

Fig. 8 Engine nacelle acoustic MORDO automation loop

Key Drivers/Recommendations for MORDO of Acoustic Liners

The key drivers to successfully develop and implement an efficient automatic methodology for multiobjective robust design optimization—that allowed improving an already hand-optimized panel, but more importantly to make it much more robust—including:

- a wise selection of uncertainties to consider and careful assessment of manufacturing tolerances to build a valid experiment database;
- an integrated and structured multicode workflow open to additions and expansions;
- tuning of mathematical and FE Actran model to make CPU cost compatible with preliminary design time constraints and UMRIDA *10 u* × *100 cores* × *10 CPUh* goal;
- rigorous step-by-step validation of tools being developed with intermediate tests on simplified models;
- efficient UQ thanks to *Adaptive Sparse Polynomial Chaos Expansion;*
- efficient RDO with *Reliability–Based* approach that by minimizing the percentile of global OASPL minimizes also mean and standard deviation of single OASPLs;
- choice of different optimization algorithm based on opportunity, e.g., SIMPLEX best is only 0.2 dB worse than MOGAII but with less than half CPU cost.

Geometrical Uncertainties—Accuracy of Parametrization and Its Influence on UQ and RDO Results

Dishi Liu, Daigo Maruyama and Stefan Görtz

Introduction

Random variations in the geometry of an aircraft (e.g., due to manufacturing tolerance) may introduce a large number of variables if uncertainty quantification (UQ) of its aerodynamic performance is to be performed. Due to the "curse of dimensionality," this makes the numerical integration of the performance statistics computationally expensive. However, since the geometry variations at different locations are by nature correlated, the number of variables can be greatly reduced if the correlation is exploited. This reduced-order modeling can be furnished by a truncated Karhunen–Loève expansion (KLE) of the random field that models the variation. Details of truncated KLE approximation can be found, e.g., in [1] for a similar application.

Assuming that the geometry is discretized for the purpose of computational fluid dynamics (CFD) simulations, a full KLE representation of the random field defined on a given surface mesh yields M variables, where M is the number of surface grid points. Model reduction is achieved by discarding the variables corresponding to "small" eigenvalues of the correlation matrix of the random field such that only $K < M$ variables are retained.

There are different choices for the threshold to delimit "small" eigenvalues. The first choice is based on the machine precision, i.e., $M \cdot eps(1.0)$ with $eps(r)$ the floating-point precision of a real number r (the elements of the correlation matrix are positive and ≤ 1.0). We denote the corresponding value of K as K^*. The precision of the computer that we used for this study is about $M \cdot$ 2.22e−16. One could also set the threshold based on the accuracy of the flow solutions used to

D. Liu · D. Maruyama · S. Görtz (✉)
German Aerospace Center (DLR), Institute of Aerodynamics and Flow Technology, Braunschweig, Germany
e-mail: Stefan.Goertz@dlr.de

C. Hirsch et al. (eds.), *Uncertainty Management for Robust Industrial Design in Aeronautics*, Notes on Numerical Fluid Mechanics and Multidisciplinary Design 140, https://doi.org/10.1007/978-3-319-77767-2_51

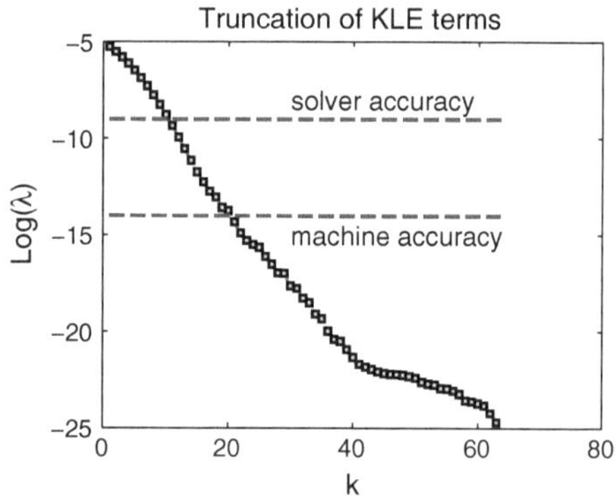

Fig. 1 Logarithm of eigenvalues of correlation matrix and truncation thresholds of K

estimate the statistic of aerodynamic quantities; e.g., make it equal to the minimum residual that is used as the CFD solver convergence threshold (in our case, 1.0e−8). The argument behind these choices is the assumption that small geometry variations lead to small perturbations in aerodynamic performance. If the perturbation is so small that is not distinguishable, it needs not to be imposed. Other choices for the threshold based on practical needs could also be adopted, e.g., some limitation in the number of variables that can be considered. Figure 1 displays the declining eigenvalues in one of the test cases in section "Test Case" and the two thresholds for truncation which lead to $K^* = 19$ and $K = 10$, respectively.

Two questions about the variable reduction are of our interest:

1. How much is the loss of accuracy of the statistics if "smaller" K values are used, i.e., if fewer KLE terms are retained?
2. Can smaller values of K help improve the efficiency of surrogate-based integration of the statistics?

In the first question, by "smaller" K values we mean values of K smaller than the K^* that corresponds to the machine accuracy. The second question is related to the common belief that for a given number of samples, surrogate models with more variables are less accurate (the so-called "curse of dimensionality").

In this work, we make some numerical experiments (tabulated in Table 1) based on a 2D Euler test case in the hope to shed some light on these two questions.

Table 1 Test cases with different number of uncertain variables K and two different correlation lengths ℓ(percentage of total perturbation variance kept in brackets)

Correlation length ℓ	0.2	0.1
Num. of variables K^*	**38**	**60**
Num. of variables K	**16** (99.9%) **10** (98.1%) **4** (74.7%)	**28** (99.9%) **20** (98.5%) **10** (83.6%) **4** (48.6%)

Test Case

This study is carried out based on the RAE2822 airfoil at a Mach number of 0.73 and an angle of attack of 2.0°. Figure 2 shows the structured grid used for the inviscid flow computations with the Euler solver of DLR's CFD code TAU [2–4].

The upper and lower surfaces of the airfoil are assumed subject to two independent random fields discretized on the surface grid (Fig. 3 left) featuring a Gaussian-type correlation with correlation length ℓ as described in [1]. The standard deviation of the zero-mean perturbation is assumed a function of x, $\sigma_p(x) =$

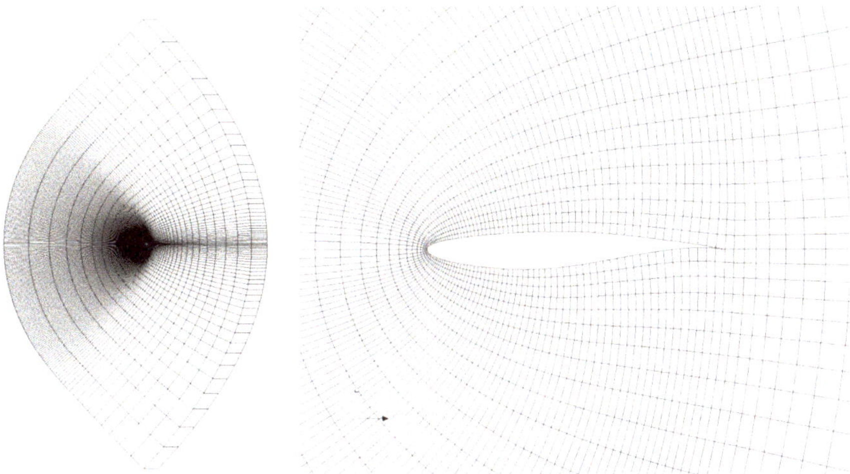

Fig. 2 Structured grid for the RAE2822 airfoil: far-field grid (left) and detailed view around the airfoil (right)

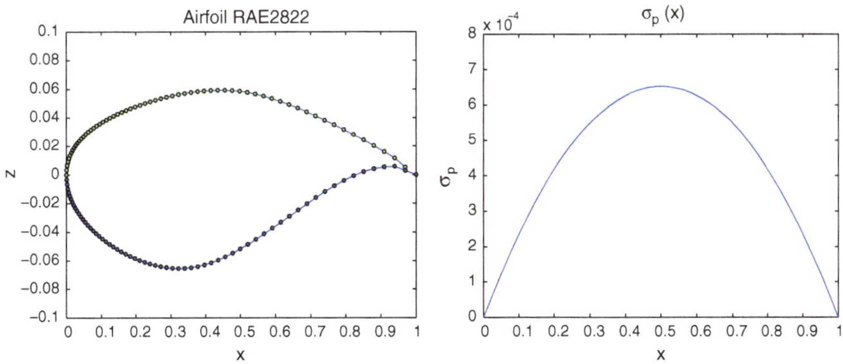

Fig. 3 Surface grid points of RAE2822 airfoil (left) and distribution of standard deviation of the random perturbation (right)

$0.01 \cdot 4 \cdot Z_{max} \cdot x \cdot (1 - x)$, with Z_{max} the maximum thickness of the airfoil, as graphed in the right part of Fig. 3. The $\sigma_p(x)$ features diminishing perturbations toward the leading and training edges of the airfoil to guarantee convergence of the CFD solver and has its maximum at the mid-cord.

We try two different correlation lengths, $\ell = 0.2$ and 0.1 for the random field. The respective K^* for the upper and lower random fields are $K_u^* = K_b^* = 19$ and $K_u^* = K_b^* = 30$. For the case $\ell = 0.2$, we consider three test cases with smaller K values, i.e., $K_u = K_b = 2$, 5, and 8, respectively, and for $\ell = 0.1$ four test cases with $K_u = K_b = 2$, 5, 10, 14, respectively. Table 1 lists the test cases in which $K = K_u + K_b$ and $K^* = K_u^* + K_b^*$ are the total number of variables that parameterize the random geometry perturbation. The two cases with $K = 16$ and $K = 28$ are based on the truncation threshold given by the convergence threshold of our CFD solver of 1e−8. The number in bracket in Table 1 is the percentage of total perturbation variance kept (average of upper and lower fields). The target statistics are mean and standard deviation of lift and drag coefficients (C_L and C_D), denoted as μ_L, σ_L, μ_D, and σ_D.

Impact of KLE Truncation on the Statistics

For every test case, we compute the target statistics by a direct integration of a very large number (N) of quasi-Monte Carlo (QMC) samples of the CFD model. The N numbers are listed in Table 2. We call the statistics obtained from the k-variable test cases k-statistics and those from k^*-variable test cases k^*-statistics. The impact of the KLE truncation is evidenced by the discrepancy between the K- and K^*-statistics, as tabulated in Tables 3 and 4.

The reliability of these discrepancies is verified through an accurate estimation of the statistics, for which we use Snyder's multi-partition method [5] since the theoretical error bound of QMC integration is not a practical accuracy indicator. Suppose S denotes any of the above statistics, by this method one makes an equal-size m-partition of all the N samples and computes m estimates S'_i, $i = 1, \ldots, m$ by integrating on each partition only. Then, compute their sample standard deviation ς_m which is an estimate of standard deviation of the QMC estimation of S with sample number N/m. To extrapolate ς_m to ς_1, one computes ς_m for four values of m, i.e., $m = \{128, 64, 32, 16\}$, and fit a line across the four $[\log(N/m), \log(\varsigma_m)]$ points using the weighted linear least squares method; the m values are used as weights to account for the increasing variability for smaller values of m. With the fitted slope a and intercept b, we have $\varsigma_1 = \exp(a^*\log(N) + b)$ as an estimate of the standard deviation of the QMC integral of S with N samples. This estimate is tabulated for all statistics in Table 5.

For every discrepancy $S_K - S_{K*}$ in Tables 2 and 3, if both $3^*\varsigma_1(S_K)$ and $3^*\varsigma_1(S_{K*})$ are no larger than $p(S_K - S_{K*})$, we regard that discrepancy reliable.

Table 2 Number of QMC CFD samples for each test case

	$\ell = 0.2$				$\ell = 0.1$				
	$K = 4$	$K = 10$	$K = 16$	$K^* = 38$	$K = 4$	$K = 10$	$K = 20$	$K = 28$	$K^* = 60$
N	352,000	1,038,700	1,000,000	2,393,500	316,000	594,300	1,279,500	390,700	581,700

Table 3 Discrepancy of *K*- and K^*-statistics of C_L

	$\ell = 0.2$ ($K^* = 38$)			$\ell = 0.1$ ($K^* = 60$)			
	$K = 4$	$K = 10$	$K = 16$	$K = 4$	$K = 10$	$K = 20$	$K = 28$
$\mu_{L,K} - \mu_{L,K^*}$	3.0e−4	3.5e−5	2.1e−6	1.1e−3	6.2e−4	1.2e−4	**1.0e−5**
$(\mu_{L,K} - \mu_{L,K^*})/\mu_{L,K^*}$	3.7e−4	4.2e−5	2.6e−6	1.4e−3	7.6e−4	1.6e−4	**1.4e−5**
$\sigma_{L,K} - \sigma_{L,K^*}$	−1.3e−3	−2.1e−4	−6.2e−6	−2.0e−3	−1.3e−3	−2.0e−4	**−1.7e−6**
$(\sigma_{L,K} - \sigma_{L,K^*})/\sigma_{L,K^*}$	−1.9e−1	−3.0e−2	−9.1e−4	−3.4e−1	−2.2e−1	−3.7e−2	**−1.1e−3**

Table 4 Discrepancy of *K*- and K^*-statistics of C_D

	$\ell = 0.2$ ($K^* = 38$)			$\ell = 0.1$ ($K^* = 60$)			
	$K = 4$	$K = 10$	$K = 16$	$K = 4$	$K = 10$	$K = 20$	$K = 28$
$\mu_{L,K} - \mu_{L,K^*}$	−5.3e−5	−7.4e−6	−3.6e−7	−2.0e−4	−9.7e−5	−1.4e−5	**−8.1e−7**
$(\mu_{L,K} - \mu_{L,K^*})/\mu_{L,K^*}$	−1.1e−2	−1.5e−3	−7.3e−5	−4.0e−2	−1.9e−2	−2.8e−3	**−2.2e−4**
$\sigma_{L,K} - \sigma_{L,K^*}$	−5.9e−5	−6.5e−6	−1.6e−7	−1.0e−4	**−3.8e−5**	**−4.2e−6**	**−2.6e−7**
$(\sigma_{L,K} - \sigma_{L,K^*})/\sigma_{L,K^*}$	−1.3e−1	−1.4e−2	−3.6e−4	−2.1e−1	**−7.8e−2**	**−9.8e−3**	**−1.5e−3**

Table 5 Estimated standard deviation of the statistics

	$\ell = 0.2$				$\ell = 0.1$				
	$K = 4$	$K = 10$	$K = 16$	$K^* = 38$	$K = 4$	$K = 10$	$K = 20$	$K = 28$	$K^* = 60$
$\varsigma_1(\mu_L)$	1e−7	4e−8	5e−8	4e−8	8e−8	1e−7	3e−7	3e−7	**4e−7**
$\varsigma_1(\sigma_L)$	1e−7	9e−8	1e−7	1e−7	1e−7	8e−7	1e−6	**1e−6**	**9e−7**
$\varsigma_1(\mu_D)$	1e−8	1e−8	1e−8	9e−9	8e−9	2e−8	6e−8	**1e−7**	**8e−8**
$\varsigma_1(\sigma_D)$	3e−8	2e−8	3e−8	1e−8	4e−8	3e−7	**4e−7**	**8e−7**	**5e−7**

$p(S_K - S_{K^*})$ denotes the precision of the discrepancies in Tables 2 and 3; e.g., for "−5.3e−5" the precision is 1e−6. By assuming the QMC estimated statistics are Gaussian distributed around the "true" statistics, the above inequalities guarantee a 99.73% confidence in the precision of the statistics and hence of their discrepancy. The discrepancies in bold font in Table 3 are those fail this judgment, and the corresponding ς_1 are also shown in bold font in Table 4. The failure is due to limitation in computing resource to achieve a sample number *N* that is large enough. Standard deviations are harder to estimate than means, and statistics of C_D are harder than those of C_L.

In the results that are reliable, we found that the truncation can have remarkable impact on the statistics. The impact is larger for C_D than for C_L, larger for the standard deviation than for the mean (more evidently seen in the relative discrepancy), and larger for the cases with shorter correlation length than for those with longer one. Not surprisingly, in these experiments the truncations always lead to underestimated σ_L and σ_D and always lead to overestimated μ_L and underestimated μ_D (i.e., undervalue the performance degenerating effect of the geometric uncertainties), which is the result of the reduced variance of the random perturbation imposed. Truncation that keeps more perturbation variance leads to smaller accuracy loss.

Impact of KLE Truncation on GEK Efficiency

The statistics could also be estimated by using gradient-enhanced kriging (GEK) surrogate method with much smaller number of CFD samples. The gradients of C_L and C_D with respect to all variables could be obtained by an adjoint solver at an additional cost roughly equals to that for two CFD evaluations. The GEK surrogates are constructed by using both the sampled responses and the gradients. So the statistics can be obtained by an integration of a large number ($3 * 10^5$, in our case) of QMC samples of the surrogate model.

In general, the convergence rate of surrogate models, i.e., the speed that its mean squared error is reduced by increasing samples, will slow down with more variables. For example, error bounds of approximation with Gaussian and inverse multiquadric (a popular radial basis function) kernel are shown to be $C_1 e^{-c_2/h}$ in [6], with h the "fill distance" (the largest distance between any nearest sample site neighbors). This translates to a bound in sample number N as $C_1 e^{-c_3 N^{1/d}}$ which deteriorates with increasing d, the number of variables. This favors a conclusion that for a certain sample number the accuracy of surrogate methods could be improved by reducing the number of variables, and hence so could the accuracy of the statistics based on the surrogate. This is the reason why further truncation from K^* is sometimes temptable. But the above convergence rate is based on an assumption that all the variables have the same importance; if this is not the case (just like in a KLE parameterization), the conclusion might be different.

To investigate the influence of the variable reduction in the accuracy of GEK surrogate of C_L and C_D, we compare the GEK error convergence rate with various K values. The error consists of two parts: The first can be written as $e_k = |\hat{S}_k - S_k|$ with $\hat{S}_k$ the statistics estimated by the GEK-based integration and S_k the corresponding K-statistics, and this part is the approximation error of GEK. The second is $|S_k - S_{k*}|$, the discrepancy of K- and K^*-statistics, since the GEK-based statistics is converging to S_k instead of S_{k*}. The second part is obviously larger for smaller K's, but in a certain range of smaller N the first part can be dominative. If with smaller K's the first part converges so fast that compensates the larger second part, further truncation from K^* would be beneficial in terms of efficiency in obtaining statistics in this range of N.

GEK in these experiments are implemented by using SMART toolbox [7], opting for cubic spline correlation function with hyper-parameters tuned by a maximum likelihood estimation.

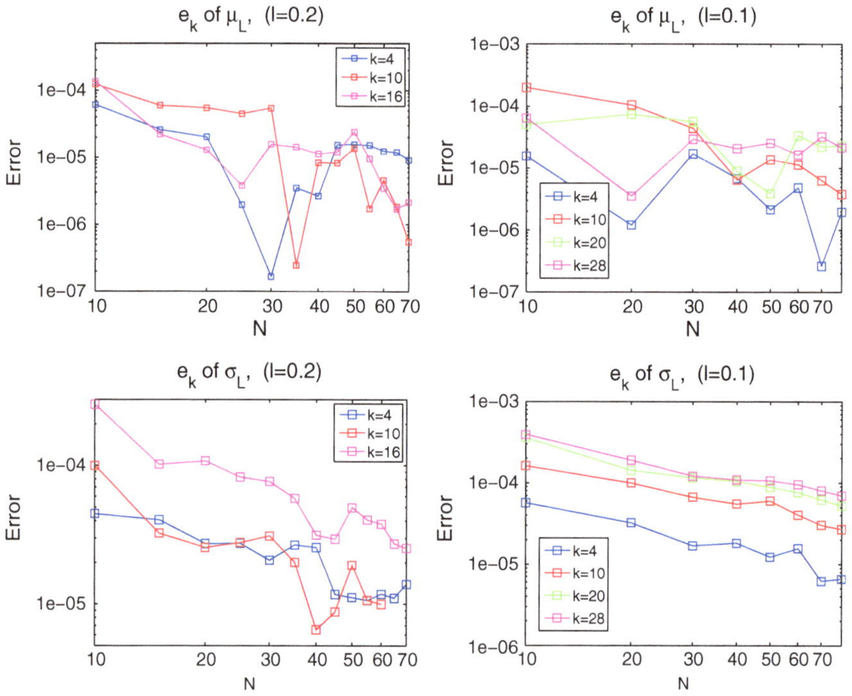

Fig. 4 e_k in GEK estimated statistics of C_L. $\ell = 0.2$ (left) and $\ell = 0.1$ (right)

Figures 4 and 5 display the convergence of e_k in estimating statistics of C_L and C_D, both axes are in logarithm scale. By these results it seems the convergence rate (the slope of the curve) is not handicapped by larger K value, if we exclude the stagnated convergence of the test case $K = 28$ due to the inaccuracy of its K-statistics. The smaller error magnitude for smaller K's is due to the relative simplicity of the surrogate with less variables (not surprisingly this is more obvious in the case with smaller correlation length because their truncations cause more loss in perturbation variance, as quantified in Table 1), but the error of the GEK surrogates with different number of variables are reduced at nearly the same speed with respect to N. This is not complying with the theoretical convergence rate.

We attribute this phenomenon to two reasons: First are the characteristics of the KLE parameterization. The variables as KLE coefficients are readily sorted in terms of importance. Variables with higher rank are less important (declining faster for larger correlation length ℓ) so that including more variables would not introduce more complexity to the surrogate in the same proportion. This can be observed in the percentage of perturbation variance retained by various K as tabulated in Table 1.

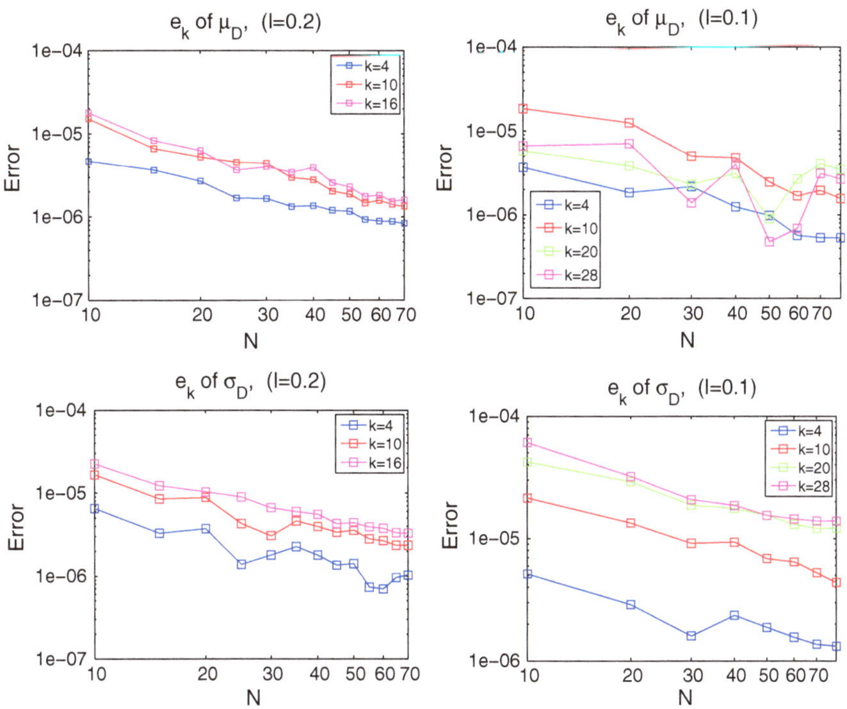

Fig. 5 e_k in GEK estimated statistics of C_D. $\ell = 0.2$ (left) and $\ell = 0.1$ (right)

The second reason is the relatively "cheap" gradient information obtained by using adjoint solver. The number of "conditions" (responses and their derivatives) utilized by GEK is $(d + 1) \cdot N$, which increases along d, but the cost remains stable as roughly that of $2N$ CFD evaluations irrelevant to d; i.e., the higher the number of variables, the cheaper the gradients. This helps to break the "curse of dimensionalities."

Now, we take the second part of the error into consideration; i.e., measure the error as $E_k = |\hat{S}_k - S_{k*}|$ (using the K^*-statistics as the reference value), and the negative effect of further truncation from K^* is revealed in Figs. 6 and 7. Due to the discrepancy of K- and K^*-statistics, E_k with K much smaller than K^* are not converging. Only E_{16} ($\ell = 0.2$) and E_{28} ($\ell = 0.1$) display a converging trend, and the associated K values keep 99.9% perturbation variance.

The result in these experiments shows reducing variables is not improving the convergence rate of GEK. Choosing $K < K^*$ will handicap the accuracy of GEK-integrated statistics unless that K keeps enough variance of the random perturbation field.

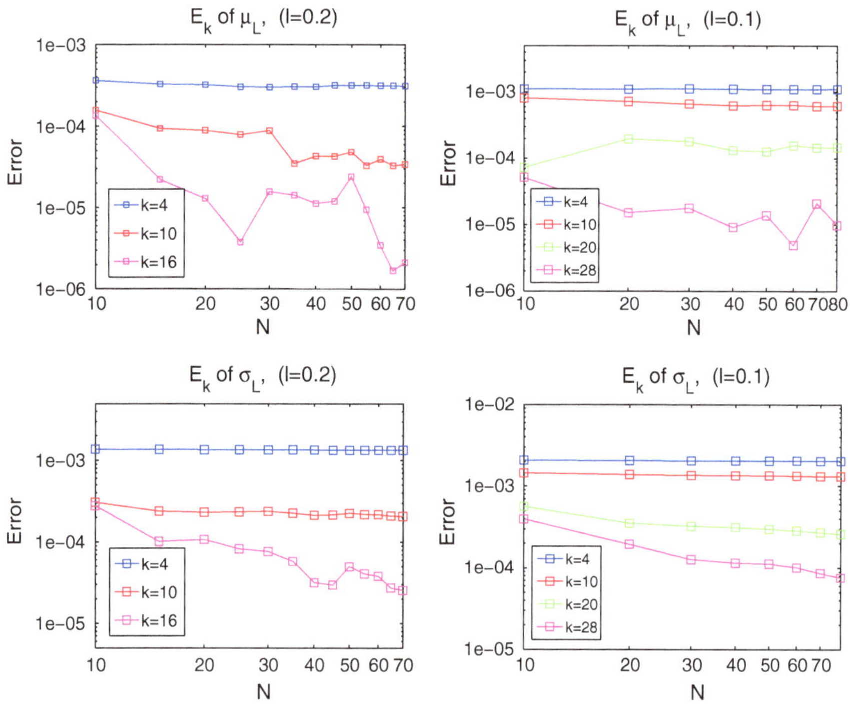

Fig. 6 E_k in GEK estimated statistics of C_L. $\ell = 0.2$ (left) and $\ell = 0.1$ (right)

Practical Applications to RDO

The influences of the accuracy of the parameterization, which was accomplished by a truncated Karhunen–Loève expansion (tKLE), on the results of robust design optimization (RDO), are finally discussed here. Under the condition that the KLE is sufficiently truncated, the accuracy of tKLE can be recovered by adding sample points, or the gradient information using GEK, and the number of the gradients also increases with K. For this reason, we concluded that GEK could efficiently compensate the errors due to lack of the sample points. One approach to investigate it was achieved by dispersion of different sample collocations. More details of this approach are discussed in [8]. Figure 8 shows mean and standard deviation of the stochastic cost function f $(f \equiv \mu_{Cd} + \sigma_{Cd})$ of the initial configuration by two different numbers of eigenvalues ($K = 10$ and 26) of tKLE. The truncated number of eigenvalues $K = 10$ satisfies neither the threshold of the flow solver nor the machine tolerance. That of $K = 26$ does the threshold of the machine tolerance. The stochastic cost function f is evaluated by GEK with a fixed number of samples as 30. Greater σ_f indicates greater dispersion (more errors) due to the lack of samples. The figure shows σ_f in $K = 26$ is less than that in $K = 10$. This can lead that the gradient information in GEK, with the help of an adjoint solver, can really

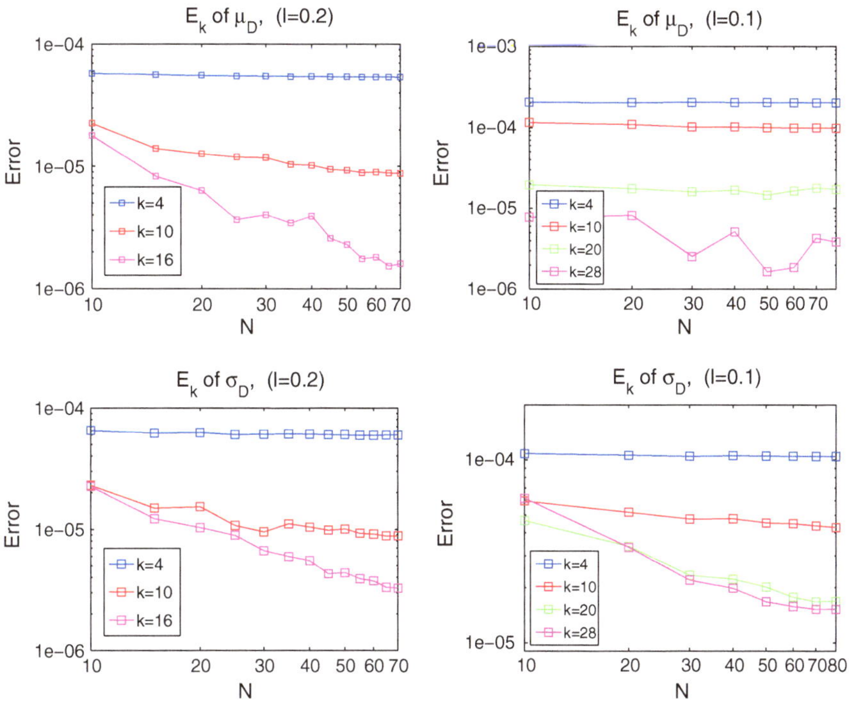

Fig. 7 E_k in GEK estimated statistics of C_D. $\ell = 0.2$ (left) and $\ell = 0.1$ (right)

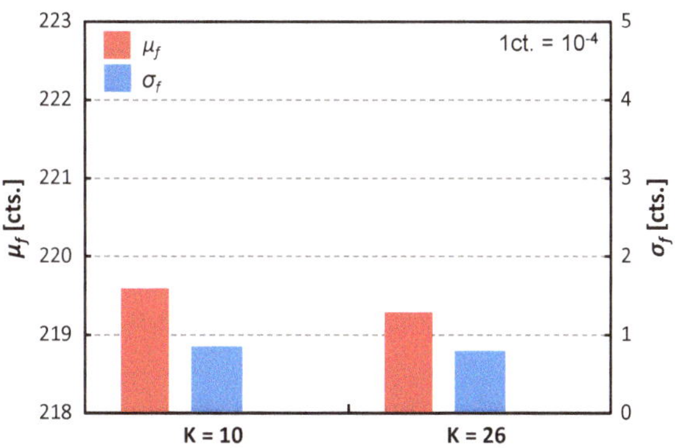

Fig. 8 Mean and standard deviation of the stochastic cost function f of the initial configuration by different numbers of eigenvalues ($K = 10$ and 26, $\ell = 0.2$) of truncated Karhunen–Loève expansion (KLE). f is evaluated by gradient-enhanced kriging (GEK) with a fixed number of samples as 30

compensate the problem of the number of sampling without any loss of the accuracy of the geometrical uncertainties unless the CPU time for construction of GEK model grows huge to become comparable to CFD computations. This should be also applied to robust design.

The result of an application to RDO is presented here. Figure 9 shows the histories of the stochastic objective function f ($f \equiv \mu_{Cd} + \sigma_{Cd}$) in RDO and the

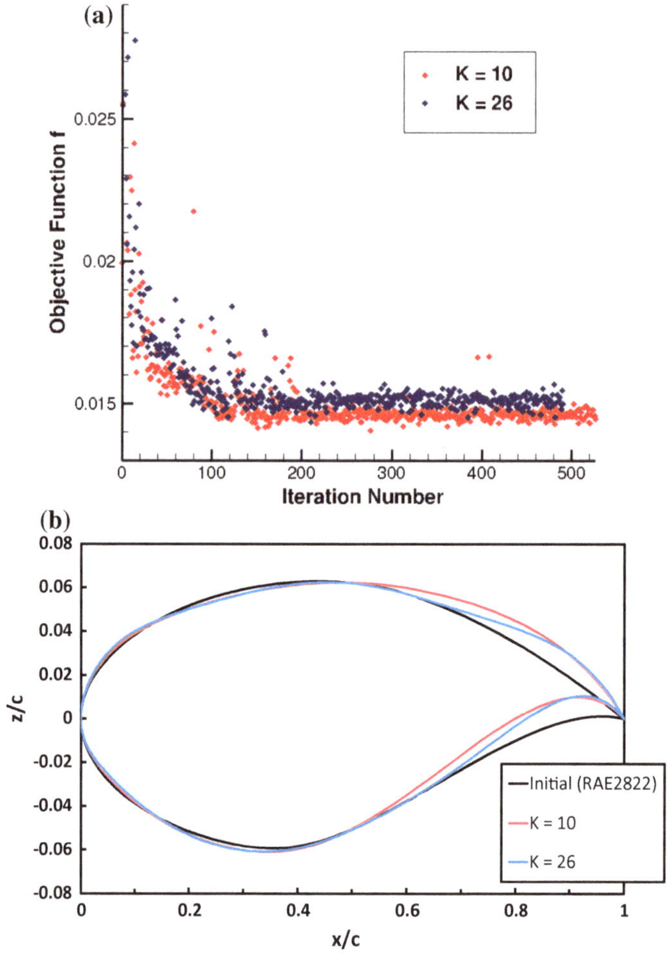

Fig. 9 Histories of objective function in application to robust design optimization (RDO) (**a**) and the designed airfoil configurations (**b**), by different numbers of eigenvalues (K = 10 and 26, ℓ = 0.2) of truncated Karhunen–Loève expansion (KLE). f is evaluated by gradient-enhanced kriging (GEK) with a fixed number of samples as 30

designed airfoil configurations, by different numbers of eigenvalues (K = 10 and 26, ℓ = 0.2) of tKLE. The stochastic objective function f is evaluated by GEK with a fixed number of samples as 30. Since there could be differences between the embodied geometrical features due to tKLE by $K = 10$ and $K = 26$, the comparison of the statistical performance between these two designed configurations is difficult (there is a slight difference between the evaluated f by $K = 10$ and $K = 26$ as can be observed by σ_f in Fig. 8). We can focus on that there is a difference in the optimized geometrical configurations in Fig. 9. Our conclusion is that if the statistics can be computed efficiently with the help of an adjoint solver to use GEK, sufficiently small eigenvalues in KLE should be maintained unless the number of the truncation becomes big in such a case of 3D aircraft/wing models.

Summary

To investigate the influence of the accuracy of random field approximation to the accuracy of the estimation of performance statistics, we conduct a series of numerical experiments with different truncation of Karhunen–Loève expansion (KLE) (hence different number of variables in the model) for random perturbation in the geometry of a RAE2822 airfoil and compare the statistics of lift and drag coefficients and efficiency of gradient-enhanced kriging (GEK) in each scenario.

Statistics obtained with different number of variables are different, and the reliability of the discrepancies is justified by an accurate estimation of the statistics. The size of discrepancy is related to the geometry perturbation variance kept by the KLE truncation.

It is observed in the experiments that the GEK's error convergence rate is not improved by reducing variables. This can be attributed to the cheaper gradient information obtained by adjoint solver in the cases with more variables, and also to the not evenly distributed "importance" of the variables as KLE coefficients.

The results of this investigation indicate that excessive KLE truncation will not improve the efficiency of surrogate-based statistics integration. On the contrast, it handicaps the accuracy of the statistics. The truncation should be made with a reasonably small threshold on the eigenvalues of the random field's covariance matrix, e.g., based on machine precision or CFD solver precision. The lack of the proper truncation could cause different results also in applications to robust design optimization.

References

1. Liu, D., Görtz, S.: Efficient quantification of aerodynamic uncertainty due to random geometry perturbations. In: New Results in Numerical and Experimental Fluid Mechanics IX, Volume 124 of the series Notes on Numerical Fluid Mechanics and Multidisciplinary Design, pp. 65–73. Springer (2014)
2. Galle, M., Gerhold, T., Evans, J.: Parallel computation of turbulent flows around complex geometries on hybrid grids with the DLR-TAU code. In: Ecer, A., Emerson, D.R. (eds.) Proceedings of the 11th Parallel CFD Conference, Williamsburg, VA, North-Holland, 23–26 May 1999
3. Gerhold, T., Hannemann, V., Schwamborn, D.: On the validation of the DLR-TAU code. In: Nitsche, W., Heinemann, H.-J., Hilbig, R. (eds.) New Results in Numerical and Experimental Fluid Mechanics. Notes on Numerical Fluid Mechanics, vol. 72, pp. 426–433. Vieweg (1999). ISBN 3-528-03122-0

4. Schwamborn, D., Gerhold, T., Heinrich, R.: The DLR TAU-code: recent applications in research and industry, invited lecture. In: Wesseling, P., Oate, E., Priaux, J. (eds.) Proceedings of the European Conference on Computational Fluid Dynamics (ECCOMAS CFD 2006), The Netherlands (2006)
5. Snyder, W.C.: Accuracy estimation for quasi-Monte Carlo simulations. Math. Comput. Simul. **54**(1–3), 131–143 (2000)
6. Wendland, H.: Scattered Data Approximation. Cambridge University Press (2005)
7. Han, Z.-H., Görtz, S., Zimmermann, R.: Improving variable-fidelity surrogate modeling via gradient-enhanced kriging and a generalized hybrid bridge function. J. Aerosp. Sci. Technol. (2012)
8. Maruyama, D., Liu, D., Görtz, S.: Surrogate model based approaches to UQ and their range of applicability. In: Hirsch, C. et al. (eds.) Uncertainty Management for Robust Industrial Design in Aeronautics, vol. 140, pp. 703–714.

Analysis and Interpretation of Probabilistic Simulation Output

Alberto Clarich and Rosario Russo

Obtaining the input/output cumulative distribution function (CDF) or probability density function (PDF) would be the main goal when dealing with uncertain input parameters. Indeed, once output CDF and output PDF are known, all the information about the uncertainties' effect of the system at hand can be monitored and then potentially controlled during the project phase.

In most of the cases unfortunately, evaluating output PDF or CDF is too expensive in terms of requested samples and the first statistical moments (mean and standard deviation) can be enough in practical applications. However, in real cases, where the outputs are complex function of uncertain input parameters, the output uncertainties cannot be modeled with a normal distribution, so a simple Six Sigma test would be not accurate to assess the stability and then the quality of the output performance.

For this reason, a way to estimate the output statistics and percentiles of the outputs, independently from the hypotheses on the CDF or PDF outputs, is required; if indeed a given percentile is known, the optimization can focus directly on the percentile value, not only optimizing the performance but also keeping at the same time the output uncertainties under control.

In this chapter, a best practice guide is therefore provided to evaluate output uncertainties, namely main output statistical moments and percentiles.

Input Parameter CDF and PDF Reconstruction

If a proper number of experimental samples are available, a way to obtain the output CDF and PDF can be achieved by a distribution fitting approach.

A. Clarich (✉) · R. Russo
ESTECO, Trieste, Italy
e-mail: clarich@esteco.com

C. Hirsch et al. (eds.), *Uncertainty Management for Robust Industrial Design in Aeronautics*, Notes on Numerical Fluid Mechanics and Multidisciplinary Design 140, https://doi.org/10.1007/978-3-319-77767-2_52

The distribution fitting tool developed within the modeFRONTIER software [1] allows to find, from a database of 11 statistical distributions (Normal, Cauchy, Logistic, Exponential, Lognormal, Weibull, Gamma, Chi Square, Beta, Student, Uniform) the one that best fits the given database (set of samples).

The tool finds the statistical distribution and parameters that best fits the samples database.

The tool uses two indexes to quantify how much the statistical distribution fits the database, the Kolmogorov–Smirnov (K-S) and the Likelihood.

The K-S test is based on the maximum distance between the discrete CDF computed by the available samples and the theoretical CDF (with given parameters) to check.

The likelihood is given by

$$\mathcal{L} = \sum_{i=1}^{n} p(x_i) \tag{1}$$

where $p(x_i)$ is the PDF corresponding to the sample x_i; since the likelihood is often small, usually it is considered minus its logarithm, the so-called log-likelihood.

Basically, an internal optimization is performed, the input parameters being the relevant parameters (i.e., mean and standard deviation for normal) and the objective function being the minimization of the log-likelihood (Fig. 1).

In particular, a genetic algorithm is used (NSGAII) as global search, and then, a gradient-based algorithm (BFGS) is used for local refinement.

As output, the 11 different statistical distributions are sorted from the one with lowest value of KS test coefficient (worst fitting) to the one with highest value of KS test (best fitting).

For all the distributions, the relative parameters are reported together with the K-S significance test index.

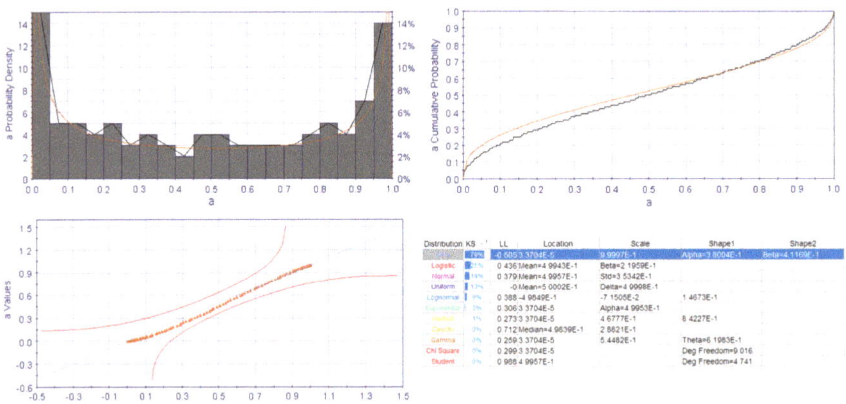

Fig. 1. Example of distribution fitting tool in modeFRONTIER

In addition, for the best-fitting distribution, a plot that compares the theoretical CDFs with the discrete CDF from the database is given.

The distribution fitting approach, however, is applicable when the number of samples is large enough (as an order of magnitude hundreds of samples is requested), otherwise the estimation of the distribution parameter would be too rough; to have an idea to have a reliable result, the K-S index should be at least 95%.

Steps and Assumptions to Assess Output Uncertainties

We can summarize in the following six steps the operations to be followed in order to asses efficiently the output uncertainties.

1. Estimation of the input uncertainties by a distribution fitting approach (see previous paragraph), following the hypothesis that the input uncertainties are uncorrelated. Especially in parametric optimization, based on, at most, few tens of parameters to be controlled and optimized, this hypothesis can be valid in many cases.
2. Consider the input uncertain parameters of the system at hand and evaluate how much each uncertain parameter affects the outputs of interest. Input parameters that do not affect the outputs beyond given thresholds should be discarded and not considered as uncertain input parameter. A tool that can be used for this kind of analysis is Smoothing Spline ANOVA [2] (described in Chapter "Screening Analysis and Adaptive Sparse Collocation Method"). A meaningful criterion is to keep only the uncertain input parameters that are able to catch up to 90% of the output variance (considering single and interaction effect).
3. The analysis above can be performed starting from a design of experiment size that, as order of magnitude, should be around the minimum number of samples to catch a second-order response, namely $\frac{(n+2)!}{2n!}$, with n number of input parameters. A Latin Hypercube DoE algorithm can be used to initialize the design population. The distribution parameters of each uncertain variable are set according to the assigned input PDF evaluated before.
4. If some uncertain parameters are also free variables of a successive optimization procedure, then the described analysis to filter out the unimportant uncertain parameters should be repeated around different nominal values of the optimization variables to check whether the behavior of the system is uniform within their range of variation.
5. Once the most important input uncertainties are identified, it is necessary to estimate the output uncertainties statistics (typically mean standard deviation and percentiles). As described in Chapter "Screening Analysis and Adaptive Sparse Collocation Method," an effective approach is the Polynomial Chaos

Expansion enhanced with the sparse approach and implemented within the modeFRONTIER software. The first goal is to assess how many samples are necessary to have a reliable estimation of the statistics. For these reasons once again is worth starting (if possible in terms of computational time) with a number of samples that is about twice the minimum number of samples to compute a second-order polynomial expansion, then the number of samples can be reduced, monitoring at the same time the R-squared Leave One Out and Adjusted R-square LOO [3] (Fig. 2).

In order to achieve reliable statistic results, it is needed to consider the minimum number of samples and the minimum polynomial degree which give adjusted R-Squared results close to 1. These values can be considered as reference values to be used for a robust design optimization.

6. The final step in the definition of the optimization problem is the definition of objective(s) function. Since the number of designs required by the optimization is directly proportional to the the number of objectives, in order to reduce the computational effort it might be important to reduce as much as possible the number of objectives. This can be achieved basically in two ways:
   (i) consider the most important criteria as one objective and keep the other criteria when possible as constraints (such as the case of airfoil drag minimization with constraints on lift and momentum), or sum together criteria which are homogenous (such as the case of acoustic performance at three different flight conditions);
   (ii) prefer the reliability-based optimization approach to the classical multi-objective approach (optimize mean performance and minimize standard deviation), since, by optimization the worst percentile (e.g., 99.9 or 0.03) of the performance distribution, the whole distribution can be optimized keeping only one objective per criteria instead of two.
   Finally, the choice of the optimization algorithm is important: If the simulation times are very expansive, a fast algorithm like Simplex

Output Variables

Name	Value	Mean	Standard Deviation	Min	Max
z	+0.0000000000E000	+5.0000000000E-002	+5.8070154616E-002	-2.6067853329E-004	+1.5945812209E-001

PCE Coefficients

Name	PCE Coefficients	Accuracy Indicators
z	Constant term: 0.05000000000000005	R-squared: 1.0
	x^2: 0.05345224838248493	R-squared LOO: 1.0
	x^1 * y^1: 0.022360679774997772	Adjusted R-squared LOO: 1.0
	x^3: 3.8723122600281566E-17	AIC: -746.746263051615
	y^3: 0.0024494897427829805	BIC: -744.9307524936506
	y^1: 0.0029999999999997893	

Fig. 2. Example of adaptive sparse PCE log report in modeFRONTIER

(single-objective) or Game Theory (multi-objective) could be used to find a compromise solution by a reduced number of design evaluations, otherwise more robust algorithms like genetic algorithm could be used as alternative.

Selection of Probabilistic Optimal Design from RDO

There are several tools which can be used to select the optimal solutions, from the Pareto frontier, once the optimization is completed.

If the number of objectives is contained, scatter or bubble charts, like the one reported in Fig. 3 from modeFRONTIER software [4], can be used: It is possible to highlight only Pareto designs in the chart, and it is possible to monitor up to four optimization criteria: ordinate and abscissas (approach and cut-back OASPL performances in the aeroacoustic application reported in Fig. 3), color of the bubbles (takeoff flight condition in the example), and eventually a four criteria can be selected as size of the bubbles.

In case of higher number of criteria, parallel chart like the one reported in Fig. 4 can finally be used: Every line represents a different design of the Pareto front, and every column reports a different objective/constraint or variable; by using the relative filter sliders, the user can easily filter out the not-desired solutions that find the optimal solution. More advanced tools, called MCDM (multi-criteria decision making), are available as well in software like modeFRONTIER, in the case the user would like to express some weights for every objective, obtaining at the end a ranking of all the Pareto solutions on the basis of the indicated preferences.

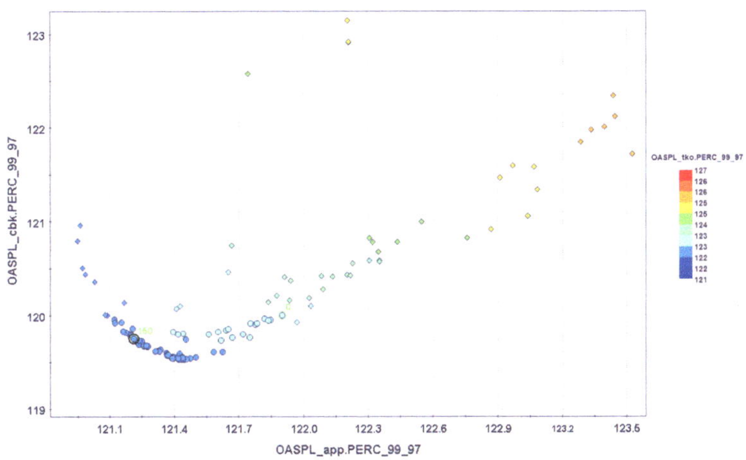

Fig. 3. Optimization results (OASPL percentiles at three flight conditions)

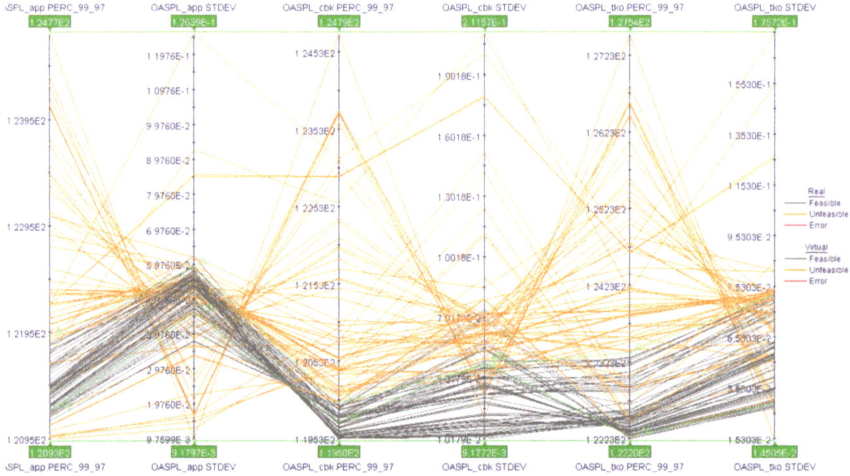

Fig. 4. Parallel chart to filter out optimization results

References

1. www.esteco.it
2. Gu, C.: Smoothing Spline ANOVA Models. Springer, New York (2002)
3. Clarich, A., Russo, R.: Innovative methodologies for robust design optimization with large number of uncertainties using modefrontier. In: Proceedings of the International conference on Evolutionary and Deterministic Methods for Design Optimization and Control with Applications to Industrial and Societal Problems, EUROGEN, pp. 233–240 (2015)
4. Clarich, A., Pediroda, V.: Robust design applications with modeFRONTIER, applying NODESIM-CFD tools. In: NODESIM-CFD Workshop on Quantification of CFD Uncertainties, Bruxelles, 29–30 October (2009)

Summary of UMRIDA Best Practices

Jordi Pons-Prats and Gabriel Bugeda

Uncertainty quantification (UQ) is becoming a strategic step in the design phase. Robust Design Optimization (RDO) is the following step. The Technological Readiness Level (TRL) of intrusive and non-intrusive methodologies is increasing rapidly, although several limitations remain. Nowadays, UQ is a major trend in research, because there is a lot of room for improvement.

It is well accepted, both by academia and industry, that uncertainty is an intrinsic issue affecting the design. This is not new at all. Methods to deal with uncertainty quantification exist from many years ago, just to mention Monte Carlo method as the main reference. The computational cost of these kinds of methods does not help to spread their use. However, not only the methods are important, but also the definition of the uncertainties, where they are applied, and the final analysis of the results are key for a good application.

The definition of new methods is a key issue and much research is focussed on it. This book is a clear example of this research; methods are presented both from the point of view of the methodology and also from the point of view of the application. The research that has been described in the previous chapters is the outcome of the UMRIDA project, but also of the NODESIM-CFD project, which was the first step. The present annex confirms the Best Practice Guide for an improved and industrial-oriented application of uncertainty quantification (UQ) and Robust Design Optimization (RDO) methods.

The Best Practice Guide, jointly with the UMRIDA book, has presented several methodologies; from Monte Carlo to Methods of moments, just to mention two of them. The present annex, as a Best Practice Guide, contains also a discussion about how to deal with UQ and RDO. The following list is a summary of methodologies:

J. Pons-Prats (✉) · G. Bugeda
Aeronautical Group, CIMNE, 08034 Barcelona, Spain
e-mail: jpons@cimne.upc.edu

G. Bugeda
Universitat Politecnica de Catalunya - BarcelonaTech, 08034 Barcelona, Spain

C. Hirsch et al. (eds.), *Uncertainty Management for Robust Industrial Design in Aeronautics*, Notes on Numerical Fluid Mechanics and Multidisciplinary Design 140, https://doi.org/10.1007/978-3-319-77767-2_53

- Non-intrusive polynomial chaos;
- Non-intrusive probabilistic collocation method;
- Adaptive sparse collocation method;
- Reduced basis methods;
- Surrogate-based methods;
- Monte Carlo-based methods;
- Intrusive perturbation methods.

Regarding the methods for RDO, the previous UQ methods have been combined with evolutionary techniques (covariance matrix-assisted, genetic algorithms) as well as gradient-based and adjoint-based techniques. For both cases, namely UQ and RDO, the book has presented the methods and their applications to industrial problems.

The Best Practice Guide is focused on the issue to take into consideration when dealing with these methodologies. It also includes the description and references to several tools, some of them are commercial software, which can be used for an as easy as possible implementation of UQ and RDO analysis. Tools like TAPENADE, a software devoted to the differentiation of a functional, developed by INRIA, or OpenTURNS, an open-source library for uncertainty and reliability analysis, whose development is funded by AIRBUS, EDF and PHIMECA, are two examples of the tools for UQ. On the other hand, RMOP, the CIMNE's optimization platform, or modeFRONTIER, scheduled by ESTECO, has been used in the framework of RDO analysis. FINE™ is a commercial solver for CFD problems, by NUMECA, which already integrated the UQ analysis as an option within its menu. It is a quite exceptional case due to the full integration of UQ methodologies into a commercial package. The standard procedure to couple the UQ analyser with the solver requires the use of PYTHON wrappers, or other type of scripting. The scripts contain the entire list of steps to be performed from the definition of the problem to its final execution. For a full automation of the analysis, it is required that all the steps can be called without user interaction. If you are dealing with a RDO analysis, this automation must be robust enough to go through an iterative process without manual and user interaction. But this is not the only issue to take into account.

The Best Practice Guide chapters have identified some key issues to bear in mind when formulating the analysis. These issues are pretty common when dealing with either a UQ or a RDO analysis. As a summary, the list of issues is as follows:

- Definition of the uncertainty: the user must have a clear idea about the uncertainty and its statistical behaviour, in order to translate that into a representative set of numerical samples, but also to accordingly select the analysis method. First chapter of this Guide is focused on the process to identify and model uncertainties. Uncertainties from the lack of knowledge of the phenomena, but also related to manufacturing tolerances, or the error derived from experimental measurements or numerical tools must be considered and modelled according to their real behaviour, in order to understand the final effect over the output of the analysis.

- Analysis of sensitivity: in order to simplify the analysis, those variables with lower effect on the variability of the output could be removed. This item links to the first one, if there is a lack of knowledge about the uncertainty, an analysis of sensitivity can help to identify the most relevant uncertain parameters to deal with.
- Sampling: all the methods require to calculate from few to hundreds of samples. Monte Carlo-based methods are the ones that demand the largest number of samples. Random selection could be an alternative option, but not the best one. So, a careful selection of the sampling technique is also relevant. There are a lot of techniques that optimize the selection of samples, from Design of Experiments to Sparse Grid Level technique (when dealing with collocation methods).
- The "Curse of Dimensionality" is an important issue when dealing with UQ and RDO. There are several methods that suffer from this problem; it can be shortly described as "the larger the number of uncertain variables is, the more costly the analysis will be". So it is more than necessary to spend time before starting the computation to save computing time.
- Truncation errors could be considered as an issue, but as described in chapter "Formulations for Robust Design and Inverse Robust Design", the efficiency of the method is not improved while the accuracy is handicapped.
- Computational cost reduction; it is well known that the computational cost of UQ and RDO could reach unaffordable levels, especially for industrial applications. Applying simplifications through the solver itself, or by using surrogate models, is a good option. The tests done by UMRIDA partners considered Reduced Order Models, definition of transfer functions as well as surrogate models like Kriging, gradient-enhanced Kriging or Differentiated Lifting Line Method (DLLM), which implemented through OpenTURNS acted as a surrogate model for the coefficient calculation, to mention three of them.
- Results analysis: it is important to highlight this item, since the full statistical information of the results could be highly expensive to obtain. The evaluation of PDF and CDF is usually computationally unaffordable due to the fact that the required number of samples increases rapidly. On the other hand, first statistical moments, namely the mean and the standard, are easy to calculate so usually used as reference. It means the assumption of a normal distribution for the output is accepted, while for complex problems could be a wrong assumption. The last chapter of this Best Practice Guide provides some recommendations about the analysis and interpretation of the results, so it is worth to read it carefully.

It is also relevant to focus our attention to the difference between an intrusive and a non-intrusive method. Some of the methods presented in this book have both implementations, but the associated analysis cost could be completely different. This is because the intrusive methods can use information directly from the calculation, adjoint information, for instance, but this benefit could be overlapped by the implementation cost. A non-intrusive method works as a black box; internal data is not available from the solver, so only the final value of the functional is available. The simple and usually fast implementation of a non-intrusive method is compensated by the additional information that an intrusive method can use. So a clear trade-off must be solved by the developer, and sometimes by the user, when selecting a method.

Although a lot of work is already done, and the TRL of the methods is increased to 6–8, there is still room for improvement. From the industrial perspective, the methods should be more attractive; reducing the computational cost, simplifying the implementation of the analysis, so industrial partners can extend daily use of UQ and RDO in their design office.

The aim of UMRIDA partnership has been not only to describe the methodologies, which are done in the main body of the book, but also to provide a guide for newcomers. The experience accumulated during the UMRIDA project, as well as during its parent project NODESIM-CFD, is intended to make UQ and RDO methods better understood, to spread the word across the engineering community so lead to an extensive use of these methodologies in any engineering and science field.

This annex to the UMRIDA book conforms the Best Practice Guide of UQ and RDO methods.

Part VI
Conclusions

Project Summary and Outlook

Charles Hirsch and Dirk Wunsch

UMRIDA Project Commitment

The UMRIDA project focused on uncertainty management and robust design methodologies at all levels of the analysis and design process. The technical objectives of UMRIDA are listed in chapter "Vision, Objectives and Research Activities". The main objectives were the development of innovative methods for UQ and RDM, the large-scale introduction of UQ methodologies into robust design, application of UQ and RDM to the created database with prescribed uncertainties by meeting the UMRIDA objective of handling at least 10 simultaneous uncertainties, in a turnover time of no more than 10 h on a 100 core parallel computer, and to facilitate the cooperation and dissemination toward European industries, research establishments, and universities.

In order to reach these objectives, a significant progress beyond the state of the art was needed. The activities of UMRIDA project partners were grouped into three main fields of activity, namely advancing methods for uncertainty quantification, characterization of the most influential uncertainties and dimension reduction, and advances in robust design methodologies.

In the following, the work performed in UMRIDA is assessed against these objectives in view of the three main fields of activity and around the following core objectives of the UMRIDA project:

- Construction of a novel and unique database of industrial relevant challenges with prescribed uncertainties

C. Hirsch (✉) · D. Wunsch
Numeca International, Chaussée de la Hulpe 189, Brussels, Belgium
e-mail: charles.hirsch@numeca.be

D. Wunsch
e-mail: dirk.wunsch@numeca.be

C. Hirsch et al. (eds.), *Uncertainty Management for Robust Industrial Design in Aeronautics*, Notes on Numerical Fluid Mechanics and Multidisciplinary Design 140, https://doi.org/10.1007/978-3-319-77767-2_54

- The UMRIDA quantifiable objective of: "*development and application of UQ methods for large number of uncertainties (>10) within an acceptable CPU return time of 10 h on no more than 100 cores parallel processors including multidisciplinary applications*"
- Application to industrial challenges and hereby increasing the Technology Readiness Level (TRL) from currently 2–3 to 5–6.

Novel and Unique Database with Prescribed Uncertainties

A unique database with prescribed uncertainties for UQ and RDM was built within the UMRIDA project as presented in chapter "UMRIDA Test Case Database with Prescribed Uncertainties". The purpose of the generic methodology description is to be (partly) independent from the specific use case, thus allowing for a common view by all partners on the different test cases. A common language of all partner's use case descriptions enables for easier collaboration and better means of comparability of different methods within the project. For this purpose, a test case description template has been used for the description of all test cases differentiating between the test case description and the definition of uncertainties. Finally, the database comprises of four basic test cases and nine industrial test cases spanning a range of applications as diverse as 2D cases to full 3D cases for exterior aerodynamics, turbomachinery design, and multidisciplinary applications such as aero- and thermo-acoustics, and fluid–structure interaction.

The database test cases with prescribed uncertainties were used throughout the project by the project partners as benchmarks for the development of various UQ and RDO methodologies and in particular for the workshops on UQ held in Delft on April 15 till 16, 2015, and on RDO held in Brussels from September 21 till 23, 2016. The database formed thus the backbone of the activities within UMRIDA and fulfills one part of objective 3 defined in chapter "Vision, Objectives and Research Activities".

Progress in Methods for Uncertainty Quantification (UQ)

The work with respect to uncertainty propagation performed within UMRIDA is based on three classes of uncertainty propagation techniques: method of moments (perturbation method) and adjoint-based methods; sampling, Monte Carlo, and multilevel Monte Carlo methods; polynomial chaos and collocation methods. These are applied to operational and geometrical uncertainties, uncertainties resulting from the manufacturing and assembly process, and modeling (epistemic) uncertainties, and by some partners combined with techniques that rely on surrogate modeling.

The general speedup reached by the developed methods is in the order of 5–10, when compared to standard approaches, such as classical regression-based polynomial chaos, tensorized sampling, or classical Monte Carlo. The standard approaches used as reference are listed in Tables 1, 2, and 3 for all methods. In some cases, the speedup compared with the situation at the start of the project can be significantly larger if, for example, sparse grid techniques are compared with full tensor grids. Tables 1, 2, and 3 provide an overview of speedups reached for the various UQ methods.

The above methodologies were applied to test cases from the UMRIDA database, showing that the UMRIDA quantifiable objective is successfully reached by several partners by applying a variety of methods (reduced basis method, method of moments, level 2 sparse grid, multilevel Monte Carlo, gradient-enhanced kriging) to deal with test cases from the UMRIDA database (BC-01, BC-02, BC-03). Industrially relevant and representative test cases (IC-03, IC-04, IC-05, IC-06, and IC-09) were also successfully handled fulfilling the UMRIDA objective. Table 4 compares the required computational effort for UQ comparing different methods on some test cases with the UMRIDA quantifiable objective in CPUh/100 cores. In these units, the UMRIDA quantifiable objective corresponds to a value of 10 and values smaller than 10 fulfill the objective. Reaching the quantifiable objective is a direct result of the above-presented speedups and was reached as a milestone after the first UMRIDA workshop.

Table 1 Speedup achieved by the consortium for polynomial chaos and collocation methods if applied to database test cases

Polynomial chaos and collocation methods					
Partner	Method	Test case	Number of uncertainties	Speedup	Reference for speedup
Airbus group innovation	Sparse PC OpenTURNS 1.4 (OT1.4)	IC-02	8 geometrical 2 operational	18	OT1.2
ESTECO	Adaptive sparse PC	BC-02	11 geometrical 2 operational	7	Classical PC
ONERA	Compressive sampling	BC-02	1 geometrical 2 operational	12.5	Tensorized sampling
NUMECA	Sparse probabilistic collocation	BC-02	10 geometrical	8.3	Classical PC with regression
		BC-01	6 geometrical 3 operational	7.2	Classical PC with regression
VUB	Reduced basis method with regression-based PC	BC-02	10 geometrical	6.7	Classical PC with regression
		BC-01	19 geometrical 2 operational	4.6	Classical PC with regression

Table 2 Speedup achieved by the consortium for Monte Carlo and sampling techniques if applied to database test cases

Monte Carlo and sampling techniques					
Partner	Method	Test case	Number of uncertainties	Speedup	Reference for speedup
CIMNE	Multilevel MC	BC-02	13	5	Class MC
DLR	Quasi-MC +gradient-enhanced surrogates (Kriging, RBF, PC)	BC-02	26 geometrical 2 operational	5-6	Class. quasi-MC
TU Dresden	(Extendible) LHS	IC-09	15 geometrical	10	Class MC (random sampling)

Table 3 Speedup achieved by the consortium for perturbation techniques if applied to database test cases

Perturbation techniques					
Partner	Method	Test case	Number of uncertainties	Speedup	Reference for speedup
Leonardo +INRIA	Automatic differentiation	BC-02	12 geometrical 2 operational	2.5	Finite diff

Table 4 Benchmarking of developed methods against the UMRIDA quantifiable objective

Test case	Partner	Method	Nature and number of uncertainties	Equations solved/ mesh size (10^6)/# of runs	CPUh/ 100 cores
BC-01	VUB	Reduced basis with regression (PC order 2)	19 geometrical+2 operational	RANS/0.77/42 (on fine grid)	0.66
BC-01	NUMECA	Level 1 sparse grid	10 operational +geometrical	RANS/2.8/21	0.56
		Level 2 sparse grid	10 operational +geometrical	RANS/2.8/241	6.43
BC-01	Saturn-NPO	Surrogate-based Monte Carlo	10 geometrical	RANS/2.8/54	1.18
BC-02	Leonardo	Second-order method of moments with auto. diff.	12 geometrical, 2 operational	RANS with auto-differentiation/ 0.04/1	1.68
BC-02 BC-03	WUT	Second-order sensitivity	40 geometrical, no operational	Euler/0.44/1	1.3
IC-03	DASSAV	First-order method of moments	10 geometrical	Linearized RANS 7.1/10	9.16
IC-03	INRIA	Norm-oriented anisotropic mesh adaptation	Estimate a point-wise error accounting for discretization errors and to drive adaptivity	RANS/1-10/20	4.8

Finally, UQ strategies were applied to non-aerodynamic cases, including thermo-acoustics, acoustics, and aeroelasticity with up to 10 operational and geometrical uncertainties. INRIA developed a set of numerical methods for the reduction and estimation of numerical errors and applied it reaching the UMRIDA quantifiable objective to the test case IC-03. This work contributes to reaching the objectives 1, 2, 3, and 4, which are listed in chapter "Vision, Objectives and Research Activities".

Progress in Characterization of Most Influential Uncertainties and Dimension Reduction

A first very important step in the quantification and propagation of uncertainties is the correct quantification of input uncertainties. TU Dresden and Airbus Group Innovation focused on measurement of production scatters. There are several problems to address in quantification of geometry variability due to production. These include the analysis of point cloud data, data reduction (as the available data is in many cases substantial), and statistical analysis. MAN Diesel and Turbo Schweiz worked on methods for identification and quantification of input experimental uncertainties. The life cycle of turbomachinery was outlined, and the major documents that deliver data for the input uncertainty quantification were identified. These documents are instructions, drawings, guidelines, and protocols. The identified uncertainties were used to analyze and improve internal manufacturing instructions for rotating parts.

Several partners worked on dimension reduction techniques in order to reduce the cost of uncertainty propagation techniques. VUB worked on reduced basis methods coupled with regression-based polynomial chaos, where coarse mesh solutions were used to sample a covariance matrix, which in return was used to reduce the dimensionality of the problem. Techniques based on Karhunen–Loeve Decomposition or Principal Component Analysis were applied by several partners to random fields representing production variability or to the handling of correlated engineering parameters. On such reduced bases, CIMNE and NUMECA used non-intrusive uncertainty quantification methods: MLMC and probabilistic collocation, respectively. Two partners (WUT and Leonardo) developed first- and second-order sensitivity methods for their in-house solvers, with the help of INRIA on the usage of their automatic differentiation software.

Surrogate modeling techniques for airfoils were used by ONERA and DLR. ONERA worked on kriging-based surrogate models for the lift-to-drag ratio, while DLR worked on adaptive gradient-enhanced surrogate models, using the adjoint TAU code, and compared the efficiency and accuracy with kriging, direct integration (quasi-MC), and full MC reference statistics.

This work contributes to reaching the objectives 1, 2, 3, and 4, which are listed in chapter "Vision, Objectives and Research Activities".

Progress in Robust Design and Optimization Methodologies

The challenge for robust design optimization lies in coupling optimization methods with the UQ approaches developed in the project. Most of the partners used the UMRIDA database test cases including industrial challenges (IC) for the assessment of their developed methods. The geometries to be optimized were described using both deterministic (nominal) design parameters and stochastic variables. In some of the applications, the number of stochastic variables exceeded 10, which was set as a threshold to be achieved to ensure the success of UMRIDA. The stochastic objective functions were optimized by using gradient-free algorithms, which are easy to implement for various types of stochastic quantities, whereas VUB formulated a gradient-based method using an adjoint solver. All uncertainty propagation techniques developed within UMRIDA were used, i.e., polynomial chaos or collocation methods, sampling-based methods and multilevel Monte Carlo, and perturbation methods. Different objectives functions were formulated to obtain robust or even reliability-based approaches. Finally, the developed robust design frameworks were demonstrated mainly by using high-fidelity methods such as RANS solvers for CFD simulations. The different partners demonstrated that methods for uncertainty quantification can successfully be integrated into design optimization frameworks to achieve a robust and reliability-based design capability based on high-fidelity CFD. The successful application of RDO to industrial challenges from the UMRIDA database was demonstrated during the final UMRIDA workshop, where RDO techniques were applied to the basic challenges BC-01 and BC-02 and to industrial challenges IC-02, IC-03, IC-04, IC-05, IC-06, and IC-07. The computational resources needed for RDO were judged acceptable in an industrial context. This work contributes to the objectives 1, 2, 3, and 4 listed in chapter "Vision, Objectives and Research Activities".

Dissemination and Exploitation

Being at the forefront of the current research in the field in UQ and RDM, UMRIDA introduced a fundamental shift in engineering practice accounting for uncertainties in the simulation process. The UMRIDA partners recognize the importance of spreading awareness on the novel results of the project, and the entire consortium has actively contributed to the diffusion of the project outcomes. UMRIDA partners produced a long list of publications in highly ranked journals, as well as proceedings publications, presentations in various conferences, seminars, workshops, and other fora.

In overall, the UMRIDA dissemination outcomes can be summarized in numbers as follows:

- **More than 60 scientific papers were published:** 20 in peer-reviewed journals and 41 in conference proceedings.

- **More than 40 presentations were realized in international conferences**.
- **Approximately, 70 presentations were realized in workshops**, including the two UMRIDA workshops, specialized in UQ and RDO.
- **Approximately, 20 presentations were realized in technical seminars, summer schools, and other events**.
- **Eight academic theses** were developed within the activities of UMRIDA.
- **Ten UMRIDA-related entries were published at newsletters of mass outreach**, including the EASN Association Newsletter and the EnginSoft Newsletter.
- **Twelve press releases, announcements, and presentations were published at various EC-supported and other media**; these include popularized articles through CORDIS, as well as information shared across the European Commission's social media channels.
- **Two UMRIDA workshops and symposia** open to external participants were realized. The final workshop counted 60 participants, where 22 participants were external to the project and saw a total of 33 presentations over its 2.5 days duration. It was rounded off by a roundtable discussion on the industrial maturity of UQ methods.
- **The current book with its Best Practice Guide** summarizing the most significant outcomes and giving advice in the use of UQ and RDM is believed to be a valuable contribution to the dissemination of UQ and RDM methods.

Developed UQ and RDM techniques and methods are already available in the design practice of UMRIDA partners or have been introduced in software packages developed and commercialized by UMRIDA partners. The further, the new findings have found their entry into the teaching activities of academic partners. The dissemination and exploitation activities fulfill objective 5 listed in chapter "Vision, Objectives and Research Activities".

Conclusions and Identified Future Challenges

The UMRIDA project reached nearly all objectives and in its whole clearly responds to all main challenges and goals that were set out to reach the start of the project. The UMRIDA database with prescribed uncertainties formed the backbone of the project, and it was used for assessment of the developed uncertainty quantification and robust design optimization methodologies. The successful application of UQ and RDO to industrial challenges from the UMRIDA database during the two workshops demonstrates that the TRL was significantly increased to a level of industrial applicability. The developed UQ and RDO methods are now in use by the industrial project partners, and software vendors integrated the findings of the project into their commercial software offer. First functionalities have been available to their clients since midterm of the project on and the availability of methods is constantly enlarged. Some industrial and academic partners contribute to in-house or open-source libraries, which equally have been extended throughout the project. The application of all types of uncertainty propagation techniques investigated in UMIRDA to industrial challenges from the database shows that they meet the quantifiable objective of 10 uncertainties in less than 10 h turnaround time on 100 core parallel computers.

The final UMRIDA workshop showed with the application to a large number of industrial challenges that the integration of these UQ techniques into robust design formulations was successful. The outcome of UMRIDA is believed to push the deployment of UQ and RDO techniques, certainly by the project partners, but also by industry and academia. The final workshop counted 33 high-level presentations on UQ and RDO and attracted a total of 60 participants, where 22 came from outside of the consortium. The workshop was concluded with a roundtable discussion of the industrial maturity of UQ and RDO methods, which gave a retrospect on the project, but also helped to identify future challenges. Four main future challenges were identified.

Industrial challenges representing multidisciplinary applications such as fluid–structure interaction, or aero- and thermo-acoustics are already part of the UMRIDA database. Based on the experience of industrial project partners, **multidisciplinary integration** in the product development cycles is identified as the weak link in the chain and must be more integrated with the robust and reliable product design process to assure that all physics are captured. This includes UQ and RDO for multi-operating-point problems, such as a performance curve for turbomachinery or a full flight envelope of an aircraft. At the example of an aircraft, preliminary design is performed by means of CFD, while structural and aerodynamic departments need to work together to analyze flutter, where CFD provides the loads and CSM provides the twist angles of a wing, for example.

This example of flutter analysis and the needed interaction between design departments sheds light on the second identified challenge, which is **uncertainty management**. Although UMRIDA carries uncertainty management in its name, the main focus of the project laid on the development of uncertainty propagation and robust design optimization techniques. Some aspects of uncertainty management were addressed in the ambitious work description of the UMRIDA project, such as including multidisciplinary test cases, work contributions on epistemic uncertainties as well as model update by Bayesian techniques. Now that an adequate industrial readiness is reached, addressing aeroelastic uncertainties in an aircraft flight envelope appears on the horizon as a next possible goal.

Currently, robust design optimization is performed on one configuration, whereas the aircraft design is a long-year process in which the RDO inputs will change with time. This shows clearly that the simulations need to be adapted during the design process and raises the question, on a design point of view, of how to design a wing section knowing that changes will come later. This again relates to the multidisciplinary integration identified as first future challenge. Such a process is likely to use tools and methodologies of different accuracy from preliminary design tools to high-fidelity CFD codes, for example. All these methodologies come with epistemic uncertainties, and it seems necessary to include both aleatory and epistemic uncertainties in the UQ approach.

Another important element relates to the complexity of organizations and their processes, such as exchange of data between different departments, different business units, or with external suppliers. It is clear that a more efficient product design process that can account for uncertainties along the design chain and that provides the necessary tools to be applicable in complex organizations bears a huge potential of increased efficiency and lower costs. All these aspects ranging from model updates during the duration of the design process combined epistemic and aleatory uncertainties, and process integrations can be summarized under the challenge of uncertainty management.

A third identified challenge relates somewhat to the organizational complexity and is concerned with the difficulties of extracting manufacturing variability from the production process in order to **accurately define the input uncertainties**. There is no standardized way of identifying or measuring uncertainties to be used in a unified way in the design process. Besides difficulties to obtain data from the production units or from external suppliers, it might be difficult to measure sensible data if the production volume is very low, such as in the case of custom built machines. One possibility is to base input uncertainty definitions on tolerances defined in technical norms, but this option will not be available in the case of additive manufacturing, where measurements of the surface variability seem necessary.

A final challenge identified is related to the complexity and **usability of the UQ and RDO methods**. It seems mandatory to reduce the user-experienced complexity of the tools as much as possible in order to make them accessible and easily understandable to a wide range of design engineers and non-technical decision takers. Some partners already addressed this aspect in UMRIDA, but not at last due to the novelty of UQ and RDO methods, many tools still require expert knowledge. This goes in hand with the second challenge identified on uncertainty management. The provided tools need to clearly state the range of applicability, need to be understandable and usable in a multidisciplinary design process, which is regularly updated with new data input, and ideally provide information on the cost savings that can be achieved.

MIX
Papier aus verantwortungsvollen Quellen
Paper from responsible sources
FSC® C105338

If you have any concerns about our products,
you can contact us on
ProductSafety@springernature.com

In case Publisher is established outside the EU,
the EU authorized representative is:
Springer Nature Customer Service Center GmbH
Europaplatz 3, 69115 Heidelberg, Germany

Printed by Libri Plureos GmbH
in Hamburg, Germany